UNIVERSES IN DELICATE BALANCE: CHEMOKINES AND THE NERVOUS SYSTEM

The cover was designed by Corinna Trebst and Richard Ransohoff (both at Dept. of Neurosciences, Lerner Research Institute, Cleveland Clinic Foundation) and executed by Dr. Trebst, with assistance from Brendan S. Jackson (Cleveland, Ohio, USA). The figure shows two micrographs, rendered as spheres, connected by an elliptical line suggesting a binary system.

One image, on the right, was reconstructed from a z-series by Dr. Grahame Kidd (Dept. of Neurosciences, Lerner Research Institute, Cleveland Clinic Foundation) showing confocal images of cells prepared and immunostained by Drs. Trebst, Zhugong Liu and Pia Kivisakk (Dept. of Neurosciences, Lerner Research Institute, Cleveland Clinic Foundation). The images demonstrate human peripheral blood mononuclear cells stained for CCR5 (red) and CD3 (green). Several CD3-positive T-cells express CCR5 and are double-positive, appearing yellow. One monocyte is CD3-negative but CCR5-positive, appearing red.

The image on the left, prepared by Dr. Deren Huang (Dept. of Neurosciences, Lerner Research Institute, Cleveland Clinic Foundation), shows a fluorescence photomicrograph of a vibratome section from the cerebral cortex of a healthy adult mouse, in which the CX3CR1 gene had been disrupted by insertion of a green fluorescence protein (GFP) gene. Microglia, which express CX3CR1 constitutively, are visualized as evenly-distributed cells with with the fluorescence delineating small somata and delicate radial processes. The mice were generously provided by Drs. Stephan Jung and Dan Littman, Howard Hughes Medical Institutes and Skirball Institute for Biomolecular Medicine, New York University, New York, NY.

UNIVERSES IN DELICATE BALANCE: CHEMOKINES AND THE NERVOUS SYSTEM

Editors:

Richard M. Ransohoff

Cleveland Clinic Foundation, The Lerner Research Institute
Mellen Center for MS Treatment and Research
Cleveland, OH 44195, USA

Kinuko Suzuki

University of North Carolina, Pathology & Lab. Medicine
Chapel Hill, NC 27599, USA

Amanda E.I. Proudfoot

Serono Pharmaceutical Research Institute
1228 Plan des Ouates, Switzerland

William F. Hickey

Dartmouth-Hitchcock Medical Center, Department of Pathology
Lebanon, NH 03756-0001, USA

Jeffrey K. Harrison

University of Florida, College of Medicine
Gainesville, FL 32610-0267, USA

2002
ELSEVIER

Amsterdam – Boston – London – New York – Oxford – Paris – San Diego
San Francisco – Singapore – Sydney – Tokyo

ELSEVIER SCIENCE B.V.
Sara Burgerhartstraat 25
P.O. Box 211, 1000 AE Amsterdam, The Netherlands

First edition 2002

Library of Congress Cataloging in Publication Data

Universes in delicate balance : chemokines and the nervous system / editors R.M. Ransohoff...[et al.].--1st ed.
p.;cm.
ISBN 0-444-51002-8 (alk. paper)
1. Neuroimmunology. 2. Chemokines. I.Ransohoff, Richard M.
[DNLM: 1. Nervous System--immunology. 2. Chemokines--physiology. 3. Inflammation--immunology. 4. Nervous System Diseases--immunology. 5. Receptors, Chemokine--physiology. WL 102 U615 2002]
QP356.47 .U54 2002
616.8′0479--dc21

2002016360

ISBN: 0 444 51002 8

⊗ The paper used in this publication meets the requirements of ANSI/NISO Z39.48-1992 (Permanence of Paper).
Printed in The Netherlands.

This book is dedicated to my wife Margaret
and children Amy and Lena;
to my parents, Sue and Jerry Ransohoff;
and to my patients, colleagues and supporters

PREFACE

It is commonly acknowledged that the nervous system and the immune system, those most complex of networks, share attributes beyond their intricacy. Elements common to the two systems include memory, connectivity, flexibility and developmental selection of cellular composition by a rigorous process involving widespread programmed cell death. There is one salient difference: the cells of the immune system are predominantly in constant motion, while post-mitotic neurons and glia are largely fixed in place. Therefore, chemokines, initially characterized as leukocyte chemoattractants, have for the last one and one-half decades been intensely and productively studied in the contexts of inflammation, immunity and hematopoietic development.

Only recently have the two fields, neurobiology and immunology, displayed mutual interests in chemokines. This convergence of the two tribes of investigators was catalyzed by the finding that SDF-1 (now known as CXCL12) and its receptor, CXCR4, exerted significant and similar functions in development of both nervous and immune systems. Indeed CXCL12 and CXCR4 were required, in an uncannily similar fashion, for retention of pre-B lymphocytes at sites of maturation in the bone marrow and of neuronal progenitors in the external granule cell layer of the developing cerebellum. Recent reports indicate that chemoattraction of cerebellar granule cells through CXCR4 can be suppressed by reverse signaling initiated by binding of soluble eph receptors to transmembrane ephrin B, thereby establishing a link between chemokine action and a cardinal patterning system of the developing nervous system. As may be anticipated when a dam breaks, a massive influx of correlative observations in the nervous and immune systems is likely to ensue.

"Universes in delicate balance: chemokines and the nervous system" represents an attempt to apprehend on the wing the state of current knowledge. To this end, introductory material for both systems is provided. Basic and advanced 'chemokinology' are presented in Chapters 2.1 (Rollins), 2.2 (Proudfoot), 2.3 (Gerard) and 2.4 (Bacon). The recipe for making a nervous system (both ingredients and instructions for preparation) is described in Chapter 1.1 (by Dr. Dupree). The roles of chemokines and their receptors in making an immune system are described in Chapter 2.5 (Farber). Given their importance and complexity, CXCL12/CXCR4 interactions are separately treated in varying contexts, in Chapters 2.4 (Bacon), 4.1 (Harrison), 4.4 (R.J. Miller) and 5.3 (Gabuzda).

Of course, the field that one might call 'neurobiology of chemokines' has not lain fallow for the past ten years. During much of this time, the principal focus has been on neuroinflammation. Here, the question most commonly posed has been "How does the nervous system use chemokines to achieve its distinctive, restricted inflammatory response?" The archetypal, basic patterns by which neural tissues react to insult are described in Chapters 1.2 and 1.3 (Suzuki). Models of neuroinflammation are addressed in Chapters 3.1 (Yong), 3.2 (Karpus), 3.3 (Silverstein), 3.4 (Boddeke), 3.5 (Lane, Buchmeier), 3.6 (Raivich), and 3.7 (Hickey). One consistent motif throughout this period of chemokine research has been generation and application of genetic models, discussed briefly in most of these chapters and addressed directly in Chapter 3.8 (Campbell). Linking the immune and nervous systems are explanations of the functions of chemokines and their receptors for resident brain macrophages, the microglia (Chapter 4.5; Biber), the unique cerebrovascular endothelium (Chapter 4.6; Pachter) and angiogenesis (Chapter 2.6; Strieter).

Understanding human disease is the goal of much of this research. The clinical neuroinflammatory diseases are depicted in Chapters 5.1 (Ransohoff), 5.4 (Fontana) and 5.5 (Hartung). Chemokines have also been implicated in the pathogenesis of certain atypical neuroinflammations. It is unsurprising to find HIV-associated dementia (Chapter 5.3; Gabuzda) on this list. Perhaps less likely, one also encounters Alzheimer's disease, in Chapter 5.2 (Xia, Hyman).

Chemokines have emerged as biologically significant peptides in neurobiology. One harbinger of this advent was the observation that there are constitutive chemokines and chemokine receptors in the nervous system, just as there are in the lymphoid organs. The constitutive neural chemokines (CXCL12, CXCR4; fractalkine/CX3CR1) are discussed in Chapter 4.1 (Harrison). Functions of chemokines and their receptors for the canonical neuroepithelial cells, oligodendroglia (Chapter 4.2; R.H. Miller), astrocytes (Chapter 4.3; Dorf) and neurons (Chapter 4.4; R.J. Miller) are each treated in detail. It is plausible that explorations of these topics provide a glimpse into the future of this field.

One measure of the success of this book will be the pace at which it becomes outdated. New discoveries are being made and reported at a gratifying rate. It is to be hoped that the efforts of the contributors to this volume will be justified by the steady production and application of useful new knowledge in this new field.

R.M. Ransohoff
Cleveland, Ohio, October 27, 2001

List of Contributors

Anuska V. Andjelkovic
University of Connecticut Health Center
Department of Pharmacology
Blood-Brain Barrier Laboratory
263 Farmington Ave
Farmington, CT 06030
USA

Douglas A. Arenberg
University of Michigan
Department of Internal Medicine
Division of Pulmonary and Critical
Care Medicine
Ann Arbor, MI 48109
USA

Valerie C. Asensio
The Scripps Research Institute
10550 North Torrey Pines Road
La Jolla, CA 92037
USA

Kevin B. Bacon
Bayer Yakuhin, Ltd.
6-5-1-3 Kunimidai
Kizu-cho, Soraku-gun
Kyoto, 619-0216
Japan

John A. Belperio
University of California
School of Medicine
Division of Pulmonary and Critical Care Medicine
Department of Medicine
900 Veteran Ave
Los Angeles, CA 90024-1922
USA

Michael A. Berman
Harvard Medical School
Department of Pathology
200 Longwood Ave
Boston, MA 02115
USA

Knut Biber
State University Groningen
Department of Medical Physiology
A. Deusinglaan 1
9700 AD Groningen
The Netherlands

H.W.G.M. Boddeke
University of Groningen
Department of Medical Physiology
A. Deusinglaan 1
Bld. 3125
9713 AV Groningen
The Netherlands

Michael J. Buchmeier
Division of Virology
Department of Neuropharmacology
The Scripps Research Institute
10550 N. Torrey Pines Road
La Jolla, CA 92037
USA

Marie D. Burdick
University of California
School of Medicine
Division of Pulmonary and Critical Care Medicine
Department of Medicine
900 Veteran Ave
Los Angeles, CA 90024-1922
USA

Iain L. Campbell
The Scripps Research Institute
10550 North Torrey Pines Road
La Jolla, CA 92037
USA

Martin E. Dorf
Harvard Medical School
Department of Pathology
200 Longwood Ave
Boston, MA 02115
USA

Jeffrey L. Dupree
University of North Carolina
Department of Pathology and Laboratory Medicine
Chapel Hill, NC 27599
USA

Joshua M. Farber
NIAID
Laboratory of Clinical Investigation
10 Center Drive
Bethesda, MD 20892-1888
USA

Falko R. Fischer
Harvard Medical School
Department of Pathology
200 Longwood Ave
Boston, MA 02115
USA

Adriano Fontana
University Hospital
Department of Internal Medicine
Section of Clinical Immunology
8044 Zurich
Switzerland

Dana Gabuzda
Dana-Farber Cancer Institute
Department of Cancer Immunology and AIDS
44 Binney Street
Boston, MA 02115
USA

John M. Galasso
University of Michigan
Medical School
Ann Arbor, MI 48109-0646
USA

Craig Gerard
Children's Hospital
Division of Respiratory Diseases
Hunnewell 2
300 Longwood Avenue
Boston, MA 02115
USA

Norma Gerard
Children's Hospital
Division of Respiratory Diseases
Hunnewell 2
300 Longwood Avenue
Boston, MA 02115
USA

Paul R. Gorry
Dana-Farber Cancer Institute
Department of Cancer Immunology and AIDS
44 Binney Street
Boston, MA 02115
USA

Jeffrey K. Harrison
University of Florida College of Medicine
Department of Pharmacology and Therapeutics
Gainesville, FL 32610-0267
USA

Hans-Peter Hartung
Karl-Franzens-University
Department of Neurology
Auenbruggerplatz 22
8036 Graz
Austria

William F. Hickey
Dartmouth Medical School
Department of Pathology
Dartmouth-Hitchcock Medical Center
Lebanon, NH 03756
USA

Bradley T. Hyman
Harvard Medical School
Massachusetts General Hospital
Alzheimer's Research, Department of Neurology
Boston, MA 02114
USA

William J. Karpus
Northwestern University Medical School
Department of Pathology
303 E Chicago Avenue
Chicago, IL 60611
USA

Michael P. Keane
University of California
School of Medicine
Division of Pulmonary and Critical Care Medicine
900 Veteran Ave
Los Angeles, CA 90024-1922
USA

Tammy Kielian
Department of Anatomy and Neurobiology
University of Arkansas for Medical Sciences
Little Rock, AR 72205
USA

Bernd C. Kieseier
Karl-Franzens-University
Department of Neurology
Auenbruggerplatz 22
8036 Graz
Austria

Thomas E. Lane
Department of Molecular Biology and Biochemistry
University of California
Irvine, CA 92697
USA

Yi Luo
Harvard Medical School
Department of Pathology
200 Longwood Avenue
Boston, MA 02115
USA

Robert H. Miller
Case Western Reserve University
School of Medicine
Department of Neurosciences
University Hospitals of Cleveland
Cleveland, OH 44106
USA

Richard J. Miller
Northwestern University Medical School
Department of Molecular Pharmacology and Biology Chemistry
303 E. Chicago Ave
Chicago, IL 60611
USA

Seog Bae Oh
Northwestern University Medical School
Department of Molecular Pharmacology and Biology Chemistry
303 E. Chicago Ave
Chicago, IL 60611
USA

Joel S. Pachter
University of Connecticut
Health Center
Department of Pharmacology
263 Farmington Ave
Farmington, CT 06030
USA

Christine A. Power
Serono Pharmaceutical Research Institute
14 Chemin des Aulx
1228 Plan les Ouates
Switzerland

Amanda E.I. Proudfoot
Serono Pharmaceutical Research Institute
14 Chemin des Aulx
1228 Plan les Ouates
Switzerland

Gennadij Raivich
Perinatal Brain Repair Group
Department of Obstetrics and Gynaecology
Department of Anatomy
University College London
Chenies Mews 86-96
WC1E 6HX London
UK

Richard M. Ransohoff
Cleveland Clinic Foundation
Department of Neurosciences of the Lerner Research Institute
NC30, 9500 Euclid Avenue
Cleveland, OH 44195
USA

Shenandoah Robinson
Case Western Reserve University
School of Medicine
Department of Neurosurgery
University Hospitals of Cleveland
Cleveland, OH 44106
USA

Barrett J. Rollins
Harvard Medical School
Brigham & Women's Hospital
Dana Farber Cancer Institute
Department of Adult Oncology
Boston, MA 02115
USA

Gretchen N. Schwartz
Hematopoietic Associates
Wheaton, MD
USA

Jeffrey P. Shaw
Serono Pharmaceutical Research Institute
14 Chemin des Aulx
1228 Plan les Ouates
Switzerland

Faye S. Silverstein
University of Michigan
Departments of Pediatrics and Neurology
8301 MSRB III
Ann Arbor, MI 48109-0646
USA

Dubhfeasa M. Slattery
Children's Hospital
Division of Respiratory Diseases
Hunnewell 2
300 Longwood Avenue
Boston, MA 02115
USA

M. Iain Smith
University of California
School of Medicine
Division of Pulmonary and Critical Care Medicine
900 Veteran Ave
Los Angeles, CA 90024-1922
USA

Katharina-Susanne Spanaus
University Hospital
Department of Internal Medicine
Section of Clinical Immunology
8044 Zurich
Switzerland

Robert M. Strieter
University of California, Los Angeles
School of Medicine
Division of Pulmonary and Critical Care Medicine
Department of Medicine
Department of Pathology and Laboratory Medicine
900 Veteran Ave
Los Angeles, CA 90024-1922
USA

Kinuko Suzuki
University of North Carolina at Chapel Hill
Department of Pathology and Laboratory Medicine
CB# 7527 Brinkhous-Bullitt Building
Chapel Hill, NC 27599-7525
USA

Corinna Trebst
Cleveland Clinic Foundation
Department of Neurosciences of the Lerner Research Institute
NC30, 9500 Euclid Avenue
Cleveland, OH 44195
USA

Jianbin Wang
Dana-Farber Cancer Institute
Department of Cancer Immunology and AIDS
44 Binney Street
Boston, MA 02115
USA

Timothy N.C. Wells
Serono Pharmaceutical Research Institute
14 Chemin des Aulx
1228 Plan les Ouates
Switzerland

Meng Qi Xia
Harvard Medical School
Department of Neurosurgery
Children's Hospital
Boston, MA 02115
USA

Haiyan Xu
University of Michigan
Department of Pediatrics
Ann Arbor, MI 48109-0646
USA

V. Wee Yong
University of Calgary
Faculty of Medicine
3330 Hospital Drive NW
Calgary, Alberta T2N 4N1
Canada

CONTENTS

Section 1

THE NERVOUS SYSTEM

Section Editor: K. Suzuki

Universes in Delicate Balance: Chemokines and the Nervous System
R.M. Ransohoff, K. Suzuki, A.E.I. Proudfoot, W.F. Hickey and J.K. Harrison, editors

Chapter 1.1

Cellular Elements, Tissue Organization, Organogenesis

Jeffrey L. Dupree

Department of Pathology and Laboratory Medicine, University of North Carolina at Chapel Hill, Chapel Hill, USA

1. INTRODUCTION

The function of the nervous system is to collect, process and integrate sensory information and ultimately to formulate and execute the appropriate motor response. Under normal circumstances this complicated series of events is regulated by the appropriate interaction of highly specialized cells that are generally divided into two groups: neurons and macroglia. In Section 1.1, I will address the development, structure and function of these cells and discuss the intricate associations that these cells maintain to ensure proper neural function. The failure of these cells to maintain their proper function can result in devastating neuropathological conditions some of which will be presented in the Section 1.2.

2. TISSUE ORGANIZATION AND DEVELOPMENT

Anatomically, the vertebrate nervous system consists of two divisions: the central nervous system (CNS) comprises the brain and spinal cord and the peripheral nervous system (PNS) is formed by the cranial and spinal nerves including their branches and the autonomic ganglia. Both the CNS and PNS originate from the ectoderm. On the dorsal side of the embryo a region of the ectoderm, known as neuroepithelium, thickens giving rise to the neural plate, which subsequently forms the neural tube (Fig. 1). Prior to closure of the neural tube, a population of cells migrates away from the edges of the closing tube and gives rise to the neural crest, the precursors of the PNS [1].

2.1. Gross Anatomic Development of the CNS

2.1.1. Neural tube

The CNS develops from the neural tube. Upon formation of the neural tube the CNS is represented by a closed tubular structure with a broad anterior end, the immature brain, and a narrow caudal region, the immature spinal cord. The immature brain consists of three dilatations known as the primary brain vesicles: the prosencephalon (forebrain), the mesencephalon (midbrain) and rhombencephalon (hindbrain) (Fig. 2A). With time the three-vesicle brain develops into a five-vesicle brain consisting of: (1) the telencephalon, precursor of the cerebral hemispheres; (2) the diencephalon, the subcortical region that contains the thalamus and hypothalamus; (3) the mesencephalon, the midbrain region that contains the motor nuclei of the third and fourth cranial nerves; (4) the metencephalon, which later forms the pons and the cerebellum; and (5) the myelencephalon, which gives rise to the medulla oblongata [2].

Caudal to the myelencephalon, the walls of the neural tube thicken effectively reducing the lumen. The neuroepithelial cells that are lumenally positioned constitute the ventricular zone which gives rise to the cells of the spinal cord. As the neural tube continues to thicken, a shallow groove, known as the sulcus limitans, develops which separates the dorsal and ventral regions. Neuroepithelial cells of the dorsal ventricular region develop into the neurons of the dorsal horns. The ventral ventricular cells are the precursors of the motor neurons of the ventral horns and are most likely the primary source of the myelin-forming cells of the spinal cord [2] (Fig. 2B).

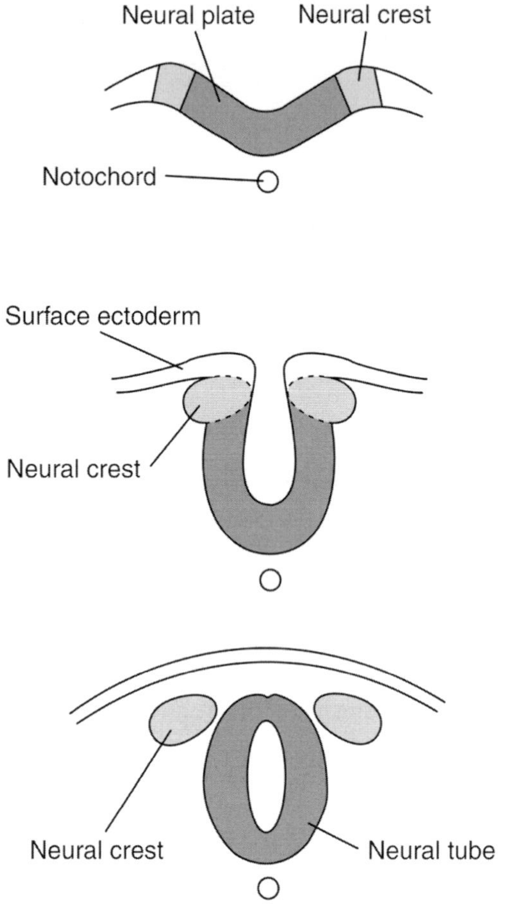

Figure 1. A thickened region of dorsal ectoderm gives rise to the neuroectoderm, which forms the neural plate. The neural plate pinches in and forms a tubular structure known as the neural tube. The neural tube is the precursor of the CNS. Prior to the closure of the neural tube, a population of cells, which originate from the lateral edges of the neural plate, migrate away from the developing tube. These cells, which constitute the neural crest, form the PNS in addition to several non-neural cell types.

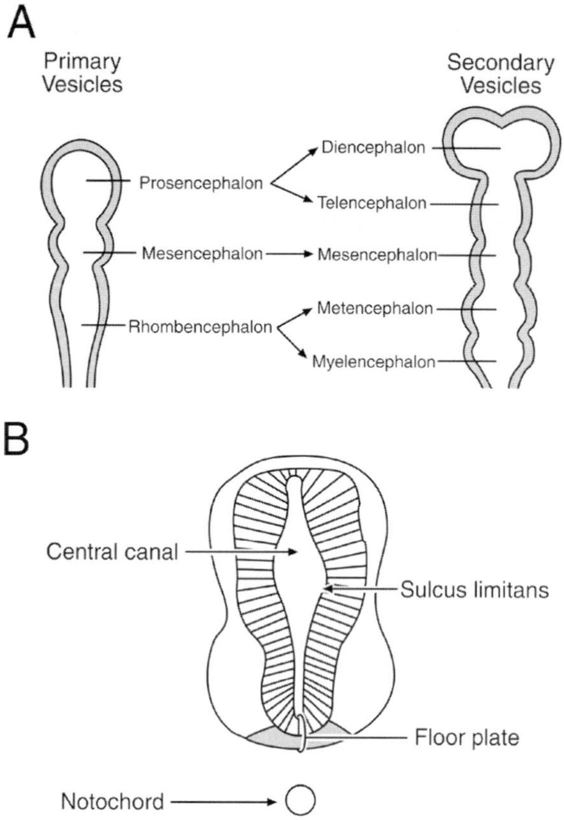

Figure 2. As the brain develops, three primary vesicles are formed that will eventually give rise to the five-vesicle brain. These five vesicles establish the five domains of the adult brain. In the developing spinal cord, the walls of the neural tube thicken producing a small groove, known as the sulcus limitans, which separates the dorsal and ventral regions of the spinal cords.

Two of the fundamental questions regarding the development of the CNS are: (1) How is the CNS patterned into functionally distinct areas? and (2) How is such great cell diversity achieved? To address the first question, investigators have employed a variety of techniques, most notably retroviral labeling. These studies have mapped the migration of individual cells and their lineages and have greatly enhanced our understanding of the mechanisms that regulate tissue and cellular organization. The second question has received a great deal of interest more recently. It is now evident that a plethora of intrinsic and extrinsic factors, whose expression is both temporally and spatially regulated, dictate patterns of cellular differentiation. Extrinsic factors are typically released from a region of specialized cells known as an organizing center. These extrinsic factors then induce, either directly or indirectly, the expression of intrinsic factors that define terminal cell fate.

At the cellular level, the CNS is composed of cell types the can be categorized as either neurons, macroglia, microglia or ependymal cells. In recent years, one of the hottest areas of neural research has been the isolation and developmental regulation of neural stem cells, cells committed to the neural lineage that are capable of self-renewal and asymmetric divisions to generate undifferentiated and differentiated neural cell populations [3]. Primarily from immature organisms, single cell populations have been repeatedly demonstrated to give rise to both neurons and macroglia, oligodendrocytes and astrocytes [4,5]. For example, when multipotent cerebral progenitors are grown in culture in the presence of platelet-derived growth factor, these cells differentiate into neurons; however, the exposure of these uncommitted cells to ciliary neurotrophic factors results in a glia phenotype [6]. The search for stem cells has not been restricted to developmentally immature organisms and several reports have shown that pluripotent cells are retained in the adult brain [7,8]. Not surprisingly, the study of stem cells has been fueled by clinical possibilities such as replacement cell therapy for the treatment of diseases such as multiple sclerosis, Parkinson's disease, Alzheimer's disease, epilepsy and stroke. Thus far, preliminary clinical applications have been extremely promising [9].

Although increasing evidence indicates that the major cell types of the CNS are of common origin and that the fate of multipotential cells is determined by environmental influences, for the sake of simplicity, the development of each of the major cell types of the CNS will be discussed independently. It must be noted that an exhaustive discussion of this rapidly changing and expanding field is outside the limitations of this chapter. Many outstanding reviews on the subject of neural tissue differentiation are available and references for some of these reviews will be given throughout the text.

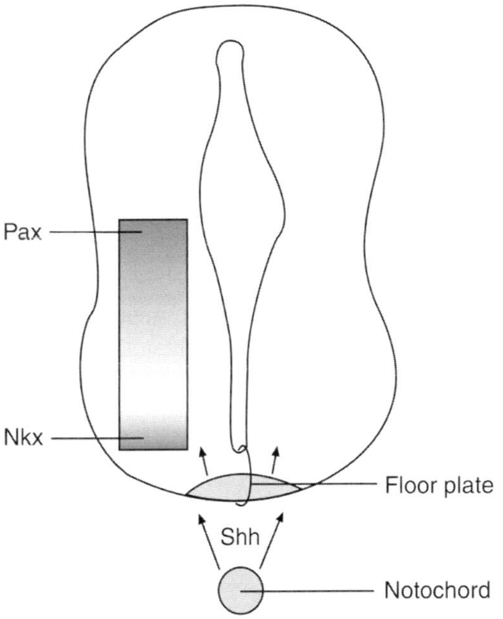

Figure 3. The notochord, an embryonic organizing center, secretes the signalling molecule Shh (arrows). Shh induces the formation of the floor plate, which in turn also secretes Shh. In the ventral region of the spinal cord, Shh establishes an activity gradient of Pax and Nkx, two homeodomain proteins that mediate the development of several classes of spinal cord neurons.

2.2. Cellular Differentiation of the CNS

2.2.1. Neuron

In general, mitotic neuronal precursors arise from the neuroectoderm of the ventricular zone of the rapidly differentiating neural tube. These precursor cells exhibit symmetric proliferation (i.e. each cell gives rise to two identical cells); however, with time the type of cellular division that dominates is asymmetric (i.e. one daughter cell initiates differentiation while the other remains a stem cell). The non-mitotic neuronal precursors then migrate to their final destination. In the cortex, neurons migrate from the ventricular zone toward the cerebral wall and establish the neuronal layers of the cortex in an inside–outside manner. Migration of the neurons to their correct cortical destination is directed by radial glial cells and the interactions between the migratory cells and the radial glia are mediated by an ever-growing list of extracellular matrix and cell adhesion molecules. Recently, evidence has also indicated that non-radial (tangential) migration occurs [10]. Additional opportunities for phenotype refining, such as non-radial migration, may be critical in achieving the cellular diversity observed in the CNS.

Compelling evidence indicates that neuronal maturation is directed by a combination of extrinsic and intrinsic factors. Perhaps one of the best characterized examples of terminal differentiation by a combination of extrinsic and intrinsic factors is the differentiation of neuronal subclasses of the ventral spinal cord and hindbrain. Briefly, the notochord, an embryonic organizing center, induces the formation of the floor plate, the ventral most region of the neural tube and a structure that is essential for the development of the spinal cord and its resident cells (Fig. 3). The signaling molecule that regulates this inductive event is sonic hedgehog (Shh). In the absence of Shh activity, either through gene disruption or antibody perturbation, both the floor plate and the spinal cord neurons fail to develop [11].

With time, floor plate cells also express Shh and are critical in establishing a ventral/dorsal Shh activity gradient in the developing cord. Based on this gradient, developing neural tube cells can be divided into several subclasses of neurons. Strong evidence indicates that the homeodomain proteins, members of the Pax and Nkx families, are intermediate signal molecules that are responsible for interpreting the Shh gradient. Both Pax and Nkx transcription factors are expressed by mitotic neural tube cells; however, the activity pattern of these factors is complementary. Pax activity, which is repressed by Shh, is highest in the dorsal region, whereas the activity of Nkx is restricted to the ventral region. Nkx activity is not directly affected by Shh but is inhibited by the Pax family. Thus Shh indirectly regulates the functional distribution of Nkx. In turn, these transcription factor families induce the expression of specific sets of proteins that define distinct neuronal populations along the ventral/dorsal axis [12–14].

2.2.2. *Oligodendrocyte*

Retroviral labeling studies in the brain have shown that oligodendrocytes originate from the subventricular zone, which lies beneath the lateral ventricles; however, the source of spinal cord oligodendrocyte precursors remains a subject of debate. The predominant view is that these cells develop from neuroepithelial cells from the ventral half of the embryonic neural tube. Elegant studies by Miller and colleagues [15,16] have demonstrated that oligodendrocytes develop in culture from ventral but not dorsal ventricular neuroepithelial cells. Further evidence was provided by in situ hybridization labeling of early glial lineage cells. Using a probe directed against an early oligodendrocyte marker, platelet-derived growth factor receptor, only cells located on the lumenal surface of the ventral half of the cord revealed expression [17,18]. In contrast, evidence has been reported that suggests that dorsal cells may also produce oligodendrocytes [19], but the possibility of an alternative source for oligodendrocytes has not been widely accepted. Recent articles have also suggested that multiple oligodendrocyte lineages are present in both the spinal cord and the forebrain. The primary evidence for this proposal is based on the early appearance of proteolipid protein (PLP)}/DM-20 positive cells in the spinal cord [20] and the forebrain [21]. An alternative explanation of these findings is that these cells are of the same lineage as the general population of oligodendrocytes, but at a different stage of development.

The majority of our understanding of the developmental lineage of oligodendrocytes has resulted from extensive in vitro analyses. Embryonic and postnatal glial precursors in culture are immunoreactive against the monoclonal antibody A2B5, which binds a specific ganglioside that is expressed by neurons and oligodendrocyte precursors. In culture, these cells are bipolar and are capable of developing into either Type 2 astrocytes or oligodendrocytes. Based on this bipotentiality, these cells have been termed oligodendrocyte-Type 2 astrocyte (O-2A) progenitors. Whether O2-A progenitors develop into oligodendrocytes or in Type 2 astrocytes depends on the culturing conditions. O-2A cells grown in the presence of ciliary neurotrophic factor will differentiate into Type 2 astrocytes whereas the default pathway is the oligodendrocyte lineage [22]. Although the presence of Type 2 astrocytes in vitro is well documented, their existence in vivo is less clear. A recent report, however, demonstrated the presence of a cell type positive for A2B5 and the astrocytic marker glial fibrillary acidic protein (GFAP) [23].

As oligodendrocytes mature, they become less motile and more process bearing. In addition to changes in their physical characteristics, the progressive states of oligodendrocyte differentiation are defined through the expression of distinct cell surface markers. A2B5+ O-2A cells also express the NG2 glycoprotein and the poorly characterized pro-oligodendrocyte antigen recognized by the monoclonal antibody known as 04. As these pro-oligodendrocytes continue to mature, they lose their immunoreactivity for these markers and differentiate into oligodendrocytes that first express galactocerebroside, the major glycolipid of myelin, and ultimately terminal myelin protein markers including myelin basic protein (MBP) and PLP.

Using stage-specific markers, oligodendrocyte differentiation has been well documented; however, the mechanisms that regulate this process have remained elusive. Several transcription factors have been identified in oligodendrocyte precursor cells, such as CREB, Gxt, SCIP/tst-1/Oct6, Brn-1 and Brn-2 and Myt-1, but their precise roles in oligodendrocyte development have not been fully elucidated

[24]. Interestingly, Shh has also been identified as a potential developmental inducer of oligodendrocytes. In support of this hypothesis, the addition of antibodies directed against Shh into the spinal cord prevents the emergence of oligodendrocyte precursors [25]. Furthermore, Shh is capable of inducing oligodendrocyte precursors from cells that do not typically enter the glial lineage [26,27]. Although it is not completely understood how a single signaling molecule induces distinct cell types from a common precursor, it is proposed that induction is concentration dependent and may rely on the presence of other factors working in synchrony with Shh. A molecule that may function in such a capacity is neuregulin. Neuregulin is required for the survival and differentiation of oligodendrocytes and its expression temporally correlates with the appearance of oligodendrocyte precursors, but not with the initial emergence of motor neurons [28–30].

2.2.3. Astrocyte

The other member of the macroglial family is the astrocyte. It appears that astrocytes originate in at least two sources. The first is the ventricular zone. Cells from this region appear to develop into astrocytes by forming an intermediate known as radial glial cells. Radial glial cells, so named because of their long processes that guide migratory neurons, first appear embryonically and largely disappear with the completion of neurogenesis. The evidence that implicates radial glial cells as astrocyte precursors is based primarily on the analysis of transitional stages between known radial glial cells and astrocytes [31]. The other possible source of astrocytes is the germinal or subventricular zone. Based on retroviral tracing studies, both astrocytes and oligodendrocytes have been demonstrated to arise from this zone. Taken together, strong evidence exists that two distinct genealogies provide the CNS sources of astrocytes.

Much of the work regarding astrocyte development has been performed in vitro. These studies have also defined two distinct cell lineages known as Type 1 and Type 2 astrocytes. Type 1 astrocytes exhibit a fibroblast-like morphology and are not immunoreactive against the A2B5 antibody. Conversely, Type 2 astrocytes have a neuron-like morphology and bind both tetanus toxin and the A2B5 antibody. Both cell types were found in cultures established from white matter, but only Type 1 astrocytes were present in cultures from gray matter [32,33].

Although these astrocyte lineages are clearly defined in culture, correlating the in vitro astrocytes with in vivo astrocytes has been problematic. Based on the timing of their appearance and antigenicity, Type 1 astrocytes may correspond to the astrocytes of the optic nerve that are derived from radial glial cells. In addition, it has also been proposed that Type 1 astrocytes correspond to protoplasmic astrocytes, an in vivo classification based on the cellular distribution and the level of GFAP expression, while the in vivo counterpart for Type 2 astrocytes is the fibrous astrocyte. Although this correlation is tempting, there is no solid evidence to corroborate this conclusion. Furthermore, attempts to identify the O-2A cell in vivo have met with little success. For example, brain sections from regions known to produce O-2A cells in culture were immunocytochemically labeled for GFAP and GD3, a marker of neuroectodermal cells that give rise to oligodendrocytes. No single cell expressing both of these epitopes was found suggesting that O-2A cells do not exist in vivo. It should be pointed out, however, that studies that rely on the co-expression of specific markers for cell-type identification are problematic since the markers may not have overlapping periods of expression. Therefore, even if a single cell expressed the correct combination of markers over time, the cell would not be identified using such this experimental approach. Regardless of the reason, current correlative data do not exist that match the in vitro astrocytes with their in vivo counterparts [31].

2.2.4. Microglia

The developmental origin of microglia remains a highly debated issue. The most widely accepted hypothesis proposes that mesodermal cells derived from the bone marrow enter the brain during embryonic development as undifferentiated monocytic ameboid cells. The evidence for this theory is the antigenic and functional similarities between brain microglia and peripheral macrophages. Also, the dramatic increase in microglia temporally correlates with the disappearance of the mesodermal ameboid cells; however, transitional forms between these two cell types have not been observed [34]. In contrast, evidence has been presented that suggests that microglia may have non-mesodermal origins. Bone marrow transplant studies have indicated that only a small portion of the resident brain microglia have mesodermal origins [35]. In addition, the number of microglia increases significantly with developmental growth of the brain. Since it is unlikely that the differentiated bone marrow-derived cells retain proliferative capabilities that could account for this increase, investigators have hypothesized that the highly proliferative neuroepithelial cells may be responsible for the microglial population increase [36]. Moreover, cultured astrocyte precursors, isolated from neonatal mouse brain, gave rise to both astrocytes and microglia. Since astrocytes originate from neuroectoderm, it was concluded that microglia also develop from neuroepithelial cells [37]. Collectively, these data suggest that brain microglial cells may arise both from mesoderm and ectoderm.

2.3. Gross Anatomic Development of the PNS

2.3.1. Neural crest

The PNS originates from a population of multipotent ectodermal cells known as the neural crest. These cells dissociate from the edges of the neural plate and reside dorsolateral to the neural tube (Fig. 2). The neural crest cells, which are highly motile, migrate in either a lateral or ventral direction. The laterally directed crest cells give rise to non-neural cells while the ventrally directed cells produce the spinal and sympathetic ganglia of the PNS. Presently, the mechanisms that regulate the migratory path of specific crest cells are not clear, but appear to involve environmental cues [1,38].

2.4. Cellular Differentiation of the PNS

2.4.1. Neuron

Similar to the CNS, functional populations of PNS neurons are established by specific environmental and intrinsic factors. This is illustrated by the differentiation of the two major nerve cell classes of the PNS: the sensory and autonomic neurons. The development of sensory neurons is dependent on neurogenin. Neurogenin is a basic helix–loop–helix transcription factor whose ectopic expression converts ectodermal cells into sensory neurons, while in the absence of neurogenin, sensory neurons fail to develop [39]. The development of the autonomic neurons is dependent on the expression of mammalian iced-scute homolog-1 (MASH-1), also a helix–loop–helix transcription factor. The expression of MASH-1, an intrinsic factor, is regulated by the extrinsic factors termed bone morphogenetic protein 2 and 4, which are secreted by dorsal aorta tissue. Such findings suggest that as undetermined neural crest cells migrate to peripheral locations, locally released factors specify their developmental fates. Although MASH-1 is required for the terminal differentiation of neural crest cells into autonomic neurons, MASH-1 alone is not sufficient for commitment to a neuronal fate. Therefore, development of PNS neurons is similar to that of their CNS counterparts as the interaction of multiple factors are required for appropriate terminal differentiation. (For more details, see Anderson and Jan [40] and Anderson et al. [41].)

2.4.2. Schwann cell

The glial cell of the PNS is the Schwann cell. Although the best known function of the Schwann cell is the formation of myelin, a large portion of Schwann cells do not form this highly specialized multilayer sheath. Therefore mature Schwann cells are typically classified as either myelinating or non-myelinating (also referred to as ensheathing) cells. Besides the differences in their myelin-forming capacity, these two populations of Schwann cells can also be distinguished by specific markers: non-myelinating cells express GFAP, p75NGF receptor and L1; myelinating Schwann cells express myelin proteins, such as MBP, protein 0 (P0) and myelin-associated glycoprotein (MAG). The mechanisms that determine which phenotype a particular cell exhibits are not clear, but substantial evidence indicates that both intrinsic and extrinsic factors are involved [42]. Surprisingly, the effect of extrinsic factors on the fate determination of Schwann cells was demonstrated almost 25 years ago. By cross-anastomosing myelinated and non-myelinated nerves, Aguayo et al. [43,44] demonstrated that a Schwann cell could switch phenotypes depending on the type of axon with which the cell associates.

As previously noted, Schwann cells, as well as the majority of the cells of the PNS, are derived from

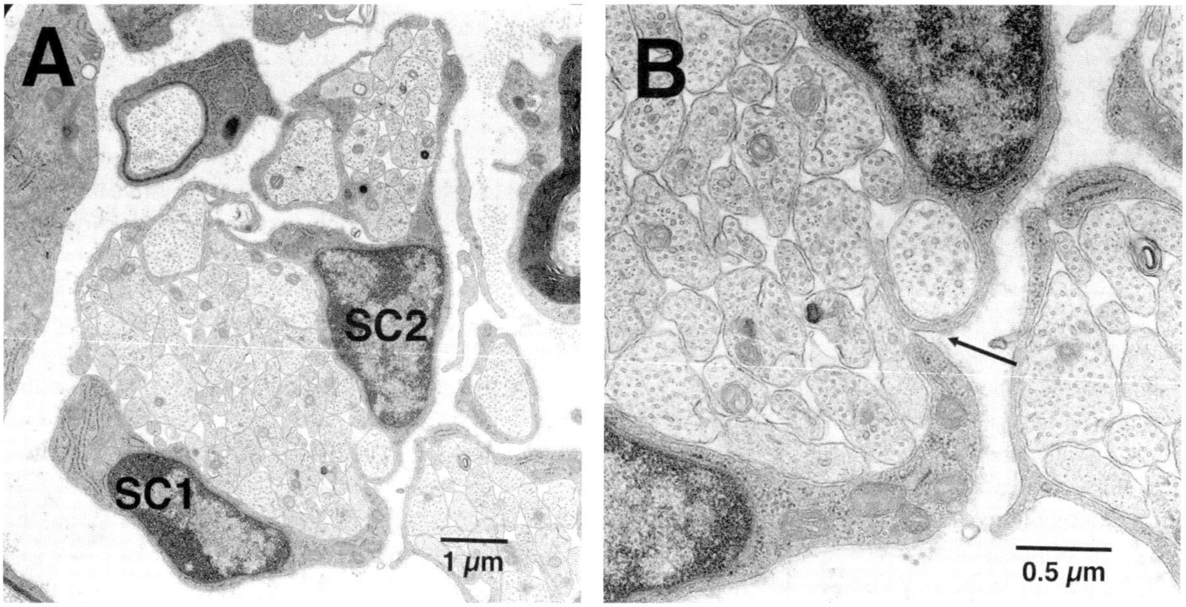

Figure 4. Early in development of the peripheral nerves, naked axons are ensheathed by multiple Schwann cells (A, SC1, SC2) in a common basal lamina (B, arrow). (B) The Schwann cells extend processes that subdivide the axon bundle into smaller bundles and eventually into isolated axons for myelination.

the neural crest. Highly motile and proliferative glial lineage cells migrate out of the neural crest along pre-existing axonal tracts and begin to invest axons. Little information is currently available concerning the differentiation of these neural crest cells into the Schwann cell precursor. The recent finding that P0, a protein previously believed to be restricted to mature myelinating Schwann cells, is an early marker for neural crest cells that have entered the glial lineage may prove to be a valuable tool in deciphering the cascade of events involved in the differentiation of Schwann cells. (For more information concerning the potential role of P0 in PNS gliogenesis, see Sommer and Suter [45]). One extrinsic factor that has been implicated in the differentiation and survival of the neural crest cell is β-neuregulin. Neuronally expressed β-neuregulin meets the temporal and spatial qualifications for neural crest differentiation and in the absence of this protein, both Schwann cells and their precursors fail to develop [42].

In addition to extrinsic factors, several intrinsic factors have been identified that regulate Schwann cell differentiation. Primarily these intrinsic signals are transcription factors, which have also served as developmental stage and phenotype markers. Immature and non-myelinating Schwann cells express Pax-3 and Krox-24, while myelinating lineage Schwann cells express Oct-6/SCIP/tst-1 and Krox-20/Egr2. Early in development, all cells of the Schwann cell lineage express Pax-3 and Krox-24. In myelinating Schwann cells, the expression of these proteins is downregulated during the periods just before active myelination and only returns following dedifferentiation (e.g. axotomy). Functionally, Pax-3 and Krox-24 have been implicated in the promoter activation of non-myelinating Schwann cell-specific genes and the repression of MBP expression. Conversely, mice that lack functional Oct-6/SCIP/tst-1 or Krox-20/Egr2 have a transient reduction or absence of myelinating Schwann cells, respectively. Although it appears evident that Oct-6/SCIP/tst-1 and Krox-20/EGR2 are important in the regulation of myelinating Schwann cells, the target genes of these transcription factors have not been identified [26,46,47].

Regardless of their terminal phenotype, all Schwann cells initially envelop numerous naked axons (Fig. 4). These axons are separated by Schwann cell processes. Interestingly, these original bundles of axons may be ensheathed by processes from multiple Schwann cells within a common basal lamina. At this immature stage of PNS development, the Schwann cells are highly mitotic resulting in a dramatic increase in the population of these cells. With more cells extending processes, the original axon bundles are further subdivided with the larger axons ultimately being segregated into a 1 : 1 relationship with a Schwann cell. These large, isolated axons will be myelinated. In contrast, the smaller axons will remain in bundles surrounded by one process from a single Schwann cell. These Schwann cells are the non-myelinating or ensheathing cells [48]. As previously stated, the mechanism that dictates which Schwann cell becomes an ensheathing cell and which becomes a myelin-forming cell has not been elucidated, but for almost a quarter of a century strong evidence has demonstrated that the axon, not the Schwann cell, regulates this decision [43,44].

Following nerve injury, Schwann cells can reverse their differentiation and regain their mitotic capabilities. The primary function of these dedifferentiated cells seems to be the formation of an environment conducive to axonal outgrowth and remyelination. In support of this role, immature Schwann cells express high levels of several neurotrophic factors, including nerve growth factor and adhesion molecules and receptors known to facilitate axonal outgrowth [47]. Furthermore, the absence of myelin and mature myelin-forming cells may enhance the permissive environment since myelin has been demonstrated to inhibit axonal outgrowth [49].

3. CELL MORPHOLOGY AND FUNCTION

3.1. Neuron

The cell type responsible for capturing, transforming and conveying neural information within the organism is the nerve cell or neuron. In order to efficiently achieve these tasks, neurons have developed into a highly specialized and diverse group of cells. Although neurons are highly diverse, all nerve cells have several traits in common. All neurons: (1) contain a cell body; (2) have at least one process that extends from the cell body; and (3) have the capacity to transmit electrical impulses. Due to the highly diverse nature of these cells, neurons are classified in a variety of ways based on function, location and neurotransmitter expression. One of the most common classification schemes is based on morphology. Using this criteria for classification, there are three types of neurons: uni- or pseudounipolar, bipolar and multipolar neurons (Fig. 5A). Uni- or pseudounipolar neurons have a single process that bifurcates. This group of cells include the neurons of dorsal root and cranial nerve ganglia and the neurons of the mesencephalic nucleus of the trigeminal nerve. Bipolar neurons are restricted to the retina, the olfactory epithelium and the ganglia of the eighth cranial nerve. The last group of neurons is the largest and includes the stereotypic stellate nerve cell. These neurons have a large cell body, multiple dendrites and an axon [50].

In general, neurons are comprised of three cellular domains known as the cell body, the axon and dendrites (Fig. 5). Each subcellular region can be distinguished both by its structural and functional specializations that are essential for maintaining the function of the nerve cell.

3.1.1. Cell body

The cell body, also known as the soma or perikaryon, contains the nucleus and many of the organelles (Fig. 6). Typically the nucleus is large, round, centrally located, contains euchromatic DNA, which is indicative of highly active cells, and displays a prominent nucleolus. The cytoplasm of the cell body contains Nissl substance, a light microscopic term. With the advent of the electron microscope, Nissl substance was identified as rough endoplasmic reticulum (RER). The dramatic amount of Nissl substance is in accordance with the high synthetic demands on the neuron. Another prominent organelle in the neuron is the Golgi apparatus, a highly heterogenous organelle that functions in the segregation and assembly of secretory proteins and the formation of lysosomes. Prominent throughout the cytoplasm are numerous mitochondria that are highly variant in size and shape. Also in the cell body cytoplasm is a network of cytoskeletal elements — intermediate filaments, known as neurofilaments, microtubules and microfilaments (Fig. 7). The neurofilaments and microtubules will be discussed in greater detail later in this chapter. Also common to many neurons is the presence of peroxisomes and lysosomes for intracellular digestion; lipofuscin granules, probable lysosomal derivatives that accumulate with age; and neuromelanin pigment granules [50].

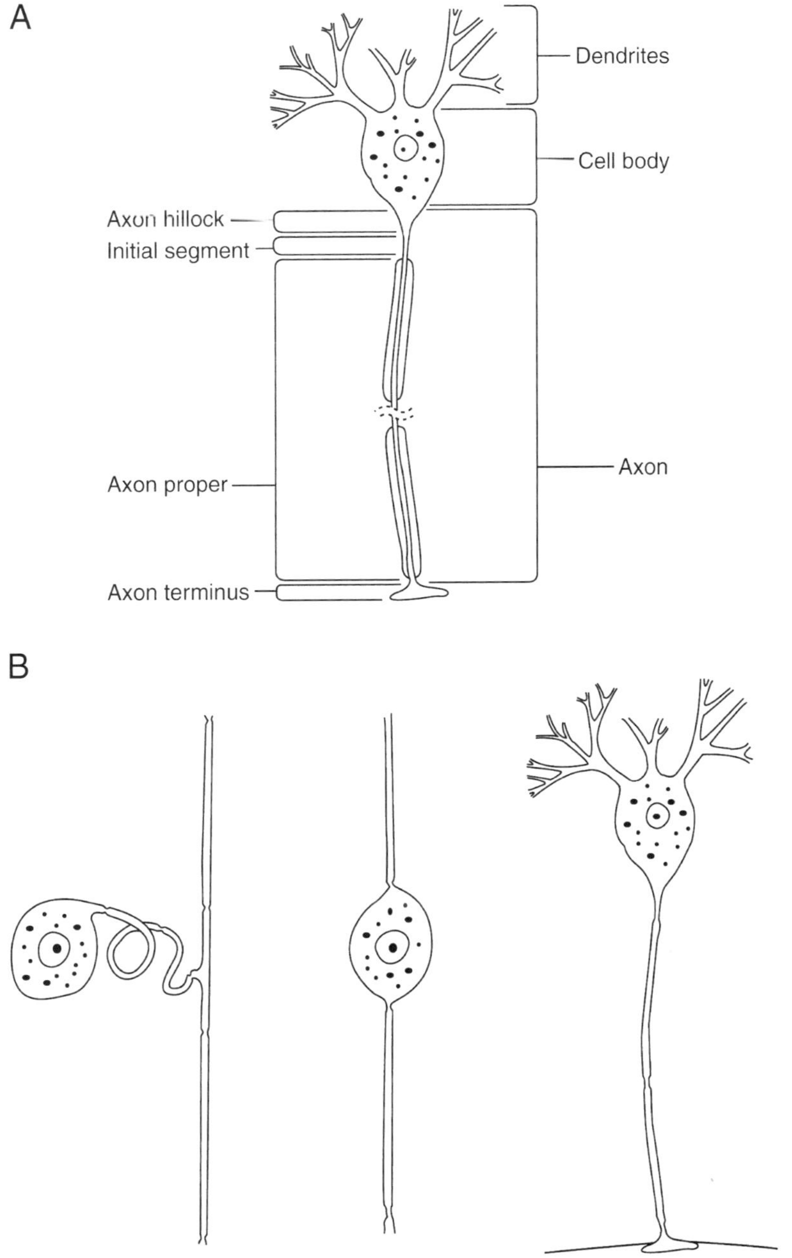

Figure 5. The typical neuron contains multiple branched dendrites, a cell body and an axon. (A) The four regions of the axon are labeled: axon hillock, initial segment, axon proper and axon terminus. (B) the three basic neuronal morphologies, pseudounipolar, bipolar, and multipolar, are illustrated.

3.1.2. Dendrite

The primary function of the dendrite is to receive signaling input from other neurons and to conduct that information to the cell body. Dendrites are highly branched with each daughter process displaying a reduction in caliber. Unlike axons, dendrites have few neurofilaments and contain some RER; however, using these classic parameters to distinguish between axons and dendrites is becoming increasingly more unreliable. Dendrites can be definitively distinguished from their efferent counterpart by the pres-

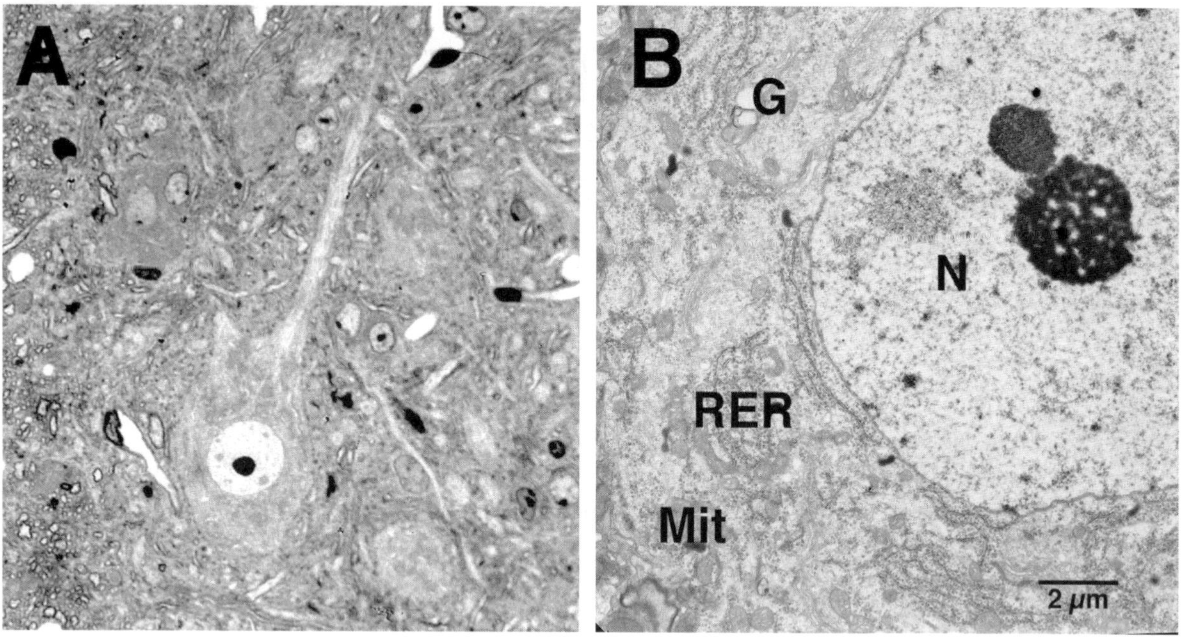

Figure 6. (A) The neuron displays a large centrally positioned nucleus with a prominent nucleolus. The dark regions of cytoplasm correspond to the Nissl substance. (B) An electron micrograph of a spinal cord neuron that displays several organelles common to neurons: rough endoplasmic reticulum (RER), mitochondria (Mit), and Golgi apparatus (G).

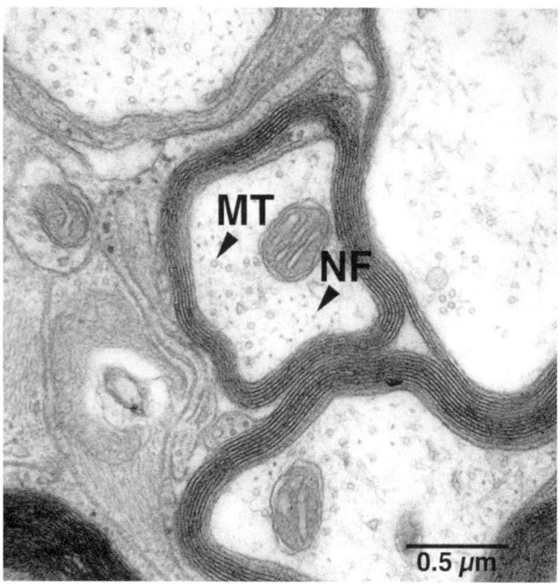

Figure 7. A high magnification electron micrograph of myelinated axons from the spinal cord displays both neurofilaments (NF) and microtubules (MT).

ence of microtubule-associated protein 2 (MAP2), a dendrite specific marker [51]. Dendrites are also covered with small protrusions known as dendritic spines. These spines significantly increase the surface area of the dendritic tree, thus increasing the potential for synaptic input. Interestingly, the number of these spines appears to directly correlate with stages of development and aging [50].

3.1.3. Axon

Nerve cells have a single axon, a cylindrical tube whose primary function is to conduct electrical impulses away from the cell body. The axon, invested by a specialized membrane known as the axolemma, can be divided into four regions: the axon hillock, the initial segment, the axon terminus and the axon proper. The axon hillock is the cone shaped portion of the cell body that contains little to no RER. In the initial segment, the axolemma contains a high concentration of sodium channels and it is in this region that the action potential is initiated. Ultrastructural analysis of the initial segment has revealed the presence of a dense granular arrangement of cytoskeletal elements under the axolemma and the presence of a few ribosomes. The most distal region of the axon is the axon terminus. This region is an unmyelinated, enlarged portion of the axon that is also called the terminal bouton [50]. The terminal bouton contains mitochondria, synaptic vesicles, microfilaments and a reduced number of neurofilaments and microtubules that do not usually extend beyond the center of the terminus [52]. Although boutons are typically located at the distal end of the axon, these specialized structures can also be positioned along the length of the axon where they are called en passant boutons. Boutons form a functional contact, or synapse, with other cells and release either chemical or electrical signals that modify the activity of the receiving cell. A more detailed discussion of the structure and function of the synapse will be presented later in the chapter.

The final region of the axon is the axon proper. Organelles in this region include mitochondria, lysosomes, smooth endoplasmic reticulum and all three members of the cytoskeleton. It is important to note that, historically, organelles involved in protein synthesis were generally reported to be excluded from the axon proper. This prevailing idea resulted from the inability to detect either RER by light microscopy or ribosomes by electron microscopy. Recent studies, however, report both biochemical and morphological data that demonstrates that this classic view of the axon needs to be re-examined [53,54].

An extremely prominent organelle of the axon proper is the neuronal cytoskeleton, which is composed primarily of microfilaments, neurofilaments and microtubules (Fig. 7). The microfilaments, of which the most prominent member is actin, are concentrated beneath the axolemma and are believed to play a role in cell shape and motility, axonal outgrowth, neurotransmitter release, intracellular trafficking and membrane protein distribution [55]. The second member of the neuronal cytoskeleton is actually a rapidly expanding family of proteins known as the neurofilaments. There are at least six members of this family but the most well known and characterized of this group is the neurofilament triplet protein. The three subunits of the triplet protein have apparent molecular masses of 212, 160 and 68 kDa and are referred to as NF-H, NF-M and NF-L, respectively [56]. Assemblage of these polypeptides results in a highly stable protein that undoubtedly is critical in maintaining axonal morphology. In addition, the neurofilament triplet protein has sidearms formed by the carboxyl termini of NF-H and NF-M. The sidearms contain phosphorylation sites and it is believed that neurofilament phosphorylation is a critical factor in determining axon caliber [57].

Microtubules, the third member of the neuronal cytoskeleton, are heterodimers of α- and β-tubulin. Microtubules are accompanied by a class of proteins known as the microtubule-associated proteins (MAPs), which assist in the assembly and disassembly of microtubules. Although MAPs are widely distributed in non-neural tissue, the expression of some MAPs is neuronally restricted. As previously mentioned, MAP2 is only found in dendrites while tau family proteins are located in the axon. Over the past decade, tau has received considerable attention since it is a major constituent of the pathological structure known as paired helical filament found in the CNS of patients suffering from Alzheimer's disease [58]. Similar to the neurofilaments, microtubules play an important role in determining axon size and morphology. Of equal importance, however, is the role that microtubules play in the specialized neuronal intracellular transport known as axonal transport.

Since the vast majority of neuronal proteins is synthesized in the cell body, a highly effective means of long distance intracellular protein transport is essential for the growth, maintenance and function of the axon. The term used to describe this intraneuronal trafficking is axonal transport. Axonal transport consists of a fast and a slow component. The fast component is primarily responsible for the movement of membrane-associated proteins and lipids and mitochondria. The movement of these macromolecules and organelles away from the cell is termed anterograde transport while movement in the opposite direction is referred to as retrograde transport. Utilizing microtubules as the conduit for transport, specific molecular motors, dyneins and kinesins, mediate retrograde and anterograde transport, respectively, at rates ranging between 50 and 400 mm/day. Whereas anterograde transport is primarily responsible for the transfer of membrane-associated macromolecules, such as metabolic enzymes, neurotransmitters and axolemmal proteins and lipids, away from the cell body to the four regions of the axon, retrograde transport shuttles lysosomal vesicles and exogenous material such as growth factors back to the cell body. In contrast, the slow component of axonal transport trafficks only in the anterograde direction. In slow component a, neurofilaments and microtubules move at a rate of 0.1–1 mm/day while actin and cytoplasmic proteins are transported in slow component b at a rate of 5–10 times faster [56].

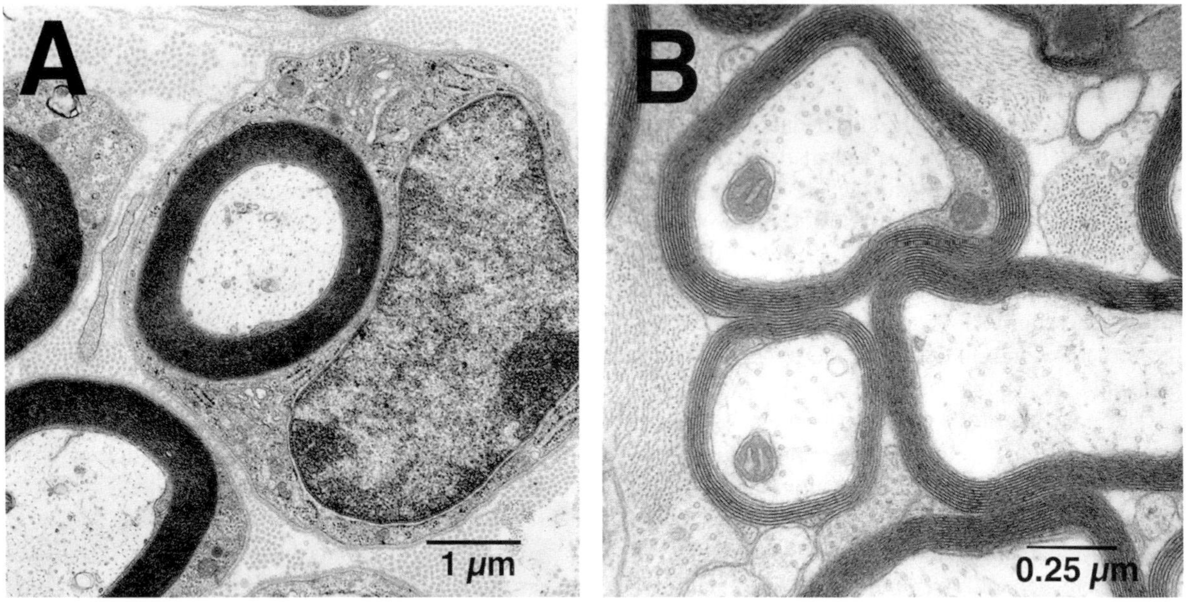

Figure 8. The axon proper is myelinated in both the PNS (A) and the CNS (B). Although PNS and CNS myelin share many simularities, PNS myelin is generally thicker and is formed by Schwann cells, whereas CNS myelin is formed by the less conspicuous oligodendrocytes.

A unique feature of the axon proper is the presence of a myelin sheath (Fig. 8), although it is important to remember that unmyelinated axons make up a large portion of the axonal population in both the peripheral and central nervous systems. Myelin, which will be addressed in greater detail later in this chapter, subdivides the axon proper into three functionally and structurally distinct domains known as the internode, the paranode and the node of Ranvier. It is at the node of Ranvier that sodium channels are highly concentrated in the axolemma, enabling the efficient and rapid transmission of electrical impulses by saltatory conduction. In the paranode, the region juxtaposed to the node, the myelin sheath and axon establish their closest apposition and it is therefore not surprising that cell adhesion molecules appear to be concentrated in this portion of the axolemma and the corresponding region of the myelin sheath [59].

Although neurons may be the cell type that classically represents the CNS, they are far outnumbered by the macroglia — astrocytes and oligodendrocytes. As previously discussed, astrocytes and oligodendrocytes may be developmentally related, but they are morphologically and functionally distinct cell types.

3.2. Astrocyte

The term astrocyte was derived from the star-like appearance of these cells in early light microscopic preparations (Fig. 9). These cells are classically categorized as either fibrous or protoplasmic astrocytes. Fibrous astrocytes are located in the white matter, and extend numerous, infrequently branched pro-

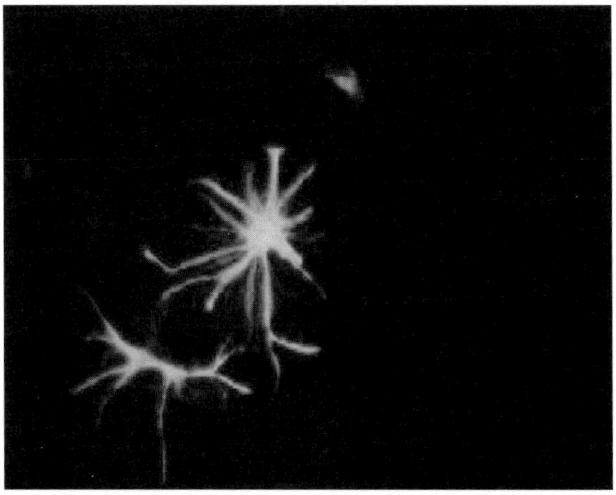

Figure 9. Cultured astrocytes, intensely labeled by a polyclonal GFAP antibody, display the stereotypic star-like shape that is responsible for the naming of this cell type.

cesses. These cells possess an irregularly shaped nucleus and the typical complement of organelles including RER, Golgi apparatus, elongated mitochondria, lysosomes and peroxisomes. In addition, fibrous astrocytes contain numerous glycogen granules and bundles of intermediate filaments. The class of intermediate filaments found in astrocytes is known as GFAP. Unlike neurofilaments, GFAP forms tightly packed bundles. GFAP is antigenically distinct from other intermediate filaments and has therefore been an extremely useful marker for astrocyte identification. Protoplasmic astrocytes are morphologically similar to their fibrous counterparts with the exception that the protoplasmic cells reside in the gray matter, possess fewer bundles of GFAP and radiate highly branched processes [31,50].

Numerous functions for astrocytes have been proposed and an in-depth discussion of these functions is out of the scope of this chapter. The main functions of astrocytes is to provide structural and trophic support for cells of the nervous system and assist in migration and axonal outgrowth of developing neurons. In addition, astrocytic processes: (1) define the limits of the CNS by forming the glia limitans; (2) contact blood vessels and may play an active role in the formation of the blood–brain barrier; and (3) surround the node of Ranvier in an attempt to prevent erroneous stimulation of closely apposed neurons [50]. Astrocytes have also been implicated in neurotransmitter uptake and ion buffering. Finally, astrocytes, in conjunction with microglia, may play a role in the recruitment of inflammatory cells, assist in the regulation of T-helper cell response and under certain rare pathological conditions participate in the presentation of antigen [60].

Under pathological conditions, whether induced by injury or disease, astrocytes respond in a complex, yet highly predictable, manner termed astrogliosis. Reactive astrocytes become mitotic and exhibit an increase in nuclear diameter, enzymatic activity and protein synthesis. Although the mechanism that triggers this response is not clearly understood, the purpose of this reaction appears to be to facilitate the astrocytic monitoring and regulation of extracellular concentrations of ions, growth factors and waste products.

3.3. Oligodendrocyte

The second member of the macroglia family is the oligodendrocyte. The term oligodendrocyte originated from early structural studies that revealed that these cells produced a few, highly delicate processes (Fig. 10). Ultrastructurally, the cell body proper of oligodendrocytes is small and contains a round heterochromatic nucleus. The nucleus is typically surrounded by a thin rim of cytoplasm that contains a well-developed RER, mitochondria and Golgi apparatus. The cytoplasm is rather dense, which has been attributed to the presence of numerous free ribosomes and other small granules whose function is currently unknown. Microtubules are present; however, intermediate filaments are absent [50]. Oligodendrocytes can also be distinguished from other cells of the CNS based on the expression of numerous cell specific markers such as MBP, PLP, MAG and galactocerebroside. The function of these and other myelin-specific molecules will be discussed later in the chapter.

Oligodendroglia, a termed used interchangeably with oligodendrocytes, make up a very heterogeneous group. Originally, del Rio Hortega [61] cat-

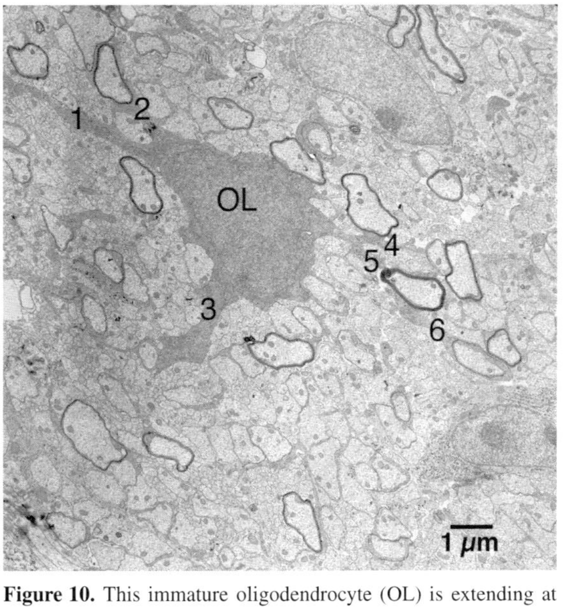

Figure 10. This immature oligodendrocyte (OL) is extending at least six (numbered 1–6) myelin-forming processes. Based on the classification of Mori and Leblon [62], this cell is most likely a large light oligodendrocyte since it displays abundant, pale staining cytoplasm is actively forming myelin.

egorized oligodendrocytes into four groups based on morphology and distribution. Currently, oligodendrocytes are more routinely referred to as either interfascicular or satellite cells. This classification is not based on morphologic differences, but strictly refers to their distribution. The interfascicular cells are found in the white matter, whereas the satellite cells are located in the gray matter. Another system of classification that should be mentioned is that of Mori and Leblond [62]. Based on cell size and electron density, these authors described three cell phenotypes: light, medium and dark. The light cells are the largest and the dark cells are the smallest. These three phenotypes probably do not define three distinct types of oligodendrocytes, but represent different stages of oligodendrocyte development with the large light cells depicting less mature cells that are actively synthesizing myelin while the small dark cells are mature, myelin-maintenance cells.

Functionally, oligodendrocytes are best known for their role in the formation of CNS myelin. Oligodendrocytes are capable of extending more than 30 membranous processes that contact and spiral around axonal segments. Oligodendrocytes also appear to play a role in neuronal nutrition and maintenance. The original evidence for such roles came from ultrastructural observations that oligodendrocytes were frequent satellites of neurons [50]. More recently, several laboratories have demonstrated the need for proper intercellular interactions between neurons and oligodendrocytes to maintain proper neuronal function [63].

3.4. Schwann Cell

Oligodendrocytes are not the only cell type that forms myelin sheaths; the other such cell type is the Schwann cell. Whereas oligodendrocytes form CNS myelin, Schwann cells form the myelin of the PNS. Although both oligodendrocytes and Schwann cells form myelin, there are differences between these cells that have significant clinical implications. First, a Schwann cell only myelinates a single axonal segment, whereas an oligodendrocyte may myelinate multiple segments on more than one axon. This distinction bears clinical significance since the loss of a few Schwann cells, perhaps due to injury or disease, would result in limited demyelination. In contrast, the loss of an equal number of oligodendrocytes could result in a dramatic increase in the loss of myelin, which could have catastrophic effects on neuronal function. Secondly, Schwann cells form a basal lamina, an extracellular matrix that surrounds the entire Schwann cell-axon unit. The basal lamina provides a permissive environment for neurite outgrowth and may facilitate axon regeneration [48,64]. Conversely, a glial basal lamina is absent in the CNS and axonal regeneration is far less successful.

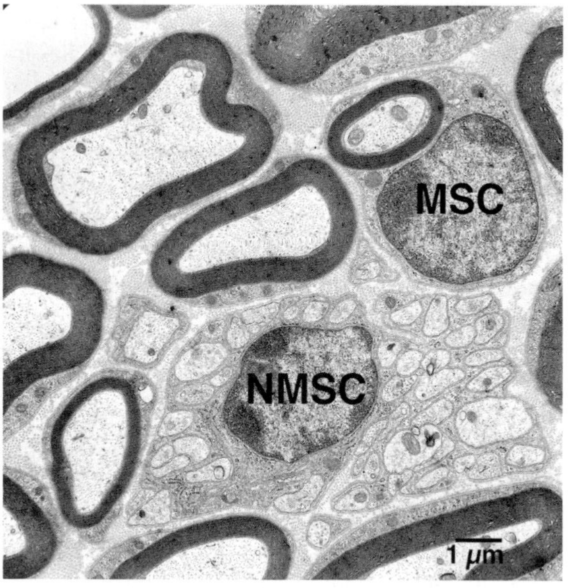

Figure 11. The two distinct phenotypes of Schwann cells are depicted in this figure: myelinating Schwann cells (MSC) and non-myelinating Schwann cells (NMSC).

Schwann cells are readily divided into two phenotypes: cells that form myelin and those that do not, commonly referred to as non-myelinating or ensheathing cells (see Fig. 11). Similar to oligodendrocytes, myelinating Schwann cells express myelin-specific molecules such as MBP, MAG, and galactocerebroside. Myelinating Schwann cells also express P0, a protein that is primarily restricted to PNS myelin [45]. Non-myelinating Schwann cells do not express most of the myelin specific markers but do express GFAP, an intermediate filament protein primarily expressed by astrocytes. Typically, myelinating Schwann cells form myelin around axons with a diameter greater than 1 μm. In contrast, ensheathing Schwann cells engulf a bundle of axons with diameters less than 1 μm.

In addition to the formation of myelin, Schwann cells synthesize collagen, which may assist in sup-

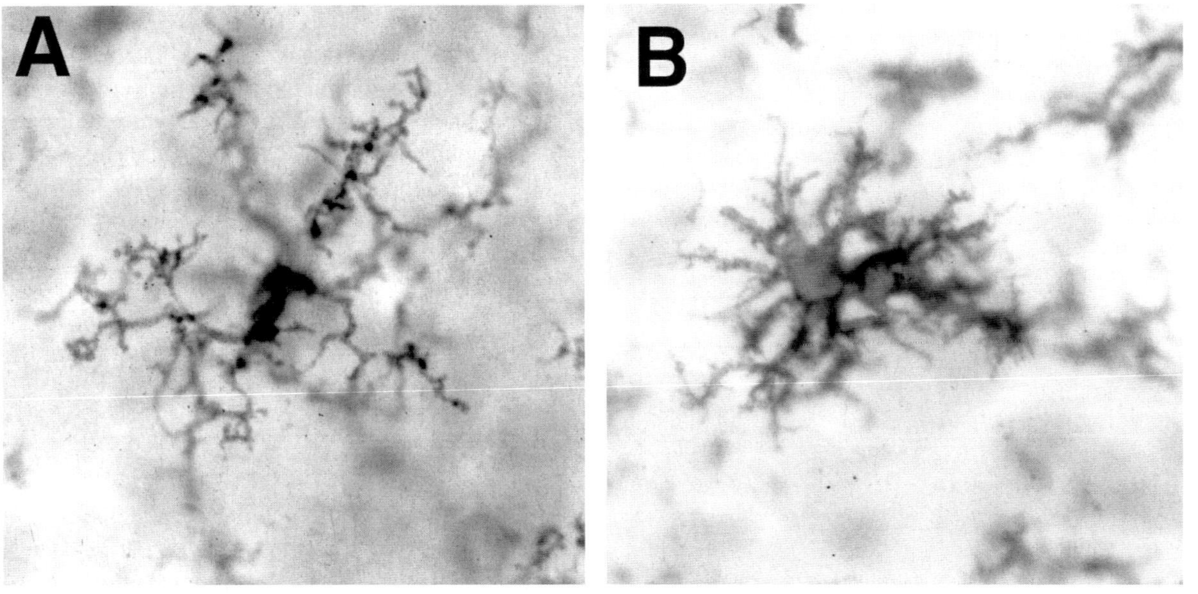

Figure 12. Although microglial are highly variable in shape, ramified microglia (A) typically have small cell bodies and thinner processes whereas reactive microglia (B) have large cell bodies and their processes are thicker and more extensively branched.

port and elasticity of the axon during movement, regulate ionic concentrations and facilitate axonal regeneration. The roles of Schwann cells during regeneration include phagocytosis of myelin and axonal debris, synthesis and release of trophic factors and formation and maintenance of the extracellular matrix that is critical to the survival of the regrowing axon [64].

3.5. Microglia

Microglia are unique from the other cells of the nervous system in that they are generally thought to be of mesodermal origin, although, as presented in Section 2, evidence to the contrary does exist [36,65]. Microglia are ubiquitously, but not uniformly, distributed throughout the CNS. The highest concentration of these cells is found in the telencephalon and they are more plentiful in gray matter than white matter. Regardless of the CNS region, microglia are often found clustered near blood vessels. Although the morphology of microglia is highly variable, Lawson et al. [66] have categorized these cells into three groups: (1) compact cells; (2) longitudinally branched cells; and (3) radially branched cells. Compact cells are small and round or slightly elongated with a few short processes and are typically found in circumventricular organs. Longitudinally branched cells are found in fiber tracts. These cells have an elongated cell body with long processes that are oriented parallel to the surrounding axons. The radially branched cells constitute the largest and most diverse group. These cells generally display a large, round cell body that extends several primary, yet highly branched, processes. Since microglia are highly variable in shape, they have routinely been misidentified. Currently, there are several antigenic markers that are used to ensure correct identification.

Regardless of their morphological categorization, the inactivated form of these cells is known as ramified microglia (Fig. 12A). With the onset of CNS injury or infection, the ramified microglia undergo a rounding and enlarging of their cell bodies and an upregulation of complement receptors. The microglia begin to proliferate, express major histocompatibility class I and II proteins and become highly migratory. Ramified microglia that have undergone these changes are termed reactive microglia (Fig. 12B). Ultimately, the microglia become phagocytic and engulf cellular debris. For example, following demyelination it is common to observe phagocytic microglia with numerous lipid inclusions in their cytoplasm (Fig. 13) [34,36].

A cell type that is commonly mislabeled as a microglia cell is the perivascular cell. Similar to microglia, perivascular cells are phagocytic and are

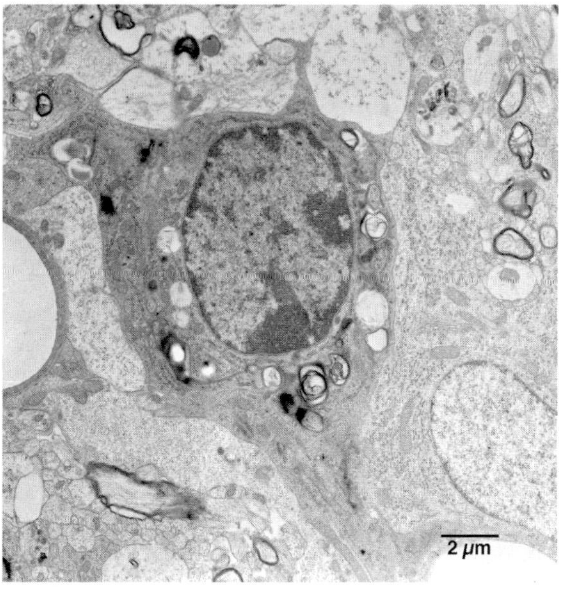

Figure 13. This microglia, which was observed in a demyelinating murine model, displays numerous cytoplasmic inclusions probably resulting from phagocytosing myelin debris.

derived from bone marrow. These cells, however, are not residents of the CNS since they are located on the endothelial side of the vascular basement membrane. Therefore, perivascular cells are not microglial, but are most likely specialized macrophages.

3.6. Ependyma

The ependyma is a simple cuboidal or columnar epithelium that lines the ventricles of the brain and the central canal of the spinal cord (Fig. 14A). The cells that form the ependyma, known as ependymal cells or ependymocytes, are a heterogeneous group. In general, ependymal cells have both cilia and microvilli on their ventricular surface (Fig. 14B). Studies of the cilia demonstrate that they beat in a coordinated manner to facilitate the flow of the cerebrospinal fluid. Other functions that have been attributed to the ependyma include neuronal guidance during development, neuroglia development and regeneration, capture of material in the cerebrospinal fluid and adhesion to inflammatory or other cells in the cerebrospinal fluid [67].

Specialized cells of the ependyma include the choroid epithelium of the choroid plexus and the tanycytes of the third ventricle. The choroid epithelium produces the cerebrospinal fluid and has been implicated in absorption. The choroid epithelial cells typically contain a round, centrally located nucleus (Fig. 15). Their lumenal surface is covered by numerous microvilli, which may assist in the function of absorption [50]. Tanycytes are highly elongated cells that extend processes to a variety of other cell types. Tanycytes are frequently found in clusters covering the circumventricular organs, areas within the CNS that lack a competent blood–brain barrier. These cells are connected by zonula occludens, or tight junctions, that may partially compensate for the absence of the barrier [68].

3.7. Pericytes

Pericytes, also known as microvascular pericytes, are a heterogeneous population of cells that surrounds endothelial cells. These cells are located on the abluminal surface of the endothelial cell and reside between two layers of the basement membrane. Numerous functions for pericytes have been proposed including: (1) induction of blood–brain barrier formation; (2) regulation of blood–brain barrier permeability; (3) vascular stability; (4) angiogenesis; (5) phagocytosis; and (6) antigen presentation and T-cell activation. The functional importance of these cells is also suggested by the correlation between pericyte degeneration and a variety of neurodegenerative diseases, such as Alzheimer's and multiple sclerosis [69].

4. SPECIALIZED STRUCTURES OF NEURAL TISSUE

4.1. Myelin

Myelin is a multilayered membrane, formed by highly specialized cell in the PNS and the CNS, that ensures the rapid and efficient propagation of electrical impulses along axons. Myelin is unique in its composition, its formation and its function, and each of these aspects will be addressed in the following sections.

4.1.1. Composition

Myelin is unlike other membranes in that it is composed of 70% lipid and 30% protein. Most mem-

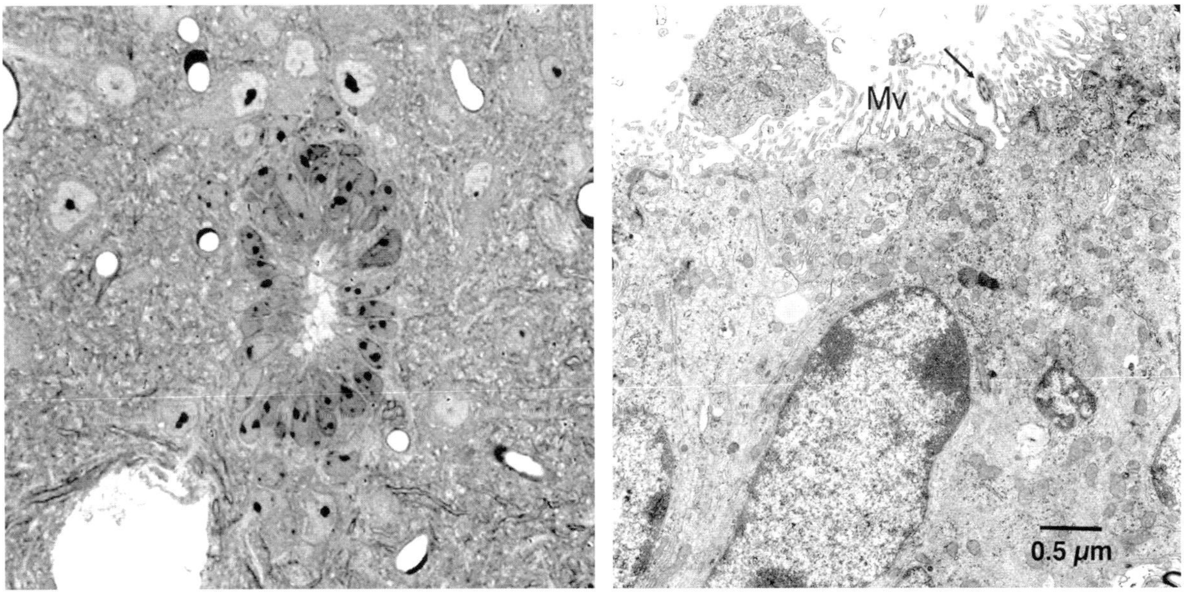

Figure 14. (A) Low magnification of the spinal cord central canal. The canal is formed by specialized cuboidal epithelial cells known as ependymal cells. (B) Electron micrograph of an ependymal cell displaying typical luminal surface specializations: microvilli (Mv) and cilia (arrow).

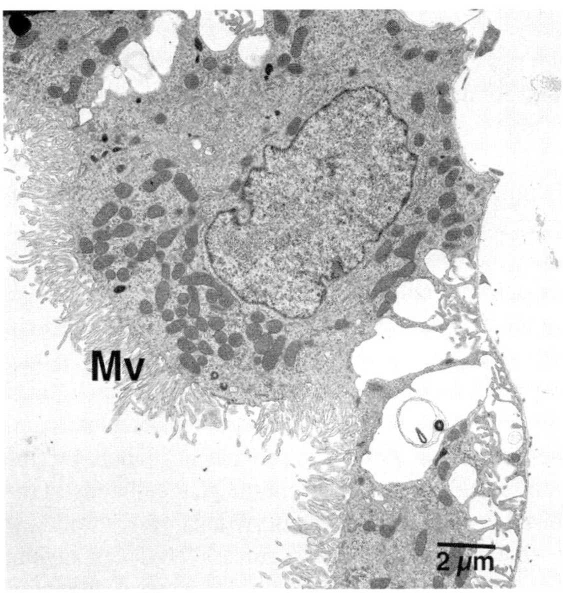

Figure 15. Choroid epithelial cells are essential for the formation of cerebrospinal fluid. In addition, microvilli (Mv), which cover the surface of these cells, may assist in absorption.

branes have a lower lipid-to-protein ration. In addition, myelin exhibits an unusually high galactolipid content. Typical of other membranes, the remaining major lipids of both PNS and CNS myelin are cholesterol and phospholipids. Although the basic lipid classes are similar between PNS and CNS myelin, quantitative differences do exist [for more details, see [70]). Similarly, PNS and CNS myelin contain many of the same proteins; however, compositional differences are prevalent. Proteins that are present in both PNS and CNS myelin are MBP, MAG and connexin 32. Also there is some evidence that PLP should be included in this list, however the presence of PLP in the PNS is controversial.

MBP, a cytoplasmic protein that is essential for the formation of the major dense line, comprises ~20% of PNS and ~30% of CNS myelin proteins. There are four isoforms that are alternatively spliced products of a single gene; however, the functions of the specific isoforms are not known. PLP is a transmembrane protein that constitutes about 40% of the protein of the CNS myelin while, if present in PNS myelin, it is in trace amounts. The function of PLP is not completely understood in spite of the availability of several PLP mutant animal models. In mice that completely lack PLP, as a result of genetic disruption of the *plp* gene, only slight abnormalities of myelin structure are observed. In contrast, animals with naturally occurring PLP mutations exhibit oligodendrocyte death and severe dysmyelination. Unlike MBP and PLP, MAG and connexin 32 are not found in compact myelin. MAG, a member of the immunoglobulin superfamily, has a periaxonal localization, and in the PNS MAG is also concentrated at the paranode and in the Schmidt–Lanterman incisures (see below). The proposed function of MAG is cell adhesion. Connexin 32, which is also concentrated in the paranodal loops and the Schmidt–Lanterman incisures, is a gap junction protein and therefore functions in inter- and intracellular communication. P0 and peripheral myelin protein 22 (PMP22) are integral membrane proteins that are restricted to the PNS. Both P0 and PMP22 appear to be structural proteins whose absence in rodent models results in dysmyelination, while mutations in the human genes result in various forms of the demyelinating disease known as Charcot–Marie–Tooth disease [71].

As previously mentioned, the galactolipids, galactocerebroside and its sulfated derivative sulfatide, comprise a significant portion, roughly 20%, of the dry weight of myelin. Numerous in vitro perturbation studies suggest that these lipids are instrumental in the development of oligodendrocytes [72]. In the absence of these lipids in vivo, oligodendrocytes appear to be arrested in a myelin-forming stage, but they express terminal differentiation markers and synthesize myelin [73]. Structurally, however, galactolipid-deficient mice exhibit myelin abnormalities in the paranodal region that suggest that these lipids function in mediating axo-glial interactions [74].

4.1.2. Structure

Myelin is a multilamellar membranous structure that is formed by the oligodendrocytes in the CNS and Schwann cells in the PNS. These cells extend cytoplasm-containing membrane processes that contact and spiral around axons. In recent years, tremendous research efforts have been dedicated to the identification of the molecules that mediate the processes of axon recognition, contact and wrapping by the glial extension. As multiple wraps are deposited, the cytoplasm is extruded; the cytoplasmic surfaces become fused forming the major dense line while the close apposition of the extracellular surfaces forms the intraperiod line (Fig. 16). Both the inner and outer borders of the sheath maintain cytoplasm and form the inner and outer cytoplasmic loops, respectively. The only other regions of

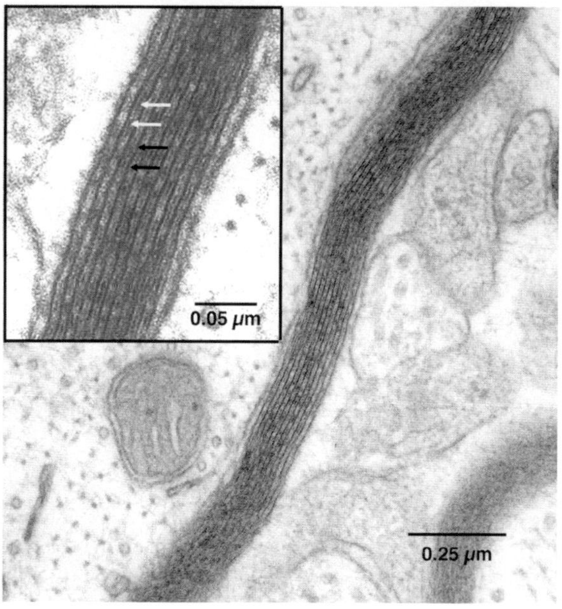

Figure 16. As multiple wraps of myelin are deposited and the cytoplasm is extruded, the cytoplasmic surfaces fuse to form the major dense line (inset, black arrows). During this compaction process, the extracellular surfaces become closely apposed and form the intraperiod line (inset, white arrows).

the compacted, mature myelin sheath that maintain cytoplasm are the Schmidt–Lanterman incisures and the paranodal loops (to be discussed below). Schmidt–Lanterman incisures are regions in compacted PNS myelin where the major dense line does not form and cytoplasm is retained. These cytoplasm-containing clefts are visible in each layer of the myelin sheath. Thus, if the myelin sheath were unrolled, the Schmidt–Lanterman incisure would form a cytoplasm-containing tube. The function of Schmidt–Lanterman incisures is not completely understood, but they may provide a conduit for the transport of molecules from the glial cell body to the distal portions of the sheath [75].

Each myelin-forming process produces myelin for only a single axonal segment known as an internode. (The formation of myelin and myelin-specific structures is illustrated in Fig. 17.) Adjacent internodes are separated by a myelin bare region of the axon known as a node of Ranvier (Fig. 18). The edges, or shoulders, of the internodal region is termed the paranode. The paranode is delineated by the paranodal loops, pockets that form as a result of the retention of cytoplasm in the lateral edges of the myelin-forming process. The paranodal re-

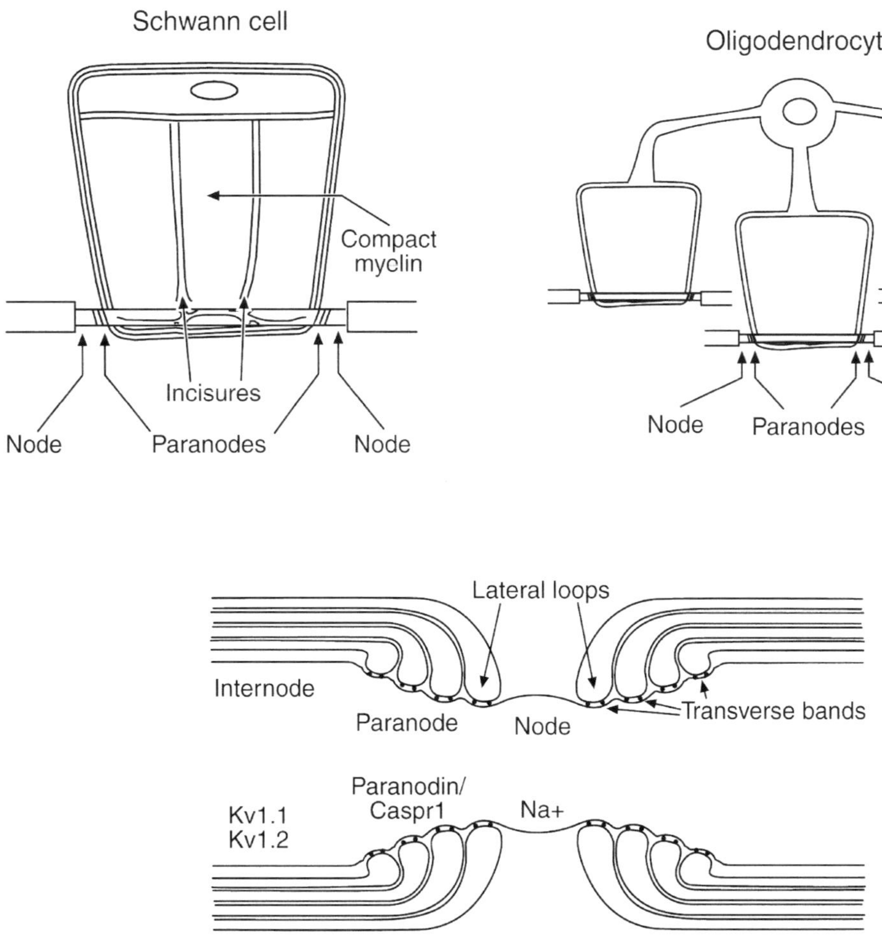

Figure 17. A cytoplasm-containing process contacts and wraps around an axonal segment. As the process continues to wrap, the cytoplasm is extruded resulting in a highly compacted, multilayered membrane sheath. In the PNS, this cytoplasm-containing process is extended by the Schwann cell. Each Schwann cell only myelinates a single axon segment. In contrast, oligodendrocytes, the myelin-forming cell of the CNS, can myelinate multiple axon segments. Although formed by different cells, PNS and CNS myelin share many common features. In both, the inner and outer borders of the sheath maintain cytoplasm and form the inner and outer cytoplasmic loops, respectively. In addition, the lateral borders also maintain cytoplasm and form the lateral loops of the paranodal regions. Junctional complexes form in the paranode and a prominent feature of these complexes is transverse bands. Schmidt–Lantermann incisures, depicted as cytoplasm containing tubes in the unrolled sheath, are common structures in the PNS, but are extremely rare in the CNS. In both the PNS and the CNS, adjacent myelin segments are separated by short unmyelinated regions of axon known as the nodes of Ranvier. Voltage-gated sodium channels are clustered at the node. Flanking the node is the paranode and lateral to the paranode is the juxtaparanode, a region that contains a concentration of potassium channels. Recently a potential cell adhesion molecule, termed paranodin/CASPR1, was shown to be abundant in the paranodal region and may be a component of the transverse bands.

gion is formed by the overlapping of slightly wider concentric membranous wraps resulting in the close apposition of the paranodal loops against the axon. Specialized septate-like junctions are formed in this region and a prominent feature of these junctions is the transverse bands, which are regularly arrayed densities that traverse the periaxonal space. Although the function of the transverse bands is not known, they may be involved in the adherence of the myelin sheath to the axon and/or instrumental in the establishment and maintenance of the axolemmal ion channel domains [50].

The prominent ion channel domains of the nodal region are the voltage-gated sodium channels that are restricted to the node and the Shaker type potassium channels, subtypes Kv 1.1 and 1.2, that are located in the paranode and juxtaparanode (Fig. 17). Interestingly, in mice that lack transverse bands the sodium

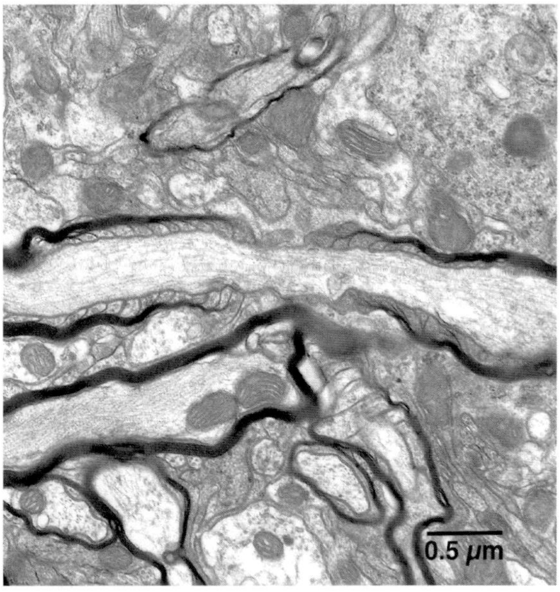

Figure 18. The node of Ranvier is the myelin-bare region of the axon that is flanked by paranodal regions of adjacent myelin sheaths.

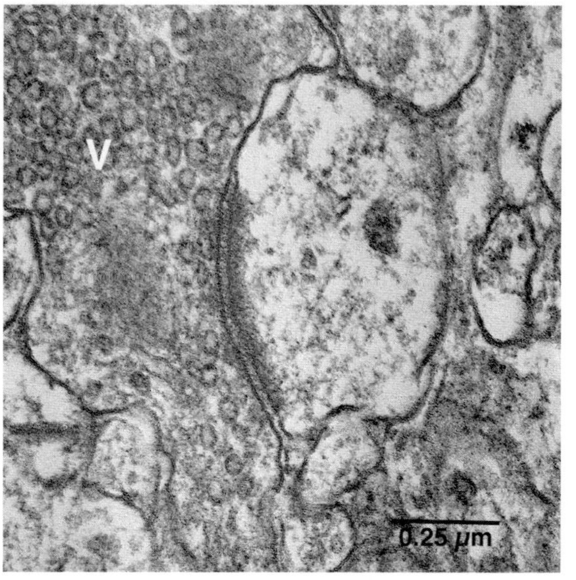

Figure 19. The presynaptic cell, which forms a terminal bouton, contains numerous neurotransmitter-containing vesicles (V), while the postsynaptic cell displays a thickened synaptic membrane.

and potassium channel domains are not clearly defined and frequently overlap [76]. Such ill-defined domains may have a dramatic effect on axonal conduction.

4.1.3. Function

The primary function of the myelin sheath is to facilitate rapid conduction of a neuronal impulse. The basis for neuronal conduction is depolarization of the axonal membrane through the opening of voltage-gated sodium channels. In unmyelinated axons, these channels are uniformly distributed along the axon. In order for an unmyelinated fiber to conduct an impulse, the entire membrane is depolarized as the impulse crawls along the axon. In myelinated axons, however, sodium channels are not uniformly distributed along the entire axon, but are concentrated at the nodes of Ranvier. Since the myelin sheath has low capacitance, the opening of only a few sodium channels is required to ensure the passage of the current through the internodal region to the adjacent node. Upon reaching the adjacent node, the current is sufficient to induce depolarization of the membrane. Since the depolarization seems to jump from node to node, this type of impulse propagation is called saltatory conduction. Importantly, since only the nodal regions are depolarized, the speed of conduction is increased with minimal energy expenditure. For further details of saltatory conduction, see Ritchie [77].

4.2. Synapse

Functionally, the synapse is one of the most important structures in the nervous system since it provides the means for interneuronal communication. There are two categories of synapses: electrical and chemical. For both types, information passed from the presynaptic neuron to the postsynaptic neuron modulates the activity of the second cell. For electrical synapses, the pre- and postsynaptic cells come close together maintaining a gap of about 2 nm and forming gap junctions, which allow the direct passage of electrical information from one cell to the next. Although this type of synapse is rare in mammals, studies have reported their presence in a variety of CNS regions including the lateral vestibular nucleus, the molecular layer of the cerebellum and in the somatosensory and motor cortices [50].

The more common type of synapse is the chemical synapse (Fig. 19). Morphologically, these synapses are generally described as being asymmetric with the postsynaptic cell displaying a prominent, electron-dense membrane. The presynaptic terminal typically contains a few cytoskeletal elements, small mitochondria and synaptic vesicles. These vesicles, which vary in size and shape, contain a chemical signal, or neurotransmitter, that will induce either an excitatory or inhibitory effect on the postsynaptic cell. Some authors have reported a correlation between vesicle shape and synaptic function while others believe that the various shapes only reflect fixation artifacts [50].

Prior to their release, the vesicles cluster near the presynaptic plasma membrane. The vesicles are linked together and these linkages are facilitated by synapsin in a phosphorylation-dependent manner [55,78,79]. A rapid influx of calcium induces fusion of the vesicles with the presynaptic membrane and the neurotransmitter in the vesicle is released. The vesicle, having fused with the plasma membrane, is retrieved through a mechanism that is not clearly understood, but is known to involve clathrin-mediated endocytosis. The neurotransmitter is released into the synaptic cleft, which is usually about 20 nm wide and filled with extracellular matrix proteins that are essential for the formation and maintenance of the cleft. Glial cells are also positioned in or near the cleft that assist in the removal of the neurotransmitter. The postsynaptic membrane is frequently highly convoluted in order to increase the surface area to maximize the number of available neurotransmitter receptors [80].

4.3. Protective Barriers

4.3.1. Blood–brain barrier

The first evidence for the existence of a blood–brain barrier came from the observation that dyes injected into the vascular system stained all organs and tissues except for the brain. For many years the structures and/or mechanisms that were responsible for this exclusion were a topic of great controversy. Now, it is generally agreed that the specialized endothelial cells of the brain vasculature are the primary regulators of the blood–brain barrier. Unlike other endothelial cells, the brain endothelial cells are joined by zonula occludens, or tight junctions, at or near their luminal surface. These junctions, formed by the fusion of the plasma membranes of adjacent cells, prevent the free diffusion of blood substances between joined cells. The intravenous injection of electron dense markers, such as lanthanum and horseradish peroxidase, has provided much of the evidence for implicating the tight junctions of the endothelial cells as the primary site of the blood–brain barrier (Fig. 20). When these markers are injected in the vasculature, they exit the blood vessel by penetrating between cells. In cases of competent tight junctions the progress of the markers is blocked, and the markers are only able to infiltrate the brain parenchyma in the absence of these junctions [81].

Brain endothelial cells also lack fenestrations and exhibit low pinocytotic activity. Transendothelial transport, however, does occur. Substances that are highly lipid soluble, such as ethanol and nicotine, are able to diffuse into the brain. This potential route of entry into the brain has clinical benefit by providing a non-invasive means for drug therapy. Another route of entry into the brain parenchyma is through carrier-mediated diffusion. The best characterized of these carriers is the glucose transporter GLUT1, a facilitative transporter located in brain endothelial membrane that carries glucose against its gradient [82].

The blood–brain barrier also provides a potential obstacle against unimpeded entry of peripheral cells. For many years it was believed that the brain was an immunologically privileged site and that entry of non-neural cells was restricted to periods of development or pathologic conditions. New data is beginning to change our understanding of cellular trafficking into and out of the CNS and the role that the blood–brain barrier plays in this process. Two examples that illustrate this point are the microglia and the T-cells. Microglia, cells that appear to be systemically derived (controversy of microglial origin was addressed previously), were believed to enter the CNS prior to the maturation of the barrier. However, increasing evidence indicates that the microglia population is supplemented in the adult in the presence of an intact blood–brain barrier [34]. Recently, Wu et al. [83] demonstrated that, in the presence of a maintained blood–brain barrier, hematogenous lineage cells entered the CNS and that this entry was mediated by cytokines or chemokines. T-cells were believed to be excluded from the CNS except during brain injury or disease. Now it is generally accepted that T-cells are readily able to cross an uncompromised barrier and are involved in im-

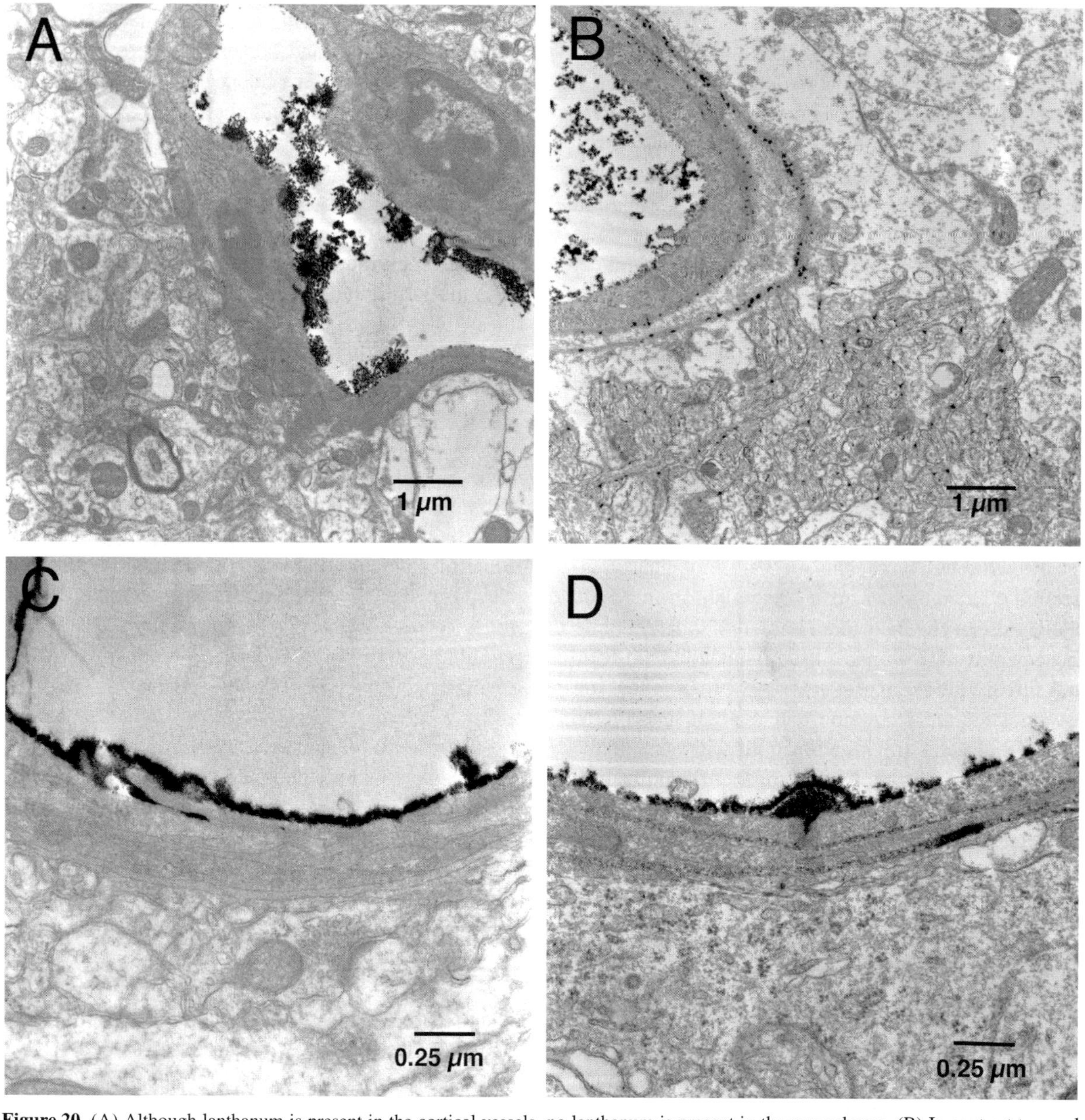

Figure 20. (A) Although lanthanum is present in the cortical vessels, no lanthanum is present in the parenchyma. (B) In contrast to panel (A), lanthanum is readily observed in the basal lamina of the endothelial cells and in the parenchyma of a mouse that has a compromised blood–brain barrier. In high magnification electron micrographs (C,D), lanthanum is observed in the intercellular spaces. In the normal brain (C), the lanthanum abruptly stops its penetration between apposing cells indicating the presence of intact tight junctions. In the blood–brain barrier compromised mouse brain (D), lanthanum penetrates beyond the endothelial cells, which is indicative of disrupted tight junction formation.

mune surveillance in the normal brain. Interestingly, chemically induced blood–brain barrier breakdown in laboratory animals does not result in a significant increase in the number of monocytes or leukocytes that enter the CNS. Apparently, factors other than the mere loss of the blood–brain barrier are required for increased cellular influx. These other factors may be a family of molecules, termed the chemokines, that are known to be secreted by activated microglia and astrocytes [84]. Chemokines are small, heparin-binding proteins with chemoattractant activity. For example, the microinjection of macrophage inflammatory protein 2 into the mammalian brain results in the recruitment of hematogenic cells into the CNS. Concomitant with the influx of these cells is the breakdown of the blood–brain barrier.

Although the brain may not be surrounded by the impenetrable barrier that was once envisioned, the blood–brain barrier still provides a formidable protective hurdle that selectively regulates entry into the CNS. Also, it should be noted that in several neurodegenerative conditions, including Alzheimer's disease, Parkinson's disease, Creutzfeldt–Jakob disease and multiple sclerosis, the blood–brain barrier is compromised and the loss of this protective agent appears to contribute to the pathologic state [84].

Interestingly, there are several regions within the brain where the endothelial cells do not exhibit the structural specializations associated with a competent blood–brain barrier. In these regions, known as the circumventricular organs, the endothelial cells lack tight junctions and are highly fenestrated. Consequently, the blood–brain barrier is compromised in these areas. The circumventricular organs include the area postrema, pineal gland, median eminence, preoptic recess, paraphysis, and pituitary [81].

Other cells that may be involved in the formation and function of the blood–brain barrier include astrocytes and pericytes. Astrocytes extend perivascular end-feet that completely surround the endothelial cells. The end-feet processes of these cells are also connected by junctions. Since these junctions are not the occluding type, it is difficult to perceive how these cells could establish a formidable physical barrier. Although astrocytes may not form the actual barrier, recent studies have implicated astrocytes, and neurons as critical inducers of barrier formation [85]. In addition, pericytes, small smooth muscle cells that are positioned on the abluminal side of the endothelial cells, may regulate endothelial proliferation and restrict entry into the CNS by phagocytosing material foreign to the nervous system [69].

4.3.2. Blood–nerve barrier

The blood–nerve barrier has received far less attention than its CNS counterpart. Presumably, the primary function of the blood–nerve barrier is to regulate the microenvironment of the peripheral nervous tissue. Similar to the blood–brain barrier, the basis for the formation of the blood–nerve barrier is the interendothelial tight junctions. In addition, the vasculature of the PNS is composed of the continuous type of vessels and share several transporters and receptors that are specifically expressed on the endothelial cells that constitute the blood–brain barrier. Although the barriers of the CNS and PNS share morphological similarities, functionally the two barriers are distinct. Peripheral nerves are substantially more permeable to sodium and chloride ions, and lanthanum, the electron dense marker commonly used to study the patency of the blood–brain barrier. Currently, the significance of this increased permeability of the blood–nerve barrier compared to the blood–brain barrier is not known [82].

4.3.3. Blood–cerebrospinal fluid barrier

Cerebrospinal fluid is primarily produced by the choroid plexus, papillary structures in the ventricles composed of simple cuboidal, specialized epithelial cells. In order for a blood-borne substance to enter the cerebrospinal fluid it must traverse the endothelial cell layer and the choroid plexus. Compounds face little resistance in entering or exiting the blood vessels of the choroid plexus. Unlike most cortical vessels, the microvasculature of the ventricles are fenestrated and are not connected by tight junctions. The most imposing obstacle for the exchange between the cerebrospinal fluid and the blood is the choroid epithelial cells. These cells are connected by tight junctions; therefore, exchange between the cerebrospinal fluid and the blood is limited by the size and lipid solubility of a compound, the two primary parameters that regulate transmembrane diffusion. Similar to the blood–brain barrier, transport mechanisms have also been identified in the cerebrospinal fluid–blood barrier for the efficient uptake and removal of certain substances. Although the presence of such mechanisms was first suggested nearly 40 years ago, the molecular identity of some of these transporters has only recently been unraveled [67].

5. AXO-GLIAL INTERACTIONS

Numerous examples of the importance of cell–cell interactions have been presented throughout this chapter. Although each of these is critical to the proper formation and function of the nervous system and merits discussion, it is beyond the scope of this chapter to treat them all. Here, the discussion will be limited to the interactions between the neuron, particularly the axon, and the cells that form myelin.

Although the most dramatic example of axo-glial interaction is the formation of the myelinated neuronal fiber, the initial interplay between these cell types far precedes myelinogenesis. From very early in development, neurons regulate the number of oligodendrocytes and Schwann cells by releasing or controlling the release of trophic factors that dictate mitotic activity and survival. In the CNS, Barres and Raff proposed that oligodendrocytes that fail to make adequate contacts do not receive sufficient axon-dependent survival signals and as a result undergo apoptosis [63,86]. Similarly, Schwann cell survival is dependent on the competitive availability of axonally derived neuregulin [87]. Schwann cells that do not contact an axon, and are thus deprived of neuregulin, will undergo apoptosis. Interestingly, Schwann cells seem to reciprocate by synthesizing a variety of growth factors (e.g. nerve growth factor, brain-derived neurotrophic factor, ciliary neurotrophic factor) that influence the survival of neurons. It now appears plausible that Schwann cells and neurons each produce trophic factors that support the survival of the other cell type; such a scenario provides a system for matching the number of neurons to myelin-forming cells during development [87,88].

Recently, reports have demonstrated that axon caliber, in both the CNS and PNS, is dependent on the interactions between the axon and the myelin-forming cell. From the analysis of several strains of dysmyelinating mice, Sanchez et al. [89] concluded that oligodendrocytic signals, but not myelin formation, were required for axonal growth through cytoskeletal phosphorylation. Similarly, Trapp and colleagues [57], exploiting a mouse deficient in MAG, reported that MAG is the glial signal that mediates axonal caliber.

In addition to modulating axon size, axo-glial interactions are also involved in the molecular organization of the axolemma. The axolemma is a highly specialized membrane that maintains distinct protein domains that ensure proper neural function. The most studied of these domains are those of the ion channels. Voltage-gated sodium channels cluster at the node of Ranvier and Shaker-type potassium channels are concentrated in a region adjacent to the paranode known as the juxtaparanode. The exact mechanism involved in establishing and maintaining these domains seem to involve specific axo-glial interactions. In a report by Kaplan et al. [90] proper sodium channel clustering in vitro was shown to be oligodendrocyte dependent. Sodium channel clustering, however, was not induced by cell–cell contact but was mediated by a soluble factor. In contrast, a growing number of reports have demonstrated that sodium channel clustering in the PNS and CNS is directed by physical contact between the myelin-forming process and axon [91–94]. Furthermore, in mice that display abnormal axo-glial contacts, the spatial organization of certain axolemmal proteins is altered in both the PNS and CNS [76]. A recently isolated and characterized axolemmal protein, named paranodin or CASPR1, was shown to be concentrated in the paranodal region of both PNS and CNS of myelinated axons. Paranodin/CASPR1 is a potential cell-adhesion molecule and is a prime candidate for mediating paranodal axo-glial interactions.

In addition to regulating cell number and axolemmal organization, axo-glial interactions mediate the formation and maintenance of the myelin sheath. Recently, efforts to unravel the mechanisms that control these processes have greatly intensified and have resulted in the identification of novel molecules and novel functions of previously known molecules. In 1994, Brophy and colleagues [95] isolated and characterized a protein termed periaxin, which may regulate very early stages of myelin formation. Utilizing genetically engineered animals, MAG has been implicated as a mediator of myelin/axon contact since mice that lack MAG have irregular periaxonal collars [96,97]. Furthermore, mice that lack galactocerebroside, a lipid that is restricted to the outer leaflet of the myelin membrane and thus optimally positioned to mediate axo-glial interactions, exhibit abnormal periaxonal spaces during development and disrupted paranodal junctions [98,99]. In addition to these few examples, numerous other candidates have been proposed as potential mediators of axo-glial interactions [59,75,94,100]. The importance of establishing and maintaining these interactions is demonstrated by diseases such as multiple sclerosis and Charcot–Marie–Tooth disease. In these pathological conditions, axo-glial interactions are compromised and dramatic demyelination results [101].

REFERENCES

1. Purves D, Litchman JW. Principles of Neural Development. Sunderland: Sinauer Associates, 1985:25–51, 81–103.
2. Sadler TW. Langman's Medical Embryology, 6th edn. Baltimore, MD: Williams and Wilkins: 1990:352–387.
3. Gage FH. Mammalian neural stem cells. Science 2000;287:1433–1438.
4. Reynolds BA, Weiss S. Generation of neurons and astrocytes from isolated cells of the adult mammalian central nervous system. Science 1992;255:1707–1710.
5. Rao MS, Mayer-Proschel M. Glia-restricted precursors are derived from multipotent neuroepithelial stem cells. Dev Biol 1997;188:48–63.
6. Williams BP, Park JK, Alberta JA, Muhlebach SG, Hwang GY, Roberts TM, Stiles CD. A PDGF-regulated immediate early response initiates neuronal differentiation in ventricular zone progenitor cells. Neuron 1997;18:553–562.
7. Svendsen CN, Smith AG. New prospects for human stem-cell therapy in the nervous system. Trends Neurosci. 1999;22:357–364.
8. Kuhn HG, Svendsen CN. Origins, functions, and potential of adults neural stem cells. BioEssays 1999;21:625–630.
9. Gage FH. Cell therapy. Nature 1998;392:18–24.
10. Parnavelas JG. The origin and migration of cortical neurones: new vistas. Trends Neurosci 2000;23:126–131.
11. Yamada T, Placzek M, Tanaka H, Dodd J, Jessell TM. Control of cell pattern in the developing nervous system: polarizing activity of the floor plate notochord. Cell 1991;64:635.
12. Richardson WD, Smith HK, Sun T, Pringle NP, Hall A, Woodruff R. Oligodendrocyte lineage and the motor neuron connection. Glia 2000;29:136–142.
13. Briscoe J, Pierani A, Jessell TM, Ericson J. A homeodomain protein code specifies progenitor cell identity and neuronal fate in the ventral neural tube. Cell 2000;101:435–445.
14. Ericson J, Briscoe J, Rashbass P, van Heyningen V, Jessell TM. Graded sonic hedgehog signaling and the specification of cell fate in the ventral neural tube. Cold Spring Harbor Symp Quant Biol 1997;62:451–466.
15. Ono K, Bansal R, Payne J, Rutishauser U, Miller RH. Early development and dispersal of oligodendrocyte precursors in the embryonic chick spinal cord. Development 1995;121:1743–1754.
16. Warf BC, Fok-Seang J, Miller RH. Evidence for the ventral origin of oligodendrocyte precursors in the rat spinal cord. J Neurosci 1991;11:2477–2488.
17. Richardson WD, Pringle NP, Yu WP, Hall AC. Origins of spinal cord oligodendrocytes: possible developmental and evolutionary relationships with motor neurons. Dev Neurosci 1997;19:58–68.
18. Goldman JE. Lineage, migration, and fate determination of postnatal subventricular zone cells in the mammalian CNS. J Neurooncol 1995;24:61–64.
19. Cameron-Curry P, Le Douarin NM. Oligodendrocyte precursors originate from both the dorsal and ventral parts of the spinal cord. Neuron 1995;15:1–20.
20. Timsit SG, Martinez S, Allinquant B, Peyron F, Puelles L, Zalc B. Oligodendrocytes originate in a restricted zone of the embryonic ventral neural tube defined by DM-20 expression. J Neurosci 1995;15:1012–1024.
21. Spassky N, Olivier C, Perez-Villegas E, Goujet-Zalc C, Martinez S, Thomas J-L, Zalc B. Single or multiple oligodendrocyte lineages: a controversy. Glia 2000;29:143–148.
22. Rogister B, Ben-Hur T, Dubois-Dalcq M. From neural stem cells to myelinating oligodendrocytes. Mol Cell Neurosci 1999;14:287–300.
23. Scolding NJ, Rayner PJ, Compston DA. Identification of A2B5-positive putative oligodendrocyte progenitor cells and A2B5-positive astrocytes in adult human white matter. Neurosci 1999;89:1–4.
24. Wegner M. Transcriptional control in myelinating glia: the basic recipe. Glia 2000;29:118–123.
25. Orentas DM, Hayes JE, Dyer KL, Miller RH. Sonic hedgehog signaling is required during the appearance of spinal cord oligodendrocyte precursors. Development 1999;126:2419–2429.
26. Pringle NP, Yu WP, Guthrie S, Roelink H, Lumsden A, Peterson AC, Richardson WD. Determination of neuroepithelial cell fate: induction of the oligodendrocyte lineage by ventral midline cells and sonic hedgehog. Dev Biol 1996;177:30–42.
27. Poncet C, Soula C, Trousse F, Kan P, Hirsinger E, Pourquie O, Durprat AM, Cochard P. Induction of oligodendrocyte progenitors in the trunk neural tube by ventralizing signals: effects of notochord and floor plate grafts, and of sonic hedgehog. Mech Dev 1996;60:13–32.
28. Adlkofer K, Lai C. Role of neuregulins in glial cell development. Glia 2000;29:104–111.
29. Canoll PD, Mussacchio JM, Hardy R, Reynolds R, Marchionni MA, Salzer JL. GGF/neuregulin is a neuronal signal that promotes the proliferation and survival and inhibits the differentiation of oligodendrocyte progenitors. Neuron 1996;17:229–243.
30. Vartanian T, Fischbach G, Miller R. Failure of spinal cord oligodendrocyte development in mice lacking neuregulin Proc Natl Acad Sci USA 1999;96: 731–735.
31. Levison SW, Goldman J. Astrocyte origins. In: Astrocytes Pharmacology and Function. In: Murphy S, ed. San Diego, CA: Academic Press, 1993:1–16.
32. Raff MC, Miller RH, Noble M. A glial progenitor cell that develops in vitro into an astrocyte or an oligodendrocyte depending on culture medium. Nature 1983;303: 390–396.
33. Raff MC, Abney ER, Cohen J, Lindsay R, Noble M. Two types of astrocytes in cultures of developing rat white matter: differences in morphology, surface gangliosides, and growth characteristics. J Neurosci 1983;6:1289–1300.
34. Stoll G, Jander S. The role of microglia and macrophages in the pathophysiology of the CNS. Prog Neurobiol 1999;58:233–247.
35. Flaris NA, Densmore TL, Molleston MC, Hickey W.F. Characterization of microglia and macrophages in the

central nervous system of rats: definition of the differential expression of molecules using standard and novel monoclonal antibodies in normal CNS and in four models of parenchymal reaction. Glia 1993;7:34–40.
36. Streit WJ, Walter SA, Pennell NA. Reactive microgliosis. Prog Neurobiol 1999;57:563–581.
37. Federoff S, Zhai R, Novak JP. Microglia and astroglia have a common progenitor cell. J Neurosci Res 1997;58:477–486.
38. Ziller C, Le Douarin NM. The neural crest in nerve development. In: Dyck PJ, Thomas PK, Griffin JW, Low PA, Poduslo JF, eds. Peripheral Neuropathy, 3rd edn. Philadelphia, PA: W.B. Saunders, 1993:230–242.
39. Ma Q, Chen Z, del Barco Barrantes I, de la Pompa JL, Anderson DJ. neurogenin is essential for the determination of neuronal precursors for proximal cranial sensory ganglia. Neuron 1998;20:469–482.
40. Anderson DJ, Jan YN. The determination of the neuronal phenotype. In: Cowan WM, Jessell TM, SL Zipursky, eds. Molecular and Cellular Approaches to Neural Development. New York: Oxford University Press, 1997:26–63.
41. Anderson DJ, Goves A, Lo L, Ma Q, Rao M, Shah NM, Sommer L. Cell Lineage determination and the control of neuronal identity in the neural crest. Cold Spring Harbor Symposia on Quantitative Biology. 1997;62:493–504.
42. Mirsky R, Jessen KR. The neurobiology of Schwann cells. Brain Pathol 1999;9:293–311.
43. Aguayo AJ, Epps J, Charron L, Bray GM. Multipotentiality of Schwann cells in cross-anastomosed and grafted myelinated and unmyelinated nerves: quantitative microscopy and radiography. Brain Res 1976;104:1–20.
44. Aguayo AJ, Charron L, Bray GM. Potential of Schwann cells from unmyelinated nerves to produce myelin: a quantitative ultrastructural and radiographic study. J Neurocytol 1976;5:565–573.
45. Sommer L, Suter U. The glycoprotein P0 in peripheral gliogenesis. Cell Tissue Res. 1998;292:11–16.
46. Zorick TS and Lemke G. Schwann cell differentiation. Curr Opin Cell Biol 1996;8:870–878.
47. Scherer SS. The biology and pathology of Schwann cells. Curr Opin Neurol 1997;10:386–397.
48. Webster HdeF. Development of peripheral nerve fibers. In: Dyck PJ, Thomas PK, Griffin JW, Low PA Poduslo JF, eds. Peripheral Neuropathy, 3rd edn. Philadelphia, PA: W.B. Saunders, 1993:243–266.
49. Filbin MT. Myelin-associated glycoprotein: a role in myelination and in the inhibition of axonal regeneration? Curr Opin Neurobiol 1995;5:588–595.
50. Peters A, Palay SL, Webster HdeF. The Fine Structure of the Nervous System, 3rd edn. New York: Oxford University Press, 1991.
51. Le Clerc N, Baas PW, Garner CC, Kosik KS. Juvenile and mature MAP2 isoforms induce distinct patterns of process outgrowth. Mol Biol Cell 1996;7:443–455.
52. Walker JH, Boustead CM, Witzmann V, Shaw G, Weber K, Osborn M. Cytoskeletal proteins at the cholinergic synapse: distribution of desmin, actin, fodrin, neurofilaments and tubulin in Torpedo electric organ. Eur J Cell Biol 1985;38:123–133.
53. Koenig E, Giuditta A. Protein-synthesizing machinery in the axon compartment. Neuroscience 1999:89;5–15.
54. Alvarez J. Protein synthesis in axons and terminals: significance for maintenance, plasticity and regulation of phenotype with a critique of slow transport theory. Prog Neurobiol 2000;62:1–62.
55. Doussau F, Augustine GJ. The actin cytoskeleton and neurotransmitter release: an overview. Biochimie 2000;82:353–363.
56. Hammerschlagg R, Cyr JL, Brady ST. Axonal transport and the neuronal cytoskeleton. In: Seigel GJ, Agronoff BW, Albers RW, Molinoff PB, eds. Basic Neurochemistry: Molecular, Cellular and Medical Aspects. New York: Raven Press, 1994:545–572.
57. Yin X, Crawford TO, Griffin JW, Tu PH, Lee VM, Li C, Roder J and Trapp BD. Myelin-associated glycoprotein is a myelin signal that modulates the caliber of myelinated axons. J Neurosci 1998;18:1953–1962.
58. Weaver CL, Espinosa M, Kress Y, Davies P. Conformational change as one of the earliest alterations of tau in Alzheimer's disease. Neurobiol Aging 2000;21:719–727.
59. Scherer SS. Molecular specialization at nodes and paranodes in peripheral nerve. Micro Res Tech 1996;34:452–461.
60. Aloisi F. The role of microglia and astrocytes in CNS immune surveillance and immunopathology. In: Matsas and Tsacopoulos, eds. The Functional Roles of Glial Cells in Health and Disease. New York: Plenum, 1999:123–133.
61. del Rio Hortega P. Microglia. In: Penfield W, ed. Cytology and Cellular Pathology of the Nervous system, Vol. 2. New York: Hoeber, 1932:483–534.
62. Mori S, Leblond CP. Electron microscopic identification of three classes of oligodendrocytes and a preliminary study of their proliferative activity in the corpus callosum of young rats. J Comp Neurol 1970;139:1–30.
63. Barres BA. Neuron–glial interactions. In: Cowan WM, Jessell TM, Zipursky SL, eds. Molecular and Cellular Approaches to Neural Development. New York: Oxford University Press, 1997;65–107.
64. Griffin JW, Hoffman PN. Degeneration and regeneration in the peripheral nervous system. In: Dyck PJ, Thomas PK, Griffin JW, Low PA Poduslo JF, eds. Peripheral Neuropathy, 3rd edn., Philadelphia, PA: W.B. Saunders, 1993:361–376.
65. Ling EA, Wong WC. The origin and nature of ramified and amoeboid microglia: a historical review and current concepts. Glia 1993;7:9–18.
66. Lawson LJ, Perry VH, Dri P, Gordon S. Heterogeneity in the distribution and morphology of microglia in the normal adult mouse brain. Neuroscience 1990;39:151–170.
67. Strazielle N, Ghersi-Egea J-F. Choroid plexus in the central nervous system: biology and physiology. J Neuropath Exp Neurol 2000;59:561–574.
68. Del Bigio MR. The ependyma: a protective barrier between brain and cerebrospinal fluid. Glia 1995;14:1–13.
69. Balabanov R, Dore-Duffy P. Role of the CNS microvascular pericyte in the blood–brain-barrier. J Neurosci Res 1998;53:637–644.
70. Norton NT, Cammer W. Isolation and characterization of

myelin. In: Morrell P, ed. Myelin, 2nd edn. New York: Plenum Press, 1984:147–195.
71. Porter BE, Tennekoon G. Myelin and disorders that affect the formation and maintenance of this sheath. Mental Retard Dev Abnorm 2000;6:47–58.
72. Dupree JL, Suzuki K, Popko B. Galactolipids in the formation and function of the myelin sheath. Micro Res Tech 1998;41:431–440.
73. Marcus J. Dupree JL. Popko B. Effects of galactolipid elimination on oligodendrocyte development and myelination. Glia 2000;30:319–328.
74. Popko B. Galactolipids: mediators of axo-glial interactions? Glia 2000;29:149–153.
75. Arroyo EJ, Scherer S. On the molecular architecture of myelinated fibers. Histochem Cell Biol 2000;113:1 18.
76. Dupree JL, Girault J-A, Popko B. Axo–glial interactions regulate the localization of axonal paranodal proteins. J Cell Biol 1999;147:1145–1151.
77. Ritchie J. Physiology of axons. In: Waxman SG, Kocsis JD, Stys PH, eds. The Axon: Structure, Function and Pathophysiology. New York: Oxford Press, 1995:68–96.
78. Augustine GJ, Burns ME, DeBello WM, Hilfiker S, Morgan JR, Schweizer FE, Tokumaru H, Umayahara K. Proteins involved in synaptic vesicle trafficking. J Physiol 1999;520:33–41.
79. Turner KM, Burgoyne RD, Morgan A. Protein phosphorylation and the regulation of synaptic membrane traffic. Trends Neurosci 1999;22:459–464.
80. Burns ME, Augustine GJ. Synaptic structure and function: dynamic organization yields architectural precision. Cell 1995;83:187–194.
81. Betz AL, Goldstein GW, Katzman R. Blood–brain–cerebrospinal fluid barriers. In: Seigel GJ, Agronoff BW, Albers RW, Molinoff PB, eds. Basic Neurochemistry: Molecular, Cellular and Medical Aspects. New York: Raven Press, 1994:681–699.
82. Allt G, Lawrenson JG. The blood–nerve barrier: enzymes, transporters and receptors — a comparison with the blood–brain barrier. Brain Res Bull 2000;52:1-12.
83. Wu Y-P, Matsuda J, Kubota A, Suzuki K, Suzuki K. Infiltration of hematogenous lineage cells into the demyelinating central nervous system of twitcher mice. J Neuropath Exp Neurol 2000;59:628–639.
84. Perry VH, Anthony DC, Bolton SJ, Brown HC. The blood–brain-barrier and the inflammatory response. Mol Med Today 1997;3:335–341.
85. Bauer H-C, Bauer H. Neural induction of the blood–brain-barrier: still an enigma. Cell Mol Neurobiol 2000;20:13–28.
86. Barres BA, Raff MC. Axonal control of oligodendrocyte development. J Cell Biol 1999;147:1123–1128.
87. Grinspan JB, Marchionni MA, Reeves M, Coulaloglou M, Scherer S. Axonal interactions regulate Schwann cell apoptosis in developing peripheral nerve: neuregulin receptors and the role of neuregulins. J Neurosci 1996;16:6107–6118.
88. Reynolds ML, Woolf CJ. Reciprocal Schwann cell–axon interactions. Curr Opin Neurobiol 1993;3:683–693.
89. Sanchez I, Hassinger L, Paskevich P, Shine D, Nixon R. Oligodendroglia regulate the regional expansion of axon caliber and local accumulation of neurofilaments during development independently of myelin formation. J Neurosci 1996;16:5095–5105.
90. Kaplan MR, Meyer-Franke A, Lamber S, Bennett V, Duncan ID, Levinson SR, Barres BA. Induction of sodium channel clustering by oligodendrocytes. Nature 1997;386:724–728.
91. Dugandzija-Novakovic S, Koszowski AG, Levinson SR, Shrager S. Clustering of Na^+ channels and node of Ranvier formation in remyelinating axons. J Neurosci 1995;15:492–503.
92. Rasband MN, Peles E, Trimmer JS, Levinson SR, Lux SE, Shrager P. Dependence of nodal sodium channel clustering on paranodal axoglial contact in the developing CNS. J Neurosci 1999;19:7516–7528.
93. Arroyo EJ, Xu YT, Zhou L, Messing A, Peles E, Chiu SY, Scherer S. Myelinating Schwann cell cells determine the internodal localization of Kv1.1, Kv1.2, Kv beta2 and Caspr. J Neurocytol 1999;28:333–347.
94. Trapp BD, Kidd GJ. Axo-glial septate junctions: the maestro of nodal formation and myelination. J Cell Biol 2000;150:F97-100.
95. Gillespie CS, Sherman DL, Blair GE, Brophy P. Periaxin, a novel protein of myelinating Schwann cells with a possible role in axonal ensheathment. Neuron 1994;12:497–508.
96. Li C, Tropak MB, Gerlai R, Clapoff S, Abramow-Newerly W, Trapp B, Peterson A, Roder J. Myelination in the absence of myelin-associated glycoprotein. Nature 1994;369:747–750.
97. Montag D, Giese KP, Bartsch U, Martini R, Lang Y, Bluthmann H, Karthigasan J, Kirschner DA, Wintergerst ES, Nave K, Zielasek J, Toyka KV, Lipp H, Shachner M. Mice deficient for the myelin-associated glycoprotein show subtle abnormalities in myelin. Neuron 1994;13:229–246.
98. Dupree JL, Coetzee T, Blight A, Suzuki K, Suzuki K, Popko B. Myelin galactolipids are essential for proper node of Ranvier formation in the CNS. J Neurosci 1998;18:1642–1649.
99. Dupree JL, Popko B. Genetic dissection of myelin galactolipid function. J Neurocytol 1998;28:271–279.
100. Tait S, Gunn-Moore F, Collison JM, Huang J, Luetzki C, Pedraza L, Sherman DL, Colman DR, Brophy P. An oligodendrocyte cell adhesion molecule at the site of assembly of the paranodal axo-glial junction. J Cell Biol 2000;150:657–666.
101. Sahenk Z. Abnormal Schwann cell–axon interactions in CMT neuropathies: the effects of mutant Schwann cells on the axon cytoskeleton and regeneration-associated myelination. Ann New York Acad Sci 1999;883:415–426.

Universes in Delicate Balance: Chemokines and the Nervous System
R.M. Ransohoff, K. Suzuki, A.E.I. Proudfoot, W.F. Hickey and J.K. Harrison, editors

Chapter 1.2

Cellular Reactions to Insult (Cellular Neuropathology)

Kinuko Suzuki

Department of Pathology and Laboratory Medicine, University of North Carolina at Chapel Hill, Chapel Hill, NC 27599-7525, USA

1. INTRODUCTION

Cells react to insult in many ways. To insults beyond the capacity of normal homeostasis of cellular function, cells die acutely or following gradual deterioration of their function. Various structural and biochemical changes take place during the process of deterioration. Some cells may be quite resistant and show only transient structural and/or biochemical alterations and survive, while others may die following a long struggle for survival. The structural alterations of neurons and glial cells that are used for the diagnosis of neurological disorders represent the on-going process or sequel of such cellular reaction. These alterations may be subtle or massive and some may appear unique in certain disorders but they are in fact the manifestation of cellular responses to the injury or insult. In this chapter, structural changes of cells in the nervous system observed in various neuropathological conditions in human will be briefly described. If appropriate, changes observed in experimental animals are also included.

2. NEURON

2.1. Cell Soma

2.1.1. Shrunken neurons

Size and shape of neurons varies considerably in the different regions even in the normal brain. In pathological conditions, neuronal soma may become shrunken or atrophic, or enlarged. In dying neurons, the soma are shrunken and deeply acidophilic. Their nuclei are hyperchromatic, often fragmented and details of the nucleoplasm are lost. These neurons are commonly seen in cerebral ischemia, hypoglycemia and hypoxic conditions, and often called hypoxic/ischemic neurons (Fig. 1) [1–3]. These neurons are ultrastructurally characterized by the flocculent mitochondria, disrupted nuclear and cytoplasmic membranes and fragmented nuclear chromatin. Shrunken soma and nuclei are also seen in neurons with generalized condensation of the soma. They are commonly called dark neurons and have affinity for both acidic and basic dyes. These neurons are considered to represent early reversible changes in metabolically perturbed neurons [4,5]. Dark neurons may be found in otherwise normal human brains and are also often seen in suboptimally perfused brains in animal experiments [6] and some may consider them as an artifact of improper fixation. Unlike dying neurons, these dark neurons show condensed but preserved cytoplasmic organelles at the ultrastructural level. Incrustation of shrunken nerve cells and processes with deeply basophilic material is found in areas of old infarction or ischemic lesions and in the brain of the patient with history of perinatal injury (Fig. 2). These neurons are called ferruginated neurons. They stain positive with Prussian blue reaction for iron and von Kossa reaction for calcium.

2.1.2. Atrophy of the soma

Small atrophic neurons are typically seen in transsynaptic (transneuronal) degeneration. Atrophy of the soma occurs in deafferented neurons when fibers afferent to them are lost. The best example is seen in the neurons in the lateral geniculate body after

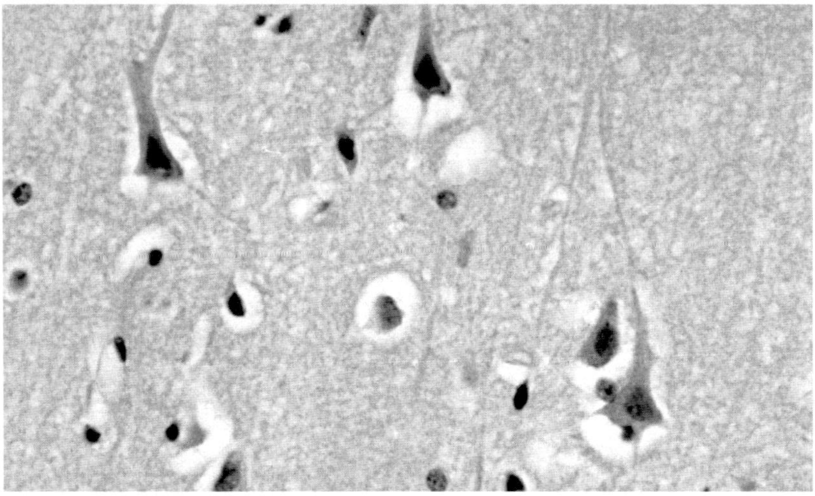

Figure 1. Hypoxic/ischemic neurons in the cerebral cortex following acute vascular occlusion.

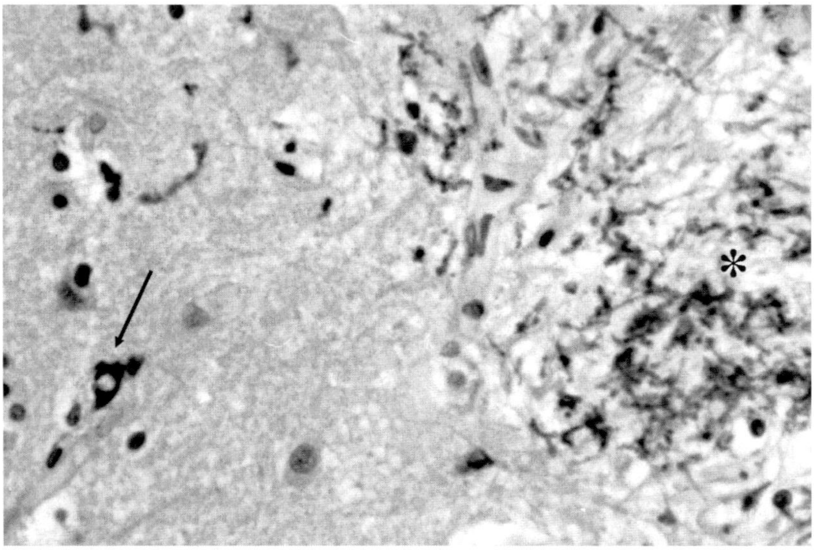

Figure 2. A ferruginated neuron (arrow) and cluster of ferruginated processes (∗).

the loss of an eye or lesions of the retina, optic nerve or tract. Other examples include neurons in the mamillary body following the damage in fornix and in pontine nuclei following frontal leukotomy. In these neurons, both nuclei and perikarya are small compared with controls but cytoplasmic organelles are preserved.

2.1.3. Central chromatolysis

Following injury to the axon, the neuronal cell body becomes rounded with central clearing and Nissl substance is displaced to the periphery (Fig. 3). This reactive change is called central chromatolysis. The nuclei of these neurons are located eccentrically and nucleoli may be enlarged. The chromatolytic central portion reveals abundant mixture of smooth endoplasmic reticulum, dense bodies, Golgi apparatus and neurofilaments at the ultrastructural level [7,8]. Chromatolysis can occur in neurons in any regions of the central nervous system. Common examples are anterior horn neurons of the spinal cord in prolonged peripheral neuropathy and cerebral cortical neurons following subcortical white matter degeneration (Fig. 3). Chromatolysis is a transient reaction of the neuronal soma to the axonal damage. Its

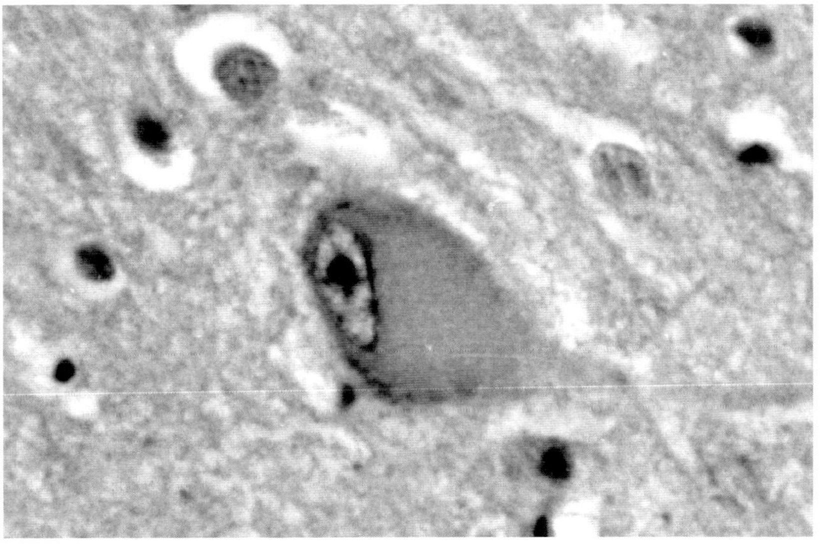

Figure 3. Central chromatolysis of a neuron in the cerebral cortex overlying degenerating white matter.

outcome differs depending on the type of neurons. In general, the neurons whose axons project out of the CNS such as motor neurons in the spinal cord and brainstem tend to regenerate and neurons recover from chromatolytic changes. However, neurons whose projection lies entirely within the CNS tend to degenerate and disappear eventually. Experimental data indicate that the neuronal changes tend to occur more rapidly in young animals than in adult animals and the axonal damage close to the soma tends to cause neurodegeneration rather than chromatolysis. In rodents, chromatolytic changes of neurons may not be apparent and facial nucleus of the mouse has been reported to show increased cytoplasmic basophilia not clearing following axotomy [9,10]. Detachment of afferent synapses and proliferation of microglia and astrocytes around axotomized neurons have been well documented [11]. Central chromatolysis has been well documented in spinal anterior horn cells, Betz cells in the cerebral cortex and neurons in pontine and other brainstem nuclei in Pellagra (niacin (nicotinic acid) deficiency) in humans [12,13]. Unlike chromatolysis observed in axotomized neurons, no associated glial changes are documented. Widespread distribution of chromatolytic neurons in Pellagra, however, suggests that these changes are more likely related to the metabolic perturbation of neuronal soma rather than secondary changes to axonal damage. Central chromatolysis is not apparent in histology sections of the rodent models of niacin deficiency. However, a mouse model of niacin deficiency generated by treatment with 6-aminonicotinamide, an antagonist of niacin, developed skin and gastrointestinal lesions similar to human Pellagra and ultrastructural features of neuronal chromatolysis in the anterior horn of the spinal cord [14].

2.1.4. Cytoplasmic inclusions

Light brown lipopigment, lipofuscin, accumulates in the perikaryal soma of neurons (Fig. 4) with age. The amounts vary but neurons in certain nuclei such as inferior olives in the medulla, lateral geniculate body and dentate nucleus in the cerebellum tend to accumulate lipofuscin even in a relatively young age (Fig. 4). As a natural aging process, numbers of neurons containing lipofuscin and amounts of lipofuscin in individual neurons increase in normal individuals. In ceroidlipofuscinosis, a genetic neurodegenerative disorders, materials having light microscopic staining characteristics of lipofuscin accumulate in neuronal soma as well as in glial cells. Their ultrastructural morphology differs significantly from that of lipofuscin accumulating in normal aging neurons, however [15]. In many lysosomal storage diseases, neuronal soma is packed with disease specific inclusions. On routine paraffin sections, the storage materials are often dissolved during the tissue preparation thus, the neuronal soma appears empty or

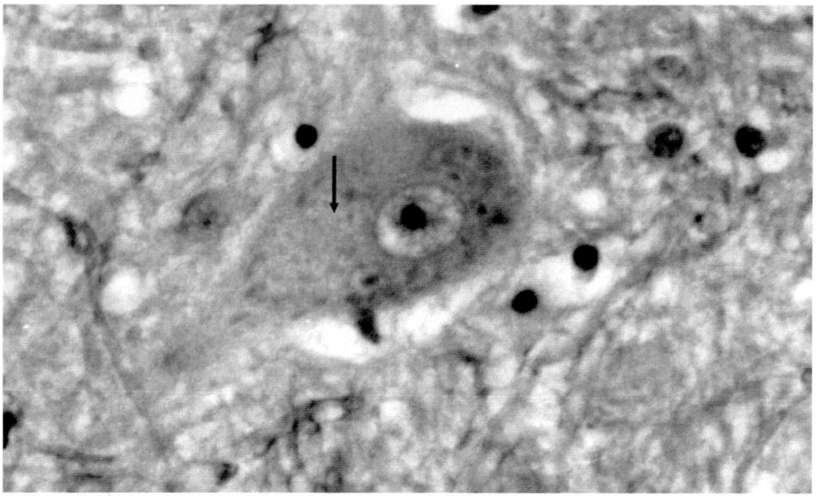

Figure 4. A neuron in the medulla containing lipofuscin in the soma (arrow).

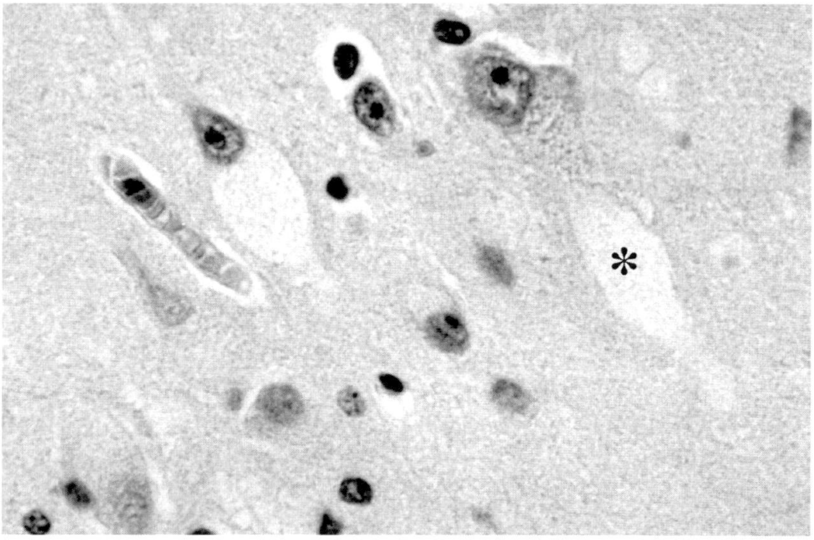

Figure 5. Neuronal soma and meganeurite (∗) that are markedly expanded with storage materials.

finely vacuolated (Fig. 5). The morphology of the inclusion depends on the nature of the stored materials resulting from the specific metabolic perturbation [16]. In addition to the perikarya, abundant inclusions are often found in the expanded axonal hillock region forming meganeurites [17]. Meganeurites can be found in routine paraffin sections (Fig. 5), but are best demonstrated by Golgi impregnation. Aberrant neurites often originating from these meganeurites and aberrant synaptic junction are found in human disease (gangliosidosis) but aberrant neurite formation has not been observed in murine models. Neuronal inclusions similar to those of lysosomal storage disease can be experimentally induced by treatment with various amphiphilic cationic drugs, such as hypocholesterolemic drug AY9944 and chloroquine [18–20]. Neuronal storage has been reported in livestock ingesting locoweed or experimental animals administered the locoweed toxin, swainsonine [21]. In some neurodegenerative disorders causing dementia, referred to as tauopathies, aggregates of hyperphosphorylated tau protein form silver-stain-positive inclusions. Tau is a microtubule-associated protein and is important in assembly and stability of microtubules. In Alzheimer's disease the major component of paired herical filaments that form neurofibril-

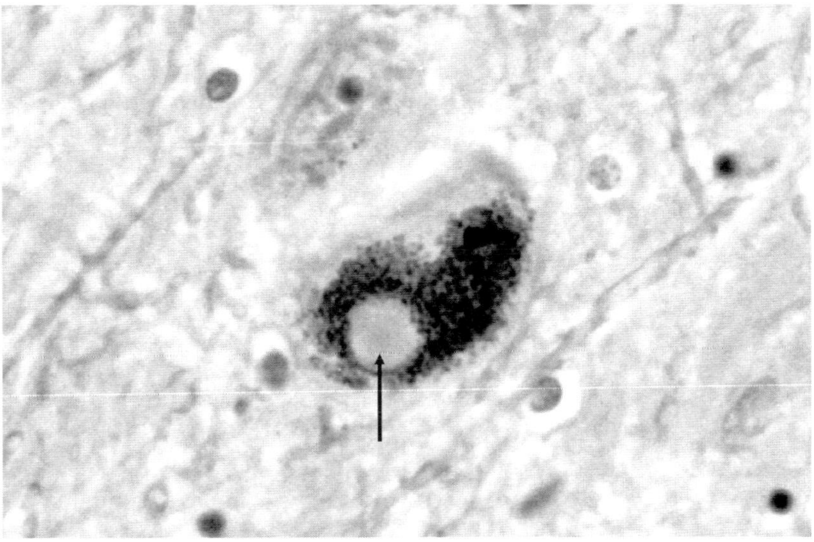

Figure 6. A Lewy body (arrow) in the pigmented neuron in the substantia nigra in Parkinson's disease.

lary tangles are tau proteins. Tau proteins have 6 isoforms and the Western-blot analysis shows different combination of tau isoforms in different types of tauopathies [22]. Lewy bodies are eosinophilic spherical intracytoplasmic inclusions (Fig. 6) observed in the neurons in the substantia nigra, locus ceruleus and autonomic ganglion neurons in patients with Parkinson's disease. Abnormal aggregates of presynaptic protein, α-synuclein, are major components of Lewy bodies, which also are immunoreactive with ubiquitin and phosphorylated neurofilament but not tau protein [23]. In certain dementia patients, Lewy bodies are found, in addition, in the cerebral cortex (dementia with Lewy bodies) [23]. Inclusion similar to Lewy body has been demonstrated in the substantia nigra in experimentally induced Parkinson's disease in non-human primate treated with 1-methyl-4-phenyl-1,2,3,6-tetrahydropyridine (MPTP), although ultrastructural features differ from that of typical Lewy body [24]. In amyotrophic lateral sclerosis (ALS), several types of neuronal inclusions have been detected by immunocytochemical staining for ubiquitin. Bunina bodies are small eosinophilic inclusions often arranged in beaded chains and found in the anterior horn of ALS. In familial type of ALS caused by the mutation of Cu–Zn superoxide dismutase (SOD1), Lewy body-like (or hyaline-like) inclusions have been reported in neurons as well as in astrocytes. These inclusions are immunoreactive to SOD1 [25,26]. Similar inclusions have been documented in both astrocytes and neurons in transgenic mice expressing mutant gene of human SOD1. These inclusions contain abundant SOD1 polypeptide [27]. Hirano body is a deeply eosinophilic round to rod like inclusion found most frequently in the pyramidal layer of the hippocampus in cases of dementia as well as elderly non-demented individuals [28]. It can be found in young normal individuals also. In young individuals, Hirano body is found mostly in the stratum lacunosum not in the pyramidal layer [29]. It is described in varieties of cells in many disorders in animals as well [30]. Deeply eosinophilic needle-like structure has been reported in the thalamus of normal aged mice or mice with chronic degenerative diseases. Light microscopically, these thalamic inclusions resemble the Hirano body, but show different ultrastructure [31]. Homogenous colloid-like inclusions may be found as an incidental finding in the hypoglossal nuclei in human. Thus, varieties of neuronal inclusions are associated with degenerative processes in neurons, indicating that they represent neuronal reactions to perturbed cellular metabolism/function. The mechanism of generating different types of inclusions, however, is not well elucidated.

2.1.5. Neurofibrillary tangles

Neurofibrillary tangles are considered to be a hallmark of Alzheimer's disease. However, they are also

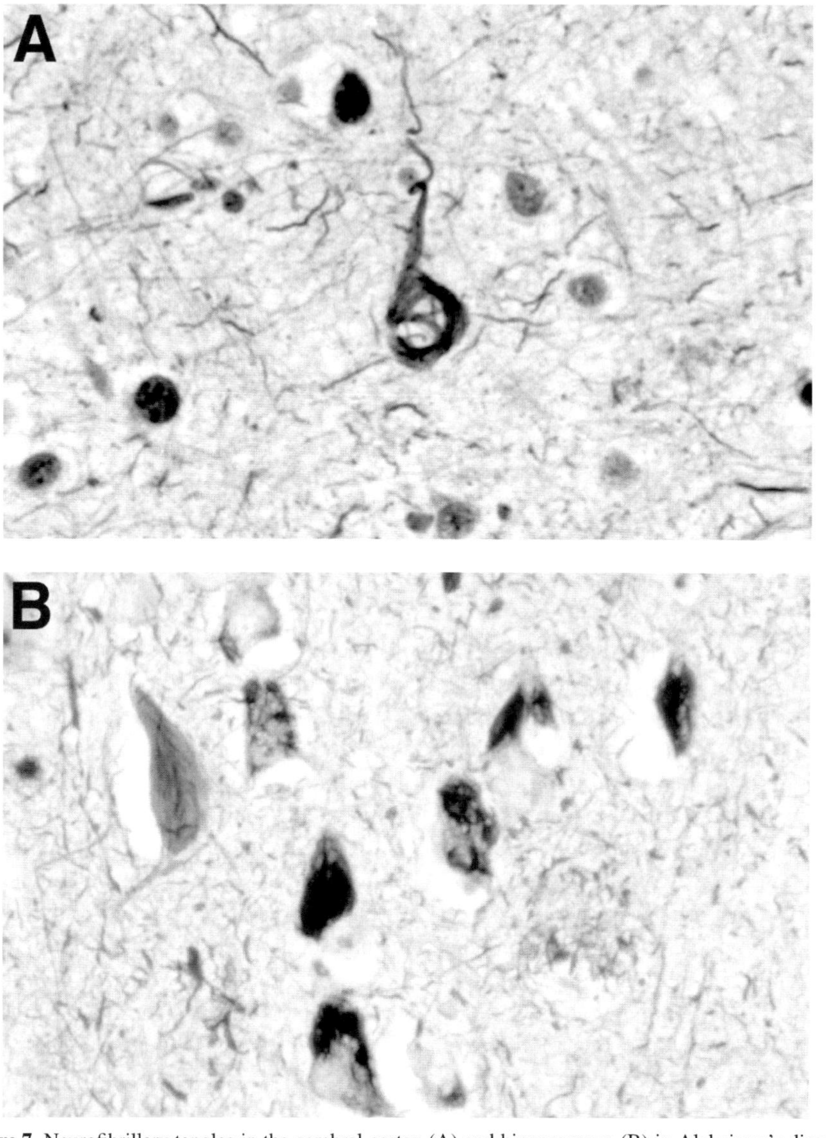

Figure 7. Neurofibrillary tangles in the cerebral cortex (A) and hippocampus (B) in Alzheimer's disease.

seen in the brain of normal elderly individuals and also have been described in varieties of neurodegenerative disorders. The tangles are seen in the brain following repeated trauma, such as dementia pugilistica, following viral infection, such as subacute sclerosing panencephalitis and postencephalitic Parkinsonism, and in various genetic neurological diseases, such as Niemann–Pick disease type C, Fukuyama disease and myotonic dystrophy [32–34]. They are constantly found in the brain of the middle-aged patient with Down's syndrome [35] and are not uncommonly found in the brain of mentally retarded adults [36]. Widespread presence in varieties of neurological disorders strongly suggests that neurofibrillary tangles are not specific neuronal change in Alzheimer's disease, but rather represent neuronal reaction to certain type of metabolic perturbation. They tend to be found in large pyramidal neurons as silver-stain-positive tortuous bundles of thick fibrous structure (Fig. 7A), but configuration may vary according to the type of neurons affected. For example, flame-shaped or torch-like tangles are often seen in hippocampus (Fig. 7B), while yarn-ball type or globose form of tangles are usually seen in the substantia nigra or locus ceruleus. In Niemann–Pick type C disease, not only perikarya but also abnormally enlarged axonal hillock regions, meganeurites, often contain neurofibrillary tangles [33]. The neurofibrillary tangles found in these different diseases are structurally similar, consisting of paired helical filaments (PHFs) at the ultrastructural level, but mixed straight tubules have also been reported. The major antigenic composition of PHFs is tau proteins and thus PHFs immunostain with ubiquitin and phosphorylated tau antibodies. In recent years, many transgenic or knockout mice have been generated as models of Alzheimer's disease. Also, there is a naturally occurring mouse model of Niemann–Pick disease type C that has the same genetic defect as human disease and show closely similar neuropathological changes. However, neurofibrillary tangles have not been seen in any of these murine models.

2.1.6. Granulovacuolar degeneration

Small inclusions consisting of a central hematoxyphilic granules surrounded by clear vacuoles are found in the pyramidal neurons in the hippocampus of patients with Alzheimer's disease and also non-demented elderly individuals. Although the nature of these inclusions is not clear, they tend to be more numerous in Alzheimer's disease. So far, there is no report of the similar structure in experimental animals.

2.2. Dendrites

2.2.1. Dendritic expansion

Focal expansion of dendritic process has been well described in the cerebellar Purkinje cells in developmental malformation, neuronal storage diseases and degeneration. Focal expansion is often seen at the branching points of the dendrites, but sometimes the terminal portion of the dendrites show cactus-like expansion with numerous fine radiating processes [37].

2.2.2. Dendritic branching and spine abnormalities

In Down's syndrome, Fragile X syndrome and other mentally retarded individuals, abnormal dendritic branching pattern, usually associated with paucity of spine or with filopodia-like configuration of spines has been reported [38–40]. Filopodia-like thin, tortuous postsynaptic spines have been also reported in the fragile X knockout mouse that expressed no fragile X mental retardation protein (FMRP) and showed mildly impaired performance on the water maze test [41]. Dendritic abnormalities have also been described in patients with AIDS dementia complex (see Chapter 1.3). In Menkes' kinky hair disease, an X-linked neurodegenerative disorder of copper homeostasis, cerebellar Purkinje cells show atrophic main dendrites with fewer, anomalously oriented dendritic branches and abnormal dendrites originating from the soma [42]. Similar dendritic abnormalities have been described in the murine model of kinky hair disease, brindled mottled mouse [43].

2.3. Axon

2.3.1. Wallerian degeneration

Transection of the nerve fibers results in degeneration of the distal segment. This process is known as Wallerian degeneration following Waller's original description of changes in the transection experiment with glossopharyngeal and hypoglossal nerves in frog [44]. The most common cause of the axonal damage in the CNS is an infarction or trauma in the cerebral white matter. In the periphery of the acutely infarcted regions, or in junctional areas of normal and infarcted necrotic tissue, numerous swollen cut-end of the severed axons, retraction balls, are conspicuous (Fig. 8). Retraction balls are eosinophilic and argyrophilic bulbs on nerve fibers. Diffuse or scattered retraction balls are also seen as result of the diffuse axonal injury in the white matter following head injury [45,46]. It is not uncommon that large numbers of retraction balls are found in one nerve fiber tract, but not in another adjacent fiber tract running in different direction. Retraction balls are acute reaction of the axonal damage 15–18 h after injury and disappear within a few days to weeks, followed by fragmentation of axons and myelin sheaths with intense glial cell reaction. Acute axonal damage can be better demonstrated immunocytochemically using antibody for β-amyloid precursor protein (APP) [47]. An early stage of axonal degeneration, accumulation of multivesicular or multilamellar dense bodies, abnormal mitochondria, microtubules and neurofilaments are identified, but later on, these structures

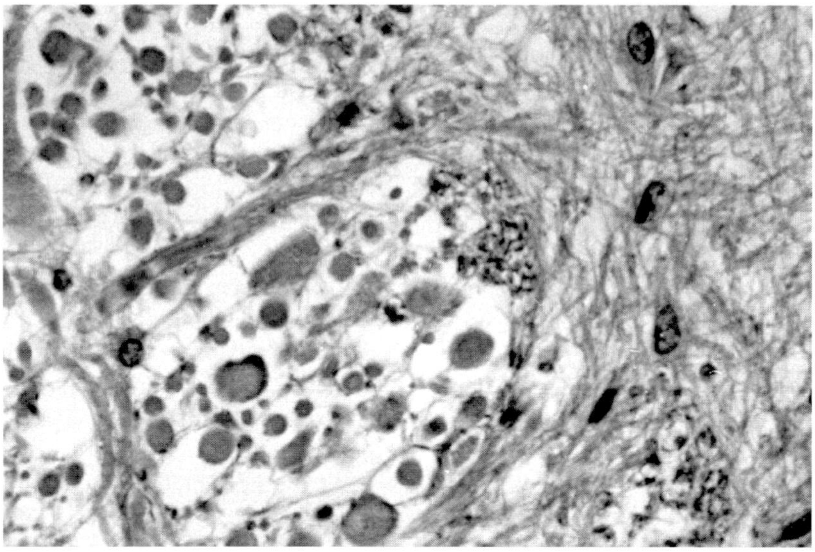

Figure 8. Clustered swollen axons (retraction balls) at the edge of acute cerebral infarction.

disintegrate and are replaced by clumps of granular debris. In the PNS, axonal damage is found in many chronic neuropathies. Teased fiber preparation of the peripheral nerve demonstrates Wallerian degeneration as linear rows of osmiophilic droplets. Wallerian degeneration has been most studied in the PNS, and structural alterations of the damaged axons share similarities in PNS and CNS. However, there is a striking difference between the CNS and PNS in the rate of myelin breakdown and removal. The rate is far quicker in the PNS than the CNS.

2.3.2. Axonal spheroid (focal axonal swelling)

The term axonal spheroid has been used interchangeably with the term neuroaxonal dystrophy, describing focal axonal swelling in certain neurodegenerative disorders, such as infantile neuroaxonal dystrophy, Hallervorden–Spatz disease and Niemann–Pick disease type C (Fig. 9) [48]. In these disorders, presence of axonal spheroids is widespread. Axonal spheroid can be found focally as normal aging process in certain regions of the CNS such as gracile and cuneate nuclei in the medulla [49]. The spheroid is usually eosinophilic and stains positive with silver-stain. Ultrastructurally, the distended axon contains mitochondria, multilamellar dense bodies and tubulo-membranous structure (Fig. 9), but generally few neurofilaments. Experimentally, axonal spheroids can be induced in vitamin E deficient rats [50] and also they are reported in mice lacking the major proteolipid of myelin [51]. Focal swelling of Purkinje cell axons has been well recog-

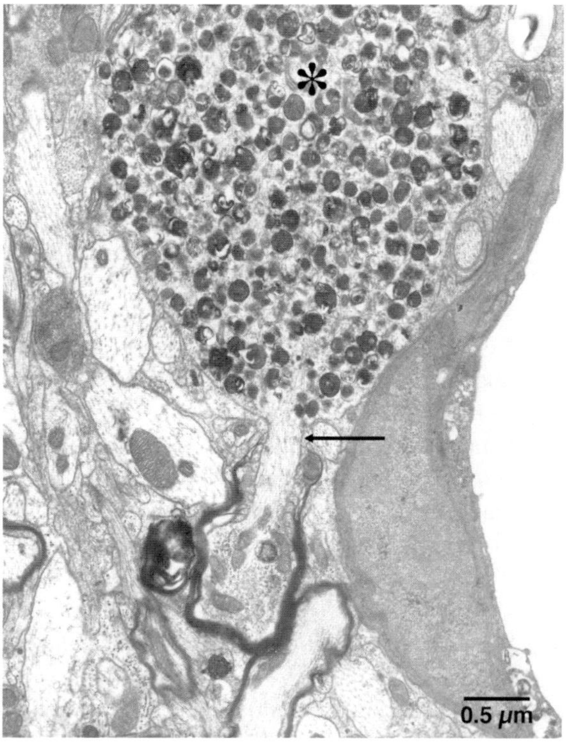

Figure 9. An electron micrograph of dystrophic axon (axonal spheroid, ∗) in the mouse model of Niemann–Pick disease type C. An arrow indicates node of Ranvier.

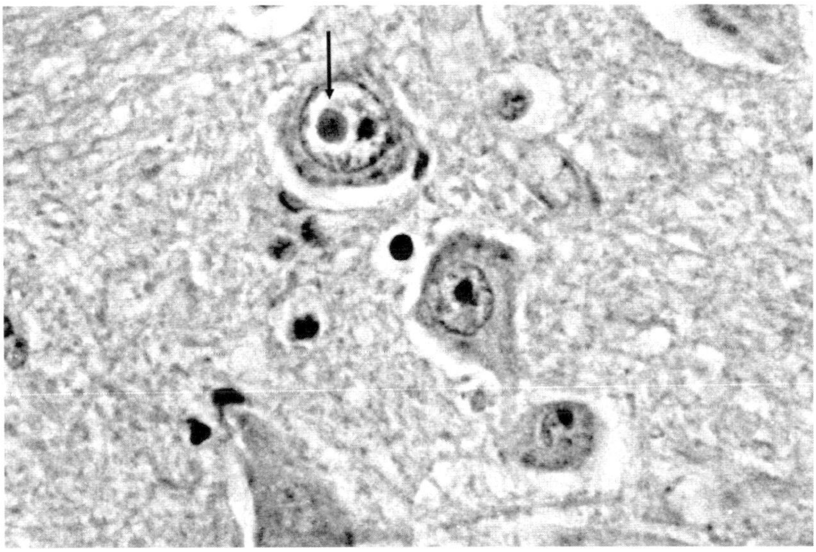

Figure 10. An intranuclear inclusion (arrow) next to the nucleolus in the neuron in the intranuclear inclusion body disease.

nized in cerebellar degeneration in humans. This is often called 'torpedo' in neuropathological literature and considered to represent dysfunctional status of the Purkinje cell. Focal axonal swelling of Purkinje cell axon has been reported in several mouse mutants with neurological disorders [51,52].

2.4. Nucleus

2.4.1. Intranuclear inclusions

Intranuclear inclusion bodies are an important pathological feature of neurons in the viral infection in the CNS and are well documented in herpes simplex encephalitis or subacute sclerosing panencephalitis. These inclusions are round to oval eosinophilic bodies and often surrounded by a clear halo. Similar but somewhat smaller intranuclear inclusion bodies are often found as an incidental finding in the substantia nigra of elderly individuals (Marinesco bodies) [53]. More recently, intranuclear eosinophilic inclusions have been reported in the neurons in intranuclear inclusion body disease and in trinucleotide repeat diseases, such as Huntington's disease, dentato-rubropallido-luisian atrophy (DRPLA) and cerebellar degeneration [54] (Fig. 10).

3. ASTROCYTES

3.1. Reactive Changes

In normal brain, astrocytes are identified by their ovoid nuclei with relatively pale nucleoplasm. Their cytoplasm and processes are not easily detected on the routine paraffin sections. With immunocytochemical stain with GFAP antibody, delicate GFAP positive filaments are identified in the soma and processes. Astrocytes respond to almost all types of insults and show remarkable structural change [55]. In acute response, the soma and processes of astrocytes become hypertrophic and packed with glial filaments that show enhanced GFAP immunoreactivity (Fig. 11). They often show abundant eosinophilic cytoplasm and are called gemistocytic astrocytes (Fig. 12). Their nuclei are slightly enlarged and displaced to the periphery. Binucleated or multinucleated reactive astrocytes may be seen on some occasion. With time, extensive formation of glial filaments leaves permanently abnormal gliotic tissue at the site of injury. Another type of diffuse astrocytic change has been well recognized in the brains of patients with hepatic encephalopathy. Their nuclei are rounded and become pale without recognizable soma at the light microscopic level. They tend to be present as a group of two or four. At the ultrastructural level, the soma becomes vacuolated and electron lucent. These changes of astrocytes are in-

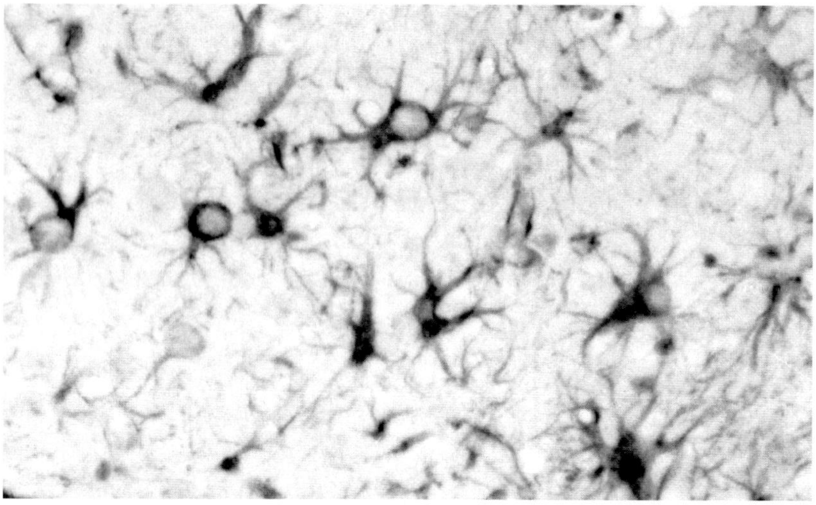

Figure 11. Reactive astrocytes with strong immunostain with GFAP antibody in the thalamus of a 40-day-old twitcher mouse. A murine model of genetic demyelinating disease/Krabbe's disease.

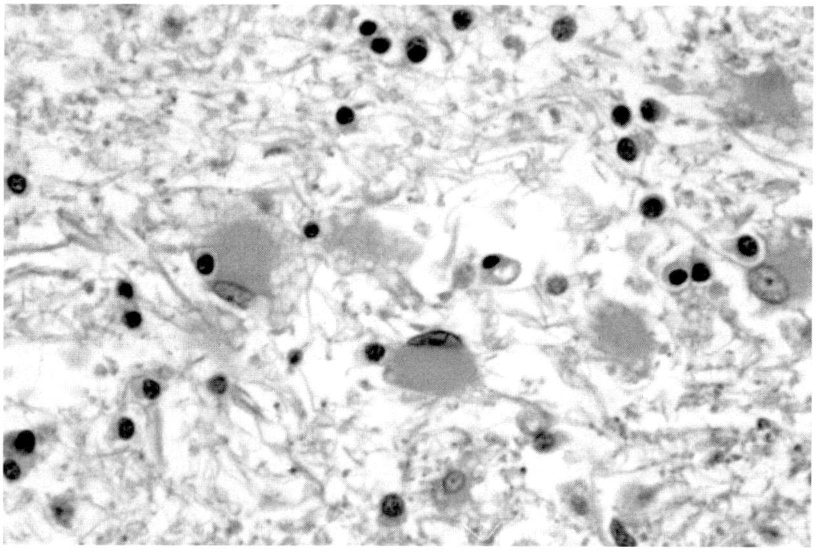

Figure 12. Gemistocytic astrocytes with abundant cytoplasm in adjacent area to radiation necrosis.

terpreted as an ammonium-induced gliopathy and named Alzheimer's type II astrocytes. Similar astrocytic changes are also seen in uremic encephalopathy as well (for review refer to [55]).

3.2. Intracytoplasmic Inclusions

Intracytoplasmic hyalin inclusions (Lewy body-like inclusions) have been reported in the astrocytes in familial amyotrophic lateral sclerosis with SOD1 mutation and also transgenic mice carrying mutant gene [24]. Astrocytic inclusions have also been reported in neurodegenerative disorders, such as progressive supranuclear palsy, corticobasal degeneration and Pick's disease [56]. The inclusions in these disorders consist of microtubule-associated protein tau.

3.3. Gliosis

Even in normal brain, there is a thick network of glial fibers in the subpial area of the cerebrum, subependymal regions of ventricles, aqueduct of

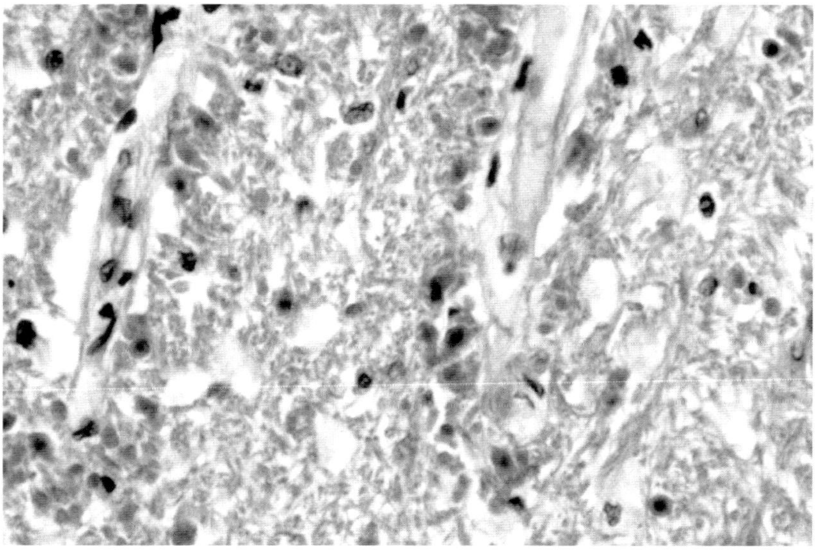

Figure 13. Rosenthal fibers around blood vessels in a plaque of multiple sclerosis.

Sylvius and around the central canal of the spinal cord. With increasing age, glial fibers accumulate more and these regions become gliotic. Gliosis is commonly seen in Ammon's horn in the hippocampal region in patients with chronic epileptic seizures. In these areas of gliosis, spherical bodies consisting of amorphous or lamellated basophilic to amphophilic material, that strongly stain with periodic acid Schiff (PAS) stain, are often found. These bodies are termed corpora amylacea, which are polyglucosan bodies accumulated within the astrocytic processes. In several strains of aging mice older than 9–10 months of age, clusters of PAS-positive granules have been reported in the hippocampus. These granules are closely associated with the processes of astrocytes and their ultrastructural feature resembles that of corpora amylacea [57]. Deeply eosinophilic structures, Rosenthal fibers (Fig. 13), which may be rounded, elongated or oval shaped, are present in areas of gliosis and also certain type of low-grade astrocytic neoplasm. Rosenthal fibers are numerous in Alexander disease and fucosidosis and tend to be accumulated around blood vessels (Fig. 13) or subpial regions. They consist of a dense mass of finely granular osmiophilic material with closely related glial fibers and contain α-B-crystallin [58]. Presence of corpora amylacea and Rosenthal fibers is apparently limited to the human brain. However, transgenic mice that carry copies of human GFAP gene generate hypertrophic astrocytes containing inclusion bodies histologically and antigenically identical to Rosenthal fibers [59,60]. Gliosis can be a sequel of acute insults such as infarct or demyelination. The pattern of gliosis is influenced by environmental factors. In gliosis resulting from destructive or necrotic lesions, glial fibers are arranged haphazardly. This pattern of gliosis is termed unisomorphic gliosis, while orderly gliosis within demyelinating fiber tracts is called isomorphic gliosis. In the latter, the preserved axons govern the orientation of the glial fibers.

4. OLIGODENDROCYTES/MYELIN

4.1. Neuronal Satellitosis

Oligodendrocytes appear normally as perineuronal satellites in the gray matter, and as interfascicular glia in the white matter and a few are located in perivascular regions as perivascular satellites. Satellite oligodendrocytes are commonly seen around medium-sized pyramidal cells of the cortex and large cells in the basal ganglia. In certain neurodegenerative disorders and also in the close vicinity of oligodendroglioma, excessive numbers of satellite oligodendrocytes are present particularly around the deep cortical neurons. Significance of neuronal satellitosis by oligodendrocytes has not been clarified.

4.2. Intracytoplasmic Inclusions

Intracytoplasmic inclusions have been reported in neurons and glial cells in multiple system atrophy. The glial cells with inclusions are immunoreactive with markers for oligodendrocytes such as carbonic anhydrase II, transferrin and myelin-associated glycoprotein (MAG). The inclusions are lightly eosinophilic and stain with silver-stains (Bodian, Bielshowsky, Gallyas stains). They are immunoreactive with ubiquitin, MAP, α-synuclein, α-B-crystallin and tubulin and consisted of loosely packed 20–30 nm filaments [61,62]. Silver-stain-positive intracytoplasmic inclusions are also reported in astrocytes as well as in oligodendrocytes in progressive supranuclear palsy, corticobasal degeneration and Pick's disease [56,63,64].

4.3. 'Dying-Back' Oligodendrogliopathy (Distal Oligodendrogliopathy)

In mice with slowly progressive demyelination following repeated treatment with cuprizone (bis-cyclohexanone oxaldihydrazone), degenerative changes were noted in the inner cytoplasmic tongue before any detectable degenerative change in the cell body of oligodendrocytes. As the inner tongue is the most distal portion of oligodendrocyte cytoplasm, Ludwin and Johnson [65] considered that this type of degenerative change in oligodendrocytes was analogous to the dying-back type of neuropathies in the peripheral nerve [66,67]. Dying-back neuropathies or axonopathies are characterized by the inability of the neuron to maintain the metabolic processes necessary to support the distal end of the axon and the distal end degenerates before any degenerative changes take place in the neuronal soma. Thus, dying back gliopathy in inner tongue can be considered as an early indication of dysfunctional oligodendrocytes. Degeneration of the inner tongue has been well documented in multiple sclerosis [68], Theiler's virus-induced demyelination, Niemann–Pick disease type C (Fig. 14) and in mice with myelin-associated glycoprotein (MAG) deficiency.

4.4. 'Emperipolesis' of Oligodendrocytes

Presence of oligodendrocytes within astrocytic cell body has been repeatedly documented in multiple sclerosis and other demyelinating conditions (Fig. 15) [69–71]. Ghatak [69] has considered that this phenomenon is a non-specific change analogous to the presence of intact lymphocytes within neoplastic cells or leukocytes within megakaryocytes. This phenomenon of internalization of apparently viable cells within other cells is known as emperipolesis [72]. Prineas and co-workers [70] have suggested that oligodendrocytes are being destroyed by astrocytes. Emperipolesis of apparently normal lymphocytes by oligodendrocytes, astrocytes and macrophages/microglia has also been reported in addition to phagocytosis of apoptotic lymphocytes in experimental autoimmune encephalomyelitis (EAE) and has been suggested to represent phagocytosis of cells committed to apoptosis [73].

4.5. Myelin Disorders

Degeneration of myelin sheaths can be a primary pathological event or secondary to axonal injury (Wallerian degeneration). In demyelinating disease and also Wallerian degeneration, myelin sheaths are fragmented and phagocytosed by macrophages. Traditionally, myelin disorders are classified into two large groups: demyelinating disorder (acquired myelin disease) and leukodystrophies (genetic myelin disease). The best known demyelinating disorder is multiple sclerosis. However, demyelination also occurs by many other causes, such as viral infection or toxic metabolic insults. In all demyelinating disorders, the patterns of myelin degeneration and phagocytosis are similar. Macrophages are foamy, containing neutral lipids, brilliantly stain with Sudan IV or oil red O. Inflammatory cell infiltration may or may not be conspicuous. Following demyelination, remyelination takes place, but unlike peripheral nerves, remyelinated fibers in the CNS never regain normal thickness. Myelin is a modified plasma membrane of oligodendrocytes and thus metabolic perturbation of oligodendrocytes results in impaired of myelin formation or maintenance. Thus, leukodystrophies may be caused by degeneration of formed myelin (demyelination) resulting from genetic defect in proper maintenance, or by poor or abnormal myelination (hypomyelination or dysmyelination) due to defective gene for normal myelination. Each leukodystrophy has its own characteristic morphological features (for review refer to [74–76]).

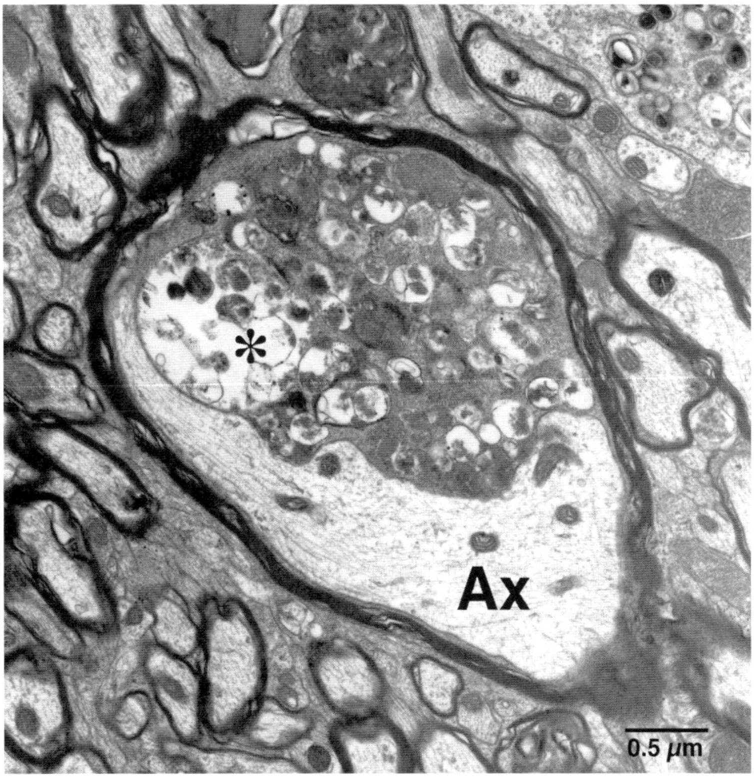

Figure 14. An electronmicrograph of distal oligodendrogliopathy in the mouse model of Niemann–Pick disease type C. The asterisk indicates degenerating cytoplasm of oligodendrocyte. Ax = axon.

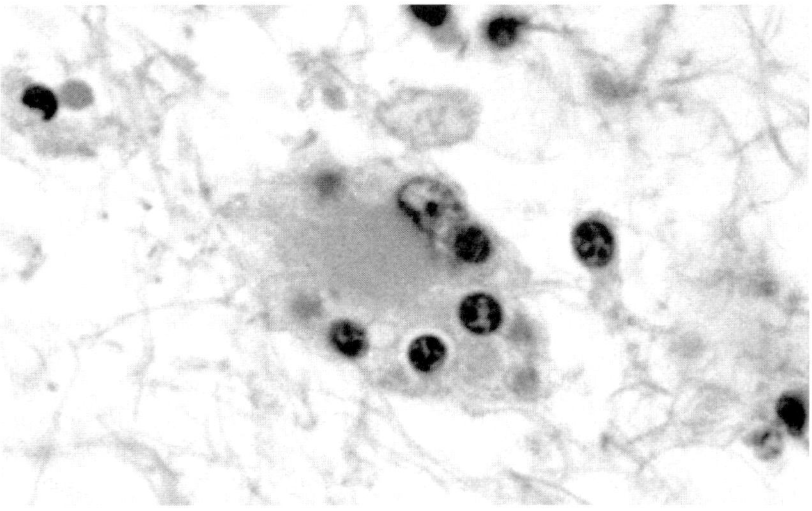

Figure 15. Several oligodendrocytes engulfed by a gemistocytic astrocyte (emperipolesis) adjacent to radiation necrosis.

4.6. Myelin Splitting (Intramyelinic Vacuoles)

In certain toxic metabolic myelin disorders, myelin lamellae split at the intraperiod line forming intramyelinic vacuoles. The affected white matter lesion appears spongy at the light microscopic level. Intramyelinic vacuoles are characteristic features in certain human disorders such as in Canavan's

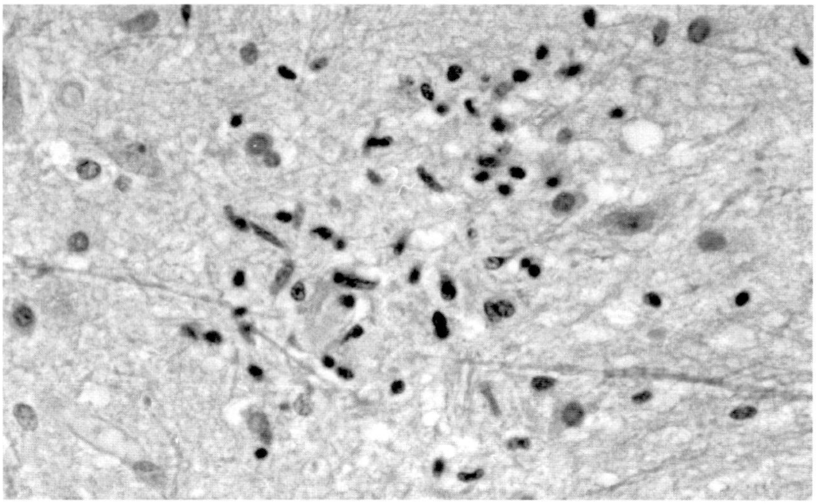

Figure 16. A microglial nodule in AIDS encephalopathy.

disease, central pontine myelinolysis and vacuolar myelopathy in AIDS [74]. Intoxication with triethyltin, hexachlorophene and cuprizone also cause intramyelinic vacuolation (for review refer to [77]). Continuous cuprizone intoxication in mice is followed by primary demyelination. This mouse model of demyelination is very useful for investigation of mechanisms of demyelination and remyelination [65,78].

5. MICROGLIA

5.1. Microglia Activation

The microglia are the most sensitive cells responding to altered environment in the CNS. In normal brain, they exist in a quiescent form, resting or ramified microglia. In response to insult, the microglia become activated and express activation markers including MHC class I and II, CD45 and CD4 and secrete cytokines and cytotoxic substances [79]. Thus, activated microglia are found in almost any pathological lesion in the CNS. From the resting ramified microglia with fine delicate processes, the activated microglia undergo morphological transformation to become almost amoeboid morphology with stout shorter processes (see Fig. 12 in Chapter 1.1). Microglia are the source of brain macrophages. They become phagocytes when the insult results in neuronal death and express a number of immunomolecules. In Alzheimer's disease, activated microglia are seen in the core as well as the periphery of senile plaques. In ALS, activated microglia are abundant in areas of degenerating motor neurons and often surrounding and phagocytosing degenerating neurons (neuronophagia). Clustered activated microglia (microglial nodules) are often seen in viral infections in the CNS (Fig. 16). Neuronophagia was one of the classical features seen in the anterior horn of the spinal cord in acute encephalomyelitis due to polioviruses. These are features of microglia acting as scavengers of damaged cells or tissues. Some morphologically activated microglia may not always express molecules typically found in macrophages and protective role of these cells to neurons from secondary injury are suggested by some investigators [80].

5.2. Perivascular Multinucleated Giant Cells

The pathological hallmark of HIV (AIDS) encephalitis is the presence of multinucleated giant cells in the perivascular region (Fig. 17). These cells express microglia/macrophage markers and are immunoreactive with HIV coat proteins such as gp120 and gp41 and contain high copy numbers of HIV nucleic acid [81]. The perivascular cells are of bone marrow origin and are infected with HIV virus relatively early in course and then, thought to be acting like the Trojan horse for the introduction of HIV into the CNS. The giant cell formation is considered to be the fusion of macrophages under the influence of HIV gp120 protein. HIV-infected perivascular

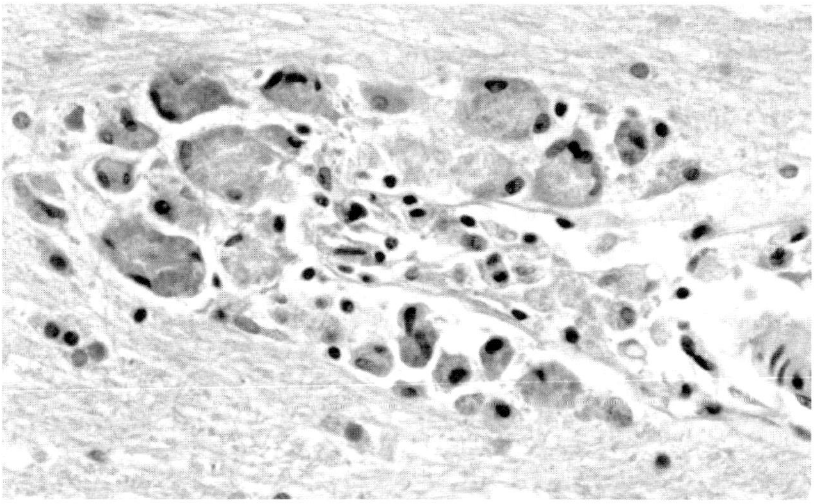

Figure 17. Perivascular multinucleated giant cells in AIDS encephalopathy.

cells (macrophages) may infect adjacent microglia and astrocytes which then secrete cytokines and influence neuronal function causing neurobehavioral problem in AIDS patients. Although very rare, multinuclear giant cells have been described in measles (paramyxo) virus infection [82].

6. EPENDYMA/CHOROID PLEXUS

6.1. Loss of Ependymal Cells/Granular Ependymitis

Ependyma is a cuboidal epithelium that lines the entire ventricular system and is connected with choroid plexus. The ependymal cells are vulnerable to a variety of injuries including infection, hemorrhage and also to simple stretching by hydrocephalus, all of which can cause ependymal cells to disappear. Ependymal cells do not regenerate, leaving denuded gliotic surface directly exposed to the cerebrospinal fluid. The gaps caused by loss of ependyma cells are soon filled with glial processes of subventricular astrocytes. Often proliferated astroglial processes form nodules bulging into the ventricle. If the nodules are large enough they can be seen grossly and the ventricular surface becomes granular. In the past, such ependymal changes were called 'granular ependymitis'. Clustered small ependymal canals (ependymal rosettes) are seen not uncommonly buried by overgrown subependymal astrocytes [83].

6.2. Cytoplasmic Inclusions in Ependyma and Choroid Plexus

In Alzheimer's disease, silver-stain-positive inclusions (Biondi inclusions) have been described in ependymal cells and choroid plexus epithelium. These inclusions contain β-amyloid protein epitopes and are immunoreactive to P component, ubiquitin, fibronectin and tau protein. Their ultrastructural features closely resemble that of neurofibrillary tangles in neurons. Highly significant correlation between the occurrence of Alzheimer's type changes in the cerebral cortex and those in the ependyma and choroid plexus has been reported [84].

7. SCHWANN CELL

The Schwann cell is a myelin-forming cell in the peripheral nervous system (PNS). Unlike oligodendrocytes which myelinate internodes of many nerve fibers, each Schwann cell myelinates an internode of a single myelinated fiber. Therefore, dysfunction or degeneration of the Schwann cell causes segmental demyelination in the peripheral nerve. The initial changes take place at the paranodal region. The lateral loops separate from the axolemma and nodal gap widens by retraction of myelin sheaths from the paranodal region. Myelin degeneration starts at the paranodal region and extends to the entire internode [85]. Remyelination tends to cause shorter internodes as more than one Schwann cell often myelinates the segment previously myelinated by a single Schwann cell. Repeated demyelination and remyelination results in concentric arrangement of many Schwann cell processes around axons and provides appearance of onion bulb formation on cross section of nerve fibers [86]. In slowly progressing axonopathies, intricate axon–Schwann cell networks develop at the paranodal regions by invagination of adaxonal Schwann cell processes. The network formation has been suggested to be a possible mechanism of sequestrating damaged or degenerated axonal organelles [87]. In many leukodystrophies, disease-specific abnormal inclusions are often detected in the Schwann cell cytoplasm [16].

Acknowledgements

The author thanks Dr. Robert C. Bagnell Jr., Microscopy Laboratory, Department of Pathology and Laboratory Medicine for his help in preparation of figures and Ms. Carol Troutner for manuscript preparation.

REFERENCES

1. Brierley JB, Meldrum BS, Brown AW. The threshold and neuropathology of cerebral 'anoxic–ischemic' cell change. Arch Neurol 1973;29:367–374.
2. Brierley JB, Brown AW, Meldrum BS. The nature and time course of the neuronal alterations resulting from oligaemia and hypoglycemia in the brain of *Macaca mulatta*. Brain Res 1971;25:483–499.
3. Auer RN, Wieloch T, Olsson Y, Siesjo BK. The distribution of hypoglycemic brain damage. Acta Neuropathol (Berl) 1984;64:177–191.
4. Soderfeldt B, Kalimo H, Olsson Y, Siesjo BK. Bicuculline-induced epileptic brain injury. Transient and persistent cell changes in rat cerebral cortex in the early recovery period. Acta Neuropathol (Berl) 1983;62:87–95.
5. Auer RN, Kalimo H, Olsson Y, Siesjo BK. The temporal evolution of hypoglycemic brain damage. 1. Light and electron microscopic findings in the rat cerebral cortex. Acta Neuropathol (Berl) 1985;67:13–24.
6. Cohen EB, Pappas GD. Dark profiles in the apparently normal central nervous system: A problem in the electron microscopic identification of early anterograde axonal degeneration. J Comp Neurol 1969;136:375–396.
7. Barron KD, Chiang TY, Daniels AC, Doolin PF. Subcellular accompaniments of axon reaction in cervical motor neurons of the cats. In: Zimmerman HM, ed. Progress in Neuropathology, Vol 1. New York: Grune and Stratton, 1971:255–280.
8. Barron KD. Comparative observations on the cytologic reactions of central and peripheral nerve cells to axotomy. In: Kao CC, Bunge RP, Reier JP, eds. Spinal Cord Reconstruction. New York: Raven Press, 1983:7–40.
9. Trovik A. Central chromatolysis and the axon reaction: a reappraisal. Neuropathol Appl Neurobiol 1976;2:423–432.
10. Trovik A, Heding A. Histological studies on the effect of actinomycin D on retrograde nerve cell reaction in the facial nucleus of mice. Acta Neuropathol (Berl) 1967;9:146–157.
11. Watson WE. Cellular responses to axotomy and to related procedures. Br Med Bull 1974;30:112–115.
12. Hauw J-J, DeBaecque C, Hausser-Hauw C, Serdaru M. Chromatolysis in alcoholic encephalopathies. Brain 1988;111:843–857.
13. Ishii N, Nishihara Y. Pellagra among chronic alcoholics: clinical and pathological study of 20 autopsy cases. J Neurol Neurosurg Psychiatry 1981;44:209–215.
14. Aikawa H, Suzuki K. Lesions in the skin, intestine, and central nervous system induced by antimetabolite of niacin. Am J Pathol 1986;122:335–342.
15. Goebel HH, Mole SE, Lake BD, eds. The Neuronal Ceroid Lipofuscinoses (Batten Disease). Amsterdam: IOS Press, 1999.
16. Suzuki K, Suzuki K. Lysosomal diseases. In: Graham DI, Lantos PL, eds. The Greenfield's Neuropathology, 7th edn. London, Arnold, in press, 2002.
17. Purpura DP, Suzuki K. Distortion of neuronal geometry and formation of aberrant synapses in neuronal storage disease. Brain Res 1976;116:1–21.
18. Gleiser CA, Bay WW, Dukes TW, Brown RS, Read WK, Pierce KR. Study of chloroquine toxicity and a drug-induced cerebrospinal lipodystrophy in swine. Am J Pathol 1968;53:27–45.
19. Suzuki K, Zagoren JC, Gonatas J, Suzuki K. Ultrastructural study of neuronal cytoplasmic inclusions produced by hypocholesterolemic drug, AY9944. Acta Neuropathol (Berl) 1973;26:185–197.
20. Lullmann H, Lullmann-Rauch, R, Wassermann O. Lipidosis induced by amphiphilic cationic drugs. Biochem Pharmacol 1978;27:1103–1108.
21. Walkley SU, James LF. Locoweed-induced neuronal storage disease characterized by meganeurite formation. Brain Res 1984;324:145–150.
22. Buee L, Delacourte A. Comparative biochemistry of tau in progressive supranuclear palsy, corticobasal degeneration, FTDP-17 and Pick's disease. Brain Pathol 1999;9:681–693.
23. Hashimoto M, Masliah E. Alpha-synuclein in Lewy body disease and Alzheimer's disease. Brain Pathol 1999;9:707–720.
24. Forno LS, Langston JW, DeLanney LE, Irwin I. An electron microscopic study of MPTP-induced inclusion bodies in an old monkey. Brain Res 1988;448:150–157.
25. Kato S, Shimono M, Watanabe Y, Nakashima K, Takahashi K, Ohama E. Familial amyotrophic lateral sclerosis with a two base pair deletion in superoxide dismutase 1 gene: multisystem degeneration with intracytoplasmic

hyaline inclusions in astrocytes. J Neuropathol Exp Neurol 1996;55:1089–1101.

26. Shibata N, Hirano A, Kobayashi M, Siddique T, Deng H-X, Hung W-Y, Kato T, Asayama K. Intense superoxide dismutase-1 immunoreactivity in intracytoplasmic hyaline inclusions of familial amyotrophic lateral sclerosis with posterior column involvement. J Neuropathol Exp Neurol 1996;55:481–490.
27. Bruijn LI, Becher MW, Lee MK, Anderson KL, Jenkins NA, Copeland NG, Sisodia SS, Rothstein JD, Borcheit DR, Price DL, Cleveland DW. ALS-linked SOD1 mutant G85R mediates damage to astrocytes and promotes rapidly progressive disease with SOD1-containing inclusions. Neuron 1997;18:327–338.
28. Gibson P, Tomlinson BE. The numbers of Hirano bodies in the hippocampus of normal and demented subjects with Alzheimer's disease. J Neurol Sci 1977;33:199–206.
29. Ogata J, Budzilovich GN, Cravioto H. A study of rod-like structure (Hirano bodies) in 240 normal and pathological brains. Acta Neuropathol 1972;21:61–67.
30. Hirano A. Hirano bodies and related neuronal inclusions Neuropathol Appl Neurobiol 1994;20:3–11.
31. Aikawa H, Suzuki K, Iwasaki Y. Ultrastructural Observations on the thalamic neuronal inclusions in the young mouse. Acta Neuropathol (Berl) 1983;59:316–318.
32. Corsellis JAN, Bruton CJ, Freeman-Browne D. The aftermath of boxing. Psychol Med 1973;3:270–303.
33. Suzuki K, Parker C, Penchev PG, Katz D, Ghetti B, D'Agostino AN. Neurofibrillary tangles in Niemann–Pick type C. Acta Neuropathol (Berl) 1995;89:227–238.
34. Wisniewski K, Jervis GS, Moretz RC, Wisniewski HM. Alzheimer neurofibrillary tangles in diseases other than senile and pre-senile dementia. Ann Neurol 1979;5:288–294.
35. Burger P, Vogel FS. The development of pathological changes of Alzheimer's disease and senile dementia in patients with Down's syndrome. Am J Pathol 1973;73:457–476.
36. Popovitch ER, Wisniewski HM, Barcikowska M, Silverman W, Bancher C, Sersen E, Wen GY. Alzheimer neuropathology in non-Down's syndrome mentally retarded adults. Acta Neuropathol 1990;80:362–367.
37. Soffer D, Grotsky HW, Rapin I, Suzuki K. Cockayne Syndrome — Unusual Neuropathological Findings and Review of the Literature. Ann Neurol 1979;6:340–348.
38. Purpura DP. Dendritic spine 'dysgenesis' and mental retardation. Science 1974;186:1126–1128.
39. Marin-Padilla M. Pyramidal cell abnormalities in the motor cortex of a child with Down's syndrome. A Golgi study. J Comp Neurol 1976;167:63–82.
40. Takashima S, Becker L, Armstrong DL, Chan F. Abnormal neuronal development in the visual cortex of the human fetus and infant with Down's syndrome. A quantitative and qualitative Golgi study. Brain Res 1981;225:1–21.
41. Comery TA, Harris JB, Willems PJ, Oostra BA, Irwin SA, Weiler IJ, Greenough WT. Abnormal dendritic spines in fragile X knockout mice: maturation and pruning deficit. Proc Natl Acad Sci USA 1997;94:5401–5404.
42. Purpura DP, Hirano A, French JH. Polydendritic Purkinje cells in X-chromosome linked copper malabsorption: A Golgi study. Brain Res 1976;117:125–129.
43. Yamano T, Suzuki K. Abnormalities in Purkinje Cell Arborization in Brindled Mouse Cerebellum. A Golgi Study. J Neuropathol Exp Neurol 1985;44:97–107.
44. Waller AV. Experiments on the section of the glossopharyngeal and hypoglossal nerves of the frog, and observations of the alterations produced thereby in the structure of their primitive fibres. Phil Trans Roy Soc Lond 1850;140:423–469.
45. Strich SJ. Shearing of nerve fibers as a cause of brain damage due to head injury. Lancet 1961;ii:443–448.
46. Adams JH, Graham DI, Murray LS, Scott G. Diffuse axonal injury due to non-missile injury in humans: an analysis of 45 cases. Ann Neurol 1982;12:557–563.
47. Sherriff FE, Bridges LR, Gentleman SM, Sivaloganathan S, Wilson S. Markers of axonal injury in post mortem human brain. Acta Neuropathol (Berl) 1994;88:433–439.
48. Seitelberger F. Neuroaxonal dystrophy: Its relation to aging and neurological diseases. In: Vinken PJ, Bruyn GW, Klawans HL, eds. Handbook of Clinical Neurology, Vol. 5: Extrapyramidal Disorders. Amsterdam: Elsevier, 1986:391–415.
49. Brannon W, McCormick W, Lampert P. Axonal dystrophy in the gracile nucleus of man. Acta Neuropathol (Berl) 1967;9:1–6.
50. Lampert P, Blumberg JM, Pentschew A. An electron microscopic study of dystrophic axons in the gracile and cuneate nuclei of vitamin E-deficient rats. Axonal dystrophy in vitamin-E deficiency. J Neuropathol Exp Neurol 1964;23:60–77.
51. Griffiths I, Klugmann M, Anderson T, Yool D, Thomson C, Schwab MH, Schneider A, Zimmermann F, McCulloch M, Nadon N, Nave K-A. Axonal swelling and degeneration in mice lacking the major proteolipid of myelin. Science 1998;280:1610–1613.
52. Suzuki K, Zagoren JC. Focal axonal swelling in cerebellum of quaking mouse: light and electron microscopic studies. Brain Res 1975;85:38–43.
53. Yuen P, Baxter DW. The morphology of Marinesco bodies (paranucleolar corpuscles) in the melanin pigmented nuclei in the brainstem. J Neurol Neurosurg Psychiatry 1963;26:178–183.
54. Becher MW, Kotzuk JA, Sharp AH, Davies SW, Bates GP, Price DL, Ross CA. Intranuclear neuronal inclusions in Huntington's disease and dentatorubral and pallidoluysian atrophy: correlation between the density of inclusions and IT15 CAG triplet repeat length. Neurobiol Dis 1998;4:387–397.
55. Norenberg MD. Astrocyte response to CNS injury. J Neuropathol Exp Neurol 1994;53:213–220.
56. Komori T. Tau-positive glial inclusions in progressive supranuclear palsy, corticobasal degeneration and Pick's disease. Brain Pathol 1999;9:663–679.
57. Jucker M, Walker LC, Kuo H, Tian M, Ingram DK. Age-related fibrillar deposits in brains of C57BL/6 mice. Mol Neurobiol 1994;9:125–133.
58. Iwaki T, Kume-Iwaki A, Leim RKH, Goldman JE. Alpha-B-crystallin is expressed in non-lenticular tissues and accumulates in Alexander's disease brain. Cell 1989;57:71–78.

59. Messing A, Head MW, Galles K, Galbreath EJ, Goldman JE, Brenner M. Fatal encephalopathy with astrocyte inclusions in GFAP transgenic mice. Am J Pathol 1998;152:391–398.
60. Eng LF, Lee YL, Kwan H, Brenner M, Messing A. Astrocytes cultured from transgenic mice carrying the added human glial fibrillary acidic protein gene contain Rosenthal fibers. J Neurosci Res 1998;53:353–360.
61. Papp MI, Kahn JE, Lantos PI. Glial cytoplasmic inclusions in the CNS of patients with multiple system atrophy (Striatonigral Degeneration, Olivopontocerebellar atrophy and Shy–Drager Syndrome). J Neurol Sci 1989;94:79–100.
62. Dickson DW, Lin W-L, Liu W-K, Yen S-H. Multiple system atrophy: A sporadic synucleinopathy. Brain Pathol 1999;9:721–732.
63. Chin SS-M, Goldman JE. Glial inclusions in CNS degenerative diseases. J Neuropathol Exp Neurol 1996;55:499–508.
64. Feany MB, Dickson DW, Widespread cytoskeletal pathology characterizes corticobasal degeneration. Am J Pathol 1995;146:1388–1396.
65. Ludwin SK, Johnson ES. Evidence for a 'Dying-Back' gliopathy in demyelinating disease. Ann Neurol 1981;9:301–305.
66. Cavanagh JB. The 'dying-back' process. A common denominator in many naturally occurring and toxic neuropathies. Arch Pathol Lab Med 1979;103:659–664.
67. Spencer PS, Schaumburg HH. Central–peripheral distal axonopathy-the pathology of dying-back polyneuropathies. In: Zimmerman HM, ed. Progress in Neuropathology, Vol. 3. New York: Grune and Stratton, 1976:253–295.
68. Rodriguez M, Scheithauer BW, Forbes G, Kelly PJ. Oligodendrocyte injury is an early event in lesions of multiple sclerosis. Mayo Clin Proc 1993;68:627–636.
69. Ghatak NR. Occurrence of oligodendrocytes within astrocytes in demyelinating lesions. J Neuropathol Exp Neurol 1992;51:40–46.
70. Prineas JW, Kwon EE, Goldenberg PZ, Cho ES, Sharer LR. Interaction of astrocytes and newly formed oligodendrocytes in resolving multiple sclerosis lesions. Lab Invest 1990;63:624–636.
71. Wu E, Raine CS. Multiple sclerosis. Interactions between oligodendrocytes and hypertrophic astrocytes and their occurrence in other non-demyelinating conditions. Lab Invest 1992;67:88–99.
72. Humble JG, Jayne WHW, Pulvertaft RJV. Biological interaction between lymphocytes and other cells. Br J Haematol 1956;2:283–294.
73. Nguyen KB, Pender MP. Phagocytosis of apoptotic lymphocytes by oligodendrocytes in experimental autoimmune encephalomyelitis. Acta Neuropathol (Berl) 1998;95:40–46.
74. Suzuki K. Demyelinating disorders and leukodystrophy. In: Galcia JH, ed. Neuropathology: The Diagnostic Approach. St Louis: Mosby-Year Book, 1997:669–715.
75. Prineas JW, McDonald WI. Demyelinating disease. In: Graham DI, Lantos PL, eds. The Greenfield's Neuropathology, 6th edn. London: Arnold, 1997:813–881.
76. Ludwin S. The pathobiology of the oligodendrocyte. J Neuropathol Exp Neurol 1997:56;111–124.
77. Spencer PS, Schaumburg HH. Experimental and Clinical Neurotoxicology. Baltimore, MD: Williams and Wilkins, 1980.
78. Hiremath MM, Saito Y, Knapp G, Ting JP-Y, Suzuki K, Matsushima GK. Macrophage/microglial accumulation during cuprizone-induced demyelination in C57BL/6 mice. J Neuroimmunol 1998;92:38–49.
79. Gehrmann J, Matsumoto Y, Kreutzberg GW. Microglia: Intrinsic immunoeffector cell of the brain. Brain Res Rev 1995:20;269–287.
80. Banati RB, Graeber MB. Surveillance, intervention and cytotoxicity: is there a protective role of microglia? Dev Neurosci 1994;16:114–127.
81. Kure K, Lyman W, Weidenheim K, Dickson DW. Cellular localization of an HIV-1 antigen in subacute AIDS encephalitis using an improved double-labeling immunohistochemical method. Am J Pathol 1992;136:1085–1092.
82. Esiri MM, Oppenheimer DR, Brownell B, Haire M. Distribution of measles antigen and immunoglobulin-containing cells in the CNS in subacute sclerosing panencephalitis (SSPE) and atypical measles encephalitis. J Neurol Sci 1981;53:29–43.
83. Sarnat HB. Ependymal reaction to injury. A review. J Neuropathol Exp Neurol 1995;54:1–15.
84. Miklossy J, Kraftsik R, Pillevutt O, Lepori D, Genton C, Bosman FT. Curly fiber and tangle-like inclusions in the ependyma and choroid plexus — A pathogenetic relationship with the cortical Alzheimer-type changes? J Neuropathol Exp Neurol 1998;57:1202–1212.
85. Webster H deF, Spiro D, Waksman B, Adams RD. Phase and electron microscopic studies of experimental demyelination. 2. Schwann cell changes in guinea pig sciatic nerves during experimental diphtheria neuritis. J Neuropathol Exp Neurol 1961;20:5–34.
86. Spencer PS, Thomas PK. Ultrastructural studies of the dying-back process. II. The sequestration and removal by Schwann cells and oligodendrocytes of organelles from normal and diseased axons. J Neurocytol 1974;3:763–783.
87. Dyck PJ. Experimental hypertrophic neuropathy. Pathogenesis of onion-bulb formations produced by repeated tourniquet applications. Arch Neurol 1969;21:73–95.

Universes in Delicate Balance: Chemokines and the Nervous System
R.M. Ransohoff, K. Suzuki, A.E.I. Proudfoot, W.F. Hickey and J.K. Harrison, editors

Chapter 1.3

Patterns of Tissue Pathology in Neurological Diseases

Kinuko Suzuki

Department of Pathology and Laboratory Medicine, University of North Carolina at Chapel Hill, Chapel Hill, NC 27599-7525, USA

1. INTRODUCTION

Cellular elements in the nervous system can respond to injuries only in certain ways. Therefore, there is inevitable overlap in the pathology at the cellular level. However, when pathology in certain diseases is expressed at the tissue level, each disease tends to follow certain disease specific or characteristic patterns. In this chapter some neuropathological patterns noted in certain neurological disorders (diseases) will be reviewed briefly.

2. Wallerian degeneration following damage of nerve fibers

Wallerian degeneration is the commonest tissue reaction that occurs when the continuity of nerve fibers is disrupted [1]. Wallerian degeneration induces changes in the axon, myelin sheaths and myelin forming cells, in the CNS and PNS, and endo- or perineurial cells in the PNS [2–8]. Damaged axons and myelin sheaths are removed by intrinsic phagocytes or infiltrated hematogenous macrophages [9]. In the clinical setting, any destructive lesions in the brain that include tumor, trauma and stroke (hemorrhage and infarction) can damage nerve fibers so that their distal portions undergo degeneration. Traumatic transection or crush is a common cause of Wallerian degeneration in the peripheral nerves. The Wallerian degeneration in the CNS and PNS shares many similarities. Earliest changes following nerve transection have been well documented at the ultrastructural level in several earlier studies [2–5]. Following the injury, the cut end of the nerve fibers is swollen, accumulating mitochondria, multivesicular bodies and lamellar electron dense bodies at nodes of Ranvier, especially on the distal side. Disappearance of microtubules, fragmentation of neurofilaments and swelling of mitochondria follow these earlier changes. Then, the damaged axon is filled with electron dense granular debris. An early change of myelin sheaths is a retraction from the node of Ranvier. The numbers of Schmidt–Lanterman incisures increase [3]. Interruption of myelin occurs immediately adjacent to the incisures and the interrupted regions seal off at either end to form the ellipsoids [10,11]. In teased fiber preparation of osmium fixed nerve fibers, segmentation of the myelin into linear rows of ellipsoids has been clearly demonstrated at the light microscopic level as characteristic feature of Wallerian (axonal) degeneration. During the process of myelin breakdown and removal in the PNS, Schwann cells proliferate within the basal lamina tubes and form continuous columns of tightly packed Schwann cells and their processes (Bungner bands). The axon sprouts from the proximal end enter the Bungner bands to be myelinated in regeneration process. Sequences of cellular events occurring in the Wallerian degeneration in the peripheral nerves have been reviewed recently [9]. Early changes in the Wallerian degeneration in the CNS studied in the spinal cord or optic nerve are similar to those reported in the PNS [12–14]. In the CNS, astrocytes and microglia proliferate in response to the axonal degeneration [15,16] and phagocytic cells remove products of degeneration. The role of debris removal by Schwann cells in the PNS or astrocytes and oligodendrocytes in the CNS is very limited [17]. The rate of removal of debris is far greater in the PNS than CNS [18] and there is a significant difference in the response of macrophages in debris removal between the CNS and PNS [19].

3. CEREBRAL ISCHEMIA AND STROKES

Cerebral ischemia is a condition in which blood supply and oxygen delivery to the brain are both inadequate. Since they have the highest demand for oxygen, the neurons are the most vulnerable cells in ischemia, followed in declining order by oligodendrocytes and astrocytes. A stroke is a commonly used term by lay people implying a sudden neurological deficit attributable to hemorrhage or ischemic injury to the brain. The most severe form of ischemic injury is a cerebral infarction, which is defined as the total necrosis of the cells and tissue elements as results of a critical reduction of blood supply. The transition from the area of infarction to surviving tissue in histological section is surprisingly sharp. Ischemia or ischemic injury in the brain is usually divided into two groups, focal cerebral ischemia (FCI) and global cerebral ischemia (GCI) [20,21]. A reduction of blood supply within the defined territory of the artery due to narrowing or occlusion of the lumen causes the former. The latter is a consequence of insufficient blood supply of the entire brain resulting from a systemic circulatory failure. If ischemia is severe enough, either may develop brain infarction. However, GCI does not result in a predictable pattern of brain lesions because of differential vulnerability of selected neuronal groups. It is well known that CA1 region of the hippocampus, cerebellar Purkinje cells, neurons in the layers 3 and 5 of the cerebral cortex and neurons in the globus pallidus are susceptible to ischemia. Also, cortical neurons in arterial border zones (watershed areas) are vulnerable to GCI. In the immediately post-ischemic period, the brain is hyperemic and markedly swollen and necrosis may not be detected. In patients who survived more than several days after GCI, neuronal necrosis in hippocampus, cerebral cortex and other susceptible regions becomes apparent [22]. The lesions of global ischemia are usually bilateral and relatively symmetrical in distribution. At sites of neuronal necrosis, reactive astrocytes with GFAP immunoreactivity are present. The sequence and time course of events following hypoxic/ischemic injury to neurons have been well defined in both human and experimental animals [20,21]. The earliest stage of ischemic changes observed in neurons is microvacuolation of the perikaryon [23,24]. This microvacuolation represents swollen mitochondria, dilated endoplasmic reticulum and other intracytoplasmic organelles. These earliest changes are detected mostly in the brains of experimental animals that are well fixed by the intracardiac perfusion but are difficult to detect in the human postmortem brain. The next stage is ischemic neuronal changes. As stated in Chapter 1.2, the affected neurons are shrunken with deeply acidophilic (eosinophilic) cytoplasm. Kalimo and co-workers [20] have estimated the natural history of evolving infarct in human brain as follows. Microvacuolation may be detected within the ischemic lesion within 1 hour. Eosinophilic shrunken neurons become visible within 4 hours after the insult and invasion of neutrophils takes place 15 to 24 hours later. Infiltration of blood-derived phagocytes occurs 2–3 days later and reactive astrocytes surround the necrotic (infarcted) tissue 1 week later. Phagocytosis of the tissue debris would take months to years depending on the size of the lesion and the end result is the cavity lined by astrocytes and filled with clear fluid. Chronological events occurring in the experimental animal models are similar, although the time course may differ depending on the type of the model used for the experiments [21]. In animal experiments of FCI, the tissue damage consisting of central core of densely ischemic tissue is usually surrounded by perifocal zone of less dense ischemia perfused by blood flow from the surrounding collateral vessels (Penumbra zone) [25]. The cells of the central core of FCI become irreversibly injured while cells of the perifocal zone are less severely damaged and thus this injury is reversible if adequate blood supply is restored.

4. NEURODEGENERATIVE DISEASES AND ASSOCIATED INFLAMMATION

The term degenerative is usually used in describing pathological changes of the brain. However, with aging process, signs of degeneration are well appreciated even in the brains of normal individuals.

When the brains of normal individuals at age 25 are compared with those at 75, differences are obvious even macroscopically. The leptomeninges covering the brain of an old individual are fibrotic while they are more transparent in a young individual. The brain weight of an older individual is less than that of a younger one. There is significant atrophy of cerebral gyri, loss of white matter volume and dilatation of the ventricles. Cerebral cortex is usually light brown in color at young age but with age, brown tinge gradually increases due to an accumulation of lipofuscin pigments in neurons and to a lesser extent in glial cells. Neurodegenerative diseases usually occur in adults whose brain already shows some evidence of degenerative changes. Microscopically several changes take place in the brain of older individuals. Apparently there is a significant loss of neurons with age and remaining neurons show changes in their dendrites and synapses and accumulate lipofuscin pigments, that are generally thought to be waste products. Accumulation of lipofuscin varies in different groups of neurons. For example, Purkinje cells show little or none, while large amounts accumulate in the inferior olive neurons in the medulla, dentate neurons in the cerebellum, neurons in the lateral geniculate bodies (nuclei) and in the cranial and spinal motor nuclei. In studies with non-human primates, lipofuscin pigments are more conspicuous in glial cells in young animals compared with older animals, raising the possibility that intraneuronal accumulation of lipofuscin with age may be a result of inefficient removal of waste products by glial cells [26]. In support of this hypothesis, Cavanagh and co-workers have shown that infusion of the protease inhibitor leupeptin inhibits the excretion of neuronal lysosomal contents to astrocytes [27]. Other structures that increase with aging process are ubiquitin immunoreactive dot-like structures in both gray and white matter which are thought to be located in dystrophic neuronal processes and/or in glial cells. As ubiquitin is a protein attached to cell protein destined to be degraded [28], the presence of ubiquitin positive structures indicates ongoing degenerative changes in aging brain. Hypertrophic astrocytes and increased numbers of reactive microglia have been well documented as normal aging process. For more details on the aging changes in the CNS, readers are advised to refer a recent review [29].

More disease-specific distribution of the pathological process and cellular changes has been recognized in neurodegenerative diseases. Neurodegenerative diseases are the diseases of neurons, primarily affecting certain systems of neurons. The diseases primarily affecting cerebral cortex present clinically as dementia such as Alzheimer disease (AD). The diseases affecting basal ganglia and related systems manifest as involuntary movement disorders. Parkinson disease (PD) and Huntington disease (HD) belong to this category. Amyotrophic lateral sclerosis (ALS) is a disease in which motor neurons degenerate systematically and thus progressive motor weakness is the major clinical symptom. Ataxia is the major symptom of diseases affecting cerebellum primarily. Thus, clinical symptoms and distribution of regional pathological processes differ in individual diseases, although the basic cellular pathology is the same, namely neuronal degeneration.

The underlying mechanisms for this selective neuronal degeneration are subject of intense investigation. Activated glial cells express many components associated with peripheral immunological and inflammatory states and these may contribute, to some extent, to the progression of the disease process [30–33]. However, there is very little evidence of inflammatory cellular infiltrates such as B or T lymphocytes, in the CNS regions in neurodegenerative diseases, as noted in the inflammatory disease process seen in visceral organs [34]. In AD, neurofibrillary tangles and senile (neuritic) plaques have been documented as histological hallmarks of the disease. Typically, neuritic plaques consist of core of β-amyloid, surrounded by dystrophic neurites containing lysosomal dense bodies and paired helical filaments. Reactive microglial cells, the CNS equivalent of monocytes/macrophages in non-neural tissues, and astrocytes are usually found as components of these plaques [35]. Neurofibrillary tangles tend to occur in the amygdala, hippocampus, parahippocampal gyrus and temporal association cortices, whereas senile plaques are more generally distributed throughout the neocortex [36]. The first brain areas affected by neurofibrillary tangles are the entorhinal cortex and hippocampal formation [37], while the first brain area developing senile plaques appears to be the temporal neocortex [38].

In ALS in which neurons in the motor system selectively degenerate, a unique lamellar pattern of astrogliosis has been noted in the pyramidal layer and deep cortical layers (gray–white junction) of the primary motor cortex. Betz cells degenerate in the primary motor cortex and the spaces occupied by Betz cells are filled by clustered reactive astrocytes and macrophages [39]. Ubiquitin immunoreactive inclusions have been documented in those neurons in the lower motor nuclei, which disappear in advanced ALS cases, suggesting that these inclusions represent structural alteration during the process of cellular degeneration [40].

In HD, neuronal loss and gliosis advance in orderly fashion along the caudo-rostral, medio-lateral and dorso-ventral axis in the striatum and earliest neuronal loss is detected in the tail of the caudate nucleus [41]. Increased endogenous complement biosynthesis by activated microglia in the striatum has been suggested to contribute to neuronal necrosis and proinflammatory activities in HD [42].

In PD, dopaminergic neurons in pars compacta of the substantia nigra gradually degenerate and disappear with progression of the disease process. The cell death is accompanied by an inflammatory reaction by activated microglial cells [43]. But apparently, unlike other neurodegenerative diseases, there is no or very little evidence of astrocytosis associated with neuronal cell death in PD [44,45].

5. VIRAL MENINGOENCEPHALITIS

The viral infection in the CNS is largely divided into two groups depending on the sites of the infection: meningitis and meningoencephalitis. In viral meningitis, infection is confined to the leptomeninges without involving underlying brain tissue. Enteroviruses are the most common cause of viral meningitis. In meningoencephalitis, both leptomeninges and underlying brain tissues are affected. Acute meningoencephalitis is usually caused by viruses of the herpes group and arboviruses. The reaction of the leptomeninges and brain to viral infection is somewhat stereotyped. Grossly the brains of the patients with acute viral infection are swollen with some vascular congestion. The exception is herpes simplex infection, in which hemorrhagic necrosis of the temporal lobe is the characteristic feature. Microscopically, inflammatory cellular infiltration is the most conspicuous histological abnormality. The cellular infiltrates are mostly of lymphocytes, plasma cells and mononuclear cells and tend to be present as perivascular cuffs. The vascular endothelial cells become reactive expressing adhesion molecules such as VCAM-1 or ICAM-1 that promote trafficking of mononuclear cells [46]. Reactive microglia (rod cells), microglial nodules and neuronophagia (see Chapter 1.2) and astrogliosis are always accompanying features. Intranuclear viral inclusions have pathognomonic diagnostic significance in certain viral infection in the CNS such as herpes simplex and cytomegalovirus infection, subacute sclerosing panencephalitis and progressive multifocal leukoencephalopathy.

In recent years, CNS infection of HIV (human immunodeficiency virus) has been the target of intense investigation. HIV is a human retrovirus responsible for subacute and chronic forms of CNS disease, AIDS dementia complex. The CT scan of the brain of the patient with AIDS dementia complex demonstrates diffuse cerebral cortical atrophy with or without abnormality in the white matter. Histologically, the white matter shows diffuse pale stain (myelin pallor) with increase of reactive microglia and astrocytic gliosis. Multinucleated giant cells (as described in Chapter 1.2) are found mostly at the perivascular regions in the white matter [47] and also in the deep cerebral nuclei. The cerebral cortex is relatively spared [48,49]. Loss of neurons and reduction of the numbers of dendritic spines and synaptic density have been reported in the cortical neurons [50–52]. These neuronal changes could explain the clinical symptom of dementia. However, viral infection is confined to monocytes/macrophages and there is no convincing evidence to suggest direct viral infection of neurons. The regional production, within the CNS, of toxic cytokines that may be augmented by circulating cytokines, has been considered to play a significant role in the pathogenesis of AIDS dementia complex [53,54]. Recent studies demonstrate the disruption of the blood–brain barrier tight junction in association with perivascular accumulation of HIV infected macrophages in the AIDS brain. This tight junction defect may contribute the influx of circulating cytokines [55,56].

Progressive multifocal leukoencephalopathy (PML) has been well recognized as one of the commonest neurological complication for the AIDS patients. It is caused by an opportunistic infection of JC virus. Histologically it is characterized by intranuclear viral inclusions in oligodendrocytes and bizarre astrocytes within multiple demyelinating foci (see Chapter 1.2). Lymphocytic cellular infiltration is relatively sparse, presumably due to underlying immunosuppressive status of patients.

6. DEMYELINATING DISEASES

Demyelinating diseases are a heterogeneous group of disorders in which formed myelin sheaths degenerate with relative preservation of axons. Demyelination can be caused by autoimmune mechanism, viral infection, toxic and metabolic perturbation of myelin forming cells [57,58]. Multiple sclerosis (MS) is generally considered to be a prototype of demyelinating disease, typically affecting young adult but MS can occur at any age including children. While the areas of demyelination may be detected anywhere in both gray and white matter, the periventricular white matter regions surrounding the lateral angle and posterior horn of the lateral ventricles are often preferentially and symmetrically affected. Active or acutely demyelinating lesions tend to be soft and ill defined from surrounding edematous white matter. The acutely demyelinating lesion is hypercellular with numerous macrophages containing myelin fragments and reactive astrocytosis. Clinically and pathologically the acute lesion can mimic brain tumor [59]. In contrast, inactive or chronic lesions have sharp margins from the adjacent normal appearing white matter. Chronic lesions may become cavitated, particularly those involving the spinal cord, resulting in secondary Wallerian degeneration of nerve fiber tracts.

Acute disseminated post-infectious encephalomyelitis (ADEM) and acute hemorrhagic leukoencephalitis (AHL) are both autoimmune demyelinating disorders. AHL is considered to be a hyperacute and fulminant form of ADEM [60]. Typically, the affected brains are swollen and multiple foci of perivascular demyelination with infiltration of neutrophils and lymphocytes characterize the histological features. In AHL, additional multiple foci of petechial hemorrhages with fibrinoid necrosis of vessels are identified. Focal demyelination may also result from iatrogenic rapid correction of chronic hyponatremia [61,62]. In this condition, the characteristic demyelinating focus is found in the middle of the pontine base and thus it has been called central pontine myelinolysis.

ACKNOWLEDGEMENTS

The author thanks Ms. Carol Troutner for manuscript preparation.

REFERENCES

1. Waller A. Experiments on the section of the glosopharyngeal and hypoglossal nerves of the frog, and observations of the alterations produced thereby in the structure of their primitive fibers. Philos Trans R Soc Lond (Biol) 1850; 140:423–429.
2. Webster H deF. Transient focal accumulation of axonal mitochondria during the early stage of Wallerian degeneration. J Cell Biol 1962;12:361–377.
3. Webster H deF. The relationship between Schmidt–Lanterman incisures and myelin segmentation during Wallerian degeneration. Ann NY Acad Sci 1965;122:29–38.
4. Ballin RHM, Thomas PK. Changes at the nodes of Ranvier during Wallerian degeneration: an electron microscope study. Acta Neuropathol 1969;14:237–249.
5. Schlaepfer WW, Micko S. Chemical and structural changes in neurofilaments in transected rat sciatic nerve. J Cell Biol 1978;78:269–278.
6. Ohara S, Takahashi H, Ikuta F. Ultrastructural alterations of perineurial cells in the early stage of Wallerian degeneration Lab Invest 1986;54:213–221.
7. Roytta M, Salonen V, Peltonen J. Reversible endoneurial changes after nerve injury. Acta Neuropathol 1987;743: 323–329.
8. Roytta M, Salonen V. Long-term endoneurial changes after nerve transection. Acta Neuropathol 1988;76:35–45.
9. Bruck, W. The role of macrophages in Wallerian degeneration. Brain Pathol 1997;7:741–752.
10. Ghabriel MN, Alt G. The role of Schmidt–Lanterman incisure in Wallerian degeneration. I. A quantitative teased fibre study. Acta Neuropathol 1979;48:83–93.
11. Ghabriel MN, Alt G. The role of Schmidt–Lanterman incisures in Wallerian degeneration. II. An electron microscope study. Act Neuropathol 1979;48:95–103.
12. Lampert PW, Cressman MR. Fine structural changes of myelin sheaths after axonal degeneration in the spinal cord of rats. Am J Pathol 1966;49:1139–1155.
13. Vaughn JE, Hinds PL, Skoff RP. Electron microscopic studies of Wallerian degeneration in rat optic nerves I. The multipotential glia. J Comp Neurol 1970;140:175–206.
14. Vaughn JE, Pease DC. Electron microscopic studies of Wallerian degeneration in rat optic nerves. II. Astrocytes, oligodendrocytes and adventitial cells. J Comp Neurol 1970;140:207–226.
15. Bignami A, Ralston HJ III. The cellular reaction to Walle-

rian degeneration in the central nervous system of the cat. Brain Res 1969;13:444–461.
16. Aldskogius H, Kozlova EN. Central neuron–glial and glial–glial interactions following axon injury. Prog Neurobiol 1998;55:1–26.
17. Ludwin SK. Phagocytosis in the rat optic nerve following Wallerian degeneration. Acta Neuropathol 1990;80:266–273.
18. Daniel PM, Strich SJ. Histological observations on Wallerian degeneration in the spinal cord of the baboon, *Papio papio*. Acta Neuropathol 1969;12:314–328.
19. Avellino AM, Hart D, Dailey AT, MacKinnon M, Ellegala D, Kliot M. Differential macrophage responses in the peripheral and central nervous system during Wallerian degeneration of axons. Exp Neurol 1995;136:183–198.
20. Kalimo H, Kaste M, Haltia M. Vascular disease. In: Graham DI, Lantos PL, eds. The Greenfield's Neuropathology, 6th edition. London: Arnold, 1997:315–381.
21. Garcia JH, Mena H. Vascular disease. In: Garcia JH, ed. Neuropathology: The Diagnostic Approach. St. Louis: Mosby-Year Book, Inc., 1997:263–320.
22. Garcia JH. Morphology of global cerebral ischemia: a review. Crit Care Med 1988;16:979–987.
23. Brown AW, Brierley JB. Evidence for early anoxic–ischemic cell damage in the rat brain. Experientia 1966; 22:546–547.
24. Brown W. Structural abnormalities in neurons. J Clin Pathol 1977;30,Suppl. (Royal College of Pathologists) 11: 155–169.
25. Astrup J, Siesjo BK, Symon L. Thresholds in cerebral ischemia — the ischemic penumbra. Stroke 1981;12:723–725.
26. ElGhazawi EF, Malaty HA. Observations on extra neuronal lipofuscin in the human brain. Cell Tissue Res 1975; 161:555–565.
27. Cavanagh JB, Nolan CC, Seville MP, Anderson VE, Leigh PN. Routes of excretion of neuronal lysosomal dense bodies after ventricular infusion of leupeptin in the rat: A study using ubiquitin and PGP9.5 immunocytochemistry. J Neurocytol 1993;22:779–791.
28. Driscoll J. The role of the proteosome in cellular protein degradation. Histol Histopathol 1994;9:197–202.
29. Dickson DW. Aging in the central nervous system. In: Markesbery, ed. Neuropathology of Dementing Disorders. London: Arnold, 1998:56–88.
30. Xia MQ, Hyman BT. Chemokines/chemokine receptors in the central nervous system and Alzheimer disease. J Neurovirol 1999;5:32–41.
31. Neuroinflammation Working Group: Akiyama H, Barger S, Barnum S, Bradt B, Bauer J, Cole GM, Cooper NR, Eikelenboom P, Emmerling M, Fiebich BL, Finch CE, Fraurschy S, Griffin WST, Hampel H, Hull M, Landreth G, Lue L-F, Mrak R, Mackenzie IR, McGeer PL, O'Banion K, Pachter J, Pasinetti G, Plata-Salaman C, Rogers J, Rydel R, Shen Y, Streit W, Strohmeyer R, Tooyama I, Van Muiswinkel FL, Veerhuis R, Walker D, Webster S, Wegrzyniak B, Wenk G, Wyss-Coray T. Inflammation and Alzheimer's disease, Neurobiol Aging 2000;21:383–421.
32. Halliday G, Robinson SR, Shepherd C, Kril J. Alzheimer's disease and inflammation: a review of cellular and therapeutic mechanisms. J Clin Exp Pharmacol Physiol 2000; 27:1–8.
33. McGeer PL, McGeer EG, Autotoxicity and Alzheimer disease. Arch Neurol 2000;57:789–790.
34. Roses AD. Causes or consequences of inflammation and pathological signs of Alzheimer disease. Neurobiol Aging 2000;21:423–425.
35. Terry RD, Gonatas NK, Weiss M. Ultrastructural studies in Alzheimer's presenile dementia. Am J Pathol 1964;44: 269–297.
36. Arnold SE, Hyman BT, Flory J, Damasio AR, Van Hoesen GW. The topographical and neuroanatomical distribution of neurofibrillary tangles and neuritic plaques in cerebral cortex of patients with Alzheimer's disease. Cerebral Cortex 1991;1:103–116.
37. Hyman BT, Van Hoesen GW, Damasio AR. Memory-related neural systems in Alzheimer's disease: An anatomical study. Neurology 1990;40:1721–1730.
38. Price JI, Davis PB, Morris JC, White DL. The distribution of tangles, plaques and related immunohistochemical markers in healthy aging and Alzheimer's disease. Neurobiol Aging 1991;12:295–312.
39. Murayama S, Inoue K, Kawakami H, Bouldin TW, Suzuki K. A unique pattern of astrocytosis in the primary motor area in amyotrophic lateral sclerosis. Acta Neuropathol 1991;82:456–461.
40. Murayama S, Mori H, Ihara Y, Bouldin TW, Suzuki K, Tomonaga M. Immunocytochemical and ultrastructural studies of lower motor neurons in amyotrophic lateral sclerosis. Ann Neurol 1990;27:137–148.
41. Vonsattel JPG, DiFiglia M. Huntington disease. J Neuropathol Exp Neurol 1998;57:369–384.
42. Singhrao SK, Neal JW, Morgan BP, Gasque P. Increased complement biosynthesis by microglia and complement activation on neurons in Huntington's disease. Exp Neurol 1999;159:362–376.
43. McGeer P, Itagaki S, Boyes BE, McGeer EG. Reactive microglia are positive for HLA-DR in the substantia nigra of Parkinson's and Alzheimer's disease brains. Neurology 1988;38:1285–1291.
44. Banati RB, Daniel SE, Blunt SB. Glial pathology but absence of apoptotic nigral neurons in long-standing Parkinson's disease. Mov Disord 1998;13:221–227.
45. Mirza B, Hadberg H, Thomsen P, Moos T. The absence of reactive astrocytosis is indicative of a unique inflammatory process in Parkinson's disease. Neuroscience 2000; 95:425–432.
46. Sobel RA, Mitchel ME, Fondren G. Intercellular adhesion molecule-1 (ICAM-1) in cellular immune reactions in the human central nervous system. Am J Pathol 1990;136: 1309–1316.
47. Price RW. The cellular basis of central nervous system HIV-infection and the AIDS dementia complex. J Neuro AIDS 1996;1:1–173.
48. Navia BA, Jordan BD, Price RW. The AIDS dementia complex. II Neuropathology. Ann Neurol 1986;19:525–535.
49. Kure K, Weidenheim KM, Lyman WD, Dickson DW. Morphology and distribution of HIV-1 gp41-positive microglia in subacute AIDS encephalitis. Acta Neuropathol 1990;80:393–400.

50. Masliah E, Heaton RK, Marcotte TD, Ellis RJ, Wiley CA, Mallory M, Achim CL, MeCutchan JA, Nelson JA, Atkinson JH, Grant I. Dendritic injury is a pathological substrate for human immunodeficiency virus-related cognitive disorders. HNRC Group. The HIV Neurobehavioral Research Center. Ann Neurol 1997;42:963–972.
51. Everall IP, Heaton RK, Marcotte TD, Ellis RJ, McCutchan JA, Atkinson JH, Grant I, Mallory M, Masliah E. Cortical synaptic density is reduced in mild to moderate human immunodeficiency virus neurocognitive disorder. HNRC Group. HIV Neurobehavioral Research Center. Brain Pathol 1999;9:209–217.
52. Everall IP, Luthert P, Lantos PA. A review of neuronal damage in human immunodeficiency virus infection: Its assessment, possible mechanism and relationship to dementia. J Neuropathol Exp Neurol 1993;52:561–566.
53. Smith TW, DeGirolami U, Henin D, Bolgert F, Hauw JJ. Human immunodeficiency virus (HIV) leukoencephalopathy and the microcirculation. J Neuropathol Exp Neurol 1990;49:357–370.
54. Power C, Kong P-A, Crawford TO, Wesselingh S, Glass JD, McArthur JC, Trapp BD. Cerebral white matter changes in acquired immunodeficiency syndrome dementia: Alterations of the blood–brain barrier. Ann Neurol 1993;34:339–350.
55. Dallasta LM, Pisarov LA, Esplen JE, Werley JV, Moses AV, Nelson JA. Blood–brain barrier tight junction disruption in human immunodeficiency virus-1 encephalitis. Am J Pathol 1999;155:1915–1927.
56. Boven LA, Middel J, Verhoef J, De Groot CJA, Nottet HSLM. Monocyte infiltration is highly associated with loss of the tight junction protein zonula occludens in HIV-1 associated dementia. Neuropathol Appl Neurobiol 2000; 26:356–360.
57. Prineas JW, McDonald WL. Demyelinating diseases. In: Graham DL, Lantos PL, eds. The Greenfield's Neuropathology, 6th edition. London: Arnold, 1997:813–896.
58. Suzuki K. Diagnosis of demyelinating disorders and leukodystrophies. In: Garcia JH, ed. Neuropathology: The Diagnostic Approach. St. Louis: Mosby-Year Book, 1997: 669–715.
59. Kepes JJ. Large focal tumor-like demyelinating lesions of the brain: intermediate entity between multiple sclerosis and acute disseminated encephalomyelitis? A study of 31 patients. Ann Neurol 1993;33:18–27.
60. Gosztonyi G. Acute hemorrhagic leukoencephalitis (Hurst disease). In: Vinken PJ, Bruyn GW, eds. Infections of the Nervous System: Part II, Handbook of Clinical Neurology, Vol. 3. Amsterdam: Elsevier, 1978:587–604.
61. Stern RH, Riggs JE, Schochet SS Jr. Osmotic demyelination syndrome following correction of hyponatremia. New Eng J Med 1986;314:1535–1542.
62. Karp BI, Laureno R. Pontine and extrapontine myelinolysis: a neurologic disorder following rapid correction of hyponatremia. Medicine 1993;729:359–373.

Section 2

THE CHEMOKINE SYSTEM

Section Editor: A.E.I. Proudfoot

Universes in Delicate Balance: Chemokines and the Nervous System
R.M. Ransohoff, K. Suzuki, A.E.I. Proudfoot, W.F. Hickey and J.K. Harrison, editors

Chapter 2.1

The Biology of Chemokines

Barrett J. Rollins

Department of Adult Oncology, Dana-Farber Cancer Institute, Department of Medicine, Brigham and Women's Hospital, Harvard Medical School, Boston, MA 02115, USA

1. INTRODUCTION

Effective defense against invasion and injury is an absolute prerequisite for an organism's evolutionary success. Obvious selective advantages accrue as multicellular entities increase in complexity, but these are neatly balanced by the disadvantage of presenting larger and more inviting targets to pathogenic microorganisms. True selective success, therefore, is accompanied by an ever more sophisticated armamentarium for battling invaders. The gradual development of leukocyte subtypes that are specialized for fighting specific threats (e.g., eosinophils against parasites) or that make specific protective agents (e.g., lymphocytes secreting antibody) are cases in point.

In order for this complex system of differentiated cells to perform their functions, they must be directed to appropriate sites of invasion at appropriate times. At the simplest level, neutrophils must be able to leave the circulation and converge specifically at sites of bacterial invasion where the body's mechanical defenses have been breached. At a more sophisticated regulatory level, naïve T lymphocytes must find their way to antigen presenting cells in lymph nodes in order to increase the probability of mounting an effective immune response. In both of these cases and countless others, the soluble factors responsible for specific leukocyte trafficking are chemokines [1].

Chemokines are phylogenetically ancient. Homologs of the chemokine receptor CXCR4 exist in *Drosophila melanogaster* and in primitive vertebrates such as lampreys [2]. As the formed elements of the blood multiplied and diversified during evolution, so did chemokines. It is likely that evolutionary pressures exerted by commensal and pathogenic microbes contributed to the panoply of chemokine ligands and receptors found in higher vertebrates today. However, as effective as this system may be in its defense of the host, maladaptive activation of any of its parts can lead to pathology. It will become apparent in later chapters that in these settings, therapies designed to disrupt chemokine action may have beneficial effects.

2. CHEMOKINES AND THEIR RECEPTORS

As of this book's publication date, more than fifty human chemokines have been described. (For the latest census, see http://cytokine.medic.kumamoto-u.ac.jp/.) Assignment of a protein to this family has been based for the past 12 years on the conserved positions of cysteines in a secreted protein (or in a domain of a larger protein) of approximately 90–100 amino acids. However, as more chemokine structures have been reported, it has become clear that they all share a nearly superimposable tertiary structure now called the chemokine fold [3,4] (see Fig. 1). This is characterized by a highly disorganized N-terminus which is tamped down by the N-terminal cysteine(s). There follows a long loop which leads into three antiparallel β-pleated sheets terminating in an α helix which overlies the sheets. All chemokine structures solved to date conform to this three-dimensional pattern. In the post-genomic proteomic world, it is likely that proteins will be designated as chemokines based on their predicted or observed assumption of the this structure.

Nonetheless, the cysteine motifs provide a useful

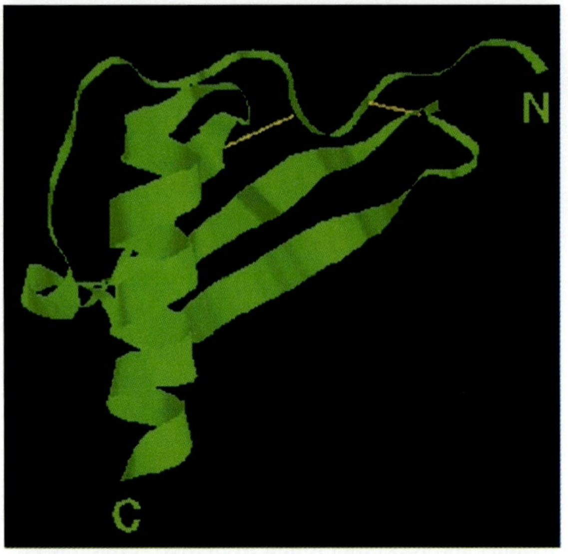

Figure 1. Ribbon diagram of the structure of monomeric IL-8 [3] (coordinates available at www.rcsb.org/pdb/cgi/explore.cgi?pdbId=1IL8). The N-terminal domain is highly disordered, and the position shown here was found in one of the structures. After the N-terminal cysteines, moving toward the C-terminus, a long loop is followed by three anti-parallel β-pleated sheets terminating in an α helix. This basic fold is shared by all of the chemokine structures solved to date.

criterion for placing chemokines into subfamilies. Specifically, there are four patterns of amino acid substitutions surrounding two cysteines in the N-terminal domain (see Table 1). A single amino acid interposed between these cysteines places the chemokine in the CXC or α family, adjacent cysteines places the chemokine in the CC or β family, a single cysteine in this domain characterizes the C or γ family, and three amino acids between two cysteines define the CX_3C or δ family. Target cell specificities do not rigidly fall within family boundaries although the most efficient neutrophil chemoattractants are all members of the CXC family and have a three amino acid motif near their N-termini comprised of glutamate–leucine–arginine (ELR) [5,6]. Most monocyte chemoattractants are members of the CC family, but many members of both families are efficient attractants for various T lymphocyte subsets. Target cells will be described for chemokines as they appear in subsequent chapters.

Membership in a subfamily is a powerful predictor for the type of receptor a chemokine activates. All functional chemokine receptors are members of the seven transmembrane-spanning, G protein-coupled family of receptors, discussed in detail below. CXC chemokines bind with high affinities to CXC chemokine receptors, CC chemokines to CC receptors, and the C and CX_3C chemokines bind to their own recep-

Table 1. Chemokine families

Family	Systematic name	Trivial name(s)
CXC	CXCL1	Gro-α, MGSA-α, MIP-2, KC
	CXCL2	Gro-β, MGSA-β, MIP-2α
	CXCL3	Gro-γ, MGSA-γ, MIP-2β
	CXCL4	Platelet factor 4
	CXCL5	ENA-78
	CXCL6	GCP-2
	CXCL7	Platelet basic protein and its products (β-TG, CTAP-III, NAP-2)
	CXCL8	IL-8
	CXCL9	Mig
	CXCL10	IP-10
	CXCL11	I-TAC
	CXCL12	SDF-1
	CXCL13	BLC, BCA-1
	CXCL14	MIP-2γ, BRAK, bolekine
	CXCL15	Lungkine
	CXCL16	
CC	CCL1	I-309 (mTCA3)
	CCL2	MCP-1
	CCL3	MIP-1α, LD78
	CCL4	MIP-1β, Act-2
	CCL5	RANTES
	CCL6	mC10
	CCL7	MCP-3, FIC, MARC
	CCL8	MCP-2
	CCL9/CCL10	MIP-1γ, MRP-2
	CCL11	Eotaxin
	CCL12	mMCP-5
	CCL13	MCP-4, CKβ10
	CCL14	HCC-1, CKβ1
	CCL15	HCC-2, MIP-5, MIP-1δ
	CCL16	HCC-4, CKβ12
	CCL17	TARC
	CCL18	DC-CK1, PARC, MIP-4, CKβ7
	CCL19	MIP-3β, ELC, exodus-3, CKβ11
	CCL20	MIP-3α, LARC, exodus-1, CKβ4
	CCL21	SLC, 6Ckine, exodus-2, TCA4
	CCL22	MDC
	CCL23	MPIF-1, CKβ8, MIP-3
	CCL24	MPIF-2, CKβ6, eotaxin-2
	CCL25	TECK, CKβ15
	CCL26	Eotaxin-3, MIP-4α
	CCL27	CTAK, ESkine, skinkine
	CCL28	MEC
C	CL1	Lymphotactin, SCM-1α
	CL2	SCM-1β
CX_3C	CX_3CL1	Fractalkine

Table 2. Chemokine receptors

Receptor	Ligand
CXCR1	IL-8, GCP-2
CXCR2	ELR$^+$ CXC chemokines (IL-8, GCP-2, Gro-α, Gro-β, Gro-γ, NAP-2, ENA-78, lungkine)
CXCR3	IP-10, Mig, I-TAC
CXCR4	SDF-1
CXCR5	BLC
CXCR6	CXCL16
CCR1	MIP-1α, RANTES, HCC-1
CCR2	MCP-1, -2, -3, -4, -5
CCR3	Eotaxin, eotaxin-2, MCP-3
CCR4	MDC, TARC, MIP-1α, RANTES
CCR5	MIP-1α, MIP-1β, RANTES
CCR6	MIP-3α
CCR7	MIP-3β
CCR8	I-309
CCR9	TECK
CCR10	CTAK
XCR1	Lymphotactin
CX_3CR1	Fractalkine

tors (see Table 2). However, within each subfamily there may be extensive promiscuity. For example, some CC chemokines bind to several CC chemokine receptors, and some CC chemokine receptors bind several chemokines. This apparent ligand–receptor promiscuity does not cross subfamily boundaries. However, there are at least two receptors that bind several chemokines from both the CXC and CC subfamilies. These are the Duffy antigen receptor for chemokines (DARC) [7,8] and D6 [9]. Although these receptors bind chemokines with high affinities, binding has not yet been shown to result in signal transduction.

Chemokines have found a mechanistic place in the multistep model of leukocyte emigration [10]. When chemokines are secreted, they diffuse toward the luminal of nearby vessels, although there are data indicating that endothelial cells transcytose chemokines through their cytoplasms [11]. Either way, chemokines are displayed on the luminal surface of endothelial cells probably in the context of cell surface glycosaminoglycans which prevent their being washed away by blood flow. If a leukocyte engaged in selectin-mediated rolling has a cognate receptor for a displayed chemokine, receptor activation will result in up-regulation of integrin affinities producing firm, flow-resistant adhesion. This is followed by diapedesis of the leukocyte into the subendothelium where it presumably follows a concentration gradient of chemokine to arrive at the source. The effects of chemokines on converting rolling interactions to firm adhesion have been demonstrated by the use of in vitro flow chambers [12,13] and by intravital microscopic examination [14].

3. CHEMOKINES IN NORMAL PHYSIOLOGY

Although chemokines were first characterized in the context of inflammatory responses, the discovery of additional members of this protein family has led to the notion that they fall into two broad categories. The first includes inflammatory chemokines. For the most part, these are not expressed constitutively, but only after cells are exposed to inflammatory stimuli such as endotoxin or inflammatory cytokines. These chemokines are induced to very high levels and are responsible for attracting cells involved in inflammatory responses. Examples include IL-8 and MCP-1.

The second category has been referred to as immune chemokines. These tend to be expressed constitutively, although higher levels of expression can be induced by specific stimuli. Constitutive expression reflects the fundamental function of these chemokines in controlling basal immune cell, usually T or B lymphocyte, trafficking [15]. For example, SLC is a CC chemokine expressed by endothelial cells of high endothelial venules (HEV) in lymph nodes and directs naïve T lymphocytes to leave the vasculature through HEV [16,17]. A second chemokine, ELC, attracts these extravasated T lymphocytes into the T cell zone [18]. In another example, skin-homing T cells are characterized by the expression of the CLA antigen and absence of the gut-specific integrin $\alpha_4\beta_7$. These cells express the chemokine receptor CCR4 and respond to one of its ligands, TARC, which is secreted by cutaneous keratinocytes [19]. Conversely, gut-homing T cells are $CLA^-\alpha_4\beta_7^+$ and express the chemokine receptor CCR9 whose ligand, TECK, is secreted by gut epithelial cells [20]. It is anticipated that several similar examples will be found in which specific chemokine/receptor pairs are responsible for tissue-specific recirculation patterns of specific leukocyte subtypes.

One of the consequences of constitutive immune chemokine expression is that persistent migration of a specific leukocyte subtype into a distinct area can result in the formation of anatomic structures. In particular, chemokines appear to be responsible for the overall and subanatomic structures of secondary lymphoid organs. This was first demonstrated in mice engineered to be deficient for the chemokine receptor CXCR5 (originally designated BLR-1) [21]. Peyer's patches and inguinal lymph nodes were nearly absent from these mice and germinal centers in spleen and lymph nodes were disrupted, consistent with the B lymphocyte expression of this receptor. Along the same lines, lymph node architecture is completely disrupted in mice deficient for CCR7, the receptor for ELC described above as being responsible for the accumulation of T lymphocytes in specific nodal zones [22]. That these mice cannot mount T cell-dependent immune responses is mute testimony to the critical importance of cell trafficking and contact in immunity. Dendritic cell contributions to the immune response depend on trafficking patterns mediated by ligand activation of CCR6 [23].

While the division of chemokines into inflammatory and immune categories may be a helpful heuristic, it should not be viewed as absolute. For example, one of the paradigmatic inflammatory chemokines is MCP-1 which, along with its receptor CCR2, is clearly involved in controlling the trafficking of monocytes and macrophages in inflammatory settings [24,25]. However, both MCP-1 and CCR2 have been demonstrated to have profound effects on the development of polarized T helper cell responses, indicating that in some circumstances, chemokines can straddle the boundary between innate inflammatory responses and adaptive immunity [24,26].

Another potential role for chemokines in normal physiology that bears mentioning is the control of stem cell homing. Blood borne stem cells for a variety of non-hematopoietic organs can be identified, but the signals that direct these cells to their intended organs remain obscure. In the case of hematopoietic stem cells, the CXC chemokine SDF-1α has been shown to play a critical role in keeping these cells in the bone marrow [27–29]. Antagonism of SDF-1α may lead to peripheral mobilization of hematopoietic stem cells. It is conceivable that this model will be recapitulated in other organs using other chemokine/chemokine receptor pairs.

4. CHEMOKINES IN DISEASE

Maladaptive engagement of host defense systems can frequently form the pathogenetic basis for disease, and chemokines and their receptors are not exempt. From the time of their discovery, chemokines have been thought to be responsible for the persistent presence of effector cells in inflammatory diseases such as asthma or arthritis. In support of this idea, chemokine expression has been easily documented in several diseases. However, it has been more difficult to demonstrate that chemokine expression is either responsible for disease pathology or that antagonism of chemokines has a therapeutic effect.

Nonetheless, the use of neutralizing antibodies and, perhaps more rigorously, gene-disrupted mice has implicated specific chemokines in a handful of diseases. For example, asthma has been approximately modeled by inhalation challenge of sensitized mice. Neutralizing antisera against chemokines have demonstrated roles for MCP-1, MCP-5, eotaxin, and RANTES in the airway hyperresponsiveness and inflammation that this model produces [30,31]. A broad spectrum antagonist known as Met-RANTES [32] also blocks inflammation and is effective in this setting [31]. While mice genetically deficient for MCP-1 show less disease in this model (Gerard and Rollins, unpublished data), CCR1- (Humbles and Gerard, unpublished), CCR4- [33], and eotaxin- [34,35] deficient mice do not. CCR3-deficient mice are only partially protected (Humbles and Gerard, manuscript submitted). These are rather surprising results given the importance of these ligands and receptors in the trafficking of eosinophils and T lymphocytes, and they suggest that much remains to be learned about the precise role of chemokines in asthma.

The effectiveness of neutralizing antibodies or a broad spectrum antagonist have been compared in two models, namely the airways inflammation model described above [31] and a murine model of glomerulonephritis [36]. While a greater effect was seen with the receptor antagonist in the airways model, it was less effective in kidney inflammation since only anti-MCP-1 treatment was able to block the fibrosis which accompanies this disease, whereas both the antagonist and the antibody reduced cellular recruitment. This result highlights the importance of

CCR2 rather than RANTES receptors in this disease model.

A somewhat stronger genetic case can be made for MCP-1 and CCR2 in atherosclerosis. As noted above, MCP-1 is a CC chemokine that attracts monocytes and macrophages through activation of its receptor, CCR2. MCP-1 is secreted by endothelial cells and arterial smooth muscle cells in response to atherogenic stimuli such as hypercholesterolemia and hypertension [37,38]. It has been thought that MCP-1 expression might be responsible for attracting circulating monocytes into the subendothelium of diseased arteries where they become macrophages and eventually the foam cells of atherosclerotic lesions. This scenario was confirmed using *MCP-1*$^{-/-}$ and *CCR2*$^{-/-}$ mice in atherosclerosis models [39,40]. Disruption either of the chemokine or its receptor resulted in substantial protection from vascular disease in the face of persistently high concentrations of plasma lipids.

The convincing demonstration of chemokine involvement in inflammatory models, adaptive immunity, and selected disease models bodes well for their having similar roles in neuroinflammation. As described elsewhere in this volume, there is already substantial evidence for the influence of chemokines on the course of rodent EAE. There is little doubt that continued application of genetic analyses to this model and to human disease will eventually demonstrate the central importance of these proteins and their receptors in inflammatory diseases of the nervous system.

REFERENCES

1. Rossi D, Zlotnik A. The biology of chemokines and their receptors. Annu Rev Immunol 2000;18:217–242.
2. Najakshin AM, Mechetina LV, Alabyev BY, Taranin AV. Identification of an IL-8 homolog in lamprey (Lampetra fluviatilis): early evolutionary divergence of chemokines. Eur J Immunol 1999;29:375–382.
3. Clore GM, Appella E, Yamada M, Matsushima K, Gronenborn AM. Three-dimensional structure of interleukin 8 in solution. Biochemistry 1990;29:1689–1696.
4. Handel TM, Domaille PJ. Heteronuclear (1H, 13C, 15N) NMR assignments and solution structure of the monocyte chemoattractant protein-1 (MCP-1) dimer. Biochemistry 1996;35:6569–6584.
5. Clark-Lewis I, Schumacher S, Baggiolini M, Moser B. Structure–activity relationships of Interleukin-8 determined using chemically synthesized analogs: critical role of NH_2-terminal residues and evidence for uncoupling of neutrophil chemotaxis, exocytosis, and receptor binding activities. J Biol Chem 1991;266:23128–23134.
6. Hebert CA, Viytangcol RV, Baker JB. Scanning mutagenesis of interleukin-8 identifies a cluster of residues required for receptor binding. J Biol Chem 1991;266: 18989–18994.
7. Peiper SC, Wang ZX, Neote K et al. The Duffy antigen/receptor for chemokines (DARC) is expressed in endothelial cells of Duffy negative individuals who lack the erythrocyte receptor. J Exp Med 1995;181:1311–1317.
8. Szabo MC, Soo KS, Zlotnik A, Schall TJ. Chemokine class differences in binding to the Duffy antigen-erythrocyte chemokine receptor. J Biol Chem 1995;270: 25348–25351.
9. Nibbs RJB, Wylie SM, Pragnell IB, Graham GJ. Cloning and characterization of a novel murine beta chemokine receptor, D6. Comparison to three other related macrophage inflammatory protein-1alpha receptors, CCR-1, CCR-3, and CCR-5. J Biol Chem 1997;272:12495–12504.
10. Springer T. Traffic signals for lymphocyte recirculation and leukocyte emigration: the multistep paradigm. Cell 1994;76:301–314.
11. Middleton J, Neil S, Wintle J et al. Transcytosis and surface presentation of IL-8 by venular endothelial cells. Cell 1997;91:385–395.
12. Campbell JJ, Qin S, Bacon KB, Mackay CR, Butcher EC. Biology of chemokine and classical chemoattractant receptors: differential requirements for adhesion-triggering versus chemotactic responses in lymphoid cells. J Cell Biol 1996;134:255–266.
13. Campbell JJ, Hedrick J, Zlotnik A, Siani MA, Thompson DA, Butcher EC. Chemokines and the arrest of lymphocytes rolling under flow conditions. Science 1998;279: 381–384.
14. Stein JV, Rot A, Luo Y et al. The CC chemokine thymus-derived chemotactic agent 4 (TCA-4, secondary lymphoid tissue chemokine, 6Ckine, exodus-2) triggers lymphocyte function-associated antigen 1-mediated arrest of rolling T lymphocytes in peripheral lymph node high endothelial venules. J Exp Med 2000;191:61–76.
15. Campbell JJ, Butcher EC. Chemokines in tissue-specific and microenvironment-specific lymphocyte homing. Curr Opin Immunol 2000;12:336–341.
16. Nagira M, Imai T, Hieshima K et al. Molecular cloning of a novel human CC chemokine secondary lymphoid-tissue chemokine that is a potent chemoattractant for lymphocytes and mapped to chromosome 9p13. J Biol Chem 1997;272:19518–19524.
17. Gunn MD, Tangemann K, Tam C, Cyster JG, Rosen SD, Williams LT. A chemokine expressed in lymphoid high endothelial venules promotes the adhesion and chemotaxis of naive T lymphocytes. Proc Natl Acad Sci USA 1998; 95:258–263.
18. Luther SA, Tang HL, Hyman PL, Farr AG, Cyster JG. Coexpression of the chemokines ELC and SLC by T zone stromal cells and deletion of the ELC gene in the plt/plt mouse. Proc Natl Acad Sci USA 2000;97:12694–12699.
19. Campbell JJ, Haraldsen G, Pan J et al. The chemokine receptor CCR4 in vascular recognition by cutaneous but not intestinal memory T cells. Nature 1999;400:776–780.

20. Zabel BA, Agace WW, Campbell JJ et al. Human G protein-coupled receptor GPR-9-6/CC chemokine receptor 9 is selectively expressed on intestinal homing T lymphocytes, mucosal lymphocytes, and thymocytes and is required for thymus-expressed chemokine-mediated chemotaxis. J Exp Med 1999;190:1241–1256.
21. Forster R, Mattis A, Kremmer E, Wolf E, Brem G, Lipp M. A putative chemokine receptor, BLR1, directs B cell migration to defined lymphoid organs and specific anatomic compartments of the spleen. Cell 1996;87:1037–1047.
22. Forster R, Schubel A, Breitfeld D et al. CCR7 coordinates the primary immune response by establishing functional microenvironments in secondary lymphoid organs. Cell 1999;99:23–33.
23. Cook DN, Prosser DM, Forster R et al. CCR6 mediates dendritic cell localization, lymphocyte homeostasis, and immune responses in mucosal tissue. Immunity 2000;12: 495–503.
24. Boring L, Gosling J, Chensue SW et al. Impaired monocyte migration and reduced type 1 (Th1) cytokine responses in C–C chemokine receptor 2 knockout mice. J Clin Invest 1997;100:2552–2561.
25. Lu B, Rutledge BJ, Gu L et al. Abnormalities in monocyte recruitment and cytokine expression in monocyte chemoattractant protein 1-deficient mice. J Exp Med 1998; 187:601–608.
26. Gu L, Tseng S, Horner RM, Tam C, Loda M, Rollins BJ. Control of TH2 polarization by the chemokine monocyte chemoattractant protein-1. Nature 2000;404:407–411.
27. Aiuti A, Webb IJ, Bleul C, Springer T, Guttierez-Ramos JC. The chemokine SDF-1 is a chemoattractant for human $CD34^+$ hematopoietic progenitor cells and provides a new mechanism to explain the mobilization of $CD34^+$ progenitors to peripheral blood. J Exp Med 1997;185: 111–120.
28. Peled A, Petit I, Kollet O et al. Dependence of human stem cell engraftment and repopulation of NOD/SCID mice on CXCR4. Science 1999;283:845–848.
29. Ma Q, Jones D, Springer TA. The chemokine receptor CXCR4 is required for the retention of B lineage and granulocytic precursors within the bone marrow microenvironment. Immunity 1999;10:463–471.
30. Jia GQ, Gonzalo JA, Lloyd C et al. Distinct expression and function of the novel mouse chemokine monocyte chemotactic protein-5 in lung allergic inflammation. J Exp Med 1996;184:1939–1951.
31. Gonzalo JA, Lloyd CM, Wen D et al. The coordinated action of CC chemokines in the lung orchestrates allergic inflammation and airway hyperresponsiveness. J Exp Med 1998;188:157–167.
32. Proudfoot AE, Power CA, Hoogewerf AJ et al. Extension of recombinant human RANTES by the retention of the initiating methionine produces a potent antagonist. J Biol Chem 1996;271:2599–2603.
33. Chvatchko Y, Hoogewerf AJ, Meyer A et al. A key role for CC chemokine receptor 4 in lipopolysaccharide-induced endotoxic shock. J Exp Med 2000;191:1755–1764.
34. Rothenberg ME, MacLean JA, Pearlman E, Luster AD, Leder P. Targeted disruption of the chemokine eotaxin partially reduces antigen-induced tissue eosinophilia. J Exp Med 1997;185:785–790.
35. Yang Y, Loy J, Ryseck RP, Carrasco D, Bravo R. Antigen-induced eosinophilic lung inflammation develops in mice deficient in chemokine eotaxin. Blood 1998;92:3912–3923.
36. Lloyd CM, Minto AW, Dorf ME et al. RANTES and monocyte chemoattractant protein-1 (MCP-1) play an important role in the inflammatory phase of crescentic nephritis, but only MCP-1 is involved in crescent formation and interstitial fibrosis. J Exp Med 1997;185:1371–1380.
37. Yu X, Dluz S, Graves DT et al. Elevated expression of monocyte chemoattractant protein 1 by vascular smooth muscle cells in hypercholesterolemic primates. Proc Natl Acad Sci USA 1992;89:6953–6957.
38. Capers Qt, Alexander RW, Lou P et al. Monocyte chemoattractant protein-1 expression in aortic tissues of hypertensive rats. Hypertension 1997;30:1397–1402.
39. Gu L, Okada Y, Clinton SK et al. Absence of monocyte chemoattractant protein-1 reduces atherosclerosis in low density lipoprotein receptor-deficient mice. Mol Cell 1998;2:275–281.
40. Boring L, Gosling J, Cleary M, Charo IF. Decreased lesion formation in $CCR2^{-/-}$ mice reveals a role for chemokines in the initiation of atherosclerosis. Nature 1998;394:894–897.

Universes in Delicate Balance: Chemokines and the Nervous System
R.M. Ransohoff, K. Suzuki, A.E.I. Proudfoot, W.F. Hickey and J.K. Harrison, editors

Chapter 2.2

Chemokines

Amanda E.I. Proudfoot, Jeffrey P. Shaw, Christine A. Power and Timothy N.C. Wells

Serono Pharmaceutical Research Institute, 14 Chemin des Aulx, 1228 Plan les Ouates, Switzerland

1. INTRODUCTION

The family of chemokines has expanded extremely rapidly since the identification of the first chemokine sequence, Platelet Factor 4/CXCL1 in 1977 [1], with IL-8/CXCL8 and the MIP-1 proteins isolated in the late 1980s. They were initially identified by isolation of the active protein from tissues or culture supernatants. The neutrophil chemoattractant IL-8 was purified from culture supernatant of stimulated blood monocytes [2] and the monocyte chemoattractants MIP-1α/CCL3 and MIP-1β/CCL4 were purified from LPS stimulated murine macrophages [3]. The primary amino acid sequence of these two chemokines rapidly led to the identification of the highly conserved four-cysteine motif that is common to almost all chemokines and also allowed their classification into two principal subclasses. The first, the CXC or α-chemokines, have a single residue between the first amino terminal Cys residues, whereas the CC, or β-chemokines have these two Cys residues adjacent. The number of chemokines then grew rapidly through homology cloning using the conserved motifs found in chemokines. Subsequently, two additional subclasses have been identified, the CX_3C subclass [4,5], which has 3 amino acids, as its name suggests, separating the amino terminal Cys pair, and the C subclass, which lacks a pair of Cys residues. The real explosion in the number of members of the chemokine family came from their identification in EST databases [6], and as is shown in Fig. 1, approximately 40 chemokines are now known to occur in humans.

Initially support for the division of chemokines into the α and β subclasses was not only structural, but also based on biological activity as it described leukocyte specificity, which held firm until the cloning of the third CXC receptor, CXCR3 [7]. Until this point, the CXC chemokines were thought to be responsible principally for neutrophil recruitment and were therefore implicated in acute inflammation, whilst CC chemokines recruited other leukocyte types and were thus involved in chronic inflammation. However CXCR3 is mainly expressed on activated T cells, and its ligands were initially identified as IFN-γ inducible polypeptides, and are therefore pivotal in chronic inflammatory disorders. The identification of CXCR4 and CXCR5 as well as several CC chemokine receptors and their respective ligands introduced another concept to chemokine biology, i.e. that chemokines could be further subdivided into two broad classes on the basis of their biology: those that are inducible and therefore involved in inflammation, and those that are constitutively expressed, and are involved in leukocyte homing.

Chemokine biology is addressed in detail in many chapters in this book. We will therefore concentrate on aspects more related to the structure of chemokines, and chemokine analogues, and how these aspects may be related to their biology.

2. CHEMOKINE GENE CHROMOSOMAL LOCATION

The genes for the majority of the CXC chemokines identified to date are co-localized on chromosome 4 with the exception of SDF-1 which is located on chromosome 10, and two of the more recently dis-

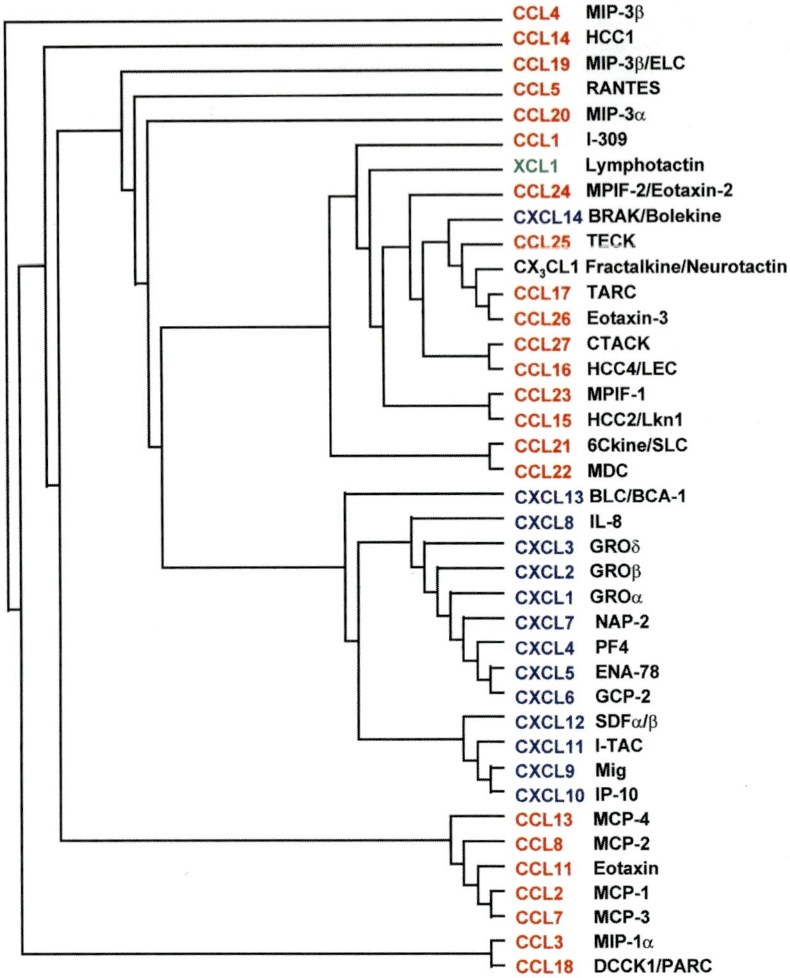

Figure 1. Dendrogram showing the relative sequence similarities between the know human chemokines. The systemic nomenclature is taken from Ref. [137]. CC chemokines are shown in red, CXC in blue, the C chemokine in green and the CX_3C chemokine in cyan. The common names that have been used to date are in black.

covered CXC chemokines BRAK (CXCL14) and CXCL16 which are located on chromosome 5 and chromosome 17 respectively [8]. Most of the genes for the CC chemokines occur on chromosome 17 although there are exceptions: MIP-3β/CCL19, SLC/CCL21 and CTACK/CCL27 occur on chromosome 9; eotaxin-2 and -3 occur on chromosome 7; MIP-3α/CCL20 occurs on chromosome 2; CCL28 occurs on chromosome 5 and the CCR4 ligands TARC/CCL17 and MDC/CCL22 occur on chromosome 16, co-localizing with the gene for fractalkine. The gene for the C chemokine lymphotactin occurs on chromosome 1. More detailed information can be obtained at http://cytokine.medic.kumamoto-u.ac.jp/CFC/CK/Chemokine.html.

In view of the similarity between the splice junctions, and numbers and sizes of introns and exons in the chemokine genes it is conceivable that the chemokine genes all arose from a common ancestral gene which has undergone chromosomal translocation and subsequent duplication giving rise to highly homologous chemokines like the MCP-1 and MIP-1α families which are generally monocyte/macrophage chemoattractants located on chromosome 17 and the IL-8 family (ELR motif containing CXC chemokines) on chromosome 4 which are mainly neutrophil chemoattractants. Interestingly the genes for the more divergent chemokines such as MIP-3α, lymphotactin and SDF-1 which are all specific for a single chemokine receptor, are found on distinct chromosomes and are not clustered with any other chemokine genes. It remains to be seen if the clustering patterns of the chemokine genes is associated with specific functional roles of chemokines.

3. RECEPTOR SPECIFICITY

Following the identification of the chemokine family of leukocyte chemoattractants, it was initially hoped that each chemokine would be responsible for directing the migration of one cell type: T- and B-lymphocytes, monocytes, eosinophils, basophils or neutrophils. This would have potentially allowed selective anti-inflammatory therapies, for example being able to block eosinophil recruitment in asthma, without altering the neutrophil response. For instance, in 1993, two neutrophil chemokine receptors for IL-8 were cloned (now called CXCR1 and CXCR2), giving two therapeutic targets for acute inflammation. At the time, the picture was even simpler for chronic inflammation where the only CC chemokine receptor known was the shared RANTES/MIP-1α receptor, CCR1 [9]. However, only three years later, in 1996, the year in which chemokine receptors were identified as an essential component for HIV-1 infectivity, the number of chemokine receptors identified had reached 9, and today it totals 19 (Fig. 2). Certain cell types express several receptors, such as activated T cells which can simultaneously express CXCR3 and CXCR4 along with certain CC chemokine receptors. The cloning of CCR3 [10,11] identified it as the major CC chemokine receptor on eosinophils, thus suggesting a very attractive target for asthma. But CCR3 expression is not restricted to eosinophils since this receptor is also expressed on basophils [12], Th2 T cells [13], neutrophils [14], monocytes [15] and microglia [16] as well as on non-leukocyte cell types such as bronchial epithelium and smooth muscle cells (C.A. Power, unpublished observations). Moreover receptor expression patterns have been shown to be under the control of pro-inflammatory cytokines [17] and thus neutrophils which normally only express the IL-8 receptors, can upregulate CC chemokine receptors upon stimulation with IFN-γ [18].

As is demonstrated in Fig. 2, the picture is further complicated in that not only are multiple receptors expressed on the same cell type, but each receptor

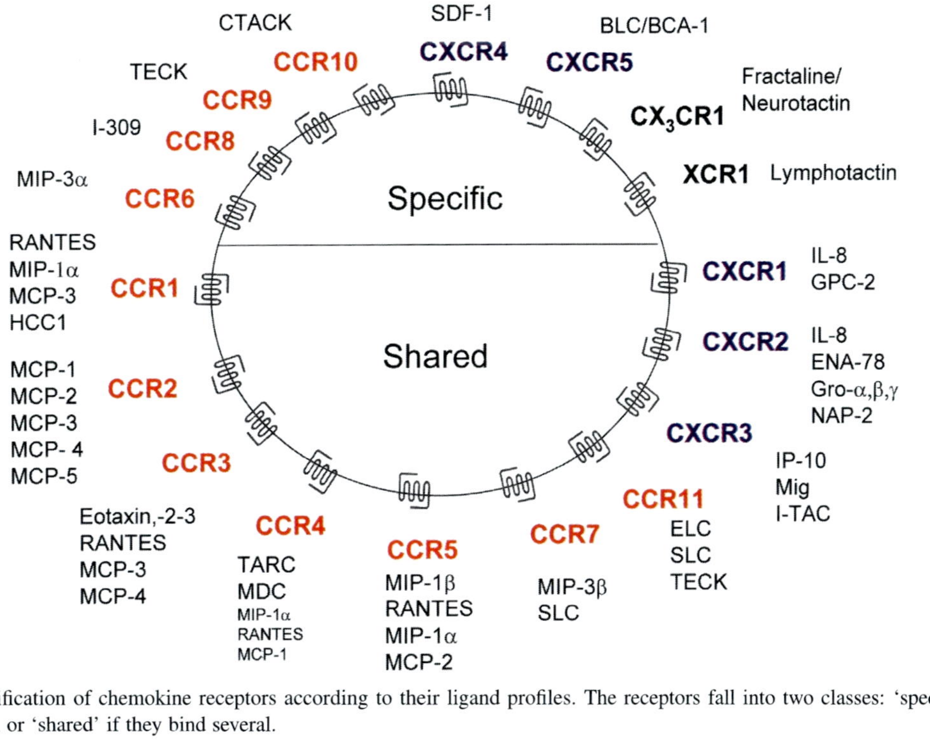

Figure 2. Classification of chemokine receptors according to their ligand profiles. The receptors fall into two classes: 'specific', if they bind one ligand, or 'shared' if they bind several.

type may bind several different chemokines — if ligand binding specificity observed in vitro holds true in vivo. There are only a few examples of highly specific chemokine receptor interactions to date: between CCR6 and MIP-3α/CCL20, CXCR4 and SDF-1/CXCL12 and CXCR5 and BLC/CXCL13. The reason that this statement is qualified by 'to date', is that as the identification of new ligands continues, absolute specificity has tended to disappear. CXCR1 was classified as a specific receptor for IL-8 for seven years, until GCP-2/CXCL6 was identified [19]. Whilst a certain degree of redundancy has been described for chemokine/receptor interactions, there appeared to be a golden rule that CXC chemokines only bind to CXC chemokine receptors and that CC chemokines only bind to CC chemokine receptors. This is still largely true but, as all rules are made to be broken, this has also been shown to be the case for the chemokine system. The duffy antigen receptor (DARC) is a promiscuous chemokine receptor expressed on erythrocytes which binds both CC and CXC chemokines [20]. Dr. Springer's group has shown that CXCR3 ligands are able to bind CC chemokines although the pharmacological properties appear to be different for the two chemokine subclasses — perhaps indicating that each class recognizes different receptor conformations [21]. Deviations from classical receptor pharmacology has also been observed for a virally expressed chemokine receptor, US28, where heterologous competition does not work as a two-way system. One of the ligands for US28, fractalkine, cannot be competed by certain CC chemokines, whereas these CC chemokines are all able to be displaced by fractalkine [22]. We believe, as is discussed below, that the study of the virally encoded members of the chemokine system will teach us a great deal about the intricacies involved in chemokine/receptor interactions, since viruses have produced chemokine ligands, for example vMIP-II, that can bind across chemokine receptor subclasses.

It is not always easy to establish which ligand is the most potent and has the highest affinity for a certain receptor. This is perhaps well illustrated by CCR5, and a comparison of the rank order potencies of its ligands that have been published. Using a calcium mobilization assay, the rank order potency of ligands on CCR5 expressed in CHO cells was described as RANTES > MIP-1β > MIP-1α [23] whereas in RBL cells stably transfected with CCR5, the rank order was RANTES > MIP-1β = MIP-1α [24] and in HEK293/CCR5 transfectants the order was different again, with MIP-1α > RANTES > MIP-1β [25]. However in the third example, the form of MIP-1α used is an allelic variant (LD78β instead of LD78α), which has a Pro instead of a Ser residue at position 2. However, MIP-1α (LD78α) was found to be the most potent ligand for IP3 accumulation when CCR5 was expressed in HEK293 cells, where the rank order was MIP-1α > RANTES > MIP-1β [26], but in the same assay using COS-7 transfectants, the most potent ligand was RANTES > MIP-1β > MIP-1α [27]. These results highlight the importance of cellular background, and implicate that care should be taken when interpreting results using transfected cell lines, where the receptors are probably expressed at far higher levels than found physiologically. In addition sequence modifications in recombinant commercially available chemokines may have profound effects on receptor interactions.

Whilst certain chemokines can bind to several different receptors, the biological activity induced may differ significantly and can even vary depending on the cell type on which the receptor is expressed. RANTES induced downregulation of three of its receptors and the ensuing recycling illustrates this phenomenon nicely. On incubation with RANTES in vitro, the surface expression of either CCR1 or CCR5 is reduced by approximately 80% from the surface of PBMC [28,29], and CCR3 from eosinophils [28,30]. However on removal of the chemokine from the culture medium, very different patterns of receptor recycling are observed. In the case of CCR5, receptor density returns to that observed initially [29]. With CCR3, only 70–80% of the initial receptor density is observed, but with CCR1, no recycling is observed [28]. Whilst the CCR3 receptors that do not recycle have been shown to traffic to the lysosomal compartment where they are degraded, the fate of CCR1 remains to be established. Therefore the apparent redundancy of a chemokine binding to more than one receptor may not be as redundant as meets the eye.

4. STRUCTURAL FEATURES

The chemokines of all three classes are composed of 70-odd amino acids, and most contain the canonical four-cysteine motif. Despite this highly conserved

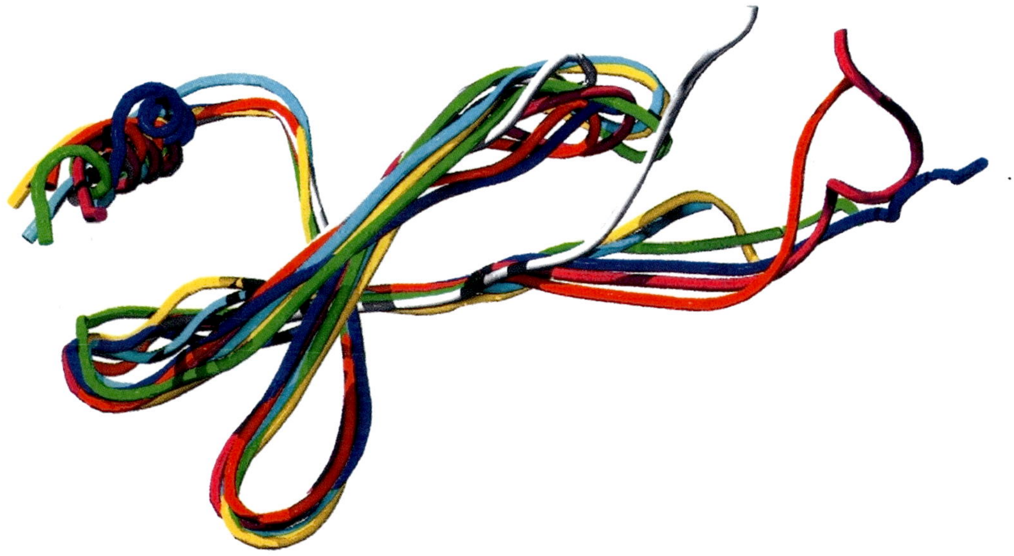

Figure 3. Overlay of the monomer structures of chemokines. The monomer structures of the following proteins are overlaid: IL-8, green; neutrophil activating protein (Nap-2), yellow; PF-4, light blue; MIP-1β, dark blue; RANTES, red; MCP-1, violet; chemokine domain of fractalkine, gray.

4-Cys motif, chemokines do not necessarily have a high level of sequence identity, which can vary from 20% to 80–90%. Alignment of the amino acid sequences of all CC chemokines for which structural information is available reveals only 10 strictly conserved residues out of the 70 or so amino acids (of which 4 are the cysteines). In the case of the CXC chemokines, the amount of identity is slightly higher (13 out of the 70 amino acids, including the 4 cysteines). This low level of identity of the primary sequence does not prevent the chemokines from having a remarkably well conserved monomeric fold consisting of an NH_2-terminal loop, three antiparallel β-strands arranged in a Greek key motif, connected by loops, and a COOH-terminal α-helix (Fig. 3). The monomer structures of all chemokines are almost superimposable; for example, the average root mean square deviation for the main chain atoms of the superimposition of the six chemokine monomers from all three classes shown in Fig. 3 is less than 1.5 Å for the 64 visible amino acids. The important structural differences that can be observed within a class usually lie at the generally floppy N- and C-termini, and are probably the result of the technique and conditions used to determine the structure. For example, several chemokine monomers contain short 3_{10} helices at the N-terminus, but many do not, and the importance of this feature has not been determined. Furthermore, the structure of the N-termini of many of these chemokines has been shown to play a very important role in their biological activity and as a consequence may only acquire their real structure when binding to their receptor. There are subtle differences in the first conserved disulfide bridge; in the case of the CC chemokines, the conformation of the bridge is a right-handed hook, while in CXC chemokines, the conformation is generally that of a left-handed spiral. The second disulfide bridge is essentially the same in all classes. Chemokines tend to have a large number of charged residues, with the majority of the chemokines having a larger number of basic residues than acidic ones, resulting in relatively high p*I* values. A comparison of their primary sequences and secondary structures does not reveal any obvious trend in charge distribution, nor does a comparison of their hydrophobic regions suggest important hydrophobic regions, though some hydrophobic amino acid positions, mainly in the β-sheet region, are conserved.

The most dramatic differences in chemokine structure lies in their quaternary structure. A few native chemokines and several modified and/or truncated analogues have only been observed as monomers as discussed below, but the majority have been observed to adopt dimer or tetramer structures. The biological significance of the quaternary structure

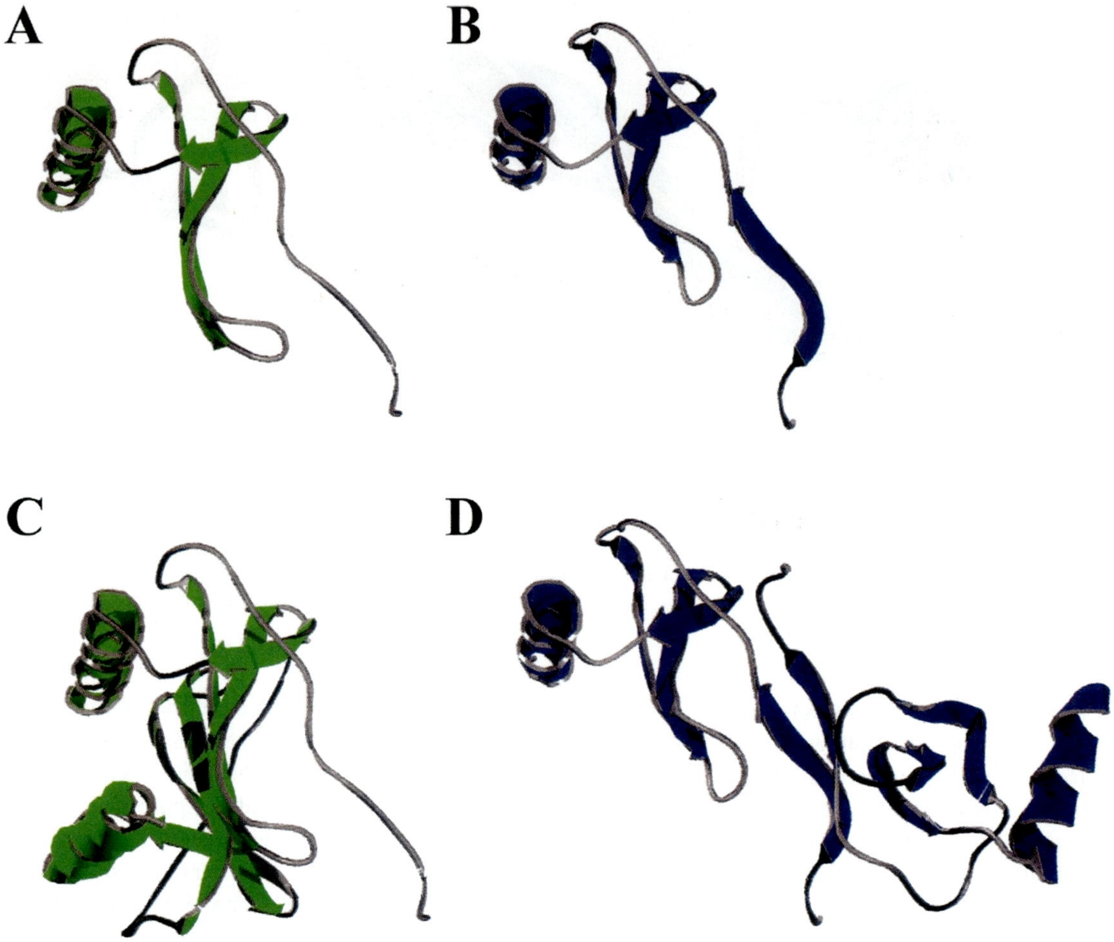

Figure 4. Comparison of the monomer and dimer structures of a CC and a CXC chemokine. Ribbon diagram of the structure of an (A) interleukin-8 monomer, (B) RANTES monomer, (C) interleukin-8 dimer, (D) RANTES dimer.

of chemokines has not been entirely elucidated, but several studies suggest that chemokines are active as monomers as well as dimers.

The most common quaternary structure observed to date is the homodimer. The vast majority of CC chemokines have been observed as dimers. Some of the structures of chemokines were determined in conditions that favored the monomeric form, although the dimer form has been observed (eotaxin-1 [31] and eotaxin-2 [32]). The only CC chemokines that have only been observed as monomers are HCC2 [33] and I-309 [34]. No native CXC chemokines have been observed as monomers with the exception of human SDF-1α, which was monomeric in solution NMR [35], but appeared dimeric in crystal form [36]. Some other CXC chemokines were purposely engineered to be obligate monomers such as IL-8 [37]. The most striking feature of the dimers is the difference in dimer structure between CC and CXC chemokines (see Fig. 4). The CXC chemokines form compact dimers reminiscent of the MHC Class II molecule with the dimer interface formed by an antiparallel β-sheet in which the first β-strand of both monomers interacts with the first β-strand of the other, thereby forming a six-stranded antiparallel β-sheet. This β-sheet is further stabilized by the two antiparallel α-helices, which lie on top of the β-sheet. The C-terminus of the α-helices interacts with the underlying β-sheet of the other monomer, thereby further stabilizing the dimer.

In contrast, the CC subclass dimers are formed by interaction of the NH_2-terminal region of one monomer with the equivalent region of the other monomer, yielding dimers with an elongated, cylindrical shape (see Fig. 4). An extended β-sheet is not formed in CC chemokines, but only a small antiparallel β-sheet consisting of the two short N-terminal strands. Despite the differences in these two dimer structures, a similar number of hydrogen bonds are created in both cases, and a similar amount of surface is buried upon dimerization. An early examination of the possible reasons for the differences in the mode of dimerization, based on a comparison of the structures of IL-8 and MIP-1β suggested the importance of certain residues in the choice of the type of dimer produced [38]. The length of the α-helix and the importance of its interaction with the other monomer were possible factors considered, since the helix was longer in IL-8 than MIP-1β. However, the determination of the structures of several CC chemokines with even longer α-helices than IL-8 (such as MCP-2) have cast doubt on this hypothesis. However, a second hypothesis would appear to have withstood the test of time, and is apparently still valid after the publication of many new CC and CXC chemokine structures. It was noticed that residues Leu25, Val27 and Leu66 in IL-8 were all hydrophobic and that this characteristic is conserved in all CXC chemokines, whereas in CC chemokines hydrophilic or charged residues replaced these residues. It was then demonstrated that these highly conserved residues play a pivotal role in selectivity since mutation of Leu25 of IL-8 to Tyr, which is highly conserved in CC chemokines at this position, introduced monocyte recruitment properties to this neutrophil chemoattractant [39]. In CC chemokines, the conserved charged residues have since been shown to play a role in dimerization and/or oligomerization, for example, Glu26 and Glu66 in RANTES [40].

The only tetramer structures observed to date are the crystallographic structures of bovine and human PF4 [41,42] and the solution NMR structure of a chimera of human PF4, in which 11 N-terminal amino acids are replaced by the 8 N-terminal amino acids of interleukin-8 [43]. This chimera appears to confirm that the tetramer is the biologically active conformation, consisting of a dimer of CXC dimers. The exact biological relevance of the quaternary structure of platelet factor-4 has not been fully investigated, but the study of the biological activity of the monomeric and dimeric forms of some chemokines has been undertaken in some cases. Interestingly, despite the fact that PF4 was the earliest chemokine to be characterized, a specific receptor for PF4 has yet to be identified.

The observation that most chemokines dimerize in solution suggests that the dimeric form is the biologically relevant form that productively binds its receptor. However it was also known that at physiological concentrations, chemokines were most likely to be monomers. The equilibrium constants for dimerization of some chemokines have been determined experimentally, and can vary from 35 μM for RANTES [44] to 40 nM for MIP-1β [45], although the experimental conditions have been shown to greatly influence these results. Many studies have been performed to determine the relative activity of the monomeric and dimeric forms of various chemokines. It has also been suggested that some chemokines (RANTES, for example) have biologically relevant multimeric oligomerization states which play an important role in their activities. RANTES has been shown to be secreted as a high molecular weight complex with glycosaminoglycans from the α-granules of platelets, and this complex has superior anti-HIV inhibitory properties than uncomplexed RANTES [46]. At high concentrations RANTES actually causes enhancement of HIV infectivity [47] and this effect has been shown to be due to the ability of RANTES to oligomerize [48].

In the case of CC chemokines, the effect of N-terminal truncation on dimer stability was studied on human MIP-1β [45]. Removal of five N-terminal residues had little effect on dimerization, but removal of six N-terminal amino acids had a marked effect on dimerization, and when eight N-terminal amino acids were removed, dimerization was abolished. This obligate monomer retains full capacity to bind to the receptor but has almost totally lost the ability to induce receptor activation [49]. However, a point mutation in the amino terminal region, P8A, prevents dimerization, but retains full binding and activation properties of CCR5. The same mutation in another CC chemokine, P8A in MCP-1, similarly produces an obligate monomer which is fully active [50]. These studies imply that CC chemokine receptor activation is mediated through the monomeric form, although care should be taken in extrapolation to all receptor/ligand pairs. However there is also evidence for functional dimerization for the CC chemokine MCP-1 based on the observations that chemically cross-linked dimers are active at nanomolar concentration [51]. The removal of residues 2–8 from the amino terminus results in an obligate monomer protein that antagonizes MCP-1, and which was hypothesized to exert its effect through dimerization with a wild type monomer [51].

In the case of CXC chemokines, evidence has been found for activation both by obligate monomers as well as obligate dimers. The obligate dimers were created using modified recombinant IL-8 molecules that were produced by joining two cDNAs coding for the monomers [52] whilst the obligate monomers were created through chemical modification (methylation of Leu25) at the dimer interface which prevents dimerization [53]. Both studies demonstrated the full activity of both monomeric and dimeric forms.

Thus the growing evidence suggests a biological activity for both chemokine monomers and dimers, though exactly what this role may be in determining biological activity and specificity in vivo remains to be elucidated. The biological relevance of oligomerization of chemokines into larger molecular oligomer states under the influence of glycosaminoglycans [54] also remains to be investigated. The continuing study of mutant forms of chemokines and the steady increase of chemokine structures can be expected to reveal some of the factors influencing their activity and hopefully lead to understanding specificity and selectivity mechanisms. However, the very large diversity of amino acid sequences, receptor and glycosaminoglycan binding characteristics, as well as the functions of these small proteins suggests that a great deal more information will be required before an understanding of their structure–function relationships is well understood.

5. RECEPTOR BINDING INTERACTIONS

Chemokines have been suggested to interact with their receptors according to a two-site model, similar to that proposed for the chemoattractant protein, C5a, which also acts on a 7TM G-protein-coupled receptor [55]. The major idea behind this model is that the main body of the ligand first reacts with an extracellular domain(s) of the receptor. This induces a conformational change allowing a flexible region to interact with a second site triggering signal transduction. In the case of C5a, the flexible triggering region is the carboxy terminal, whereas in the chemokine system, the N-terminal region is proposed to play this role. The extracellular regions of certain chemokine receptors have been mapped by chimeric receptor constructs or by site directed mutagenesis and thereby have identified the second extracellular loop (ECL2) as being the principal region of Site 1 for CCR2 and CCR5 [56], although the N-terminal loop is also important. Selectivity was found to reside in ECL2 with the N-terminal loop playing a role in stabilization of the ligand–receptor interaction. In the case of CXCR1, CXCR2 and CCR2 the N-terminal receptor loop has been shown to play an important initial role in ligand recognition [57,58]. Chimeric receptor constructs between CCR1 and CCR3 have shown that whilst the N-terminal loop is essential for ligand recognition, these receptors are activated by another site — again supporting the multiple site model [59]. However to date chimerae have not been created between receptor pairs which bind a common chemokine such as RANTES and MIP-1α, which could substantiate the differential binding effects of the N-terminal analogues RANTES and the MIP/RANTES chimera discussed below.

The evidence for the triggering role of the amino terminus of chemokines is multiple. The most conclusive is the observation that N-terminal truncation of chemokines resulted in proteins that have lost their capacity to activate their receptors, whilst retaining high affinity receptor binding. The first example of this phenomenon was demonstrated with IL-8, which was shown to have a three-amino-acid motif: Glu–Leu–Arg (ELR) preceding the CXC motif that was required for optimal binding [60,61]. This three-amino-acid motif is found on all neutrophil attracting CXC chemokines, but is absent on members such as PF4 and IP-10 that lack this activity. Interestingly, when the ELR motif was introduced into PF4, it was now able to attract neutrophils [62]. Removal of the first five amino acid residues from IL-8, although retaining Arg-6 produced a protein that was still able to bind to the receptor [63]. Furthermore this analogue could antagonize the actions of IL-8 highlighting the importance of the amino terminal in activation through Site 2, whilst not affecting Site 1. Other modifications to the amino termini of several chemokines further strengthened this argument, and their antagonist properties are discussed below.

However this two-site paradigm does not appear to hold true for all chemokine–receptor pairs.

RANTES, one of the most promiscuous chemokines, that binds to six human chemokine receptors (see Fig. 2) as well as to the functional virally encoded receptor US28, does not appear to bind to these receptors using the same molecular interactions. This was first demonstrated by the N-terminally modified RANTES protein AOP-RANTES, which retains high affinity binding, but is virtually inactive on CCR1, leading to its description as a receptor antagonist since it demonstrated no activity in a monocyte chemotaxis assay [64]. It was subsequently shown to be more effective in inducing CCR5 downmodulation than the CCR5 ligands, RANTES, MIP-1α and MIP-1β [65], which agreed well with the observation that it was also more effective in inducing phosphorylation of the carboxy terminus of this receptor [24]. In fact CCR5 appears to be relatively insensitive to amino terminal modifications of RANTES, since the truncated (3–68)RANTES is equipotent to the native protein [66]. We have recently observed that substitution of the residues preceding the N-terminal CC pair with the MIP-1α sequence, to form a chimera which we have called MIP/RANTES, retains full activity on CCR5, but only binds with low affinity and is totally inactive on CCR1 [67]. Interestingly, the truncated (9–68)RANTES analogue retains a small capacity to activate CCR5 in mobilizing calcium, and full activity in CCR5 downmodulation [68]. The addition of eight amino acids to this truncated protein, even from another chemokine with low sequence identity is sufficient to restore full activity in mobilizing calcium. This suggests that the amino terminal segment serves as a stabilizing region for optimal RANTES interactions with CCR5, but is not essential for specific receptor activation.

This chimeric molecule allowed a further distinction amongst the receptors that obeyed the two-site paradigm and those that seem to deviate. MIP/RANTES is totally unable to displace eotaxin from CCR3, whereas RANTES is able to do so. On the other hand, in HEK293/CCR4 transfectants, RANTES shows low affinity in that it can displace the high affinity ligand TARC at micromolar concentrations, and these binding properties are retained by all the N-terminally modified RANTES analogues tested, including MIP/RANTES [67]. These observations suggest that CCR1 and CCR3 obey the two-site paradigm described above, whereas CCR5 is activated through a different mechanism.

Mutagenesis studies of the CC chemokine RANTES, have revealed that distinct but overlapping epitopes are necessary for high affinity binding to CCR1, 3 and 5 [69]. An aromatic residue in the RANTES N-loop (F12) is necessary for high affinity binding to CCR3 and CCR5. Interestingly, a Phe in the same position in the 20's loop of MIP-1α, MIP-1β and MCP-1 is also crucial for binding to CCR5 and CCR2b respectively [49,50], suggesting that this aromatic side chain of the N-loop of CC chemokines may represent an essential element. However, for CCR1 binding, the crucial residue in the RANTES 20's loop is R17 [69]. This loop has also been shown to be important in IL-8 binding [70] but the most important element for receptor binding of this CXC chemokine is the ELR epitope preceding the CXC motif [60,61]. The carboxy terminal helix does not appear to play any role in receptor binding, as was demonstrated by its deletion in IL-8 ([63]) and furthermore several chemokines have been expressed as fusions of large proteins to the carboxy terminus which retain full receptor binding and activity [71].

6. GLYCOSAMINOGLYCAN (GAG) BINDING SITES

The pioneering studies on the importance of the chemokine/glycosaminoglycan (GAG) interaction were carried out several years ago where the immobilization of IL-8 on endothelial surfaces was demonstrated [72]. An aspect of selectivity in this interaction was demonstrated using electrophoresis of chemokines in the presence of different heparin fragments [73] and a solid phase heparin binding assay [74]. Interestingly, GAGs have been shown to both potentiate and inhibit chemokine activities. The potentiation of chemokine activity by GAGs has been reported in vitro by using CHO cells that are deficient in cell surface GAGs where it was demonstrated that whilst biological activities such as receptor binding and activation is independent of the presence of GAGs, the expression of cell surface GAGs helped to sequestrate the chemokines [75]. On the other hand, the activation of platelets by chemokines is inhibited by GAGs such as heparin [76]. Recently it has been shown that RANTES secreted from the alpha granules of platelets is found in large GAG associated complexes, and that the GAG

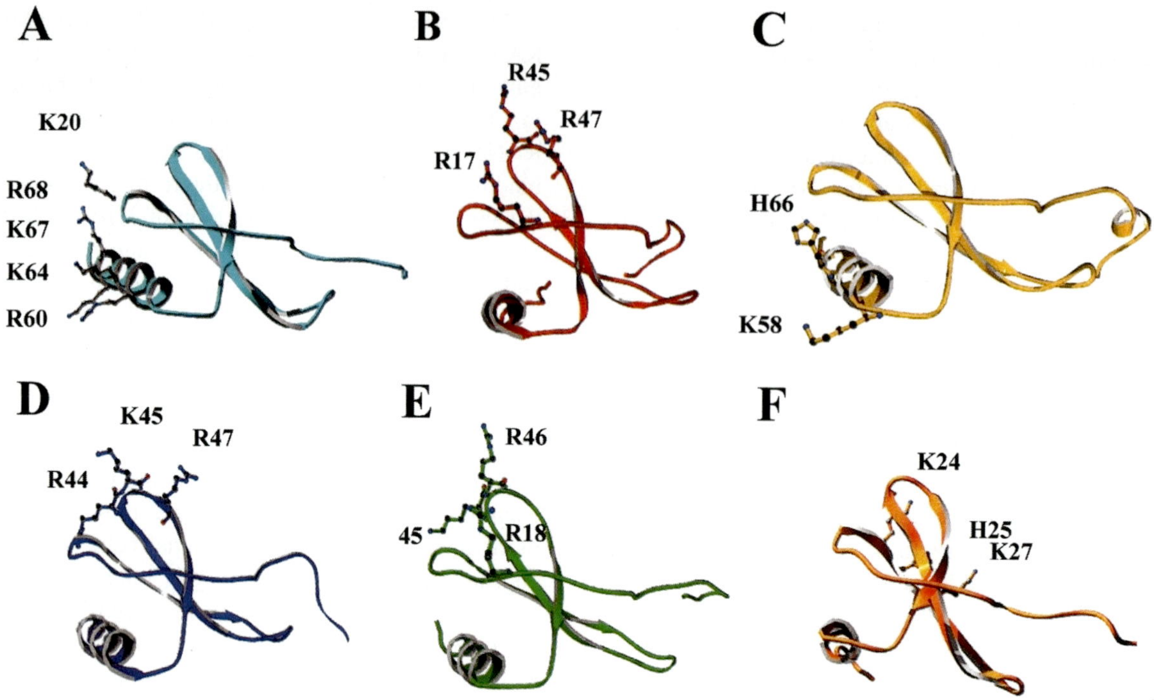

Figure 5. GAG binding residues of chemokines. The chemokine monomers with the GAG binding basic residues identified by mutagenesis studies. (A) IL-8, (B) MIP-1α, (C) MCP-1, (D) RANTES, (E) Mip-1β, (F) SDF-1α.

association is highly significant in the anti-HIV inhibitory properties of this chemokine [46]. RANTES has also been shown to induce enhanced infection of HIV-1 in macrophages [77], as well as enhancing other viral infections in vitro, albeit at micromolar concentrations, which is attributed in part to the oligomerizing property of this chemokine [47,78]. As discussed below, this property is mediated by the binding of RANTES to GAGs.

Receptor binding regions and GAG binding regions have been delineated by mutagenesis studies for certain chemokines such as IL-8 [70] and MCP-1 [79,80]. These studies led to the idea that receptor and GAG binding domains were spatially separated in that receptor binding was governed by the amino terminus and GAG binding by the carboxy terminal helix. The importance of the carboxy terminal helix in GAG binding had already been shown for PF4 [81]. However it appears that this is a generalization that does not hold true for all chemokines. In fact, in IL-8 the 20's loop contains residues important both for receptor binding and GAG binding [70]. The GAG binding sites have now been mapped for three more chemokines, and we have recently extended these studies to RANTES. The residues on chemokines that interact with GAGs identified to date are represented in Fig. 5. Heparin binding regions are often a cluster of basic residues, forming either a BBXB or a BBXXB motif. Inspection of the primary sequence of chemokines shows that the BBXB motif is in fact present on several chemokines. These residues have been mutated to Ala in the CCR5 ligands MIP-1α [82,83], MIP-1β [84] and recently in RANTES in our laboratory [85], showing that this region of these chemokines appears to be the principal heparin binding site. Whilst no effect on receptor binding was observed for the R46A mutant of MIP-1β to its receptor CCR5, the mutation R46A in MIP-1α abolished binding to CCR1. Our results with RANTES have shown a similar phenomenon. The mutations of the residues R44, K45 and R47 individually to Ala had no significant effect on all the parameters tested — binding to heparin, CCR1, CCR5 or on in vitro chemotaxis. However, the concomitant mutation of the basic residues to Ala in the ^{44}RKNR^{47} motif had a profound effect on heparin binding and a significant effect on binding to CCR1, whilst wild type activity was retained on CCR5. RANTES has a second cluster of basic residues immediately preceding the carboxy terminal helix, ^{55}KKWVR^{59} but mutation of these residues either individually or as a triple mutation had no effect in all of the above assays.

The GAG binding residues on MCP-1 are located towards the carboxy terminus, and do not form a cluster. In this chemokine, the two residues involved are K58 which is located immediately before the carboxy terminal helix, and H66 which is found at the end of the helix, whereas the other basic residues in this region, K56, K69 and K75 are not involved in GAG binding [79]. The carboxy terminal helix of IL-8 presents four basic residues which interact with GAGs: K64 and R68 playing the major role, but R60 and K67 also contribute [70]. K20, in the 20s loop which has been shown to be involved in receptor binding, is also involved in GAG interactions. However, in the IL-8 protein the main receptor interface is spatially separated from the carboxy terminal helix which is the principal GAG binding region. SDF-1 has introduced an entirely new region that is implicated in GAG binding [86]. In this chemokine the GAG binding is mediated by a BBXB motif, but which is located in the 20's loop, since it consists of the residues ^{24}KHLK^{27}. However these basic amino acids play no role in CXCR4 activation.

Chemokine/GAG binding has been measured by several different methods in different laboratories. These include heparin sepharose chromatography [70,83–85], electrophoresis in the presence of different GAG species [73], binding assays to radiolabelled heparin [79] and binding of radiolabelled chemokines to heparin immobilized on beads [74,85]. The most common procedure is the analysis of the NaCl concentration required to elute the chemokine from heparin sepharose columns. In these analyses, certain mutants are no longer able to bind to the column as was observed for MIP-1α [83] and MIP-1β, where the binding was abrogated in the presence of the physiological concentration of NaCl [84]. However, in the chemokines that bind much more strongly to heparin, such as IL-8, SDF-1 and RANTES, the mutation of the heparin binding sites does not totally abolish the capacity to bind to heparin, indicating either that other specific sites exist, or that there is a certain amount of non-specific electrostatic interaction. Our analyses of the RANTES triple ^{44}RKNR^{47} Ala mutant indicate the former. The residual 20% binding capacity that this mutant retained for heparin, had the same affinity for heparin as WT RANTES however, but lost most of the selectivity for the four GAG families, heparin, heparin sulfate, dermatan sulfate and chondroitin sulfate [85].

7. CHEMOKINE ANALOGUES

Alterations in the N-termini of certain chemokines have been shown to result in profound changes on their activity. The pioneering work in this area was carried out by two laboratories on IL-8. Hebert et al. carried out an alanine scanning mutagenesis study and identified that the ELR motif present in several CXC chemokines was essential for receptor activation [61]. Using total chemical synthesis, amino terminally truncated forms of IL-8 revealed that removal of the first five amino acids producing (6–68)-IL-8 did not have a major impact on binding, but the removal of the N-terminal tail had abolished receptor activation properties [63]. This validated the two-site hypothesis described above, and furthermore was the first example of a functional chemokine receptor antagonist.

Similar modifications of CC chemokines have also been shown to produce antagonists. N-terminally truncated RANTES, MCP-1 and MCP-3 proteins have been shown to antagonize the effect of their parent ligands in vitro [87]. Interestingly, removal of the first eight residues from the N-terminal of RANTES changed its specificity since the truncated protein was now able to bind to CCR2, to which the full-length protein does not bind. The removal of seven residues from the N-terminal of MCP-1 forms an antagonist protein that has been proposed to act through an interesting mechanism by preventing the formation of the active MCP-1 dimer, analogous to dominant negative mutations [51]. However there is only one report of the activity of such truncated chemokines in vivo, where the administration of the truncated (9–68)-MCP-1 protein into mice that spontaneously develop arthritic symptoms was very effective in reducing the inflammatory symptoms, when administered in a therapeutic protocol [88].

The extension of the amino terminal of certain chemokines can also have a profound effect on the biological properties. In the case of RANTES, when the recombinant human protein is produced in

Escherichia coli, despite being correctly folded as assessed by NMR spectroscopy [89], the chemokine had no activity [90]. It was established that the retention of the initiating methionine was responsible for this effect, and that the protein was a potent antagonist — an entirely serendipitous finding. This analogue, Met-RANTES, retains high affinity binding for human CCR1 [90] and CCR5 [64] but only moderate affinity for CCR3 [28,91,92], and we have recently shown that it only binds to CCR1 and CCR5 in the murine system (R. Buser and A. Proudfoot, unpublished observations). Met-RANTES has been tested in several rodent inflammatory models which have served to highlight the efficacy of chemokine receptor antagonism in preventing inflammation. Thus Met-RANTES has been reported to reduce inflammation in models of arthritis [93], nephritis [94], organ transplant [95], colitis [96] and lastly asthma [97]. Since this analogue has high affinity for mCCR1 and mCCR5, chemokine receptors expressed on leukocytes implicated in many inflammatory disorders such as T-lymphocytes and monocytes, these results indicate that these two receptors are good therapeutic targets. It is however interesting to note that Met-RANTES is also effective in reducing airways inflammation in the murine ovalbumin sensitization model. We have recently analyzed the murine receptor pharmacology for this analogue and found that it has virtually no affinity for mCCR3. Since CCR3 is the predominant CC chemokine receptor expressed on eosinophils and that this cell type is widely thought to be the main pathogenic culprit in asthma, this receptor was widely believed to be a prime target for this disease. However the use of this analogue demonstrates that antagonism of receptors other than CCR3 is effective in reducing airway inflammation.

Our initial studies characterizing Met-RANTES were carried out on the pro-monocytic cell line, THP-1, where we found no activity. However, on eosinophils from certain patients, weak partial agonist activity was observed [92]. With the aim of producing a derivative that was devoid of this partial agonist activity, we subsequently carried out a study through chemical modifications of the amino terminus, and produced a series of aminooxy alkyl derivatives with different chain lengths. This led to the identification of an analogue, AOP-RANTES (aminooxy pentane RANTES) that had higher affinity for RANTES receptors than Met-RANTES [64,15]. Again the amino terminally extended RANTES protein was inactive on monocytes, and antagonized the effects of RANTES, MIP-1α and MIP-1β on this cell type. This analogue has since taught us new mechanistic pathways that play a role in disease.

AOP-RANTES was found to be a particularly potent inhibitor of HIV-1 infection, being significantly more potent than RANTES itself [64]. The fact that CCR5 ligands are able to inhibit HIV-1 infection [98] helped to identify CCR5 as the macrophage tropic receptor shortly after the revelation that the long sought after co-receptor for HIV infection was a chemokine receptor (reviewed in [99]). The inhibition of HIV-1 infection by chemokines could be mediated by two mechanisms — first simply by steric hindrance, or secondly by inducing downregulation of the receptor from the cell surface. Although AOP-RANTES was initially thought to be an antagonist, subsequent studies demonstrated that it also retained partial agonist activity on CCR1 and CCR3, but was a full agonist of CCR5 [92]. In fact in certain activities, it was even more potent than RANTES itself [29,100]. Thus its potent HIV-1 inhibitory properties were attributed to its greater effectiveness in downregulating CCR5, and furthermore in preventing the recycling of the receptor to the cell surface [29]. However AOP-RANTES did not target the receptor to the lysosomal compartment where it is degraded, as has been shown to occur for other chemokine receptors such as CCR3 [30]. It has subsequently been demonstrated that CCR5 does in fact recycle to the cell surface after AOP-RANTES treatment, but through a mechanism that remains to be elucidated, prevents the functionality of CCR5 [101]. Thus modification of the receptor endocytotic pathways is an effective mechanism of receptor antagonism.

To date there are not much data available on receptor antagonism in inflammatory neurological disorders, but there is accumulating evidence that chemokines and their receptors play a role, and this is treated in depth in other chapters in this book. Neutralization of MIP-1α using a monoclonal antibody has a profound effect on the acute phase of an animal model of multiple sclerosis, T cell mediated experimental autoimmune encephalomyelitis (EAE) [102] whereas anti-MCP-1 treatment was effective in preventing relapses [103]. Examination of human autopsy samples, has shown high levels of expression of two receptors which are found on Th1 T cells, CXCR3 and CCR5, the latter being also upregulated on activated macrophages [104,105]. CCR1 has also been observed in certain samples. The role of CCR1 in this disease is further supported by two lines of evidence. First, it was reported by Dr. C. Gerard (Pasteur Institute Chemokine Meeting, Paris, 1999) that CCR1-deficient mice are resistant to murine EAE, and recently a small-molecule CCR1 receptor antagonist has been shown to have a dose-related effect in inhibiting a chronic EAE model in the rat, when administered prophylactically [106]. Recent unpublished studies with Met-RANTES in Dr. R. Ransohoff's laboratory have shown that whilst this analogue has little effect in the chronic MOG-induced model of EAE, it had significant effects in the relapsing, remitting chronic EAE model in SWXJ mice. Met-RANTES treatment resulted in significant reductions in severity during the chronic period of disease. Concomitant experiments in EAN (experimental autoimmune neuritis) show that the Met-RANTES treated animals have a less severe disease course and weight-loss is minor (B. Kieseier, personal communication). Statistical and histological analyses of these two experiments are still ongoing at the time of writing, and will hopefully provide evidence for new therapeutical approaches to inflammatory neurological disorders.

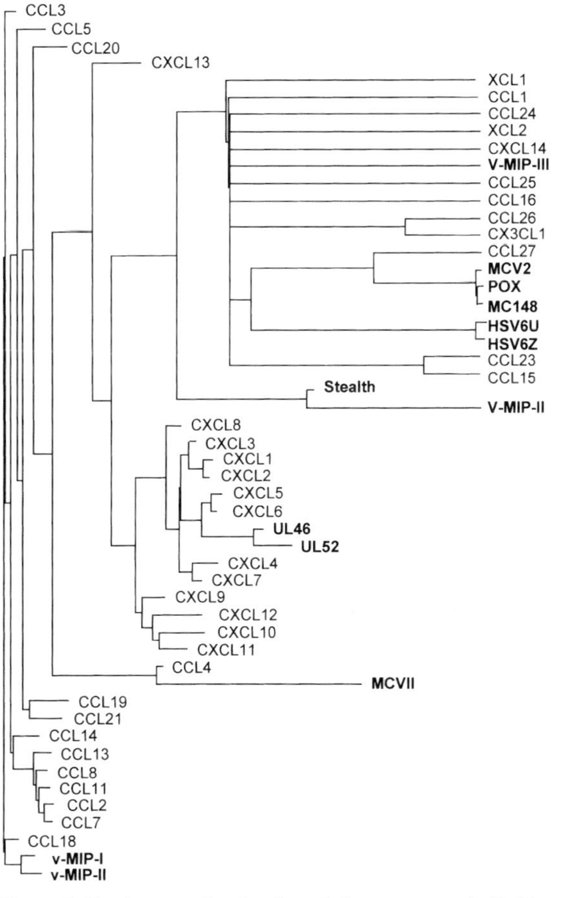

Figure 6. Dendrogram showing the relative sequence similarities between human and viral chemokines. The viral chemokines are shown in bold.

8. VIRAL PIRACY OF THE CHEMOKINE SYSTEM — WHAT CAN WE LEARN?

Viruses have evolved throughout the millennia in order to devise ways to combat the host's defense system. During this process they have captured genes from the immune system and mutated them in order to best serve their purposes. Fig. 6 shows their evolutionary proximity to the human chemokines. The early observation that many poxviruses and herpesviruses expressed chemokines, receptors or chemokine binding proteins, highlighted the subtlety of the host response to infection and the ability of viruses to evade that response. Since the principal aim of a virus is to outwit the host's immune response, continual mutation will result in molecules well equipped to do so. Therefore, we can probably learn a great deal from these virally encoded chemokines, their receptors and the proteins that interfere with chemokine activities. The field of virally encoded chemokines, receptors and chemokine binding proteins exceeds the scope of this chapter, and the reader is referred to specialist reviews [107–109].

Since those early observations, many new members of the chemokine family have been identified from viral sources, and characterized for their ligand binding and biological activities. The predominant actions of these proteins are the control of the host immune system, and downregulation of inflammation, to create an environment more suited to the replication of the virus. In addition, effects on cellular replication have been noted, providing an in vitro link between some of the viruses and their observed enhancement of cellular proliferation in disease.

The earliest virally encoded chemokines to be studied come from the human herpesvirus 8 (which was originally called the Kaposi's sarcoma herpesvirus) [110]. Three open reading frames encoding potential chemokines were identified in the initial sequencing of the HHV8 genome, and these were named vMIP-I, vMIP-II and BCK (for beta chemokine) — otherwise known as vMIP-III. The assignment of the name vMIP (for viral macrophage inhibitory protein) was based only on the observed sequence identity between the virally encoded molecules and the human MIP-1α sequence. This identity level is low — 42% for vMIP-I and II, and tenuous for vMIP-III. The best studied of these three proteins is vMIP-II. Initial studies showed that this protein has a very interesting receptor binding profile, which was unique at the time, in that it is capable of binding to both CXC and the CC chemokine receptors. Specifically, vMIP-II binds CXCR4, and CCR5 (the two key co-receptors for HIV infection), and prevents infection of cell lines and primary cells by the appropriate strain of HIV. Thus it is the only molecule identified to date that is able to inhibit HIV-1 infection of both R5 and X4 viral strains — the elucidation of the binding properties of this chemokine should surely help us to design a pan-chemokine receptor anti-HIV inhibitor.

It was therefore suggested that it had a role in the interplay between the HHV8 and HIV infections [111]. The protein is also highly angiogenic in a chorioallantoic assay, indicating a role in the neovascularization of the sarcoma [110]. Beyond the biological role that the protein plays in viral infections, several studies have been carried out attempting to exploit its anti-inflammatory activity in a therapeutic setting. In a model of glomerulonephritis, vMIP-II was able to significantly reduce cell infiltration in the glomeruli, and also attenuate proteinurea [112]. Gene transfer experiments using plasmid DNA have shown that vMIP-II can enhance the survival of cardiac allografts [113]. This effect can be enhanced if the gene is co-administered with the virally encoded analogue of interleukin-10 (vIL-10).

Studies of vMIP-I have shown that it is a highly selective agonist for CCR8 [114,115]. Functionally, it has been shown to induce calcium flux and chemotaxis in cell lines — effects that can be blocked by vMIP-II. The interplay between vMIP-I and v-MIP-II in vivo is still not clearly understood, and the current interest in the vMIP-I/CCR8 interaction seems to focus somewhat on the Th1/Th2 balance, since CCR8 is specifically expressed in Th2 T-lymphocytes. The third chemokine, vMIP-III has been shown to be a CCR4 agonist, and is incapable of blocking HIV infection [116]. The presence of the protein in Kaposi sarcoma lesions, and the receptor binding profile suggests that there may be a Th2 component to the inflammation. However, from studies of CCR4-deficient mice, the in vivo role of CCR4 in Th2 pathologies is still debatable [117].

Other viral chemokines have so far been studied in less detail, but the functional characterization of several have started to appear in the literature. The human poxvirus, molluscum contagiosum virus (MCV) encodes an open reading frame called MC148, with significant identity to the chemokines, and with conservation of the two disulfide bridges. Pharmacological analysis of the receptor binding profile [118] showed that it bound selectively to human CCR8, and was able to antagonize I-309 mediated calcium flux. Since human CCR8 was originally identified as a dendritic cell expressed chemokine receptor, it has been suggested that the protein plays a role in the control of antigen processing/presentation. As with most of this field, the functional role of such proteins comes from an extrapolation of the in vitro cell biology, since there is a relative paucity of data in a pathological setting on the functions in vivo.

Viral chemokines are continually being identified and a lot remains to be learned. Human Herpesvirus 6 open reading frame U83, encodes a functional chemokine, which is capable of mobilizing calcium in THP-1 cells, and also is a chemotactic agent [119]. The receptor has not been identified to date, nor has binding of the ligand to THP-1 cells been demonstrated — data required for formal proof that the protein is a member of the chemokine superfamily. Given that the spacing of the two disulfide bridges in U83 is atypical, it will be interesting to see what effect this has on receptor selectivity. A CXC chemokine has been found in the full length sequence of the Marek's disease virus [120]. The biological significance of this has not been determined and since this is an avian virus, this elucidation will probably not be straightforward.

Whilst the scope of this chapter is devoted to the chemokine ligands, it is worthwhile discussing the virally encoded receptors to see what clues they can give us to understanding chemokine/receptor interactions. The concept that viruses could encode proteins similar to the seven transmembrane G-protein coupled chemokine receptors similar to those found in mammals came from the sequencing of a genomic fragment of human cytomegalovirus. Subsequent pharmacology showed that one of these sequences, known as US28 was capable of binding to CC chemokines [121]. The pharmacology of this receptor became even more interesting with a series of observations made about its ability to bind the soluble domain of the CX3C chemokine fractalkine [22]. The interaction shows a sub-nanomolar affinity, and this interaction could only be competed by other US28 binding CC chemokines if they were added at relatively high concentrations. The explanation of this relative difficulty to compete the receptor binding with heterologous ligands is most likely due to a slow conformational change in the US28/fractalkine complex. Functionally, it implies that the interaction between the fractalkine and cell surface US28 will be a highly stable interaction. Since fractalkine is normally found tethered to the cell surface by a mucin stalk, the US28/fractalkine interaction may be involved in the cell/cell contact with infected cells, and in the transfer of the virus from cell to cell.

The seven transmembrane receptor encoded by Herpesvirus 8, (known as ORF74) is a CXC chemokine receptor which is constitutively active [122]. When expressed in mice under the control of a CD2 promoter, it produces a tumor in mice which is pathologically very similar to the Kaposi lesion [123]. In addition, if the human CXCR2 (the closest human homologue of ORF74) is mutated at position 138 (the DRY sequence) — to introduce the VRY seen in ORF74, then the human receptor becomes constitutively active, and has transforming activity in NIH-3T3 cells [124].

The other virally encoded seven transmembrane receptors have been studied in less detail. The equine herpesvirus 2 E21 open reading frame encodes a CC chemokine receptor that shows functional responses to human eotaxin [125]. The pathological consequences of this are not clear, and it cannot be ruled out that the true ligand will only be identified when a range of equine chemokines are tested in binding assays. Human Herpesvirus 6 contains open reading frames for two receptors, U12 and U51. The U51 gene product has been shown to be a CC chemokine receptor, binding RANTES. RANTES can be competed with a range of other CC chemokines such as MCP-1, -3 and -4, eotaxin and the HHV8 chemokine, vMIP-II [126]. In epithelial cells the expression of U51 has been shown to reduce the secretion of RANTES into the cell culture medium, suggesting an ability of the virus to downregulate the host protective inflammatory response. The U12 gene product is also a CC chemokine receptor, having been shown to direct calcium flux in response to RANTES and MIP-1α, but not IL-8 [127].

Viruses have been shown to produce a number of proteins that can neutralize inflammatory mediators, namely cytokines and chemokines. We believe that they can thus teach us how to neutralize chemokine activities, and also give us important insights into the structure–function modalities of these apparently highly related small proteins. Chemokine binding proteins were initially found in the poxviruses. They are characterized based on their ability to bind chemokines in solid phase assays. So far, no mammalian homologues of these binding proteins have been identified by primary sequence alignments. In addition, since these proteins are secreted and soluble, no direct role in signaling has been proposed. One of the best characterized is the myxoma poxvirus M-T7 protein which binds promiscuously to a variety of CXC, CC and C chemokines as well as to interferon gamma [128]. In vivo, this protein has been shown to inhibit intimal hyperplasia following restenosis in the rabbit showing that these proteins can be used as anti-inflammatory agents [129]. A second protein, M-T1 (also confusingly known as CBP-II), is a member of a family of chemokine binding proteins produced by the orthopoxviruses [130], and has only a 40% sequence identity with the M-T7 protein. In rabbit models of myxoma virus infection, it has been shown that deletion of the M-T1 from the poxvirus genome results in a dramatic increase in infiltration of leukocytes around the sites of infection, pointing to an anti-inflammatory role of this protein in the rabbit disease. The vaccinia virus derived chemokine binding protein, p35, which is closely related to the T1 proteins [131], has been shown to selectively bind to CC chemokines and not CXC or C chemokines. In vivo this has been used to block the recruitment of eosinophils in a skin model of inflammation. Mechanistically, what is most interesting in these two classes of chemokine binding proteins is that they appear to use entirely different modes of action. The T7 protein is thought to bind to the glycosaminoglycan binding domain of the chemokine, and does not inhibit chemokine activity in vitro [128] whilst the T1/p35 class are able to block in vitro activities [132] and therefore probably bind to the amino terminal region. Whilst the structure of a member of the T1/p35 class has been solved [133], despite intensive efforts by several laboratories, the structure of a complex of a virally encoded chemokine binding protein with its ligand(s) which would provide precious structure/function information, has so far proved elusive. More recently, a broad spectrum chemokine binding protein, encoded by the gene M3 has been identified in the murine gamma herpesvirus. Once again, this has no clear sequence identity with the known human chemokine receptors [134].

The initial excitement following the characterization of chemokine binding proteins and viral chemokines with antagonist activities led to many discussions about their therapeutic possibilities. Whilst there are several reports that such proteins are active in vivo, in animal models of inflammatory disease, these results should be treated with caution. Since there is a very low sequence identity level between the virally encoded chemokines and their human homologues, these proteins could be antigenic in humans. It could justifiably be argued that the viral chemokines and chemokine binding proteins will have evolved to evade the host immune system — both by stealth approaches (i.e. that their surfaces contain few novel epitopes) and by direct suppression of the host antigen presentation process in some cases. These are both highly plausible arguments, but in the absence of published data on the antibody response to exogenously applied viral chemokines or chemokine binding proteins, their therapeutic use should be viewed with caution, at least for treatment of chronic inflammatory disorders.

The other potential application of the chemokine binding proteins that is yet to be explored is their use in the rational design of chemokine receptor antagonists. Since the chemokine binding proteins are not classical seven transmembrane receptor coupled G-proteins, it should be possible to elucidate the three dimensional structure of the p35/chemokine complex using X-ray crystallography. However, as mentioned above, we have not yet managed to achieve the solution of a chemokine/chemokine binding protein complex. We believe though that this will be accomplished, and depending on the precise geometry of the interaction, and the distribution of key contact residues, it may be possible to use this structural information to rationally design molecules which bind to chemokines, and prevent them from binding to their conventional G-protein coupled receptors.

9. CONCLUSIONS

The chemokine family of ligands appears rather daunting at first glance due to their ever-increasing number and the apparent redundancy of the system. However, as studies of their biology continues, and in particular as more structure–function investigations are carried out, these apparently very similar polypeptides are beginning to show rather amazing diversity. Moreover, with the availability of receptor antagonists as proof of concept, and recently with the development of small molecule chemokine receptor antagonists (for a review see [135,136], intervention with this system appears more and more promising as a therapeutic approach. Hopefully we will learn a great deal more on how to combat the nefarious side of chemokines by studying how viruses have evolved to evade the inflammatory responses mediated by the chemokine family.

REFERENCES

1. Deuel TF, Keim PS, Farmer M, Heinrikson RL. Amino acid sequence of human platelet factor 4. Proc Natl Acad Sci USA 1977;74(6):2256–2258.
2. Yoshimura T, Matsushima K, Tanaka S, Robinson EA, Appella E, Oppenheim JJ et al. Purification of a human monocyte-derived neutrophil chemotactic factor that has peptide sequence similarity to other host defense cytokines. Proc Natl Acad Sci USA 1987;84(24):9233–9237.
3. Wolpe SD, Davatelis G, Sherry B, Beutler B, Hesse DG, Nguyen HT et al. Macrophages secrete a novel heparin-binding protein with inflammatory and neutrophil chemokinetic properties. J Exp Med 1988;167(2): 570–581.
4. Pan Y, Lloyd C, Zhou H, Dolich S, Deeds J, Gonzalo JA et al. Neurotactin, a membrane-anchored chemokine upregulated in brain inflammation. Nature 1997;387(6633): 611–617.
5. Bazan JF, Bacon KB, Hardiman G, Wang W, Soo K, Rossi D et al. A new class of membrane-bound chemokine with a CX_3C motif. Nature 1997;385:640–644.

6. Wells TNC, Peitsch MC. The chemokine information source: identification and characterization of novel chemokines using the WorldWideWeb and expressed sequence tag databases. J Leukoc Biol 1997;61(5):545–550.
7. Loetscher M, Gerber B, Loetscher P, Jones SA, Piali L, Clark-Lewis I et al. Chemokine receptor specific for IP10 and mig: structure, function, and expression in activated T-lymphocytes. J Exp Med 1996;184(3):963–969.
8. Zlotnik A, Morales J, Hedrick JA. Recent advances in chemokines and chemokine receptors. Crit Rev Immunol 1999;19(1):1–47.
9. Neote K, DiGregorio D, Mak JY, Horuk R, Schall TJ. Molecular cloning, functional expression, and signaling characteristics of a C-C chemokine receptor. Cell 1993; 72(3):415–425.
10. Ponath PD, Qin S, Ringler DJ, Clark-Lewis I, Wang J, Kassam N et al. Cloning of the human eosinophil chemoattractant, eotaxin. Expression, receptor binding, and functional properties suggest a mechanism for the selective recruitment of eosinophils. J Clin Invest 1996; 97(3):604–612.
11. Daugherty BL, Siciliano SJ, DeMartino JA, Malkowitz L, Sirotina A, Springer MS. Cloning, expression, and characterization of the human eosinophil eotaxin receptor. J Exp Med 1996;183(5):2349–2354.
12. Uguccioni M, Mackay CR, Ochensberger B, Loetscher P, Rhis S, LaRosa GJ et al. High expression of the chemokine receptor CCR3 in human blood basophils — role in activation by eotaxin, MCP-4, and other chemokines. J Clin Invest 1997;100(5):1137–1143.
13. Qin S, Rottman JB, Myers P, Kassam N, Weinblatt M, Loetscher M et al. The chemokine receptors CXCR3 and CCR5 mark subsets of T cells associated with certain inflammatory reactions. J Clin Invest 1998;101(4):746–754.
14. Bonecchi R, Polentarutti N, Luini W, Borsatti A, Bernasconi S, Locati M et al. Up-regulation of CCR1 and CCR3 and induction of chemotaxis to CC chemokines by IFN-gamma in human neutrophils. J Immunol 1999;162(1):474–479.
15. Proudfoot AE, Buser R, Borlat F, Alouani S, Soler D, Offord RE et al. Amino-terminally modified RANTES analogues demonstrate differential effects on RANTES receptors. J Biol Chem 1999;274(45):32478–32485.
16. He J, Chen Y, Farzan M, Choe H, Ohagen A, Gartner S et al. CCR3 and CCR5 are co-receptors for HIV-1 infection of microglia. Nature 1997;385(6617):645–649.
17. Loetscher P, Seitz M, Baggiolini M, Moser B. Interleukin-2 regulates CC chemokine receptor expression and chemotactic responsiveness in T lymphocytes [see comments]. J Exp Med 1996;184(2):569–577.
18. Bonecchi R, Polentarutti N, Luini W, Borsatti A, Bernasconi S, Locati M et al. Up-regulation of CCR1 and CCR3 and induction of chemotaxis to CC chemokines by IFN-gamma in human neutrophils. J Immunol 1999;162(1):474–479.
19. Wuyts A, Van Osselaer N, Haelens A, Samson I, Herdewijn P, Ben Baruch A et al. Characterization of synthetic human granulocyte chemotactic protein 2: usage of chemokine receptors CXCR1 and CXCR2 and in vivo inflammatory properties. Biochemistry 1997;36(9): 2716–2723.
20. Neote K, Darbonne W, Ogez J, Horuk R, Schall TJ. Identification of a promiscuous inflammatory peptide receptor on the surface of red blood cells. J Biol Chem 1993;268(17):12247–12249.
21. Weng YM, Siciliano SJ, Waldburger KE, Sirotinameisher A, Staruch MJD, Gould SL et al. Binding and functional properties of recombinant and endogenous CXCR3 chemokine receptors. J Biol Chem 1998;273(29):18288–18291.
22. Kledal TN, Rosenkilde MM, Schwartz TW. Selective recognition of the membrane-bound CX3C chemokine, fractalkine, by the human cytomegalovirus-encoded broad-spectrum receptor US28. FEBS Lett 1998;441(2): 209–214.
23. Blanpain C, Migeotte I, Lee B, Vakili J, Doranz BJ, Govaerts C et al. CCR5 binds multiple CC-chemokines: MCP-3 acts as a natural antagonist. Blood 1999;94(6): 1899–1905.
24. Oppermann M, Mack M, Proudfoot AE, Olbrich H. Differential effects of CC chemokines on CC chemokine receptor 5 (CCR5) phosphorylation and identification of phosphorylation sites on the CCR5 carboxyl terminus. J Biol Chem 1999;274(13):8875–8885.
25. Nibbs RJ, Yang J, Landau NR, Mao JH, Graham GJ. LD78beta, a non-allelic variant of human MIP-1alpha (LD78alpha), has enhanced receptor interactions and potent HIV suppressive activity. J Biol Chem 1999;274(25): 17478–17483.
26. Gosling J, Monteclaro FS, Atchison RE, Arai H, Tsou CL, Goldsmith et al. Molecular uncoupling of C-C chemokine receptor 5-induced chemotaxis and signal transduction from HIV-1 coreceptor activity. Proc Natl Acad Sci USA 1997;94(10):5061–5066.
27. Raport CJ, Gosling J, Schweickart VL, Gray PW, Charo IF. Molecular cloning and functional characterization of a novel human CC chemokine receptor (CCR5) for RANTES, MIP-1beta, and MIP- 1alpha. J Biol Chem 1996;271(29):17161–17166.
28. Elsner J, Mack M, Bruhl H, Dulkys Y, Kimmig D, Simmons G et al. Differential activation of CC chemokine receptors by AOP-RANTES. J Biol Chem 2000;275(11): 7787–7794.
29. Mack M, Luckow B, Nelson PJ, Cihak J, Simmons G, Clapham PR et al. Aminooxypentane-RANTES induces CCR5 internalization but inhibits recycling: a novel inhibitory mechanism of HIV infectivity. J Exp Med 1998; 187(8):1215–1224.
30. Zimmermann N, Conkright JJ, Rothenberg ME. CC chemokine receptor-3 undergoes prolonged ligand-induced internalization. J Biol Chem 1999;274(18):12611–12618.
31. Crump MP, Rajarathnam K, Kim KS, Clark-Lewis I, Sykes BD. Solution structure of eotaxin, a chemokine that selectively recruits eosinophils in allergic inflammation. J Biol Chem 1998;273(35):22471–22479.
32. Mayer KL, Stone MJ. NMR solution structure and receptor peptide binding of the CC chemokine eotaxin-2. Biochemistry 2000;39(29):8382–8395.
33. Sticht H, Escher SE, Schweimer K, Forssmann WG, Rosch P, Adermann K. Solution structure of the hu-

man CC chemokine 2: A monomeric representative of the CC chemokine subtype. Biochemistry 1999;38(19): 5995–6002.

34. Keizer DW, Crump MP, Lee TW, Slupsky CM, Clark-Lewis I, Sykes BD. Human CC chemokine I-309, structural consequences of the additional disulfide bond. Biochemistry 2000;39(20):6053–6059.
35. Crump MP, Gong JH, Loetscher P, Rajarathnam K, Amara A, Arenzana-Seisdedos F et al. Solution structure and basis for functional activity of stromal cell-derived factor-1; dissociation of CXCR4 activation from binding and inhibition of HIV-1. EMBO J 1997;16(23):6996–7007.
36. Dealwis C, Fernandez EJ, Thompson DA, Simon RJ, Siani MA, Lolis E. Crystal structure of chemically synthesized [N33A] stromal cell-derived factor 1alpha, a potent ligand for the HIV-1 'fusin' coreceptor. Proc Natl Acad Sci USA 1998;95(12):6941–6946.
37. Rajarathnam K, Clark-Lewis I, Sykes BD. 1H NMR solution structure of an active monomeric interleukin-8. Biochemistry 1995;34(40):12983–12990.
38. Clore GM, Gronenborn AM. Three-dimensional structures of alpha and beta chemokines. [Review] [26 refs]. FASEB J 1995;9(1):57–62.
39. Lusti-Narasimhan M, Power CA, Allet B, Alouani S, Bacon KB, Mermod JJ et al. Mutation of Leu25 and Val27 introduces CC chemokine activity into interleukin-8. J Biol Chem 1995;270(6):2716–2721.
40. Czaplewski LG, McKeating J, Craven CJ, Higgins LD, Appay V, Brown A et al. Identification of amino acid residues critical for aggregation of human CC chemokines macrophage inflammatory protein (MIP)-1alpha, MIP-1beta, and RANTES. Characterization of active disaggregated chemokine variants. J Biol Chem 1999; 274(23):16077–16084.
41. Zhang X, Chen L, Bancroft DP, Lai CK, Maione TE. Crystal structure of recombinant human platelet factor 4. Biochemistry 1994;33(27):8361–8366.
42. St Charles R, Walz DA, Edwards BF. The three-dimensional structure of bovine platelet factor 4 at 3.0-Å resolution. J Biol Chem 1989;264(4):2092–2099.
43. Mayo KH, Roongta V, Ilyina E, Milius R, Barker S, Quinlan C et al. NMR solution structure of the 32-kDa platelet factor 4 ELR-motif N-terminal chimera: a symmetric tetramer. Biochemistry 1995;34(36):11399–11409.
44. Skelton NJ, Aspiras F, Ogez J, Schall TJ. Proton NMR assignments and solution conformation of RANTES, a chemokine of the C-C type. Biochemistry 1995;34(16): 5329–5342.
45. Laurence JS, Liwang AC, Liwang PJ. Effect of N-terminal truncation and solution conditions on chemokine dimer stability: nuclear magnetic resonance structural analysis of macrophage inflammatory protein 1 beta mutants. Biochemistry 1998;37(26):9346–9354.
46. Wagner L, Yang OO, Garcia-Zepeda EA, Ge Y, Kalams SA, Walker BD et al. b-Chemokines are released from HIV-1-specific cytolytic granules complexed to proteoglycans. Nature 1998;391:908–911.
47. Gordon CJ, Muesing MA, Proudfoot AEI, Power CA, Moore JP, Trkola A. Enhancement of human immunodeficiency virus type 1 infection by the CC-chemokine RANTES is independent of the mechanism of virus–cell fusion. J Virol 1999;73(1):684–694.
48. Trkola A, Gordon C, Matthews J, Maxwell E, Ketas T, Czaplewski L et al. The CC-chemokine RANTES increases the attachment of human immunodeficiency virus type 1 to target cells via glycosaminoglycans and also activates a signal transduction pathway that enhances viral infectivity. J Virol 1999;73(8):6370–6379.
49. Laurence JS, Blanpain C, Burgner JW, Parmentier M, Liwang PJ. CC chemokine MIP-1 beta can function as a monomer and depends on Phe13 for receptor binding. Biochemistry 2000;39(12):3401–3409.
50. Paavola CD, Hemmerich S, Grunberger D, Polsky I, Bloom A, Freedman R et al. Monomeric monocyte chemoattractant protein-1 (MCP-1) binds and activates the MCP-1 receptor CCR2B. J Biol Chem 1998;273(50): 33157–33165.
51. Zhang Y, Rollins BJ. A dominant negative inhibitor indicates that monocyte chemoattractant protein 1 functions as a dimer. Mol Cell Biol 1995;15(9):4851–4855.
52. Leong SR, Lowman HB, Liu J, Shire S, Deforge LE, Gillece-Castro BL et al. IL-8 single-chain homodimers and heterodimers: interactions with chemokine receptors CXCR1, CXCR2, and DARC. Protein Sci 1997;6(3): 609–617.
53. Rajarathnam K, Sykes BD, Kay CM, Dewald B, Geiser T, Baggiolini M et al. Neutrophil activation by monomeric interleukin-8. Science 1994;264(5155):90–92.
54. Hoogewerf AJ, Kuschert GS, Proudfoot AE, Borlat F, Clark-Lewis I, Power CA et al. Glycosaminoglycans mediate cell surface oligomerization of chemokines. Biochemistry 1997;36 (44):13570–13578.
55. Siciliano SJ, Rollins TE, DeMartino J, Konteatis Z, Malkowitz L, van Riper G et al. Two-site binding of C5a by its receptor: an alternative binding paradigm for G protein-coupled receptors. Proc Natl Acad Sci USA 1994;91(4):1214–1218.
56. Rucker J, Samson M, Doranz BJ, Libert F, Berson JF, Yi Y et al. Regions in beta-chemokine receptors CCR5 and CCR2b that determine HIV-1 cofactor specificity. Cell 1996;87(3):437–446.
57. Ahuja SK, Lee JC, Murphy PM. CXC chemokines bind to unique sets of selectivity determinants that can function independently and are broadly distributed on multiple domains of human interleukin-8 receptor B. Determinants of high affinity binding and receptor activation are distinct. J Biol Chem 1996;271(1):225–232.
58. Monteclaro FS, Charo IF. The amino-terminal domain of CCR2 is both necessary and sufficient for high affinity binding of monocyte chemoattractant protein 1. Receptor activation by a pseudo-tethered ligand. J Biol Chem 1997;272(37):23186–23190.
59. Pease JE, Wang J, Ponath PD, Murphy PM. The N-terminal extracellular segments of the chemokine receptors CCR1 and CCR3 are determinants for MIP-1alpha and eotaxin binding, respectively, but a second domain is essential for efficient receptor activation. J Biol Chem 1998;273(32):19972–19976.

60. Clark-Lewis I, Schumacher C, Baggiolini M, Moser B. Structure–activity relationships of interleukin-8 determined using chemically synthesized analogs. Critical role of NH2-terminal residues and evidence for uncoupling of neutrophil chemotaxis, exocytosis, and receptor binding activities. J Biol Chem 1991;266(34):23128–23134.
61. Hebert CA, Vitangcol RV, Baker JB. Scanning mutagenesis of interleukin-8 identifies a cluster of residues required for receptor binding. J Biol Chem 1991;266(28):18989–18994.
62. Clark-Lewis I, Dewald B, Gciser T, Moser B, Baggiolini M. Platelet factor 4 binds to interleukin 8 receptors and activates neutrophils when its N terminus is modified with Glu–Leu–Arg. Proc Natl Acad Sci USA 1993; 90(8):3574–3577.
63. Moser B, Dewald B, Barella L, Schumacher C, Baggiolini M, Clark-Lewis I. Interleukin-8 antagonists generated by N-terminal modification. J Biol Chem 1993; 268(10):7125–7128.
64. Simmons G, Clapham PR, Picard L, Offord RE, Rosenkilde MM, Schwartz TW et al. Potent inhibition of HIV-1 infectivity in macrophages and lymphocytes by a novel CCR5 antagonist. Science 1997;276(5310):276–279.
65. Mack M, Luckow B, Nelson PJ, Cihak J, Simmons G, Clapham PR et al. Aminooxypentane-RANTES induces CCR5 internalization but inhibits recycling: a novel inhibitory mechanism of HIV infectivity. J Exp Med 1998; 187(8):1215–1224.
66. Proost P, De Meester I, Schols D, Struyf S, Lambeir AM, Wuyts A et al. Amino-terminal truncation of chemokines by CD26/dipeptidyl-peptidase IV. Conversion of RANTES into a potent inhibitor of monocyte chemotaxis and HIV-1-infection. J Biol Chem 1998;273(13): 7222–7227.
67. Blanpain C, Buser R, Power CA, Edgerton MD, Buchanan C, Mack M et al. A chimeric MIP-1a/RANTES protein demonstrates the use of different regions of the RANTES protein to bind and activate its receptors. J Leukoc Biol 2001;69(6):977–985.
68. Amara A, Gall SL, Schwartz O, Salamero J, Montes M, Loetscher P et al. HIV coreceptor downregulation as antiviral principle: SDF-1alpha-dependent internalization of the chemokine receptor CXCR4 contributes to inhibition of HIV replication. J Exp Med 1997;186(1):139–146.
69. Pakianathan DR, Kuta EG, Artis DR, Skelton NJ, Hebert CA. Distinct but overlapping epitopes for the interaction of a CC chemokine with CCR1, CCR3, and CCR5. Biochemistry 1997;36(N32):9642–9648.
70. Kuschert GS, Hoogewerf AJ, Proudfoot AE, Chung CW, Cooke RM, Hubbard RE et al. Identification of a glycosaminoglycan binding surface on human interleukin-8. Biochemistry 1998;37(32):11193–11201.
71. Hieshima K, Imai T, Baba M, Shoudai K, Ishizuka K, Nakagawa T et al. A novel human CC chemokine PARC that is most homologous to macrophage-inflammatory protein-1 alpha/LD78 alpha and chemotactic for T lymphocytes, but not for monocytes. J Immunol 1997; 159(3):1140–1149.
72. Rot A. Endothelial cell binding of NAP-1/IL-8: role in neutrophil emigration. [Review] [47 refs]. Immunol Today 1992;13(8):291–294.
73. Witt DP, Lander AD. Differential binding of chemokines to glycosaminoglycan subpopulations. Curr Biol 1994; 4(5):394–400.
74. Kuschert GS, Coulin F, Power CA, Proudfoot AE, Hubbard, RE et al. Glycosaminoglycans interact selectively with chemokines and modulate receptor binding and cellular responses. Biochemistry 1999;38(39): 12959–12968.
75. Ali S, Palmer ACV, Banerjee B, Fritchley SJ, Kirby JA. Examination of the function of RANTES, MIP-1 alpha, and MIP-1 beta following interaction with heparin-like glycosaminoglycans. J Biol Chem 2000;275(16):11721–11727.
76. Clemetson KJ, Clemetson JM, Proudfoot AE, Power CA, Baggiolini M, Wells TN. Functional expression of CCR1, CCR3, CCR4, and CXCR4 chemokine receptors on human platelets. Blood 2000;96(13):4046–4054 (December 15).
77. Schmidtmayerova H, Sherry B, Bukrinsky M. Chemokines and HIV replication. Nature 1996;382(6594):767.
78. Trkola A, Gordon C, Matthews J, Maxwell E, Ketas T, Czaplewski L et al. The CC-chemokine RANTES increases the attachment of human immunodeficiency virus type 1 to target cells via glycosaminoglycans and also activates a signal transduction pathway that enhances viral infectivity. J Virol 1999;73(8):6370–6379.
79. Chakravarty L, Rogers L, Quach T, Breckenridge S, Kolattukudy PE. Lysine 58 and histidine 66 at the C-terminal alpha-helix of monocyte chemoattractant protein-1 are essential for glycosaminoglycan binding. J Biol Chem 1998;273(45):29641–29647.
80. Jarnagin K, Grunberger D, Mulkins M, Wong B, Hemmerich S, Paavola C et al. Identification of surface residues of the monocyte chemotactic protein 1 that affect signaling through the receptor CCR2. Biochemistry 1999;38(49):16167–16177.
81. Stringer SE, Gallagher JT. Specific binding of the chemokine platelet factor 4 to heparan sulfate. J Biol Chem 1997;272(N33):20508–20514.
82. Koopmann W, Krangel MS. Identification of a glycosaminoglycan-binding site in chemokine macrophage inflammatory protein-1alpha. J Biol Chem 1997;272(15): 10103–10109.
83. Graham GJ, Wilkinson PC, Nibbs RJ, Lowe S, Kolset SO, Parker A et al. Uncoupling of stem cell inhibition from monocyte chemoattraction in MIP-1alpha by mutagenesis of the proteoglycan binding site. EMBO J 1996; 15(23):6506–6515.
84. Koopmann W, Ediriwickrema C, Krangel MS. Structure and function of the glycosaminoglycan binding site of chemokine macrophage-inflammatory protein-1 beta. J Immunol 1999;163(4):2120–2127.
85. Proudfoot AE, Fritchley S, Borlat F, Shaw JP, Vilbois F, Zwahlen C, Trkola A et al. The BBXB motif of RANTES is the principal site for heparin binding, and controls receptor selectivity. J Biol Chem 2001; 276:10620–10626.

86. Amara A, Lorthioir O, Valenzuela A, Magerus A, Thelen M, Montes M et al. Stromal cell-derived factor-1alpha associates with heparan sulfates through the first beta-strand of the chemokine. J Biol Chem 1999;274(34): 23916–23925.
87. Gong JH, Uguccioni M, Dewald B, Baggiolini M, Clark-Lewis I. RANTES and MCP-3 antagonists bind multiple chemokine receptors. J Biol Chem 1996;271(18):10521–10527.
88. Gong JH, Ratkay LG, Waterfield JD, Clark LI. An antagonist of monocyte chemoattractant protein 1 (MCP-1) inhibits arthritis in the MRL-lpr mouse model. J Exp Med 1997;186(1):131–137.
89. Chung CW, Cooke RM, Proudfoot AEI, Wells TNC. The three-dimensional solution structure of RANTES. Biochemistry 1995;34(29):9307–9314.
90. Proudfoot AEI, Power CA, Hoogewerf AJ, Montjovent MO, Borlat F, Offord RE et al. Extension of recombinant human RANTES by the retention of the initiating methionine produces a potent antagonist. J Biol Chem 1996;271(5):2599–2603.
91. Elsner J, Petering H, Hochstetter R, Kimmig D, Wells TNC, Kapp A et al. The CC chemokine antagonist Met-RANTES inhibits eosinophil effector functions through the chemokine receptors CCR1 and CCR3. Eur J Immunol 1997;27(11):2892–2898.
92. Proudfoot AE, Buser R, Borlat F, Alouani S, Soler D, Offord RE et al. Amino-terminally modified RANTES analogues demonstrate differential effects on RANTES receptors. J Biol Chem 1999;274(45):32478–32485.
93. Plater-Zyberk C, Hoogewerf AJ, Proudfoot AE, Power CA, Wells TN. Effect of a CC chemokine receptor antagonist on collagen induced arthritis in DBA/1 mice. Immunol Lett 1997;57(1–3):117–120.
94. Lloyd CM, Minto AW, Dorf ME, Proudfoot A, Wells TN, Salant DJ et al. RANTES and monocyte chemoattractant protein-1 (MCP-1) play an important role in the inflammatory phase of crescentic nephritis, but only MCP-1 is involved in crescent formation and interstitial fibrosis. J Exp Med 1997;185(7):1371–1380.
95. Grone HJ, Weber C, Weber KC, Grone EF, Rabelink T, Klier CM et al. Met-RANTES reduces vascular and tubular damage during acute renal transplant rejection: blocking monocyte arrest and recruitment. FASEB J 1999; 13(11):1371–1383.
96. Ajuebor MN, Hogaboam CM, Kunkel SL, Proudfoot AE, Wallace JL. The chemokine RANTES is a crucial mediator of the progression from acute to chronic colitis in the rat. J Immunol 2001;166(1):552–558.
97. Gonzalo JA, Lloyd CM, Wen D, Albar JP, Wells TN, Proudfoot A et al. The coordinated action of CC chemokines in the lung orchestrates allergic inflammation and airway hyperresponsiveness. J Exp Med 1998;188(1): 157–167.
98. Cocchi F, DeVico AL, Garzino-Demo A, Arya SK, Gallo RC, Lusso P. Identification of RANTES, MIP-1 alpha, and MIP-1 beta as the major HIV-suppressive factors produced by CD8+ T cells. Science 1995;270(5243): 1811–1815.
99. Berger EA. HIV entra and tropism: the chemokine receptor connection. AIDS 1997;11:S3–S16.
100. Oppermann M, Mack M, Proudfoot AE, Olbrich H. Differential effects of CC chemokines on CC chemokine receptor 5 (CCR5) phosphorylation and identification of phosphorylation sites on the CCR5 carboxyl terminus. J Biol Chem 1999;274(13):8875–8885.
101. Signoret N, Pelchen-Matthews A, Mack M, Proudfoot AEI, Marsh M. Endocytosis and recycling of the HIV coreceptor CCR5. J Cell Biol 2000;151(6):1281–1293.
102. Karpus WJ, Lukacs NW, McRae BL, Strieter RM, Kunkel SL, Miller SD. An important role for the chemokine macrophage inflammatory protein-1 alpha in the pathogenesis of the T cell-mediated autoimmune disease, experimental autoimmune encephalomyelitis. J Immunol 1995;155(10):5003–5010.
103. Kennedy KJ, Strieter RM, Kunkel SL, Lukacs NW, Karpus WJ. Acute and relapsing experimental autoimmune encephalomyelitis are regulated by differential expression of the CC chemokines macrophage inflammatory protein-1 alpha and monocyte chemotactic protein-1. J Neuroimmunol 1998;92(1–2):98–108.
104. Sorensen TL, Tani M, Jensen J, Pierce V, Lucchinetti C, Folcik VA et al. Expression of specific chemokines and chemokine receptors in the central nervous system of multiple sclerosis patients. J Clin Invest 1999;103(6): 807–815.
105. Balashov KE, Rottman JB, Weiner HL, Hancock WW. CCR5(+) and CXCR3(+) T cells are increased in multiple sclerosis and their ligands MIP-1 alpha and IP-10 are expressed in demyelinating brain lesions. Proc Natl Acad Sci USA 1999;96(12):6873–6878.
106. Liang M, Rosser M, Ng HP, May K, Bauman JG, Islam I et al. Species selectivity of a small molecule antagonist for the CCR1 chemokine receptor. Eur J Pharmacol 2000;389(1):41–49.
107. Wells TNC, Schwartz TW. Plagiarism of the host immune system — lessons about chemokine immunology from viruses [Review]. Curr Opin Biotechnol 1997;8(6): 741–748.
108. McFadden G, Kelvin D. New strategies for chemokine inhibition and modulation: you take the high road and I'll take the low road. Biochem Pharmacol 1997;54(12): 1271–1280.
109. Smith GL, Dunn W. Poxvirus chemokine-binding proteins. Eur Cytokine Netw 1999;10(2):288–289.
110. Boshoff C, Endo Y, Collins PD, Takeuchi Y, Reeves JD, Schweickart VL et al. Angiogenic and HIV-inhibitory functions of KSHV-encoded chemokines. Science 1997; 278(5336):290–294.
111. Kledal TN, Rosenkilde MM, Coulin F, Simmons G, Johnsen AH, Alouani S et al. A broad-spectrum chemokine antagonist encoded by kaposis sarcoma-associated herpesvirus. Science 1997;277(N5332):1656–1659.
112. Chen S, Bacon KB, Li L, Garcia GE, Xia Y, Lo D et al. In vivo inhibition of CC and CX3C chemokine-induced leukocyte infiltration and attenuation of glomerulonephritis in Wistar–Kyoto (WKY) rats by vMIP-II. J Exp Med 1998;188(1):193–198.
113. DeBruyne LA, Li K, Bishop DK, Bromberg JS. Gene transfer of virally encoded chemokine antagonists vMIP-II and MC148 prolongs cardiac allograft survival and

inhibits donor-specific immunity. Gene Ther 2000;7(7): 575–582.

114. Dairaghi DJ, Fan RA, McMaster BE, Hanley MR, Schall TJ. HHV8-encoded vMIP-I selectively engages chemokine receptor CCR8. Agonist and antagonist profiles of viral chemokines. J Biol Chem 1999;274(31):21569–21574.
115. Endres MJ, Garlisi CG, Xiao H, Shan LX, Hedrick JA. The Kaposi's sarcoma-related herpesvirus (KSHV)-encoded chemokine nu. J Exp Med 1999;189(12):1993–1998.
116. Stine JT, Wood C, Hill M, Epp A, Raport CJ, Schweickart VL et al. KSHV-encoded CC chemokine vMIP-III is a CCR4 agonist, stimulates angiogenesis, and selectively chemoattracts TH2 cells. Blood 2000;95(4):1151–1157.
117. Chvatchko Y, Hoogewerf AJ, Meyer A, Alouani S, Juillard P, Buser R et al. A key role for CC chemokine receptor 4 in lipopolysaccharide-induced endotoxic shock. J Exp Med 2000;191(10):1755–1763.
118. Luttichau HR, Stine J, Boesen TP, Johnsen AH, Chantry D, Gerstoft J et al. A highly selective CC chemokine receptor (CCR)8 antagonist encoded by the poxvirus molluscum contagiosum. J Exp Med 2000;191(1):171–180.
119. Zou P, Isegawa Y, Nakano K, Haque M, Horiguchi Y, Yamanishi K. Human herpesvirus 6 open reading frame U83 encodes a functional chemokine. J Virol 1999;73(7): 5926–5933.
120. Lee LF, Wu P, Sui D, Ren D, Kamil J, Kung HJ et al. The complete unique long sequence and the overall genomic organization of the GA strain of Marek's disease virus. Proc Natl Acad Sci USA 2000;97(11):6091–6096.
121. Ahuja SK, Gao JL, Murphy PM. Chemokine receptors and molecular mimicry. [Review] [42 refs]. Immunol Today 1994;15(6):281–287.
122. Rosenkilde MM, Schwartz TW. Potency of ligands correlates with affinity measured against agonist and inverse agonists but not against neutral ligand in constitutively active chemokine receptor. Mol Pharmacol 2000;57(3): 602–609.
123. Rosenkilde MM, Kledal TN, Holst PJ, Schwartz TW. Selective elimination of high constitutive activity or chemokine binding in the human herpesvirus 8 encoded seven transmembrane oncogene ORF74. J Biol Chem 2000;275(34):26309–26315.
124. Burger M, Burger JA, Hoch RC, Oades Z, Takamori H, Schraufstatter IU. Point mutation causing constitutive signaling of CXCR2 leads to transforming activity similar to Kaposi's sarcoma herpesvirus-G protein-coupled receptor. J Immunol 1999;163(4):2017–2022.
125. Camarda G, Spinetti G, Bernardini G, Mair C, Davis-Poynter N, Capogrossi MC et al. The equine herpesvirus 2 E1 open reading frame encodes a functional chemokine receptor. J Virol 1999;73(12):9843–9848.
126. Milne RS, Mattick C, Nicholson L, Devaraj P, Alcami A, Gompels UA. RANTES binding and down-regulation by a novel human herpesvirus-6 beta chemokine receptor. J Immunol 2000;164 (5):2396–2404.
127. Isegawa Y, Ping Z, Nakano K, Sugimoto N, Yamanishi K. Human herpesvirus 6 open reading frame U12 encodes a functional beta-chemokine receptor. J Virol 1998;72(7):6104–6112.
128. Lalani AS, Graham K, Mossman K, Rajarathnam K, Clark LI, Kelvin D et al. The purified myxoma virus gamma interferon receptor homolog M-T7 interacts with the heparin-binding domains of chemokines. J Virol 1997;71(6):4356–4363.
129. Liu L, Lalani A, Dai E, Seet B, Macauley C, Singh R et al. The viral anti-inflammatory chemokine-binding protein M-T7 reduces intimal hyperplasia after vascular injury [see comments]. J Clin Invest 2000;105(11):1613–1621.
130. Lalani AS, McFadden G. Secreted poxvirus chemokine binding proteins. J Leukoc Biol 1997;62(5):570–576.
131. Alcami A, Symons JA, Collins PD, Williams TJ, Smith GL. Blockade of chemokine activity by a soluble chemokine binding protein from vaccinia virus. J Immunol 1998;160(2):624–633.
132. Lalani AS, Ness TL, Singh R, Harrison JK, Seet BT, Kelvin DJ et al. Functional comparisons among members of the poxvirus T1/35 kDa family of soluble CC-chemokine inhibitor glycoproteins. Virology 1998; 250(1):173–184.
133. Carfi A, Smith CA, Smolak PJ, McGrew J, Wiley DC. Structure of a soluble secreted chemokine inhibitor vCCI (p35) from cowpox virus. Proc Natl Acad Sci USA 1999; 96(22):12379–12383.
134. Parry CM, Simas JP, Smith VP, Stewart CA, Minson AC, Efstathiou S et al. A broad spectrum secreted chemokine binding protein encoded by a herpesvirus. J Exp Med 2000;191(3):573–578.
135. Ponath PD. Chemokine receptor antagonists: novel therapeutics for inflammation and AIDS. Exp Opin Invest Drugs 1998;7:1–18.
136. Schwarz MK, Wells TNC. Recent developments in modulating chemokine networks. Exp Opin Ther Patents 2000;9(11):1471–1490.
137. Zlotnik A, Yoshie O. Chemokines: a new classification system and their role in immunity. Immunity 2000;12(2): 121–127.

Universes in Delicate Balance: Chemokines and the Nervous System
R.M. Ransohoff, K. Suzuki, A.E.I. Proudfoot, W.F. Hickey and J.K. Harrison, editors

Chapter 2.3

Chemokine Receptors

Dubhfeasa M. Slattery, Norma Gerard and Craig Gerard

[1] *Ina Sue Perlmutter Laboratory, Boston, MA, USA;* [2] *Division of Respiratory Diseases, Hunnewell 2, Children's Hospital, 300 Longwood Avenue, Boston, MA 02115, USA;* [3] *Harvard Medical School, 25 Shattuck Street, Boston, MA 02115-6092, USA*

1. INTRODUCTION

Since the discovery of the first chemokine, platelet factor 4 (CXC chemokine ligand 1: CXCL1) in 1977 and the first chemokine receptors, CXC chemokine receptor 1 (CXCR1) and CXC chemokine receptor 2 (CXCR2) ten years ago it is now clear that the chemokine system is the most important regulator of white cell movement both constitutively and under inducible terms. The seven transmembrane domain receptors are coupled to heterotrimeric Gi proteins and potentially to downstream effectors. Here we give an overview of chemokine receptor classification, genomic location, structure, expression, binding, function, viral mimics, therapeutic implications and potential 'decoy receptors'.

1.1. Evolution of Chemokine Receptors

The chemokine system consists of at least 50 structurally related peptide agonists and 20 G protein-coupled receptors. Chemokine receptors are similar to other chemoattractant receptors, e.g. the anaphylatoxin, complement component 5a (C5a) and formyl peptide receptors which in turn resemble neuropeptide receptors. As vertebrates developed a more sophisticated host defense system the number and complexity of chemokines and their receptors increased. It has been postulated that CXC ligand 12 (CXCL12 previously referred to as stromal cell-derived factor 1 or SDF-1) and its receptor, CXCR4, are the primordial chemokine and receptor [1]. Comparative phylogenetic analyses suggest the presence of a rudimentary chemokine system in Drosophila and primitive vertebrates.

The rapid identification and description of chemokines and their receptors has been facilitated by DNA database mining with the establishment of expressed sequence tag (EST) databases and by bioinformatics (computer-assisted sequence analysis). The characteristic small size and conserved motifs of chemokines both made these techniques very useful in gene identification because the probability that their entire coding regions are represented as ESTs in databases is very high. Understanding the multifaceted roles of chemokines and their receptors in vivo has been facilitated by the development of transgenic mice and targeted deletion/mutagenesis of chemokines and their receptors. Full details of this ever expanding field of science are available at http://cytokine.medic.kumamoto-u.ac.jp and at the GPCR database http://www.gpcr.org.7tm/

1.2. Chemokine Gene Clusters

Genes encoding chemokines cluster at specific loci. Generally, the CC chemokine genes, many of which act on lymphoid cells, are grouped at 17q11.2-12 and the CXC (e.g. human CXCL1–8) chemokine genes, many of which act on myeloid cells, cluster at 4q12-q13, though multiple exceptions exist. These clusters suggest that chemokines arose by gene duplication and divergence from a primordial chemokine gene with an early split into two loci. This rapid duplication leading to the formation of a cluster is postulated to explain why their function remains related.

Table 1. Chemokine receptors

Name	Alternative name	Ligands	Chromosome	Expression
CCR-1	CMKBR1, CC CKR1, MIP-1α R, RANTES R	MIP-1α, RANTES, MCP-3, MCP-1, HCC-1, HCC-4, MPIF-1, MIP-1δ	3p21	Mo, T, B, Eo, Ba
CCR-2A/CCR-2B	CMKBR2, CC CKR2, MCP-1R	MCP-1, MCP-2, MCP-3, MCP-4, MCP-5	3p21	Mo, T
CCR-3	CMKBR3, CC CKR3	Eot, Eot-2, MCP-4	3p21	Eo
	Eotaxin (Eot) R	Eot-3, RANTES, MCP-3, MCP-2, MIP-1δ		
CCR-4	CMKBR4, CC CKR4, K5-5	MDC, TARC	3p21	act T, Th1>2
CCR-5	CMKBR5, CC CKR5, ChemR13	MIP-1α, MIP-1β, RANTES	3p21	T, Mo, Th1>2
CCR-6	CMKBR6, CC CKR6, GPR29, CKR-L3, GPRCY4, STRL22, DRY6, LARC receptor	MIP-3α	6q27	T, B, Mo
CCR-7	CMKBR7, CC CKR7, EBI-1, BLR2	MIP-3β, 6CKine	17q12-q21.2	T, B
CCR-8	CKMBR-8, CC CKR8, TER1, CY6, ChemR1, CKR-L1	I-309, vMIP-I	3p22-p23	T
CCR-9A/CCR-9B	CC CKR9, GPR-9-6	TECK	3pter-qter	T, epith
CCR-10		CTACK	17q21	T, epith, fibrobl
CCR-11			3q22	

Mo, monocytes; B, B cells; Eo, eosinophils; Ba, basophils; act. T, activated T cells; epith, epithelial cells; fibrobl, fibroblasts.

1.3. Chemokine Receptor Classification and Nomenclature

Chemokines are chemotactic cytokines which are 8–10 kDa proteins with 20–70% homology in amino acid sequences. They are subdivided into classes based on the relative positions of their N-terminal cysteine residues. The alpha and beta chemokine classes, both of which contain four cysteines are the two largest families. In the alpha family, one amino acid separates the first two cysteine residues, CXC, while in the beta family, the first two cysteine residues are adjacent to each other, CC. Fractalkine, comprises a third family and is a membrane-bound glycoprotein in which the first two cysteine residues are separated by three amino acids (CX_3C) and the chemokine domain sits on a mucin-like stalk [2]. Lymphotactin (C) lacking cysteine one and three of the typical chemokine structure, [3] comprises the fourth class. The alpha chemokine class is further subdivided into those that contain a three amino acid motif, ELR (glutamic acid–leucine–arginine) close to the N-terminal (preceding the CXC sequence) and those that do not. Those with the ELR sequence are chemotactic for neutrophils and those without the sequence act on lymphocytes. Similar to the CXC family, the N-terminal amino acids preceding the CC residues of beta chemokines are a critical component of the biologic activity and leukocyte selectivity of these chemokines. The chemokine nomenclature is based on the above chemokine receptor nomenclature followed by 'R' for receptor, replacing 'L' for ligand, followed by the number which designates the gene encoding each chemokine, e.g. CCR1–9. Detailed reviews of the widely accepted chemokine receptor nomenclature system as ratified by the International Union of Pharmacology have recently been published and are beyond the scope of this chapter [4] (Tables 1 and 2).

2. CHEMOKINE RECEPTOR STRUCTURE

In 1991, two independent groups identified the genes that encode human CXCL8 receptors (interleukin-8 receptors IL-8): CXCR1 [5] and CXCR2 [6]. Receptors have a molecular size of approx. 40 kDa and like their ligands, most chemokine receptor genes

Table 2. CXC, CX_3C and XC chemokine receptors

Name	Alternative name	Ligands	Chromosome	Expression
CXCR-1	CMKBRL1, V28	Fractalkine, VMIP-II	3p21	T, Mo
CXCR-1	IL-8RA, IL-8R1, IL-8Rα	IL-8, GCP-2, NAP-2	2q33-q36	
CXCR-2	IL-8RB, IL-8 R2, IL-8Rβ	IL-8, GCP-2, ENA-78, NAP-2, GROα, GROβ, GROγ	2q33-q35	T, Mo
CXCR-3	IP-10/MIG R, GPR9	IP-10, MIG, I-TAC	8p12-p11.2	T, Eo, platelets
CXCR-4	LESTR, HUMSTR, fusin, HM89	SDF-1α/SDF-1β	2q21	T, B
CXCR-5	Burkitt lymphoma receptor 1 (BLR-1)	BCA-1	11	B, T, neurons
CXCR-6	Bonzo, STRL33, TYMSTR	SexCkine	3	T, dendritic
CX3CR1	V28, GPR13, CMKBRL1	fractalkine	3	neurons
XCR1	GPR5, SCM-1R, lymphotactin R	SCM-1β, SCM-1α, vMIP-II	3p21.3-p21.1	T

Mo, monocytes; eo, eosinophils.

are found in clusters located on chromosomes 2 and 3. Chemokine receptors have a structure similar to that of other G protein-coupled receptors. Existing as integral membrane proteins containing seven transmembrane-spanning helices, chemokine receptors have both an extracellular NH_2-terminus and an intracellular, cytoplasmic COOH-terminus.

Their structure is characteristic of receptors such as rhodopsin and the muscarinic acetylcholine receptor, which are coupled to guanine nucleotide binding proteins (G proteins). Chemokines also activate G proteins. Chemokine receptors, heptahelical structure, initially predicted from the two-dimensional crystals of bacteriorhodopsin have since been confirmed by the discovery of the three-dimensional crystal structure of rhodopsin [7]. Through spreading of the intracellular loops along the surface of the inner leaflet of the plasma membrane and the lateral orientation of the COOH-terminus, these receptors provide a larger surface than previously predicted for the coupling with effector molecules. In this way receptors can connect to intracellular signal transduction pathways and can interact with both the alpha (α), beta (β) and gamma (γ) subunits of G proteins and probably additional effectors. In the case of CCR5 it has been shown that the N-terminus is important for binding while the COOH-terminus important in phosphorylation leading to receptor internalization. These characteristics, although likely, have not been shown for all chemokine receptors.

3. CHEMOKINE RECEPTOR EXPRESSION

Chemokine receptor expression varies with cell type and receptor. Some receptors are restricted to certain cell types while others are expressed on many cell types. CXCR1's expression predominately on neutrophils is an example of the former while CCR2's expression on monocytes, T, natural killer (NK), dendritic cells and basophils, exemplifies the latter. Receptor expression can be constitutive or inducible: CCR1 and CCR2 are constitutively expressed on monocytes but are expressed on lymphocytes only after stimulation by interleukin-2 (IL-2) [8]. In a subgroup, constitutive expression can be down-regulated: e.g. CCR2 by lipopolysaccharide, making the cells unresponsive to monocyte chemoattractant protein (MCP)-1 and -3 [9]. With others, receptor expression is dependent on the state of activation or differentiation of a cell, e.g. CXCR3 is expressed on activated lymphocytes of the T helper type 1 phenotype (Th1), whereas CCR3 is preferentially expressed on T helper type 2 (Th2) phenotype, among other cells. A further group are expressed on non-hematopoietic cells, such as neurons, astrocytes, epithelial cells and endothelial cells.

3.1. Functions of Chemokines

The chemokine system has extensive functions. It exerts its influence on a wide range of cell types including neutrophils, monocytes, lymphocytes, esoinophils, endothelial and dendritic cells and its functions have implications for immunology, virology, neurobiology, developmental biology and oncology. Redundancy and promiscuity are two attributes frequently ascribed to the chemokine system. The former may be more of a reflection of our lack of knowledge than a biological characteristic. However, the latter is 'real' because it involves high affinity interactions. Thus a single chemokine may bind several chemokine receptors whereas a single chemokine receptor may transduce signals for several chemokines.

One of the major roles of the chemokine system is to govern leukocyte movement across the areas of homeostasis, inflammation and development. Chemokines provide signals that convert the low affinity selectin-mediated interaction into the higher affinity integrin-mediated interaction which leads to extravasation of leukocytes from blood to the tissues.

Regarding inflammatory diseases, the subgroup of chemokines expressed in diseased tissue partially dictates the type of inflammatory infiltrate which characterizes a specific disease. In viral meningitis, CXCL10/IP-10 and CCL2/MCP-1 concentrations are increased in the cerebrospinal fluid in proportion to the influx of monocytes and lymphocytes recruited, cells on which the chemokines are expressed [10]. Chemokines have been documented to inhibit stem cell proliferation, a characteristic which may be beneficial in cancer therapy [11]. In addition, platelet factor 4 [12] and IP-10 [13,14] inhibit neovascularization, tumor growth and metastasis [15]. Chemokines play a role in cellular and developmental biology and recently RANTES has been shown to produce different effects on developing human astrocytes based on their age causing proliferation at 5 weeks but maturation at 10 weeks gestation [16]. Chemokine receptor mediated signals vary from 'essential' (SDF/CXCR4 deletion is responsible for early neonatal death) to apparently 'redundant' as in CCR5 where CCR5-deficient humans appear immunologically intact.

3.2. Function of Chemokine Receptors

Chemokine receptors play a major role in two infectious diseases which cause significant morbidity and mortality worldwide, human immunodeficiency virus (HIV) and malaria. *Plasmodium vivax* binds to the Duffy antigen receptor for chemokines (DARC) and Duffy-negative human erythrocytes are resistant to invasion by *P. vivax* and the related monkey malaria, *P. knowlesi* [17]. Regarding HIV pathophysiology, CCR5 and CXCR4 have been identified as obligate co-receptors with CD4 in viral envelope fusion and entry [18]. This sequential two-step process involves the opening of a cryptic binding site on CD4 for the chemokine receptor secondary to the interaction of gp120 with CD4 while the binding energy contributes to the conformational rearrangement that exposes the fusion peptide gp41 [19,20]. The importance of CCR5 in HIV pathogenesis is demonstrated by the fact that people homozygous for a 32 base pair deletion in the gene for CCR5, despite being in a high risk category remain uninfected with HIV while heterozygotes have a slower rate of progression of the disease than expected. Mouse genetics and molecular biology have helped develop techniques of transgenic mice and targeted mutagenesis of chemokine and their receptors such that the importance of individual chemokines and their receptors can be seen in vivo.

Using different models, CCR1 knock out mice have been shown to be more susceptible to aspergillus infection (neutrophil mediated) and sustain more severe glomerulonephritis and have an enhanced Th1 response, have prolonged cardiac allograft survival and incur less acute lung injury in a caerulin-induced pancreatitis model when compared to their wild type littermates. CCR2 knock out mice demonstrate reduced atheroma formation, reduced macrophage recruitment and 'protection' in an influenza A model, but appear less effective at clearing Listeria in comparison to their wild type counterparts. Both CCR4 and CCR5 knock out mice exhibit increased resistance to lipopolysaccharide (LPS) induced shock and CCR4 demonstrates reduced peritoneal macrophages and tumor necrosis factor compared to wild type. CCR5 knock out mice display an increased mortality when exposed to influenza A but protection from colitis in the dextran sodium sulphate colitis model in comparison to their wild type counterparts. CCR6 has a role in systemic and mucosal immunity as evidenced by impaired humoral response to oral antigen and rotavirus in the knock out mice together with an increase in T cell subsets in the mucosa. CCR7 plays an important role in the development of secondary lymphoid organs and migration of naive T, B and antigen bearing dendritic cells. CXCR2 deficiency slows early production of IgG antibodies and atherosclerosis progression while CXCR4 deficiency results in perinatal death, congenital heart disease and organ specific vascular abnormalities. CXCR5 has important, immunological functions and deficiency results in a lack of Peyer's patches, defective formation of primary follicles in the spleen and altered B/T cell ration in the blood. Thus chemokine receptors have multiple and varied functions which span many organ systems. This topic is covered in detail in a recent review [21].

4. BINDING OF CHEMOKINES TO CHEMOKINE RECEPTORS

Small molecules such as retinal and catecholamines are known to bind to their G protein-coupled receptors at a single interhelical site. However, one of the accepted models for chemokine/chemokine receptor interaction is the two-step binding and signaling model [22–24]. In this model the chemokine binds to the initial binding site and the ensuing conformational change allows the pharmacophore (generally located at the N-terminus of the ligand), to interact with the transmembrane helices to trigger signal transduction. Many controversies have surrounded the identity and location of the initial binding site: is it the N-terminus or the second extracellular loop (ECL2)? Multiple reports exist proposing the ECL2 as being the major or sole chemokine binding site [25–28]. However, most of these studies were performed using chimeras made between two closely related molecules, and depending on the degree of homology of the two domains which were switched results were ambiguous and open to varied interpretation. In addition, antibodies used in some of these studies lead to steric hinderance due to their physical bulk and again results may be misleading. More recently, studies involving the more precise methods of site-directed mutagenesis or deletion mutation point to the N-terminus of the receptor as being the initial binding site for many chemokines.

That the N-terminus of the receptor is the initial binding site has been well studied in the case of the seven transmembrane receptor C5aR [29–31]. This hypothesis is supported by studies on N-terminal tyrosine sulfation of many 7 transmembrane (7TMS) receptors. In the case of CCR5, it has been demonstrated that tyrosines in the N-terminus of the receptor are post-translationally modified by sulfation and that mutations of these sulfated tyrosines are detrimental to CCR5 function as a co-receptor for HIV-1 infection [32]. Treatment of CCR5 expressing calls with the sulfation inhibitor chlorate reduces significantly MIP-1 alpha and MIP-1 beta binding to these cells [32], demonstrating the importance of sulfated tyrosines for the binding of the receptors' natural ligands. Following this initial observation, others have shown that N-terminal tyrosine mutations of chemokine receptors and related 7TMS receptors affect ligand binding [33–36]. This hypothesis certainly does not hold true for all 7TMS or chemokine receptors. In the case of formylmethionylleucylphenylalanine (fMLP) receptor FPR [37] the ligand is a small peptide which binds directly to the signaling site, bypassing the N-terminus. Keeping exceptions in mind, the two-site paradigm with the N-terminus of the receptor acting as the initial binding site for macromolecular ligands has been shown for CCR5, CCR2 and CXCR4 [37–40].

It is suspected that chemokine receptors couple not only to G proteins but to other downstream effectors. This theory is based on the fact that diverse signal cascades are triggered depending on the chemokine and receptor which bind. It has been shown that other seven transmembrane receptors can bind at their COOH-terminus scaffolding proteins containing PDZ domains [41–43]. This may explain why CXCL8's two receptors (CXCR1 and CXCR2) although having similar affinity for their common ligand and similar binding selectivity for G proteins, activate different signal transduction pathways. CXCR1 stimulates phospholipase D activation and the formation of superoxide by NADPH oxidase whereas CXCR2 does not trigger either response [44]. Because both receptors possess a similar affinity for CXCL8 and bind with the same selectivity to G proteins, additional receptor-specific signal transmission mechanisms are assumed but their existence remains to be proven in the case of chemokine receptors.

4.1. Viral Mimicry

In excess of 30 virally encoded chemokine and chemokine receptors mimics have been discovered within the poxvirus, herpes virus and retrovirus families. Viral mimicry of chemokines and their receptors provides us with new insights into viral immunopathogenesis, new approaches to drug therapy and possible vaccine development. The six known viral chemokine receptors all bind multiple chemokine ligands but differ from each other with respect to specific signaling properties and chemokine speci-

Table 3. Viral encoded chemokine and chemokine receptor mimics

In excess of 30 in total but all in the herpesvirus, poxvirus and retrovirus families, of which 6 are medically important. – human immunodeficiency virus (HIV) – human cytomegalovirus (CMV) – human herpes virus 6, 7 and 8 (Kaposi's sarcoma-associated herpes virus KSHV) – molluscum contagiosum virus

ficity. They have between 1/4 to over 1/2 amino acid sequence identity to mammalian chemokine receptors and share an acidic NH_2-terminus and a short basic third-intracellular loop [45].

Examples of chemokine mimicry include the human herpesvirus 8, HHV8 ORF 74 (also called Kaposi's sarcoma-associated herpesvirus G protein-coupled receptor or KSHV GPCR and vGPCR). This is a constitutively active chemokine receptor [46] which has been shown to induce lesions histologically similar to kaposi sarcoma when expressed in immunocompetent transgenic mice [47]. The most famous among the chemokine mimics is gp120 of the HIV-1 envelope glycoprotein which exploits HIV co-receptors and enables infections [48,49]. Chemokine mimicry has taught us that viruses can exploit the immune system to facilitate viral dissemination or directly infect leukocyte target cells (HIV-1) and can block chemokine action via chemokine mimics as in the cases of herpesvirus and poxvirus (Table 3).

5. DRUG TARGETING OF CHEMOKINE RECEPTORS

Chemokines and their receptors are attractive therapeutic targets by virtue of their specificity. Antagonizing a single ligand or receptor offers a more selective anti-inflammatory effect than can be exerted by corticosteroids. Much work has been performed on the HIV-1 co-receptors CCR5 and CXCR4. Small-molecule CCR5 antagonists could serve to slow disease progression, prevent infection of leukocytes or act as adjuvant therapy to highly active antiretroviral therapy (HAART) and permit 'drug holidays' from HAART. A concern is that blockade of the CCR5 receptor may cause the virus to adapt and use another co-receptor either CXCR4 or one of at least 13 or so minor co-receptors. Some reports exist that people who are heterozygous for delta 32 have an increased frequency of CXCR4 utilizing viral isolates [50]. It remains unclear whether it is more effective to antagonize several receptors as in the case of viral chemokines or to be selective, targeting only one for increased precision. The answer lies in the particular disease and specifically associated chemokines and receptors. In airway hyperresponsiveness, MCP-5, eotaxin, RANTES and MCP-1 have all been shown to contribute to the disease process [50]. A pan beta chemokine antagonist protein from the poxvirus family was shown to block inflammation in this model [51], as did Met-RANTES [52] (a CCR1 and CCR5 antagonist). Genetic deletion of chemokine receptors in murine models has demonstrated that antibodies to single chemokines or receptors may have a role. CCR1 is important in allograft rejection and CCR2 deficiency decreases atherosclerosis by reducing macrophage recruitment [53–55]. Similarly, RANTES, a natural ligand for CCR5, can inhibit HIV-1 entry by receptor down-regulation as well as receptor blockade.

6. DECOY RECEPTORS

Decoy receptors recognize the agonist but are structurally incapable of signaling or presenting the agonist to signaling receptor components. The interleukin-1 type II receptor (IL-1RII) was the first pure decoy to be identified. The Duffy antigen receptor for chemokines (DARC) and D6 both bind their ligands with high affinity but no detectable signaling or chemotactic activity ensues. This suggests that they can act as chemokine decoy receptors (Table 4). Functional decoy receptors exist in addition to structural decoys. Although structurally identical to signaling receptors, functional receptors can be generated by the concomitant exposure to pro-inflammatory signals and the anti-inflammatory interleukin-10 (IL-10). This double stimulation uncouples the receptors which are then unable to elicit cell migration and act as molecular sinks for inflammatory chemokines. CCR1, CCR2 and CCR5 in monocytes and dendritic cells were transformed into decoy receptors by the simultaneous stimulation with IL-10 and LPS and were unable to migrate in response to CCL3, CCL4, CCL5 or CL2 [56].

Table 4. Decoy receptors

Structural decoy receptors		
Name	Mutation	Expression
HCR	DRY sequence mutated to QRY TM2 aspartic acid mutated to asparagine	spleen, BM, fetal liver, LN
D6	DRY sequence mutated to DKY TM2 aspartic acid mutated to asparagines	placenta, lymphatic, endothelial cells
CCR-bp	Possible deletion at N-terminus and extension of hydrophobic residues at C-terminus of the predicted TM4	heart, lung

HCR, human chemokine receptor; TM, transmembrane segment; BM, bone marrow; LN, lymph node.

6.1. Enigmas in the Chemokine System: Decoy Receptors

In the past decade, a tremendous amount of information has been gathered invoking the chemokine system in the basal trafficking of leukocytes as well as host defense. Inappropriate activation of the chemokine system, however, has been postulated to be at the root of inflammatory diseases. Here we will focus on a hypothesis that the chemokine system, like the IL-1 pathway, has endogenous 'decoy receptors' in addition to the Duffy antigen.

In the case of the pro-inflammatory and host defense cytokine families IL-1 and TNF alpha, it has been well established that high affinity, non-signaling decoy receptors exist and that these receptors modulate biological responses. For example, certain cells may be protected from TNF-cytokine-mediated apoptosis by decoy receptors [57]. Since the chemokine system is similarly poised to have potentially deleterious effects, it seems reasonable to assume that a similar decoy system might exist.

In fact, in 1995 the Duffy blood group antigen, which is a heptihelical receptor found on erythrocytes and post-capillary venular endothelium, was described as the 'sink' for CXC as well as CC chemokines which had been detected previously on red blood cells [16,58,59]. Africans and their descendants were shown to be deficient in the Duffy antigen, which confers a resistance to *Plasmodium vivax* malaria infection. This, in fact, was the first demonstration of a serpentine receptor being co-opted by an infectious agent, a scenario replayed with CCR5 and HIV infection [1,60]. Excitement over the role of Duffy as a decoy receptor was enhanced when the mutation for the African allele was shown to be a promoter defect restricted to the erythrocyte lineage. Thus, these individuals retain expression of Duffy antigen on their post-capillary venular endothelium and CNS sites in the cerebellum and on neurons.

However, subsequent analyses of several non-African Duffy-negative individuals revealed that these alleles encoded premature stop codons which resulted in truncated proteins [61]. Since these individuals possessed strong anti-Duffy antibodies, it is likely that the mutant proteins are not expressed at the cell surface. However, since the chemokine binding domain exists upstream from the predicted stop codon, it remains formally possible that some functional expression of a chemokine binding moiety occurs. Notwithstanding, the fact that there were no apparent illnesses in these Duffy-negative individuals led the authors to conclude that Duffy was functionally redundant.

7. GENETIC DISRUPTION OF MURINE DUFFY ANTIGEN

In order to examine the function of the Duffy antigen, hereafter referred to as DARC (Duffy antigen receptor for chemokines), Luo et al. [62] and Dawson et al. [63] independently deleted murine DARC. Both groups found no differences in any developmental or immunologic parameters. However, they reached opposite conclusions as regards inflammatory responses following LPS challenge. Luo et al reported a statistically significant *decrease* in lung and peritoneal PMN infiltration in DARC-deficient mice compared to wild type, while Dawson et al. found an *increase* in PMN infiltration in lung and liver following intraperitoneal LPS. The doses used were 10 and 30 mg/kg respectively, which should not create such vast differences. The likely major experimental variable was time of autopsy, which was 24 hours in Luo et al but only 2 hours in Dawson et al. following LPS injection. Taken together, these findings suggest that DARC, acting as a decoy, may dramatically change the kinetics of leukocyte trafficking.Thus, the effect of DARC deficiency in man could be very subtle, and might include increased susceptibility to certain pathogens. As should be appreciated, over evolutionary time very small advantages in survival may be sufficient to retain an allele. However, it is also possible that the selection for a decoy receptor is a *polygenic trait.* And it is precisely redundancy which is selected for over several genes. Therefore, it becomes interesting to ask the question 'do alternatives to the Duffy receptor exist?'

8. NON-SIGNALING CHEMOKINE RECEPTORS?

A review of the current literature in fact reveals not one but three chemokine receptors which have binding function without typical signal transduction capabilities. These sequences are depicted in Fig. 1. The first of these is published as human chemokine receptor, HCR [64]. HCR was initially found in the public databases as a putative CCR6 (U95626), which was found on the HUGO sequencing project to be in the same loci as CCR2 and CCR5. However, examination of the sequence reveals alteration of two amino acids which, in other GPCRs, are essential for coupling to G proteins. The first of these is an aspartic acid in transmembrane segment 2, which in this receptor is an asparagine. Mutation of this Asp to Asn in adrenergic receptors, C5a receptor, FSH/LH receptor and CCR5 leads to loss of signaling with

HCR

1 manytlaped eydvliegel esdeaeqcdk ydaqalsaql vpslcsavfv igvldnllvv
61 lilvkykglk rveniyllnl **avsnlcfllt** lpfwahaggd pmckiligly fvglysetff
121 nclltvqryl vflhkgnffs arrrvpcgii tsvlawvtai latlpeyvvy kpqmedqkyk
181 cafsrtpflp adetfwkhfl tlkmnisvlv lplfiftfly vqmrktlrfr eqryslfklv
241 faimvvfllm wapyniaffl stfkehfsls dckssynldk svhitkliat thccinplly
301 afldgtfsky lcrcfhlrsn tplqprgqsa qgtsreepdh stev

D6

1 maataspqpl atedadsens sfyyydylde vafmlcrkda vvsfgkvflp vfyslifvlg
61 lsgnllllmv llryvprrrm veiyllnlai **snllflvtlp** fwgisvawhw vfgsflckmv
121 stlytinfys giffiscmsl dkyleivhaq pyhrlrtrak slllativwa vslavsipdm
181 vfvqthenpk gvwnchadfg ghgtiwklfl rfqqnllgfl lpllamiffy srigcvlvrl
241 rpagqgralk iaaalvvaff vlwfpynltl flhtlldlqv fgncevsqhl dyalqvtesi
301 aflhccfspi lyafsshrfr qylkaflaav lgwhlapgta qaslsscses siltaqeemt
361 gmndlgerqs enypnkedvg nksa

CCR BP

1 maleqnqstd yyyeenemng tydysqyeli cikedvrefa kvflpvflti vfviglagns
61 mvvaiyayyk kqrtktdvyi lnlavadlll lftlpfwavn avhgwvlgki mckitsalyt
121 lnfvsgmqfl **acisidryva vtkvpsqsgv** gkpcwiicfc vwmaaillsi pqlvfytvnd
181 narcipifpr ylgtsmkali qmleicigfv vpflimgvcy fitartlmkm pnikisrplk
241 vlltvvivfi vtqlpynivk fcraidiiys litscnmskr mdiaiqvtes ialfhsclnp
301 ilyvfmgasf knyvmkvakk ygswrrqrqs veefpfdseg pteptstfsi

C5aR2

1 mgndsvsyey gdysdlsdrp vdcldgacla idplrvaplp lyaaiflvgv pgnamvawva
61 gkvarrrvga twllhlavad llcclslpil avpiargghw pygavgcral psiilltmya
121 svlllaalsa **dlcflalgpa** wwstvqracg vqvacgaawt lallltvpsa iyrrlhqehf
181 parlqcvvdy ggssstenav tairflfgfl gplvavasch sallcwaarr crplgtaivv
241 gffvcwapyh llglvltvaa pnsallaral raeplivgla lahsclnpml flyfgraqlr
301 rslpaachwa lresqgqdes vdskkstshd lvsemev

Figure 1. Decoy chemoattractant receptors. In boldface are regions in each receptor which contain mutations known to interfere with G protein coupling, but which do not affect binding. In HCR, note the asparagines (N) replacing the canonical aspartic acid (D) found in all other signaling receptors; the same residue is mutated in D6. In the CCR BP, the canonical DRYLAIV box found in most chemokine receptors is not easily recognizable. In the C5aR2 (formerly C5L2), the essential arginine (R) in the DRY box is mutated to leucine. This loss of function mutation can be partially restored by back-mutagenesis to arginine (D.M. Slattery, unpublished observation).

retention of high affinity ligand binding. The second substitution occurs in the canonical 'DRY box' which occurs at the terminus of transmembrane segment 3. In HCR, the DRY sequence occurs as QRY. This replacement of glutamine for aspartic acid results in loss of a negative charge. In the rhodopsin structure, the Glu and Arg in TM3 form salt bridges with Glu and Thr residues in TM6. Thus, stabilization of these structures is speculated to be essential in maintaining a conformation that is, indirectly critical for G protein coupling. Data have been presented by Biber and colleagues that HCR serves as a high affinity receptor for MCP-1 (Keystone conference on Chemokines, Taos, New Mexico, February, 2001), but other investigators have not yet confirmed this finding.

The second chemokine receptor which is a broad-spectrum high affinity chemokine receptor with no apparent coupling to G proteins (as assessed by calcium flux and chemotaxis readouts) was identified by Graham and colleagues and called D6 [65]. This receptor binds CCL3, CCL4 and CCL5, as well as MCP1-5, CCL11/eotaxin CCL11 and CCL14/HCC with nanomolar affinity. Here again, the same two critical regions in TM2 and TM3 involving the aspartate and DRY boxes are found substituted. The aspartate in TM2 is again changed to asparagine, and the DRY sequence is found as DKY. While the lysine/arginine substitution maintains overall charge, the unique hydrogen bonding structure seen in rhodopsin suggests that lysine might not maintain such a structure.

The third receptor in the chemokine family which has been identified to bind without evidence of G protein-mediated signaling was recently identified by two groups [66,67]. While there was initial confusion over nomenclature (one group called the molecule CCR10 while the other called it CCR11), as well as discrepancies in ligand binding, these have now been resolved. The molecule is neither CCR10 [66] nor CCR11, [67] but rather a CCR-binding protein (CCR-bp). (Chemokine nomenclature for receptors has required both a binding and signaling function to be given a designation.) This protein binds the CCR7 ligands CCL19/ELC and CCL21/SLC, as well as the CCR9 ligand CCL25/TECK. As with other presumed decoys, there is no apparent chemotaxis or calcium flux in response to the ligands in cells bearing this binding protein. Analyses of the primary structure reveals its closest relatives in the chemokine family are CCR6, 7 and 9, which is in good agreement with the binding of canonical ligands of CCR7 and CCR9. However, substitutions are not seen in the TM2 and DRY boxes as was discussed above. Thus, the reason for lack of G protein signaling is unclear from simple inspection of the sequence. One possibility suggested, by an alignment of the structures, is that there is a deletion at the N-terminus of CCR-bp relative to CCR7 and CCR9, and an extension of hydrophobic residues at the C-terminus of the predicted TM4. This could result in TM4 of CCR-bp adopting a different 'depth' in the membrane, with resulting conformational implications for both intra- and extracellular domains.

A second interesting feature of these 'non-signaling' chemokine receptors is their expression. Interestingly, all three of these receptors have their most abundant expression outside the leukocyte series. In the case of HCR, expression is highest in spleen, fetal liver, bone marrow, and lymph nodes — but not detected in peripheral leukocytes.

Similarly, D6 is highly expressed in the placenta, and also in *lymphatic* (as opposed to vascular) endothelial cells. Highest expression by northern analyses for CCR-bp is in heart, followed by lung. PCR evidence (not quantitative) also suggested some expression in immature dendritic cells and certain T cells [68]. This pattern of expression is similar to that seen with the DARC, which is found extensively on endothelia, Purkinje neurons, but not leukocytes.

9. HYPOTHESES REGARDING THE NON-SIGNALING CHEMOKINE RECEPTORS

We are considering two alternative but not mutually exclusive hypotheses for HCR, D6 and CCR-bp. First, that these receptors are localized to particular solid organs and endothelial surfaces and placenta to limit local concentrations of chemokines, and thus effectively limit a local inflammatory response. Variations on this theme might be that these receptors can locally sequester chemokines and then present them to passing cells which might have a higher affinity for a particular ligand. Further, expression of these particular receptors might be permissive for some chemokines to function while sequestering others. For example, D6 might prevent leukocytes in the lymphatics (which may be homing to or from lymph nodes) to 'ignore' beta chemokines other than CCR7 ligands, or SDF.

A second hypothesis we are entertaining is that these receptors *do signal* but not by classical G protein-mediated pathways. An interesting precedent for this is the case made for beta arrestin-mediated activation of the MAP kinase pathways [69]. In this system, ligand drives phosphorylation of the receptors by G protein receptor kinases. The phosphorylated receptor binds an arrestin, which then engages the endocytic pathway via clathrin to endosomes. However, in addition the beta arrestin serves as a scaffold for the recruitment of proteins in the MAP kinase cascade leading to activation of ERKs. This has been not only demonstrated for the beta adrenergic receptor, but also for the substance p receptor and the IL-8 receptor CXCR1 [70]. Since these pathways in these systems occur in *parallel* to G protein-mediated signaling, we cannot predict the ultimate result of activation of this pathway alone. However, we may predict that if this hypothesis is correct, it would require that ligand drives phosphorylation of the C-terminal domain. Experiments to address this issue are underway in our laboratory.

REFERENCES

1. Mackay CR. Chemokines: immunology's high impact factors. Nat Immunol 20012(2):95–101. Review.
2. Bazan JF, Bacon KB, Hardiman G, Wang W, Soo K, Rossi D, Greaves DR, Zlotnik A, Schall TJ. A new class of membrane-bound chemokine with a CX3C motif. Nature 1997;385:640–644.
3. Kelner GS, Kennedy J, Bacon KB, Kleyensteuber S, Largaespada DA, Jenkins NA, Copeland NG, Bazan JF, Moore KW, Schall TJ et al. Lymphotactin: a cytokine that represents a new class of chemokine. Science 1994;266: 1395–1399.
4. Murphy PM, Bagglioni M, Charo IF, Hebert CA, Horuk R, Matsushima K, Miller LH, Oppenheim JJ, Power CA. International union of pharmacology. XXII. Nomenclature for chemokine receptors. Pharmacol Rev 2000;52(1):145–176.
5. Holmes WE, Lee J, Kuang WJ, Rice GC, Wood WI. Structure and functional expression of a human interleukin-8 receptor. Science 1991;253:1278–1280.
6. Murphy PM, Tiffany HL. Cloning of complementary DNA encoding a functional human interleukin-8 receptor. Science 1991;253:1280–1283.
7. Palczewski K, Kumasaka T, Hori T, Behnke CA, Motoshima H, Fox BA, Le Trong I, Teller DC, Okada T, Stenkamp RE, Yamamoto M, Miyano M. Crystal structure of rhodopsin: A G protein-coupled receptor. Science 2000; 289(5480):739–745.
8. Loetscher P, Seitz M, Baggiolini M, Moser B. Interleukin-2 regulates CC chemokine receptor expression and chemotactic responsiveness in T lymphocytes. J Exp Med 1996; 184:569–577.
9. Sica A, Saccani A, Borsatti A, Power CA, Wells TN, Luini W, Polentarutti N, Sozzani S, Mantovani A. Bacterial lipopolysaccharide rapidly inhibits expression of C-C chemokine receptors in human monocytes. J Exp Med 1997;185(5):969–974.
10. Lahrtz F, Piali L, Nadal D, Pfister HW, Spanaus KS, Baggiolini M, Fontana A. Chemotactic activity on mononuclear cells in the cerebrospinal fluid of patients with viral meningitis is mediated by interferon-gamma inducible protein-10 and monocyte chemotactic protein-1. Eur J Immunol 1997;27(10):2484–2489.
11. Graham GJ, Wright EG, Hewick R, Wolpe SD, Wilkie NM, Donaldson D, Lorimore S, Pragnell IB. Identification and characterization of an inhibitor of haemopoietic stem cell proliferation. Nature 1990;344(6265):442–444.
12. Sharpe RJ, Byers HR, Scott CF, Bauer SI, Maione TE. Growth inhibition of murine melanoma and human colon carcinoma by recombinant human platelet factor 4. J Natl Cancer Inst 1990;82:848–853.
13. Luster AD, Greenberg SM, Leder P. The IP-10 chemokine binds to a specific cell surface heparan sulfate site shared with platelet factor 4 and inhibits endothelial cell proliferation. J Exp Med 1995;182:219–231.
14. Luster AD, Leder P. IP-10, a C-X-C chemokine, elicits a potent thymus dependent antitumor response in vivo. J Exp Med 1993;178:1057–1065.
15. Strieter RM, Polverini PJ, Arenberg DA, Walz A, Opdenakker G, Van Damme J, Kunkel SL. Role of C-X-C chemokines as regulators of angiogenesis in lung cancer. J Leukoc Biol 1995;57(5):752–762. Review.
16. Bakhiet M, Tjernlund A, Mousa A, Gad A, Stromblad S, Kuziel WA, Seiger A, Andersson J. RANTES promotes growth and survival of human first-trimester forebrain astrocytes. Nat Cell Biol 2001;3(2):150–157.
17. Horuk R, Chitnis CE, Darbonne WC, Colby TJ, Rybicki A, Hadley TJ, Miller LH. A receptor for the malarial parasite Plasmodium vivax: the erythrocyte chemokine receptor. Science 1993;261(5125):1182–1184.
18. Berger EA, Murphy PM, Farber JM. Chemokine receptors as HIV-1 co-receptors: Roles in viral entry, tropism and disease. Ann Rev Immunol 1999;17:657–700.
19. Wu L, Gerard NP, Wyatt R, Choe H, Parolin C, Ruffing N, Borsetti A, Cardoso AA, Desjardin E, Newman W, Gerard C, Sodroski J. CD4-induced interaction of primary HIV-1 gp120 glycoproteins with the chemokine receptor CCR-5. Nature 1996;384(6605):179–183.
20. Trkola A, Dragic T, Arthos J, Binley JM, Olson WC, Allaway GP, Cheng-Mayer C, Robinson J, Maddon PJ, Moore JP. CD4-dependent, antibody-sensitive interactions between HIV-1 and its co-receptor CCR-5. Nature 1996; 384(6605):184–187.
21. Slattery DM, Gerard N, Gerard C. Gene targeting of chemokines and their receptors. Springer Semin Im-

munopathol 2000;22(4):417–432. Review. No abstract available.
22. Yancey KB, Lawley TJ, Dersookian M, Harvath L. Analysis of the interaction of human C5a and C5a des Arg with human monocytes and neutrophils: flow cytometric and chemotaxis studies. J Invest Dermatol 1989;92(2): 184–189.
23. Siciliano SJ, Rollins TE, De Martino J, Konteatis Z, Malkowitz L, van Riper G et al. Two site binding of C5a by its receptor: an alternative binding paradigm for G protein-coupled receptors. Proc Natl Acad Sci USA 1994; 91(4):1214–1218.
24. Kolakowski LF Jr, Lu B, Gerard C, Gerard NP. Probing the 'message: address' sites for chemoattractant binding to the C5a receptor. Mutagenesis of hydrophilic and proline residues within the transmembrane segments. J Biol Chem 1995;270(30):18077–18082.
25. Rucker J, Samson M, Doranz BJ, Libert F, Berson JF, Yi Y et al. Regions in beta-chemokine receptors CCR5 and CCR2b that determine HIV-1 cofactor specificity. Cell 1996;87(3):437–446.
26. Samson M, LaRosa G, Libert F, Paindavoine P, Detheux M, Vassart G, Parmentier M. The second extracellular loop of CCR5 is the major determinant of ligand specificity. J Biol Chem 1997;272(40):24934–24941.
27. Lee B, Sharron M, Blanpain C, Doranz BJ, Vakili J, Setoh P, Berg E, Liu G, Guy HR, Durell SR, Parmentier M, Chang CN, Price K, Tsang M, Doms RW. Epitope mapping of CCR5 reveals multiple conformational states and distinct but overlapping structures involved in chemokine and coreceptor function. J Biol Chem 1999;274(14): 9617–9626.
28. Han KH, Green SR, Tangirala RK, Tanaka S, Quehenberger O. Role of the first extracellular loop in the functional activation of CCR2. The first extracellular loop contains distinct domains necessary for both agonist binding and transmembrane signaling. J Biol Chem 1999;274(45): 32055–32062.
29. Gerard NP, Gerard C. Construction and expression of a novel recombinant anaphylatoxin, C5a-N19, as a probe for the human C5a receptor. Biochemistry 1990;29(39): 9274–9281.
30. Opperman M, Raedt U, Hebell T, Schmidt B, Zimmermann B, Gotze O. Probing the human receptor for C5a anaphylatoxin with site directed antibodies. Identification of a potential ligand binding site on the NH2-terminal domain. J Immunol 1993;151(7):3785–3794.
31. DeMartino JA, Van Riper G, Siciliano SJ, Molineaux CJ, Konteatis ZD, Rosen H, Springer MS. The amino terminus of the human C5a receptor is required for high affinity C5a binding and for receptor activation by C5a but not C5a analogs. J Biol Chem 1994;269(20):14446–14450.
32. Farzan M, Mirzabekov T, Kolchinsky P, Wyatt R, Cayabyab M, Gerard NP, Gerard C, Sodroski J, Choe H. Tyrosine sulfation of the amino terminus of CCR5 facilitates HIV-1 entry. Cell 1999;96(5):667–676.
33. Preobrazhensky AA, Dragan S, Kawano T, Gavrilin MA, Gulina IV, Chakravarty L, Kolattukudy PE. Monocyte chemoattractic protein-1 receptor CCR2B is a glycoprotein that has tyrosine sulfation in a conserved extracellular N-terminal region. J Immunol 2000;165(9):5295–5303.
34. Cormier EG, Persuh M, Thompson DA, Lin SW, Sakmar TP, Olson WC, Dragic T. Specific interaction of CCR5 amino-terminal domain peptides containing sulfotyrosines with HIV-1 envelope glycoprotein gp120. Proc Natl Acad Sci 2000;97(11):5762–5767.
35. Farzan M, Vasilieva N, Schnitzler CE, Chung S, Robinson J, Gerard NP, Gerard C, Choe H, Sodroski J. A tyrosine-sulfated peptide based on the N terminus of CCR5 interacts with a CD4-enhanced epitope of the HIV-1 gp120 envelope glycoprotein and inhibits HIV-1 entry. J Biol Chem 2000;257(43):33516–33521.
36. Farzan M, Schnitzler CE, Vasilieva N, Leung D, Kuhn J, Gerard C, Gerard NP, Choe H. Sulfated tyrosines contribute to the formation of the C5a docking site of the human C5a anaphylatoxin receptor. J Exp Med 2001; 193(9):1059–1066.
37. Mery L, Boulay F. The NH2-terminal region of C5aR but not that of FPR is critical for both protein transport and ligand binding. J Biol Chem 1994;269(5):3457–3463.
38. Monteclaro FS, Charo IF. The amino-terminal domain of CCR2 in both necessary and sufficient for high affinity binding of monocyte chemoattractant protein1. Receptor activation by a pseudo-tethered ligand. J Biol Chem 1997; 272(37):23186–23190.
39. Blanpain C, Doranz BJ, Vakili J, Rucker J, Govaerts C, Baik SS, Lorthioir O, Migeotte I, Libert F, Baleux F, Vassart G, Doms RW, Parmentier M. Multiple charged and aromatic residues in CCR5 amino-terminal domain are involved in high affinity binding of both chemokines and HIV-1 Env protein. J Biol Chem 1999;274(49):34719–34727.
40. Doranz BJ, Orsini MJ, Turner JD, Hoffman TL, Berson JF, Hoxie JA, Peiper SC, Brass LF, Doms RW. Identification of CXCR4 domains that support coreceptor and chemokine receptor functions. J Virol 1999;73(4):2752–2761.
41. Fanning AS, Anderson JM. PDZ domains: fundamental building blocks in the organization of protein complexes at the plasma membrane. J Clin Invest 1999;103:767–772.
42. Hall RA, Premont RT, Chow CW, Blitzer JT, Pitcher JA, Claing A, Stoffel RH, Barak LS, Shenolikar S, Weinman EJ, Grinstein S, Lefkowitz RJ. The beta2-adrenergic receptor interacts with the Na^+/H^+-exchanger regulatory factor to control Na^+/H^+ exchange. Nature 1998; 392(6676):626–630.
43. Cao TT, Deacon HW, Reczek D, Bretscher A, von Zastrow M. A kinase-regulated PDZ-domain interaction controls endocytic sorting of the beta2-adrenergic receptor. Nature 1999;401(6750):286–290.
44. Jones SA, Wolf M, Qin S, Mackay CR, Baggiolini M. Different functions for the interleukin 8 receptors (IL-8R) of human neutrophil leukocytes: NADPH oxidase and phospholipase D are activated through IL-8R1 but not IL-8R2. Proc Natl Acad Sci USA 1996;93(13):6682–6686.
45. Murphy PM. Viral exploitation and subversion of the immune system through chemokinemimicry. Nature Immunol 2001;2:116–122.
46. Arvanitakis L, Geras-Raaka E, Varma A, Gershengorn MC, Cesarman E. Human herpesvirus KSHV encodes a

constitutively active G-protein-coupled receptor linked to cell proliferation. Nature 1997;385(6614):347–350.
47. Yang B et al. Transgenic expression of the chemokine receptor encoded by human herpesvirus 8 induces an angioproliferative disease resembling Kaposi's Sarcoma. J Exp Med 2000;191:445–454.
48. Berger EA, Murphy PM, Farber JM. Chemokine receptors as HIV-1 coreceptors: roles in viral entry, tropism and disease. Annu Rev Immunol 1999;17:657–700.
49. Littman DR. Chemokine receptors: key to AIDS pathogenesis? Cell 1998;93:677–680.
50. Tuttle DL, Harrison JK, Anders C, Sleasman JW, Goodenow MM. Expression of CCR5 increases during monocyte differentiation and directly mediates macrophage susceptibility to infection by human immunodeficiency virus type 1. J Virol 1998;72(6):4962–4969.
51. Elsner J et al. The CC chemokine antagonist Met-RANTES inhibits eosinophil effector functions through the chemokine receptors CCR1 and CCR3. Eur J Immunol 1997;27:2892–2898.
52. Dabbagh K et al. Local blockage of allergic airway hyperreactivity and inflammation by the poxvirus derived pan-CC-chemokine inhibitor vCC1. J Immunol 2000;165: 3418–3422.
53. Gao W, Topham PS, King JA, Smiley ST, Csizmadia V, Lu B, Gerard CJ, Hancock WW. Targeting of the chemokine receptor CCr1 suppresses development of acute and chronic allograft rejection. J Clin Invest 2000;105(1): 35–44.
54. Gu L et al. Absence of monocyte chemoattractant protein-1 reduces atherosclerosis in low density lipoprotein receptor-deficient mice. Mol Cell 1998;2:275–281.
55. Gosling J, Slaymaker S, Gu L, Tseng S, Zlot CH, Young SG, Rollins BJ, Charo IF. MCP-1 deficiency reduces susceptibility to atherosclerosis in mice that over express human apolipoprotein. J Clin Invest 1999;103(6):773–778.
56. Mantovani A, Massimo L, Vecchi A, Sozzani S, Allavena P. Decoy receptors: a strategy to regulate inflammatory cytokines and chemokines. Trends Immunol 2001;22(6): 328–336.
57. Dinarello CA. The role of the interleukin-1-receptor antagonist in blocking inflammation mediated by interleukin-1. N Engl J Med 2000;343(10):732–734.
58. Chaudhuri A, Zbrzezna V, Polyakova J, Pogo AO, Hesselgesser J, Horuk R. Expression of the Duffy antigen in K562 cells. Evidence that it is the human erythrocyte chemokine receptor. J Biol Chem 1994;269(11):7835–7838.
59. Neote K, Mak JY, Kolakowski LF Jr, Schall TJ. Functional and biochemical analysis of the cloned Duffy antigen: identity with the red blood cell chemokine receptor. Blood 1994;84(1):44–52.
60. Gerard C, Rollins BJ. Chemokines and disease. Nat Immunol 2001;2(2):108–115. Review.
61. Rios M, Chaudhuri A, Mallinson G, Sausais L, Gomensoro-Garcia AE, Hannon J, Rosenberger S, Poole J, Burgess G, Pogo O, Reid M. New genotypes in Fy(a-b-) individuals: nonsense mutations (Trp to stop) in the coding sequence of either FY A or Fy B. Br J Haematol 2000;108(2):448–454.
62. Luo H, Chaudhuri A, Zbrzezna V, He Y, Pogo AO. Deletion of the murine Duffy gene (Dfy) reveals that the Duffy receptor is functionally redundant. Mol Cell Biol 2000; 20(9):3097–3101.
63. Dawson TC, Lentsch AB, Wang Z, Cowhig JE, Rot A, Maeda N, Peiper SC. Exaggerated response to endotoxin in mice lacking the Duffy antigen/receptor for chemokines (DARC). Blood 2000;96(5):1681–1684.
64. Fan P, Kyaw H, Su K, Zeng Z, Augustus M, Carter KC, Li Y. Cloning and characterization of a novel human chemokine receptor. Biochem Biophys Res Commun 1998; 243(1):264–268.
65. Nibbs RJ, Wylie SM, Yang J, Landau NR, Graham GJ. Cloning and characterization of a novel promiscuous human beta-chemokine receptor D6. J Biol Chem 1997; 272(51):32078–32083.
66. Gosling J, Dairaghi DJ, Wang Y, Hanley M, Talbot D, Miao Z, Schall TJ. Cutting edge: identification of a novel chemokine receptor that binds dendritic cell- and T cell-active chemokines including ELC, SLC, and TECK. J Immunol 2000;164(6):2851–2856.
67. Schweickart VL, Epp A, Raport CJ, Gray PW. CCR11 is a functional receptor for the monocyte chemoattractant protein family of chemokines. J Biol Chem 2000;275(13): 9550–9556.
68. McDonald PH, Chow CW, Miller WE, Laporte SA, Field ME, Lin FT, Davis RJ, Lefkowitz RJ. Beta-arrestin 2: a receptor-regulated MAPK scaffold for the activation of JNK3. Science 2000;290(5496):1574–1577.
69. DeFea KA, Vaughn ZD, O'Bryan EM, Nishijima D, Dery O, Bunnett NW. The proliferative and antiapoptotic effects of substance P are facilitated by formation of a beta-arrestin-dependent scaffolding complex. Proc Natl Acad Sci USA 2000;97(20):11086–11091.
70. Barlic J, Andrews JD, Kelvin AA, Bosinger SE, DeVries ME, Xu L, Dobransky T, Feldman RD, Ferguson SS, Kelvin DJ. Regulation of tyrosine kinase activation and granule release through beta-arrestin by CXCR1. Nat Immunol 2000;1(3):227–233.

Universes in Delicate Balance: Chemokines and the Nervous System
R.M. Ransohoff, K. Suzuki, A.E.I. Proudfoot, W.F. Hickey and J.K. Harrison, editors

Chapter 2.4

Chemokine Receptor Signal Transduction

K.B. Bacon

Bayer Yakuhin, Ltd., 6-5-1-3 Kunimidai, Kizu-cho, Soraku-gun, Kyoto, 619-0216 Japan

1. CHEMOKINES AND CHEMOKINE RECEPTORS

The chemokines are a superfamily of proteins known to be vital for leukocyte migration and activation in both physiological and pathological contexts [1,2]. Initially classified based on their roles in leukocyte trafficking and adhesion, chemokines have been implicated in the modulation of numerous biological functions, including phagocytosis, cytokine secretion, T cell activation, apoptosis, angiogenesis, proliferation, and viral pathogenesis, both in the peripheral immune system and the Central Nervous System (CNS). The majority of the superfamily members are small molecular mass proteins (8–14 kDa) although the CX_3C family member, fractalkine, exists as a membrane-anchored form, and as a 95 kDa secreted protein [3,4].

The chemokines bind to 7-transmembrane (7-TM) G-protein-coupled receptors (GPCRs) and it is now appreciated that a large majority of functions are elicited through complex signal transduction cascades similar, if not identical to those shown for more extensively studied GPCRs such as the fMLP chemoattractant, β-adrenergic, and M2-muscarinic 7-TMs [5].

This chapter will focus on post-receptor signaling of chemokine receptors and attempts to focus as much as possible on known functions of this class of mediators. Since the original characterizations have centered around migration-inducing events, the greatest focus will be on this functionality. It must be stated, however, that in a neurobiological context outside neuroinflammatory pathologies, the relevance of cell migration will be limited, thus signaling that outcomes are still to be defined. What is of relevance nonetheless, is that many of the ultrastructural changes transduced in response to these signals, can be applied in diverse functions including CNS-relevant systems such as axonal positioning, dendrite extension/retraction and nerve repair.

2. CHEMOKINE SIGNALING FOR CELL MIGRATION AND POLARIZATION

Directed migration is one of the most important functions exhibited by numerous cell types during development and embryogenesis, immune surveillance, neuronal positioning and axon guidance, wound healing and metastases. Additionally, in inflammatory pathologies, the uncontrolled infiltration of leukocytes to sites of trauma remains one of the most important characteristics during disease progression.

Migration of cells is a complex process involving not only the intricate mechanics of cytoplasmic structural proteins within the cells, but also their molecular interactions with the cells' surface receptors. In turn, the interactive role of surface receptors with the extracellular milieu, both in two and three dimensions, is highly regulated. Following polarization, the directed migration of a cell involves the balance between surface receptor redistribution (chemoattractant receptors and adhesion molecules) and the directional persistence of the cell. Purchase on the substratum is afforded by the redistribution of adhesion receptors through lamellae and lamellipodia. Once this polarization has been established, the interaction at the back edge must be terminated through detachment. Through a transfer of receptors and adhesion molecules, and the polymerization–depolymerization of actin at the leading edge, the cell executes a continual process of migration, changing according to agonist concentration fluctuations in their micro-environment. In the migrating cell, membrane extension at the leading edge is associated with actin polymerization. There are potentially three mechanisms whereby the cell can regulate the polymerization–depolymerization of actin: increasing the concentration of monomeric G-actin available at the F-actin growth (barbed) ends; uncapping, severing and re-polymerization of existent F-actin bundles; and de novo actin nucleation (for a detailed review, see [6]). There appears to be a tight and complex regulation by numerous actin-binding proteins and signal transduction cascades including the uncapping and severing of F-actin which is regulated by gelsolin. Severing activity is predominant when the concentration of intracellular calcium is at high micromolar levels and low concentration of phosphoinositides exist — a situation clearly elicited by chemoattractants in leukocytes, for example. At lower intracellular calcium concentrations, there is a dissociation of gelsolin from actin, leaving barbed ends free for re-polymerization. Regulation of free G-actin is brought about by binding proteins such as the family of profilins and ADFs/cofilin, which serve to promote the severing and elongation of actin filaments, allowing a constant dynamic or turnover at the leading edge [7,8].

3. SEVEN-TRANSMEMBRANE CHEMOATTRACTANT RECEPTORS (7-TM) AND HETEROTRIMERIC G-PROTEIN-MEDIATED SIGNALING

Heterotrimeric G-protein-coupled 7-TMs constitute one of the most thoroughly researched receptor–ligand interactions (Fig. 1). The organization of the heterotrimeric G-protein superfamily and their biological actions will not be covered here (see [9] for review). A model system for investigation of signal transduction by chemoattractant receptors has

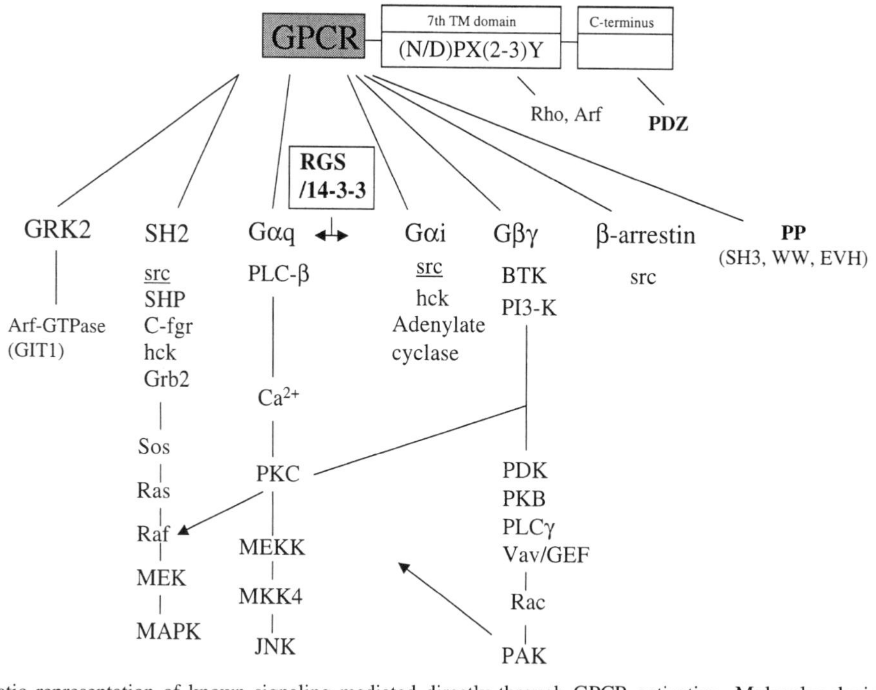

Figure 1. Schematic representation of known signaling mediated directly through GPCR activation. Molecules depicted represent the most well-characterized signaling entities to date and thus constitute the major signaling pathways. Abbreviations: BTK, Bruton's tyrosine kinase; GEF, guanine nucleotide exchange factor; GRK, G-protein receptor kinase; JNK, c-jun N-terminal kinase; MAPK, mitogen-activated protein kinase; MEK, MKK, MAPK kinase; MEKK, MEK kinase; PAK, p21-activated kinase; PDK, phosphoinositide-dependent kinase; PI3K, phosphoinositide 3-kinase; PKB, PKC, protein kinase B/C; PLC, phospholipase C; RGS, regulator of G-protein signaling.

been that of the bacterial formyl peptide f-Met-Leu-Phe (fMLP) receptor. This 7-TM receptor has been shown to be coupled to the heterotrimeric G-proteins Gα16 or Gαi2 [10] via the first and second intracellular loops of the receptor, respectively [11,12]. This is clearly different from the situation with certain chemokine receptors. Mutagenesis studies have revealed that the first intracellular loop does not bind the G-proteins in signaling through CXCR1, CXCR2, CCR2b, CCR5 and CXCR4, but that this function occurs through the second and third intracellular loops [13–16].

In general, activation of heterotrimeric G-proteins leads to exchange of GDP for GTP on the α subunit and dissociation of the α and βγ subunits for interactions with downstream effectors [17,18]. This highly dynamic and continual process of GDP–GTP exchange is effected by subsequent hydrolysis of the GTP, by the RGS (negative regulators of G-protein signaling) family of proteins which function as GTPase-activating proteins (GAPs) [19]. GAPs in turn are kept in check by phosphorylation-dependent binding of 14-3-3 proteins at the Gα-binding domain of the RGS (RGS7 and RGS3) [20]. Indeed, RGS1 and RGS3 have been implicated in the regulation of B-cell migration induced by SDF-1α and BLC, while Gαi-binding RGS14 regulates IL-8-induced MAPK activation [21,22]. Further as detailed below, the migration of neuronal precursors in the developing cerebellum is regulated in part by SDF-1/CXCL12 and CXCR4 [23,24]. Interaction with ephrin receptors during this process utilizes a novel PDZ/RGS protein [25]. PLC-β is activated mainly by the G-protein βγ subunits resulting in the generation of IP_3 which induces a transient release of calcium from intracellular stores, and DAG, which in conjunction with Ca^{2+}, stimulates PKC translocation and activation [26].

Generally, the role of the Gαi G-proteins in the actions of most of the chemoattractants which bind 7-TMs was shown to be a prominent one through use of pertussis toxin (*Bortedella pertussis*) which is known to ADP ribosylate and inactivate the Gα subunits of this subfamily. Numerous reports have now demonstrated that heterotrimeric G-proteins of the Gαq subfamily (pertussis-toxin-insensitive) can also functionally couple to 7-TM chemoattractant receptors [11,14,27–30]. However, the vast majority of chemokines appear to preferentially couple through Gαi since migration and immediate intracellular calcium transients are completely antagonized by pertussis toxin.

3.1. Adenylate Cyclase

Activation of adenylate cyclase by 7-TM receptors signaling through Gsα members of the heterotrimeric G-proteins, has also been widely investigated. In general however, this system counteracts that induced by members of the Gαi family, being negative regulators of leukocyte chemotaxis and activation [31–33]. There are three examples which do provide compelling evidence for the induction of cAMP in response to chemokines, however [34–36]. At present, it is unclear as to why there is such discrepancy compared with other results using chemokines.

3.2. Regulation of GPCR Function — Arrestins and GRKS

7-TM receptors are phosphorylated by G-protein-coupled receptor kinases (GRKs) following ligand binding. Following phosphorylation, arrestin family proteins bind to the 7-TM GPCR at the third intracellular loop and membrane-proximal region of the carboxy tail, preventing further interaction of the receptor with G-proteins. This process serves to desensitize the receptor and contributes to the process of internalization into clathrin-coated pits. β-arrestin has been shown to regulate the signaling and internalization of CCR5, CCR2b, CXCR4 and CXCR1 [16,37,38]. RANTES induces the association of GRK-2, β-arrestin and clathrin with CCR5 [39]. In another study, binding of β-arrestin to the C-terminal portion of CXCR4 did not appear to be critical for ligand-induced internalization, since over-expression of arrestin with a C-terminal-truncated CXCR4 led to efficient internalization. There is some question as to the relevance of the C-terminal residues for G-protein coupling and signaling. Cheng and colleagues [16] have shown the importance of this domain, while Brelot et al. [40] demonstrated almost complete loss of signaling through deletion of the third intracellular loop. It is thus likely that residues in both domains contribute significantly to the binding of the G-proteins.

3.3. SH2 Domain-Containing Proteins

While a large volume of literature suggests the negative regulatory role of arrestins and GRKs in GPCR signaling, there does exist significant evidence demonstrating an initiating role of these molecules in recruiting other signaling components to the receptor, including SH2 domain-containing proteins. β-arrestin has been shown to directly bind *Src*, suggestive of a role in the regulation of MAPK signaling — which has been shown repeatedly for chemokine receptors (see below). Further support for such a switch from classical signaling to kinase pathways was recently provided by the evidence for a direct interaction between Gαi and *Src* [41]. Eotaxin binding to CCR3 in eosinophils has been shown to stimulate the phosphorylation of multiple intracellular signaling proteins and the association of the Src-family members *Hck* and *c-fgr* [42]. Evidence favoring a function for arrestins in coupling CXCR1 to these Src-family kinases was recently provided [43]. Further, this pathway was linked to granule exocytosis in IL-8-stimulated cells [42]. CCR5 signaling induced by MIP-1β in transformed cells has also been shown to involve *syk* activation and binding of SHP1 and Grb2 [44]. Ptasznik et al. [45] demonstrated that a tyrosine kinase signaling pathway mediates PIP_3 formation in the neutrophil following fMLP stimulation, as well as the activation of the *Src*-like kinase, *lyn*. The role of PTKs relevant to the chemokine system was expanded to include *Pyk2-H* which was induced by RANTES and MIP-1α [46]. It will be interesting to understand if such activity is related to direct Gα binding of *Src* members [41] or the direct association of these SH2-domain-containing proteins with arrestins.

As detailed above, an immunoprecipitated CCR5-SHP-*Syk*-Grb2 complex has been observed. It has been reported that the β-adrenergic receptor can directly bind Grb2 following phosphorylation of Tyr350/354 in the receptor [47] suggesting that there is a possibility that GPCRs can directly interact with SH2-containing proteins in addition to the role arrestin may play. It has been demonstrated that direct *Src*-mediated phosphorylation of the AT1 receptor, following angiotensin II stimulation results in activation of Jak2 tyrosine kinase in the presence of SHP (a SH2-domain-containing tyrosine phosphatase [48,49]. Vila-Coro et al. [50] have shown that SDF-1α activation of CXCR4 induces dimerization and the activation of the JAK/STAT pathway. In addition, RANTES and MIP-1α have been shown to activate Stats 1 and 3 resulting in transcriptional activation of c-fos [51]. While it has been demonstrated that phosphorylation of CCR5 and CXCR4 occurs, chemokine receptors do not contain obvious SH2-binding domains, thus it is unclear as to how these cascades were initiated and not unlikely that other proteins may be playing a role in this signaling.

Recently, the SH2-domain-containing adapter Crk and substrate p130Cas have been shown to be phosphorylated by SDF-1α [52]. In addition, phosphorylation of associated kinases, RAFTK, p125FAK and paxillin was observed. The action of these SH2/SH3-domain-containing adapters and kinases are intricately linked to the function of integrins and integrin-mediated cell trafficking (as will be discussed below). More recently, SLAP-130/FYB (SH2-domain-containing 76-kDa [SLP-76]-associated phosphoprotein of 130 kDa; also known as $p59^{fyn}$-binding protein (FYB)) has been implicated in SDF-1-induced signaling through CXCR4. How this adapter protein functions with CXCR4 is still unclear; however, it has been shown to enhance SDF-1α- and β1-integrin-mediated migration of T cells by 2- to 3-fold [53].

4. CHEMOKINE RECEPTORS/ HETEROTRIMERIC G-PROTEIN COUPLING: FURTHER COMPLEXITY

The interaction of chemokine receptors with Gαi, Gα16 or Gαq proteins leading to hydrolysis of PLCβ, as the main regulatory mechanism, has recently been called into question through a number of basic observations. It has been demonstrated that certain chemokines, including RANTES, MIP-1α, MIP-1β, MCP-1, MIP-3α and β, eotaxin, SDF1α, fractalkine, IL-8, Gro-α, MDC, TARC, SLC, and I-309 can induce tyrosine phosphorylation of intracellular substrates [42,44,54–59]. There is also interest in the ability of chemokines to induce homo- or hetero-dimerization of their receptors, a topic that has recently been reviewed [60]. Of these, two major kinase pathways have been investigated: mitogen-activated protein kinase (MAPK) and phosphoinositide-3-kinase (PI3K) activation.

4.1. MAPK Signaling

In response to growth factors and mitogens, MAPK (p38, p42, p44) is activated by MAPK kinase (MEK), which in turn is activated by the ras-induced activation of raf family members [61,62]. Activated Ras (Ras-GTP) binds inactive Raf1, releasing the inhibition of the catalytic domain by the N-terminal regulatory domain, and translocates it to the membrane where it can be phosphorylated by PKC (on serine 621 [63]). Raf1 is also bound by the 14-3-3 protein, retaining its catalytic activity, and subsequently activates MEK and MAPK. Initial observations using an MCP-1-stimulated murine T-cell hybrid, demonstrated clear activation of PTKs including MAPK [56]. Thompson et al. [64] have demonstrated two induced forms of MEK in response to IL-8 as well as other chemoattractants, and further analysis of this has been detailed by Grinstein et al. [65] and Downey et al. [66]. Knall et al. [67] demonstrated IL-8-induced activation of Ras/raf/MAPK in neutrophils, which appeared to be dependent on PI3K. Knall et al. [68] further demonstrated that in IL-8-stimulated PMN, where PI3K and MAPK enzymes were activated, the migration was unaffected by the activation of ERK/MAPK or the p38MAPK enzymes, implying an independent role for PI3K in the chemotactic event.

The functional consequences following the activation of MAPK family members are not completely clear at present. Whether these components exert a direct role, or simply function as intermediaries in other signaling cascades needs to be investigated. MAPK activation is highly implicated in growth factor receptor function, and transcriptional regulation, as detailed above. There have been some more recent reports implicating a role in myosin phosphorylation hence contractility (see below); however, these functions have to be further clarified for chemoattractant receptors.

4.2. Phosphoinositide 3-Kinase (PI3K) Signaling

Mechanistically, there is differential regulation of the α, β and δ isoforms of PI3K, and PI3Kγ which is known to be activated by βγ G-protein subunits [69]. Stephens et al. [70] identified two Gβγ-dependent PI3K family members in neutrophils, characterized by two subunits, one homologous to the p110 isoform but whose activity is highly regulated by an adapter of 101 kDa. There is no binding of p110γ to the p85 subunits as occurs with the other isoforms. PI3Kγ is directly activated by Gβγ subunits, resulting in the formation of phosphatidylinositol 3,4,5-triphosphate and the activation of PDK, protein kinase B (PKB), $p70^{S6kinase}$ and p42/p44 MAPK. Investigations of PI3Kγ null mice have now revealed the central role this kinase plays in the chemoattractant-induced migratory response of leukocytes in inflammation [71–73]. In a series of studies analyzing neutrophil and monocyte migration in response to standard chemoattractants, it was revealed that PLCβ2 was dispensable, while PI3Kγ was important but not essential for neutrophil migration in response to IL-8 and MIP-1α, even though it was the sole PI3K isoform coupled to the chemoattractant receptors. In contrast, however, the macrophage/monocyte response to chemokines was significantly reduced (MIP-5, 52%; SDF-1α, 85%; MDC, 70%; and RANTES, 80%). That the defect was linked to receptor signaling as opposed to intrinsic problems associated with the locomotory capacity, was evidenced by the unaffected increase in intracellular Ca^{2+} in response to these agonists. Thus, it is clear that the migratory machinery is inextricably linked to PI3Kγ activation in this leukocyte population. Indeed, long before the evidence was obtained from the knock-out studies, chemokine-induced activation of PI3Kγ (including activation of PKB as a substrate) in numerous cell types (neutrophils, monocytes, T cells, eosinophils, microglia) had been shown, in response to IL-8 [68,72–74], SDF-1α [52,75–78], RANTES [79–81], MIP-3α and β (K.B.B., unpublished), MCP-1 [81–83], lymphotactin and IP-10 [81] and fractalkine [58,84].

It is interesting that coupling of PI3Kγ, but not the classical PI3Kα to 7-TM muscarinic M2 receptors via G-protein βγ subunits was also demonstrated [85–88]. This linkage induced the activation of MAPK through binding of PI3K to a tyrosine kinase, which ultimately stimulated the cascade of Shc, Grb2, Sos, Ras and Raf. Although the TK involved in PI3K stimulation was uncharacterized and the receptor was not a classical chemoattractant receptor, similar mechanisms may be operating to induce cross-talk between the two kinase pathways. While this adds to our understanding of upstream signals potentially involved in chemoattractant receptor signaling, we are still some way from clarifying the link to the actin cytoskeleton. One hypothesis further implicates PI3K. The major metabolite of PI3K-induced hydrolysis, PIP_3, induces the release of Rac-GDP from Rho-GDI, or similar GDI, allowing GDP–GTP exchange and the activity of Rac in actin polymerization [89] potentially providing one mechanism for chemotaxis induction (see below). Recent evidence for the fMLP receptor revealed Gβγ-dependent PI3K activation, linkage of the Rac-GEF *vav*, and binding and activation of Rac [90].

5. CALCIUM MOBILIZATION AND CYTOSKELETAL REARRANGEMENT

For some time now, intracellular calcium gradients have been perceived as being important in the control of cell motility, the suggestion being that higher concentrations of $[Ca^{2+}]_i$ at the rear of the cell promotes actin de-polymerization, while lower concentrations at the front, would promote actin polymerization [91]. This theory, while potentially useful for initial polarization of the cell, is complicated by two issues. First, the polarization of the cell and re-distribution of chemoattractant and adhesion receptors to the leading edge, would, in theory, induce greater concentrations of transient Ca^{2+} release at the front of the cell. Secondly, in calcium-buffering conditions (e.g. the use of BAPTA) or Ca^{2+}-free medium, neutrophils and other leukocytes have been observed to continue migration during in vitro assays, albeit in a time-limited fashion [92]. It is now widely appreciated that one of the main requirements for Ca^{2+} during migration is for the role this divalent cation plays in adhesion molecule avidity. Ca^{2+}-dependent integrin binding has been demonstrated for the β1 and β2 integrins, molecules known to be vital in the trafficking of leukocytes and other cell types [93].

In order to further understand the role(s) Ca^{2+} plays in the motility of the cell, in vitro observations of the patterns and waves of intracellular Ca^{2+} release, using fluorimetry, has been useful. 7-TM ligation by agonist classically induces a rapid transient of calcium release from intracellular stores in numerous leukocyte subtypes. This immediate release (representing an increase in cytoplasmic-free Ca^{2+}) may be followed by more sustained levels of Ca^{2+}i, due to the opening of membrane ion channels [94]. Unfortunately, this gross representation of the Ca^{2+} patterns tells us little of the regulatory role of Ca^{2+}. Single cell imaging has, however, provided the most compelling evidence of Ca^{2+}-dependency for migration and supported the view that a process which by nature involves continual change, both at the cell surface and intracytoplasmically, requires dynamic concentration changes of Ca^{2+}. It is evident that over a period of chemoattractant stimulation and migration of neutrophils, the ligation of 7-TMs by chemoattractant (assuming a concentration gradient) will continually stimulate Ca^{2+} transients and oscillations over a period determined by the migrating cell. These temporary increases in Ca^{2+} allow intracellular processes, such as gelsolin severing of actin filaments. Once the concentration decreases, the nucleation site is available for actin re-polymerization. MARCKS protein (myristoylated, alanine-rich C-kinase substrate) is a specific PKC substrate which binds actin and calmodulin [95]. Following chemoattractant stimulation of PKC, MARCKS cycles between the membrane and the cytosol. Upon phosphorylation by calcium-dependent PKC, MARCKS is released from focal contacts where it interacts with vinculin and talin and cross-links actin. The cross-linking of actin filaments by fodrin or MARCKS protein is inhibited by Ca^{2+}-calmodulin, further emphasizing the importance of Ca^{2+}. The Ca^{2+}-calmodulin axis is responsible for activation of MLCK, a function also contributed to by the LMWG protein Rho [89]. Thus the continual dynamic of actin polymerization leading to cytoskeletal re-distribution is tightly regulated by the dynamic of calcium release from stores and its binding to downstream effectors such as calmodulin and calcineurin.

6. LOW MOLECULAR WEIGHT G-PROTEINS (LMWG) AND CYTOSKELETAL RE-ARRANGEMENT

By far some of the most substantive advances in the understanding of chemoattractant-induced signal transduction facilitating cytoskeletal re-organization and migration have been as a result of the analysis of the LMWG proteins of the Ras superfamily (Ras, Rho, Arf, Ran, Rab). For some years now, it has been appreciated that contact points of migrating cells (focal adhesions) contain numerous signal transduction molecules in close proximity to the intracellular domains of the adhesion molecule as well as proteins involved in the regulation of the actin cytoskeleton, including vinculin, α-actinin and talin [89,96]. The formation and activation of focal adhesion complexes will be discussed in detail below.

It has been known for a number of years that members of the Ras superfamily of LMWG proteins, specifically Ran, Rab and ARF, are vital to processes involved in vesicular transport, protein trafficking and endocytosis — all processes involving membrane re-shaping and cytoskeletal re-arrangements [97,98]. The LMWG proteins are tightly regulated by factors which stimulate the cyclical exchange of GDP for GTP. In the GDP-bound (inactive) form, the actions of GEFs stimulate the exchange of GTP for GDP, while GAPs and GDIs (guanine nucleotide dissociation inhibitors) revert or maintain the protein in the GDP-bound (inactive) state, respectively. Of potential relevance is that ARF and Rho bind to NP-X-X-Y motifs in the TM7 region of GPCRs. Many chemokine GPCRs contain this motif. Laudanna et al. [99] demonstrated chemokine-induced Rho activation in transfected cell systems, and both ARF and Rho were shown to be active in RANTES-mediated Jurkat T cell signaling [100]. Recently, a role for Rho in fractalkine-mediated microglial cell signaling was demonstrated; we have observed similar events in MIP-3β-induced dendritic cell activation ([58]; D. Maciejewski-Lenoir and K.B. Bacon, unpublished). Whether this cascade is involved solely in vesicular transport (for receptor internalization) or migration (actin re-organization) is presently undetermined. However, pre-treatment of dendritic cells or microglia with the specific Rho inhibitor, Botulinum. C3 exotoxin, effectively inhibited these cells from spreading in culture (D. Maciejewski-Lenoir and K.B. Bacon, unpublished; S. Sullivan and K.B. Bacon, unpublished).

6.1. Rho, Rac, Cdc42Hs and Cytoskeletal Organization

The Rho family of LMWG proteins consists of six distinct members in mammalian cells: Rho (A, B, C isoforms), Rho G, Rho E, Rac (1 and 2 isoforms), Cdc42 (Cdc42Hs, G25K isoforms) and T10 [89]. These LMWG proteins exist as either GDP-bound (inactive) or GTP-bound (active) forms, the GTP/GDP exchange being controlled by GEFs (including DBL, vav and OST [101–103]), GAPs and GDIs [62].

There has been extensively detailed research into this family of LMWG and their role in mediating organization of actin filaments, and the regulation of and by integrin–focal adhesion complexes. Early studies identified the role of Rho in formation of actin stress fibers. Interestingly, this was in response to selective stimulation by LPA and bombesin working through their 7-TM receptors in fibroblasts [104,105]. PDGF or EGF stimulation of Rac leads to lamellipodia formation and membrane ruffling [106], while Cdc42Hs stimulation by bradykinin (again through a 7-TM) results in filopodial formations [107]. Rho is also involved in focal adhesion generation and activation. These findings have led to the suggestion that this family of LMWG proteins is intricately involved in cell migration and adhesion in response to a variety of extracellular stimuli.

The downstream targets of these LMWG proteins may provide us with interesting clues as to their mechanism of action in stimulating cell migration and adhesion. In early 1996, targets of Rho were identified by yeast two-hybrid and affinity isolation systems. Two serine/threonine kinases, PKN (p128) and Rho-kinase (p164), as well as a related molecule, termed rhophilin, were shown to be directly activated by LPA-induced Rho stimulation [108–110]. It was later demonstrated that Rho-kinase phosphorylates the myosin binding subunit of myosin phosphatase (MBS), regulating (increasing) the phosphorylation of myosin light chain [111]. It is this regulation, thus contractility induced by MLC which controls the formation of stress fibers and re-distribution of adhesion molecules to form focal adhesions [89]. Two other targets, Rhotekin and citron, have also been isolated and while Rhotekin has no known function, citron is implicated in cytokinesis. Another Rho substrate is p140mDia, which has sequence similarity to the yeast Bni1, can bind profilin and may be involved in actin polymerization. Finally, GTP-bound Rho can associate and modulate current activity through the Kv1.2 ion channel, a channel shown to be involved in O_2 chemosensitivity in cells, as well as being widely expressed throughout the substantia nigra in the basal ganglia. Investigations of MIP-3β-induced dendritic cell migration have revealed PKN phosphorylation with rapid kinetics (D. Maciejewski-Lenoir and K.B. Bacon, unpublished). This PKN phosphorylation was dependent on Rho activation and PI3K (as evidenced using C3 exotoxin, wortmannin and LY294002); however, it is unclear as to the relevance of this activity.

The elongation of F-actin plays a central role in numerous processes regulating cell morphology [6,7]. Polymerization and cross-linking, followed by de-polymerization, regulates the elongation of filaments, hence directional placement of the cells. Increases in phosphoinositide (PI) turnover correlate with F-actin levels, and the concentration of free barbed (high affinity) ends at the leading edge of the migrating cell. In part, this is regulated by inhibition of gelsolin 'capping' of the barbed ends. Capping of barbed ends restricts the levels of F-actin available for cytoskeletal re-structuring. The actin-de-polymerizing factors (ADF), including cofilin [8], are responsible for the severing of filamentous actin, and their activity is restricted by phosphorylation, possibly via LIM-kinase [89]. Stimulation of phosphatase activity by chemoattractants activates ADFs. Thus a constant dynamic is established. Rho is also thought to activate PI-5K, regulating the level of PIP_2 intracellularly available for binding to profilin, gelsolin, α-actinin, talin, tensin and vinculin. Binding of PIP_2 unmasks the talin- and actin-binding domains of vinculin allowing integrin clustering and formations of focal adhesions [112].

Additional RhoA targets include adducin, moesin, PLD, PI3K, p38 MAPK, p42/p4 MAPK, JNK and $p125^{FAK}$ [113,114]. Phosphatidic acid (PA), the metabolite of PLD activity, stimulates PI-5K, hence PIP_2 formation, which is ultimately involved in the regulation of actin-binding proteins including gelsolin and profilin. In addition PA directly activates PLCγ1, elevating intracellular Ca^{2+}, and activating PKC, as well as having a direct activating role on PKC itself. PKC-dependent PLD activation has been demonstrated in neutrophils and T cells in response to IL-8, RANTES and other chemokines (including lymphotactin, MIP-1α, MIP-1β, and MCP-1; ([115–117] and K.B. Bacon, unpublished). Cross et al. [118] demonstrated that LPA stimulated the activity of PLD via a TK mechanism, and that the generation of PA correlated with actin stress-fiber formation. This is one of the first reports demonstrating a physiological role for PLD, and it is clearly implicated in cytoskeletal rearrangement. The interesting feature was that this PA-dependent activation was upstream of Rho, in contrast to numerous reports analyzing Rho-mediated activation of PLD in purified extracts. Colley et al. [119] cloned a novel PLD isozyme, PLD2, which exhibited distribution characteristics unique from that of PLD1. PLD2 is localized mainly at the plasma membrane, is constitutively active and independent of ARF, Rho and PKC, and promotes signal-induced cytoskeletal rearrangements. The regulation of this PLD enzyme in vivo, must therefore occur via inhibitory mechanisms. Further characterization is required to elucidate the mechanism of action of PLD2.

Although not clear as to whether it is a direct target, the activation of Rho leads to formation of PIP_2 and this in turn also regulates the binding of the ERM/CD44 complex [120]. ERM (ezrin, radixin, moesin) proteins cross-link actin filaments with the plasma membrane via integral membrane glycoproteins [121]. Several studies analyzing ERM functions with chemokine stimulation have been performed by the group of Sanchez-Madrid [122–124]. In SDF-1α- and RANTES-stimulated migrating lymphocytes, the ERM proteins moesin and ezrin localize to the uropod at the trailing edge, while radixin localizes in the uropod 'neck'. The polarization of the ERM proteins mirrored that of ICAM-1, ICAM-3, CD43 and CD44 adhesion receptors, while the chemokine receptors were polarized at the anterior, leading edge of the cell, suggesting a functional interaction of the adhesion receptors and ERM proteins in mediating migration.

Downstream targets of Rac and Cdc42Hs include the serine/threonine protein kinase $p65^{PAK}$, and Cdc42 has also been shown to bind a non-receptor PTK in rat, $p120^{ACK}$ [125]. Mammalian cells contain at least three isoforms of these enzymes: PAK1 (PAKα), highly expressed in brain, muscle and spleen; PAK2 (PAKγ), which is ubiquitously expressed and PAK3 (PAKβ), which is also highly expressed in brain. PAK activation appears to be regulated by the p21 GTPases, whose activity is in turn regulated by activation of specific guanosine exchange factors (GEFs) stimulated by cytokines and growth factors such as PDGF, EGF, LPA, bradykinin and angiotensin. Candidate exchange factors for the PAKs include members of the Cool/PIX/$p85^{SPR}$ family which associate with PAKs through their SH3 domains [126,127].

Since the p21 GTPases appear to be central to the induction of actin remodeling, it has been hypothesized that PAKs are in some way involved in the signaling cascades determining cell motility. PAKs can directly phosphorylate serine residues in the enzyme $p47^{phox}$, an enzyme vital to superoxide generation and cell migration in neutrophils [128]. PAKs localize to Rac-induced membrane ruffles and mutationally active PAKs can induce cytoskeletal rearrangement [129,130]. Actin remodeling is mediated by interactions of myosin and actin via regulation by myosin light chain kinase (MLCK). PAKs have been shown to directly phosphorylate myosin heavy chains in yeast [131] and MLCK [132], as well as negatively regulate MLCK activation [133], a mechanism which may lead to de-stabilization of focal contacts [130]. PAKs have also been shown to be regulated by phosphoinositide 3-kinase (PI3K) and to play a role in the regulation of MAPK pathways (JNK, p42/p44 and p38 MAPK) in response to cytokines and stress stimuli [134–136]. In turn, MAPK activation has been linked to MLCK phosphorylation and subsequent phosphorylation on MLC [136–139]. Phosphorylation of MLC on serine-19 stimulates the actin-activated ATPase activity of myosin-II and promotes myosin binding to filamentous actin, which drives contraction of the actin cytoskeleton [140,141] and may contribute to force generation for cell migration.

The product of the X-linked gene encoding WASP (Wiskott–Aldrich syndrome protein) was shown to bind Cdc42Hs in a PIP2-dependent manner, and related members of the WASP/Scar/WAVE family bind Rac-GTP. Following interaction of the LMWG with an SH3 domain containing adapter such as Nck or Grb21, the actin regulatory protein Arp2/3 can function to assemble actin filaments through polymerization of G-actin/profilin subunits, a function antagonized by capping proteins [142,143]. WASP, which is only expressed in hematopoietic cells (as opposed to N-WASP which is ubiquitously expressed and abundant in the brain), is the site of mutation in the Wiskott–Aldrich syndrome, their T cells showing altered morphological characteristics compared to normal T cells. Over-expression of WASP in Jurkat cells induces clusters of polymerized actin, co-localized with the gene product. Co-expression of dominant negative Cdc42 attenuates the clustering and polymerization, suggesting that this gene, or closely related homologs, may be vital regulators of the cytoskeleton [89,144]. N-WASP appears to play a critical role in filopodia formation. Preliminary studies undertaken in our laboratory have revealed that fractalkine stimulation of microglial cells induces a rapid serine phosphorylation of Nck in vitro (S. Sullivan and K.B. Bacon, unpublished results). In addition there is some evidence for PAK activation in this model although the significance to migration is still unclear. Nonetheless, there is clearly a precedent for the stimulation of this pathway in chemokine-stimulated cells.

$p65^{PAK}$ is thought to be involved in Rac-induced activation stimulated by G-protein-coupled receptors. The mechanism of activation may involve PAK binding to Rac in conjunction with an SH3-domain containing molecule, analogous to the situation in budding yeast, where Ste20p (a PAK homolog) is linked to F-actin through binding SH3-domain-containing Bem1p [145]. The adapter protein Nck contains 3 SH3 domains and the amino-terminal PXXP domain of PAK has been shown to directly interact with the second SH3 domain of Nck, possible functions being in targeting these kinases to the membrane or sites of focal adhesions. Su et al. [146] identified Nck Interacting Kinase (NIK) which directly binds the SH3 domains of Nck and through its C-terminal regulatory domain, can bind MEKK1 to regulate the SAPK/JNK pathway. The significance of the MEKK activation is however questionable, since deletion mutants of PAK, rendering it kinase deficient, fail to demonstrate impaired ability to induce cytoskeletal re-arrangements [145,147]. Por1 and p140Sra-1 (specific targets of Rac) may be involved in lamellipodia formation and membrane ruffling, while IQGAP1 (a target of both Rac and CdC42) appears to be involved in cross-linking actin filaments [89].

Of major interest is the fact that these LMWG can function in a cascade, activating each other, such that Cdc42Hs activates Rac which in turn activates Rho, through the actions of the exchange factor OST [102]. $p190^{RhoGAP}$ may also be responsible for Ras-induced de-activation of RhoA, Rac and CDC42-GTPs, while the GTPase $p120^{RasGAP}$ (acting on Ras-GTP) can be switched on by Rho, Rac and CDC42 [148]. Cdc42 has been shown to be directly activated by G-protein $\beta\gamma$ subunits and to stimulate Rac [90,114]. Therefore, 7-TM chemoattractant

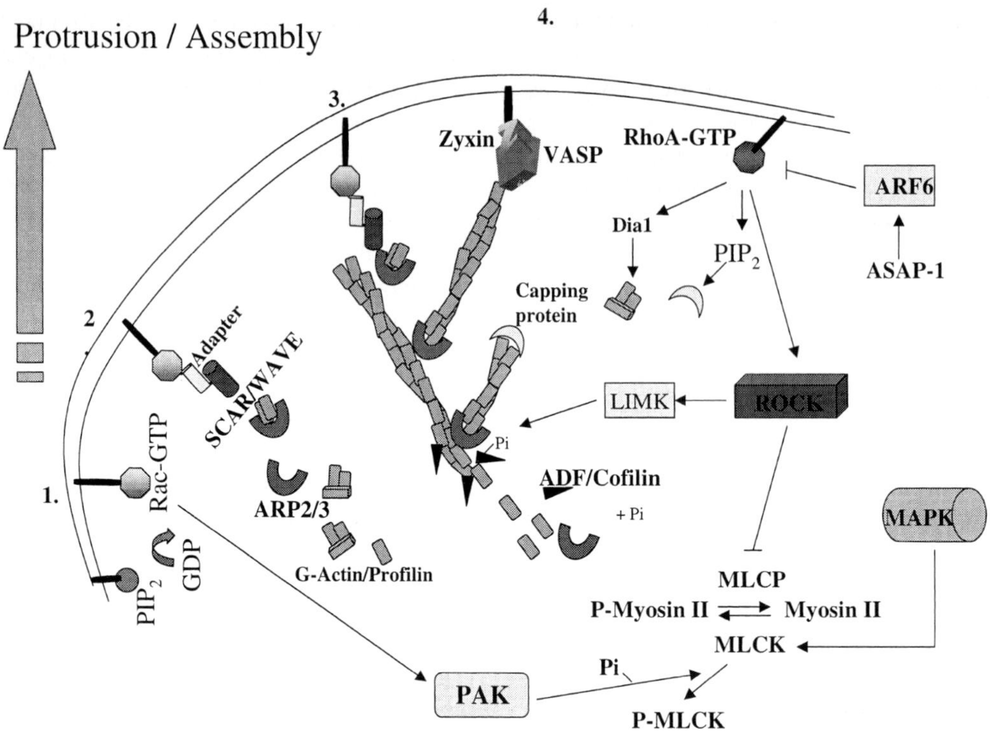

Figure 2. Adapters and signaling cascade implicated in the polymerization and severing of actin filaments, allowing directional migration or positioning of cell processes and cells. 1. Stimulation of PIP2 possibly via PI3K activation leads to Rac-GDP–GTP exchange. 2. Adapters (SH3-containing proteins) can bind Rac-GTP and trigger a sequence of assembly through Scar/WAVE, Arp2/3 and G-actin/profilin. 3. Branching of filaments on existing actin bundles in competition with capping proteins, and tethering to the membrane with Zyxin/VASP heterodimers. Rho-GTP acts as an antagonist through PIP2 and capping proteins. In balance ROCK activation stimulates LIMK which prevents the actions of the severing proteins ADF/cofilin by phosphorylation. Abbreviations: ADF, actin depolymerizing factor; ARF, ADP-ribosylating factor; ARP, actin regulatory protein; mDia, Drosophila homolog, diaphanous; LIMK, LIM kinase; MLCK, myosin light chain kinase; MLCP, myosin light chain phosphatase; PIP2, phosphoinositol 4,5-bisphosphate; ROCK, Rho-associated coiled-coil containing protein kinase; Scar/WAVE, WASP-like verprolin-homologous protein.

receptors which transduce Ras-related signals may activate the LMWG proteins and facilitate migration in a complex cascade of signaling intermediaries (Fig. 2).

In a different context, Rho is associated with LPA-induced neurite contraction through a myosin-II-dependent mechanism, while Rac and CdC42 are implicated in neurite extension and growth cone formation [89]. It is thus likely that LMWG activation in response to chemokines, if it occurs in CNS tissue, may regulate functions associated with the cytoskeleton unrelated to migration, such as axonal and dendrite positioning, neurotransmitter transport to synapses, and so forth. It is especially interesting to investigate the similarities and differences in signaling cascades for a single chemokine receptor expressed in functionally diverse tissue backgrounds such as a neuron and a leukocyte in order to determine the points of divergence within each cascade.

7. GPCR-INTEGRIN SIGNALING

Numerous experiments have demonstrated increased integrin avidity following treatments of cells with phorbol esters (for review, see [93]). Studies by Woldemarr Carr et al. [149] and Weber et al. [150] revealed the effect of chemokines on the avidity of β integrins in T lymphocytes and eosinophils, respectively. In these studies, MCP-1 induced transient increases in β_1 avidity in T cells, while RANTES and MCP-3 transiently regulated β_1 integrins in eosinophils, suggesting important cross-talk between the molecular entities (chemokines/7-TM interactions and adhesion receptors) in cell transmigration. It will now be interesting to catalog a comprehensive list of chemoattractant factors and their effects on adhesion machinery in order to characterize the preferential ligand–receptor–adhesion receptor groups involved in leukocyte migration, as well as integrin-mediated non-migratory events such as axonal positioning.

The mechanisms involved in this cross-talk between chemokine receptors and integrins have not been defined. It is widely appreciated that the coordination between cell adhesion and migration is regulated (at least in part) by the assembly of signaling molecules at the focal adhesion site. Focal adhesions are comprised of adhesion molecules, integral membrane glycoproteins extracellularly, and intracellularly of talin, vinculin, α-actinin, tensin and paxillin, among others [96]. Certain of these molecules (α-actinin and talin) have been shown to directly bind the cytoplasmic domain of β1 integrins [151]. One of the first intracellular signaling components to be recruited to the focal adhesion is p125 Focal Adhesion Kinase (FAK). FAK is capable of binding numerous adapter and signaling molecules (Grb2/Sos, src, PI3K, $p130^{cas}$, $p40^{crk}$ and ZAP-70 [55,152,153]), as well as important structural proteins (paxillin, talin [96]). It is the activity of individual members of this complex, acting either alone or in conjunction with each other, which triggers the formation and de-formation of the focal adhesion, allowing cell spreading and adhesion, as well as, potentially, cell activation and proliferation. Interestingly, however, targeted gene disruption of FAK does not prevent the formation of stress fiber and focal adhesion complexes [154] calling into question the exact role this molecule has to play in focal adhesion formation. However, there is a marked disruption of cell migratory potential, indicating that the signaling dynamic associated with this kinase must play a fairly important role in the migratory process. Importantly, it has been hypothesized that activation of cells following adhesion is due to the action of Grb2 binding to FAK, leading to activation of the Ras-GEF mSos, hence activation of MAPK [155]. SDF1α, RANTES and MIP-3β have all been shown to increase FAK phosphorylation [52,55,61] and SDF1α and RANTES increase RAFTK activity [52]. It is likely that chemokine-induced migration is the result of multiple interacting cascades involved in the re-organization of the cytoskeletal machinery and the activation/deactivation of adhesion molecules.

8. CHEMOKINE SIGNALING IN THE CENTRAL AND PERIPHERAL NERVOUS SYSTEMS

A number of studies have documented both chemokine and chemokine receptor localization in normal and pathological nervous tissue, including their relationship to HIV-1 infectivity and neuroAIDS [156,157]. While a number of signaling mechanisms have also been documented, the functional consequence remains unclear.

Chemokine receptors that have been demonstrated to be present in the CNS include CXCR2, CXCR3, CXCR4, CXCR-5 (BLR-1), DARC (Duffy antigen receptor for chemokines), CCR1, CCR2, CCR3, CCR4, CCR5, CCR7 (K.B. Bacon, unpublished), CCR8, and CX_3CR1 [158–164]. These genes are expressed on neurons and glia, as well as infiltrating leukocytes, however, with the exception of CXCR2, DARC, CCR3, CCR5, CXCR4, and CX_3CR1, the expression has been detected exclusively in vitro in culture, and/or from pathological tissue samples. CCR3, CCR5 and CXCR4 are expressed in subpopulations of large hippocampal neurons, neocortical pyramidal neurons and glial cells [165], while CCR3 is expressed in isolated microglia [159]. DARC expression has been localized exclusively to Purkinje cells in the cerebellum. CXCR2, the receptor shared by numerous ELR-CXC chemokines, including IL-8, Gro-α and NAP-2, demonstrates widespread expression in projection neurons in brain and spinal cord [158].

Since classic functions of chemokines relate to the migration-promoting effects on leukocytes and other cells, this area has been widely investigated [156,157] and will not be repeated here.

8.1. Chemokine Signaling in the PNS and CNS: New Paradigms

During neuropoiesis, there is a requirement for migration of progenitors. Our studies have demonstrated that RANTES signals through tyrosine kinase activation and calcium mobilization and stimulates migration and differentiation of dorsal root ganglion cells [166]. RANTES also mediated survival of human forebrain astrocytes, through a mechanism that requires interferon gamma [167]. While studies are required to determine the potential for chemokine-induced axon and dendrite positioning of sensory afferents from these cells, in the periphery, as well as any regulatory role at the level of the synapse, there is clearly a potential for functional consequence of chemokine signaling in this paradigm.

Studies with IL-8 and Gro-α reveal a direct effect of these chemokines on Purkinje cell activation and signal transduction [168]. Interestingly, immediate increases in cytosolic calcium and IP_3 levels point to direct receptor-mediated activation of the Purkinje cell, and receptor presence was confirmed by molecular analyses. More interesting however was the enhancement of neurotransmitter release and impairment of long-term depression of synaptic strength — the first real suggestion that chemokines can play a role in this vitally important neuronal function (see also Chapter 4.4).

8.2. Chemokines as Antagonists of Apoptosis and Cell Death

A critical role for SDF-1α and CXCR4 in neuronal development, and MAPK signaling in neuronal progenitors and adult neurons has been implicated from studies of knock-out mice, as aberrant migration of external granule layer cells into the internal granule layer was evident [169,23,24]. SDF-1α has been one of the most extensively studied chemokines, and with perhaps greater relevance to the CNS arena, its signaling properties in HIV-induced pathology has revealed interesting cross-talk with 'classical' neuronal signaling. CXCR4 is increased in cases of dementia and induces cell death following gp120 binding, through calmodulin kinase II, protein kinase A, p38 MAPK, caspase 3 activation and increased synaptic transmission [170–172]. In contrast RANTES and MIP-1β are protective in this case [171].

While cytopathic effects of SDF-1α have been observed with HIV pathology, there is a growing volume of evidence which suggests that chemokines as a class confer protection in CNS tissue. A role for GRO-α in oligodendrocyte precursor proliferation is also indicated from experimental data [173] and the inclusion of IL-8 into cultures of mixed hippocampal neurons resulted in an increase in cell survival [174]. Under precisely defined conditions, it was observed that murine MDC (the ligand for CCR4 and an as yet undescribed receptor) and soluble fractalkine could inhibit spontaneous glutamatergic excitatory post-synaptic currents, and high doses of fractalkine inhibited voltage-sensitive calcium currents and reduced neuronal cell death either upon removal of glial feeder layers or treatment with gp120 [163]. Coupled with a reduction in calcium oscillations in these neurons, it was suggested that chemokines can regulate synaptic transmission, as well as enhance survival.

One of the most commonly studied models is that of NMDA-induced cytotoxicity. SDF-1α can protect neurons if administered during or immediately after this agent, and IL-8, RANTES and MCP-1 are also protective [175]. RANTES can also protect neurons from the effects of β-amyloid (25–35 peptide) [175], and Groβ has been shown to protect cerebellar granule cells through indirect activation of AMPA-type glutamate channels [176]. In addition, fractalkine has been shown to prolong survival and protect against FAS-induced microglial death [84]. The mechanisms through which chemokines induce their protective effects appear to be diverse and dependent on the cell type. Groβ in this case, was shown to increase post-synaptic currents in an ERK-dependent manner [176]. There appeared to be no effect on glutamate release, induction of current or gating of the channel, but it was suggested that AMPA-type channels, in their interaction with PDZ-containing proteins may forge a link between the GPCR allowing the indirect activation. In cerebellar granule cells and Purkinje cells, SDF-1α increases intracellular calcium in the cell and induces spontaneous synaptic activity with a slow inward current [177]. This is thought to be mediated through extrasynaptic glutamate since the effect is blocked by inotropic glutamate channel blockers. There is also evidence that SDF-1 ligation of CXCR4 on astrocytes can mediate glutamate release and thus augment neurotoxicity [178]. One of the most prominent mechanisms of survival induction and protection can be linked to one of the most prominent chemokine-induced signaling cascades — the PI3K cascade. A major substrate of PI3K is PKB/Akt. PKB is widely reported as a survival inducer and investigations with fractalkine revealed a strong induction of this kinase in microglia [58,84]. In addition to stimulation of PKB

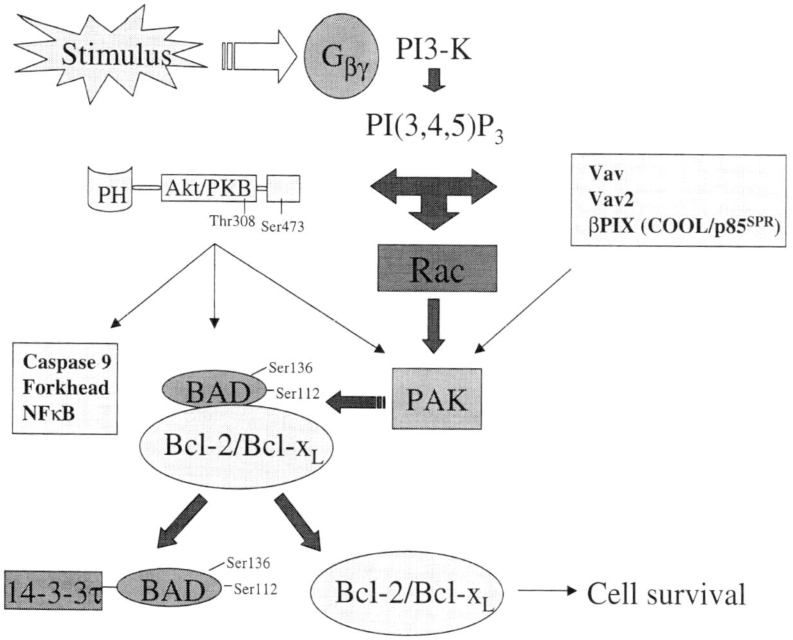

Figure 3. Stimulation of cell survival through PI3K-induced activation of protein kinase B (PKB). Active PKB or PAK converge on the death-inducer BAD, phosphorylating it and releasing its inhibition of survival inducing Bcl-2. BAD is stabilized in phosphorylated form by 14-3-3 proteins preventing its re-association with Bcl-2 family proteins.

activity, there was a concomitant increase in Bcl-2 and Bcl-$_{XL}$ with a decrease in FAS-mediated induction of BAD and BID activity. PAKs have also been shown to have direct capacity to phosphorylate BAD, releasing it from its complex with Bcl-2, thus providing a link between signaling for cytoskeletal re-organization and survival [179] (Fig. 3). We have recently demonstrated the ability of fractalkine to stimulate PAK activation (S. Sullivan and K.B. Bacon, submitted for publication), clearly providing further evidence for the cross-talk of these systems within chemokine signaling cascades. This was the first report demonstrating chemokine-induced induction of these death agonist and antagonist signaling pathways, and clearly expand the scope for further investigation into more relevant roles for chemokines in the CNS. While interesting, mice lacking the receptor for fractalkine (CX_3CR-1) do not display any overt CNS pathology, thus it is unclear as to the exact significance of this finding [180]. Whether or not such survival phenomena can significantly contribute to the regulation of neuronal turnover (and repair during nerve injury) remains to be thoroughly investigated, as does the precise mechanism of action; nonetheless, the nervous system does represent a unique paradigm with which to investigate chemokine signaling unrelated to the migratory response.

9. SUMMARY AND PERSPECTIVES

It is clear from the wealth of information amassed that the signal transduction components triggered by chemokines are complex, and as yet, not fully understood. What is clear, however, is that there are apparently numerous points of specificity control. These vary from, but are not mutually exclusive to, ligand affinity, receptor expression, and G-protein coupling. Downstream signaling can then involve numerous cascades, acting in isolation, or in concert (thus amplifying) with one another. In addition, it is apparent that there maybe a central role in the signal transduction cascade(s) for PI3K. While the numerous signaling cascades may appear a redundant scheme, one must remain aware that the migratory event, for example, is a constant dynamic, mechanistically dependent on numerous intracellular and extracellular components, each requiring precise spatio-temporal control. What is also likely is that there remain significant levels of specificity, convergence and divergence of signaling intermediaries, including those as yet unidentified or undefined, which may be involved in signal amplification in response to receptor ligation by this large and diverse family of proteins, the chemokines.

REFERENCES

1. Rollins B. Chemokines. Blood 1997;90:909–928.
2. Murphy PM, Baggiolini M, Charo IF et al. International Union of Pharmacology, XXII. Nomenclature for chemokine receptors. Pharm Rev 2000;52:145–176.
3. Bazan JF, Bacon KB, Hardiman G et al. A new class of membrane-bound chemokine with a CX3C motif. Nature 1997;385:640–644.
4. Pan Y, Lloyd C, Zhou H et al. Neurotactin, a membrane-anchored chemokine upregulated in brain inflammation. Nature 1997;387:611–617.
5. Bacon KB. Chemokines and chemokine receptors:Therapeutic targets in inflammation and infectious disease. Drug News Perspect 1997;10:133–143.
6. Lauffenburger DA, Horwitz AF. Cell migration: a physically integrated molecular process. Cell 1996;84:359–369.
7. Zigmond SH. Signal transduction and actin filament organization. Curr Opin Cell Biol 1996;8:66–73.
8. Carlier M-F, Laurent V, Santolini J et al. Actin depolymerizing factor (ADF/cofilin) enhances the rate of filament turnover: implication in actin-based motility. J Cell Biol 1997;136:1307–1323.
9. Watson S, Arkinstall S. The G-Protein Linked Receptor Facts Book. London: Academic Press, 1994.
10. Gierschik P, Sidiropoulos D, Jakobs KH et al. Two distinct Gi proteins mediate formyl peptide receptor signal transduction in human leukemia (HL-60) cells. J Biol Chem 1989;264:21470–21473.
11. Amatruda TT, Dragas-Graonic S, Holmes R et al. Signal transduction by the formyl peptide receptor. J Biol Chem 1995;270:28010–28013.
12. Schreiber RE, Prossnitz ER, Ye RD et al. Domains of the human N-formyl peptide receptor involved in G protein coupling. J Biol Chem 1994;269:326–331.
13. Damaj BB, Mcoll SR, Neote K et al. Identification of G-protein binding sites of the human interleukin-8 receptors by functional mapping of the intracellular loops. FASEB J 1996;10:1426–1434.
14. Arai H, Charo IF. Differential regulation of G-protein-mediated signaling by chemokine receptors. J. Biol Chem 1996;271:21814–21819.
15. Ling K, Wang P, Zhao J et al. Five-transmembrane domains appear sufficient for a G-protein-coupled receptor: funtional five-transmembrane domain chemokine receptors. Proc Natl Acad Sci USA 1999;96:7922–7927.
16. Cheng Z-J, Zhao J, Sun Y et al. b-Arrestin differentially regulates the chemokine receptor CXCR4-mediated signaling and receptor internalization, and this implicates multiple interaction sites between b-arrestin and CXCR4. J Biol Chem 2000;275:2479–2485.
17. Sondek J, Bohm A, Lambright DG et al. Crystal structure of a G_A protein βγ dimer at 2.1 A resolution. Nature 1996;379:369–374.
18. Lambright DG, Sondek J, Bohm A et al. The 2.0 A crystal structure of a heterotrimeric G-protein. Nature 1996;379:311–319.
19. Wilkie TM. GTPase-activating proteins for heterotrimeric G-proteins: regulators of G-protein signaling (RGS) and RGS-like proteins. Annu Rev Biochem 2000; 69:795–827.
20. Benzing T, Yaffe MB, Arnould T et al. 14-3-3 interacts with regulator of G protein signaling proteins and modulates their activity. J Biol Chem 2000;275:28167–28172.
21. Reif K, Cyster JG. RGS molecule expression in murine B lymphocytes and ability to down-regulate chemotaxis to lymphoid chemokines. J Immunol 2000;164:4720–4729.
22. Cho H, Kozasa T, Takekoshi K et al. RGS14, a GTPase-activating protein for Gialpha, attenuates Gialpha- and G13alpha-mediated signaling pathways. Mol Pharmacol 2000;58:569–576.
23. Ma Q, Jones D, Borghesani PR, Segal RA, Nagasawa T, Kishimoto T, Broson RT, Springer TA. Impaired B-lymphopoiesis, myelopoiesis, and derailed cerebellar neuron migration in CXCR4- and SDF-1-deficient mice. Proc Natl Acad Sci USA 1998;95:9448–9453.
24. Zou Y-R, Kottmann AH, Kuroda M et al. Function of the chemokine receptor CXCR4 in haematopoiesis and in cerebellar development. Nature 1998;393:595–599.
25. Lu Q, Sun EE, Klein RS, Flanagan JG. Ephrin-B reverse signaling is mediated by a novel PDZ-RGS protein and selectively inhibits G protein-coupled chemoattraction. Cell 2001;105:69–79.
26. Nishizuka Y. Protein kinase C and lipid signaling for sustained cellular responses. FASEB J 1995;9:484–496.
27. Wu D, LaRosa GL, Simon MI. G-protein-coupled signal transduction pathways for interleukin-8. Science 1993; 261:101–104.
28. Buhl AM, Eisfelder BJ, Worthen GS et al. Selective coupling of the human anaphylatoxin C5a receptor and 16 in human kidney 293 cells. FEBS Lett 1993;323:132–134.
29. Amatruda III TT, Gerard NP, Gerard C et al. Specific interactions of chemoattractant receptors with G-proteins. J Biol Chem 1993;268:10139–10144.
30. Kuang Y, Wu Y, Jiang H et al. Selective G protein coupling by C-C chemokine receptors. J Biol Chem 1996;271:3975–3978.
31. Hidi R, Timmermans S, Liu E et al. Phosphodiesterase and cyclic adenosine monophosphate-dependent inhibition of T-lymphocyte chemotaxis. Eur Respir J 2000;15: 342–349.
32. Mirzadegan T, Diehl F, Ebi B et al. Identification of the binding site for a novel class of CCR2b chemokine receptor antagonists: binding to a common chemokine receptor motif within the helical bundle. J Biol Chem 2000;275:25562–25571.
33. Poznansky MC, Olszak IT, Foxall R et al. Active movement of T cells away from a chemokine. Nat Med 2000; 6:543–548.

34. Jinquan T, Jing C, Jacobi HH et al. CXCR3 expression and activation of eosinophils: role of IFN-γ-inducible protein-10 and monokine induced by IFN-γ. J Immunol 2000;165:1548–1556.
35. Richardson RM, Ali H, Prigden BC et al. Multiple signaling pathways of human interleukin-8 receptor A. Independent regulation by phosphorylation. J Biol Chem 1998;273:10690–10695.
36. Mantel C, Aronica S, Luo Z et al. Macrophage inflammatory protein-1a enhances growth factor-stimulated phosphatidylcholine metabolism and increases cAMP levels in the human growth factor-dependent cell line M7e, events associated with growth suppression. J Immunol 1995;154:2342–2350.
37. Aramori I, Zhang J, Ferguson SSG et al. Molecular mechanism of desensitization of the chemokine receptor CCR5: receptor signaling and internalization are dissociable from its role as an HIV-1 co-receptor. EMBO J 1997;16:4606–4616.
38. Aragay AM, Mellado M, Frade JMR et al. Monocyte chemoattractant protein-1-induced CCR2b receptor desensitization mediated by the G protein-coupled receptor kinase 2. Proc Natl Acad Sci USA 1998;95:2985–2990.
39. Vila-Coro AJ, Mellado M, Martin de Ana A et al. Characterization of RANTES- and aminooxypentane-RANTES-triggered desensitization signals reveals differences in recruitment of the G protein-coupled receptor complex. J Immunol 1999;163:3037–3044.
40. Brelot A, Heveker N, Montes M et al. Identification of residues of CXCR4 critical for human immunodeficiency virus coreceptor and chemokine receptor activities. J Biol Chem 2000;275:23736–23744.
41. Ma Y-C, Huang J, Ali S et al. Src tyrosine kinase is a novel direct effector of G proteins. Cell 2000;102:635–646.
42. El-Shazly A, Yamaguchi N, Masuyama K et al. Novel association of the src family kinases, hck and c-fgr, with CCR3 receptor stimulation: a possible mechanism for eotaxin-induced human eosinophil chemotaxis. Biochem Biophys Res Commun 1999;264:163–170.
43. Barlic J, Andrews JD, Kelvin AA, Bosinger SE, De Vries ME, Xu L, Dobransky T, Feldman RD, Ferguson SS, Kelvin DJ. Regulation of tyrosine kinase activation and granule release through beta-arrestin by CXCRI. Nat Immunol 2000;1:227–233.
44. Ganju RK, Brubaker SA, Chernock RD et al. Beta-chemokine receptor CCR5 signals through SHP1, SHP2 and Syk. J Biol Chem 2000;275:17263–17268.
45. Ptasznik A, Traynor-Kaplan A, Bokoch GM. G protein-coupled chemoattractant receptors regulate Lyn tyrosine kinase. Shc adapter protein signaling complexes. J Biol Chem 1995;270:19969–19973.
46. Dikic I, Dikic I, Schlessinger J. Identification of a new Pyk2 isoform implicated in chemokine and antigen receptor signaling. J Biol Chem 1998;273:14301–14308.
47. Hall RA, Premont RT, Lefkowitz RJ. Heptahelical receptor signaling: beyond the G protein paradigm. J Cell Biol 1999;145:927–932.
48. Marrero MB, Scheiffer B, Paxton WG et al. Direct stimulation of Jak/STAT pathway by the angiotensin II AT1 receptor. Nature 1995;375:247–250.
49. Marrero MB, Venema VJ, Ju H et al. Regulation of angiotensin II-induced Jak2 tyrosine phosphorylation: roles of SHP-1 and SHP-2. Am J Physiol 1998;275:C1216–C1223.
50. Vila-Coro AJ, Rodriguez-Frade JM, Martin de Ana A et al. The chemokine SDF-1α triggers CXCR4 receptor dimerization and activates the JAK/STAT pathway. FASEB J 1999;13:1699–1710.
51. Wong M, Fish EN. RANTES and MIP-1α activate stats in T cells. J Biol Chem 1998;273:309–314.
52. Wang JF, Park IW, Groopman JE. Stromal cell-derived factor-1a stimulates tyrosine phosphorylation of multiple focal adhesion proteins and induces migration of hematopoietic progenitor cells: roles of phosphoinositide-3 kinase and protein kinase C. Blood 2000;95:2505–2513.
53. Hunter AJ, Ottoson N, Boerth N et al. A novel function for the SLAP-130/FYB adapter protein in β1 integrin signaling and T lymphocyte migration. J Immunol 2000; 164:1143–1147.
54. Gaudry M, Caon AC, Gilbert C et al. Evidence for the involvement of tyrosine kinases in the locomotory responses of human neutrophils. J Leuk Biol 1992;51: 103–108.
55. Bacon KB, Szabo MC, Yssel H et al. RANTES induces tyrosine kinase activity of stably complexed p125FAK and ZAP-70 in human T cells. J Exp Med 1996;184: 873–882.
56. Dubois PM, Palmer D, Webb ML et al. Early signal transduction by the receptor to the chemokine monocyte chemotactic protein-1 in a murine T cell hybrid. J Immunol 1996;156:1356–1361.
57. Dairaghi DJ, Franz-Bacon K, Callas E et al. Macrophage inflammatory protein-1β induces migration and activation of human thymocytes. Blood 1998;91:2905–2913.
58. Maciejewski-Lenoir D, Chen S, Feng L et al. Characterization of fractalkine in rat brain cells: migratory and activation signals for CX3CR-1-expressing microglia. J Immunol 1999;163:1628–1635.
59. Sullivan SK, McGrath DA, Grigoriadis D et al. Pharmacological and signaling analysis of human chemokine receptor CCR-7 stably expressed in HEK-293 cells: high-affinity binding of recombinant ligands MIP-3β and SLC stimulates multiple signaling cascades. Biochem Biophys Res Commun 1999;263:685–690.
60. Mellado M, Rodriguez-Frade JM, Manes S, Martinez AC. Chemokine signaling and functional responses: the role of receptor dimerization and TK pathway activation. Annu Rev Immunol 2001;19:397–421.
61. Seger R, Krebs EG. The MAPK signaling cascade. FASEB J 1995;9:726–735.
62. Denhardt DT. Signal-transducing protein phosphorylation cascades mediated by Ras/Rho proteins in the mammalian cell: the potential for multiplex signalling. Biochem J 1996;318:729–747.
63. Yip-Schneider MT, Miao W, Lin A et al. Regulation of the Raf-1 kinase domain by phosphorylation and 14-3-3 association. Biochem J 2000;351:151–159.
64. Thompson HL, Marshall CJ, Saklatvala J. Characterization of two different forms of mitogen-activated protein

kinase induced in polymor-phonuclear leukocytes following stimulation by N-formylmethionyl-leucyl-phenylalanine or granulocyte-macrophage colony-stimulating factor. J Biol Chem 1994;269:9486–9492.
65. Grinstein S, Butler JR, Furuya W et al. Chemotactic peptides induce phosphorylation and activation of MEK-1 in human neutrophils. J Biol Chem 1994;269:19313–19320.
66. Downey GP, Butler JR, Brumell J et al. Chemotactic peptide-induced activation of MEK-2, the predominant isoform in human neutrophils. J Biol Chem 1996;271: 21005–21011.
67. Knall C, Young S, Nick JA et al. Interleukin-8 regulation of the Ras/Raf/mitogen-activated protein kinase pathway in human neutrophils. J Biol Chem 1996;271: 2832–2838.
68. Knall C, Worthen GS, Johnson GL. Interleukin 8-stimulated phosphoinositol-3-kinase activity regulates the migration of human neutrophils independent of extracellular signal-regulated kinase and p38 mitogen-activated protein kinases. Proc Natl Acad Sci USA 1997;94:3052–3057.
69. Stoyanov B, Volinia S, Hanck T et al. Cloning and characterization of a G protein-activated human phosphoinositide-3 kinase. Science 1995;269:690–693.
70. Stephens LR, Equinoa A, Erdjument-Bromage H et al. The G$\beta\gamma$ sensitivity of a PI3K is dependent upon a tightly associated adaptor, p101. Cell 1997;89:105–114.
71. Sasaki T, Irie-sasaki J, Jones RG et al. Function of PI3Kγ in thymocyte development, T cell activation and meutrophil migration. Science 2000;287:1040–1046.
72. Li Z, Jiang H, Xie W et al. Roles of PLC-β2 and -β3 and PI3Kγ in chemoattractant-mediated signal transduction. Science 2000;287:1046–1049.
73. Hirsch E, Katanaev VL, Garlanda C et al. Central role for G-protein coupled phosphoinositide 3-kinase γ in inflammation. Science 2000;287:1049–1053.
74. Naccache PH, Levasseur S, Lachance G et al. Stimulation of human neutrophils by chemotactic factors is associated with the activation of phosphatidylinositol 3-kinase gamma. J Biol Chem 2000;275:23636–23641.
75. Sotsios Y, Whittaker GC, Westwick J et al. The CXC chemokine stromal cell-derived factor activates a Gi-coupled phosphoinositide 3-kinase in T lymphocytes. J Immunol 1999;163:5954–5963.
76. Vicente-Manzanares M, Rey M, Jones DR et al. Involvement of phosphatidylinositol 3-kinase in stromal cell-derived factor-1α-induced lymphocyte polarization and chemotaxis. J Immunol 1999;163:4001–4012.
77. Majka M, Ratajczak J, Kowalska MA et al. Binding of stromal cell-derived factor-1α (SDF-1α) to CXCR4 chemokine receptor in normal human megakaryoblasts but not in platelets induces phosphorylation of mitogen-activated protein kinase p42/44 (MAPK), ELK-1 transcription factor and serine threonine kinase AKT. Eur J Haematol 2000;64:164–172.
78. Tilton B, Ho L, Oberlin E et al. Signal transduction by CXC chemokine receptor 4. Stromal cell-derived factor 1 stimulates prolonged protein kinase B and extracellular signal-regulated kinase 2 activation in T lymphocytes. J Exp Med 2000;192:313–324.
79. Turner L, Ward SG, Westwick J. RANTES-activated human T lymphocytes. A role for phosphoinositide 3-kinase. J Immunol 1995;155:2437–2444.
80. Coffer PJ, Schweizer RC, Dubois GR et al. Analysis of signal transduction pathways in human eosinophils activated by chemoattractants and the T-helper 2-derived cytokines interleukin-4 and interleukin-5. Blood 1998; 91:2547–2557.
81. Al-Aoukaty A, Rolstad B, Maghazachi AA. Recruitment of pleckstrin and phosphoinositide 3-kinase γ into cell membranes, and their association with G-$\beta\gamma$ after activation of NK cells with chemokines. J Immunol 1999;162: 3249–3255.
82. Turner SJ, Domin J, Waterfield MD et al. The CC chemokine monocyte chemotactic peptide-1 activates both the class I p85/p110 phosphatidylinositol 3-kinase and the class II PI3KC2α. J Biol Chem 1998;273:25987–25995.
83. Marra F, Romanelli RG, Giannini C et al. Monocyte chemotactic protein-1 as a chemoattractant for human hepatic stellate cells. Hepatology 1999;29:140–148.
84. Boehme SA, Lio FM, Maciejewski-Lenoir D et al. The chemokine fractalkine inhibits Fas-mediated cell death of brain microglia. J Immunol 2000;165:397–403.
85. Crespo P, Xu N, Simonds WF et al. Ras-dependent activation of MAP kinase pathway mediated by G-protein $\beta\gamma$ subunits. Nature 1994;369:418–420.
86. Van Biesen T, Hawes BE, Luttrell DK et al. Receptor-tyrosine-kinase- and G$\beta\gamma$-mediated MAP kinase activation by a common signalling pathway. Nature 1995;376: 781–784.
87. Wan Y, Kurosaki T, Huang X-Y. Tyrosine kinases in activation of the MAP kinase cascade by G-protein-coupled receptors. Nature 1996;380:541–546.
88. Lopez-llasaca M, Crespo P, Pellici PG et al. Linkage of G protein-coupled receptors to the MAPK signaling pathway through PI 3-kinase γ. Science 1997;275:394–397.
89. Kaibuchi K, Kuroda S, Amano M. Regulation of the cytoskeleton and cell adhesion by the Rho family GTPases in mammalian cells. Annu Rev Biochem 1999;68:459–486.
90. Ma AD, Metjian A, Bagrodia S et al. Cytoskeletal reorganization by G protein-coupled receptors is dependent on phosphoinositide 3-kinase γ, a Rac guanosine exchange factor and Rac. Mol Cell Biol 1998;18:4744–4751.
91. Gilbert SH, Perry K, Fay FS. Mediation of chemoattractant-induced changes in $[Ca^{2+}]$i and cell shape, polarity, and locomotion by InsP3, DAG, and protein kinase C in newt eosinophils. J Cell Biol 1994;127:489–503.
92. Marks PW, Maxfield FR. Transient increases in cytosolic free calcium appear to be required for the migration of adherent neutrophils. J Cell Biol 1990;110:43–52.
93. Springer TA. Traffick signals for lymphocyte recirculation and leukocyte transmigration: the multistep paradigm. Cell 1994;76:301–314.
94. Clapham DE. Calcium signaling. Cell 1995;80:259–268.
95. Maxfield FR. Regulation of leukocyte locomotion by Ca^{2+}. Trends Cell Biol 1993;3:386–391.

96. Bishop AL, Hall A. Rho GTPases and their effector proteins. Biochem J 2000;348:241–255.
97. Frohman MA, Morris AJ. Phospholipid signaling: Rho is only ARF the story. Curr Biol 1996;6:945–947.
98. Pryer NK, Wuestehube LJ, Schekman R. Vesicle-mediated protein sorting. Annu Rev Biochem 1992;61: 471–516.
99. Laudanna C, Campbell JJ, Butcher EC. Role of Rho in chemoattractant-activated leukocyte adhesion through integrins. Science 1996;271:981–983.
100. Bacon KB, Schall TJ, Dairaghi DJ. RANTES activation of phospholipase D in Jurkat T cells: requirement of GTP-binding proteins ARF and RhoA. J Immunol 1998; 160:1894–1900.
101. Ridley AJ. Membrane ruffling and signal transduction. Bioessays 1994;16:321–327.
102. Hori Y, Beeler JF, Sakaguchi K et al. A novel oncogene, ost, encodes a guanine nucleotide exchange factor that potentially links Rho and rac signaling pathways. EMBO J 1994;13:4776–4786.
103. Han J, Das B, Wei W et al. Lck regulates the Vav activation of members of the Rho family of GTPases. Mol Cell Biol 1997;17:1346–1353.
104. Ridley AJ, Hall A. The small GTP-binding protein rho regulates the assembly of focal adhesions and actin stress fibres in response to growth factors. Cell 1992;70:389–399.
105. Ridley AJ, Paterson HF, Johnston CL et al. The small GTP-binding protein rac regulates growth factor-induced membrane ruffling. Cell 1992;70:401–410.
106. Nobes CD, Hall A. Rho, rac and Cdc42 GTPases regulate the assembly of multimolecular focal complexes associated with actin stress fibres, lamellipodia and filopodia. Cell 1995;81:1–20.
107. Kozma R, Ahmed S, Best A et al. The ras-related protein Cdc42Hs and bradykinin promote formation of peripheral actin microspikes and filopodia in Swiss 3T3 fibroblasts. Mol Cell Biol 1995;15:1942–1952.
108. Matsui T, Amano M, Yamamoto T et al. Rho-associated kinase, a novel serine/threonine kinase, as a putative target for the small GTP binding protein Rho. EMBO J 1996;15:2208–2216.
109. Watanabe G, Saito Y, Madaule P et al. Protein kinase N (PKN) and PKN-related protein rhophilin as targets of small GTPase Rho. Science 1996;271:645–648.
110. Amano M, Mukai H, Ono Y et al. Identification of a putative target for Rho as the serine–threonine kinase protein kinase N. Science 1996;271:648–650.
111. Kimura K, Ito M, Amano M et al. Regulation of myosin phosphatase by Rho and Rho-associated kinase (Rho-kinase). Science 1996;273:245–248.
112. Gilmore AP, Burridge K. Regulation of vinculin binding to talin and actin by phosphatidylinositol-4,5-bisphosphate. Nature 1996;381:531–535.
113. Frost JA, Xu S, Hutchinson MR et al. Actions of Rho family small G proteins and p21-activated protein kinases on mitogen-activated protein kinase family members. Mol Cell Biol 1996;17:3707–3713.
114. Craig SW, Johnson RP. Assembly of focal adhesions: progress, paradigms, and portents. Curr Opin Cell Biol 1996;8:74–85.
115. L'Hereux GP, Bourgoin S, Jean N et al. Diverging signal transduction pathways activated by interleukin-8 and related chemokines in human neutrophils: interleukin-8 but not NAP-2 or Gro-α, stimulates phospholipase D activity. Blood 1995;85:522–531.
116. Sozzani S, Agwu DE, Ellenburg MD et al. Activation of phospholipase D by interleukin-8 in human neutrophils. Blood 1994;84:3895–3901.
117. Bacon KB, Flores-Romo L, Life PF et al. Interleukin-8-induced signal transduction in T lymphocytes involves receptor-mediated activation of phospholipases C and D. J Immunol 1995;154:3654–3666.
118. Cross MJ, Roberts S, Ridley AJ et al. Stimulation of actin stress fibre formation mediated by activation of phospholipase D. Curr Biol 1996;6:588–597.
119. Colley WC, Sung T-C, Roll R et al. Phospholipase D2, a distinct phospholipase D isoform with novel regulatory properties that provokes cytoskeletal reorganization. Curr Biol 1997;7:191–201.
120. Hirao M, Sato N, Kondo T et al. Regulation mechanism of ERM (ezrin/radixin/moesin) protein/plasma membrane association: possible involvement of phosphatidylinositol turnover and Rho-dependent signaling pathway. J Cell Biol 1996;135:37–51.
121. Tsukita S, Yonemura S, Tsukita S. ERM (ezrin/radixin/moesin) family: from cytoskeleton to signal transduction. Curr Opin Cell Biol 1997;9:70–75.
122. Serrador JM, Alonso-Lebrero JL, del Pozo MA et al. Moesin interacts with the cytoplasmic region of intercellular adhesion molecule-3 and is redistributed to the uropod of T lymphocytes during cell polarization. J Cell Biol 1997;138:1409–1423.
123. Serrador JM, Nieto M, Alonso-Lebrero JL et al. CD43 interacts with moesin and ezrin and regulates its redistribution to the uropods of T lymphocytes at the cell–cell contacts. Blood 1998;91:4632–4644.
124. Del Pozo MA, Vicente-Manzanares M, Tejedor R et al. Rho GTPases control migration and polarization of adhesion molecules and cytoskeletal ERM components in T lymphocytes. Eur J Immunol 1999;29:3609–3620.
125. Daniels RH, Bokoch GM. p21-activated protein kinase: a crucial component of morphological signaling. Trends Biochem Sci 1999;24:350–355.
126. Crespo P, Schuebel KE, Ostrom AA et al. PhosphoTyr-dependent activation of Rac-1 GDP/GTP exchange by the *vav* protooncogene product. Nature 1997;385:169–171.
127. Han J, Luby-Phelps K, Das B et al. Role of substrates and products of PI 3-kinase in regulating activation of Rac-related guanosine triphosphatases by *vav*. Science 1998;279:558–560.
128. Ding J, Badwey JA. Stimulation of neutrophils with a chemoattractant activates several novel protein kinases that can catalyze the phosphorylation of peptides derived from $p47^{phox}$ and MARKS. J Biol Chem 1993;268: 17326–17333.
129. Nobes CD, Hall A. Rho, Rac, and Cdc42 GTPases regulate the assembly of multimolecular focal adhesions and actin stress fibres, lamellipodia and filopodia. Cell 1995; 81:53–62.

130. Manser E, Huang H-Y, Loo T-H et al. Expression of constitutively-active a-PAK reveals effects of the kinase on actin and focal complexes. Mol Cell Biol 1997;17: 1129–1143.
131. Wu C, Lee S-F, Furmaniak-Kazmierczak E et al. Activation of myosin-I by members of the Ste20p protein kinase family. J Biol Chem 1996;271:31787–31790.
132. Goeckeler ZM, Masaracchia RA, Zeng Q et al. Phosphorylation of myosin light chain kinase by p21-activated kinase PAK2. J Biol Chem 2000;275:18366–18374.
133. Sanders LC, Matsumura F, Bokoch GM et al. Inhibition of myosin light chain kinase by p21-activated kinase. Science 1999;283:2083–2085.
134. Ding J, Vlahos CJ, Liu R et al. Antagonists of phosphatidylinositol 3-kinase block activation of several novel protein kinases in neutrophils. J Biol Chem 1995; 270:11684–11691.
135. Zhang S, Han J, Sells MA et al. Rho family GTPases regulate p38 MAP kinase through the downstream mediator Pak1. J Biol Chem 1995;270:23934–23936.
136. Frost JA, Steen H, Shapiro P et al. Cross-cascade activation of ERKs and ternary complex factors by Rho family proteins. EMBO J 1997;16:6426–6438.
137. Klemke RL, Cai S, Giannini AL et al. Regulation of cell motility by mitogen-activated protein kinase. J Cell Biol 1997;137:481–492.
138. King AJ, Sun H, Diaz B et al. The protein kinase Pak3 positively regulates Raf-1 activity through phosphorylation of serine 338. Nature 1998;396:180–183.
139. Nguyen DHD, Catling AD, Webb DJ et al. Myosin light chain kinase functions downstream of Ras/ERK to promote migration of urokinase-type plasminogen activator-stimulated cells in an integrin-dependent manner. J Cell Biol 1999;146:149–164.
140. Wilson AK, Gorgas W, Claypool WD et al. An increase or a decrease in myosin II phosphorylation inhibits macrophage motility. J Cell Biol 1991;114:277–283.
141. Matsumura F, Ono S, Yoshihiko Y et al. Specific localization of serine 19 phosphorylate myosin II during cell locomotion and mitosis of cultured cells. J Cell Biol 1998;140:119–129.
142. Moreau V, Frischknecht F, Reckmann I et al. A complex of N-WASP and WIP integrates signaling cascades that lead to actin polymerization. Nat Cell Biol 2000;2:441–448.
143. Pantaloni D, Boujemaa R, Didry D et al. The Arp2/3 complex branches filament barbed ends: functional antagonism with capping proteins. Nat Cell Biol 2000;2: 385–391.
144. Symons M, Derry JMJ, Karlak B et al. Wiskott–Aldrich syndrome protein, a novel effector for the GTPase cdc42Hs, is implicated inactin polymerization. Cell 1996; 84:723–734.
145. Sells MA, Knaus UG, Bagrodia S et al. Human p21-activated kinase (Pak1) regulates actin organization in mammalian cells. Curr Biol 1997;7:202–210.
146. Su Y-C, Han J, Xu S et al. NIK is a new Ste20-related kinase that binds NCK and MEKK1 and activates the SAPK/JNK cascade via a conserved regulatory domain. EMBO J 1997;16:1279–1290.
147. Lamarche N, Tapon N, Stowers L et al. Rac and Cdc42 induce actin polymerization and G1 cell cycle progression independently of p65PAK and the JNK/SAPK MAP kinase cascade. Cell 1996;87:519–529.
148. Boguski MS, McCormick F. Proteins regulating Ras and its relatives. Nature 1993;366:643–654.
149. Carr MW, Alon R, Springer TA. The C-C chemokine MCP-1 differentially modulates the avidity of β1 and β2 integrins on T lymphocytes. Immunity 1996;4:179–187.
150. Weber C, Kitayama J, Springer TA. Differential regulation of 1 and 2 integrin avidity by chemoattractants in eosinophils. Proc Natl Acad Sci USA 1996;93:10939–10944.
151. Yamada KM, Geiger B. Molecular interactions in cell adhesion complexes. Curr Opin Cell Biol 1997;9:76–85.
152. Kanner SB, Aruffo A, Chan PY. Lymphocyte antigen receptor activation of a focal adhesion kinase-related tyrosine kinase substrate. Proc Natl Acad Sci USA 1994; 91:10484–10487.
153. Rabinowich H, Manciuba M, Herberman RB et al. b1 Integrin-mediated activation of focal adhesion kinase and its association with fyn and ZAP-70 in human NK cells. J Immunol 1996;157:3860–3868.
154. Llic D, Furuta Y, Kanazawa S et al. Reduced cell motility and enhanced focal adhesion contact formation in cells from FAK-deficient mice. Nature 1995;377:539–544.
155. Schlaepfer DD, Hanks S, Hunter T et al. Integrin-mediated signal transduction linked to Ras pathway by GRB2 binding to focal adhesion kinase. Nature 1994; 372:786–791.
156. Ransohoff RM, Tani, M. Do chemokines mediate leukocyte recruitment in post-traumatic CNS inflammation? Trends Neurosci 1998;21:154–159.
157. Glabinski AR, Ransohoff RM. Chemokine and chemokine receptors in CNS pathology. J Neurovirol 1999;5: 3–12.
158. Horuk R, Martin AW, Wang Z-X et al. Expression of chemokine receptors by subsets of neurons in the central nervous system. J Immunol 1997;158:2882–2890.
159. He J, Chen Y, Farzan M et al. CCR3 and CCR5 are coreceptors for HIV-1 infection of microglia. Nature 1997; 385:645–649.
160. Jiang Y, Salafranca MN, Adhikari S et al. Chemokine receptor expression in cultured glia and rat experimental allergic encephalomyelitis. J Neuroimmunol 1998;86:1–12.
161. Jinno A, Shimizu N, Soda Y et al. Identification of the chemokine receptor TER1/CCR8 expressed in brain-derived cells and T cells as a new coreceptor for HIV-1 infection. Biochem Biophys Res Commun 1998;243: 497–502.
162. Harrison JK, Barber CM, Lynch KR. cDNA cloning of a G-protein coupled receptor expressed in rat spinal cord and brain related to chemokine receptors. Neurosci Lett 1994;169:85–89.
163. Meucci O, Fatatis A, Simen AA et al. Chemokines regulate hippocampal neuronal signaling and gp120 neurotoxicity. Proc Natl Acad Sci USA 1998;95:14500–14505.
164. Sorenson TL, Tani M, Jensen J et al. Expression of specific chemokines and chemokine receptors in the cen-

tral nervous system of multiple sclerosis patients. J Clin Invest 1999;103:807–815.

165. Westmoreland SV, Rottman JB, Williams KC et al. Chemokine receptor expression on resident and inflammatory cells in the brain of macaques with simian immunodeficiency virus encephalitis. Am J Pathol 1998;152:659–665.

166. Bolin LM, Murray R, Lukacs NW et al. Primary sensory neurons migrate in response to the chemokine RANTES. J Neuroimmunol 1998;81:49–57.

167. Bakhiet M, Tjernlun A, Mousa A, Gad A, Stromblad S, Kuziel WA, Seiger A, Anderson J. RANTES promotes growth and survival of human first-trimester forebrain astrocytes. Nac Cell Biol 2001;3:150–157.

168. Giovannelli A, Limatola C, Ragozzino D et al. CXC chemokines interleukin-8 (IL-8) and growth-related gene product alpha (Gro-α) modulate Purkinje neuron activity in mouse cerebellum. J Neuroimmunol 1998;92:122–132.

169. Nagasawa T, Hirota S, Tachibana K et al. Defects of B-cell lymphopoiesis and bone-marrow myelopoiesis in mice lacking the CXC chemokine PBSF/SDF-1. Nature 1996;382:635–638.

170. Zheng J, Ghorpade A, Niemann D et al. Lymphotropic virions affect chemokine receptor-mediated neural signaling and apoptosis: implications for human immunodeficiency virus type 1-associated dementia. J Virol 1999; 73:8256–8267.

171. Kaul M, Lipton SA. Chemokines and activated macrophages in HIV gp120-induced neuronal apoptosis. Proc Natl Acad Sci USA 1999;96:8212–8216.

172. Zheng J, Thylin MR, Ghorpade A et al. Intracellular CXCR4 signaling, neuronal apoptosis and neuropathogenic mechanisms of HIV-1-associated dementia. J Neuroimmunol 1999;98:185–200.

173. Robinson S, Tani M, Strieter RM et al. The chemokine growth-regulated oncogene-a promotes spinal cord oligodendrocyte precursor proliferation. J Neurosci 1998;18: 10457–10463.

174. Araujo DM, Cotman CW. Trophic effect of interleukin-4, -7 and -8 on hippocampal neuronal cultures: potential involvement of glial-derived factors. Brain Res 1993; 600:49–55.

175. Bruno V, Copani A, Besong G et al. Neuroprotective activity of chemokines against N-methyl-D-aspartate or beta-amyloid-induced toxicity in culture. Eur J Pharmacol 2000;399:117–121.

176. Limatola C, Ciotti MT, Mercanti D et al. The chemokine growth-related gene product β protects rat cerebellar granule cells from apoptotic cell death through α-amino-3-hydroxy-5-methyl-4-isoxazolepropionate receptors. Proc Natl Acad Sci USA 2000;97:6197–6201.

177. Limatola C, Giovannelli A, Maggi L et al. SDF-1α-mediated modulation of synaptic transmission in rat cerebellum. Eur J Neurosci 2000;12:2497–2504.

178. Bezzi P, Domercq M, Brambilla L, Galli R, Schold D, De Clerq E, Vescovi A, Bagetta G, Kollias G, Meldolesi J et al. CXCR4-activated astrocyte glutamate release via TNF alpha: amplification by microglia triggers neurotoxicity. Nat Neurosci 2001;4:702–710.

179. Schurmann A, Mooney AF, Sandres LC et al. p21-activated kinase 1 phosphorylates the death agonist BAD and protects cells from apoptosis. Mol Cell Biol 2000; 20:453–461.

180. Jung S, Aliberti J, Graemmel P et al. Analysis of fractalkine receptor CX_3CR1 function by targeted deletion and green fluorescent protein reporter gene insertion. Mol Cell Biol 2000;20:4106–4114.

Universes in Delicate Balance: Chemokines and the Nervous System
R.M. Ransohoff, K. Suzuki, A.E.I. Proudfoot, W.F. Hickey and J.K. Harrison, editors

Chapter 2.5

Development and Function of the Hemato-lymphopoietic System

Gretchen N. Schwartz [1] and Joshua M. Farber [2]

[1] Hematopoietic Associates, Wheaton, MD, USA; [2] Laboratory of Clinical Investigation, National Institutes of Allergy and Infectious Diseases, Building 10, Room 11N228, 10 Center Drive, MSC 1888, Bethesda, MD, 20892-1888, USA

1. CHEMOKINES AND CHEMOKINE RECEPTORS IN REGULATION OF MARROW PROGENITOR CELL RESPONSES

Blood cell production is provided by a heterogeneous population of committed and primitive progenitors normally found in the marrow or fetal liver [1,2]. Long-term repopulating cells, or pluripotential progenitors, proliferate to produce cells that reconstitute the lymphoid and hematopoietic system in lethally irradiated animals. In irradiated mice, some progenitors in the transplanted cells home to the spleen and proliferate to form colonies of hematopoietic cells. Each colony is derived from a single cell called a colony-forming unit-spleen (CFU-S). In response to stimulatory factors, some progenitors form colonies of differentiated cells in a semisolid matrix such as methylcellulose or agar. There are colonies derived from progenitors with characteristics of CFU-S called CFU-A [3,4]. There are also CFU for single lineage progenitors (e.g., CFU-Erythroid, CFU-Granulocyte, CFU-Macrophage), multilineage progenitors (e.g., CFU-GM and CFU-GEM), and more primitive progenitors that form clusters of colonies called burst-forming units (e.g., BFU-E). CFU and BFU are generated in liquid cultures from more primitive progenitors called long-term initiating cells (LTC-IC). Human progenitors are found among cells expressing the CD34 cell surface antigen and murine progenitors are enriched in cells expressing Sca-1, c-kit, and Thy-1. Progenitor cell assays along with FACS analyses for cell surface antigens identify pathways regulating blood cell production [1,2].

The role of chemokines in hematopoiesis was first recognized with the discovery of a factor produced by macrophages that inhibited in vivo CFU-S and in vitro CFU-A colony formation [4]. This factor and the previously described stem cell inhibitory factor isolated from bone marrow cultures were identified as MIP-1α (CCL3). In other studies, MIP-1α inhibited proliferation of CFU-Meg [5] and BFU-E [6,7], prevented stem cell growth factor (SCF)-dependent increases in GM-CSF-stimulated colony numbers [8,9], and abrogated insulin-like growth factor II-dependent increases in IL-3-stimulated CD34$^+$ cell proliferation [10]. In other studies, MIP-1α promoted an increase in colony numbers from murine and human marrow [11–14]. In vitro studies demonstrated that MIP-1α alone did not stimulate CFU colony formation, but promoted approximately a two-fold increase in the number of colonies formed in response to GM-CSF [11,14], IL-3 [11], or M-CSF [14]. Further work revealed that the effect of MIP-1α on progenitor cell proliferation was dependent upon which factors were used to stimulate proliferation and on the stages of progenitor cell differentiation or maturation (Table 1) [5–8,10–15].

The mechanisms for the effects of chemokine on progenitor cell proliferation are not well defined. CCR1 and CCR5, receptors for MIP-1α, are expressed on subsets of CD34$^+$ cells [6,16,17]. For example, CCR5 was detectable on 0.5–3% of adult and fetal marrow Lin$^-$CD34$^+$ cells [16] and CD34$^+$ and CD34$^+$AC133$^+$ (a subset of cells enriched for primitive progenitors) marrow cells (unpublished data, GNS). CCR1 expression on CD34$^+$ cells ranged from minimal [17] (unpublished data, GNS) to some

Table 1. Effects of MIP-1α (CCL3) on progenitor cell proliferation

	Source of cells and stimulatory factors	Inhibitory effects on proliferation	Stimulatory or no effects on proliferation
1	Murine marrow cells stimulated with GM-CSF, IL-3, or CSF-1 [11]	↓ Proliferation of Lin^-Thy^+ cells	↑ Proliferation of mature Lin^-Thy^- cells
2	Murine marrow stimulated with GM-CSF ± SCF [8,13]	↓ SCF-dependent increase in colony numbers	↑ Colony numbers in absence of SCF
3	Subsets of human marrow $CD34^+$ cells stimulated with IL-3 + IL-6 + IL-1 + SCF [12]	↓ Number of CFU and cells generated from primitive $CD34^{2+}CD38^-$ cells	↑ Number of CFU and cells generated from mature $CD34^+CD38^+$ cells
4	Human marrow $CD34^+$ cells stimulated with IL-3 + GM-CSF + aplastic anemia sera [a] [5]	↓ CFU-Meg colony numbers	No effect on CFU-GM colony numbers
5	Human marrow $CD34^+$ cells stimulated with erythropoietin and SCF [6]	↓ BFU-E colony numbers	No effect on CFU-E colony numbers
6	Human marrow CFU long-term cultures [b] [7]	↓ % primitive BFU-E and CFU-GM in S-phase	No effect on mature CFU-GM
7	Liquid cultures of $CD34^+$ marrow cells stimulated with IL-3 ± IGF-II [c] [10]	↓ IGF-II-dependent increase in CFU-GM colony numbers	No effect on CFU-GM in absence of IGF-II
8	Human marrow or cord blood transplanted into NOD/SCID mice [d] [15]	↓ % CFU-G and primitive BFU-E in S-phase	No effect on LTC-IC or primitive CFU-GM

[a] MIP-1β, PF4, NAP-2, IL-8, C10 had similar inhibitory effects on CFU-Meg colony numbers and no effect on CFU-GM colony numbers.

[b] MCP-1 had similar inhibitory effects. IL-8, PF4, IP-10, RANTES, MCP-2, or MCP-3 were also evaluated, but did not have any effect on proliferation.

[c] Mig and IP-10 also abrogated IGF-II-dependent increased numbers of CFU-GM colonies and Mig inhibited IGF-II-dependent increased numbers of LTC-IC in liquid cultures of $CD34^+$ human marrow cells.

[d] In vivo administration of MCP-1 inhibited the cycling of both primitive BFU-E and CFU-M, but had no effect on LTC-IC.

expression on most cells [18]. Several studies investigated the possible role of CCR1 in the effects of MIP-1α on progenitor cell proliferation [3,6,19]. In vitro results demonstrated that almost all BFU-E and CFU-E from human marrow were in fractions of $CCR1^+$ selected cells [6]. In the presence of MIP-1α, BFU-E colony numbers from $CD34^+$ cells were reduced by 30%, but only by 15% when antibody to CCR1 was also added to the cultures. Even though CFU-E were also $CCR1^+$, MIP-1α had no effect on CFU-E colony numbers, so that expression of CCR1 did not necessarily result in inhibition by MIP-1α.

In murine studies, cell and CFU numbers and percentages of CFU in S-phase were similar in marrow from wild-type and CCR1-deficient mice [19]. Similar to earlier results [13,14], in cultures of marrow from wild-type mice the number of colonies formed in response to GM-CSF or M-CSF was two fold higher in the presence of MIP-1α. In contrast, no enhancing effect of MIP-1α was observed on CFU colony numbers in marrow from CCR1-deficient mice. The results indicated that the stimulatory effects of MIP-1α on CFU colony formation were dependent on signaling through CCR1. Loss of CCR1 expression did not block the inhibitory effects of MIP-1α on SCF-dependent increases in colony numbers from marrow stimulated with either GM-CSF or M-CSF. Other experiments in which MIP-1α was modified to prevent its binding to CCR1 demonstrated that signaling through CCR1 was not required for the inhibitory effects of MIP-1α on CFU-A colony formation, but was necessary for monocyte chemotactic responses to MIP-1α [3]. Taken together, the results indicate that signaling through CCR1 promoted the stimulatory effects of MIP-1α, but was not essential for the inhibitory effects of MIP-1α on progenitor cell proliferation.

In addition to MIP-1α, many other chemokines have been reported to inhibit hematopoietic progenitor cell proliferation [5,7–10,20,21]. Not only were the particular responses sensitive to stimulatory factors and differences among subsets of progenitors as observed for MIP-1α, but also specific chemokines had different effects on progenitor cell proliferation. Similar to MIP-1α, MCP-1 (CCL2) prevented SCF-dependent increases in CFU colony numbers

[8,22] and when added to marrow long-term cultures both chemokines inhibited proliferation of primitive BFU-E and CFU-GM (Table 1) [7]. On the other hand, MCP-3 (CCL7), which, along with MCP-1, is a ligand for CCR2, had no inhibitory effects on CFU proliferation in either culture system [7,8,22]. MCP-1 did not inhibit the proliferation of CFU in marrow from CCR2-deficient mice [22]. The inhibitory effects of MCP-4 (CCL13) and MCP-5 (CCL12), also ligands for CCR2, on CFU proliferation were also absent. In vivo, CFU numbers in the marrow of wild-type and CCR2-deficient mice were similar. However, in marrow from CCR2-deficient mice, there was a significant elevation in the percentage of CFU in S-phase and in the percentage of progenitors undergoing apoptosis [22]. It is not clear whether these effects on proliferation and/or survival of hematopoietic progenitors are a consequence of the loss of CCR2 signaling on the progenitors or the loss of signaling on accessory cells of the marrow microenvironment that support hematopoiesis.

CXCR4, the receptor for SDF-1 (CXCL12), is the most widely expressed chemokine receptor on $CD34^+$ cells. Most cord blood $CD34^+$ cells [16,23] and approximately 50% of marrow $CD34^+$ cells express CXCR4 [17,23,24]. In the marrow, CXCR4 has been observed on 5–20% of $CD34^+$ cells enriched for primitive progenitors [16] and on a subset of $CD34^+$ cells enriched for B- and T-cell progenitors [23]. SDF-1 has either no or only slightly inhibitory effects on CFU proliferation [6,8]. It is, however, a potent chemoattractant for human and murine primitive and committed hematopoietic and lymphopoietic progenitor cells [17,25,26]. Thus, CXCR4 and SDF-1 may be important in regulating movement and localization of progenitor cells during developmental stages in the marrow.

Marrow B-cell maturation is accompanied by changes in CXCR4 expression. In adult marrow, CXCR4 was expressed on approximately 30% pro-B-cells ($CD34^+CD19^+$) and on 3-fold more pre-B-cells ($CD34^-CD19^+$), but only on 30% of immature B-cells ($CD19^+IgD^-$). A high percentage of $CXCR4^+$ cells was again observed on stages of B-cells ($CD19^+IgD^+$) that emigrate from the marrow [26]. CXCR4 is important in regulating B-lymphopoiesis and myelopoiesis during fetal development. CXCR4- and SDF-1-deficient mice die in utero or shortly after birth and have marrow that is almost completely devoid of B- and myeloid cells [27–29]. Pro B-cells were also reduced in the blood of CXCR4-deficient mice, but myeloid precursors were 5-fold higher than in the controls. Abnormalities in B-cell and granulocyte reconstitution were also observed in irradiated mice transplanted with fetal liver cells from CXCR4-deficient mice [27]. In the marrow of recipients of fetal liver cells from CXCR4-deficient mice, the number of B-cells and granulocytes, particularly the later immature stages, usually restricted to marrow, was lower than in recipients cells from wild-type mice. Associated with the reduced number in the marrow, the immature stages of B-cells and granulocytes were prevalent in the blood. Results indicate that CXCR4 expression is important for retention of immature B-cells and granulocytes in the marrow until maturation is complete before their release into the periphery. CXCR4 may also have a role in localization of hematopoietic progenitors to specific marrow microenvironments that regulate proliferation and differentiation.

2. CHEMOKINES AND CHEMOKINE RECEPTORS IN REGULATION OF T-CELL DEVELOPMENT IN THE THYMUS

Generation of immature thymocytes from T-cell progenitors and the successive stages in development to mature T-cells occur in the thymus [1]. T-cell development depends on the recruitment of progenitors from the blood into the thymus and then on the migration of the progeny from the thymic cortex to the medulla. In each region, distinct microenvironments regulate the processes of positive and negative selection and development of the progeny into mature T-cells that exit the thymus into the periphery [1,30]. There is indirect, but convincing evidence that signals through chemokine and/or other chemoattractant receptors are necessary for migration within and emigration from the thymus. For example, thymocyte distribution in mice treated with pertussis toxin (PT) to block signaling through chemokine and other receptors linked to G_i proteins was abnormal [31]. Total thymic cellularity was reduced in the treated mice and instead of a relative restriction of $CD4^+$ and $CD8^+$ single positive (SP) thymocytes to the thymic medulla, as observed in control mice, many SP cells were also found in the thymic cortex. In

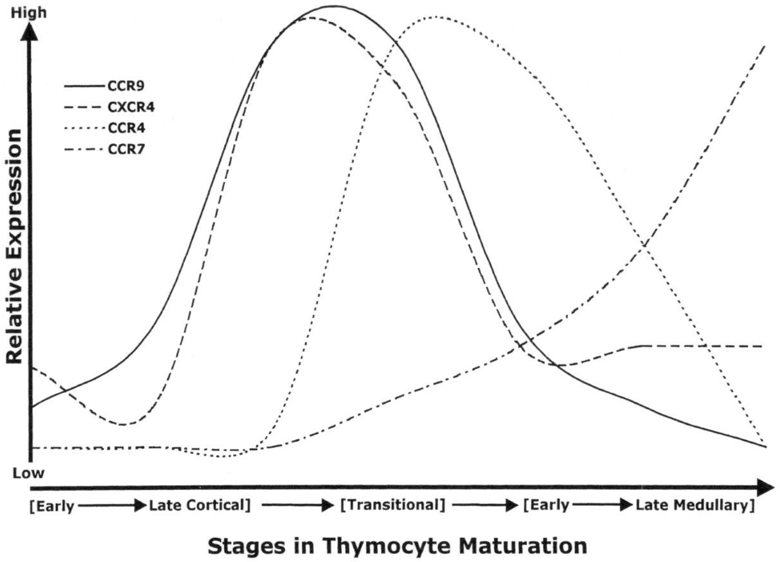

Figure 1. Patterns of chemokine receptor expression during thymocyte maturation. Early cortical stage thymocytes are $CD4^{-}CD8^{-}$ (DN) with the T-cell receptor (TCR) in the germ-line configuration (TCR^{-}). Early cortical events involve immigration of progenitors (mouse Sca-1^{+}, human $CD34^{+}$), commitment to the T-cell lineage, TCRβ gene rearrangements and selection. Late cortical events include rearrangement of TCRα genes, formation of TCRαβ complex ($TCR\alpha\beta^{lo}$), and expression of CD4 plus CD8 (DP). Transitional stage cells are DP and express CD69. Early medullary thymocytes are $CD4^{+}CD8^{-}$ or $CD4^{-}CD8^{+}$ (SP) and $CD69^{+}CD62L^{lo/-}$. As maturation progresses, CD69 expression is lost and late medullary stage thymocytes that emigrate from the thymus are $CD69^{-}CD62L^{+}$SP. Relative levels were estimated from receptor analyses and chemotactic responses.

earlier studies, in transgenic mice where PT was expressed under the *lck* promotor, T-cells accumulated in the thymus and did not enter the periphery, demonstrating a requirement for G_i protein signals for emigration from the thymus [32].

Different patterns of chemokine receptor expression are associated with distinct stages of thymocyte maturation (Fig. 1). CXCR4 and CCR9 are predominantly expressed on stages of thymocytes found in the cortex, also the region where high levels of SDF-1 (CXCL12) and TECK (CCL25) are expressed. CCR4 is predominantly expressed by $CD4^{+}CD8^{+}$ double positive (DP) transitional thymocytes that migrate from the thymic cortex into the medulla and on early stage $CD4^{+}$ or $CD8^{+}$ single positive (SP) cells in the medulla. For MDC (CCL22), a ligand for CCR4, expression is localized to areas near the corticomedullary junction. For CCR7, a receptor for SLC (CCL21) and ELC (CCL19) expression increases as SP thymocytes mature to final stages that leave the thymus. Changes in the pattern of chemokine receptor expression may have a role in recruitment of progenitors into the thymus and in regulating migration as thymocytes mature.

It is still not clear what role, if any, CXCR4 has in regulating the migration of T-cell progenitors from the blood into the thymus [23,33]. In adult human marrow, $CD34^{+}$ progenitors restricted to T- and/or B-cell production expressed CXCR4 [23]. Results in other studies demonstrated that levels of calcium flux in response to SDF-1 were several times higher in fractions of human pediatric early stage thymocytes ($TCR^{-}CD4^{-}CD8^{-}$, double negative, DN) expressing CD34 than in fractions of CD34 negative early stage thymocytes [33]. On the other hand, even though SDF-1- and CXCR4-deficient mice as well as recipients of cells from CXCR4-deficient mice all had profound abnormalities in the B-cell and granulocyte maturation pathways, thymopoiesis and T-cell function were normal [27–29]. Those results demonstrated that CXCR4 expression was not essential for the homing of fetal T-cell progenitors to the thymus or their subsequent maturation. It is possible that in the absence of responses to SDF-1, T-cell progenitor recruitment is regulated by other known or unknown thymus expressed chemokines.

As observed during marrow B-cell maturation, stages of thymocyte maturation differ in expression of CXCR4 or in responses to SDF-1 [24,31,33–35]. In human and murine studies, more than 70% of cortical thymocytes expressed CXCR4 or migrated in response to SDF-1 [24,31,35], but there were differences among subsets of thymocytes [24,31, 33,35]. In murine studies, most $CD44^+CD25^+$DN, $CD44^-CD25^+$DN, and DP thymocytes expressed CXCR4, but no expression was detectable on the more immature $CD44^+CD25^{-/lo}$DN cells [31]. SP human and murine thymocytes found in the medulla had the lowest levels of CXCR4 expression or responses to SDF-1 (Fig. 1) [24,31,33,35,36]. The potential role of CXCR4 in adult thymopoiesis was investigated by reconstituting irradiated mice with marrow cells transduced with an SDF-1-intrakine gene, which encodes SDF-1 fused to an endoplasmic reticulum retention sequence and sequesters CXCR4, preventing cell surface expression of the receptor [37]. Thymus recovery in those mice was abnormal and few donor thymocytes were produced. In contrast to observations in fetal development, CXCR4 expression may have a significant role in regulating thymocyte migration within the adult thymus.

Similar to observations with CXCR4 and SDF-1, stages of thymocyte maturation differ in expression of CCR9 or in chemotactic responses to TECK. Results in murine studies demonstrated that diffusible factors produced by day 14 fetal thymic lobes depleted of lymphoid cells induced the migration of early thymocyte precursors from intact fetal thymic lobes. Even though some thymocyte precursors migrated in response to TECK, treatment of the alymphoid lobes with neutralizing antibody to TECK did not prevent precursor migration from intact lobes [38]. Those results suggest that CCR9 is not necessary for recruitment of T-cell progenitors into the thymus, although it is possible that CCR9 has a yet unidentified thymus-expressed ligand. In later stages of thymocyte maturation, few DN thymocytes and more than 80% of later stage DP thymocytes migrated in response to TECK [39]. There was a similar pattern for expression of CCR9 during thymocyte maturation in Rag-$2^{-/-}$ mice [40]. In Rag-$2^{-/-}$ mice, thymocyte development is blocked at $CD44^-CD25^+$DN stage, and anti-CD3ε treatment promotes a synchronous maturation to DP cells. Although CCR9 mRNA levels were low in DN thymocytes before treatment, expression increased many fold as cells matured to the DP stage [40], when thymocytes migrate to inner cortical microenvironments that regulate the next stages of development. As thymocytes matured from DN to SP cells, there was a loss of CCR9 expression [39]. In other experiments, early stages of SP thymocytes migrated in response to TECK, but this response was lost on SP thymocytes expressing the adhesion molecule, CD62L, the stage where response of SP cells to SDF-1 was also reduced [36]. It is possible that loss of CCR9 expression may allow for the migration of thymocytes from the cortex into the medulla.

MDC, SLC, and ELC are expressed at high levels in thymic medullary regions, but expression is absent in the thymic cortex [41,42]. All three chemokines are chemotactic for SP and late stage DP thymocytes [36,40,42]. Subsets of SP thymocytes differ in responses to these chemokines. For example, both early stage $CD69^+CD62L^{-/low}$ and late stage $CD69^-CD62L^{high}$ SP thymocytes migrated in response to the CCR7 ligands, SLC and ELC. However, only the early stage SP thymocytes migrated in response to MDC, a ligand for CCR4. Recent studies demonstrated that MDC and ELC production was localized to different areas in the human thymic medulla [42]. MDC-producing cells were concentrated near the outer wall of Hassal's corpuscles and ELC-producing cells were found near corticomedullary junctions and dispersed throughout the inner medullary regions. MDC-responsive, but not ELC-responsive cells, expressed CD30, a molecule that may be important in negative selection of DP thymocytes that do not recognize self-MHC [42,43]. Results from the above studies indicate that there is a progressive loss of CCR4 expression as thymocytes mature from late DP to mature SP stages and a progressive increase in CCR7 expression during maturation from DP to SP stages that emigrate from the thymus into the periphery.

Taken together, studies of chemokines that their receptors in thymus suggest that patterns of chemokine receptor expression have a role in regulating the proper sequence essential for thymocyte maturation. Additional functional studies will be necessary to delineate the roles of individual chemokine/receptor groups in T-cell development.

3. CHEMOKINES AND CHEMOKINE RECEPTORS IN REGULATION OF T- AND B-CELL DIFFERENTIATION IN SECONDARY LYMPHOID TISSUES

Antigen-dependent stages of T- and B-lymphocyte activation and differentiation occur in secondary lymphoid tissues that include the spleen, lymph nodes, and mucosa-associated lymphoid tissues (e.g., tonsils and Peyer's patches) [1]. Development of antigen-dependent immune responses is dependent on the recruitment of naive (i.e., antigen inexperienced) T-cells, naive B-cells, and antigen-presenting dendritic cells into the lymphoid tissues. Events within lymphoid tissues are determined by specific interactions among these three types of cells. Recruitment of lymphocytes and dendritic cells into T- and B-cell regions of lymphoid tissues depends on the receptors CCR7 and CXCR5.

Most human ($CD45RA^+CD62L^{hi}$) and murine ($CD44^{lo}CD62L^{hi}$) naive $CD4^+$ T-cells express CCR7 and migrate in response to SLC and ELC [44–47]. High endothelial venules (HEV), through which cells from the blood traffic into lymph nodes, express high levels of SLC, which has been shown to induce adhesion of naive T-cells to ICAM-1-coated glass [47]. Critical roles for CCR7 and its ligands in recruitment of cells into the T-cell zones of spleen and lymph nodes have been revealed by studies of naturally arising mutant *plt* (for paucity of lymph node T-cells) mice and CCR7 knockout mouse. The *plt/plt* mice do not express either ELC or the lymphoid tissue form of SLC [48,49], resulting in a deficiency of CCR7 ligands in lymphoid tissues. Lymphocytes from wild-type and *plt/plt* mice were recruited into lymphoid tissues of wild-type mice, but not into lymphoid tissues of *plt/plt* mice [50–52]. Results from in vitro studies demonstrated that HEV from wild-type mice supported the adhesion of T-cells from both wild-type and *plt/plt* mice. HEV from *plt/plt* mice, however, did not support T-cell adhesion [50]. In CCR7-deficient mice, reduced numbers of T-cells in the T-cell zones within lymphoid tissues were associated with elevated numbers of T-cells in the blood and the splenic red pulp; and clusters of T-cells were observed in B-cell follicles and in the germinal centers [53]. Abnormal T-cell distribution was also observed in wild-type recipients of lymphocytes from CCR7-deficient mice [53]. Similarly, few dendritic cells were observed in T-cell zones of lymphoid tissues from *plt/plt* mice and from CCR7-deficient mice [52,53]. There was also a lack of dendritic cell recruitment into the lymphoid tissues of CCR7-deficient mice even after treatments shown to increase their numbers in lymph nodes of wild-type mice [53].

Recruitment of naive B-cells into secondary lymphoid tissues of *plt/plt* mice and CCR7-deficient mice was not as compromised as it was for T-cells and dendritic cells [51,53]. This is not unexpected since studies have shown that only 10–20% of naive B-cells migrate in response to SLC or ELC [46,47] and that B-cell migration from the blood through HEV into Peyer's patches is not dependent on responses to SLC or ELC [50]. However, some impairment in B-cell migration within the tissues of CCR7-deficient mice was observed. Adoptively transferred donor B-cells from CCR7-deficient mice were almost absent in the T-cell regions of wild-type mice, but were predominately localized to B-cell follicles [53]. One speculation was that $CCR7^-$ B-cells were not retained normally in the T-cell zones on their way to the follicles. Finally, analysis of human effector/memory T-cells from peripheral blood suggested that in contrast to memory T-cells that mediate effector functions in peripheral tissue, memory T-cells important for providing T-cell help within lymphoid tissues, so called central memory cells (T_{CM}) express CCR7 [45]. Together these results suggest that CCR7 and its ligands are best thought of as a receptor/ligand group that functions in a compartment-specific fashion to recruit or retain cells within the T-cell compartments of secondary lymphoid tissues.

CXCR5 and its ligand BLC (CXCL13) function analogously for the B-cell compartment. Most naive (IgM^+IgD^+) recirculating and follicular B-cells express CXCR5 [54,55] and after recruitment into the lymphoid tissues the cells migrate through the T-cell zone to B-cell follicles where high levels of BLC are expressed [56–59]. CXCR5-deficient mice lacked inguinal lymph nodes and had reduced numbers of Peyer's patches [60,61]. Studies in these mice demonstrated that although the recruitment of B-cells into lymphoid tissues was not dependent upon signaling through CXCR5, CXCR5 was required for homing to and formation of B-cell follicles. B-cells in the spleen surrounded the T-cell zones rather than being localized to one pole as in wild-type mice, and PNA^+ cells, a marker for germinal center cells were scattered in the red pulp instead of being localized to germinal center light zones.

Experiments analyzing the expression and function of CXCR5 on T-cells also suggested that the receptor is involved in recruitment of T-cells to the B-cell compartment. In adoptive transfer studies in mice, $CCR7^-$ memory/effector T-cells transduced to express CXCR5 homed to B-cell follicles [44]. Also in murine studies, CXCR5 mRNA expression increased after stimulation of naive T-cells with the B-cell activation antigen OX40L [62] and following activation in vivo [56]. In humans, CXCR5 is expressed on a subset of $CD4^+$ memory cells and is found on cells expressing CD154 (CD40 ligand) [55,57], a protein on the surface of activated T-cells that is critical for B-cell responses to T-cell-dependent antigen. $CXCR5^+$ T-cells in secondary lymphoid tissues have been detected primarily in T-cell zones near the B-cell follicles and in the mantle and germinal center light zones [55–58]. $CXCR5^+$ T-cells are thus localized to BLC-expressing sites where T-cell support for cognate activation of B-cells and maturation of antibody responses occur. Properties of $CXCR5^+$ T-cells in secondary lymphoid tissues indicate their role in supporting those functions. $CXCR5^+$ T-cells cocultured with B-cells support 5- to 10-fold increases in IgA and IgG antibody production and $CXCR5^+CCR7^-CD62L^{lo/-}$ T-cells supported B-cell affinity maturation [57,58].

The profound effects of CXCR5 and BLC on B- and T-cell trafficking in lymphoid tissues is reflected, as noted above, in a requirement for CXCR5 function for normal lymph node architecture. This requirement was also revealed in studies of TNF-, TNFR1-, lymphotoxin α(LTα)- or LTβ-deficient mice. Although B-cells from TNF- or TNFR1-deficient mice, which have defective B-cell follicle and germinal center formation, expressed CXCR5, BLC mRNA levels in spleens from the mutant mice were only 40% of levels in wild-type mice [59]. BLC expression was almost absent in spleens from LTα- or LTβ-deficient mice, which show a more severe disruption of secondary lymphoid tissue structure, including a lack of follicular dendritic cells. Further analyses demonstrated that BLC expression was partially regulated by membrane LTα/β on B-cells. Not only has CXCR5/BLC been found to be necessary for establishing basic lymph node architecture, but BLC has been found to be sufficient for establishing basic lymph node structure. In transgenic mice generated to express BLC in pancreatic islet β-cells $CXCR5^+$ B- and T-cells as well as some $CXCR5^{lo/-}$ T-cells and $CD11c^+$ dendritic cells infiltrated the BLC-producing islets [63]. As more cells migrated into the islets, distinct B-cell and T-cell regions were observed and HEV-like structures and stromal networks developed with SLC being expressed in the T-cell regions. Thus the ectopic expression of a single chemokine was able to produce organized lymphocyte compartments.

Production and differentiation of hemato-lymphopoietic cells and the organization of these cells into a functional immune system depend critically on the anatomy of bone marrow, thymus, and secondary lymphoid tissues. The chemokine system is being recognized as an essential element in recruiting and retaining cells within the anatomic compartments where the cellular interactions occur that are necessary for immune cell development and for mounting an effective immune response. Also, the positioning of cells in response to chemokines can serve as the event that initiates lymphoid organ formation. Although in our discussion we have focused on the roles of chemokines in cell migration, it is also possible that chemokines serve directly to enhance cell–cell adhesion and activation important for immune cell differentiation and/or function. Increasing attention is now being paid to how multiple components cooperate at the level of the immune system and the organism as a whole to provide for host defense. As indicated by the data reviewed above, the chemokine system serves not only to orchestrate the effector responses during inflammation, but also to support immune cell production and organize the structures on which the immune response depends.

REFERENCES

1. Bondurant MC, Koury MJ, Dessypris EN, Telen MJ, Kaufman RE, Lee GR, et al. Part III. The normal hematologic system. In: Lee GR, Foerster J, Lukens J, Paraskevas F, Greer JP, Rodgers GM, eds. Wintrobe's Clinical Hematology. Baltimore, MD: Williams and Wilkins, 1999:145–614.
2. Charbord P. Hemopoietic stem cells: analysis of some parameters critical for engraftment. Stem Cells 1994;12:

545–562.

3. Graham GJ, Wilkinson PC, Nibbs RJ, Lowe S, Kolset SO, Parker A et al. Uncoupling of stem cell inhibition from monocyte chemoattraction in MIP-1alpha by mutagenesis of the proteoglycan binding site. EMBO J 1996;15:6506–6515.
4. Graham GJ, Wright EG, Hewick R, Wolpe SD, Wilkie NM, Donaldson D et al. Identification and characterization of an inhibitor of haemopoietic stem cell proliferation. Nature 1990;344:442–444.
5. Gewirtz AM, Zhang J, Ratajczak J, Ratajczak M, Park KS, Li C et al. Chemokine regulation of human megakaryocytopoiesis. Blood 1995;86:2559–2567.
6. Su S, Mukaida N, Wang J, Zhang Y, Takami A, Nakao S et al. Inhibition of immature erythroid progenitor cell proliferation by macrophage inflammatory protein-1alpha by interacting mainly with a C–C chemokine receptor, CCR1. Blood 1997;90:605–611.
7. Cashman JD, Eaves CJ, Sarris AH, Eaves AC. MCP–1, not MIP-1alpha, is the endogenous chemokine that cooperates with TGF-beta to inhibit the cycling of primitive normal but not leukemic (CML) progenitors in long-term human marrow cultures. Blood 1998;92:2338–2344.
8. Broxmeyer HE, Kim CH. Regulation of hematopoiesis in a sea of chemokine family members with a plethora of redundant activities. Exp Hematol 1999;27:1113–1123.
9. Broxmeyer HE, Sherry B, Cooper S, Lu L, Maze R, Beckmann MP et al. Comparative analysis of the human macrophage inflammatory protein family of cytokines (chemokines) on proliferation of human myeloid progenitor cells. Interacting effects involving suppression, synergistic suppression, and blocking of suppression. J Immunol 1993;150:3448–3458.
10. Schwartz GN, Liao F, Gress RE, Farber JM. Suppressive effects of recombinant human monokine induced by IFN-gamma (rHuMig) chemokine on the number of committed and primitive hemopoietic progenitors in liquid cultures of $CD34^+$ human bone marrow cells. J Immunol 1997;159: 895–904.
11. Keller JR, Bartelmez SH, Sitnicka E, Ruscetti FW, Ortiz M, Gooya JM et al. Distinct and overlapping direct effects of macrophage inflammatory protein-1 alpha and transforming growth factor beta on hematopoietic progenitor/stem cell growth. Blood 1994;84:2175–2181.
12. Van Ranst PC, Snoeck HW, Lardon F, Lenjou M, Nijs G, Weekx SF et al. TGF-beta and MIP-1 alpha exert their main inhibitory activity on very primitive $CD34^{2+}$. Exp Hematol 1996;24:1509–1515.
13. Broxmeyer HE, Sherry B, Lu L, Cooper S, Oh KO, Tekamp-Olson P et al. Enhancing and suppressing effects of recombinant murine macrophage inflammatory proteins on colony formation in vitro by bone marrow myeloid progenitor cells. Blood 1990;76:1110–1116.
14. Broxmeyer HE, Sherry B, Lu L, Cooper S, Carow C, Wolpe SD et al. Myelopoietic enhancing effects of murine macrophage inflammatory proteins 1 and 2 on colony formation in vitro by murine and human bone marrow granulocyte/macrophage progenitor cells. J Exp Med 1989;170:1583–1594.
15. Cashman JD, Clark-Lewis I, Eaves AC, Eaves CJ. Differentiation stage-specific regulation of primitive human hematopoietic progenitor cycling by exogenous and endogenous inhibitors in an in vivo model. Blood 1999;94: 3722–3729.
16. Rosu-Myles M, Khandaker M, Wu DM, Keeney M, Foley SR, Howson-Jan K et al. Characterization of chemokine receptors expressed in primitive blood cells during human hematopoietic ontogeny. Stem Cells 2000;18:374–381.
17. Majka M, Ratajczak J, Lee B, Honczarenko M, Douglas R, Kowalska MA et al. The role of HIV-related chemokine receptors and chemokines in human erythropoiesis in vitro. Stem Cells 2000;18:128–138.
18. Su SB, Mukaida N, Wang J, Nomura H, Matsushima K. Preparation of specific polyclonal antibodies to a C–C chemokine receptor, CCR1, and determination of CCR1 expression on various types of leukocytes. J Leukocyte Biol 1996;60:658–666.
19. Broxmeyer HE, Cooper S, Hangoc G, Gao JL, Murphy PM. Dominant myelopoietic effector functions mediated by chemokine receptor CCR1. J Exp Med 1999;189: 1987–1992.
20. Patel VP, Kreider BL, Li Y, Li H, Leung K, Salcedo T et al. Molecular and functional characterization of two novel human C–C chemokines as inhibitors of two distinct classes of myeloid progenitors. J Exp Med 1997;185: 1163–1172.
21. Durig J, Testa NG, Heyworth CM. Distinct biological effects of macrophage inflammatory protein-1alpha and stroma-derived factor-1alpha on $CD34^+$ hemopoietic cells. Stem Cells 1999;17:62–71.
22. Reid S, Ritchie A, Boring L, Gosling J, Cooper S, Hangoc G et al. Enhanced myeloid progenitor cell cycling and apoptosis in mice lacking the chemokine receptor, CCR2. Blood 1999;93:1524–1533.
23. Ishii T, Nishihara M, Ma F, Ebihara Y, Tsuji K, Asano S et al. Expression of stromal cell-derived factor-1/pre-B cell growth- stimulating factor receptor, CXC chemokine receptor 4, on $CD34^+$ human bone marrow cells is a phenotypic alteration for committed lymphoid progenitors. J Immunol 1999;163:3612–3620.
24. Aiuti A, Tavian M, Cipponi A, Ficara F, Zappone E, Hoxie J et al. Expression of CXCR4, the receptor for stromal cell-derived factor-1 on fetal and adult human lympho-hematopoietic progenitors. Eur J Immunol 1999; 29:1823–1831.
25. Jo DY, Rafii S, Hamada T, Moore MA. Chemotaxis of primitive hematopoietic cells in response to stromal cell-derived factor-1. J Clin Invest 2000;105:101–111.
26. Honczarenko M, Douglas RS, Mathias C, Lee B, Ratajczak MZ, Silberstein LE. SDF-1 responsiveness does not correlate with CXCR4 expression levels of developing human bone marrow B cells. Blood 1999;94:2990–2998.
27. Ma Q, Jones D, Springer TA. The chemokine receptor CXCR4 is required for the retention of B lineage and granulocytic precursors within the bone marrow microenvironment. Immunity 1999;10:463–471.
28. Ma Q, Jones D, Borghesani PR, Segal RA, Nagasawa T, Kishimoto T et al. Impaired B-lymphopoiesis, myelopoiesis, and derailed cerebellar neuron migration in CXCR4- and SDF-1-deficient mice. Proc Natl Acad Sci USA 1998;95:9448–9453.

29. Zou YR, Kottmann AH, Kuroda M, Taniuchi I, Littman DR. Function of the chemokine receptor CXCR4 in haematopoiesis and in cerebellar development [see comments]. Nature 1998;393:595–599.
30. Anderson G, Moore NC, Owen JJ, Jenkinson EJ. Cellular interactions in thymocyte development. Annu Rev Immunol 1996;14:73–99.
31. Suzuki G, Sawa H, Kobayashi Y, Nakata Y, Nakagawa K, Uzawa A et al. Pertussis toxin-sensitive signal controls the trafficking of thymocytes across the corticomedullary junction in the thymus. J Immunol 1999;162:5981–5985.
32. Chaffin KE, Perlmutter RM. A pertussis toxin-sensitive process controls thymocyte emigration. Eur J Immunol 1991;21:2565–2573.
33. Zaitseva MB, Lee S, Rabin RL, Tiffany HL, Farber JM, Peden KW et al. CXCR4 and CCR5 on human thymocytes: biological function and role in HIV-1 infection. J Immunol 1998;161:3103–3113.
34. Youn BS, Kim CH, Smith FO, Broxmeyer HE. TECK, an efficacious chemoattractant for human thymocytes, uses GPR-9- 6/CCR9 as a specific receptor. Blood 1999;94: 2533–2536.
35. Kitchen SG, Zack JA. CXCR4 expression during lymphopoiesis: implications for human immunodeficiency virus type 1 infection of the thymus. J Virol 1997;71: 6928–6934.
36. Campbell JJ, Pan J, Butcher EC. Cutting edge: developmental switches in chemokine responses during T cell maturation. J Immunol 1999;163:2353–2357.
37. Onai N, Zhang Y, Yoneyama H, Kitamura T, Ishikawa S, Matsushima K. Impairment of lymphopoiesis and myelopoiesis in mice reconstituted with bone marrow-hematopoietic progenitor cells expressing SDF-1-intrakine. Blood 2000;96:2074–2080.
38. Wilkinson B, Owen JJ, Jenkinson EJ. Factors regulating stem cell recruitment to the fetal thymus. J Immunol 1999; 162:3873–3881.
39. Wurbel MA, Philippe JM, Nguyen C, Victorero G, Freeman T, Wooding P et al. The chemokine TECK is expressed by thymic and intestinal epithelial cells and attracts double- and single-positive thymocytes expressing the TECK receptor CCR9. Eur J Immunol 2000;30:262–271.
40. Norment AM, Bogatzki LY, Gantner BN, Bevan MJ. Murine CCR9, a chemokine receptor for thymus-expressed chemokine that is up-regulated following pre-TCR signaling. J Immunol 2000;164:639–648.
41. Annunziato F, Romagnani P, Cosmi L, Beltrame C, Steiner BH, Lazzeri E et al. Macrophage-derived chemokine and EBI1-ligand chemokine attract human thymocytes in different stage of development and are produced by distinct subsets of medullary epithelial cells: possible implications for negative selection. J Immunol 2000;165: 238–246.
42. Wu L, Shortman K. Isolation and characterization of murine early intrathymic precursor populations. Methods Mol Biol 2000;134:25–35.
43. Romagnani P, Annunziato F, Manetti R, Mavilia C, Lasagni L, Manuelli C et al. High CD30 ligand expression by epithelial cells and Hassal's corpuscles in the medulla of human thymus. Blood 1998;91:3323–3332.
44. Randolph DA, Huang G, Carruthers CJ, Bromley LE, Chaplin DD. The role of CCR7 in TH1 and TH2 cell localization and delivery of B cell help in vivo. Science 1999;286:2159–2162.
45. Sallusto F, Lenig D, Forster R, Lipp M, Lanzavecchia A. Two subsets of memory T lymphocytes with distinct homing potentials and effector functions [see comments]. Nature 1999;401:708–712.
46. Ngo VN, Tang HL, Cyster JG. Epstein–Barr virus-induced molecule 1 ligand chemokine is expressed by dendritic cells in lymphoid tissues and strongly attracts naive T cells and activated B cells. J Exp Med 1998;188:181–191.
47. Gunn MD, Tangemann K, Tam C, Cyster JG, Rosen SD, Williams LT. A chemokine expressed in lymphoid high endothelial venules promotes the adhesion and chemotaxis of naive T lymphocytes. Proc Natl Acad Sci USA 1998; 95:258–263.
48. Nakano H, Gunn MD. Gene Duplications at the Chemokine Locus on Mouse Chromosome 4: Multiple Strain-Specific Haplotypes and the Deletion of Secondary Lymphoid-Organ Chemokine and EBI-1 Ligand Chemokine Genes in the *plt* Mutation. J Immunol 2001;166:361–369.
49. Vassileva G, Soto H, Zlotnik A, Nakano H, Kakiuchi T, Hedrick JA et al. The reduced expression of 6Ckine in the *plt* mouse results from the deletion of one of two 6Ckine genes. J Exp Med 1999;190:1183–1188.
50. Warnock RA, Campbell JJ, Dorf ME, Matsuzawa A, McEvoy LM, Butcher EC. The role of chemokines in the microenvironmental control of T versus B cell arrest in Peyer's patch high endothelial venules. J Exp Med 2000;191:77–88.
51. Nakano H, Mori S, Yonekawa H, Nariuchi H, Matsuzawa A, Kakiuchi T. A novel mutant gene involved in T-lymphocyte-specific homing into peripheral lymphoid organs on mouse chromosome 4. Blood 1998;91:2886–2895.
52. Gunn MD, Kyuwa S, Tam C, Kakiuchi T, Matsuzawa A, Williams LT et al. Mice lacking expression of secondary lymphoid organ chemokine have defects in lymphocyte homing and dendritic cell localization [see comments]. J Exp Med 1999;189:451–460.
53. Forster R, Schubel A, Breitfeld D, Kremmer E, Renner-Muller I, Wolf E et al. CCR7 coordinates the primary immune response by establishing functional microenvironments in secondary lymphoid organs. Cell 1999;99: 23–33.
54. Bowman EP, Campbell JJ, Soler D, Dong Z, Manlongat N, Picarella D et al. Developmental switches in chemokine response profiles during B cell differentiation and maturation. J Exp Med 2000;191:1303–1318.
55. Forster R, Emrich T, Kremmer E, Lipp M. Expression of the G-protein-coupled receptor BLR1 defines mature, recirculating B cells and a subset of T-helper memory cells. Blood 1994;84:830–840.
56. Ansel KM, McHeyzer-Williams LJ, Ngo VN, McHeyzer-Williams MG, Cyster JG. In vivo-activated CD4 T cells upregulate CXC chemokine receptor 5 and reprogram their response to lymphoid chemokines. J Exp Med 1999; 190:1123–1134.

57. Breitfeld D, Ohl L, Kremmer E, Ellwart J, Sallusto F, Lipp M et al. Follicular B Helper T Cells Express CXC Chemokine Receptor 5, Localize to B Cell Follicles, and Support Immunoglobulin Production. J Exp Med 2000; 192:1545–1552.
58. Schaerli P, Willimann K, Lang AB, Lipp M, Loetscher P, Moser B. CXC Chemokine Receptor 5 Expression Defines Follicular Homing T Cells with B Cell Helper Function. J Exp Med 2000;192:1553–1562.
59. Ngo VN, Korner H, Gunn MD, Schmidt KN, Riminton DS, Cooper MD et al. Lymphotoxin alpha/beta and tumor necrosis factor are required for stromal cell expression of homing chemokines in B and T cell areas of the spleen. J Exp Med 1999;189:403–412.
60. Voigt I, Camacho SA, de Boer BA, Lipp M, Forster R, Berek C. CXCR5-deficient mice develop functional germinal centers in the splenic T cell zone. Eur J Immunol 2000;30:560–567.
61. Forster R, Mattis AE, Kremmer E, Wolf E, Brem G, Lipp M. A putative chemokine receptor, BLR1, directs B cell migration to defined lymphoid organs and specific anatomic compartments of the spleen. Cell 1996;87:1037–1047.
62. Flynn S, Toellner KM, Raykundalia C, Goodall M, Lane P. CD4 T cell cytokine differentiation: the B cell activation molecule, OX40 ligand, instructs CD4 T cells to express interleukin 4 and upregulates expression of the chemokine receptor, Blr-1. J Exp Med 1998;188:297–304.
63. Luther SA, Lopez T, Bai W, Hanahan D, Cyster JG. BLC expression in pancreatic islets causes B cell recruitment and lymphotoxin-dependent lymphoid neogenesis. Immunity 2000;12:471–481.

Universes in Delicate Balance: Chemokines and the Nervous System
R.M. Ransohoff, K. Suzuki, A.E.I. Proudfoot, W.F. Hickey and J.K. Harrison, editors

Chapter 2.6

CXC Chemokines in Angiogenesis

Robert M. Strieter [1], John A. Belperio [1], Douglas A. Arenberg [2], M. Iain Smith [1], Marie D. Burdick [1] and Michael P. Keane [1]

[1] *Department of Medicine, Division of Pulmonary and Critical Care Medicine, UCLA School of Medicine, Los Angeles, CA 90024-1922, USA;* [2] *Department of Internal Medicine, Division of Pulmonary and Critical Care Medicine, University of Michigan, Ann Arbor, MI 48109, USA*

1. INTRODUCTION

Angiogenesis, defined as the growth of new capillaries from pre-existing vessels, is a pervasive biological phenomenon that is at the core of many physiologic and pathologic processes. Examples of physiologic processes which depend upon angiogenesis include embryogenesis, wound repair, and the ovarian/menstrual cycle. In contrast, chronic inflammation associated with chronic fibroproliferative disorders as well as growth and metastasis of solid tumors are associated with aberrant angiogenesis. Angiogenesis is similar to but distinct from vasculogenesis, which describes the de novo formation of the vascular system from precursor blood islands during embryogenesis. Development of the heart and great vessels occurs via vasculogenesis, whereas organs which require invasion of blood vessels for development (brain, lung, kidney) are supplied by angiogenesis [1]. Neovascularization is a term which can be used interchangeably with angiogenesis, but may be more appropriately reserved for describing aberrant angiogenesis which accompanies pathologic processes, such as tumorigenesis or chronic inflammation.

Inflammation and angiogenesis, while distinct and separable, are closely related processes [2]. The histologic appearance of chronic inflammation includes the presence of 'granulation-like' tissue, a prominent feature of which is neovascularization. The marked increase in the metabolic demands of tissue which proliferates, repairs, or hypertrophies must be accompanied by a proportional increase in capillary blood supply. This form of aberrant angiogenesis is analogous to angiogenesis associated with tumorigenesis. This absolute dependence on neovascularization suggests several important characteristics of angiogenesis. First, the vascular system must be able to rapidly respond to increased tissue metabolic needs with increased microvasculature. Second, because of the high metabolic cost of angiogenesis, under basal conditions, the process must be tightly controlled, occurring only when necessary. Indeed, angiogenesis occurs rarely in the adult organism. Finally, in the absence of such strict control, pathophysiology, or disease is likely to result. To understand pathologic angiogenesis, it is important to understand the sequential events during physiologic angiogenesis.

2. SEQUENCE OF EVENTS DURING PHYSIOLOGIC ANGIOGENESIS

While endothelial cells are normally quiescent, during the angiogenic response, they become activated. The rate of normal capillary endothelial cell turnover is typically measured in months or years [3,4]. However, when endothelial cells of the microvasculature are stimulated, they will degrade their basement membrane and proximal extracellular matrix, migrate directionally, divide, and organize into functioning capillaries invested by a new basal lamina all within a matter of days. These steps are not sequential. Rather, they represent an orchestration of overlapping events necessary to return injured tissue to homeostasis. In each of the steps of angiogenesis, there are gaps in our knowledge of the exact mechanisms. Nevertheless, significant advances in cell and molecular biology have led to a greater understanding of angiogenesis (Fig. 1).

2.1. The Angiogenic Signal

The signal(s) which initiate angiogenesis vary with the conditions requiring angiogenesis, and may be organ specific [5]. During the wound response, angiogenic factors may be released through platelet degranulation [6] or proteolytic digestion of extracellular matrix [7]. The importance of these latter mechanisms may lie in the fact that they occur rapidly in response to tissue injury since they do not require new protein synthesis [8]. However, many cells may be the source of angiogenic signals, including tumor cells [9], fibroblasts [10], endothelial cells [11,12], epithelial cells [13] or activated macrophages [14]. Embryonic angiogenesis is activated by genes which are transcribed in response to hypoxia and hypoglycemia in the developing embryonic tissue [15]. Importantly, the signal for angiogenesis may also be initiated by a decrease in the presence of an inhibitory signal, rather than simply requiring a positive stimulus [16].

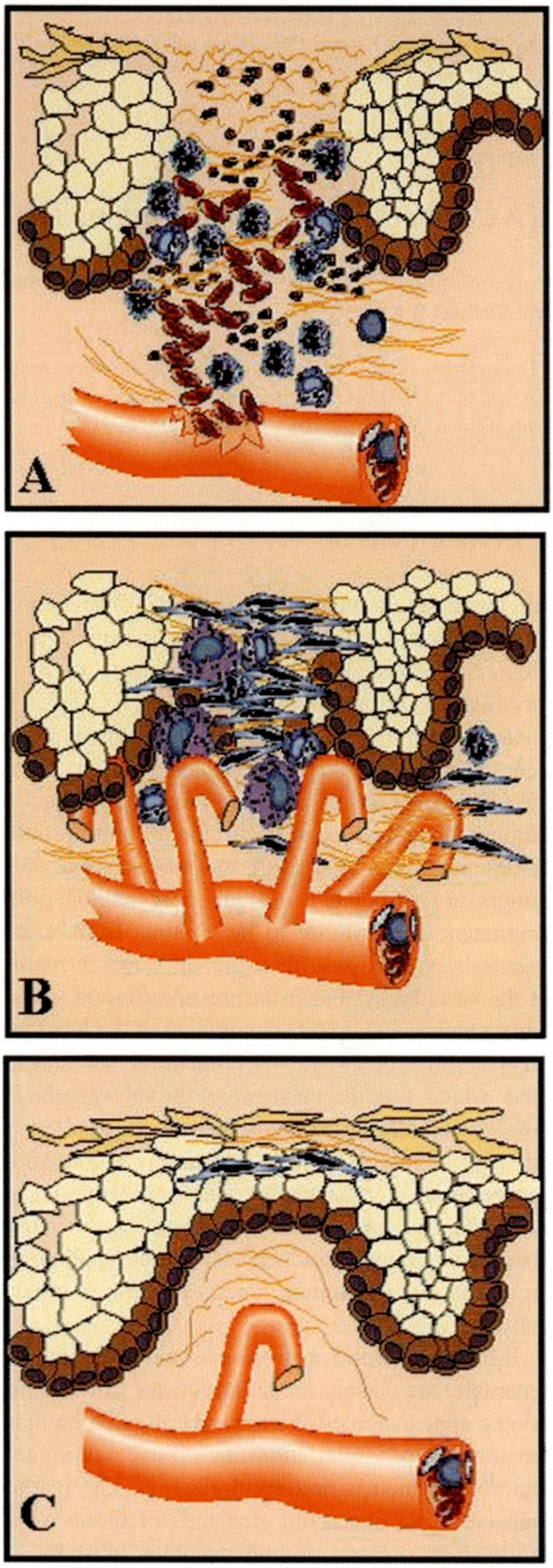

Figure 1. The angiogenic response to a wound. Initially there is coagulation due to leakage of plasma proteins, which leads to deposition of a primordial matrix. Many events, including platelet degranulation, macrophage activation, and tissue injury itself can release other angiogenic stimuli, endothelial cells begin to detach from their neighboring cells by altering adherens junctions. In order to maintain the integrity of the circulation, the step involved in the angiogenic response must occur in a complex and well orchestrated manner. (A) Degradation of basement the membrane requires the action of matrix metalloproteases. Proteolysis of extracellular matrix occurs by urokinase and tissue type plasminogen activators. (B) Adhesion to matrix components via $\alpha v\beta 3$ and $\alpha v\beta 5$ integrins allows endothelial cell haptotaxis through the matrix. Cell proliferation and DNA synthesis is necessary for continuing the angiogenic response. Tube formation requires tight cell–cell adhesions via both cadherin 5, and PECAM-1 (CD31). Collagen synthesis is required for deposition of new basement membrane. (C) Regression of newly formed vessels occurs when the expression of angiogenic factors is overcome by expression of angiostatic factors. When this occurs, the wound will regress to the prior steady-state level. This process by which this occurs probably involves apoptosis of endothelial cells and subsequent phagocytosis of cellular debris.

2.2. Endothelial De-differentiation

Once the endothelial cell receives an angiogenic stimulus, a process of 'de-differentiation' is initiated. For endothelium to invade into the surrounding matrix, cells must first detach from their tight association with neighboring endothelial cells. These cell–cell appositions, called adherens junctions, are composed of proteins of the cadherin family. Vascular endothelial cadherin (VE-cadherin, or cadherin 5) is highly specific for endothelial cells [17], and is associated with the cytoskeleton through other intermediate proteins, β-catenin, and plakoglobin [18,19]. One of the earliest events in the angiogenic response is alteration of the adherens junction complexes, leading to increased pericellular permeability, and detachment of the endothelial cell from its neighboring cells [20].

2.3. Proteolysis and Cell Migration

In order to form new vessels, the existing basement membrane must be degraded. Proteinase activity is necessary to digest the basement membrane, and early in the angiogenic response metalloproteinases are expressed [21]. Inhibitors of matrix metalloproteinases are capable of inhibiting angiogenesis [22]. This proteolytic activity also results in the further release of growth factors and angiogenic factors which are sequestered in the basement membrane [11,23].

After proteolysis of the basement membrane, angiogenic endothelial cells must migrate through the extracellular matrix. The structural constituents of extracellular matrix consist of a variety of components, including fibrin, fibronectin, vitronectin, and hyaluronan as well as other glycosaminoglycans [24]. Endothelial cells express both urokinase, and tissue-type plasminogen activators during angiogenesis which allow invasion into the surrounding matrix [25]. Motility through this environment requires cell–matrix adhesion which occurs through cell surface associated integrins.

In vitro endothelial cell chemotaxis, is typically assessed in an aqueous environment. However, endothelial cell migration in vivo occurs in a 'solid-phase' environment, referred to as haptotaxis. Reversible integrin-mediated binding to matrix components allows haptotaxis of endothelial cells along a chemotactic gradient through the extracellular matrix. Since endothelial cells must respond to injury in all organs and tissues of the body, they must be capable of adherence to a variety of matrix components. The $a_v\beta_3$ and $a_v\beta_5$ integrins are important endothelial cell adhesion molecules which display appropriately promiscuous binding profiles, and are both involved in angiogenesis [26,27]. The importance of this cell–matrix interaction in angiogenesis is demonstrated by work showing that specific inhibition of integrin binding in angiogenic endothelium leads to apoptosis of the endothelial cells [26,27]. Thus, integrin binding not only facilitates adhesive and locomotive functions, but also inhibits apoptotic cell death.

2.4. Cell Proliferation

While DNA synthesis occurs early in the angiogenic response, it has been shown both in vitro [28] and in vivo [29] that vascular sprouting can occur in the absence of endothelial cell proliferation. However, when proliferation is inhibited, the angiogenic response does not progress beyond vascular sprouting, the earliest stage of neovascularization [29]. Maintenance of the angiogenic response requires an increase in the number of endothelial cells to provide adequate capillary perfusion to the proliferative and repair processes of wound repair. While some angiogenic factors are only chemotactic for endothelial cells [30], most are also endothelial cell mitogens [28,30].

The signaling pathways which control cell growth are separable from those which lead to other aspects of the angiogenic response, and may be dependent on the degree of cell–matrix adhesion [31]. For example, in vitro studies demonstrate that endothelial cell proliferation occurs in conditions of increased cell–matrix adhesiveness, whereas growth arrest and loss of viability occur in cells plated on poorly adhesive substrates [31]. Intermediate levels of adhesiveness promote differentiation into tube-like structures [31]. One might hypothesize that these in vitro findings have an in vivo correlate during wound repair. The early primordial matrix of a wound is rich in plasma proteins which provide an abundant source of extracellular matrix to which endothelial cells may adhere, thus promoting cell migration and proliferation. As a wound matures, however, the primordial matrix is altered. Fibroblasts deposit type III collagen, and phagocytic cells remove debris, perhaps leading to reduced adhesiveness of the matrix and promoting capillary tube formation [8,14,32–35].

Additional evidence relating cell adhesion to control of proliferation is seen in studies of cadherin 5, the main component of the endothelial cell adherens junction. Homophilic cell–cell interaction through cadherin 5 inhibits endothelial cell proliferation and motility [36]. Similarly, cadherin 5 expressing cells have impaired proliferation when cultured on a cadherin 5-coated substrate [36]. Thus, cell–cell contact through the adherens junction and cadherin 5 may impair cell cycle progression. Early alteration of endothelial cell adherens junctions during the angiogenic response allows not only for vascular permeability and motility, but also permits endothelial cell proliferation.

2.5. Tube Formation

Once endothelial cells have invaded injured, inflamed, or hypoxic tissue, the integrity of the circulatory system must then be maintained. This critical process requires the formation of functioning capillaries with tight cell–cell adhesion. Investigators are beginning to define the nature of the cellular interactions underlying tube formation. In addition to the importance of cadherin 5 and the adherens junction, tube formation requires the function of CD31 (PECAM-1) [37]. PECAM-1 is a 130-kDa membrane glycoprotein and a member of the immunoglobulin supergene family that can mediate both heterotypic and homotypic adhesion [38,39]. The importance of this interaction has been demonstrated by neutralizing PECAM-1, which inhibited tube formation in vitro and angiogenesis in vivo [39]. However, in another model system, the combination of blocking antibodies to both CD31 and cadherin 5 were necessary to fully inhibit tube formation [37].

Once the capillary tube is established, the final step in forming a new blood vessel is the deposition of a basement membrane. Similar to other steps in the sequence of events which occur during angiogenesis, inhibition of collagen biosynthesis necessary for basement membrane formation prevents the in vitro formation of capillary-like tubes and inhibits angiogenesis in vivo [40]. The alteration and attenuation of any one of the steps involved in angiogenesis appears to have a profound effect on inhibiting the process (Fig. 2).

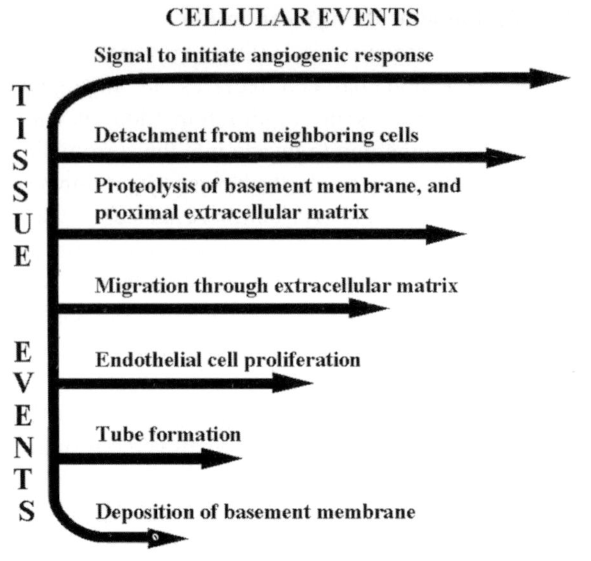

Figure 2. Cellular and tissue events related to the sequential events of angiogenesis.

2.6. Maintenance of the Angiogenic Response

Maintenance of neovascularization requires the persistence of a pro-angiogenic environment. This implies a local imbalance in the production of angiogenic factors that outweighs the expression of angiostatic factors. The events which dictate this imbalance have not been fully defined. Since the number of factors capable of regulating angiogenesis is extensive, it is not possible to discuss all the interactions which could affect this balance. However, there are several potential factors which may be important. First, early in the wound response, tissue injury itself leads to release of angiogenic factors from the extracellular matrix. Platelet degranulation releases both angiostatic [41] and angiogenic factors [42]. Infiltrating and activated macrophages release their own angiogenic factors [14,30,43–47], while macrophage-derived proteinases are probably important in the generation of angiostatic factors such as angiostatin and endostatin [48,49]. Other cells which can affect this balance include fibroblasts [10], epithelial cells [50], and endothelial cells [11,12]. These counterbalancing events that occur simultaneously during a wound response suggest that strict control is inherent in 'physiologic' angiogenesis.

2.7. Termination of the Angiogenic Response

Once the imbalance of angiogenic and angiostatic factors favors angiostasis and returns to homeostasis, the newly formed blood vessels will regress. The process of vascular regression has been studied in the cornea of the rabbit, and the sequence of events is similar whether vessels form in response to a wound or tumor-derived factors [51]. Regression of capillaries begins with morphologic changes of endothelial cells including mitochondrial swelling and development of cytoplasmic projections. This is followed by platelet adherence to vessel walls and platelet degranulation [51]. This process is eventually accompanied by degeneration of endothelial cells and removal of debris by mononuclear cells migrating into the area. Regression first occurs in the newest vessels and proceeds proximally [51]. These morphologic changes of regressing vessels resemble apoptosis [52].

In contrast to the precise regulation of physiologic angiogenesis, aberrant angiogenesis can lead to an imbalance in the relationship of angiogenic and angiostatic factors that favors persistent and unregulated angiogenesis. This type of environment can contribute to the pathogenesis of tumor growth and metastases, and the promotion of chronic fibroproliferative disorders. The complement of angiogenic and angiostatic factors and their imbalance may vary among different physiologic and pathologic settings. However, the recognition of this dual mechanism of regulatory control is critical in order to gain insight into this complex process and understand aberrant neovascularization associated with a variety of pathological angiogenesis-dependent conditions (Fig. 3). The CXC chemokines are a unique family of cytokines that can regulate angiogenesis in a disparate manner and will be the focus of this chapter.

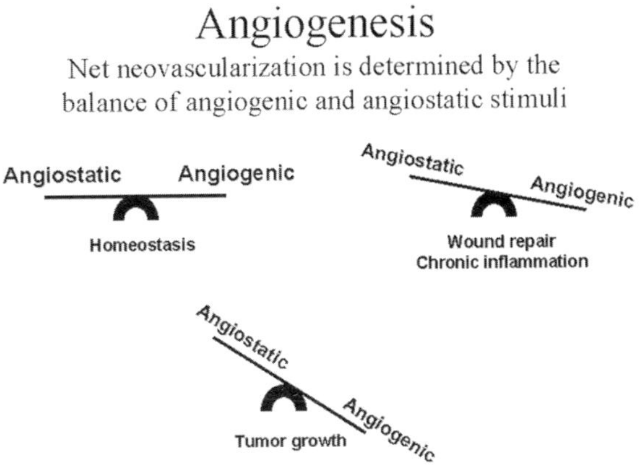

Figure 3. Dual mechanism of regulatory control of angiogenesis.

3. CXC CHEMOKINES, CXC CHEMOKINE RECEPTORS, AND ANGIOGENESIS

A variety of factors have been described that promote angiogenesis [5,53–66]. In parallel, a growing list of angiogenesis inhibitors (i.e., angiostatin, endostatin, vasostatin, EMAP-II, METH-1, and METH-2) have been identified [48,49,67–72]. While these factors are important in the regulation of angiogenesis, these molecules may not fully account for all of the modulation of the neovascular response in pathological conditions, such as tumorigenesis or chronic fibroproliferative disorders. CXC chemokines are characteristically heparin-binding proteins. On a structural level, they have four highly conserved cysteine amino acid residues, with the first two cysteines separated by one non-conserved amino acid residue, hence the name CXC [73–85]. Although the CXC motif distinguishes this family from other chemokine families, a second structural domain within this family dictates their angiogenic potential. The NH_2-terminus of the majority of the CXC chemokines contains three amino acid residues (Glu-Leu-Arg: the 'ELR' motif) which precedes the first cysteine amino acid residue of the primary structure of these cytokines [73–85]. The family members that contain the 'ELR' motif (ELR^+) are potent promoters of angiogenesis in physiological concentrations of 1–100 nM [81]. In contrast, members of the family that lack the ELR motif (ELR^-) and, in general interferon-inducible, are potent inhibitors of angiogenesis in physiological concentrations of 500 pM to 100 nM [81,86–88]. On a structural/functional level, this suggests that the CXC chemokine family is an unique family of cytokines due to their ability to behave in a disparate manner in the promotion and inhibition of angiogenesis.

3.1. Angiogenic (ELR^+) CXC Chemokines

The angiogenic members of the CXC chemokine family include interleukin-8 (IL-8), epithelial neutrophil activating protein-78 (ENA-78), growth-

Table 1. The ELR$^+$ and ELR$^-$ CXC chemokines are angiogenic and angiostatic factors, respectively

Angiogenic CXC chemokines containing the ELR motif (ELR$^+$)
Interleukin-8 (IL-8)/CXCL8
Epithelial neutrophil activating protein-78 (ENA-78)/CXCL5
Growth-related gene alpha (GRO-α)/CXCL1
Growth-related gene beta (GRO-β)/CXCL2
Growth-related gene gamma (GRO-γ)/CXCL3
Granulocyte chemotactic protein-2 (GCP-2)/CXCL6
Platelet basic protein (PBP)
Connective tissue activating protein-III (CTAP-III)
Beta-thromboglobulin (β-TG)
Neutrophil activating protein-2 (NAP-2)/CXCL7
Angiostatic CXC chemokines that lack the ELR motif (ELR$^-$)
Platelet factor-4 (PF4)
Interferon-γ-inducible protein (IP-10)/CXCL10
Monokine induced by interferon-γ (MIG)/CXCL9
Interferon-inducible T-cell α-chemoattractant/CXCL11

related genes (GRO-α, -β, and -γ), granulocyte chemotactic protein-2 (GCP-2), and NH_2-terminal truncated forms of platelet basic protein (PBP), which include connective tissue activating protein-III (CTAP-III), beta-thromboglobulin (β-TG), and neutrophil activating protein-2 (NAP-2) [44,81,89,90] (Table 1; systematic nomenclature is indicated in the table). ELR$^+$ CXC chemokines have been shown to mediate both in vitro endothelial cell chemotactic and proliferative activity, as well as in vivo angiogenesis in a direct manner using bioassays of angiogenesis [44,81,89–92]. These experiments prove that ELR$^+$ CXC chemokines have a direct effect on the endothelial cell, and that this angiogenic activity is distinct from their ability to induce inflammation. Furthermore, ELR$^+$ CXC chemokines have been found to induce the expression of the metalloproteinases, MMP-2 and MMP-9, by tumor cells during tumorigenesis [93–95]. The production of MMP-2 and MMP-9 are not only important for promoting endothelial cell migration in ECM during angiogenesis, but are also involved in enhancing tumor cell migration in ECM leading to metastasis. Therefore, ELR$^+$ chemokines not only have a direct effect on endothelial cell chemotaxis and proliferation, but also have an indirect effect in mediating their migration through ECM via the local production of MMPs.

Although stromal cell-derived factor (SDF-1) is a non-ELR CXC chemokine, it remains unclear whether this ELR$^-$ CXC chemokine inhibits or promotes angiogenesis. SDF-1 has been found to induce in vitro migration of human umbilical vein endothelial cells [96,97]. Mice with targeted disruption of the SDF-1 gene perinatally die [98]. This appears to be multi-factorial and includes defects in B-cell and myeloid progenitors suggesting that SDF-1 is involved in lymphopoiesis and myelopoiesis. In addition, these mice demonstrate cardiac ventricular septal defects [98]. Recently, targeted disruption of the receptor for SDF-1, CXCR4, has demonstrated that this CXC chemokine receptor is essential for vascularization of the gastrointestinal tract, hematopoiesis, and cerebellar development in these mice [99,100]. In contrast to these findings, SDF-1 can attenuate the angiogenic activity of either ELR$^+$ CXC chemokines, bFGF, or VEGF using the in vivo cornea micropocket assay of angiogenesis [101]. Thus, the role of SDF-1 in modulating angiogenesis in the context of tumorigenesis or chronic fibroproliferative disorders awaits further study as to whether SDF-1 directly contributes to the pathogenesis of angiogenesis-dependent disorders.

3.2. CXCR2 is the Putative Receptor for Angiogenic (ELR$^+$) CXC Chemokine-Mediated Angiogenesis

The fact that all ELR$^+$ CXC chemokines mediate angiogenesis highlights the importance of identifying a common receptor. This would provide the unique opportunity to target a putative receptor for ELR$^+$ CXC chemokine-induced angiogenesis in the context of tumorigenesis and chronic fibroproliferative disorders. The candidate CXC chemokine receptors are CXCR1 and/or CXCR2. Only IL-8 and GCP-2 specifically bind to CXCR1, whereas, all ELR$^+$ CXC chemokines bind to CXCR2 [73–85]. The ability of all ELR$^+$ CXC chemokine ligands to bind to CXCR2 supports the notion that this receptor mediates the angiogenic activity of ELR$^+$ CXC chemokines.

Evidence to support this contention is the following: (1) ELR$^+$ CXC chemokines bind endothelial cells [102]. (2) CXCR2 is the receptor used by all ELR$^+$ CXC chemokines [73,74,83,103,104], and all ELR$^+$ CXC chemokines are potent angiogenic factors [81]. (3) The expression of CXCR2 in human burn wound granulation tissue, melanoma, breast cancer, and head and neck squamous cell carcinoma is uniquely found in association with microvascular endothelial cells in areas of neovascularization, and not on normal adjacent endothelium [105–108]. (4) CXCR2 is a member of the G-protein-coupled seven-transmembrane receptor (GPCR) family [109]. GPCR activation leads to dissociation of the heterotrimeric protein complex ($G_{\alpha\beta\gamma}$) to α and $\beta\gamma$ subunits that mediate downstream regulation of several intracellular signaling pathways (i.e., cAMP/protein kinase A (PKA), protein kinase C (PKC), phospholipase C (PLC), phosphoinositide 3-kinase (PI3-K), Ras, Raf, and mitogen-activated protein kinases (MAPK)) [110–112]. Some of these signaling pathways are identical to signal transduction by receptor protein tyrosine kinases that are important for cellular proliferation, migration, and regulation of apoptosis [110–112]. (5) The human Kaposi's sarcoma herpes virus encoded GPCR (KSHV-GPCR) is homologous to CXCR2, is constitutively active and further augmented with IL-8 and/or GROα binding, and linked to angiogenesis of Kaposi's sarcoma [113–116]. In fact, transgenic mice expressing KSHV-GPCR within hematopoietic cells develop angioproliferative lesions in multiple organs that morphologically resemble KS lesion [117]. These findings suggest that the expression of only one viral chemokine receptor-like gene can lead to the histopathological recapitulation of KS. This supports the notion that a CXCR2-like receptor facilitates angiogenesis and tumorigenesis of this angiosarcoma. Moreover, the function of the KSHV-GPCR for inducing KS lesions requires ELR^+ CXC chemokine ligand signal-coupling, as mutation of the binding domain of this receptor followed by transgenic expression under the same promoter leads to *no* lesion formation in mice (personal communication, Sergio Lira). (6) A point mutation of CXCR2, not CXCR1, results in constitutive signaling and cellular transformation similar to KSHV-GPCR [118]. In fact, persistent activation of CXCR2 by an angiogenic ELR^+ CXC chemokine leads to similar cellular transformation as seen with either the point mutation of CXCR2 or KSHV-GPCR [118]. Thus, the potential expression of CXCR2 on endothelial cells in the presence of persistent juxtacrine (i.e., pericyte, smooth muscle cell, fibroblast or tumor cell) and paracrine (i.e., smooth muscle cell, fibroblast, epithelial cell, or tumor cell) stimulation with ELR^+ CXC chemokines has important implications in promoting an angiogenic environment of the endothelium.

While studies have suggested that both CXCR1 and CXCR2 may be expressed on human endothelial cells [119,120], other studies have suggested that CXCR2 is the putative receptor for ELR^+ CXC chemokine-mediated angiogenic activity [121,122]. Addison and colleagues have demonstrated that CXCR2 is detected in human microvascular endothelial cells at both the mRNA and protein levels. In addition, the expression of CXCR2, not CXCR1, was found to be functional in mediating endothelial cell chemotaxis. Moreover, this response was sensitive to Pertussis toxin, suggesting a role for G-protein linked receptor mechanisms in mediating endothelial cell chemotaxis [121]. Furthermore, the importance of CXCR2 in mediating ELR^+ CXC chemokine-induced angiogenesis was demonstrated in vivo using the cornea micropocket assay of angiogenesis in $CXCR2^{+/+}$ and $CXCR^{-/-}$ animals. ELR^+ CXC chemokine-mediated angiogenesis was inhibited in the corneas of $CXCR2^{-/-}$ mice, and in the presence of neutralizing antibodies to CXCR2 in the rat corneal micropocket assay. These studies have been further substantiated using $CXCR2^{-/-}$ mice in a wound repair model system [122]. Devalaraja and associates have examined the significance of CXC chemokines in wound healing [122]. In this study, full excisional wounds were created on CXCR2 wild-type (+/+), heterozygous (+/−), or knockout (−/−) mice. Significant delays in wound healing parameters were found in $CXCR^{-/-}$ mice, including decreased neovascularization. These in vitro and in vivo studies establish that CXCR2 is the receptor that mediates ELR^+ CXC chemokine-dependent angiogenic activity.

3.3. Interferon (IFN)-inducible (ELR^-) CXC Chemokines are Inhibitors of Angiogenesis

The angiostatic members of the CXC chemokine family include PF4, monokine induced by interferon-γ (MIG), and interferon-γ-inducible protein (IP-10) [123–128] (Table 1). IP-10 can be induced by all three interferons (IFN-α, -β, and -γ) [73,123–128]. MIG is unique in that it is only induced by IFN-γ [73,123–128]. Recently, a new ELR^- member of the CXC chemokine family, IFN-inducible T-cell α-chemoattractant (I-TAC), has been cloned and its expression appears to be induced primarily by IFN-γ [129]. I-TAC, similar to IP-10 and MIG, inhibits neovascularization in the CMP assay in response to

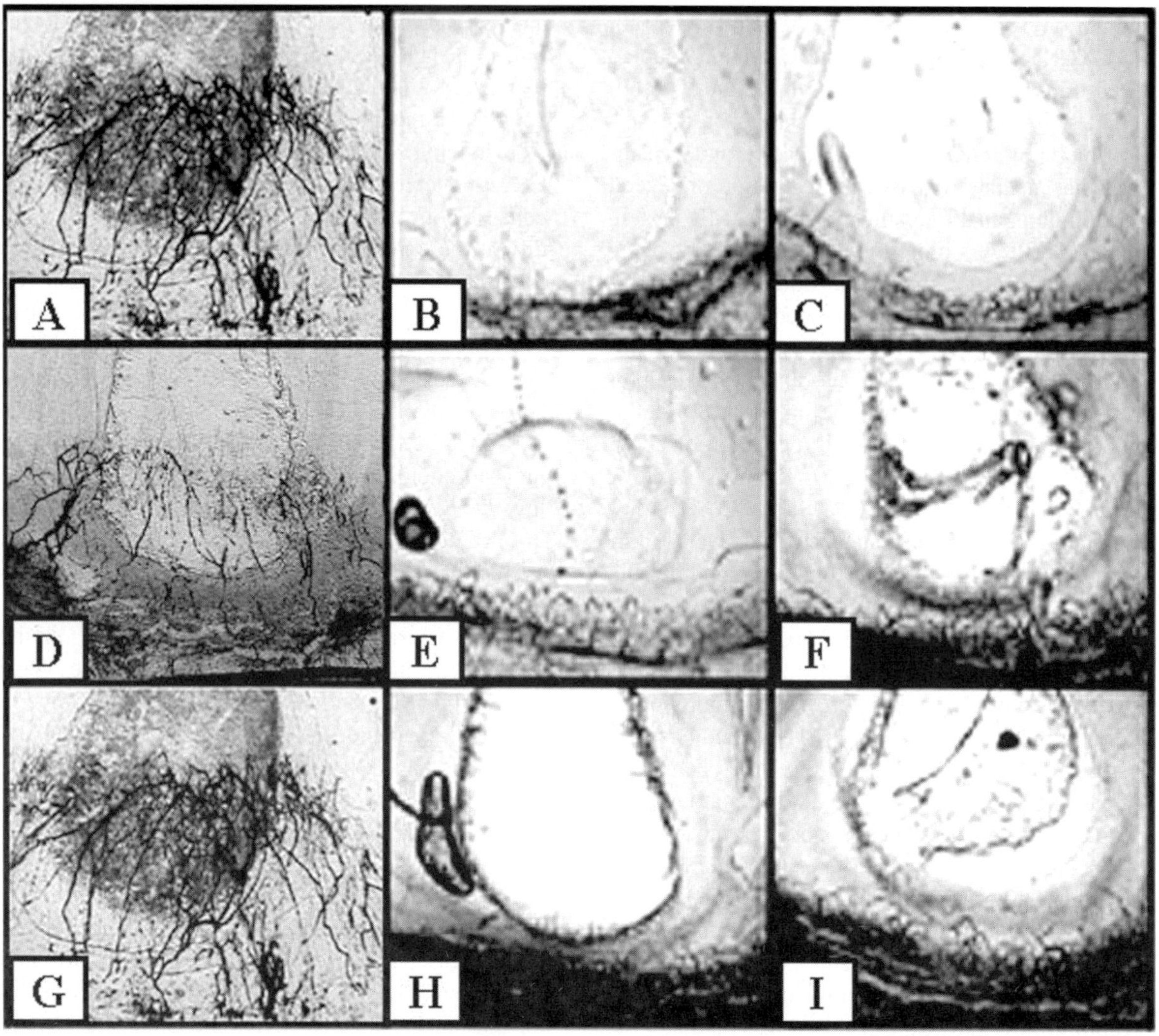

Figure 4. Cornea micropocket assay of neovascularization demonstrating the ability of interferon-inducible ELR$^-$ CXC chemokine inhibition of angiogenesis. (A) IL-8 (80 ng). (B) IL-8 (80 ng) + IP-10 (80 ng). (C) IL-8 (80 ng) + I-TAC (80 ng). (D) bFGF (50 ng). (E) bFGF (50 ng) + IP-10 (80 ng). (F) bFGF (50 ng) + I-TAC (80 ng). (G) VEGF (50 ng). (H) VEGF (50 ng) + IP-10 (80 ng). (I) VEGF (50 ng) + I-TAC (80 ng).

either ELR$^+$ CXC chemokines or VEGF (Fig. 4). These findings suggest that all IFN-inducible ELR$^-$ CXC chemokines are potent inhibitors of angiogenesis. Moreover, this interrelationship of interferon and IFN-inducible CXC chemokines and their biological function are directly relevant to the function of IL-18 and IL-12, or other molecules that stimulate the expression of IFN. The capability of IL-18 and IL-12 to induce IFN-γ and subsequently IFN-inducible CXC chemokines may explain their ability to inhibit angiogenesis. Therefore, IL-12 and IL-18, via the induction of IFN-γ, will have a profound effect on the production of IP-10, MIG, and I-TAC. The subsequent expression of IFN-inducible CXC chemokines may represent the final common pathway and explain the mechanism for the attenuation of angiogenesis related to interferons.

All three IFN-inducible ELR$^-$ CXC chemokines specifically bind to the CXC chemokine receptor, CXCR3 [129,130], and the expression of CXCR3 mRNA has been associated with endothelial cells [131]. Salcedo and colleagues have recently demonstrated that CXCR3 is expressed and functional on microvascular endothelial cells [120]. Although IP-

10, MIG, and I-TAC at high concentrations were found to induce endothelial cell chemotaxis, they were found to inhibit the endothelial cell chemotactic response to IL-8 [120]. Further studies will be necessary to determine whether CXCR3 is the putative receptor for the biological affect of interferon-inducible ELR^- CXC chemokines and inhibition of angiogenesis.

3.4. Mechanisms Related to the Angiostatic Activity of PF4 that may be Applicable to Interferon-Inducible ELR^- CXC Chemokines

Eloquent studies have delineated potential mechanisms for the ELR^- CXC chemokine, PF4, and its ability to inhibit of angiogenesis that may be relevant to interferon-inducible ELR^- CXC chemokines. The ability of PF4 to bind to glycosaminoglycans (GAG; heparin and heparan sulfate) with high affinity, appears to be important to several of its biological functions. PF4 inhibits endothelial cell migration, proliferation, and in vivo angiogenesis in response to bFGF or VEGF [132,133]. Moreover, FITC-labeled PF4 injected systemically, selectively binds to the endothelium in only areas of active angiogenesis [41,134]. This suggests that the microvasculature is the major target for the biological effects of PF4 during angiogenesis. PF4 has been shown to inhibit bFGF and $VEGF_{165}$ binding to their respective receptors [135–137]. One mechanism for this effect is related to the generation of PF4–bFGF or PF4–$VEGF_{165}$ heterodimeric complexes, which impairs bFGF or $VEGF_{165}$ binding to their respective receptors [136–138]. Basic FGF must undergo dimerization in the presence of endogenous heparin in order to bind to its receptor [137,138]. PF4 complexes to bFGF and prevents bFGF dimerization followed by impaired receptor binding and internalization [137]. $VEGF_{165}$ possesses heparin binding ability similar to bFGF. PF4 impairs $VEGF_{165}$ binding to its receptors on endothelium via a mechanism similar to what has been reported for its ability to inhibit bFGF [136]. Although, the ability of PF4 to form heterodimers with bFGF and $VEGF_{165}$ is one potential mechanism to inhibit bFGF and $VEGF_{165}$ biological activity, it appears that PF4 may inhibit angiogenesis through additional mechanisms.

While PF4 inhibits specific $VEGF_{165}$ binding, it does not inhibit $VEGF_{121}$ binding to VEGF receptors on endothelial cells. In contrast to $VEGF_{165}$, $VEGF_{121}$ is not a heparin-binding protein [136,139,140]. PF4 neither forms heterodimers with $VEGF_{121}$ nor competitively interferes with $VEGF_{121}$ binding to its receptor. However, PF4 directly inhibits $VEGF_{121}$-induced endothelial cell proliferation [136]. These findings suggest that other mechanisms must be operative for PF4 inhibition of mitogen stimulation of endothelial cells, perhaps mediated through its own independent biological signal. Although a specific receptor for PF4 on endothelium to-date has not been discovered, studies have suggested that PF4 inhibits endothelial cell cycle by preventing cell entry into S-phase [133]. In a model system of endothelial cell stimulation independent of interaction with cell-surface GAGs, PF4 inhibits epidermal growth factor (EGF)-stimulated endothelial cell proliferation by causing a decrease in cyclin E-cyclin-dependent kinase 2 (cdk2) activity that results in attenuation of retinoblastoma protein (pRb) phosphorylation [141]. The mechanism is related to PF4-dependent sustained increase in the levels and binding of the cyclin-dependent kinase inhibitor (CKI), $p21^{Cip1/WAF1}$, to the cyclin E-cdk2 complex. This inhibits cell cycle progression by preventing the down-regulation of $p21^{Cip1/WAF1}$ leading to inhibition of both cyclin E-cdk2 activity and phosphorylation of pRb [141]. These studies suggest that PF4 can inhibit a variety of endothelial cell mitogens at multiple levels. These events may be relevant to interferon-inducible ELR^- CXC chemokines, as IP-10 has been shown to compete with PF4 for binding, and inhibition of endothelial cell proliferation that may be related to inhibition of the cell cycle [87]. This supports the notion that interferon-inducible ELR^- CXC chemokines may have similar mechanisms for their inhibition of bFGF, VEGF, EGF, and ELR^+ CXC chemokine-induced angiogenesis.

4. ELR^+ CXC CHEMOKINES PROMOTE ANGIOGENESIS ASSOCIATED WITH TUMORIGENESIS

The ELR^+ CXC chemokines are important mediators of tumorigenesis related to their angiogenic properties. Although GRO-β has been recently reported to inhibit angiogenesis [142], the concentration used in this study was 1000-fold higher (1–10 μM) than what was found for its angiogenic activity (1–10 nM) [81,101]. This would suggest that super-physiological concentrations of GRO-β can 'desensitize' the angiogenic response. Moreover, studies in melanoma tumors support that all GROs play a significant role in mediating tumorigenesis related to both their mitogenic and angiogenic activities. For example, GRO-α, -β, and -γ have all been found to be highly expressed in human melanoma [106]. To determine the biological significance of the presence of these ELR^+ CXC chemokines in melanoma, human GRO-α, -β, and -γ genes have been transfected into immortalized murine melanocytes [106,143]. The persistent expression of GROs in these cells transforms their phenotype to one with anchorage-independent growth in vitro and the ability to form tumors in vivo in nude and SCID mice [106,143]. The tumors were highly vascular and similar to the vascularity of B16 melanoma controls [106,143]. When tumors were depleted of GROs there was a marked reduction of tumor-derived angiogenesis directly related to inhibition of tumor growth [106,143]. In addition to GROs, the expression of another ELR^+ CXC chemokine, IL-8, by human melanoma cells correlates with their metastatic potential in vivo [95]. Transfection of non-metastatic and IL-8-negative melanoma cells with the IL-8 gene rendered them highly tumorigenic and increase their metastatic potential in nude mice [95]. The IL-8-transfected cells demonstrate an up-regulation of MMP-2 expression and increased invasiveness in ECM [95]. IL-8 induced MMP-2 promotes the invasion of tumor cells through the host stroma, and concomitantly increases angiogenesis leading to enhanced metastasis [95]. These findings support the notion that the ELR^+ CXC chemokines, such as IL-8, GRO-α, -β, and -γ, have the ability to act both as autocrine growth factors for melanoma and as potent paracrine mediators of angiogenesis to promote tumorigenesis and metastases.

The progression and growth of ovarian carcinoma is also dependent on successful angiogenesis, and IL-8 has been determined to play a significant role in mediating human ovarian carcinoma-derived angiogenesis and tumorigenesis [144]. The expression of IL-8, bFGF, and VEGF was examined in five different human ovarian carcinoma cell lines [144]. All cell lines in vitro expressed similar levels of bFGF; however, these cells expressed either high or low levels of IL-8 or VEGF. When implanted into the peritoneum of nude mice, the high expressing IL-8 tumors were associated with all animals dying <51 days [144]. The expression of IL-8 was directly correlated with neovascularization and inversely correlated with survival, whereas VEGF expression was only correlated with production of ascites [144]. No correlation was found for bFGF with either tumor neovascularization or survival [144]. This study has been substantiated in patients with ovarian cancer, where ascites fluid demonstrates angiogenic activity directly correlated to IL-8 [145]. These findings support the notion that antigenic ELR^+ CXC chemokines play a greater role than bFGF and VEGF in mediating angiogenesis associated with ovarian cancer.

IL-8 is markedly elevated and contributes to the overall angiogenic activity of non-small cell lung cancer (NSCLC) [146]. Extending these studies to an in vivo model system of human tumorigenesis (i.e., human NSCLC/SCID mouse chimera) [147], tumor-derived IL-8 was found to be directly correlated with tumorigenesis [147]. Tumor-bearing animals depleted of IL-8 demonstrated a $>40\%$ reduction in tumor growth and a reduction in spontaneous metastases [147]. The attenuation of tumor growth and metastases was directly correlated to reduced angiogenesis. These findings have been further corroborated using several human NSCLC cell lines grown in nude mice. NSCLC cell lines that constitutively express IL-8 display greater tumorigenicity that is directly correlated to angiogenesis [148].

While IL-8 was the first angiogenic CXC chemokine to be discovered in NSCLC, ENA-78 has now been determined to have a higher degree of correlation with NSCLC-derived angiogenesis [149]. Surgical specimens of NSCLC tumors demonstrate a direct correlation of ENA-78 with tumor angiogenesis. These studies were extended to an SCID mouse model of human NSCLC tumorigenesis. ENA-78 expression was directly correlated with tumor growth. Moreover, when NSCLC tumor-bearing animals were depleted of ENA-78, both tumor growth and spontaneous metastases were markedly attenuated [149]. The reduction of angiogenesis is also accompanied by an increase in tumor cell apoptosis, consistent with the previous observation that inhibition of tumor-derived angiogenesis is associated with increased tumor cell apoptosis [49,68]. Similarly, in vivo and in vitro proliferation of NSCLC cells was unaffected by the presence of ENA-78. While a significant correlation of ENA-78 exists with tumor-derived angiogenesis, tumor growth, and metastases, ENA-78 depletion does not completely inhibit tumor growth. This reflects that the angiogenic activity of NSCLC tumors is related to many overlapping and potentially redundant factors acting in a parallel or serial manner.

Prostate cancer tumorigenesis and metastases is dependent on angiogenesis [150,151]. Serum levels of IL-8 have been found to be markedly elevated in patients with prostate cancer. These levels are highly correlated with the stage of the disease and have been determined to be an independent variable from the ratio of free/total prostate specific antigen (PSA) [152]. In fact, the combined use of f/t PSA and IL-8 levels were more effective in distinguishing prostate cancer from benign prostatic hypertrophy. This suggests that an ELR$^+$ CXC chemokine may be playing an important role in mediating prostate cancer-derived angiogenesis in support of tumorigenesis and metastases. This observation in patients has been substantiated in human/SCID mice chimeras of human prostate cancer tumorigenesis [153]. Three human prostate cancer cell lines were examined for constitutive production of angiogenic ELR$^+$ CXC chemokines [153]. Tumorigenesis of the human prostate cancer cell line, PC-3, was shown to be attributable, in part, to the production of the angiogenic CXC chemokine, IL-8. Depletion of endogenous IL-8 inhibited PC-3 tumor growth in SCID mice, that was entirely attributable to inhibition of PC-3 tumor-derived angiogenesis. In contrast, the human prostate cancer cell line, Du145, was found to utilize a different angiogenic CXC chemokine, GRO-α, to mediate tumor-derived angiogenesis. Depletion of endogenous GRO-α, but not anti-IL-8, reduced tumor growth that was directly related to attenuated angiogenic activity. Thus, prostate cancer cell lines can utilize distinct CXC chemokines to mediate their tumorigenic potential. These findings have been further substantiated by Inoue and colleagues, who found that the highly metastatic PC-3M-LN4 cell line over-expresses IL-8 relative to the poorly metastatic PC-3P cell line [93]. They evaluated whether IL-8 expression by human prostate cancer growing within the prostate of immunodeficient mice regulates tumor angiogenesis, growth, and metastasis. PC-3P cells were transfected with the full-length sense IL-8 cDNA, whereas PC-3M-LN4 cells (high expressing IL-8 cells) were transfected with the full-sequence anti-sense IL-8 cDNA. After orthotopic implantation, the sense-transfected PC-3P cells were highly tumorigenic and metastatic, with significantly increased neovascularity and IL-8 expression compared with either PC-3P cells or controls [93]. Antisense transfection significantly reduced the expression of IL-8 and MMP-9 and tumor-induced neovascularity, resulting in inhibition of tumorigenicity and metastasis [93]. These results demonstrate that IL-8 expression regulates angiogenesis in prostate cancer, in part by induction of MMP-9 expression, and subsequently regulates the growth and metastasis of human prostate cancer. Similar findings have been shown in gastric carcinoma, breast, and head and neck cancer [107,108,154–157]. The findings for the unique use of one or the redundancy of ELR$^+$ CXC chemokines in human tumors provides the unique opportunity to target a putative receptor for ELR$^+$ CXC chemokine-mediated angiogenesis.

5. ELR$^-$ CXC CHEMOKINES ATTENUATE ANGIOGENESIS ASSOCIATED WITH TUMORIGENESIS

ELR$^-$ CXC chemokines have been shown to inhibit angiogenesis in several model systems, for example, Burkitt's lymphoma cell lines form tumors in nude mice [158]. Angiogenesis is essential for tumorigenesis of these lymphomas, analogous to carcinomas. The expression of IP-10 and MIG was found to be higher in tumors that demonstrated spontaneous regression, and was directly related to impaired angiogenesis [159]. To determine whether this effect was attributable to IP-10 or MIG, more virulent Burkitt's lymphoma cell lines were grown in nude mice and subjected to intra-tumor inoculation with either IP-10 or MIG. Both conditions resulted in marked reduction in tumor-associated angiogenesis [88,160]. Although both IP-10 and MIG have been demonstrated to induce mononuclear cell recruitment via the interaction with their putative CXC chemokine receptor (CXCR3) [73–76,124], the ability of both of these ELR$^-$ CXC chemokines to inhibit angiogenesis and induce lymphoma regression in nude mice support that these chemokines mediate their effects in a T-cell-independent manner.

To examine the role of IP-10 in the regulation of angiogenesis in a carcinoma, the level of IP-10 from human surgical NSCLC tumor specimens was examined, and found to be significantly higher in the tumor specimens than in normal adjacent lung tissue [161]. The increase in IP-10 from human NSCLC tissue was entirely attributable to the higher levels of IP-10 present in squamous cell carcinoma (SCCA), as compared to adenocarcinoma. Moreover, depletion of IP-10 from SCCA surgical specimens resulted in augmented angiogenic activity [161]. The marked difference in the levels and bioactivity of IP-10 in SCCA and adenocarcinoma is clinically and pathophysiologically relevant, and represents a possible mechanism for the biologic differences of these two cell-types of NSCLC. Patient survival is lower, metastatic potential is higher, and evidence of angiogenesis is greater for adenocarcinoma, as compared to SCCA of the lung [162–164]. These studies were extended to a SCID mouse system to examine the effect of IP-10 on human NSCLC cell line tumor growth in a T- and B-cell independent manner. SCID mice were inoculated with either adenocarcinoma or SCCA cell lines [161]. The production of IP-10 from adenocarcinoma and SCCA tumors was inversely correlated with tumor growth [161]. However, IP-10 levels were significantly higher in the SCCA, as compared to adenocarcinoma tumors. The appearance of spontaneous lung metastases in SCID mice bearing adenocarcinoma tumors occurred after IP-10 levels from either the primary tumor or plasma had reached a nadir. In subsequent experiments, SCID mice bearing SCCA tumors were treated with either neutralizing anti-IP-10 antibodies, whereas, animals bearing adenocarcinoma tumors were treated with intra-tumor IP-10. Depletion of IP-10 in SCCA tumors resulted in a 2-fold increase in their size. In contrast, reconstitution of intra-tumor IP-10 in adenocarcinoma tumors reduced both their size and metastatic potential, that was unrelated to infiltrating neutrophils or mononuclear cells (i.e., macrophages or NK cells) and directly attributable to a reduction in tumor-associated angiogenesis.

Similar to IP-10, the interferon-inducible ELR^- CXC chemokine, MIG, also plays a significant role in regulating angiogenesis of NSCLC. Addison and associates have found that MIG levels in human specimens of NSCLC was not significantly different from that found in normal lung tissue [153]. However, these results suggested that the increased expression of ELR^+ CXC chemokines found in these tumor samples was not counter-regulated by a concomitant increase in the expression of the angiostatic interferon-inducible ELR^- CXC chemokine MIG. This imbalance would promote neovascularization. To alter this imbalance, they demonstrated that over-expression of the interferon-inducible ELR^- CXC chemokine, MIG, by three different strategies including gene transfer, resulted in the inhibition of NSCLC tumor growth and metastasis via a decrease in tumor-derived vessel density. These findings support the importance of the interferon-inducible ELR^- CXC chemokines in inhibiting NSCLC tumor growth by attenuation of tumor-derived angiogenesis. Furthermore, the above study demonstrates the potential efficacy of gene therapy as an alternative means to deliver and over-express a potent angiostatic CXC chemokine.

6. THE ROLE OF ANGIOGENIC (ELR^+) AND ANGIOSTATIC INTERFERON-INDUCIBLE (ELR^-) CXC CHEMOKINES IN THE REGULATION OF ANGIOGENESIS ASSOCIATED WITH CHRONIC FIBROPROLIFERATIVE DISORDERS

Angiogenesis is increasingly being recognized for its role in promoting the pathogenesis of chronic inflammatory/fibroproliferative disorders. For example, rheumatoid arthritis is associated with the unrestrained proliferation of fibroblasts and capillary blood vessels that leads to the formation of the pannus and destruction of joint spaces. Macrophages isolated from rheumatoid synovium produce pro-angiogenic factors [43]. Recently, Koch and associates have found that IL-8 and ENA-78 represent major angiogenic factors in the synovium of rheumatoid arthritis [165]. Psoriasis, is a well known angiogenesis-dependent skin disorder that is characterized by marked dermal neovascularization. Keratinocytes isolated from psoriatic plaques demonstrate a greater production of angiogenic activity. Interestingly, this angiogenic phenotype is due, in part, to a combined defect in the over-expression of the angiogenic cytokine IL-8, and a deficiency in the production of the angiogenesis inhibitor, thrombospondin-1, resulting in a pro-angiogenic environment [50].

Coronary artery atherosclerosis continues to be the leading cause of morbidity and mortality in the United States [166]. The pathogenesis of coronary atherosclerotic plaque formation is a complex process that demonstrates features of exaggerated injury and repair including recruitment of mononuclear cells, fibroproliferation, deposition of extracellular matrix, and angiogenesis, which lead to progressive fibrosis, calcification, and eventual luminal occlusion [167–169]. Aberrant angiogenesis has also been demonstrated in atherosclerosis [8,43,170,171]. Simonini and colleagues have demonstrated that IL-8 is a significant angiogenic factor in coronary atherectomy specimens [172]. These investigators found IL-8 levels to be markedly greater in coronary atherectomy specimens, as compared to control samples from the internal mammary arteries [172]. IL-8 expression by immunohistochemistry was highly correlated with localization of Factor VIII-related antigen expression on endothelial cells in the coronary atherectomy specimens. The contribution of IL-8 to net angiogenic activity from coronary atherectomy specimens was assessed using the rat cornea micropocket assay, and was found to represent the majority of the angiogenic activity. These findings suggest that, in human coronary atherosclerosis, IL-8 is an important mediator of angiogenesis and may be contributory to plaque formation via its angiogenic properties.

Evidence exists for angiogenesis in the lung. The angiogenic response of the bronchial circulation/systemic circulation in the lung is a fundamental response related to alterations in the pulmonary vascular resistance [173–177]. Compensatory neovascularization of up to 30% of the original pulmonary blood flow can occur in the bronchial circulation in all mammals in response to marked increases in pulmonary vascular resistance [177]. In fact, Mitzner and colleagues [173], have recently demonstrated in the mouse that neovascularization from the systemic circulation can supply up to 15% of the normal pulmonary flow within 5–6 days postligation, and this angiogenic response can occur in the absence of any hypoxic stimulus. This response is an attempt to maintain blood flow to the metabolic pulmonary tissue, especially in reaction to injury and repair.

Idiopathic pulmonary fibrosis (IPF) is a chronic and often fatal pulmonary fibroproliferative disorder. The pathogenesis of IPF that ultimately leads to end-stage fibrosis demonstrates features of dysregulated/abnormal repair with exaggerated neovascularization/vascular remodeling, fibroproliferation, and deposition of extracellular matrix, leading to progressive fibrosis and loss of lung function. While numerous eloquent studies have examined the biology of fibroblast proliferation and deposition of ECM in interstitial lung disease, few studies have examined the role of angiogenesis/vascular remodeling that promotes fibrogenesis in these disorders.

The existence of neovascularization in IPF was originally identified by Turner-Warwick, who examined the lungs of patients with widespread interstitial fibrosis and demonstrated neovascularization leading to anastomoses between the systemic and pulmonary microvasculatures and evidence of extensive vascular remodeling in areas of fibrosis [178]. These findings have been further substantiated with evidence of extensive neovascularization during the pathogenesis of pulmonary fibrosis in bleomycin-induced pulmonary fibrosis [179].

Recently, studies have corroborated the findings of Turner-Warwick, and have shown that the BALF and lung tissue from patients with IPF have marked angiogenic activity that is almost entirely attributable to the imbalance in the over-expression of the angiogenic ELR^+ CXC chemokine, IL-8, as compared to the relative down-regulation of the angiostatic interferon-inducible ELR^- CXC chemokine, IP-10 [10]. To determine whether the imbalance in the expression of these CXC chemokines were relevant to the pathogenesis of pulmonary fibrosis, studies were extended to a murine model system of bleomycin-induced pulmonary fibrosis. In this model system, the expression and biological activity of murine macrophage inflammatory protein-2 (MIP-2; an angiogenic ELR^+ CXC chemokine homologous to human GRO-β/γ) and the angiostatic CXC chemokine, IP-10, were correlated to the magnitude of lung fibrosis during bleomycin-induced pulmonary fibrosis [180,181]. MIP-2 and IP-10 were measured during bleomycin-induced pulmonary fibrosis from BAL and whole lung tissue homogenates, and were found to be directly and inversely correlated, respectively, with total lung hydroxyproline levels, a measure of lung collagen deposition [180,181]. Moreover, if either endogenous MIP-2 was depleted or exogenous IP-10 was administered to the animals during bleomycin exposure, both treatment strategies resulted in marked attenuation of pulmonary fibrosis that was entirely attributable to a reduction in angiogenesis in the lung [180,181]. These findings support the notion that angiogenesis is critical to promote fibroplasia and deposition of ECM during pulmonary fibrosis, and that angiogenic and angiostatic factors, such as ELR^+ and interferon-inducible ELR^- CXC chemokines play an important role in the pathogenesis of this process. Furthermore, with the recent demonstration of the efficacy of IFN-γ treatment of IPF patients [182,183], the above studies substantiate that IFN-γ treatment of IPF may mediate its effect, in part, by shifting the imbalance of the expression of angiogenic ELR^+ and angiostatic interferon-inducible ELR^- CXC chemokines to favor an angiostatic environment leading to inhibition of dysregulated neovascularization/vascular remodeling, fibroproliferation and deposition of extracellular matrix in IPF patients.

7. CONCLUSION

Angiogenesis is regulated by an opposing balance of angiogenic and angiostatic factors. CXC chemokines are an unique cytokine family that contain members that exhibit on a structural/functional basis either angiogenic or angiostatic biological activity. The above studies have demonstrated that as a family, the CXC chemokines appear to be important in the regulation of angiogenesis associated with both tumorigenesis and the pathogenesis of chronic inflammatory/fibroproliferative disorders. These findings support the notion that therapy directed at either inhibition of angiogenic or augmentation of angiostatic CXC chemokines may be a novel approach in the treatment of solid tumors, and chronic fibroproliferative disorders.

ACKNOWLEDGEMENTS

This work was supported, in part, by NIH Grants HL66027, CA87879, P50 HL67665 and P50 CA90388 (R.M.S.), CA72543 (D.A.A.), HL04493 (J.A.B), and HL03906 and P50 HL67665 (M.P.K.).

REFERENCES

1. Risau W. Vasculogenesis, angiogenesis and endothelial cell differentiation during embryonic development. In: Feinberg RN, Sherer GK, Auerbach R, eds. The development of the vascular system. Basel: Karger, 1991:58–68.
2. Jackson JR, Seed MP, Kircher CH, Willoughby DA, Winkler JD. The codependence of angiogenesis and chronic inflammation. FASEB J 1997;11:457–465.
3. Engerman RL, Pfaffenenbach D, Davis MD. Cell turnover of capillaries. Lab Invest 1967;17:738–743.
4. Tannock IF, Hayashi HS. The proliferation of capillary and endothelial cells. Cancer Res 1972;32:77–82.
5. Auerbach R, Auerbach W, Polakowski I. Assays for angiogenesis: a review. Pharmacol Ther 1991;51:1–11.
6. Sato N, Beitz JG, Kato J, Yamamoto M, Clark JW, Calabresi P et al. Platelet-derived growth factor indirectly stimulates angiogenesis in vitro. Am J Pathol 1993;142:1119–1130.
7. Vlodavsky I, Fuks Z, Ishai-Michaeli R, Bashkin P, Levi E, Korner G et al. Extracellular matrix-resident basic fibroblast growth factor: implication for the control of angiogenesis. J Cell Biochem 1991;45:167–176.
8. Clark RA. Basics of cutaneous wound repair. J Dermatol Surg Oncol 1993;19:693–706.
9. Gimbrone MA, Leapman SB, Cotran RS, Folkman J. Tumor dormancy in vivo by prevention of neovascularization. J Exp Med 1972;136:261–276.
10. Keane MP, Arenberg DA, Lynch JP, III, Whyte RI, Iannettoni MD, Burdick MD et al. The CXC chemokines, IL-8 and IP-10, regulate angiogenic activity in idiopathic pulmonary fibrosis. J Immunol 1997;159:1437–1443.
11. Vlodavski I, Folkman J, Sullivan R, Fridman R, Ishai-Michaeli R, Sasse J et al. Endothelial cell-derived basic fibroblast growth factor: synthesis and deposition into subendothelial extracellular matrix. Proc Natl Acad Sci USA 1987;84:2292–2296.
12. Strieter RM, Kunkel SL, Showell HJ, Marks RM. Monokine-induced gene expression of human endothelial cell-derived neutrophil chemotactic factor. Biochem Biophys Res Commun 1988;156:1340–1345.
13. Nickoloff BJ, Karabin GD, Barker JWCN, Griffiths CE, Sarma V, Mitra R et al. Cellular localization of interleukin-8 and its inducer, tumor necrosis factor-alpha in psoriasis. Am J Pathol 1991;138:129–140.
14. Leibovich SJ, Wiseman DM. Macrophages, wound repair and angiogenesis. Prog Clin Biol Res 1988;266:131–145.
15. Maltepe E, Schmidt JV, Baunoch D, Bradfield CA, Simon MC. Abnormal angiogenesis and responses to glucose and oxygen deprivation in mice lacking the protein ARNT. Nature 1997;386:403–407.
16. Rastinejad F, Polverini PJ, Bouck NP. Regulation of the activity of a new inhibitor of angiogenesis by a cancer suppressor gene. Cell 1989;56:345–355.
17. Lampugnani MG, Resnati M, Raiteri M, Pigott R, Pisacane A, Houen G et al. A novel endothelial-specific membrane protein is a marker of cell– cell contacts. J Cell Biol 1992;118:1511–1522.
18. Lampugnani MG, Corada M, Caveda L, Breviario F,

Ayalon O, Geiger B et al. The molecular organization of endothelial cell to cell junctions: differential association of plakoglobin, beta-catenin, and alpha-catenin with vascular endothelial cadherin (VE-cadherin). J Cell Biol 1995;129:203–217.

19. Tanihara H, Kido M, Obata S, Heimark RL, Davidson M, St John T et al. Characterization of cadherin-4 and cadherin-5 reveals new aspects of cadherins. J Cell Sci 1994;107:1697–1704.
20. Dejana E. Endothelial adherens junctions: implications in the control of vascular permeability and angiogenesis. J Clin Invest 1997;100:S7–10.
21. Pepper MS, Vassaui JD, Orci L, Montesano R. Proteolytic balance and capillary morphogenesis in vitro. In: Steiner R, Weisz PB, Langer R, eds. Angiogenesis: Key Principles. Basel: Birkhauser, 1992:137–145.
22. Takigawa M, Nishida Y, Suzuki F, Kishi J, Yamashita K, Hayakawa T. Induction of angiogenesis in chick yolk-sac membrane by polyamines and its inhibition by tissue inhibitors of metalloproteinases (TIMP and TIMP-2). Biochem Biophys Res Commun 1990;171:1264–1271.
23. Vlodavsky I, Korner G, Ishai-Michaeli R, Bashkin P, Bar-Shavit R, Fuks Z. Extracellular matrix-resident growth factors and enzymes: possible involvement in tumor metastases and angiogenesis. Cancer Metast Rev 1990;9:203–226.
24. Arnold F, West DC. Angiogenesis in wound healing. Pharmacol Ther 1991;52:407–422.
25. Gross JL, Moscatelli D, Rifkin DB. Increased capillary endothelial cell protease activity in response to angiogenic stimuli in vitro. Proc Natl Acad Sci USA 1983;80:2623–2627.
26. Brooks PC, Montgomery AM, Rosenfeld M, Reisfeld RA, Hu T, Klier G et al. Integrin alpha v beta 3 antagonists promote tumor regression by inducing apoptosis of angiogenic blood vessels. Cell 1994;79:1157–1164.
27. Brooks PC, Clark RA, Cheresh DA. Requirement of vascular integrin alpha v beta 3 for angiogenesis. Science 1994;264:569–571.
28. Koolwijk P, van Erck MG, de Vree WJ, Vermeer MA, Weich HA, Hanemaaijer R et al. Cooperative effect of TNFalpha, bFGF, and VEGF on the formation of tubular structures of human microvascular endothelial cells in a fibrin matrix. Role of urokinase activity. J Cell Biol 1996;132:1177–1188.
29. Sholly MM, Fergusen GP, Seibel HR, Montour JL, Wilson JD. Mechanisms of neovascularization: vascular sprouting can occur without proliferation of endothelial cells. Lab Invest 1984;51:624–634.
30. Sunderkotter C, Goebeler M, Schulze-Osthoff K, Bhardwaj R, Sorg C. Macrophage-derived angiogenesis factors. Pharmacol Ther 1991;51:195–216.
31. Ingber DE, Folkman J. Mechanochemical switching between growth and differentiation during fibroblast growth factor-stimulated angiogenesis in vitro: role of extracellular matrix. J Cell Biol 1989;109:317–330.
32. Clark RA. Cutaneous tissue repair:Basic biologic considerations. I. Am Acad Dermatol. 1985;13:701–725.
33. Davidson JM. Wound repair. In: Gallin JI, Goldstein IM, Snyderman R, eds. Inflammation: Basic Principles and Clinical Correlates. New York: Raven Press, 1992.
34. French-Constant C, Van DWL, Dvorak HF, Hynes RO. Reappearance of an embryonic pattern of fibronectin splicing during wound healing in the adult rat. J Cell Biol 1989;109:903–914.
35. Kurkinen M, Vaheri A, Roberts PJ, Stenan S. Sequential appearance of fibronectin and collagen in experimental granulation tissue. Lab Invest 1980;43:47–51.
36. Caveda L, Martin-Padura I, Navarro P, Breviario F, Corada M, Gulino D et al. Inhibition of cultured cell growth by vascular endothelial cadherin (cadherin-5/VE-cadherin). J Clin Invest 1996;98:886–893.
37. Matsumura T, Wolff K, Petzelbauer P. Endothelial cell tube formation depends on cadherin 5 and CD31 interactions with filamentous actin. J Immunol 1997;158:3408–3416.
38. DeLisser HM, Newman PJ, Albelda SM. Platelet endothelial cell adhesion molecule (CD31). Curr Top Microbiol Immunol 1993;184:37–45.
39. Delisser HM, Christofidou-Solomidou M, Strieter RM, Burdick MD, Robinson CS, Wexler RS et al. Involvement if endothelial PECAM-1/CD31 in angiogenesis. Am J Pathol 1997;151:671–677.
40. Haralabopoulos GC, Grant DS, Kleinman HK, Lelkes PI, Papaioannou SP, Maragoudakis ME. Inhibitors of basement membrane collagen synthesis prevent endothelial cell alignment in matrigel in vitro and angiogenesis in vivo. Lab Invest 1994;71:575–582.
41. Hansell P, Maione TE, Borgstrom P. Selective binding of platelet factor 4 to regions of active angiogenesis in vivo. Am J Physiol 1995;269:H829–836.
42. Walz A, Baggiolini M. Generation of the neutrophil-activating peptide NAP-2 from platelet basic protein or connective tissue-activating peptide III through monocyte proteases. J Exp Med 1990;171:449–454.
43. Koch AE, Leibovich SJ, Polverini PJ. Stimulation of neovascularization by human rheumatoid synovial tissue macrophages. Arthritis Rheum 1989;29:471–479.
44. Koch AE, Polverini PJ, Kunkel SL, Harlow LA, DiPietro LA, Elner VM et al. Interleukin-8 as a macrophage-derived mediator of angiogenesis [see comments]. Science 1992;258:1798–1801.
45. Leibovich SJ, Polverini PJ, Shepard HM, Wiseman DM, Shively V, Nuseir N. Macrophage-induced angiogenesis is mediated by tumour necrosis factor-alpha. Nature 1987;329:630–632.
46. Polverini PJ, Cotran PS, Gimbrone MA, Unanue ER. Activated macrophages induce vascular proliferation. Nature 1977;269:804–806.
47. Polverini PJ. How the extracellular matrix and macrophages contribute to angiogenesis- dependent diseases. Eur J Cancer 1996;32A:2430–2437.
48. O'Reilly MS, Holmgren L, Shing Y, Chen C, Rosenthal RA, Cao Y et al. Angiostatin: a circulating endothelial cell inhibitor that suppresses angiogenesis and tumor growth. Cold Spring Harb Symp Quant Biol 1994;59:471–482.
49. O'Reilly MS, Boehm T, Shing Y, Fukai N, Vasios G, Lane WS et al. Endostatin: an endogenous inhibitor of angiogenesis and tumor growth. Cell 1997;88:277–285.
50. Nickoloff BJ, Mitra RS, Varani J, Dixit VM, Polverini PJ. Aberrant production of interleukin-8 and

thrombospondin-1 by psoriatic keratinocytes mediates angiogenesis. Am J Pathol 1994;144:820–828.
51. Ausprunk DH, Folkman J. The sequence of events in the regression of corneal capillaries. Lab Invest 1978;38: 284–296.
52. Robaye B, Mosselmans R, Fiers W, Dumont JE, Galand P. Tumor necrosis factor induces apoptosis (programmed cell death) in normal endothelial cells in vitro. Am J Pathol 1991;138:447–453.
53. Auerbach W, Auerbach R. Angiogenesis inhibition: a review. Pharmacol Ther 1994;63:265–311.
54. Ziche M, Morbidelli L, Donnini S. Angiogenesis. Exp Nephrol 1996;4:1–14.
55. Pluda JM. Tumor-associated angiogenesis: mechanisms, clinical implications, and therapeutic strategies. Semin Oncol 1997;24:203–218.
56. Pluda JM, Parkinson DR. Clinical implications of tumor-associated neovascularization and current antiangiogenic strategies for the treatment of malignancies of pancreas. Cancer 1996;78:680–687.
57. Gastl G, Hermann T, Steurer M, Zmija J, Gunsilius E, Unger C et al. Angiogenesis as a target for tumor treatment. Oncology 1997;54:177–184.
58. Risau W. Angiogenesis is coming of age [editorial; comment]. Circ Res 1998;82:926–928.
59. Risau W. Mechanisms of angiogenesis. Nature 1997;386: 671–674.
60. Hotfilder M, Nowak-Gottl U, Wolff JE. Tumorangiogenesis: a network of cytokines. Klin Padiatr 1997;209:265–270.
61. Hui YF, Ignoffo RJ. Angiogenesis inhibitors. A promising role in cancer therapy. Cancer Pract 1998;6:60–62.
62. Kumar R, Fidler IJ. Angiogenic molecules and cancer metastasis. In Vivo 1998;12:27–34.
63. Zetter BR. Angiogenesis. State of the art. Chest 1988;93: 159S–166S.
64. Zetter BR. Angiogenesis and tumor metastasis. Annu Rev Med 1998;49:407–424.
65. Lund EL, Spang-Thomsen M, Skovgaard-Poulsen H, Kristjansen PE. Tumor angiogenesis — a new therapeutic target in gliomas. Acta Neurol Scand 1998;97: 52–62.
66. Lund EL, Kristjansen PE. [Neovascularization of tumors. New therapeutic possibilities]. Ugeskr Laeger 1999;161: 2929–2933.
67. O'Reilly MS, Holmgren L, Shing Y, Chen C, Rosenthal RA, Moses M et al. Angiostatin: a novel angiogenesis inhibitor that mediates the suppression of metastases by a Lewis lung carcinoma [see comments]. Cell 1994;79: 315–328.
68. O'Reilly MS, Holmgren L, Chen C, Folkman J. Angiostatin induces and sustains dormancy of human primary tumors in mice. Nat Med 1996;2:689–692.
69. O'Reilly MS. Angiostatin: an endogenous inhibitor of angiogenesis and of tumor growth. Exs 1997;79:273–294.
70. Schwarz MA, Kandel J, Brett J, Li J, Hayward J, Schwarz RE et al. Endothelial-monocyte activating polypeptide II, a novel antitumor cytokine that suppresses primary and metastatic tumor growth and induces apoptosis in growing endothelial cells. J Exp Med 1999;190: 341–354.
71. Vazquez F, Hastings G, Ortega MA, Lane TF, Oikemus S, Lombardo M et al. METH-1, a human ortholog of ADAMTS-1, and METH-2 are members of a new family of proteins with angio-inhibitory activity. J Biol Chem 1999;274:23349–23357.
72. Pike SE, Yao L, Jones KD, Cherney B, Appella E, Sakaguchi K et al. Vasostatin, a calreticulin fragment, inhibits angiogenesis and suppresses tumor growth. J Exp Med 1998;188:2349–2356.
73. Luster AD. Chemokines–chemotactic cytokines that mediate inflammation. New Engl J Med 1998;338:436–445.
74. Rollins BJ. Chemokines. Blood 1997;90:909–928.
75. Baggiolini M. Chemokines and leukocyte traffic. Nature 1998;392:565–568.
76. Baggiolini M, Dewald B, Moser B. Human chemokines: an update. Annu Rev Immunol 1997;15:675–705.
77. Baggiolini M, Dewald B, Walz A. Interleukin-8 and related chemotactic cytokines. In: Gallin JI, Goldstein IM, Snyderman R, eds. Inflammation: Basic Principles and Clinical Correlates,. New York: Raven Press, 1992.
78. Adams DH, Lloyd AR. Chemokines: leucocyte recruitment and activation cytokines. Lancet 1997;349:490–495.
79. Strieter RM, Kunkel SL, Shanafelt AB, Arenberg DA, Koch AE, Polverini PJ. CXC chemokines in regulation of angiogenesis. In: Koch AE, Strieter RM, eds. Chemokines in Disease. Austin, TX: R.G Landes, 1996:195–210.
80. Strieter RM, Polverini PJ, Arenberg DA, Kunkel SL. The role of CXC chemokines as regulators of angiogenesis. Shock 1995;4:155–160.
81. Strieter RM, Polverini PJ, Kunkel SL, Arenberg DA, Burdick MD, Kasper J et al. The functional role of the ELR motif in CXC chemokine-mediated angiogenesis. J Biol Chem 1995;270:27348–27357.
82. Strieter RM, Kunkel SL. Chemokines and the lung. In: Crystal R, West J, Weibel EPPB, eds. Lung: Scientific Foundations, 2nd edn. New York: Raven Press, 1997: 155–186.
83. Taub DD, Oppenheim JJ. Chemokines, inflammation and the immune system. Ther Immunol 1994;1:229–246.
84. Walz A, Kunkel SL, Strieter RM. CXC chemokines — an overview. In: Strieter RM, ed. Chemokines in Disease. Austin, TX: R.G. Landes, 1996:1–26.
85. Balkwill F. The molecular and cellular biology of the chemokines. J Viral Hepat 1998;5:1–14.
86. Angiolillo AL, Sgadari C, Taub DD, Liao F, Farber JM, Maheshwari S et al. Human interferon-inducible protein 10 is a potent inhibitor of angiogenesis in vivo. J Exp Med 1995;182:155–162.
87. Luster AD, Greenberg SM, Leder P. The IP-10 chemokine binds to a specific cell surface heparan sulfate site shared with platelet factor 4 and inhibits endothelial cell proliferation. J Exp Med 1995;182:219–231.
88. Sgadari C, Farber JM, Angiolillo AL, Liao F, Teruya-Feldstein J, Burd PR et al. Mig, the monokine induced by interferon-gamma, promotes tumor necrosis in vivo. Blood 1997;89:2635–2643.

89. Hu DE, Hori Y, Fan TPD. Interleukin-8 stimulates angiogenesis in rats. Inflammation 1993;17:135–143.
90. Strieter RM, Kunkel SL, Elner VM, Martonyi CL, Koch AE, Polverini PJ et al. Interleukin-8. A corneal factor that induces neovascularization. Am J Pathol 1992;141: 1279–1284.
91. Norrby K. Interleukin-8 and de novo mammalian angiogenesis. Cell Prolif 1996;29:315–323.
92. Yoshida S, Ono M, Shono T, Izumi H, Ishibashi T, Suzuki H et al. Involvement of interleukin-8, vascular endothelial growth factor, and basic fibroblast growth factor in tumor necrosis factor alpha-dependent angiogenesis. Mol Cell Biol 1997;17:4015–4023.
93. Inoue K, Slaton JW, Eve BY, Kim SJ, Perrotte P, Balbay MD et al. Interleukin 8 expression regulates tumorigenicity and metastases in androgen-independent prostate cancer. Clin Cancer Res 2000;6:2104–2119.
94. Luca M, Huang S, Gershenwald JE, Singh RK, Reich R, Bar-Eli M. Expression of interleukin-8 by human melanoma cells up-regulates MMP-2 activity and increases tumor growth and metastasis. Am J Pathol 1997; 151:1105–1113.
95. Bar-Eli M. Role of interleukin-8 in tumor growth and metastasis of human melanoma. Pathobiology 1999;67: 12–18.
96. Gupta SK, Lysko PG, Pillarisetti K, Ohlstein E, Stadel JM. Chemokine receptors in human endothelial cells. Functional expression of CXCR4 and its transcriptional regulation by inflammatory cytokines. J Biol Chem 1998; 273:4282–4287.
97. Salcedo R, Wasserman K, Young HA, Grimm MC, Howard OM, Anver MR et al. Vascular endothelial growth factor and basic fibroblast growth factor induce expression of CXCR4 on human endothelial cells: In vivo neovascularization induced by stromal-derived factor-1alpha. Am J Pathol 1999;154:1125–1135.
98. Nagasawa T, Hirota S, Tachibana K, Takakura N, Nishikawa S, Kitamura Y et al. Defects of B-cell lymphopoiesis and bone-marrow myelopoiesis in mice lacking the CXC chemokine PBSF/SDF-1. Nature 1996;382: 635–638.
99. Tachibana K, Hirota S, Iizasa H, Yoshida H, Kawabata K, Kataoka Y et al. The chemokine receptor CXCR4 is essential for vascularization of the gastrointestinal tract [see comments]. Nature 1998;393:591–594.
100. Zou YR, Kottmann AH, Kuroda M, Taniuchi I, Littman DR. Function of the chemokine receptor CXCR4 in haematopoiesis and in cerebellar development [see comments]. Nature 1998;393:595–599.
101. Arenberg DA, Polverini PJ, Kunkel SL, Shanafelt A, Strieter RM. In vitro and in vivo systems to assess role of C–X–C chemokines in regulation of angiogenesis. Methods Enzymol 1997;288:190–220.
102. Schönbeck U, Brandt E, Petersen F, Flad H, Loppnow H. IL-8 specifically binds to endothelial but not to smooth muscle cells. J Immunol 1995;154:2375–2383.
103. Ahuja SK, Murphy PM. The CXC chemokines growth-regulated oncogene (GRO) alpha, GRObeta, GROgamma, neutrophil-activating peptide-2, and epithelial cell-derived neutrophil-activating peptide-78 are potent agonists for the type B, but not the type A, human interleukin-8 receptor. J Biol Chem 1996;271:20545–20550.
104. Lee J, Horuk R, Rice GC, Bennett GL, Camerato T, Wood WI. Characterization of two high affinity human interleukin-8 receptors. J Biol Chem 1992;267:16283–16287.
105. Nanney LB, Mueller SG, Bueno R, Peiper SC, Richmond A. Distributions of melanoma growth stimulatory activity or growth-related gene and the interleukin-8 receptor type B in human wound repair. Am J Pathol 1995;147: 1248–1260.
106. Luan J, Shattuck-Brandt R, Haghnegahdar H, Owen JD, Strieter R, Burdick M et al. Mechanism and biological significance of constitutive expression of MGSA/GRO chemokines in malignant melanoma tumor progression. J Leukocyte Biol 1997;62:588–597.
107. Miller LJ, Kurtzman SH, Wang Y, Anderson KH, Lindquist RR, Kreutzer DL. Expression of interleukin-8 receptors on tumor cells and vascular endothelial cells in human breast cancer tissue. Anticancer Res 1998;18: 77–81.
108. Richards BL, Eisma RJ, Spiro JD, Lindquist RL, Kreutzer DL. Coexpression of interleukin-8 receptors in head and neck squamous cell carcinoma. Am J Surg 1997;174:507–512.
109. Murphy PM. The molecular biology of leukocyte chemoattractant receptors. Annu Rev Immunol 1994;12: 593–633.
110. Sugden PH, Clerk A. Regulation of the ERK subgroup of MAP kinase cascades through G protein-coupled receptors. Cell Signal 1997;9:337–351.
111. Pawson T, Scott JD. Signaling through scaffold, anchoring, and adaptor proteins. Science 1997;278:2075–2080.
112. Shyamala V, Khoja H. Interleukin-8 receptors R1 and R2 activate mitogen-activated protein kinases and induce c-fos, independent of Ras and Raf-1 in Chinese hamster ovary cells. Biochemistry 1998;37:15918–15924.
113. Arvanitakis L, Geras-Raaka E, Varma A, Gershengorn MC, Cesarman E. Human herpesvirus KSHV encodes a constitutively active G-protein- coupled receptor linked to cell proliferation [see comments]. Nature 1997;385: 347–350.
114. Bais C, Santomasso B, Coso O, Arvanitakis L, Raaka EG, Gutkind JS et al. G-protein-coupled receptor of Kaposi's sarcoma-associated herpesvirus is a viral oncogene and angiogenesis activator [see comments] [published erratum appears in Nature 1998 Mar 12;392(6672):210]. Nature 1998;391:86–89.
115. Geras-Raaka E, Arvanitakis L, Bais C, Cesarman E, Mesri EA, Gershengorn MC. Inhibition of constitutive signaling of Kaposi's sarcoma-associated herpesvirus G protein-coupled receptor by protein kinases in mammalian cells in culture. J Exp Med 1998;187:801–806.
116. Geras-Raaka E, Varma A, Ho H, Clark-Lewis I, Gershengorn MC. Human interferon-gamma-inducible protein 10 (IP-10) inhibits constitutive signaling of Kaposi's sarcoma-associated herpesvirus G protein-coupled receptor. J Exp Med 1998;188:405–408.
117. Yang TY, Chen SC, Leach MW, Manfra D, Homey B, Wiekowski M et al. Transgenic expression of the chemokine receptor encoded by human herpesvirus 8 induces

an angioproliferative disease resembling Kaposi's sarcoma [see comments]. J Exp Med 2000;191:445–454.
118. Burger M, Burger JA, Hoch RC, Oades Z, Takamori H, Schraufstatter IU. Point mutation causing constitutive signaling of CXCR2 leads to transforming activity similar to Kaposi's sarcoma herpesvirus-G protein-coupled receptor. J Immunol 1999;163:2017–2022.
119. Murdoch C, Monk PN, Finn A. Cxc chemokine receptor expression on human endothelial cells. Cytokine 1999; 11:704–712.
120. Salcedo R, Resau JH, Halverson D, Hudson EA, Dambach M, Powell D et al. Differential expression and responsiveness of chemokine receptors (CXCR1–3) by human microvascular endothelial cells and umbilical vein endothelial cells [In process citation]. FASEB J 2000;14:2055–2064.
121. Addison CL, Daniel TO, Burdick MD, Liu H, Ehlert JE, Xue YY et al. The CXC Chemokine Receptor 2, CXCR2, Is the Putative Receptor for ELR($^+$) CXC Chemokine-Induced Angiogenic Activity. J Immunol 2000;165:5269–5277.
122. Devalaraja RM, Nanney LB, Qian Q, Du J, Yu Y, Devalaraja MN et al. Delayed wound healing in CXCR2 knockout mice. J Invest Dermatol 2000;115:234–244.
123. Farber JM. HuMIG: a new member of the chemokine family of cytokines. Biochem Biophys Res Commun 1993;192:223–230.
124. Farber JM. Mig and IP-10: CXC chemokines that target lymphocytes. J Leukocyte Biol 1997;61:246–257.
125. Farber JM. A macrophage mRNA selectively induced by gamma-interferon encodes a member of the platelet factor 4 family of cytokines. Proc Natl Acad Sci USA 1990;87:5238–5242.
126. Farber JM. A collection of mRNA species that are inducible in the RAW 264.7 mouse macrophage cell line by gamma interferon and other agents. Mol Cell Biol 1992;12:1535–1545.
127. Luster AD, Unkeless JC, Ravetch JV. Gamma-interferon transcriptionally regulates an early-response gene containing homology to platelet proteins. Nature 1985;315: 672–676.
128. Luster AD, Ravetch JV. Biochemical characterization of a gamma interferon-inducible cytokine (IP-10). J Exp Med 1987;166:1084–1097.
129. Cole KE, Strick CA, Paradis TJ, Ogborne KT, Loetscher M, Gladue RP et al. Interferon-inducible T cell alpha chemoattractant (I-TAC): a novel non-ELR CXC chemokine with potent activity on activated T cells through selective high affinity binding to CXCR3. J Exp Med 1998;187:2009–2021.
130. Loetscher M, Gerber B, Loetscher P, Jones SA, Piali L, Clark-Lewis I et al. Chemokine receptor specific for IP10 and mig: structure, function, and expression in activated T-lymphocytes [see comments]. J Exp Med 1996;184: 963–969.
131. Soto H, Wang W, Strieter RM, Copeland NG, Gilbert DJ, Jenkins NA et al. The CC chemokine 6Ckine binds the CXC chemokine receptor CXCR3. Proc Natl Acad Sci USA 1998;95:8205–8210.
132. Maione TE, Gray GS, Petro J, Hunt AJ, Donner AL, Bauer SI et al. Inhibition of angiogenesis by recombinant human platelet factor-4 and related peptides. Science 1990;247:77–79.
133. Gupta SK, Singh JP. Inhibition of endothelial cell proliferation by platelet factor-4 involves a unique action on S-phase progression. J Cell Biol 1994;127:1121–1127.
134. Borgstrom P, Discipio R, Maione TE. Recombinant platelet factor 4, an angiogenic marker for human breast carcinoma. Anticancer Res 1998;18:4035–4041.
135. Sato Y, Abe M, Takaki R. Platelet factor 4 blocks the binding of basic fibroblast growth factor to the receptor and inhibits the spontaneous migration of vascular endothelial cells. Biochem Biophys Res Commun 1990; 172:595–600.
136. Gengrinovitch S, Greenberg SM, Cohen T, Gitay-Goren H, Rockwell P, Maione TE et al. Platelet factor-4 inhibits the mitogenic activity of VEGF121 and VEGF165 using several concurrent mechanisms. J Biol Chem 1995;270: 15059–15065.
137. Perollet C, Han ZC, Savona C, Caen JP, Bikfalvi A. Platelet factor 4 modulates fibroblast growth factor 2 (FGF-2) activity and inhibits FGF-2 dimerization. Blood 1998;91:3289–3299.
138. Jouan V, Canron X, Alemany M, Caen JP, Quentin G, Plouet J et al. Inhibition of in vitro angiogenesis by platelet factor-4-derived peptides and mechanism of action. Blood 1999;94:984–993.
139. Houck KA, Leung DW, Rowland AM, Winer J, Ferrara N. Dual regulation of vascular endothelial growth factor bioavailability by genetic and proteolytic mechanisms. J Biol Chem 1992;267:26031–26037.
140. Houck KA, Ferrara N, Winer J, Cachianes G, Li B, Leung DW. The vascular endothelial growth factor family: identification of a fourth molecular species and characterization of alternative splicing of RNA. Mol Endocrinol 1991;5:1806–1814.
141. Gentilini G, Kirschbaum NE, Augustine JA, Aster RH, Visentin GP. Inhibition of human umbilical vein endothelial cell proliferation by the CXC chemokine, platelet factor 4 (PF4), is associated with impaired downregulation of p21(Cip1/WAF1). Blood 1999;93:25–33.
142. Cao Y, Chen C, Weatherbee JA, Tsang M, Folkman J. gro-beta, a –C–X–C– chemokine, is an angiogenesis inhibitor that suppresses the growth of Lewis lung carcinoma in mice. J Exp Med 1995;182:2069–2077.
143. Owen JD, Strieter R, Burdick M, Haghnegahdar H, Nanney L, Shattuck-Brandt R et al. Enhanced tumor-forming capacity for immortalized melanocytes expressing melanoma growth stimulatory activity/growth-regulated cytokine beta and gamma proteins. Int J Cancer 1997;73:94–103.
144. Yoneda J, Kuniyasu H, Crispens MA, Price JE, Bucana CD, Fidler IJ. Expression of angiogenesis-related genes and progression of human ovarian carcinomas in nude mice. J Natl Cancer Inst 1998;90:447–454.
145. Gawrychowski K, Skopinska-Rozewska E, Barcz E, Sommer E, Szaniawska B, Roszkowska-Purska K et al. Angiogenic activity and interleukin-8 content of human ovarian cancer ascites. Eur J Gynaecol Oncol 1998;19: 262–264.
146. Smith DR, Polverini PJ, Kunkel SL, Orringer MB, Whyte RI, Burdick MD et al. Inhibition of IL-8 at-

tenuates angiogenesis in bronchogenic carcinoma. J Exp Med 1994;179:1409–1415.
147. Arenberg DA, Kunkel SL, Polverini PJ, Glass M, Burdick MD, Strieter RM. Inhibition of interleukin-8 reduces tumorigenesis of human non-small cell lung cancer in SCID mice. J Clin Invest 1996;97:2792–2802.
148. Yatsunami J, Tsuruta N, Ogata K, Wakamatsu K, Takayama K, Kawasaki M et al. Interleukin-8 participates in angiogenesis in non-small cell, but not small cell carcinoma of the lung. Cancer Lett 1997;120:101–108.
149. Arenberg DA, Keane MP, DiGiovine B, Kunkel SL, Morris SB, Xue YY et al. Epithelial-neutrophil activating peptide (ENA-78) is an important angiogenic factor in non-small cell lung cancer. J Clin Invest 1998;102:465–472.
150. Bostwick DG, Iczkowski KA. Microvessel density in prostate cancer: prognostic and therapeutic utility. Semin Urol Oncol 1998;16:118–123.
151. Fregene TA, Khanuja PS, Noto AC, Gehani SK, Van Egmont EM, Luz DA et al. Tumor-associated angiogenesis in prostate cancer. Anticancer Res 1993;13:2377–2381.
152. Veltri RW, Miller MC, Zhao G, Ng A, Marley GM, Wright GL, Jr. et al. Interleukin-8 serum levels in patients with benign prostatic hyperplasia and prostate cancer. Urology 1999;53:139–147.
153. Addison CL, Arenberg DA, Morris SB, Xue YY, Burdick MD, Mulligan MS, Iannettoni MD, Strieter RM. The CXC chemokine, monokine induced by interferon-gamma, inhibits non-small cell lung carcinoma tumor growth and metastasis. Hum Gene Ther 2000;11(2):247–261.
154. Kitadai Y, Haruma K, Sumii K, Yamamoto S, Ue T, Yokozaki H et al. Expression of interleukin-8 correlates with vascularity in human gastric carcinomas. Am J Pathol 1998;152:93–100.
155. Singh RK, Gutman M, Radinsky R, Bucana CD, Fidler IJ. Expression of interleukin 8 correlates with the metastatic potential of human melanoma cells in nude mice. Cancer Res 1994;54:3242–3247.
156. Cohen RF, Contrino J, Spiro JD, Mann EA, Chen LL, Kreutzer DL. Interleukin-8 expression by head and neck squamous cell carcinoma. Arch Otolaryngol Head Neck Surg 1995;121:202–209.
157. Chen Z, Malhotra PS, Thomas GR, Ondrey FG, Duffey DC, Smith CW et al. Expression of proinflammatory and proangiogenic cytokines in patients with head and neck cancer. Clin Cancer Res 1999;5:1369–1379.
158. Gurtsevitch VE, O'Conor GT, Lenoir GM. Burkitt's lymphoma cell lines reveal different degrees of tumorigenicity in nude mice. Int J Cancer 1988;41:87–95.
159. Sgadari C, Angiolillo AL, Cherney BW, Pike SE, Farber JM, Koniaris LG et al. Interferon-inducible protein-10 identified as a mediator of tumor necrosis in vivo. Proc Natl Acad Sci USA 1996;93:13791–13796.
160. Teruya-Feldstein J, Jaffe ES, Burd PR, Kanegane H, Kingma DW, Wilson WH et al. The role of Mig, the monokine induced by interferon-gamma, and IP-10, the interferon-gamma-inducible protein-10, in tissue necrosis and vascular damage associated with Epstein–Barr virus-positive lymphoproliferative disease. Blood 1997; 90:4099–4105.
161. Arenberg DA, Kunkel SL, Polverini PJ, Morris SB, Burdick MD, Glass MC et al. Interferon-gamma-inducible protein 10 (IP-10) is an angiostatic factor that inhibits human non-small cell lung cancer (NSCLC) tumorigenesis and spontaneous metastases. J Exp Med 1996;184: 981–992.
162. Minna JD. Neoplasms in the lung. In: Isselbacher KJ, ed. Principles of Internal Medicine, 12th edn. New York: McGraw-Hill, 1991:1102–1110.
163. Carney DN. Cancers of the lungs. In: Fishman AP, ed. Pulmonary Diseases and Disorders, 2nd edn. New York: McGraw-Hill, 1988:1885–2068.
164. Yuan A, Pan-Chyr Y, Chong-Jen Y, Lee Y, Yu-Tuang Y, Chi-Long C et al. Tumor angiogenesis correlates with histologic type and metastasis in non-small cell lung cancer. Am J Resp Crit Care Med 1995;152:2157–2162.
165. Koch AE, Volin MV, Woods JM, Kunkel SL, Connors MA, Harlow LA, Woodruff DC, Burdick MD, Strieter RM. Regulation of angiogenesis by the C–X–C chemokines interleukin-8 and epithelial neutrophil activating peptide 78 in the rheumatoid joint. Arthritis Rheum 2001;44(1):31–40.
166. Nicholls ES, Peruga A, Restrepo HE. Cardiovascular disease mortality in the Americas. World Health Stat Q 1993;46:134–150.
167. Fowler S, Berberian PA, Shio H, Goldfischer S, Wolinsky H. Characterization of cell populations isolated from aortas of rhesus monkeys with experimental atherosclerosis. Circ Res 1980;46:520–530.
168. Joris I, Zand T, Nunnari JJ, Krolikowski FJ, Majno G. Studies on the pathogenesis of atherosclerosis. I. Adhesion and emigration of mononuclear cells in the aorta of hypercholesterolemic rats. Am J Pathol 1983; 113:341–358.
169. Munro JM, Cotran RS. The pathogenesis of atherosclerosis: atherogenesis and inflammation. Lab Invest 1988; 58:249–261.
170. Folkman J. Angiogenesis in cancer, vascular, rheumatoid and other disease. Nat Med 1995;1:27–31.
171. Folkman J, Shing Y. Angiogenesis. J Biol Chem 1992; 267:10931–10934.
172. Simonini A, Moscucci M, Muller DW, Bates ER, Pagani FD, Burdick MD et al. IL-8 is an angiogenic factor in human coronary atherectomy tissue. Circulation 2000; 101:1519–1526.
173. Mitzner W, Lee W, Georgakopoulos D, Wagner E. Angiogenesis in the mouse lung. Am J Pathol 2000;157: 93–101.
174. Charan NB, Baile EM, Pare PD. Bronchial vascular congestion and angiogenesis. Eur Respir J 1997;10:1173–1180.
175. Charan NB, Carvalho P. Angiogenesis in bronchial circulatory system after unilateral pulmonary artery obstruction. J Appl Physiol 1997;82:284–291.
176. Schlaepfer K. Ligation of the pulmonary artery of one lung with and without resection of the phrenic nerve. Arch Surg 1924;9:25–94.
177. Michel RP, Hakim TS, Petsikas D. Segmental vascular

resistance in postobstructive pulmonary vasculopathy. J Appl Physiol 1990;69:1022–1032.
178. Turner-Warwick M. Precapillary systemic-pulmonary anastomoses. Thorax 1963;18:225–237.
179. Peao MND, Aguas AP, DeSa CM, Grande NR. Neoformation of blood vessels in association with rat lung fibrosis induced by bleomycin. Anat Rec 1994;238:57–67.
180. Keane MP, Belperio JA, Arenberg DA, Burdick MD, Xu ZJ, Xue YY et al. IFN-gamma-inducible protein-10 attenuates bleomycin-induced pulmonary fibrosis via inhibition of angiogenesis [In process citation]. J Immunol 1999;163:5686–5692.
181. Keane MP, Belperio JA, Moore TA, Moore BB, Arenberg DA, Smith RE et al. Neutralization of the CXC chemokine, macrophage inflammatory protein-2, attenuates bleomycin-induced pulmonary fibrosis. J Immunol 1999;162:5511–5518.
182. Ziesche R, Hofbauer E, Wittmann K, Petkov V, Block LH. A preliminary study of long-term treatment with interferon gamma-1b and low-dose prednisolone in patients with idiopathic pulmonary fibrosis [see comments]. New Engl J Med 1999;341:1264–1269.
183. Du Bois RM. Interferon gamma-1b for the treatment of idiopathic pulmonary fibrosis [editorial comment]. New Engl J Med 1999;341:1302–1304.

Section 3

CHEMOKINES AND NEURAL INFLAMMATION IN MODEL SYSTEMS

Section Editor: W.F. Hickey

Universes in Delicate Balance: Chemokines and the Nervous System
R.M. Ransohoff, K. Suzuki, A.E.I. Proudfoot, W.F. Hickey and J.K. Harrison, editors

Chapter 3.1

Expression, Functions and Interactions of Chemokines in CNS Trauma

V. Wee Yong

Departments of Oncology and Clinical Neurosciences, University of Calgary, 3330 Hospital Drive NW, Calgary, AB T2N 4N1, Canada

1. INTRODUCTION

Virtually all types of insults to the adult CNS lead to the infiltration of leukocytes into the CNS parenchyma. Since chemokines regulate the passage of leukocytes into tissues, it is reasonable to predict that chemokines are up-regulated in CNS insults. This chapter describes the chemokine expression that occurs in CNS trauma, such as that inflicted by a penetrating stab wound, corticectomy or contusion insults. It then reviews the stimulus for chemokine expression within the CNS and discusses the possible roles of chemokines in traumatic CNS injuries. Finally, the interactions between cytokines, chemokines and matrix metalloproteinases (MMPs) are emphasized.

2. CHEMOKINE EXPRESSION IN CNS TRAUMA

As with classical neuro-inflammatory conditions of the CNS such as multiple sclerosis, a traumatic injury to the CNS results in the induction or up-regulation of chemokine expression (reviewed in [1]). A chemokine commonly found to be up-regulated in trauma is macrophage chemoattractant protein (MCP)-1/CCL-2. In a brain stab wound model in neonatal rats, MCP-1 was elevated by 12 h whereas IP-10/CXCL10, MIP-1α/CCL3 and RANTES/CCL5 were not detected [2]. In situ hybridization combined with immunohistochemistry showed that astrocytes were the source of MCP-1 at the early time points, before any prominent infiltrates could be found. In brain stab wound injury to adult animals, an increase in MCP-1 mRNA was detected as early as at 3 h (but not at 1 h), peaked between 6 and 12 h, and then declined [3]. An increase of MCP-1 was also documented at 12–24 h following a contusion injury to the spinal cord of rats [4]. Similarly, a cryogenic lesion to the cortex resulted in an increase of MCP-1 that peaked at 6 h [5].

Elevations of other chemokines are also observed following CNS trauma. The presence of RANTES (regulated upon activation normal T expressed and secreted) and MIP (macrophage inflammatory protein)-1β/CCL4 is detected by immunohistochemistry by 1d of brain stab wound injury. Double labeling studies show that MIP-1β, but not RANTES, was expressed by reactive astrocytes near the lesion site [6]. Using RT-PCR and dot blot hybridization assays of the rat spinal cord after contusion injury, gro-α/CXCL1 was 30 times higher than control values at 6 h and decayed thereafter. IP-10/CXCL10 was elevated at 6 and 12 h while MIP-1α and RANTES were minimally affected [4].

Although the above discussion would suggest that chemokines are expressed only hours following CNS trauma, an early rise in chemokine expression has indeed been noted after spinal cord injury in rats. Thus, MIP-1α and MIP-1β were up-regulated in the first hour of injury, in cells surrounding the lesion area (probably CNS resident cells). Around 24 h, the mRNA for MIP-1α and -β was expressed in a generalized way in the gray matter of the entire spinal cord [7]. Muessel et al. [8] detected the elevation of MCP-1 mRNA by 1 h of an aspiration lesion of the visual cortex. Protein content of MCP-1 was also measured, and this was found to be significantly elevated at 6 h, peaked at 12 h, and remained above baseline levels for at least 1 week after lesion [8]. In another study involving cortical lesions, particularly with the further addition of lipopolysaccharide, all chemokines studied (MCP-1, MIP-1α, MIP-1β, RANTES, IP-10 and N51/KC/CXCL1) were elevated by 1–2 h after injury [9]. Recently, Lee et al. [10] reported that MCP-1 mRNA was increased by 1 h after spinal cord contusion injury in rats.

Chemokines are also expressed following traumatic brain injuries in humans. In children with severe head injury, interleukin (IL)-8/CXCL8 was found to be elevated in the cerebrospinal fluid; multivariate analysis demonstrated a strong association of cerebrospinal fluid IL-8 with mortality [11].

In summary, a number of chemokines are elevated in CNS trauma. Cellular sources at early time points include cells intrinsic to the CNS, and, indeed, all neural cell types can produce chemokines in various situations [12,13]. Later sources of chemokines in CNS trauma include the leukocytes that infiltrate the CNS parenchyma following the initial burst of CNS-intrinsic chemokines. While results may differ in the spectrum of chemokines that are found to be up-regulated as a result of the insult, the rise of chemokines occurs early in the course of traumatic brain injuries.

3. STIMULI FOR CHEMOKINE PRODUCTION WITHIN THE CNS

Reported stimuli for chemokine expression in a variety of cell types are the inflammatory cytokines, bacterial products, viruses and other chemotactic factors (e.g. C5a). These cause similar effects in CNS-derived cells. Several tissue-culture-based studies have demonstrated that a variety of chemokines are expressed by astrocytes and microglia, and that these are up-regulated by inflammatory cytokines that include IL-1, IL-6, tumor necrosis factor (TNF)-α and colony stimulating factor (CSF)-1 [14–21]. MIP-2 and KC induce their own production in astrocytes [22]; similarly, MIP-1α and MIP-1β increase their own levels in cultured astrocytes and microglia [23], thereby providing an amplification cascade for chemokine activity.

Cytokines may also down-regulate chemokine expression by glial cells. In this regard, the anti-inflammatory cytokines, IL-10 and transforming growth factor (TGF)-β were found to inhibit the production of RANTES by microglia in culture [20].

What might be the in vivo stimulus for chemokine expression in CNS trauma? If chemokines are expressed by the first hour of injury, then the initiator of chemokine expression must precede this. Since cytokines are potent regulators of chemokine transcription, these are viable candidates as regulators of chemokine expression. However, it has become clear that leukocytes, which are rich sources of inflammatory cytokines, infiltrate the CNS parenchyma only after a substantial delay. Neutrophils, the first leukocyte to transit into the CNS parenchyma following trauma, is detected only after 3 h [24,25]. Recently, we described that after a corticectomy injury to the adult mouse brain, IL-1β is increased within 15 min and that the cellular source is microglia [26]. Thus, this CNS-initiated inflammation, which triggers the production of ciliary neurotrophic factor [26], could be a stimulus to initiate chemokine expression. It would be interesting to evaluate mice genetically deficient for IL-1β, in order to ascertain whether chemokines are expressed in a delayed time course compared to wild type controls, if at all.

When blood is leaked into the CNS parenchyma, as occurs in traumatic insults or hemorrhagic strokes, the complement cascade can become activated. Thus, complements, including C5a which itself is a chemotactic source, represent another potential stimulus to increase chemokine expression within the CNS.

Adhesive interaction between cells with extracellular matrix (ECM) or other cells serves as another means to regulate chemokine production [27]. With respect to leukocytes, the adherence of peripheral blood monocytes onto plastic up-regulated IL-8 [28]. Adhesion of leukocytes onto endothelial cells also promotes chemokine expression, by both cell types [27]. Thus, it is possible that CNS trauma alters cell–cell adhesion and cell–ECM interaction within the CNS, resulting in chemokine expression.

4. FUNCTIONS OF CHEMOKINES IN CNS TRAUMA

The function of chemokines following CNS trauma is likely to be multi-dimensional (Fig. 1). The most

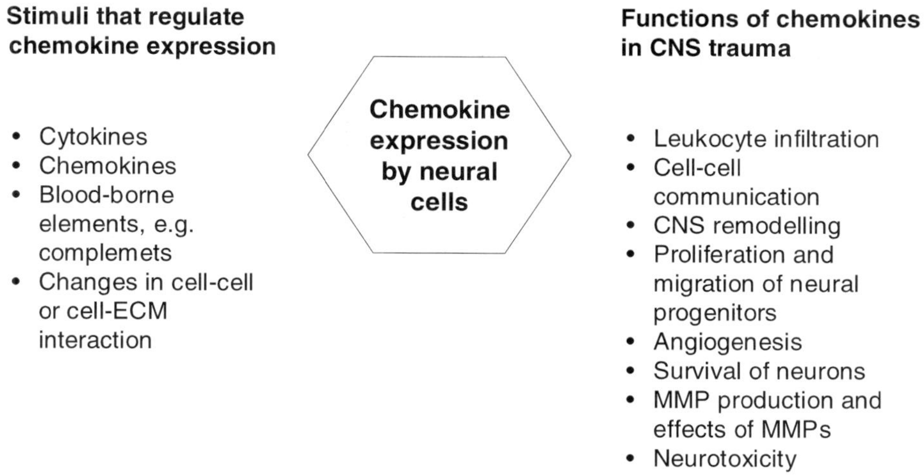

Figure 1. Expression and functions of chemokines in CNS trauma.

obvious role is the recruitment of leukocytes into the CNS. Indeed, in brain stab wound injured rats infused with antisense oligonucleotides to MCP-1, the number of infiltrating macrophages was reduced by 30% [29]. The vMIPII molecule, which blocks the activity of several CC and CXC chemokine receptors [30], reduced the infiltration of neutrophils and monocytes following spinal cord contusion injury in rats [31].

It is difficult to evaluate whether the chemokine-mediated leukocyte trafficking is generally detrimental or beneficial to the CNS. Infiltrating neutrophils and macrophages are phagocytic cells that can clear the cellular debris that accumulate in CNS trauma. Nonetheless, these cells are also the source of numerous molecules, including proteases, radicals and inflammatory cytokines, which can injure the constituents of the CNS. Similarly, infiltrating T lymphocytes and NK (natural killer) cells are potential amplifiers of an immune cascade within the CNS. However, recent evidence suggests that T and NK cells are also the source of neurotrophic factors and neurotrophins that prevent the secondary degeneration of neurons that follows traumatic CNS injuries [32,33]. On balance, it remains unclear whether the infiltration of leukocytes into the CNS has a general negative or positive role for recovery. Both roles probably exist and the challenge is to discover means to tip the balance towards a reparative role for infiltrating leukocytes.

Chemokines themselves can also have direct activities on CNS constituents unrelated to their recruitment of leukocytes. CNS cells including neurons have receptors for chemokines [34]. Indeed, most, if not all, cells of the CNS can produce chemokines and have receptors for chemokines, suggesting that chemokines may be critical for intercellular communication within the CNS [12]. Experimentally, rat hippocampal neurons in culture produce calcium fluxes in response to SDF-1, RANTES and fractalkine [35]. Increased calcium mobilization was also observed in cultured astrocytes and microglia exposed to SDF-1α [36]. An interesting example of chemokines in intercellular communication is provided by fractalkine/CX3CL1 (or its murine analog, neurotactin) interacting with its receptor, CX3CR1. Harrison et al. [37] described the widespread distribution of fractalkine in the rat brain, primarily on neurons. After facial motor nerve axotomy, the number of perineuronal microglia expressing CX3CR1 increased. As fractalkine can stimulate the chemotaxis of microglia in culture, these authors suggested that neuronally derived fractalkine mediates the interaction between neurons and microglia in vivo [37]. Recent studies in CX3CR1-null mice challenge this hypothesis [38]; however, in neonatal rat brains subjected to excitotoxic injury, the concordant rise of MCP-1 and the expression of CCR2 on microglia invites the possibility that MCP-1 may regulate the microglia response to acute brain injury [39]. On balance, it appears likely that communication between neurons and other neural cell types in CNS trauma utilize chemokines as messengers.

Increasing evidence implicates chemokines in

CNS development and remodeling. Mice genetically deficient for CXCR4 or its ligand, SDF-1/CXCL12, have abnormalities in cerebellar development associated with the derailed migration of granule cells [40]. While chemokines are unlikely to affect the migration of neurons in the normal adult CNS, it is possible that the chemokines that are expressed following trauma may help to remodel the CNS. MCP-1 regulates the migration of astrocytes while SDF-1α alters microglia chemotaxis [36]. Chemokines may also regulate the proliferation and migration of neural progenitor cells to the sites of CNS trauma in an attempt to replenish lost cells. In this regard, it is noteworthy that the chemokine, gro-α, has been shown to be a potent mitogen for spinal cord oligodendrocyte precursors by enhancing the effect of the major mitogen for oligodendrocyte precursors, platelet-derived growth factor [41].

In addition to promoting CNS remodeling following trauma, chemokines may also confer trophic activity on neurons that survive the initial insult. Although this concept has yet to be proven in vivo, the chemokine IL-8/CXCL8 was noted to produce a pronounced survival of hippocampal neurons in culture [42].

Yet another potential activity of chemokines following CNS trauma is that of angiogenesis. CXC chemokines with an ELR (glutamate–leucine–arginine) motif at their N-terminus (e.g. gro-α and IL-8) have a positive role in angiogenesis [43], and may help to recruit new blood vessels to areas of repair. However, this function has to be tempered by the finding that the non-ELR CXC chemokines (e.g. IP-10 and MIG) are angiostatic.

While the above discussions allude to a favorable role for chemokines in CNS trauma, it should be noted that when the vMIPII chemokine receptor antagonist was used to attenuate the influx of neutrophils and monocytes in spinal cord injury, the degree of neuronal loss in vMIPII-treated animals was reduced compared to controls [31]. This highlights the view that chemokines in CNS trauma may have detrimental effects. Whether this is a direct consequence of toxicity of chemokines on neurons, or whether it is the indirect result of vMIPII limiting the neurotoxic actions of chemokine-recruited leukocytes, remains to be determined. Furthermore, how chemokines interact with cytokines and matrix metalloproteinases (next section) in the context of CNS trauma, may also determine the outcome of the injury.

In summary, the chemokines that are expressed in CNS trauma appear to have multiple functions. Much remains to be done to define precisely which functions of chemokines are critical for CNS recovery, if at all.

5. INTERACTIONS OF CHEMOKINES, CYTOKINES AND MATRIX METALLOPROTEINASES

The family of matrix metalloproteinases (MMPs) are proteolytic enzymes that can degrade all components of the ECM. MMPs are implicated in a variety of physiological and pathological processes that involve some degree of ECM turnover (reviewed in [44]). It has become evident that MMPs are elevated in CNS trauma [45], where their functions remain elusive. Nonetheless, possible beneficial roles of MMPs in CNS trauma include the remodeling of the post-injury ECM to modulate cell migration, signaling and survival. This contention has its basis in the multiple neurotrophic roles of MMPs. Thus, MMPs regulate growth factor availability, by virtue of releasing ECM- or cell surface-bound growth factors [46–48]. MMPs may also act as an intermediary in growth factor signaling. The activation of G-protein-coupled receptors by ligands including bombesin or carbachol induced the transactivation of the epidermal growth factor receptor and this occurred through a metalloproteinase intermediary [49]. When active MMP-9 is focussed on the cell membrane by its interaction with CD44 (hyaluronan receptor), angiogenesis is promoted through the conversion of pro-TGFβ to TGF-β [50]. MMPs are involved in growth cone activity since MMP inhibitors attenuate this process [51]. Furthermore, MMP-2 produced by dorsal root ganglion neurons inactivates inhibitory chondroitin sulfate proteoglycans and expose the permissive laminin substrate to facilitate the elongation of neurites [52]. Despite these observations, however, MMPs also have detrimental properties in CNS trauma. Indeed, mice deficient for MMP-9 have less behavioral deficits and smaller lesion volumes following spinal cord injury when compared to wild-type controls [53].

The discussion of MMPs is relevant to this re-

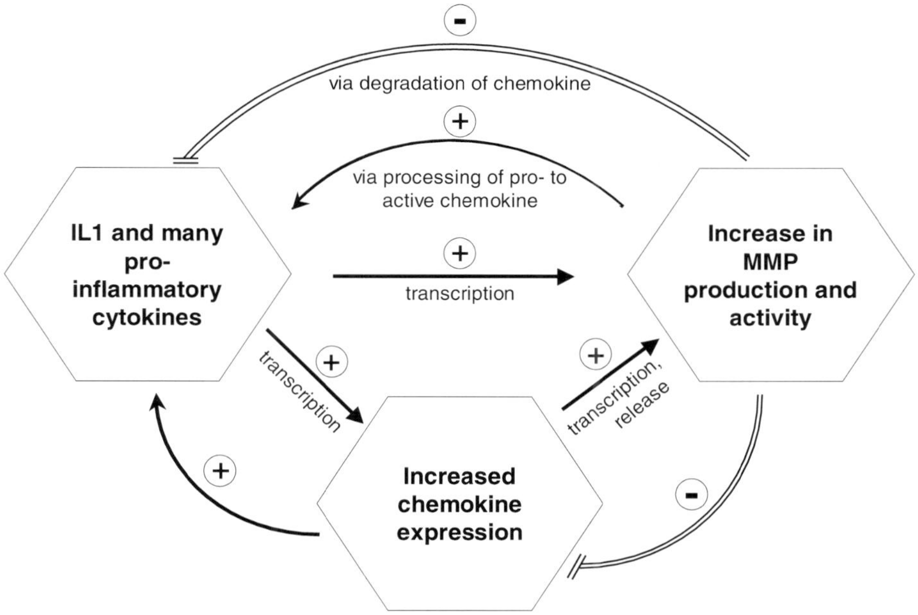

Figure 2. Interplay between chemokines, MMPs and cytokines.

view since the production of MMPs can be regulated by chemokines. Thus, MCP-1, -2 and -3 have been shown to induce the production of MMP-9 [54–56]; MCP-1, MIP-1β, RANTES, IL-8 and fractalkine increased MMP-9 in microglia [57]. IL-8 has been found to cause the release of MMP-9 from neutrophils [58]. In addition, metalloproteinases remove CXCR1 and CXCR2 from the surface of monocytoid cells, and thus negate their response to chemokines [59]. In contrast, the chemokine-induced recruitment of neutrophils is impaired in MMP-9 null mice [60], suggesting that MMP-9 aids chemokine functions.

Given that MMPs are upregulated by 6 h after a corticectomy injury to the adult mouse brain [45], there is tremendous potential for chemokines and MMPs to interact in the post-trauma CNS (Fig. 2). It is likely that chemokines may regulate the production and activity of MMPs in CNS trauma, and produce a larger spectrum of effects attributable to MMP activity. These interactions deserve further investigations.

It has been noted previously that truncated forms of chemokines may serve to block the actions of the full length chemokine. Thus, a naturally occurring human RANTES, lacking 2 N-terminal residues {RANTES(3-68)}, blocks the effects of RANTES, MIP or MCP-3 in stimulating monocyte chemotaxis [61]. Deletions of amino acids 2–8 of MCP-1 results in a polypeptide that acts as its inhibitor [62]. Recently, MMPs have been implicated in the N-terminal processing of chemokines [63]. Using a yeast 2-hybrid system, MCP-3 (but not MCP-1) was found to be a physiological substrate of MMP-2 (but not MMP-7, -8 or -9). The resultant cleaved MCP-3 (first 4 N-terminal amino acids removed) bound to CCR-1, -2 and -3 as an antagonist, decreased calcium mobilization in THP-1 cells in response to MCP-1 or -3, and reduced migration of these cells. In an in vivo model of zymosan A-induced peritonitis, inflammation was reduced by ip injection of the cleaved MCP-3 peptide. In this manner, the interaction between MMPs and chemokines dampens the inflammatory response [63]. It is conceivable that such interactions may occur in the CNS post-trauma.

While the interaction between MMPs and chemokines may lead to an attenuation of the inflammatory response, the converse can also occur. In this regard, IL-8 is processed by MMP-9 at the aminoterminus to result in a truncation variant [IL-8(7-77)] with increased activity [64]. The MMP-9 potentiation of IL-8 response is selective in that MMP-9 degrades CTAP-III, PF-4 and gro-α slowly, and does not affect RANTES and MCP-2 [64].

Finally, the interactions between chemokines, cytokines and MMPs deserve mention. As noted earlier, cytokines can regulate chemokine production by neural cells. A large literature also demonstrates that inflammatory cytokines, and particularly IL-1β, are potent transcriptional activators of several MMPs [65]. Furthermore, MMPs can process several procytokines, including pro-TNF-α [66] or pro-IL-1β [67], into their active forms. Alternatively, MMPs may degrade mature IL-1β into inactive fragments [67]. Ito et al. [68] also demonstrated that prolonged incubation with MMP-1, -2, -3 and -9 resulted in IL-1β (but not IL-1α) degradation and loss of its biological activity. Interestingly, the primary site of cleavage of IL-1β by MMP-2 was identified as the Glu^{25}–Leu^{26} bond, not unlike the glutamate–leucine bond of chemokines with the ELR motif. Finally, MCP-1 was recently found to induce MMP-1 in primary skin fibroblasts but this was mediated through an IL-1α intermediary [69].

On balance, a complex and intricate interaction exists between chemokines, cytokines and MMPs and the individual or sum effect of this on CNS injury and recovery has yet to be elucidated.

6. FINAL CONSIDERATIONS

It is clear that chemokines have multiple functions in the CNS following trauma, other than being mere chemoattractants for leukocytes. Many of these functions may contribute to limit CNS destruction or may enhance repair, although detrimental activities also exist for chemokines. Clearly, there is a need to establish the many functions of chemokines that are up-regulated following trauma, so that the CNS environment can be manipulated towards enabling recovery from CNS insults.

REFERENCES

1. Glabinski AR, Ransohoff RM. Chemokines and chemokine receptors in CNS pathology. J Neurovirol 1999;5:3–12.
2. Glabinski AR, Balasingam V, Tani M, Kunkel SL, Strieter RM, Yong VW, Ransohoff RM. Chemokine monocyte chemoattractant protein 1 is expressed by astrocytes after mechanical injury to the brain. J Immunol 1996;156:4363–4368.
3. Berman JW, Guida MP, Warren J, Amat J, Brosnan CF. Localization of monocyte chemoattractant peptide-1 expression in the central nervous system in experimental autoimmune encephalomyelitis and trauma in the rat. J Immunol 1996;156:3017–3023.
4. McTigue DM, Tani M, Krivacic K, Chernosky A, Kelner GS, Maciejewski D, Maki R, Ransohoff RM, Stokes BT. Selective chemokine mRNA accumulation in the rat spinal cord after contusion injury. J Neurosci Res 1998;53:368–376.
5. Grzybicki D, Moore SA, Schelper R, Glabinski AR, Ransohoff RM, Murphy S. Expression of monocyte chemoattractant protein (MCP-1) and nitric oxide synthase-2 following cerebral trauma. Acta Neuropathol 1998;95:98–103.
6. Ghirnikar RS, Lee YL, He TR, Eng LF. Chemokine expression in rat stab wound brain injury. J Neurosci Res 1996;46:727–733.
7. Bartholdi D, Schwab ME. Expression of pro-inflammatory cytokine and chemokine mRNA upon experimental spinal cord injury in mouse: an in situ hybridization study. Eur J Neurosci 1997;7:1422–1438.
8. Muessel MJ, Berman NE, Klein RM. Early and specific expression of monocyte chemoattractant protein-1 in the thalamus induced by cortical injury. Brain Res 2000;870:211–221.
9. Hausmann EH, Berman NE, Wang YY, Meara JB, Wood GW, Klein RM. Selective chemokine mRNA expression following brain injury. Brain Res 1998;788:49–59.
10. Lee YL, Shih K, Bao P, Ghirnikar RS, Eng LF. Cytokine chemokine expression in contused rat spinal cord. Neurochem Int 2000;36:417–425.
11. Whalen MJ, Carlos TM, Kochanek PM, Wisnewski SR, Bell MJ, Clark RS, DeKosky ST, Marion DW, Adelson PD. Interleukin-8 is increased in cerebrospinal fluid of children with severe head injury. Crit Care Med 2000;28:929–934.
12. Arsenio VC, Campbell IL. Chemokines in the CNS: plurifunctional mediators in diverse states. Trends Neurosci 1999;22:504–512.
13. Hesselgesser J, Horuk R. Chemokine and chemokine receptor expression in the central nervous system. J Neurovirol 1999;5:13–26.
14. Barna BP, Pettay J, Barnett GH, Zhou P, Iwasaki K, Estes ML. Regulation of monocyte chemoattractant protein-1 expression in adult human non-neoplastic astrocytes is sensitive to tumor necrosis factor (TNF) or antibody to 55-kDa TNF receptor. J Neuroimmunol 1994;50:101–107.
15. Hurwitz AA, Lyman WD, Berman JW. Tumor necrosis factor α and transforming growth factor β upregulate astrocyte expression of monocyte chemoattractant protein-1. J Neuroimmunol 1995;57:193–198.
16. Hayashi M, Luo Y, Laning J, Strieter RM, Dorf ME. Production and function of monocyte chemoattractant protein-1 and other β chemokines in murine glial cells. J Neuroimmunol 1995;60:143–150.
17. Majumder S, Zhou LZH, Ransohoff RM. Transcriptional gene regulation of chemokine gene expression in astrocytes. J Neurosci Res 1996;45:758–769.
18. Calvo CF, Yoshimura T, Gelman M, Mallat M. Production of monocyte chemotactic protein-1 by rat brain macrophages. Eur J Neurosci 1996;8:1725–1734.

19. McManus CM, Brosnan CF, Berman JW. Cytokine-induced MIP-1α and MIP-1β expression in human fetal microglia. J Immunol 1998;160:1449–1455.
20. Hu S, Chao CC, Ehrlich LC, Sheng WS, Sutton RL, Rockswold GL, Peterson PK. Inhibition of microglial cell RANTES production by IL-10 and TGF-β. J Leukoc Biol 1999;65:815–821.
21. Miyamoto Y, Kim SU. Cytokine-induced production of macrophage inflammatory protein-1α (MIP-1α) in cultured human astrocytes. J Neurosci Res 1999;55:245–251.
22. Luo Y, Fischer FR, Hancock WW, Dorf ME. Macrophage inflammatory protein-2 and KC induce chemokine production by mouse astrocytes. J Immunol 2000;165:4015–4023.
23. McManus CM, Weidenheim K, Woodman SE, Nunez J, Hesselgesser J, Nath A, Berman JW. Chemokine and chemokine-receptor expression in human glial elements. Am J Pathol 2000;156:1441–1452.
24. Dusart I, Schwab ME. 1994. Secondary cell death and the inflammatory reaction after dorsal hemisection of the rat spinal cord. Eur J Neurosci 1994;6:712–724.
25. Schnell L, Schneider R, Berman MA, Perry VH, Schwab ME. Lymphocyte recruitment following spinal cord injury in mice is altered by prior viral exposure. Eur J Neurosci 1997;9:1000–1007.
26. Herx LM, Rivest S, Yong VW. Central nervous system-initiated inflammation and neurotrophism in trauma: IL-1β is required for the production of ciliary neurotrophic factor. J Immunol 2000;165:2232–2239.
27. Smith RE, Hogaboam CM, Strieter RM, Lukacs NW, Kunkel SL. Cell-to-cell and cell-to-matrix interactions mediate chemokine expression: an important component of the inflammatory lesion. J Leukoc Biol 1997;62:612–619.
28. Kasahara K, Strieter RM, Chensue SW, Sandiford TJ, Kunkel SL. Mononuclear cell adherence induces neutrophil chemotactic factor/interleukin-8 gene expression. J Leukoc Biol 1991;50:287–295.
29. Ghirnikar RS, Lee YL, Li JD, Eng LF. Chemokine inhibition in rat stab wound injury using antisense oligonucleotides. Neurosci Lett 1998;247:21–24.
30. Kledal TN, Rosenkilde MM, Coulin F, Simmons G, Johnsen AH, Alouani S, Power CA, Luttichau HR, Gerstoft J, Clapham PR, Clark-Lewis I, Wells TNC, Schwartz TW. A broad spectrum chemokine antagonist encoded by Kaposi's sarcoma-associated herpesvirus. Science 1997;277:1656–1659.
31. Ghirnikar RS, Lee YL, Eng LF. Chemokine antagonist infusion attenuates cellular infiltration following spinal cord contusion injury in rat. J Neurosci Res 2000;59:63–73.
32. Hauben E, Butovsky O, Nevo U, Yoles E, Moalem G, Agranov E, Mor F, Leibowitz-Amit R, Pevsner E, Akselrod S, Neeman M, Cohen IR, Schwartz M. Passive or active immunization with myelin basic protein promotes recovery from spinal cord contusion. J Neurosci 2000;20:6421–6430.
33. Hammarberg H, Lidman O, Lundberg C, Eltayeb SY, Gielen AW, Muhallab S, Svenningsson A, Linda H, van der Meide PH, Cullheim S, Olsson T, Piehl F. Neuroprotection by encephalomyelitis and neurotrophin production by CNS-infiltrating T and natural killer cells. J Neurosci 2000;20:5283–5291.
34. Horuk R, Martin AW, Wang Z, Schweitzer L, Gerassimides A, Guo H, Lu Z, Hesselgesser J, Perez HD, Kim J, Parker J, Hadley TJ, Peiper SC. Expression of chemokine receptors by subsets of neurons in the central nervous system. J Immunol 1997;158:2882–2890.
35. Meucci O, Fatatis A, Simen AA, Bushell TJ, Gray PW, Miller RJ. Chemokines mediate hippocampal neuronal signaling and gp120 neurotoxicity. Proc Natl Acad Sci USA 1998;95:14500–14505.
36. Tanabe S, Heesen M, Yoshizawa I, Berman MA, Luo Y, Bleul CC, Springer TA, Okuda K, Gerard N, Dorf ME. Functional expression of the CXC-chemokine receptor-4/fusin on mouse microglial cells and astrocytes. J Immunol 1997;159:905–911.
37. Harrison JK, Jiang Y, Chen S, Xia Y, Maciejewski D, McNamara RK, Streit WJ, Salafranca MN, Adhikari S, Thompson DA, Botti P, Bacon KD, Feng L. Role for neuronally derived fractalkine in mediating interactions between neurons and CX3CR1-expressing microglia. Proc Natl Acad Sci USA 1998;95:10896–10901.
38. Jung et al. Mol Cell Biol 2000;20:4106–4114.
39. Galasso JM, Miller MJ, Cowell RM, Harrison JK, Warren JS, Silverstein FS. Acute excitotoxic injury induces expression of monocyte chemoattractant protein-1 and its receptor, CCR2, in neonatal rat brain. Exp Neurol 2000;165:295–305.
40. Ma Q, Jones D, Borghesani PR, Segal RA, Nagasawa T, Kishimoto T, Bronson RT, Springer TA. Impaired B-lymphopoiesis, myelopoiesis, and derailed cerebellar neuronal migration in CXCR4- and SDF-1-deficient mice. Proc Natl Acad Sci USA 1998;95:9448–9453.
41. Robinson S, Tani M, Streiter RM, Ransohoff RM, Miller RH. The chemokine growth-regulated oncogene-α promotes spinal cord oligodendrocyte precursor proliferation. J Neurosci 1998;18:10457–10463.
42. Araujo DM, Cotman CW. Trophic effects of interleukin-4, -7 and -8 on hippocampal neuronal cultures: potential involvement of glial-derived factors. Brain Res 1993;600:49–55.
43. Belperio JA, Keane MP, Arenberg DA, Addison CL, Ehlert JE, Burdick MD, Strieter RM. CXC chemokines in angiogenesis. J Leukoc Biol 2000;68:1–8.
44. Yong VW, Power C, Forsyth P, Edwards DR. Metalloproteinases in biology and pathology of the nervous system. Nat Rev Neurosci 2001;2:502–511.
45. Vecil GG, Larsen PH, Herx LM, Besson A, Goodyer CG, Yong VW. Interleukin-1 is a key regulator of matrix metalloproteinase-9 in human neurons in culture and following mouse brain trauma in vivo. J Neurosci Res 2000;61:212–224.
46. Levi E, Fridman R, Miao HQ, Ma YS, Yayon A, Vlodavsky I. Matrix metalloproteinase-2 releases active soluble ectodomain of fibroblast growth factor receptor 1. Proc Natl Acad Sci 1996;93:7069–7074.
47. Whitelock JM, Murdoch AD, Iozzo RV, Underwood PA. The degradation of human endothelial cell-derived perlecan and release of bound basic fibroblast growth factor by stromelysin, collagenase, plasmin, and heparanases. J Biol Chem 1996;271:10079–10086.

48. Suzuki M, Raab G, Moses MA, Fernandez CA, Klagsbrun M. Matrix metalloproteinase-3 releases active heparin-binding EGF-like growth factor by cleavage of a specific juxtamembrane site. J Biol Chem 1997;272:31730–31737.
49. Prenzel N, Zwick E, Daub H, Leserer M, Abraham R, Wallasch C, Ullrich A. EGF receptor transactivation by G-protein-coupled receptors requires metalloproteinase cleavage of proHB-EGF. Nature 1999;402:884–888.
50. Yu Q, Stamenkovic I. Cell surface-localized matrix metalloproteinase-9 proteolytically activates TGF-beta and promotes tumor invasion and angiogenesis. Genes Dev 2000;14:163–176.
51. Sheffield JB, Krasnopolsky V, Dehlinger E. Inhibition of retinal growth cone activity by specific metalloproteinase inhibitors in vitro. Dev Dyn 1994;200:79–88.
52. Zuo J, Ferguson TA, Hernandez YJ, Stetler-Stevenson WG, Muir D. Neuronal matrix metalloproteinase-2 degrades and inactivates a neurite-inhibiting chondroitin sulfate proteoglycan. J Neurosci 1998;18:5203–5211.
53. Wang X, Jung J, Asahi M, Chwang W, Russo L, Moskowitz MA, Dixon CE, Fini ME, Lo EH. Effects of matrix metalloproteinase-9 gene knock-out on morphological and motor outcomes after traumatic brain injury. J Neurosci 2000;20:7037–7042.
54. Van Damme J, Proost P, Lanaerts JP, Opdenakker G. Structural and functional identification of two human, tumor-derived monocyte chemotactic proteins (MCP-2 and MCP-3) belonging to the chemokine family. J Exp Med 1992;176:59–65.
55. Montovani A, Sozzani S, Bottazzi B, Peri G, Sciacca FL, Locati M, Colotta F. Monocyte chemotactic protein-1 (MCP-1): signal transduction and involvement in the regulation of macrophage traffic in normal and neoplastic tissues. Adv Exp Med Biol 1993;351:47–54.
56. Stuve O, Chabot S, Jung SS, Williams G, Yong VW, Chemokine-enhanced migration of T lymphocytes is antagonized by interferon beta-1b through an effect on matrix metalloproteinase-9. J Neuroimmunol 1997;80:38–46.
57. Cross AK, Woodroofe MN. Chemokine modulation of matrix metalloproteinase and TIMP production in adult rat brain microglia and a human microglial cell line in vitro. GLIA 1999;28:183–189.
58. Pugin J, Widmer MC, Kossodo S, Liang CM, Preas HLN, Suffredini AF. Human neutrophils secrete gelatinase B in vitro and in vivo in response to endotoxin and proinflammatory mediators. Am J Respir Cell Mol Biol 1999;20:458–464.
59. Khandaker MS, Mitchell G, Xu L, Andrews JD, Singh R, Leung H, Madrenas J, Ferguson SSG, Feldman RD, Kelvin DJ. Metalloproteinases are involved in lipopolysaccharide- and tumor necrosis factor-α-mediated regulation of CXCR1 and CXCR2 chemokine receptor expression. Blood 1999;93:2173–2185.
60. D'Haese A, Wuyts A, Dillen C, Dubois B, Billau A, Heremans H, Van Damme J, Arnold B, Opdenakker G. In vivo neutrophil recruitment by granulocyte chemotactic protein-2 is assisted by gelatinase B/MMP-9 in the mouse. J Interferon Cytokine Res 2000;20:667–674.
61. Struyl S, De Meester I, Scharpe S, Lenaerts JP, Manten P, Wang JM, Proost P, Van Damme J. Natural truncation of RANTES abolishes signaling through the CC chemokine receptors CCR1 and CCR3, impairs its chemotactic potency and generates a CC chemokine inhibitor. Eur J Immunol 1998;28:1262–1271.
62. Gong JH, Ratkay LG, Waterfield JD, Clark-Lewis I. An antagonist of monocyte chemoattractant protein 1 (MCP-1) inhibits arthritis in the MRL-lpr mouse model. J Exp Med 1997;186:131–137.
63. McQuibban GA, Gong JH, Tam EM, McCulloch CA, Clark-Lewis I, Overall CM. Inflammation dampened by gelatinase A cleavage of monocyte chemoattractant protein-3. Science 2000;289:1202–1206.
64. Van den Steen PE, Proost P, Wuyts A, Van Damme J, Opdenakker G. Neutrophil gelatinase B potentiates interleukin-8 tenfold by aminoterminal processing, whereas it degrades CTAP-III, PF-4, and GRO-alpha and leaves RANTES and MCP-2 intact. Blood 2000;96:2673–2681.
65. Yong VW. The potential use of MMP inhibitors to treat CNS diseases. Expert Opin Invest Drugs 1999;8:255–268.
66. English WR, Puente XS, Freije JM, Knauper V, Amour A, Merryweather A, Lopez-Otin C, Murphy G. Membrane type 4 matrix metalloproteinase (MMP17) has tumor necrosis factor-alpha convertase activity but does not activate pro-MMP2. J Biol Chem 2000;275:14046–14055.
67. Schonbeck U, Mach F, Libby P. Generation of biologically active IL-1 beta by matrix metalloproteinases: a novel caspase-1-independent pathway of IL-1β processing. J Immunol 1998;161:3340–3346.
68. Ito A, Mukaiyama A, Itoh Y, Nagase H, Thogersen IB, Enghild JJ, Sasaguri Y, Mori Y. Degradation of interleukin 1β by matrix metalloproteinases. J Biol Chem 1996;271:14657–14660.
69. Yamamoto T, Eckes B, Mauch C, Hartmann K, Krieg T. Monocyte chemoattractant protein-1 enhances gene expression and synthesis of matrix metalloproteinase-1 in human fibroblasts by an autocrine IL-α loop. J Immunol 2000;164:6174–6179.

Universes in Delicate Balance: Chemokines and the Nervous System
R.M. Ransohoff, K. Suzuki, A.E.I. Proudfoot, W.F. Hickey and J.K. Harrison, editors

Chapter 3.2

Animal Models of Multiple Sclerosis

William J. Karpus

Department of Pathology, Northwestern University Medical School, Chicago, IL 60611, USA

Experimental autoimmune encephalomyelitis (EAE) has been extensively studied as a model for multiple sclerosis (MS) and is characterized by progressive ascending clinical paralysis resulting from central nervous system (CNS) mononuclear cell infiltration that consists of antigen specific and nonspecific $CD4^+$ and $CD8^+$ T-cells as well as macrophages, but no demonstrable polymorphonuclear cell infiltration [1,2]. The mechanism of immune-mediated inflammation in this animal model for MS has been extensively studied in an effort to develop therapeutic modalities for tissue-specific autoimmune inflammatory diseases. Despite this extensive investigation, the mechanism by which mononuclear cells traffic to the CNS and accumulate prior to and during clinical CNS disease is not well understood. Experimental evidence has suggested that following peripheral activation T-cells rapidly traffic to the CNS within 24 h and begin the process of accumulation [3,4]. It is widely believed that activation of antigen-specific Th1 lymphocytes results in a population of cells capable of activating additional cell types such as macrophages and microglia that act as the effector cells in the demyelinating process [5,6]. Entry of T-cells into the CNS requires an interaction with endothelial-expressed P-selectin as an early migration event [7], $\alpha_4\beta_1$ (VLA-4) integrin expression on the T-cells [8] for interaction with VCAM-1 expressed on cerebrovascular endothelial cells. Treatment of mice with anti-VLA-4 monoclonal antibodies at the initiation of disease has been shown to prevent development of CNS inflammation and subsequent clinical disease [9]. The importance of $\alpha_L\beta_2$ (LFA-1) [10,11] and $\alpha_M\beta_2$ (Mac-1) [12] in the disease process was similarly demonstrated using antibody therapy. Whether LFA-1 or Mac-1 are required for the migration of T-cells and monocytes into the CNS or whether they serve as adhesion molecules during antigen presentation and T-cell activation has not been conclusively determined. A role for selectins in the initial T-cell migration into the CNS has not been demonstrated although L-selectin may function during later steps of disease pathogenesis [13] perhaps by effector cells in the demyelinating process [14]. Our current concept of CNS mononuclear cell infiltration occurs as a regulated, step-wise process beginning with antigen-specific T-cell infiltration and proceeding with infiltration of monocytes, antigen-nonspecific T-cells, and spreading epitope-specific T-cells. The accumulation of these cells in the CNS is the prelude to inflammation and self tissue damage and subsequent clinical manifestation.

The function of chemokines in the entry of T-cells and monocytes into the CNS during disease pathogenesis has been the subject of intense investigation. According to the three-step model of leukocyte migration, chemokines are thought to be required for leukocyte integrin activation and subsequent transendothelial migration [15]. However, in the absence of CNS inflammation, most chemokines are not expressed [16]. The exceptions to this are fractalkine (CX3CL1) which is constitutively expressed by neurons and endothelium in the brain [17] and SDF-1 (CXCL12) which is both constitutively expressed as well as induced by a variety of cell types in the CNS [18]. These two chemokines may play a role in the initial endothelial arrest and/or migration of T-cells and monocytes into the CNS for the initiation of inflammation; however, it is also conceivable that initial T-cell entry to the CNS does not require a chemokine signal.

There are two reasonable postulates that help to explain the role of chemokines in the initial CNS T-cell infiltration. The first is that neuroantigen-specific T-cells enter the CNS perivascular space in the absence of a chemokine signal from endogenous tissue and induce the recruitment of additional antigen-specific and antigen-non-specific T-cells as well as macrophages by secreting chemokines. Kuchroo et al. [19] demonstrated that PLP139-151-specific T-cell clones expressing the chemokines MIP-1α (CCL3) and TCA-3 (CCL1) in response to neuroantigen-specific stimulation were able to elicit clinical disease following adoptive transfer. Those clones that did not express chemokines in response to neuroantigen-specific stimulation did not induce EAE. Furthermore, Godiska et al. [20] also demonstrated that an encephalitogenic T-cell line was capable of expressing CCL3, MIP-1β (CCL4), RANTES (CCL5), and CCL1 following neuroantigen stimulation. The temporal decline in the ability of this T-cell line to induce EAE following antigen stimulation was related to a decline in the levels of CCL3, CCL4, and CCL1 mRNA expression. The second possibility is that encephalitogenic T-cells, through expression of cytokines such as IFN-γ, TNF-α, and/or LT, induce the cerebrovascular endothelium to secrete chemokines which in turn may increase the avidity of β-integrins expressed on T-cells [21] to mediate arrest and firm adhesion. Although chemokine production by cerebrovascular endothelium in EAE has not been described, several candidate chemokines, such as MCP-1 (CCL2), GROα (CXCL1), and CCL5 have been shown to be expressed in a variety of cultured endothelial model systems [22–24]. In all of these examples, endothelial chemokine expression was augmented by proinflammatory cytokine stimulation, thereby making it possible that encephalitogenic Th1 cells producing IFN-γ and/or TNF-α have the ability to stimulate chemokine expression by cerebrovascular endothelium. However, murine EAE does not require IFN-γ expression as mice deficient for this gene product are able to develop disease [25] suggesting that multiple inflammatory factors may have the capacity to induce endothelial chemokine expression.

The first demonstration that chemokine expression was associated with acute clinical EAE disease signs was performed by Hulkower et al [26]. Using a rat model of acute EAE where animals develop a single disease episode followed by spontaneous remission, they demonstrated that CCL2 mRNA was temporally expressed in the CNS of rats with disease. Furthermore, they noted that when the animals entered remission, CCL2 mRNA could no longer be detected. In both the active and adoptive SJL mouse models of EAE, Godiska et al. [20] demonstrated CNS chemokine mRNA expression (CCL3, CCL4, CCL5, CCL1, IP-10 (CXCL10), CCL2, and CXCL12) prior to the onset of clinical disease and that it remained elevated throughout the course of acute clinical disease signs. Using a MOG peptide-induced model of acute EAE in the C57Bl/6 mouse, it has been recently shown that CCL2 and CCL5 expression correlated with clinical disease severity [27], whereas CCL3 expression does not appear to play a role in this particular model of EAE [28]. These examples underscore the variability of chemokine expression patterns in different strains of mice induced to develop EAE with different antigen priming and cell transfer regimens [29]. Collectively, these data state that there is a correlation between chemokine expression and CNS disease; however, they raise the question: does initial T-cell migration occur in response to CNS tissue chemokine expression?

The question of whether CNS parenchymal chemokine expression precedes encephalitogenic T-cell migration or whether expression is necessary to induce T-cell migration into the CNS was addressed by sensitive chemokine mRNA detection. Chemokine expression was not detected in the absence of an inflammatory infiltrate [16]. Therefore, it appears that activated neuroantigen-specific T-cells migrate to the CNS perivascular sites and secrete chemokines that act to amplify the subsequent inflammatory process, including accumulation of mononuclear cells. Furthermore, these encephalitogenic T-cells secrete proinflammatory cytokines, such as TNF-α and IFN-γ that have the ability to induce astrocytes to express CCL5 [30] as well as CXCL10 [16], CCL3, and CCL2. Therefore, it appears that infiltrating T-cells can drive astrocyte expression of chemokines which in turn can induce the recruitment of additional mononuclear cells. This parenchymal expression of chemokines may contribute to the progression of CNS inflammatory disease by directing infiltration of effector T-cells and macrophages from the perivascular space into the tissue where ensuing cellular damage results in clinical disease manifestation.

The biological significance of chemokine expression in the pathogenesis of acute EAE was demonstrated using in vivo antibody treatment experiments [31]. CNS CCL3 and CXCL10, but not CCL2, protein expression was shown to correlate with increasing EAE severity and anti-CCL3 [32] and anti-CXCL10 [33], but not anti-CCL2 or anti-CCL5, treatment prevented acute clinical EAE. Furthermore, in vivo anti-CCL3 or anti-CXCL10 treatment reduced accumulation of disease inducing Th1 cells, host-derived bystander T-cells, and monocytes in the CNS [32,33]. These studies demonstrated that CCL3 and CXCL10 are important factors in the pathogenesis of acute EAE in SJL mice by their ability to directly induce CNS T-cell and monocyte accumulation.

The clinically relevant question arises as to what chemokines are produced in the CNS during the relapses of EAE. Does the CNS expression pattern remain the same or change with the evolving disease presentation? The CNS of (SJL $\times$ SWR)F1 mice with chronic relapsing EAE was examined for the expression of chemokines as a function of time after disease induction [34]. There was a dramatic increase in CCL2 mRNA and protein expression in the brain and an increase in CCL2 mRNA expression in the spinal cord during the relapsing phase of disease. CCL3 mRNA expression in the spinal cord remained elevated from the acute EAE episode through the relapsing phase of disease. Furthermore, CCL2 expression was localized to astrocytes while CCL3 expression was localized to the perivascular mononuclear cell infiltrate. That CCL2 expression plays a biologically relevant role in the relapsing EAE disease process was demonstrated by the ability to ameliorate relapsing clinical disease with anti-CCL2 treatment by inhibiting the accumulation of monocytes in the CNS [31].

The role of CNS expressed chemokines for the accumulation of monocytes has recently been studied using mice deficient for CCL2 [35] and its receptor CCR2 [36]. CCR2-deficient mice were immunized with MOG peptide and monitored for development of active clinical EAE [37]. The results of these experiments demonstrated that the CCR2-deficient mice did not develop clinical EAE, showed no CNS mononuclear cell infiltration, and had indistinguishable T-cell cytokine responses from control mice. This latter point was an important consideration as chemokines have been shown to be involved in T-cell costimulation [38], enhancement of cytokine expression [39], and T-cell differentiation [40]. Disruption of CCR2 has also been shown to result in a decrease in both Th1 and Th2 associated cytokine production [41]. However, CCR2-deficient mice fail to develop EAE, not as a result in decreased inflammatory cytokine production, rather, there is reduced accumulation of CD4 and CD8 T-cells as well as $CD45^{high}F4/80^{+}$ monocytes [42] with no change in endogenous $CD45^{low}F4/80^{+}$ microglia. Similar results were obtained using CCL2-deficient mice and demonstrating a lack of CNS monocyte accumulation [43]. The results of these two studies indicate a very critical role for CNS MCP-1 expression and corresponding CCR2 expression on monocytes in the development of clinical and histological EAE.

CCR1 and CCR5 knockout mice have also been used to study the pathogenesis of EAE. Mice deficient for CCR1 show a partial decrease in the severity of acute EAE [44], while mice deficient for CCR5 show no change in the ability to induce EAE [28]. Both CCR1 and CCR5 are receptors for CCL3 and CCL5 [45] and at least one would be predicted to have a role in disease development given the expression of its ligand during EAE onset [27].

Collectively, the studies examining the role of chemokines in EAE have shown that this class of inflammatory cytokines are important inflammatory mediators involved in the regulation of CNS inflammatory autoimmune diseases. It is attractive to postulate that differential spatial and temporal chemokine production by specific cell types in the CNS, from infiltrating mononuclear cells and resident perivascular cells to parenchymal astrocytes and microglia, serves as an important regulatory mechanism in the pathogenesis of EAE and similar immune-mediated inflammatory disease. Experiments demonstrating early CCL3 and CCL1 expression by encephalitogenic T-cells and macrophages associated with the perivascular mononuclear cell infiltrate suggest that the these chemokines are early regulators for the subsequent mononuclear infiltration and/or accumulation. Delayed CCL2 expression by activated astrocytes may be a later regulatory factor in the induction of further mononuclear cell infiltration including antigen-specific T-cells responsible for epitope spreading and episodes of relapsing disease as well as effector macrophages. Discovering the cellular sources, expression patterns, and temporal regulation of chemokine production during CNS autoimmune demyelinating disease opens up new potential targets for therapeutic intervention.

REFERENCES

1. Hickey WF, Gonatas NK, Kimura H, Wilson DB. Identification and quantitation of T lymphocyte subsets found in the spinal cord of the Lewis rat during acute experimental allergic encephalomyelitis. J Immunol 1983;131: 2805–2809.
2. Cross AH, Cannella B, Brosnan CF, Raine CS. Homing to central nervous system vasculature by antigen-specific lymphocytes. I. Localization of ^{14}C-labeled cells during acute, chronic, and relapsing experimental allergic encephalomyelitis. Lab Invest 1990;63:162–170.
3. Hickey WF, Hsu BL, Kimura H. T-lymphocyte entry into the central nervous system. J Neurosci Res 1991;28:254–260.
4. Hickey WF. Migration of hematogenous cells through the blood–brain barrier and the initiation of CNS inflammation. Brain Pathol 1991;1:97–105.
5. Brosnan CF, Bornstein MB, Bloom BR. The effects of macrophage depletion on the clinical and pathologic expression of experimental allergic encephalomyelitis. J Immunol 1981;126:614–620.
6. Huitinga I, Van Rooijen N, De Groot CJA, Uitdehaag BMJ, Dijkstra CD. Elimination of macrophages infiltrating into the central nervous system suppresses experimental allergic encephalomyelitis in Lewis rats. Ann New York Acad Sci 1990;594:458–460.
7. Carrithers MD, Visintin I, Kang SJ, Janeway CA, Jr. Differential adhesion molecule requirements for immune surveillance and inflammatory recruitment. Brain 2000; 123:1092–1101.
8. Yednock TA, Cannon C, Fritz LC, Sanchez-Madrid F, Steinman L, Karin N. Prevention of experimental autoimmune encephalomyelitis by antibodies against $\alpha 4\beta 1$ integrin. Nature 1992;356:63–66.
9. Baron JL, Madri JA, Ruddle NH, Hashim G, Janeway CA, Jr. Surface expression of alpha 4 integrin by CD4 T cells is required for their entry into brain parenchyma. J Exp Med 1993;177:57–68.
10. Kawai K, Kobayashi Y, Shiratori M, Sobue G, Tamatani T, Miyasaka M, Yoshikai Y. Intrathecal administration of antibodies against LFA-1 and against ICAM-1 suppresses experimental allergic encephalomyelitis in rats. Cell Immunol 1996;171:262–268.
11. Willenborg DO, Staykova MA, Miyasaka M. Short term treatment with soluble neuroantigen and anti-CD11a (LFA-1) protects rats against autoimmune encephalomyelitis — Treatment abrogates autoimmune disease but not autoimmunity. J Immunol 1996;157:1973–1980.
12. Gordon EJ, Myers KJ, Dougherty JP, Rosen H, Ron Y. Both anti-CD11a (LFA-I) and anti-CD11b (MAC-1) therapy delay the onset and diminish the severity of experimental autoimmune encephalomyelitis. J Neuroimmunol 1995;62:153–160.
13. Huang K, Kikuta A, Rosen SD. Myelin localization of a central nervous system ligand for L-selectin. J Neuroimmunol 1994;53:133–141.
14. Grewal IS, Foellmer HG, Grewal KD, Wang H, Lee WP, Tumas D, Janeway CA, Jr, Flavell RA. CD62L is required on effector cells for local interactions in the CNS to cause myelin damage in experimental allergic encephalomyelitis. Immunity 2001;14:291–302.
15. Springer TA. Traffic signals for lymphocyte recirculation and leukocyte emigration: the multistep paradigm. Cell 1994;76:301–314.
16. Glabinski AR, Tani M, Tuohy VK, Tuthill RJ, Ransohoff RM. Central nervous system chemokine mRNA accumulation follows initial leukocyte entry at the onset of acute murine experimental autoimmune encephalomyelitis. Brain Behav Immun 1995;9:315–330.
17. Schwaeble WJ, Stover CM, Schall TJ, Dairaghi DJ, Trinder PK, Linington C, Iglesias A, Schubart A, Lynch NJ, Weihe E, Schafer MK. Neuronal expression of fractalkine in the presence and absence of inflammation. FEBS Lett 1998;439:203–207.
18. Bajetto A, Bonavia R, Barbero S, Piccioli P, Costa A, Florio T, Schettini G. Glial and neuronal cells express functional chemokine receptor CXCR4 and its natural ligand stromal cell-derived factor 1. J Neurochem 1999; 73:2348–2357.
19. Kuchroo VK, Martin CA, Greer JM, Ju S-T, Sobel RA, Dorf ME. Cytokines and adhesion molecules contribute to the ability of myelin proteolipid protein-specific T cell clones to mediate experimental allergic encephalomyelitis. J Immunol 1993;151:4371–4382.
20. Godiska R, Chantry D, Dietsch GN, Gray PW. Chemokine expression in murine experimental allergic encephalomyelitis. J Neuroimmunol 1995;58:167–176.
21. Carr MW, Alon R, Springer TA. The C–C chemokine MCP-1 differentially modulates the avidity of $\beta 1$ and $\beta 2$ integrins on T lymphocytes. Immunity 1996;4:179–187.
22. Goebeler M, Yoshimura T, Toksoy A, Ritter U, Bröcker EB, Gillitzer R. The chemokine repertoire of human dermal microvascular endothelial cells and its regulation by inflammatory cytokines. J Invest Dermatol 1997;108:445–451.
23. Lukacs NW, Strieter RM, Elner V, Evanoff HL, Burdick MD, Kunkel SL. Production of chemokines, interleukin-8 and monocyte chemoattractant protein-1, during monocyte endothelial cell interactions. Blood 1995;86:2767–2773.
24. Zach O, Bauer HC, Richter K, Webersinke G, Tontsch S, Bauer H. Expression of a chemotactic cytokine (MCP-1) in cerebral capillary endothelial cells in vitro. Endothelium 1997;5:143–153.
25. Ferber IA, Brocke S, Taylor-Edwards C, Ridgway W, Dinisco C, Steinman L, Dalton D, Fathman CG. Mice with a disrupted IFN-gamma gene are susceptible to the

induction of experimental autoimmune encephalolmyelitis (EAE). J Immunol 1996;156:5–7.
26. Hulkower K, Brosnan CF, Aquino DA, Cammer W, Kulshrestha S, Guida MP, Rapoport DA, Berman JW. Expression of CSF-1, c-fms, and MCP-1 in the central nervous system of rats with experimental allergic encephalomyelitis. J Immunol 1993;150:2525–2533.
27. Juedes AE, Hjelmstrom P, Bergman CM, Neild AL, Ruddle NH. Kinetics and cellular origin of cytokines in the central nervous system: insight into mechanisms of myelin oligodendrocyte glycoprotein-induced experimental autoimmune encephalomyelitis. J Immunol 2000;164: 419–426.
28. Tran EH, Kuziel WA, Owens T. Induction of experimental autoimmune encephalomyelitis in C57BL/6 mice deficient in either the chemokine macrophage inflammatory protein- 1alpha or its CCR5 receptor. Eur J Immunol 2000;30:1410–1415.
29. Karpus WJ, Ransohoff RM. Chemokine regulation of experimental autoimmune encephalomyelitis: temporal and spatial expression patterns govern disease pathogenesis. J Immunol 1998;161:2667–2671.
30. Barnes DA, Huston M, Holmes R, Benveniste EN, Yong VW, Scholz P, Perez HD. Induction of RANTES expression by astrocytes and astrocytoma cell lines. J Neuroimmunol 1996;71:207–214.
31. Kennedy KJ, Strieter RM, Kunkel SL, Lukacs NW, Karpus WJ. Acute and relapsing experimental autoimmune encephalomyelitis are regulated by differential expression of the CC chemokines macrophage inflammatory protein-1α and monocyte chemotactic protein-1. J Neuroimmunol 1998;92:98–108.
32. Karpus WJ, Lukacs NW, McRae BL, Strieter RM, Kunkel SL, Miller SD. An important role for the chemokine macrophage inflammatory protein-1α in the pathogenesis of the T cell-mediated autoimmune disease, experimental autoimmune encephalomyelitis. J Immunol 1995;155: 5003–5010.
33. Fife BT, Kennedy KJ, Paniagua MC, Lukacs NW, Kunkel SL, Luster AD, Karpus WJ. CXCL10 (IFN-gamma-inducible protein-10) control of encephalitogenic $CD4^+$ T cell accumulation in the central nervous system during experimental autoimmune encephalomyelitis. J Immunol 2001;166:7617–7624.
34. Glabinski AR, Tani M, Strieter RM, Tuohy VK, Ransohoff RM. Synchronous synthesis of α- and β-chemokines by cells of diverse lineage in the central nervous system of mice with relapses of chronic experimental autoimmune encephalomyelitis. Am J Pathol 1997;150:617–630.
35. Gu L, Okada Y, Clinton SK, Gerard C, Sukhova GK, Libby P, Rollins BJ. Absence of monocyte chemoattractant protein-1 reduces atherosclerosis in low density lipoprotein receptor-deficient mice. Mol Cell 1998;2:275–281.
36. Kuziel WA, Morgan SJ, Dawson TC, Griffin S, Smithies O, Ley K, Maeda N. Severe reduction in leukocyte adhesion and monocyte extravasation in mice deficient in CC chemokine receptor 2. Proc Natl Acad Sci USA 1997;94: 12053–12058.
37. Fife BT, Huffnagle GB, Kuziel WA, Karpus WJ. CC Chemokine Receptor 2 Is Critical for Induction of Experimental Autoimmune Encephalomyelitis. J Exp Med 2000; 192:899–906.
38. Taub DD, Ortaldo JR, Turcovski-Corrales SM, Key ML, Longo DL, Murphy WJ. β Chemokines costimulate lymphocyte cytolysis, proliferation, and lymphokine production. J Leukocyte Biol 1996;59:81–89.
39. Lukacs NW, Chensue SW, Karpus WJ, Lincoln P, Keefer C, Strieter RM, Kunkel SL. C–C chemokines differentially alter interleukin-4 production from lymphocytes. Am J Pathol 1997;150:1861–1868.
40. Karpus WJ, Lukacs NW, Kennedy KJ, Smith WS, Hurst SD, Barrett TA. Differential CC chemokine-induced enhancement of T helper cell cytokine production. J Imunol 1997;158:4129–4136.
41. Boring L, Gosling J, Chensue SW, Kunkel SL, Farese RVJ, Broxmeyer HE, Charo IF. Impaired monocyte migration and reduced type 1 (Th1) cytokine responses in C–C chemokine receptor 2 knockout mice. J Clin Invest 1997;100:2552–2561.
42. Ford AL, Goodsall AL, Hickey WF, Sedgwick JD. Normal adult ramified microglia separated from other central nervous system macrophages by flow cytometric sorting. Phenotypic differences defined and direct ex vivo antigen presentation to myelin basic protein-reactive $CD4^+$ T cells compared. J Immunol 1995;154:4309–4321.
43. Huang DR, Wang J, Kivisakk P, Rollins BJ, Ransohoff RM. Absence of monocyte chemoattractant protein 1 in mice leads to decreased local macrophage recruitment and antigen-specific T helper cell type 1 immune response in experimental autoimmune encephalomyelitis. J Exp Med 2001;193:713–726.
44. Rottman JB, Slavin AJ, Silva R, Weiner HL, Gerard CG, Hancock WW. Leukocyte recruitment during onset of experimental allergic encephalomyelitis is CCR1 dependent. Eur J Immunol 2000;30:2372–2377.
45. Murphy PM, Baggiolini M, Charo IF, Hebert CA, Horuk R, Matsushima K, Miller LH, Oppenheim JJ, Power CA. International union of pharmacology. XXII. Nomenclature for chemokine receptors. Pharmacol Rev 2000;52:145–176.

Universes in Delicate Balance: Chemokines and the Nervous System
R.M. Ransohoff, K. Suzuki, A.E.I. Proudfoot, W.F. Hickey and J.K. Harrison, editors

Chapter 3.3

Chemokines and Neonatal Excitotoxic Brain Injury

John M. Galasso, Haiyan Xu and Faye S. Silverstein

Medical School and Departments of Pediatrics and Neurology, University of Michigan, Ann Arbor, MI 48109-0646, USA

1. INTRODUCTION

Perinatal brain injury is a major cause of long-term neurological disability. Recent research has heightened awareness of the complexity of the pathophysiological mechanisms that contribute to brain injury in the perinatal period. Prenatal events contribute substantially to neurological abnormalities of infancy and childhood, and prenatal factors may also increase susceptibility to asphyxial complications of labor and delivery. There is growing recognition that inflammation contributes to the pathogenesis of perinatal brain injury [1]. Although etiological links between certain maternal infections during pregnancy and adverse neurodevelopmental outcome were first suggested in the 1970s [2], the underlying pathophysiological mechanisms received little attention up until the past five years. The recent convergence of supportive epidemiologic and experimental data has stimulated many broad-ranging studies of the roles of inflammatory mediators in the pathogenesis of perinatal brain injury.

Epidemiologic data implicating inflammation in the pathogenesis of neonatal brain injury include evidence of the association of chorioamnionitis, preterm rupture of membranes, or prolonged preterm rupture of membranes with increased risk of cerebral palsy; these associations have been documented in several independent studies, both in preterm and in term infants [3–5]. Moreover, in a cohort of infants with cerebral palsy, retrospective analysis of blood samples obtained in the neonatal period revealed remarkably strong associations between increased levels of a broad range of pro-inflammatory cytokines and a subsequent diagnosis of spastic diplegia, a common form of cerebral palsy [6]. Of note, all β-chemokines assayed in these neonatal blood samples had sensitivities and specificities close to or greater than 90% in identification of infants who developed spastic diplegia.

The best-defined neuroanatomic substrate for spastic diplegia is periventricular white matter injury. Premature infants are at increased risk for the development of a distinctive pattern of perinatal brain injury characterized by gliosis and cystic degeneration of the deep white matter surrounding the lateral ventricles. The mechanisms underlying the vulnerability of immature oligodendroglia are incompletely understood; impaired perfusion in end-arterial border zones has been implicated as a major factor. Immature oligodendroglia have distinctive patterns of responsiveness to pro-inflammatory cytokines, and these properties may contribute to their vulnerability to injury in the neonatal period [7].

Recent experimental evidence also strongly implicates inflammatory mediators in the pathogenesis of neonatal brain injury. Hypoxic–ischemic and excitotoxic injury elicit rapid and intense inflammatory responses in neonatal rodent brain [8–11]. Of particular interest, acute neonatal brain injury stimulates production of monocyte chemoattractant protein-1 (MCP-1/CCL2) in multiple cell types, including injured neurons [9]. Systemic treatment with an IL-1 receptor antagonist can limit neonatal hypoxic–ischemic brain injury [12], but to date, no studies have shown that chemokines contribute directly to the pathogenesis of ischemic neuronal damage. The challenges of understanding the influences of specific inflammatory mediators in the pathogenesis of neonatal brain injury are illustrated by the results of a recent study in which we found that treatment with a tumor necrosis factor (TNF)-α antagonist had opposite effects on the progression of excitotoxic brain damage in two different brain regions in neonatal rats (the antagonist was protective vs. striatal excitotoxic damage, but amplified hippocampal excitotoxic injury) [13].

A major rationale underlying our studies of the roles of chemokines (and other inflammatory mediators) in neonatal models of neuropathology is that pathophysiological mechanisms of brain injury differ substantially between the immature and the adult brain.

2. ANIMAL MODELS

2.1. Maturational Stage Influences Pathogenetic Mechanisms of Brain Injury

Basic mechanisms of cell death are shared across all age groups. Yet, there are many compelling examples of maturational stage-related differences in the pathophysiology of brain injury. For example, hyperglycemia may worsen cerebral ischemic injury in the adult, but is protective in the neonate [14]; neonatal transgenic mice that over-express copper/zinc superoxide dismutase are rendered more vulnerable to hypoxic–ischemic brain injury, whereas in adults, injury is attenuated [15]; treatment with brain-derived neurotrophic factor attenuates hypoxic–ischemic brain injury in postnatal day 7 (P7) rats; however, 2 weeks later, at P21, the beneficial effects are substantially diminished [16]. Susceptibility to the deleterious effects of intracerebral injection of the excitatory amino agonist *N*-methyl-D-aspartate (NMDA) peaks in the early postnatal period in rodent brain [17].

The influence of maturational stage on susceptibility to brain pathology has also been documented in a model of chronic inflammatory neuropathology. Results of a recent study of the role of matrix metalloproteinases in the pathogenesis of experimental allergic encephalitis (EAE) revealed that in genetically modified mice lacking matrix metalloprotease-2 (gelatinase B), young animals (< one month old) were resistant to induction of EAE, whereas adult animals became susceptible; the specific mechanisms underlying this ontogenetic trend were not identified [18]. Together, such distinctive maturation-dependent features of brain injury provide a compelling rationale for the study of developmental stage appropriate models.

Our in vivo studies are performed in immature rodents, P7 rats and P9–10 mice. Precise correspondence of the maturational stage of the immature rodent brain with that of the human neonate is difficult to establish. Based on consideration of a broad range of indices of maturation, the P7–14 rat brain corresponds overall with features of the late preterm to term human brain.

2.2. Neonatal Hypoxic–Ischemic Brain Injury

One of the best-characterized and widely used animal models to study cerebral ischemic injury in neonatal rodents was developed by Vannucci and colleagues [14]; to elicit unilateral forebrain ischemic injury, P7 rats undergo unilateral carotid ligation, followed by 1.5–3 h exposure to 8% oxygen (O_2). Only the combination of ligation and hypoxia results in forebrain ischemia and infarction; neither ligation nor 8% O_2 exposure alone elicit overt injury. The duration of hypoxia (beyond a threshold at 1.5 h) determines the severity of tissue damage. The distribution of neuronal injury differs substantially from that observed in adult middle cerebral artery occlusion models; in the neonate, in addition to ipsilateral cortical and striatal infarction, there is widespread hippocampal damage. The carotid ligation/hypoxia model has been adapted to mice [15,19]; reproducible injury is elicited in P9–10 mice that undergo right carotid ligation, followed by moderate hypoxia (10% O_2, 70–120 min). In neonatal rats, hypoxic–ischemic brain injury elicits an acute inflammatory response, characterized by increased expression of pro-inflammatory cytokines and chemokines [9,10,20], a rapid microglial–monocyte response [8], and gliosis [21]. Neonatal mice that lack interleukin (IL)-1 converting enzyme and consequently do not produce mature IL-1β are relatively resistant to hypoxic–ischemic lesioning [19]; in these animals, there is attenuation of hypoxia–ischemia induced MCP-1 production [11].

2.3. Excitotoxic Lesioning

Overactivation of excitatory amino acid (EAA) receptors contributes to irreversible neuronal injury in a variety of neurologic disorders. Brain injury, elicited by intracerebral injections of the EAA agonists, NMDA and AMPA in immature rat brain, is dose-related, reproducible, and quantifiable. Susceptibility to NMDA-mediated injury peaks in the early postnatal period [17]. Major strengths of in vivo models of excitotoxicity include the opportunities to study the impact of the excitotoxin on many interacting cell types and to systematically evaluate developmental-stage specific pathogenetic mechanisms.

3. CHEMOKINES AND NEONATAL BRAIN INJURY

3.1. Microglia

In the context of our studies of chemokines and neonatal brain injury, we have been particularly interested in microglia as cellular targets of chemokines and as important cellular effectors of the inflammatory response. In the immature rodent central nervous system (CNS), both resting and activated microglia can be identified using lectin histochemistry; activated microglia can also be identified with the ED-1 antibody, which labels an uncharacterized cell surface antigen selectively expressed on activated microglia and monocytes [8]. In fact, fully activated microglia and blood-derived monocytes are indistinguishable using conventional immunohistochemical methods or based upon morphological criteria [22]. In the acutely injured neonatal brain, it is not readily feasible to determine the relative contributions of microglia and blood-derived monocytes to the inflammatory response. Subtle morphological differences in cell surface spines have been used to distinguish microglia and monocytes with electron microscopy [23], but these methods have not been widely applied.

In contrast to normal adult rat CNS, where no activated microglia are typically present, activated microglia are concentrated in white matter tracts of the immature rat brain [8,9]. Physiologically activated microglia may participate in remodeling of neuronal cytoarchitecture during development; these cells may also act as sentinels for pathophysiological stimuli and respond rapidly to acute insults.

In the neonatal rat brain, there is a rapid and robust microglia–monocyte response to acute brain injury. Direct intracerebral injection of NMDA into P7 rat hippocampus results in relatively rapid activation and recruitment of microglia; Fig. 1 illustrates microglia infiltrating the pyramidal cell layer of the lesioned hippocampus. This response to acute excitotoxic brain injury is readily discerned as early as 8 h and peaks at 24–48 h after lesioning [24]. Using lectin histochemistry to delineate the transition from resting to activated microglia, Acarin et al. [25] have shown, in a similar neonatal excitotoxicity model, that a substantial fraction of infiltrating inflammatory cells are activated microglia. There is a similar microglia–monocyte reaction in response to cerebral hypoxia–ischemia in neonatal rats [8]; the relative contribution of blood-derived monocytes to this inflammatory response is uncertain.

Of note, no corresponding population of physiologically activated microglia has been identified in human neonatal brain. However, human fetal microglia have been well studied in vitro, and in culture, these cells produce and respond to a broad range of cytokines and chemokines [25,26]. Of interest, postmortem evaluation of hippocampal tissue from a large group of infants revealed that in samples obtained from those up to the age of 8–9 months with an antecedent history of hypoxic or hypotensive episodes due to perinatal asphyxia, congenital heart defects, or chronic lung disease, there was often a dense infiltrate of microglial cells in the dentate gyrus. A comparable microglial infiltrate was seen in most children who died under circumstances consistent with sudden infant death syndrome. This histopathologic pattern was never found in older children. These neuropathological data were interpreted as providing evidence of a maturational stage-specific inflammatory response to chronic hypoxic–ischemic brain injury that was more pronounced in younger infants [27].

Microglia could contribute to the progression of brain injury through the release of inflammatory mediators and other soluble neurotoxins; yet, microglia are also potential sources of neurotrophins and may contribute to neuronal recovery after injury [28,29]. Microglia may concurrently exert both beneficial and

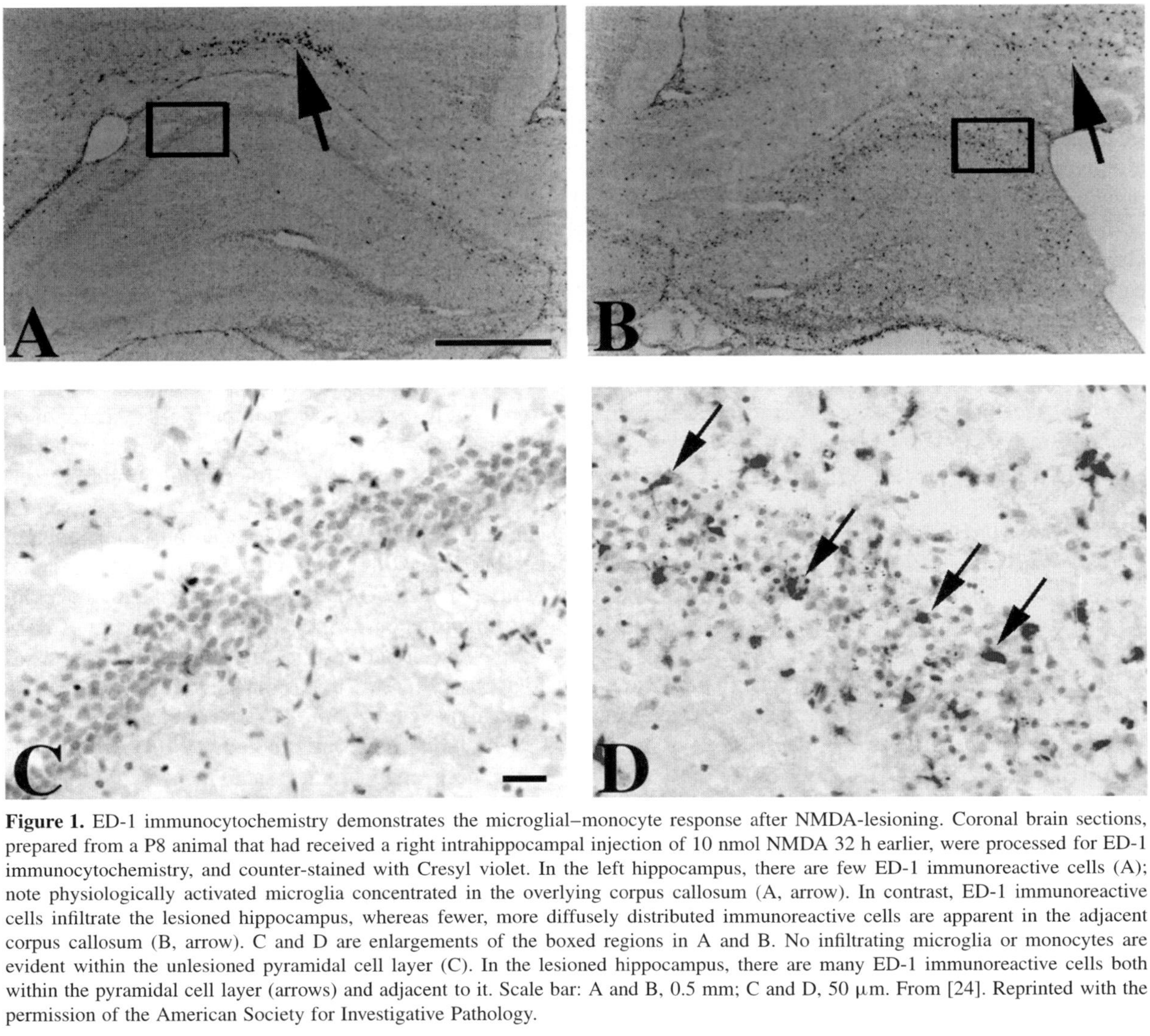

Figure 1. ED-1 immunocytochemistry demonstrates the microglial–monocyte response after NMDA-lesioning. Coronal brain sections, prepared from a P8 animal that had received a right intrahippocampal injection of 10 nmol NMDA 32 h earlier, were processed for ED-1 immunocytochemistry, and counter-stained with Cresyl violet. In the left hippocampus, there are few ED-1 immunoreactive cells (A); note physiologically activated microglia concentrated in the overlying corpus callosum (A, arrow). In contrast, ED-1 immunoreactive cells infiltrate the lesioned hippocampus, whereas fewer, more diffusely distributed immunoreactive cells are apparent in the adjacent corpus callosum (B, arrow). C and D are enlargements of the boxed regions in A and B. No infiltrating microglia or monocytes are evident within the unlesioned pyramidal cell layer (C). In the lesioned hippocampus, there are many ED-1 immunoreactive cells both within the pyramidal cell layer (arrows) and adjacent to it. Scale bar: A and B, 0.5 mm; C and D, 50 μm. From [24]. Reprinted with the permission of the American Society for Investigative Pathology.

detrimental effects upon distinct cell populations in the acutely injured brain and these differential effects may greatly complicate delineation of chemokine signaling mechanisms and functional activity in vivo.

There is currently no information about the influence of developmental stage, or of specific signaling mechanisms (physiologic vs. pathologic) on the responsiveness of activated microglia, and their secretory products. Based on the findings that acute neonatal brain injury elicited such a prominent microglial–monocyte reaction, our laboratory became interested in identifying the molecular signals that regulated this inflammatory response. This led us to focus upon analysis of specific chemokines in the acutely injured brain.

3.2. Chemokines

Based on the converging evidence, highlighted in other chapters of this volume, we hypothesized that monocyte chemoattractant protein-1 (MCP-1/CCL2) could be a key regulator of the microglial–monocyte response in the injured neonatal brain. MCP-1 has a critical regulatory role in a variety of physiological and pathological monocyte-dependent processes. In the CNS, MCP-1 could activate resident microglia and/or contribute to recruitment of blood-derived monocytes into the acutely injured brain. Of interest, MCP-1 may also play a distinctive role in normal brain development. Neuronal MCP-1 expression was recently reported in human fetal brain tissue; in cerebellar neurons, MCP-1 immunoreactivity was present at 27 weeks gestation, peaked at 36 weeks, expression decreased in early postnatal life, and was absent in tissue samples from older children and adults [30]. Similarly, in vitro studies showed that NT2 cells, which were derived from a human teratoma and differentiated to express a neuronal phenotype, express both MCP-1 and a broad range of chemokine receptors [31].

Increased MCP-1 mRNA and protein expression had been documented in experimental models of acute brain injury in adult animals [32–36]. In general, in adult brain injury models, the most commonly identified cellular source of MCP-1 has been reactive astrocytes, and the major cellular targets implicated have been blood-derived monocytes. Our initial studies in two neonatal brain injury models highlighted major differences from trends in the mature CNS. One of the most surprising initial findings was that injured neurons were a major source of MCP-1 [9,37]. Specifically, in P7 rats, using in situ hybridization, we found that both hypoxic–ischemic and excitotoxic injury induced marked stimulation of MCP-1 mRNA in injured neurons throughout the injured hemisphere 4–16 h after lesioning; other cell types, in particular ependyma, also expressed MCP-1 mRNA. Another characteristic finding was that MCP-1 mRNA expression waned quite rapidly and was no longer discerned at 24 h.

We incorporated two approaches to evaluate MCP-1 protein expression in the acutely injured brain. In P7–9 rat brain, immunocytochemistry assays revealed that an acute hypoxic–ischemic insult induced MCP-1 expression 4–48 h later in multiple cell types in the lesioned hemisphere [9]. We subsequently developed an ELISA to measure MCP-1 in brain homogenates [38]. Table 1 shows results of a preliminary experiment in which we evaluated whether the amount of MCP-1 produced was influenced by the severity of the ischemic insult. The neonatal 'unilateral carotid ligation + global hypoxia' model enables us to regulate the severity of the ischemic insult (and resulting severity of tissue damage) predictably over a fairly broad range by altering the duration of hypoxia exposure (from 1.5 to 3 h in P7 rats). As early as 6 h post-hypoxia–ischemia, the duration of hypoxia–ischemia clearly influenced the magnitude of stimulation of MCP-1; trends were similar in complementary experiments in the neonatal mouse [11]. These findings implicate MCP-1 as a potential pivotal mediator of the inflammatory response to ischemic brain injury.

Table 1. Hypoxic–ischemic injury stimulates production of MCP-1 in neonatal rat brain

	Left	Right
Normal	2.7 ± 0.3	3.5 ± 0.2
2 h	2.5 ± 1.2	61.3 ± 54.3
2.5 h	12.0 ± 3.8	94.0 ± 42.6
3 h	39.8 ± 20.1 [a]	156.6 ± 57.2 [b]

P7 rats underwent right carotid artery ligation, followed by exposure to 8% oxygen for 2, 2.5, or 3 h ($n = 3$/group). After a 6-h recovery period in room air, tissue homogenates were prepared from the left and right cerebral hemispheres, and MCP-1 concentrations were determined, using a rat MCP-1 ELISA [38]. Values are expressed as pg/mg protein ± SEM.

[a] $P < 0.05$, Kruskal–Wallis test, comparing MCP-1 concentrations in all left hemisphere samples.

[b] $P < 0.05$, Kruskal–Wallis test, comparing MCP-1 concentrations in all right hemisphere samples.

Nonetheless, stronger evidence was needed to demonstrate that MCP-1 played an important role in the pathogenesis of brain injury. To further study MCP-1 functional activity in the acutely injured brain, we decided to focus upon excitotoxin lesioning for several reasons, including the greater reproducibility of the lesions, the ability to compare inflammatory responses in discrete anatomic regions (of particular interest to us, striatum and hippocampus) [13], and the ability to readily modulate injury pharmacologically with a broad range of EAA antagonists.

We will highlight results of several complementary groups of experiments that together provide compelling evidence that MCP-1 is a pathophysiologically important modulator of neonatal brain injury.

3.2.1. *Semi-quantitative RT-PCR assays of MCP-1 and CCR2 mRNA*

In P7 rats, NMDA was stereotaxically injected into the right hippocampus; RNA was then isolated from left and right hippocampal samples, collected at

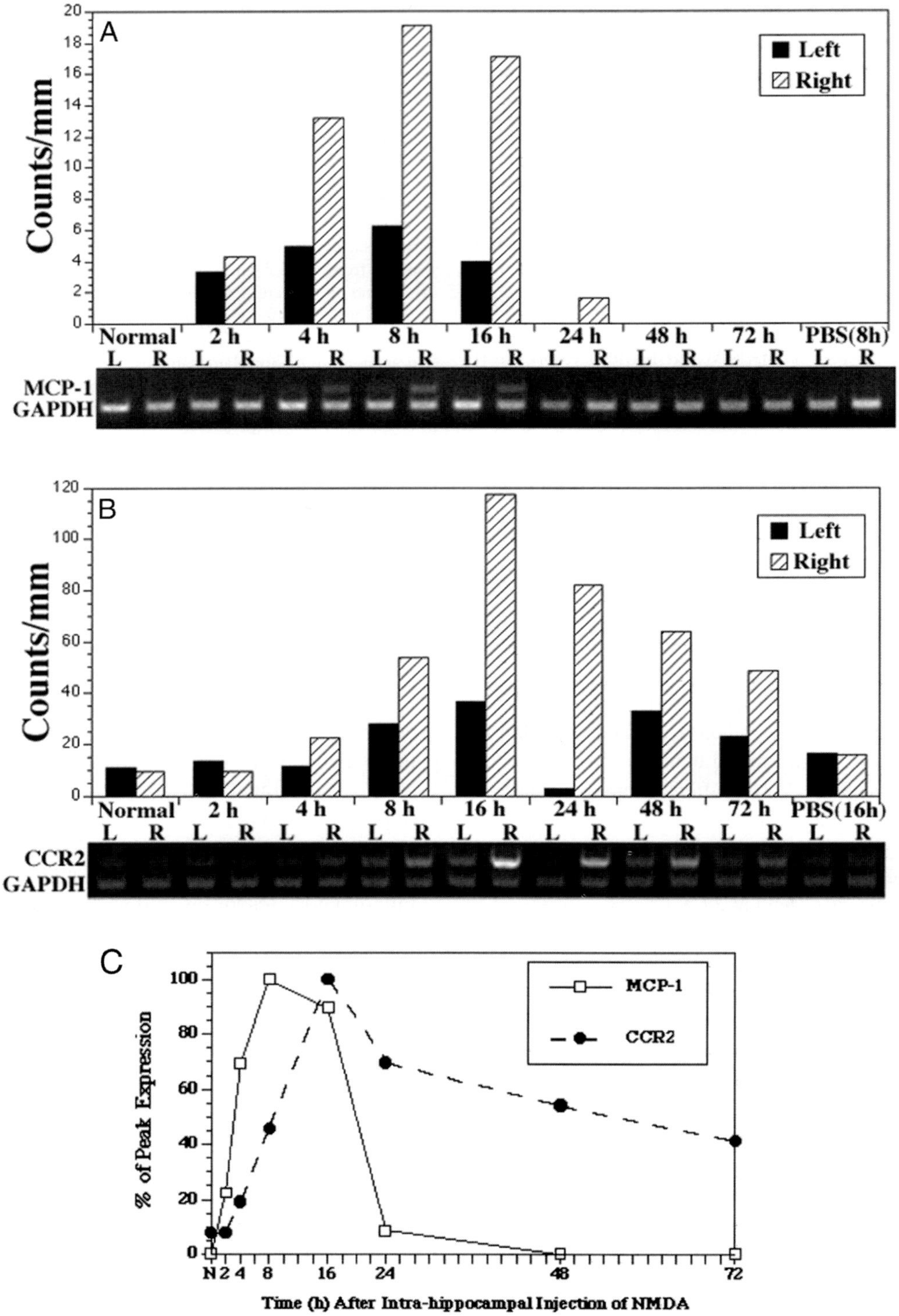
A
Left
Right
Counts/mm
Normal
2 h
4 h
8 h
16 h
24 h
48 h
72 h
PBS(8h)
L R L R L R L R L R L R L R L R L R
MCP-1
GAPDH
B
Left
Right
Counts/mm
Normal
2 h
4 h
8 h
16 h
24 h
48 h
72 h
PBS(16h)
L R L R L R L R L R L R L R L R L R
CCR2
GAPDH
C
MCP-1
CCR2
% of Peak Expression
N 2 4 8 16 24 48 72
Time (h) After Intra-hippocampal Injection of NMDA

specified postinjection time points. Semi-quantitative RT-PCR assays were developed to evaluate changes in gene expression of both MCP-1 and the MCP-1 receptor CCR2 (detailed methods are reported in [38]). Fig. 2 summarizes the major trends. In the right hippocampus, there was a substantial increase in MCP-1 mRNA as early as 2 h after NMDA administration (Fig. 2A); MCP-1 gene expression peaked at 8 h, when there was a greater than 3-fold increase in the lesioned hippocampus. MCP-1 mRNA declined to baseline levels by 24 h after lesioning. There was no increase in MCP-1 mRNA in control animals injected with PBS. Of note, MCP-1 gene expression also increased, to a lesser degree, in the contralateral hippocampus 4–16 h after lesioning; no overt injury evolves in the contralateral hippocampus with this lesioning method. Assays of CCR2 mRNA, in aliquots from the same RNA samples, demonstrated that excitotoxic lesioning also induced expression of CCR2 mRNA (Fig. 2B); expression peaked later, at 16 h, and high levels of CCR2 mRNA were detected up to 72 h after the administration of NMDA. Fig. 2C graphically illustrates the distinct temporal sequences of NMDA-induced MCP-1 and CCR2 gene expression; these disparate trends suggest that the regulatory mechanisms (and probably also their cellular sources) differ.

3.2.2. Quantitative analysis of NMDA-induced MCP-1 production

In order to study the regulation of MCP-1 in the injured neonatal brain, we developed a rat MCP-1 ELISA [38]. Of interest, MCP-1 was detectable, at very low levels, in unlesioned brain (Fig. 3A; also see Table 1). NMDA lesioning induced a marked increased in MCP-1 production throughout the lesioned cerebral hemisphere, in samples obtained at 4–48 h after lesioning; concentrations peaked at 8–12 h and returned to normal at 96 h after lesioning. MCP-1 protein expression also increased in the contralateral hemisphere 8–24 h after lesioning. In addition, mechanical injury (saline injection) stimulated a minimal increase in MCP-1. Treatment with the competitive NMDA antagonist, CPP, blocked the peak in NMDA-induced increase in MCP-1 expression (Fig. 3B), suggesting that the increase in MCP-1 is directly linked to NMDA receptor over-activation. We relied on whole hemisphere assays, rather than regional dissections of hippocampus, because in preliminary experiments, we found that concentrations attained were similar with both tissue sources.

The specific regulatory mechanisms controlling in vivo MCP-1 expression are incompletely understood. Observations in diverse experimental models of tissue injury indicate that pro-inflammatory cytokines, such as IL-1β can upregulate MCP-1 expression and that inhibition of IL-1β and TNF-α can reduce both the extent of injury and MCP-1 expression [39]. In human fetal astrocytes, MCP-1 expression is selectively upregulated by IL-1β and TNF-α [40]. Interestingly, in this neonatal excitotoxicity model, increased stimulation of IL-1β production precedes peak MCP-1 expression; and pharmacologic inhibition of IL-1β markedly attenuates injury [41]. Whether treatment with IL-1β antagonists can block MCP-1 production in the injured brain remains to be determined.

3.2.3. Does MCP-1 regulate the microglial response to excitotoxic injury?

We hypothesized that MCP-1 is an important regulator of the microglial response to acute neonatal brain injury and incorporated several experimental approaches to address this question.

Figure 2. NMDA-induced changes in MCP-1 and CCR2 gene expression. RT-PCR assays were performed to estimate changes in tissue content of MCP-1 and CCR2 mRNA, relative to GAPDH mRNA. Samples from left (L) and right (R) hippocampus of P8 normals, PBS-injected controls, and from animals that received right intrahippocampal NMDA (10 nmol) injections (evaluated at 2, 4, 8, 16, 24, 48, 72 h post-injection) were assayed. PCR products were visualized in ethidium bromide-stained agarose gels, and scanned fluorometrically. In A and B, the histograms provide quantitative estimates of MCP-1 and CCR2 mRNA, respectively, based on normalization to GAPDH mRNA/sample (counts = optical density units). C compares the temporal patterns of NMDA-induced changes in MCP-1 and CCR2 mRNA in the lesioned hippocampus; values are expressed as % peak expression. (A) In the right hippocampus, note a 3-fold increase in MCP-1 at 8 h, compared to the left hippocampus. (B) Note that CCR2 mRNA was detected in all samples. In the NMDA-injected right hippocampus, CCR2 gene expression peaked at 16 h (>11-fold increase compared to normal hippocampus). (C) In the lesioned hippocampus, MCP-1 mRNA expression peaked at 8 h and returned to baseline levels at 24 h. CCR2 mRNA expression peaked at 16 h and increased expression was sustained up until 72 h. From [38]. Reprinted with the permission of Academic Press.

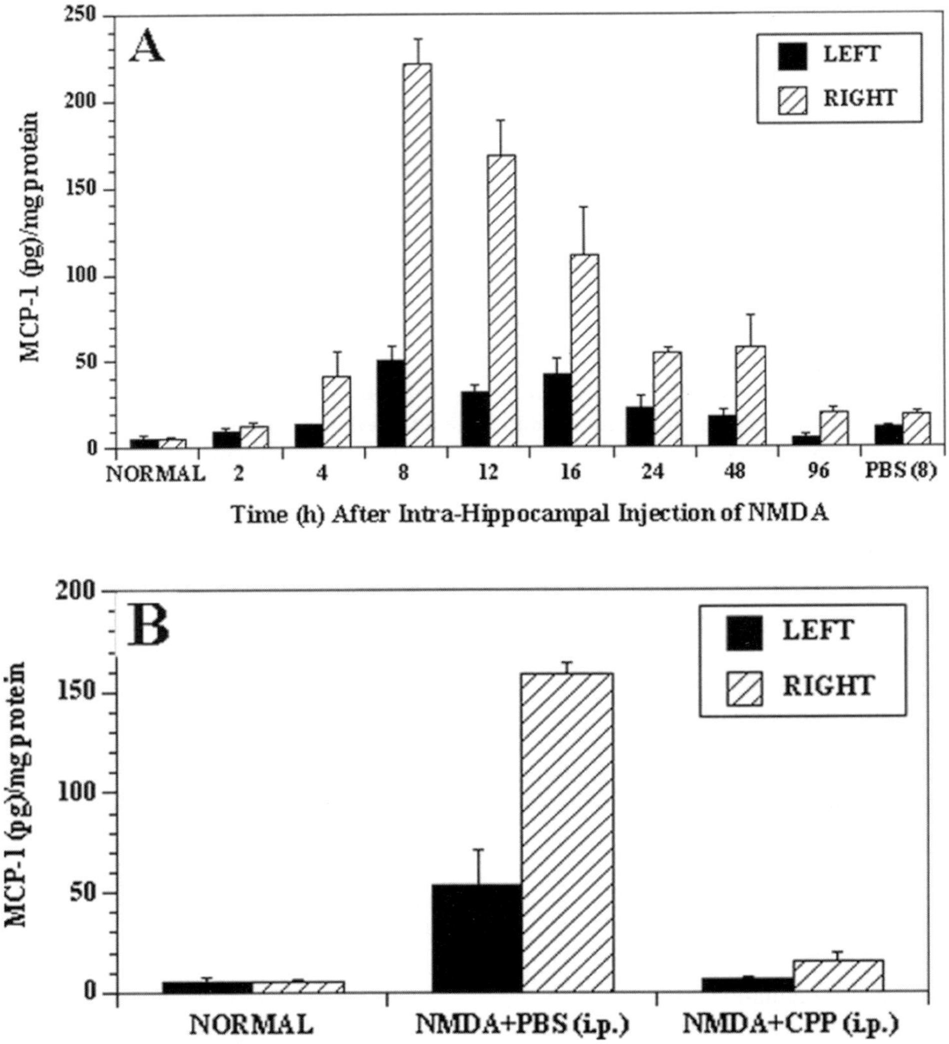

Figure 3. NMDA-induced increases in MCP-1 protein in acutely lesioned brain tissue. P7 animals received right intrahippocampal injections of 10 nmol NMDA, and tissue extracts were prepared from the left and right cerebral hemispheres at 2, 4, 8, 12, 16, 24, 48, and 96 h post-injection. Control samples were collected from normal P8 and intrahippocampal PBS-injected animals. Samples were also obtained (at 12 h postlesioning) from animals that had received injections of NMDA, followed by intraperitoneal injections of PBS or the competitive NMDA antagonist, CPP. (A) At 8–12 h, there was a >30-fold increase compared to normals; increases in MCP-1 were detected in all lesioned hemisphere samples at 4–48 h ($P < 0.03$; ANOVA). Contralaterally, there were corresponding increases of lesser magnitude ($P < 0.003$; ANOVA). (B) In NMDA-lesioned animals that received CPP, MCP-1 content did not differ from normal P8 rat brain. From [38]. Reprinted with the permission of Academic Press.

We first attempted a direct cell-counting approach, despite awareness of the many limitations of this strategy. Direct intracerebral injection of recombinant MCP-1 (rMCP-1) into P7 rat brain did not elicit any discernible histopathological changes (compared with saline injection). We then evaluated whether co-injection of rMCP-1 with NMDA could amplify the microglial response. P7 animals received intrastriatal co-injections of a dose of NMDA at the threshold of toxicity (5 nmol) and 50 ng rMCP-1; controls received injections of 5 nmol NMDA. ED-1 immunoassays were performed at 48 h after lesioning (see Table 2), and ED-1 immunoreactive cells were counted in the lesioned and contralateral striatum, and adjacent white matter tracts. In the controls, equal numbers of ED-1 reactive cells were counted

Table 2. Intrastriatal co-injection of 5 nmol NMDA and rMCP-1 induces microglial infiltration

	Number of ED-1 immunoreactive cells		
	Left	Right	R–L difference
Control	1506 (±338)	1545 (±339)	39 (±116)
rMCP-1	1357 (±779)	1832 (±968)	475 (±222) [a]

P7 rats received a right intrastriatal co-injection of 5 nmol NMDA and 50 ng rMCP-1; littermate controls received NMDA and heat-treated rMCP-1 ($n = 5$/group). Tissue samples were obtained 48 h after lesioning; ED-1 immunocytochemistry (see [8] for methods) assays were done on three corresponding sections through the striatum/brain. ED-1 cell counting was performed (using a semi-automated particle counting routine, NIH Image) bilaterally in fields that included both striatum and corpus callosum; cell counting was performed in >7 fields/hemisphere at 10× magnification. Sections from both groups were assayed concurrently, and the analysis was done by an individual unaware of treatment group assignment. Mean number of ED-1 immunopositive cells (±SD) in the lesioned and unlesioned hemispheres and the Right–Left difference in the number of ED-1 immunopositive cells were compared. In the r-MCP1 + NMDA group, more ED-1 immunoreactive cells were identified in the lesioned than in the contralateral hemisphere ([a] $P < 0.005$, t-test), whereas in the control group, there was no interhemispheric difference.

in each hemisphere; this finding was somewhat surprising, and likely reflected the limited sensitivity of the assay. In contrast, in the rMCP-1-NMDA group, there was a substantial increase in the number of ED-1 reactive cells concentrated in the right striatum and adjacent white matter. These results provided some preliminary suggestive evidence that MCP-1 could influence the microglial response. Further support for the hypothesis that MCP-1 regulates the microglial response would be provided, if we could document that infiltrating microglia expressed MCP-1 receptors.

CCR2 is the best-characterized MCP-1 receptor. CCR2 expression has been reported in rat microglia in vitro [42], and NMDA-lesioning induced expression of CCR2 mRNA in neonatal rat brain (Fig. 2). In fact, several recent studies have identified multiple chemokine receptors in the CNS (both in human and in rodent tissue) [43–45]. Surprisingly, chemokine receptors are expressed by many cell types in the brain, including, macrophages, astrocytes, endothelium, and neurons.

In order to determine which cells expressed CCR2 and were potentially responsive to MCP-1 in the acutely injured brain, we developed immunocytochemistry assays using fluorescence-labeled secondary antibodies and analyzed the distribution of signal with confocal microscopy. Fig. 4 illustrates our major findings (details are reported in [38]); immunoassays were performed in coronal brain sections of P8 animals, that had received a right intrahippocampal injection of NMDA (10 nmol) 16 or 32 h earlier. ED-1 immunoreactive activated microglia were detected in the corpus callosum bilaterally (Fig. 4A,B); yet, their distributions differed markedly. In the left corpus callosum, activated microglia are densely concentrated in the white matter (Fig. 4A). In contrast, these cells are more widely dispersed in the corpus callosum adjacent to the lesioned hippocampus (Fig. 4B); this pattern suggests that microglia are being recruited from this site into the lesion territory. Although ED-1 reactive cells were present bilaterally, CCR2 immunoreactive cells were only detected in the right corpus callosum (Fig. 4C,D); these CCR2 immunoreactive cells were consistently co-localized with ED-1 immunopositive cells (Fig. 4E–G). CCR2 and ED-1 immunoreactive cells were also detected in the injured pyramidal cell layer (Fig. 4H–I). These results suggest that 'physiologically' and 'pathologically' activated microglia may have distinct properties, including differential responsiveness to MCP-1. These results are consistent with the hypothesis that MCP-1 signals the initial microglial response to excitotoxic injury, but do not provide any evidence that chemokines influence the progression of tissue injury.

3.2.4. Effect of rMCP-1 on excitotoxic injury

To address the question of whether MCP-1 could directly influence the progression of brain injury in vivo, we examined the impact of co-administration of rMCP-1 and a dose of NMDA at the threshold of toxicity (5 nmol) on the extent of NMDA-induced striatal and hippocampal injury. In preliminary experiments where rMCP-1 alone (50 ng) was injected into the striatum, there was no evidence of histopathologic damage or microglial or monocyte recruitment [46]. This differs from reported findings in the adult rodent brain, where intracerebral administration of rMCP-1 results in the infiltration of mononuclear phagocytes [47]. The same amount of rMCP-1 was co-injected with NMDA (5 nmol) into the striatum or hippocampus of P7 rats. Controls

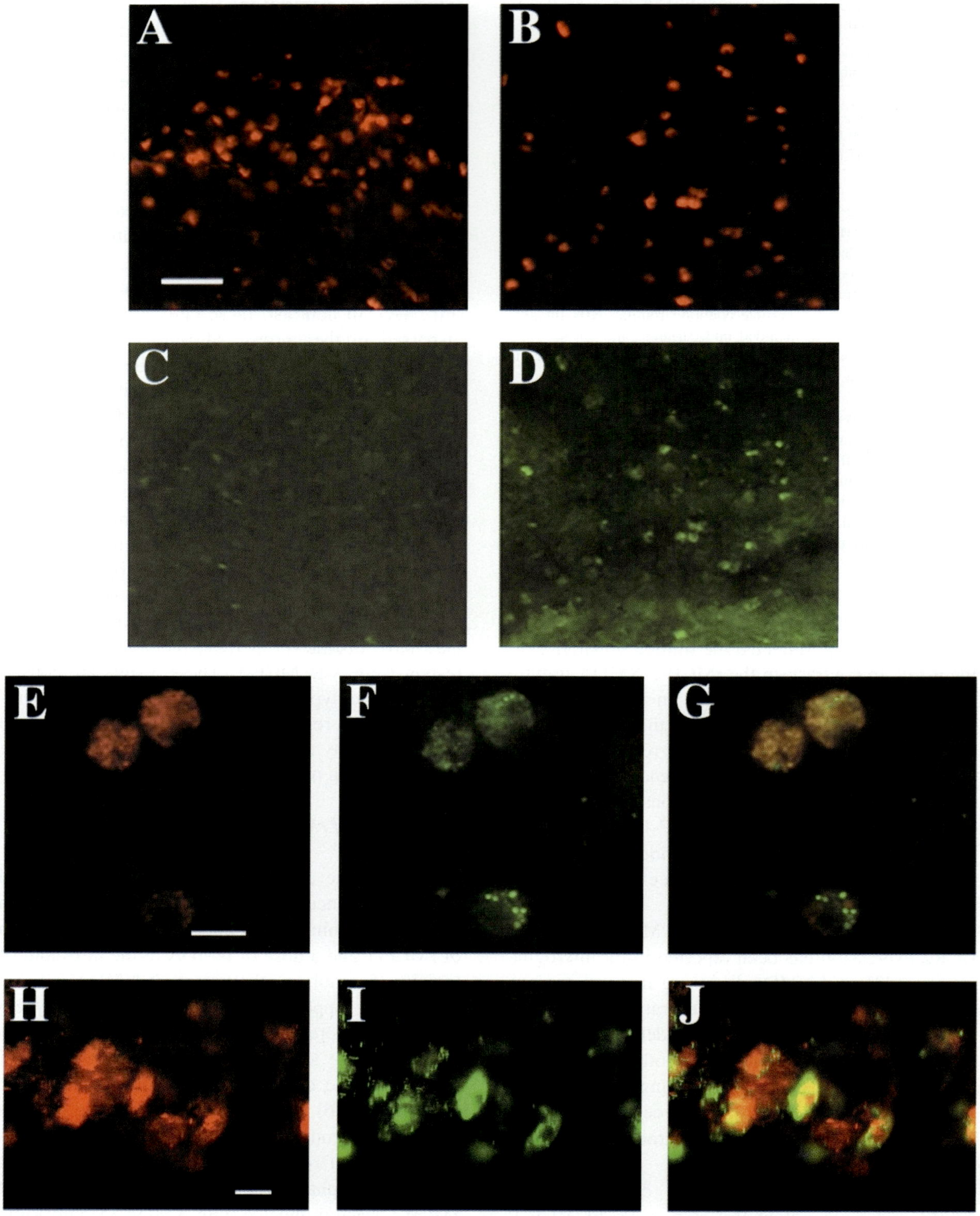
A
B
C
D
E
F
G
H
I
J

received co-injections of heat-inactivated rMCP-1 (to control for any non-specific inflammation resulting from administration of a foreign protein). Outcome was assessed by morphometric measurement of regional cross-sectional areas 5 days after lesioning; based on our previous experience with this animal model, this analytical method provides a reliable measure of the extent of irreversible tissue injury. In both the striatum and the hippocampus, co-injection of rMCP-1 with NMDA markedly amplified NMDA-induced injury (by 55% in the striatum and by 167% in the hippocampus) (results reported in detail in [46]). The histopathological features of the lesions elicited by the combination of rMCP-1+ NMDA were indistinguishable from the lesions elicited by higher doses of NMDA (7.5–10 nmol) injected alone, into the same brain regions.

3.2.5. Effects of treatment with an MCP-1 neutralizing antibody on excitotoxic injury

To complement these data, we wanted to document that functional antagonism of MCP-1 would attenuate injury. Initially, we evaluated the efficacy of systemic treatment with an MCP-1 neutralizing antibody, and results were inconsistent; this variability appeared to be attributable, in large part, to differences in neutralizing activity among lots of antibody. In our neonatal rat models, the amount of a reagent that can be reliably administered is always a limiting factor; maximum intraperitoneal injection volume is about 0.1 ml, maximum intracerebro-ventricular injection volume is about 5 μl, and maximum intracerebral injection volume is about 1.5 μl. Taking this methodological issue into account, we maximally concentrated our available neutralizing Ab. To evaluate the effects of functional inhibition of MCP-1 in vivo, this anti-MCP-1 antibody was co-injected with 10 nmol NMDA (a dose that elicits substantial tissue injury) into the striatum and hippocampus of P7 rats; controls were co-injected with an equal amount of rabbit immunoglobulin (IgG). Outcome was evaluated in the same way as in the preceding experiments, morphometric measurements of regional cross-sectional areas 5 days after lesioning. Treatment with the MCP-1 neutralizing antibody attenuated NMDA-induced striatal and hippocampal injury by 57 and 39%, respectively [46]; Fig. 5 illustrates these trends, and demonstrates the differences in the severity of hippocampal injury between antibody-treated and control animals.

These findings complement results of a recent study in which treatment with an MCP-1 neutralizing antibody decreased both the severity of clinical symptoms during relapsing EAE and was also associated with a reduction in macrophage accumulation [48]. We did not attempt to evaluate whether treatment with the MCP-1 antibody influenced the microglial–monocyte response, in part, because results of such studies would be very difficult to interpret mechanistically. Yet, together, these findings strongly implicate MCP-1 as a pathophysiologically significant mediator of neonatal excitotoxic brain injury. It will be critically important to extend these studies of chemokine biology to more clinically relevant hypoxia–ischemia models as well, and these studies are in progress in our laboratory.

3.3. Other Chemokines may also Contribute to Neonatal Brain Injury

Related chemokines may also contribute to the pathogenesis of neonatal brain injury. In P7 rat brain, hypoxic–ischemic lesioning induces expres-

Figure 4. Immunofluorescence labeling to co-localize CCR2 and ED-1 immunoreactive cells. Coronal brain sections, prepared from animals that had received a right intrahippocampal injections of 10 nmol NMDA at 16 h (A–G) or 32 h (H–J) earlier, were double-labeled to detect CCR2 and ED-1. Confocal microscopy was used to visualize the distributions of fluorescent-tagged 2° Abs; an Alexa 488 donkey anti-goat 2° Ab was used to detect CCR2 Ab and an Alexa 568 rabbit anti-mouse 2° Ab was used to detect ED-1. The distributions of ED-1 and CCR2 immunoreactive cells in the left corpus callosum are illustrated in A and C, respectively. B and D show the distributions of ED-1 and CCR2 immunoreactive cells in the right corpus callosum. ED-1 immunoreactive cells were detected bilaterally (A,B) but their distributions differed; ipsilaterally, these cells were more widely dispersed (B); CCR2 immunoreactive cells were evident only ipsilaterally (D). E and F show ED-1 and CCR2 immunoreactive cells in the ipsilateral corpus callosum; G, a composite image, shows co-localization of ED-1 and CCR2 immunoreactivity. At 32 h after lesioning, ED-1 (H) and CCR2 (I) immunoreactive cells were identified in the hippocampus; J illustrates CCR2/ED-1 immunoreactive cells infiltrating the CA3 pyramidal cell layer. Scale bars: A–D, 100 μM; E–J, 20 μM. From [38]. Reprinted with the permission of Academic Press.

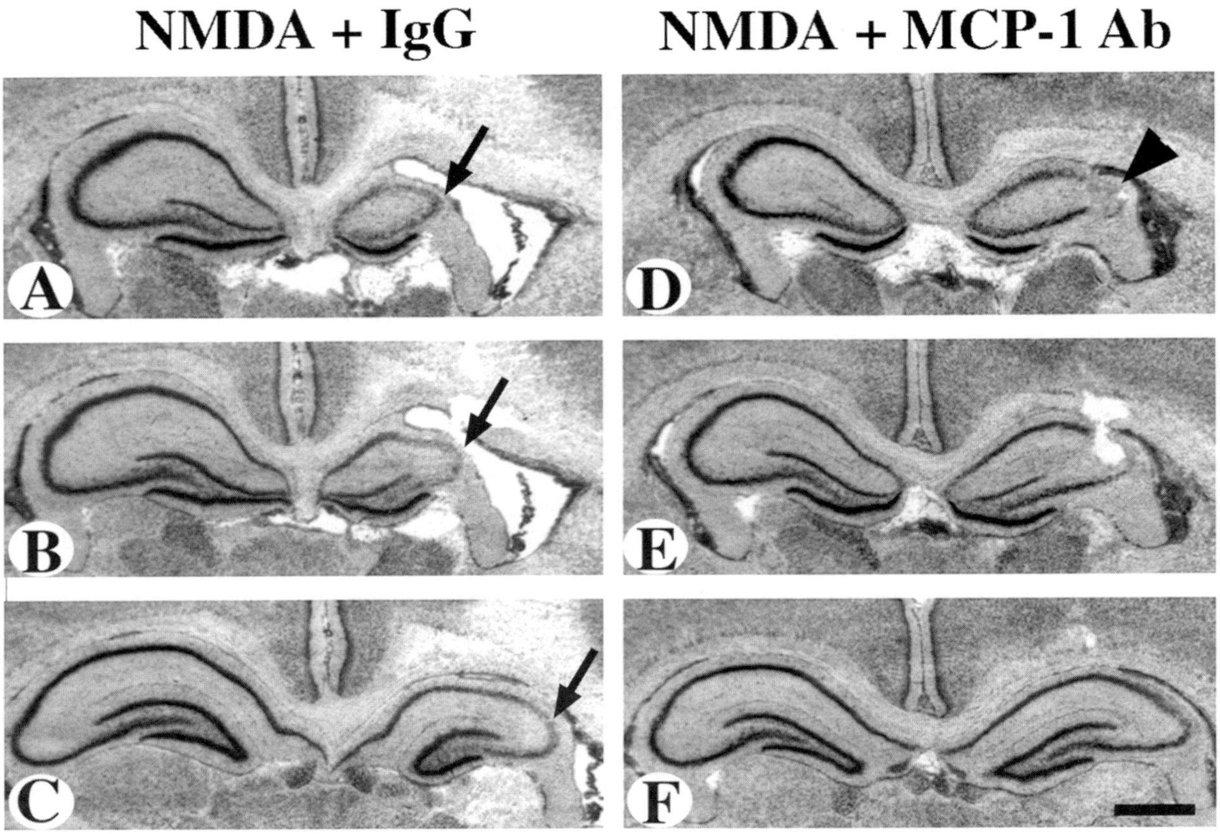

Figure 5. Treatment with MCP-1 neutralizing antibody attenuates NMDA-induced hippocampal injury. This montage compares representative Cresyl-violet-stained coronal sections through the hippocampus, at comparable anatomic levels, from one P12 control animal that received an intrahippocampal co-injection of 10 nmol NMDA and 6 μg rabbit IgG (A–C) and from a P12 animal that received an intrahippocampal co-injection of 10 nmol NMDA and 6 μg anti-MCP-1 Ab (D–F). In the control, there is focal loss of Nissl staining in the pyramidal cell layer, particularly in CA3 (A–C, arrows) and hippocampal atrophy at all levels. In contrast, in the animal in which MCP-1 Ab was co-administered with NMDA, there is minimal loss of Nissl staining in the pyramidal cell layer (D, arrowhead) and relative preservation of hippocampal tissue. Scale bar: 2 mm. From [46]. Reprinted with the permission of Elsevier Science.

sion of MIP-1α/CCL3 [49]; its functional role has not yet been evaluated in detail, but it is clear that its cellular sources and duration of expression differ substantially from those of MCP-1.

CCR5, one of the best-characterized MIP-1α receptors, is also expressed in neonatal rat brain, and its expression is altered by injury [35]. In P7 rat brain, intrahippocampal injection of NMDA induces neuronal CCR5 expression in pyramidal CA3 neurons that are highly vulnerable to injury [35]; a potential link with mechanisms of neurodegeneration is provided by data demonstrating that chemokines stimulate rises in intracellular calcium in cultured hippocampal neurons [50]. Moreover, the timing and neuroanatomic distribution of CCR2 and CCR5 expression in the NMDA-lesioned hippocampus differ substantially, indicating that distinct regulatory mechanisms are involved, and suggesting that CCR5 and CCR2 play different roles in the acutely injured brain.

4. UNANSWERED QUESTIONS ABOUT CHEMOKINES AND NEONATAL BRAIN INJURY

Our current research focuses on understanding how inflammation, initiated by acute brain injury, influences the ultimate expression of tissue damage, and how specific components of this inflammatory cascade can be modulated to protect the neonatal brain from irreversible injury. The degree to which inflammatory mediators cause injury and/or contribute to ultimate tissue repair is currently uncertain; one factor that may particularly complicate analysis of inflammatory mechanisms in the acutely injured brain is that specific mediators may have different effects on specific cell types (e.g. neurons and oligodendroglia [51]), and on cells at different maturational stages (e.g. oligodendroglial precursors vs. mature oligodendroglia).

There are significant limitations to reliance on neonatal rodent models of cerebral hypoxia–ischemia and excitotoxicity to the understanding of clinical disorders in human neonates. No available animal models precisely replicate the pathophysiology of the human neonatal neuropathology. Experimental animals are generally otherwise healthy; in contrast, neonates who are the most vulnerable to cerebral ischemia, often have multi-system disease. In addition, the epidemiological data cited earlier suggests that factors such as maternal infection and other unidentified perinatal stressors can induce a pro-inflammatory milieu, which could then amplify intrinsic inflammatory responses to brain injury. These antecedent stresses are difficult to reproducibly model experimentally, although some recent progress has been made in achieving this goal.

Future studies will need to determine whether chemokines derived from the maternal circulation, the placenta, or the fetus, in response to infection (or other stressors) reach the developing brain (perhaps, at least in part, because of immaturity of the blood–brain barrier). It will also be essential to determine whether chemokines exert direct effects in the immature brain and/or alter susceptibility of the developing brain to other agents or insults.

Our long-term goal is to identify more effective treatment interventions to limit the adverse impact of acute brain injury in the neonate and infant. Whether chemokines represent an appropriate molecular target for therapeutic intervention in infants who have incurred brain injury remains to be determined. A greater understanding of the mechanisms that regulate chemokine and chemokine receptor functional activity has the potential to enable therapeutic modulation after the onset of brain injury. Concurrently, it will be essential to identify any adverse effects of agents that alter chemokine function on brain development.

ACKNOWLEDGEMENTS

The research highlighted in this chapter was supported by USPHS Grants NS35059 and NS31504 (to F.S.S.).

REFERENCES

1. Dammann O, Leviton A. Maternal intrauterine infection, cytokines, and brain damage in the preterm infant. Pediatr Res 1997;42:1–8.
2. Gilles FH, Averill DR Jr, Kerr CS. Neonatal endotoxin encephalopathy. Ann Neurol 1977;2:49–56.
3. Grether JK, Nelson KB. Maternal infection and cerebral palsy in infants of normal birth weight. J Am Med Assoc 1997;278:207–211.
4. Yoon BH, Jun JK, Romero R, Park KH, Gomez R, Choi JH, Kim IO. Amniotic fluid inflammatory cytokines (interleukin-6, interleukin-1beta, and tumor necrosis factor-alpha), neonatal brain white matter lesions, and cerebral palsy. Am J Obstet Gynecol 1997;177:19–26.
5. Yoon BH, Romero R, Park JS, Kim CJ, Kim SH, Choi JH, Han TR. Fetal exposure to intra-amniotic inflammation and the development of cerebral palsy at the age of three years. Am J Obstet Gynecol 2000;182:675–681.
6. Nelson KB, Dambrosia JM, Grether JK, Phillips TM. Neonatal cytokines and coagulation factors in children with cerebral palsy. Ann Neurol 1998;44:665–675.
7. Baerwald KD, Popko B. Developing and mature oligodendrocytes respond differently to the immune cytokine interferon-γ. J Neurosci Res 1998;52:230–239.
8. Ivacko JA, Sun R, Silverstein FS. Hypoxic–ischemic brain injury induces an acute microglial reaction in perinatal rats. Pediatr Res 1996;39:1–9.
9. Ivacko J, Szaflarski J, Malinak C, Flory C, Warren JS, Silverstein FS. Hypoxic–ischemic injury induces monocyte chemoattractant protein-1 expression in neonatal rat brain. J Cerebr Blood Flow Metab 1997;17:759–770.
10. Bona E, Andersson AL, Blomgren K, Gilland E, Puka-Sundvall M, Gustafson K, Hagberg H. Chemokine and inflammatory cell response to hypoxia–ischemia in immature rats. Pediatr Res 1999;45:500–509.
11. Xu H, Barks JDE, Schielke GP, Silverstein FS. Attenuation of hypoxia–ischemia-induced monocyte chemoattractant protein-1 expression in brain of neonatal mice deficient in interleukin-1 converting enzyme. Mol Brain Res 2001;90:57–67.
12. Martin D, Chinookoswong FN, Miller G. The interleukin-1 receptor antagonist (rhIL-1ra) protects against cerebral infarction in a rat model of hypoxia-ischemia. Exp Neurol 1994;130:362–367.
13. Galasso JM, Wang P, Martin D, Silverstein FS. Inhibition of TNF-α can attenuate or exacerbate excitotoxic injury in neonatal rat brain. NeuroReport 2000;11:231–235.
14. Vannucci RC, Brucklacher RM, Vannucci SJ. The effect of hyperglycemia on cerebral metabolism during hypoxia-ischemia in the immature rat. J Cereb Blood Flow Metab 1996;16:1026–1033.

15. Ditelberg JS, Sheldon RA, Epstein CJ, Ferriero DM. Brain injury after perinatal hypoxia–ischemia is exacerbated in copper/zinc superoxide dismutase transgenic mice. Pediatr Res 1996;39:204–208.
16. Cheng Y, Gidday JM, Yan Q, Shah AR, Holtzman DM. Marked age-dependent neuroprotection by brain-derived neurotrophic factor against neonatal hypoxic–ischemic brain injury. Ann Neurol 1997;41:521–529.
17. McDonald JW, Silverstein FS, Johnston MV. Neurotoxicity of NMDA is markedly enhanced in developing rat central nervous system. Brain Res 1988;459:200–203.
18. Dubois B, Masure S, Hurtenbach U, Paemen L, Heremans H, van den Oord J, Sciot R, Meinhardt T, Hammerling G, Opdenakker G, Arnold B. Resistance of young gelatinase B-deficient mice to experimental autoimmune encephalomyelitis and necrotizing tail lesions. J Clin Invest 1999;104:1507–1515.
19. Liu XH, Kwon D, Schielke G, Yang GY, Silverstein FS, Barks JDE. Mice deficient in interleukin-1 converting enzyme are resistant to neonatal hypoxic–ischemic brain damage. J Cereb Blood Flow Metab 1999;19:1099–1108.
20. Szaflarski J, Burtrum D, Silverstein FS. Cerebral hypoxia–ischemia stimulates cytokine gene expression in perinatal rats. Stroke 1995;26:1093–1100.
21. Burtrum D, Silverstein FS. Hypoxic–ischemic injury stimulates glial fibrillary acidic protein mRNA expression in perinatal rat brain. Exp Neurol 1994;126:112–118.
22. Flaris NA, Densmore TL, Molleston MC, Hickey WF. Characterization of microglia and macrophages in the central nervous system of rats: definition of the differential expression of molecules using standard and novel monoclonal antibodies in normal CNS and in four models of parenchymal reaction. Glia 1993;7:34–40.
23. Giulian D, Li J, Broker J, Li X, Kirkpatrick JB. Cell surface morphology identifies microglia as a distinct class of mononuclear phagocyte. J Neurosci 1995;15:7712–7726.
24. Galasso JM, Harrison JK, Silverstein FS. Excitotoxic brain injury stimulates expression of the chemokine receptor CCR5 in neonatal rats. Am J Pathol 1998;153:1631–1640.
25. Acarin L, Gonzalez B, Castellano B, Castro AJ. Microglial response to *N*-methyl-D-aspartate-mediated excitotoxicity in the immature rat brain. J Comp Neurol 1996;367:361–374.
26. McManus CM, Brosnan CF, Berman JW. Cytokine induction of MIP-1alpha and MIP-1beta in human fetal microglia. J Immunol 1998;160:1449–1455.
27. Del Bigio MR, Becker LE. Microglial aggregation in the dentate gyrus: a marker of mild hypoxic–ischaemic brain insult in human infants. Neuropathol Appl Neurobiol 1994;20:144–151.
28. Gonzalez-Scarano F, Baltuch G. Microglia as mediators of inflammatory and degenerative diseases. Annu Rev Neurosci 1999;22:219–240.
29. Elkabes S, DiCicco-Bloom EM, Black IB. Brain microglia/macrophages express neurotrophins that selectively regulate microglial proliferation and function. J Neurosci 1996;16:2508–2521.
30. Meng SZ, Oka A, Takashima S. Developmental expression of monocyte chemoattractant protein-1 in the human cerebellum and brainstem. Brain Dev 1999;21:30–35.
31. Coughlan CM, McManus CM, Sharron M, Gao ZY, Murphy D, Jaffer S, Choe W, Chen W, Hesselgesser J, Gaylord H, Kalyuzhny A, Lee VMY, Wolf B, Doms RW, Kolson DL. Expression of multiple functional chemokine receptors and monocyte chemoattractant protein-1 in human neurons. Neuroscience 2000;97:591–600.
32. Wang X, Yue T, Barone FC, Feurstein GZ. Monocyte chemoattractant protein-1 messenger RNA expression in rat ischemic cortex. Stroke 1995;26:661–666.
33. Kim JS, Gautam SC, Chopp M, Zaloga C, Jones ML, Ward PA, Welch KMA. Expression of monocyte chemoattractant protein-1 and macrophage inflammatory protein-1 after focal cerebral ischemia in the rat. J Neuroimmunol 1995;56:127–134.
34. Berman JW, Guida MP, Warren J, Amat J, Brosnan CF. Localization of monocyte chemoattractant peptide-1 expression in the central nervous system in experimental autoimmune encephalomyelitis and trauma in the rat. J Immunol 1996;156:3017–3023.
35. Glabinski AR, Balasingam V, Tani M, Kunkel SL, Strieter RM, Yong VW, Ranshoff RM. Chemokine monocyte chemoattractant protein-1 is expressed by astrocytes after mechanical injury to the brain. J Immunol 1996;156:4363–4368.
36. Gourmala NG, Buttini M, Limonta S, Sauter A, Boddeke HW. Differential and time-dependent expression of monocyte chemoattractant protein-1 mRNA by astrocytes and macrophages in rat brain: effects of ischemia and peripheral lipopolysaccharide administration. J Neuroimmunol 1997;74:35–44.
37. Szaflarski J, Ivacko J, Liu XH, Warren JS, Silverstein FS. Excitotoxic injury induces monocyte chemoattractant protein-1 expression in neonatal rat brain. Mol Brain Res 1998;55:306–314.
38. Galasso JM, Miller MJ, Cowell RM, Harrison JK, Warren JS, Silverstein FS. Acute excitotoxic injury induces expression of monocyte chemoattractant protein-1 and its receptor, CCR2, in neonatal rat brain. Exp Neurol 2000;165:295–305.
39. Flory CM, Jones ML, Miller BF, Warren JS. Regulatory roles of tumor necrosis factor-α and interleukin-1β in monocyte chemoattractant protein-1-mediated pulmonary granuloma formation in the rat. Am J Pathol 1995;146:450–462.
40. Weiss JM, Downie SA, Lyman WD, Berman JW. Astrocyte-derived monocyte chemoattractant protein-1 directs the transmigration of leukocytes across a model of the human blood–brain barrier. J Immunol 1998;161:6896–6903.
41. Hagan P, Barks JDE, Yabut M, Davidson BL, Roessler B, Silverstein FS. Adenovirus-mediated over-expression of interleukin-1 receptor antagonist reduces susceptibility to excitotoxic brain injury in perinatal rats. Neuroscience 1996;75:1033–1045.
42. Boddeke EWGM, Meigel I, Frentzel S, Gourmala NG, Harrison JK, Buttini M, Spleiss O, Gebicke-Harter P. Cultured rat microglia express functional β-chemokine receptors. J Neuroimmunol 1999;98:176–184.
43. Horuk R, Martin AW, Wang Z, Schweitzer L, Gerassimides A, Guo H, Lu Z, Hesselgesser J, Perez HD, Kim

J, Parker J, Hadley TJ, Peiper SC. Expression of chemokine receptors by subsets of neurons in the central nervous system. J Immunol 1997;158:2882–2890.
44. Vallat AV, de Girolami U, He J, Mhashilkar A, Marasco W, Shi B, Gray F, Bell J, Keohane C, Smith TW, Gabuzda D. Localization of HIV-1 co-receptors CCR5 and CXCR4 in the brain of children with AIDS. Am J Pathol 1998;152:167–178.
45. Galasso JM, Stegman L, Blaivis M, Ross B, Silverstein FS. Experimental gliosarcoma induces chemokine receptor expression in rat brain. Exp Neurol 2000;161:85–95.
46. Galasso JM, Szaflarski J, Warren JS, Silverstein FS. Monocyte chemoattractant protein-1 is a mediator of acute excitotoxic injury in neonatal rat brain. Neuroscience 2000;101:737–744.
47. Bell MD, Taub DD, Perry VH. Overriding the brain's intrinsic resistance to leukocyte recruitment with intraparenchymal injections of recombinant chemokines. Neuroscience 1996;74:283–292.
48. Kennedy KJ, Strieter RM, Kunkel SL, Lukacs NW, Karpus WJ. Acute and relapsing experimental autoimmune encephalomyelitis are regulated by differential expression of the CC chemokines macrophage inflammatory protein-1 alpha and monocyte chemotactic protein-1. J Neuroimmunol 1998;92:98–108.
49. Cowell RM, Galasso JM, Silverstein FS. The chemokine macrophage inflammatory protein 1-alpha (MIP-1) and its receptor are upregulated in a neonatal rat stroke model. Soc Neurosci Abstr 1999;738:11.
50. Meucci O, Fatatis A, Simen AA, Bushell TJ, Gray PW, Miller RJ. Chemokines regulate hippocampal neuronal signaling and gp120 neurotoxicity. Proc Natl Acad Sci USA 1998;95:14500–14505.
51. Debillon T, Gras-Leguen C, Verielle V, Winer N, Caillon J, Roze JC, Gressens P. Intrauterine infection induces programmed cell death in rabbit periventricular white matter. Pediatr Res 2000;47:736–742.

Universes in Delicate Balance: Chemokines and the Nervous System
R.M. Ransohoff, K. Suzuki, A.E.I. Proudfoot, W.F. Hickey and J.K. Harrison, editors

Chapter 3.4

Stroke: Chemokine-Induced Infiltration of Immune Cells

H.W.G.M. Boddeke

Department of Medical Physiology, University of Groningen, Deusinglaan 1, Building 3125, 9713 AV Groningen, Netherlands

1. INTRODUCTION

The pathological mechanisms involved in brain ischemia are only partly understood. It is assumed that, in addition to a phase of rapid necrotic cell death, a subsequent degeneration process is caused by a prolonged delayed phase of neurodegeneration in the penumbra zone of the ischemic lesion. One of the mechanisms presumably involved in delayed ischemic neurodegeneration is neuroinflammation. In this review an important aspect of this neuroinflammatory process, namely the infiltration of both resident and various types of hematogenous immune cells that invade the ischemic area at different stages of the pathological process, is discussed.

2. BRAIN ISCHEMIA AND NEUROINFLAMMATION

Most cases of stroke are caused by interruption of the brain arterial blood flow by a thrombus, leading to ischemia. Fewer cases of stroke are caused by blood vessel rupture leading to hemorrhage. Many of the pathological processes involved in ischemia have been investigated using animal models such as middle cerebral artery occlusion (MCAO) and experimental hypoxic ischemia [1,2]. During and after stroke insufficient supply of oxygen and nutrients results in loss of ATP and neuronal depolarization [3]. This leads to massive release of excitotoxic concentrations of glutamate causing excessive calcium influx via NMDA receptors and additional activation of AMPA receptors and voltage gated calcium channels [4–6]. The resulting elevated intracellular calcium concentrations lead to lethal metabolic dysfunction. This type of excitotoxicity causes necrosis [7] but also apoptosis [8,9]. Particularly the necrotic process is a major starting point for neuroinflammatory processes. Presumably as a result of necrosis in the infarct region microglia are activated within hours and transform into phagocytes and start to produce proinflammatory cytokines, chemokines and radical species and these contribute to infarct progression.

It is thus clear now that cytokines are involved in the pathophysiology of focal cerebral ischemia [10]. Within hours after induction of MCAO prominent expression of the proinflammatory cytokine interleukin-1β (IL-1β) has been observed [11–13]. Furthermore, it has been shown that exogenous administration of interleukin-1 receptor antagonist (IL1ra) or the use of recombinant adenovirus vectors carrying human IL1ra cDNA reduced focal ischemic brain damage [14,15]. In addition to IL-1β, induction of tumor necrosis factor-α (TNF-α) expression was observed at the same time [13,16]. Conflicting results concerning putative harmful [13,17] and protective effects of TNF-α [18,19] have been reported and await further clarification.

In addition, a distinct group of small inflammatory cytokines, called chemokines, plays an important role in the infiltration of hematogenous immune cells and the attraction of local microglia in the ischemic area [20]. The time-related expression of chemokines in the ischemic area and their specific chemoattractant activity are the subject of this review.

3. THE BLOOD–BRAIN BARRIER DURING ISCHEMIA

The blood–brain barrier is a complex system consisting of endothelial cells, astroglia, pericytes, perivascular macrophages and a basal lamina [21]. Astrocytes project their endfeet close to the cerebral endothelial cell layer and contribute to the barrier function of the cerebral endothelial cells. The cerebral endothelial cells are colocalized with pericytes and perivascular macrophages in the basal lamina [22]. Compared to peripheral endothelium the cerebral endothelium is much tighter and contains narrow intercellular tight junctional structures. These tight junctions prohibit paracellular transport across the cerebral endothelium. Under physiological circumstances the blood–brain barrier maintains the homeostasis of the central nervous system and protects against factors from outside. Thus, although a certain degree of immune surveillance of the brain has been reported [23], under healthy conditions, infiltration of the brain by hematogenous immune cells is low. When inflammation occurs, the isolating capacity of the blood–brain barrier can decrease dramatically and infiltration of hematogenous immune cells occurs together with an influx of serum proteins. This increased permeability of the blood–brain barrier results from opening of tight junctions, enhanced pinocytosis and the formation of transendothelial channels [24]. Increased endothelial cell permeability and degradation of matrix proteins of the vascular basal lamina are observed after ischemia [25].

Increased infiltration of eosinophils and at a later time point of monocytes, macrophages and lymphocytes into the ischemic area can lead to perivascular inflammatory reactions, in which the blood–brain barrier may play an active role by production of inflammatory mediators and chemotactic substances. It has indeed been suggested that a degradation of the blood–brain barrier contributes to the neuroimmune response by production of inflammatory mediators and the expression of adhesion molecules [25,26].

4. INFILTRATION OF BLOOD CELLS DURING BRAIN ISCHEMIA

Shortly after the local release of inflammatory cytokines early infiltration of granulocytes in the ischemic lesion is observed. Granulocyte infiltration is followed at a later time point by blood macrophages and T cells [27]. The recruitment of inflammatory cells to the ischemic area is facilitated by endothelial expression of adhesion molecules [28,29] and by the establishment of a chemotactic gradient in the ischemic area [20]. Thus infiltration of immune cells in the ischemic zone involves rolling, adhesion and transendothelial migration [30]. The infiltration process is facilitated by various types of adhesion molecules in the blood vessel endothelium. The rolling of immune cells along the endothelium is mediated by selectins like E-selectin and P-selectin [31,32]. Expression of the receptors for P-selectin and E-selectin is induced by inflammatory cytokines. The subsequent tighter adhesion is mediated by integrin type adhesion molecules such as CD11b/CD18, leukocyte associated antigen-1 (LFA-1) and very late antigen-4 complex (VLA-4). The matching endothelial receptors for these adhesion molecules are intercellular adhesion molecule-1 and -2 (ICAM-1 and -2) for CD11b/CD18 and LFA-1 and vascular cell adhesion molecule-1 (VCAM-1) for VLA-4 [33,34]. Thus various adhesion molecule receptors recognize their immune cell ligand and support adhesion and migration of immune cells from the blood vessel.

During ischemia local factors such as inflammatory cytokines induce enhanced expression of adhesion molecules like ICAM-1, and VCAM-1 and P-selectin and E-selectin in brain endothelial cells [27].

The first hematogenous cells that infiltrate the ischemic brain area are granulocytes. Within hours after MCAO granulocytes accumulate in cerebral vessels [35,36] and then invade the ischemic lesion. The maximal granulocyte infiltration occurs approximately after 24 h. During ischemia granulocyte infiltration is presumably initiated by increased expression of chemokines like cytokine-induced neutrophil attractant (CINC) and interleukin-8 (IL-8; CXCL8) [37,38]. The pathophysiological consequences of the granulocyte invasion are presently, however, unclear. The immune cells that most numerously invade the infarct region in an MCAO model, at a later stage, are monocytes/macrophages [27,39]. Monocytes/macrophages start to infiltrate approximately 12 h after ischemia and this process continues up to 14 days after ischemia [35,36,40]. Within weeks the tissue damaged by ischemia is completely cleared by these macrophages. Interestingly, there are no surface markers that distinguish between hematogenous macrophages and resident activated microglia. Accordingly, the accumulation of these macrophages, found in the ischemic area, may stem from resident microglia as well as from hematogenous macrophages. The abundance of macrophages/microglia fits well with the pronounced expression of CC-chemokines like monocyte chemoattractant protein-1 (MCP-1/CCL2), macrophage inflammatory protein-1α and -1β (MIP-1α/CCL3 and MIP-1β/CCL4) in ischemic brain tissue [2,42–45].

In addition, it has been shown that 12 h after induction of hypoxia–ischemia natural killer cells start to invade the ischemic area [2]. These cells kill target cells rapidly by an antigen-nonspecific mechanism and could be attracted by ischemia-associated chemokines such as regulated on activation normal T cell expressed and secreted (RANTES/CCL5) or CCL3.

Furthermore, T cells enter the infarct region [39, 40]. Specific chemokine signals involved in T cell attraction could be chemokines induced in ischemic area such as CCL3, CCL4, CCL5 and interferon-inducible protein-10 (IP-10/CXCL10).

In addition, local brain immune cells, astrocytes and microglia respond to the pathological events during ischemia. Based on only partly understood ischemic signals normally quiescent ramified microglia are immune activated and transform into phagocytes [46]. Presumably, activated microglia migrate along the chemotactic gradient in the ischemic zone. In the ischemic area microglia and astrocytes are involved in the production of inflammatory cytokines such as IL-1β and TNFα [10,12].

5. CHEMOKINE SIGNALING IN ISCHEMIC BRAIN TISSUE

Chemokines are small chemoattractant cytokines involved in the recruitment of immune cells towards sites of inflammation and tissue damage. Furthermore, chemokines are involved in processes such as the upregulation of expression of adhesion molecules [47] and vascular neogenesis [48]. Importantly, it has been reported that application of α-chemokines induces a widespread destruction of the blood–brain barrier and thus strongly compromises the resistance to infiltration by immune cells; this effect may be very important for ischemia [49]. Most of these chemokine-induced effects are mediated by seven transmembrane spanning, G-protein coupled receptors [50]. In conjunction with disruption of the blood–brain barrier, expression of various chemokines is involved in the infiltration of hematogenic blood cells as well as attraction of local immune cells to the ischemic area. Based on the presence of conserved cysteine residues chemokines are divided into four subclasses: CXC-, CC-, and C- and CX3C-chemokines [50]. These subclasses display different specific cell target selectivities; thus CXC chemokines are chemotactic to granulocytes and CC chemokines mainly attract mononuclear cells such as monocytes and lymphocytes [50]. C-chemokines are chemotactic to lymphocytes [51] and CX3CL1 is chemotactic to neutrophils, T cells and monocytes [52,53]. Specific chemokines are induced during/after ischemia and their time-dependent induction seems in accordance with the temporal profile of infiltration of granulocytes, leukocytes and lymphocytes. It is likely that the expression of chemokines is induced by proinflammatory cytokines which are produced in the ischemic area. For example, TNF-α and IL-1β activate transcription factors such as nuclear factor κB [54] that regulate the transcription of chemokines indicating that these proinflammatory cytokines activate the production of chemokines. Indeed expression of CCL2 mRNA in monocytes and astrocytes is induced by cytokines like IL-1β and TNFα and TGFβ [55–58]. Furthermore, expression of CXCL10 in the brain is induced by interferon-γ [59] and presumably also by factors such as IL-1β and TNFα. These results suggest that cytokines produced in the ischemic area are involved in the regulation of chemokine expression during the ischemic process.

The time course of chemokine expression seems clearly associated with the time-dependent infiltration of different immune cells in the ischemic brain area. Early expression of CXC chemokines like macrophage inflammatory protein-2 (CXCL2/CXCL3), growth-related oncogene (gro; CXCL1) and CINC has been observed [2,60] in a hypoxia–ischemia model and precedes infiltration of neutrophils. Liu et al. [61] observed expression of CINC 6 h after induction of MCAO. In a rat model of reperfusion after ischemia also elevated concentrations of CINC were found [37]. In a rat model of transient MCAO Yamagami et al. [60] also observed early production of CINC which peaked 6 h after induction of MCAO. The early expression of CINC clearly matched that of rapid neutrophil infiltration. Subsequent expression of CC-chemokines (predominantly CCL4) preceded the appearance of lymphocytes, microglia/macrophages and NK cells; the rapid induction of CCL2 and CCL4 could match the early activation of microglia [35,62].

Thus regarding the time-course of infiltration it is clear that CXC chemokines are expressed rapidly after the onset of ischemia. For example in a model of hypoxic ischemia in young rats early expression of the chemokines CXCL1 and CXCL2/CXCL3 occurred already 1 h after induction of ischemia [2]. A rapid but more complex ischemia-induced expression pattern was observed for CXCL10. This chemokine displays strong chemoattractant activity towards lymphocytes. In a rat model of focal ischemia (MCAO) Wang et al. [63] found that CXCL10 is expressed at two distinct time points during ischemia. Early expression of CXCL10 mRNA was found 3 h after induction of ischemia and maximal expression was observed 6 h after induction of MCAO. A second phase of CXCL10 mRNA expression was observed 10–15 days after ischemic injury. Interestingly, immunohistochemical analysis revealed that CXCL10 protein expression in the early phase of ischemia occurred in neurons whereas the late phase expression of CXCL10 was performed in astrocytes. These experiments clearly show the remarkable feature that neurons already in an early phase of ischemia send signals to the immediate environment presumably astrocytes and microglia, thereby starting an early ischemia-induced local immune response. The chemokine receptor involved in CXCL10 signaling in ischemia, most likely is CXCR3, which is time-dependently upregulated in rat ischemic cortex [63].

CXCL8 is a potent neutrophil chemotactic and activating factor and most likely plays an important role in neutrophil-mediated acute inflammation [64]. In addition to attracting neutrophils, CXCL8 also stimulates a number of processes which are relevant for ischemia such as the release of neutrophil granules, mediation of the respiratory burst and upregulation of complement receptors and integrins in neutrophils [65–68]. Kostulas et al. [69] reported that they found high numbers of CXCL8 mRNA expressing blood monocytic cells in patients with ischemic stroke. Furthermore, the CXCL8 levels in blood plasma correlated positively with CXCL8 mRNA expression in patients with stroke and could be reminiscent of a chemotactic gradient originating in the ischemic brain area. They suggest that CXCL8 could be involved in recruiting polymorphonuclear leukocytes to the ischemic area.

Matsumoto et al. [38] observed increasing brain tissue levels of interleukin-8 within 6 h after reperfusion. Furthermore, they detected CXCL8 protein immunohistochemically in the vascular wall, suggesting local production of CXCL8 in reperfused brain tissues. Matsumoto et al. [38] also found that a neutralizing anti-CXCL8 antibody reduced the infarct size and accordingly suggested that neutrophils are involved in reperfusion damage. It has also been reported that CXCL8 is produced by endothelial cells. In an in vitro hypoxia model Zhang et al. [70] showed that both cultured fetal human astrocytes and human cerebral endothelial cells release CXCL8 and CCL2. Under hypoxic conditions fetal human astrocytes, however, produce much more CCL2 and CXCL8 than human cerebral endothelial cells.

The early infiltration of the ischemic area by neutrophils is followed by subsequent infiltration by monocytes and accordingly, the expression of CXC chemokines is followed by expression of CC chemokines.

Transient increases in mRNA expression of the CC-chemokine CCL2 were detected in temporary and permanent cerebral ischemia models in adult rats using Northern analysis. CCL2 protein expression was detected in ischemic brain tissue and localized in endothelial cells and macrophage-like cells [41,42].

In a recent study concerning permanent MCAO in rats the cellular source of CCL2 mRNA was further investigated [45]. A biphasic pattern of CCL2 expression was observed: 6 h to 2 days after MCAO strong expression of CCL2 mRNA was observed in astrocytes surrounding the penumbra; 4–6 days after MCAO CCL2 mRNA was predominantly observed in macrophages/microglia in the infarcted tissue. This temporal profile correlated with that determined by Northern analysis in previous studies [41,42]. The early expression of CCL2 mRNA in this study precedes the peak of emergence of monocyte infiltrates [35] suggesting involvement of CCL2 in the invasion of monocytes. The later phase of CCL2 expression by recruited monocytes most likely serves as a signal which further potentiates monocyte accumulation and activation [60,71]. The expression of CCL2 mRNA in astrocytes is at variance with the study of Kim et al. [41] who attributed CCL2 protein expression to endothelial and macrophage-like cells. This discrepancy could possibly be due to differences in translation efficiency or binding of CCL2, secreted by astrocytes to endothelial and macrophage-like cells.

In an hypoxic–ischemic injury model Ivacko et al. [72] found expression of CCL2 mRNA 1 h after induction of ischemia and peak expression of CCL2 was found within 8–24 h. In the same study CCL2 protein expression was observed 2.5 up to 48 h after induction of ischemia. Interestingly, Ivaco et al. suggested that the CCL2 protein is localized in neurons. This is in contrast with the afore-mentioned studies which suggest expression in glial cells and endothelial cells.

Yamagami et al. [60] reported CCL2 protein expression within the same time window in a transient MCAO model. The maximal concentration of CCL2 detected within the ischemic area in this study was 89.2 ± 28.2 ng/g tissue ww; this concentration corresponds to 6.4 nM, within the range of biological activity of CCL2. From these studies it is thus clear that pronounced expression of CCL2 both on mRNA and protein level occurs and that particularly the early expression presumably is involved in chemoattraction of hematogenous macrophages. Whereas the chemokine receptor subtype involved in CCL2-induced chemotaxis in blood leukocytes most likely is CCR2, involvement of CCR2 in CCL2-induced signaling in resident astrocytes and microglia, is currently ambiguous. There are conflicting reports on the expression of CCR2 in cultured astrocytes and microglia cells. Low CCR2 mRNA expression was found in cultured microglia by Boddeke et al. [73] and McManus et al. [74], whereas no CCR2 mRNA in cultured microglia was found by others [75,76]. Further studies concerning CCL2-induced signaling via CCR2 or other putative chemokine receptors are required to resolve this issue.

Expression of the CC-chemokines CCL3 and CCL4 is prominently induced during ischemia. Induction of CCL3 mRNA and protein was observed by Kim et al. [41] in an MCAO model in rats. In this study CCL3 was localized in astrocytes in the penumbra zone; however, no co-localization experiments were performed. In contrast, expression of CCL3 and CCL4 after induction of permanent and transient MCAO in rats has been found by Takami et al. [44] rather in microglia/macrophages and not in astrocytes. Peak expression of CCL3/CCL4 mRNAs occurred 4–6 h after the induction ischemia. Furthermore, expression of CCL3/CCL4 mRNAs was observed first in the ischaemic core later in the penumbra. In agreement with Takami et al. [44], Gourmala et al. [43] reported expression of CCL3 protein in microglia/macrophages. Furthermore, in both studies CCL3/CCL4 mRNA expression is generally stronger in transient than in permanent MCAO. This higher expression level of chemokines observed after transient MCAO may be attributed to a larger rate of infiltration of polymorphonuclear cells and macrophages under transient ischemic conditions. In the study of Gourmala et al. [43] already 1 h after ischemia expression of CCL3 and CCL4 mRNA was observed and the expression of both CCL3/CCL4 mRNAs was peaking 8–16 h after MCAO. This early expression of CCL3/CCL4 mRNAs suggests that microglia are an important source of CCL3/CCL4 mRNAs.

In an hypoxia–ischemia model applied in neonatal rats also expression of CCL3 and CCL4 mRNA was demonstrated using RT-PCR analysis [2]. CCL3 rose slightly 6 h after ischemia, CCL4 much more so 1–6 h after ischemia.

CCL3 binds the chemokine receptors CCR1 and 5 and CCL4 more selectively binds CCR5 [50]. This makes both CCL3 and CCL4 realistic chemoattractants for macrophages and T cells invading the ischemic area.

The biological activity of CCL3 and CCL4, however, is different. Whereas CCL3 stimulates secretion of TNF and interleukin-1α and interleukin-6, CCL4 does not [77]. Furthermore, CCL3 and CCL4 act on different populations of activated T cells, CCL3 specifically attracts CD-8$^+$ cells and CCL4 acts on CD-4$^+$ T cells [78].

Recently, it has been shown that secondary lymphoid-tissue chemokine (SLC), which controls homing of dendritic cells and T cells to secondary lymphoid organs is also expressed by neurons in ischemic mouse cortex [79]. The ischemia-induced expression of SLC in neurons acting on microglia suggests that SLC, like CXCL10 [80], is involved in a neuron–microglia signaling system which may be relevant for brain ischemia. The receptor for SLC on microglia is not yet well defined. Since no expression of CCR7, the original receptor for SLC in brain tissue, has been found, another candidate receptor could be CXCR3, which has been described as a receptor for SLC as well [81].

Whereas the involvement of the above-mentioned CXC and CC chemokines in chemotaxis of immune cells seems plausible, the regulation and function of CCL5 in ischemia is not well understood. In a recent study Spleiss et al. [82] observed that a distinct basal mRNA expression level of CCL5 was present in control rat brains. Interestingly, after induction of MCAO the mRNA expression of CCL5 in the infarcted hemisphere was not affected, but the mRNA level in the contralateral hemisphere decreased after induction of MCAO. In the same study an ischemia-induced increase of mRNA expression of the chemokines CCL3 and CCL4 was demonstrated. Also Bona et al. [2] applying hypoxic ischemia in neonatal rats observed a basal mRNA expression level of CCL5 in control brain. At a later stage, 24 h to 14 days after ischemia increased mRNA expression of CCL5 was observed, which could coincide with the $CD4^+$ cell response in ischemic tissue [2].

Using a differential display assay Utans-Schneitz et al. [83] demonstrated expression of a new chemokine factor ST38 induced after MCAO in rats. The nucleotide sequence of ST38 is identical to that of macrophage inflammatory protein-3α (MIP-3α/CCL20, the accession number identified in a blast search was RNU 90447) and this is the first time that involvement of CCL20 in ischemia-mediated neuroinflammation has been demonstrated. The specific chemokine receptor for CCL20 is CCR6. Up to now the expression of CCR6 in the brain has not been reported; on the other hand, CCR6 has been found in $CD4^+$ and $CD8^+$ T cells [84]. Accordingly, these cells could be attracted by a chemotactic gradient of CCL20 mediated by ischemia. This possibility awaits further research.

6. CONCLUSION

Infiltration by blood cells seems an important phenomenon which takes place at all pathological stages of ischemia. Accordingly, both local and hematogenous immune cells seem involved in the pathology of ischemia. Chemokines clearly play an important role in the time-dependent attraction of immune cells to the ischemic area. The current knowledge concerning chemoattraction of immune cells to the ischemic area, however, is still far from complete. Furthermore, it remains to be established whether antagonism of chemokines could be a therapeutic target for brain ischemia.

REFERENCES

1. Sauter A, Rudin M. Strain-dependent drug effects in rat middle cerebral artery occlusion model of stroke. J Pharmacol Exp Ther 1995;274:1008–1013.
2. Bona E, Andersson A-L, Blomgren K, Gilland E, Puka-Sundvall M, Gustafson K, Hagberg H. Chemokine and inflammatory cell response to hypoxia–ischemia in immature rats. Pediatr Res 1999;45:500–509.
3. Nicholls D, Attwell D. The release and uptake of excitatory amino acids. Trends Pharmacol Sci 1990;11:462–468.
4. Rothman SM, Olney JW. Glutamate and pathophysiology of hypoxic–ischemic brain damage. Ann Neurol 1986;19: 105–111.
5. Choi DW. Calcium: still center-stage in hypoxic ischemic neuronal death. Trends Neurosci 1995;18:68–60.
6. Choi DW. Calcium-mediated neurotoxicity: relationship to specific channel types and role in ischemic damage. Trends Neurosci 1988;1:465–469.
7. Gwag BJ. Slowly triggered excitotoxicity occurs by necrosis in cortical cultures. Neuroscience 1997;77:393–401.
8. Linnik MD, Zobrist RH, Hatfield MD. Evidence supporting a role for programmed cell death in focal cerebral ischemia in rats. Stroke 1993;24:2002–2008.
9. MacManus JP, Buchan AM, Hill IE, Rasquinha I, Preston E. Global ischemia can cause DNA fragmentation indicative of apoptosis in rat brain. Neurosci Lett 1993;164: 89–92.
10. Feuerstein GZ, Wang X, Barone FC. Inflammatory gene expression in cerebral ischemia and trauma. Potential new therapeutic targets. Ann NY Acad Sci 1997;825:179–193.
11. Liu T, McDonnel PC, Young PR, White RF, Barone FC, Feuerstein GZ. Interleukin-1β mRNA expression in ischemic rat cortex. Stroke 1993;24:1746–1751.
12. Buttini M, Sauter A, Boddeke HWGM. Induction of interleukin-1 beta mRNA after focal cerebral ischemia in rat brain. Brain Res 1994;23:126–134.
13. Arvin B, Neville LF, Barone FC, Feuerstein GZ. The role of inflammation and cytokines in brain injury. Neurosci Biobehav Rev 1999;20:445–452.
14. Relton JK, Rothwell NJ. Interleukin-1 receptor antagonist inhibits ischemic and excitotoxic neuronal damage in the rat. Brain Res Bull 1992;29:243–246.
15. Betz AL, Yang GY, Davidson BL. Attenuation of stroke

size in rats using an adenoviral vector to induce overexpression of interleukin-1 receptor antagonist in brain. J Cereb Blood Flow Metab 1995;15:547–551.
16. Liu T, Clark RK, McDonnel PC, Young PR, White RF, Barone FC, Feuerstein GZ. Tumor necrosis factor-α expression in ischemic neurons. Stroke 1994;25:1481–1488.
17. Barone FC, Arvin B, White RF, Miller A, Web CL, Willette RN, Lysko PG, Feuerstein GZ. Tumor necrosis factor-alpha. A mediator of focal ischemic brain injury. Stroke 1997;28:1233–1244.
18. Bruce AJ, Boling W, Kindy MS, Peschon J, Kramer PJ, Carpenter MK, Holtsberg FW, Mattson MP. Altered neuronal and microglial responses to excitotoxic and ischemic mice lacking TNF receptors. Nat Med 1996;2:788–794.
19. Rothwell NJ, Luheshi GN. Brain TNF: damage limitation or damaged reputation? Nat Med 1996;7:746–747.
20. Ransohoff RM, Tani M. Do chemokines mediate leukocyte recruitment in post-traumatic CNS inflammation? Trends Neurosci 1998;21:54–159.
21. Bradbury MWB. The blood–brain barrier: transport across the cerebral endothelium. Circ Res 1985;57:213–222.
22. Graeber MB, Streit WJ, Kreuzberg GW. Identity of ED-2 positive perivascular cells in rat brain. J Neurosci Res 1989;22:103–106.
23. Bradl M. Immune control of the brain. Springer Semin Immunopathol 1996;18:35–49.
24. Juhler M, Blasberg RG, Fenstermacher JD, Patlak CS, Paulson OB. A spatial analysis of the blood–brain barrier damage in experimental allergic encephalomyelitis. J Cereb Blood Flow Metab 1985;5:545–553.
25. Del Zoppo GJ, Hallenbeck JM. Advances in the vascular pathophysiology of ischemic stroke. Thromb Res 2000; 98:V73–V81.
26. De Vries H, Kuiper J, De Boer AG, Van Berkel JC, Breimer DD. The blood–brain barrier in neuroinflammatory diseases. Pharm Rev 1997;49:143–152.
27. Stoll G, Jander S, Schroeter M. Inflammation and glial responses in ischemic brain lesions. Prog Neurobiol 1998; 56:149–171.
28. Kishimoto TK, Rothlein R. Integrins, ICAMs, and selectins: role and regulation of adhesion molecules in neutrophil recruitment to inflammatory sites. Adv Pharmacol 1994;25:117–169.
29. Kim JS. Cytokines and adhesion molecules in stroke and related diseases. J Neurol Sci 1996;137:69–78.
30. Springer TA. Traffic signals for lymphocyte recirculation and leukocyte emigration: the multistep paradigm. Cell 1994;76:301–314.
31. Okada Y, Copeland BR, Mori E, Tung MM, Thomas SW, del Zoppo GJ. P-selectin and intracellular adhesion molecule-1 expression after focal brain ischemia and reperfusion. Stroke 1994;25:202–211.
32. Suzuki H, Abe K, Tojo S, Morooka S, Kimura K, Mizugaki M, Itoyama Y. Expression of P-Selectin and HSP72-like immunoreactivities in rat brain after transient middle cerebral artery occlusion. Brain Res 1997;759:321–329.
33. Osborn L. Leucocyte adhesion to endothelium in inflammation. Cell 1990;62:3–6.
34. Springer TA. Adhesion receptors of the immune system. Nature 1990;346:425–434.
35. Clark R, Lee EV, Fish CJ, White RF, Price WJ, Jonak ZL, Feuerstein GZ, Barone FC. Development of tissue damage, inflammation and resolution following stroke: an immunohistochemical and quantitativeplanimetric study. Brain Res Bull 1993;31:565–572.
36. Garcia JH, Liu KF, Yoshida Y, Lian J, Chen S, Del Zoppo G. Influx of leukocytes and platelets in an evolving brain infarct. Am J Pathol 1994;144:188–199.
37. Yamasaki Y, Matsuo Y, Matsuura N, Onodera H, Itoyama Y, Kogure K. Transient increase of cytokine-induced neutrophil chemoattractant, a member of the interleukin-8 family, in ischemic brain areas after focal ischemia in rats. Stroke 1995;26:318–322.
38. Matsumoto T, Ikeda K, Mukaida N, Harada A, Matsumoto Y, Yamashita J, Matsushima K. Prevention of cerebral edema and infarct in cerebral reperfusion injury by an antibody to interleukin-8. Lab Invest 1997;77:119–125.
39. Jander S, Schroeter M, D'Urso D, Gillen C, Witte OW, Stoll G. Focal cerebral ischemia of the rat brain illicits an unusual inflammatory response: early appearance of CD8+ macrophages/microglia. Eur J Neurosci 1998;10: 680–688.
40. Schroeter M, Jander S, Witte OW, Stoll G. Local immune responses in the rat cerebral cortex after middle cerebral artery occlusion. J Neuroimmunol 1994;55:195–203.
41. Kim JS, Gautam SC, Chopp M, Zaloga C, Jones ML, Ward PA, Welch KM. Expression of monocyte chemoattractant protein-1 and macrophage inflammatory protein-1 after focal cerebral ischemia in the rat. J Neuroimmunol 1995;56:127–134.
42. Wang X, Yue TL, Barone FC, Feuerstein GZ. Monocyte chemoattractant protein-1 messenger RNA expression in rat ischemic cortex. Stroke 1995;26:661–665.
43. Gourmala NG, Limonta S, Bochelen D, Sauter A, Boddeke HWGM. Localization of macrophage inflammatory protein: macrophage inflammatory protein-1 expression in rat brain after peripheral administration of lipopolysaccharide and focal cerebral ischemia. Neuroscience 1999;88: 1255–1266.
44. Takami S, Nishikawa H, Minami M, Nishiyori A, Sato M, Akaike A, Satoh M. Induction of inflammatory protein MIP-1α mRNA on glial cells after focal cerebral ischemia in the rat. Neurosci Lett 1997;227:173–176.
45. Gourmala NG, Buttini M, Limonta S, Sauter A, Boddeke HWGM. Differential and time-dependent expression of monocyte chemoattractant protein-1 mRNA by astrocytes and macrophages in rat brain: effects of ischemia and peripheral lipopolysaccharide administration. J Neuroimmunol 1997;74:35–44.
46. Kreuzberg GW. Microglia: a sensor for pathological events in the CNS. Trends Neurosci 1996;19:312–318.
47. Jiang Y, Beller DI, Frendl G, Graves DT. Monocyte chemoattractant protein-1 regulates adhesion molecule expression and cytokine production in human monocytes. J Immunol 1992;148:2423–2428.
48. Baggiolini M. Chemokines and leukocyte traffic. Nature 1998;392:565–568.
49. Baggiolini M, Imboden P, Detmers P. Neutrophil activation and the effects of interleukin-8/neutrophil-activating peptide 1 (CXCL8/NAP-1). Cytokines 1992;4:1–17.

50. Murphy PM, Baggiolini M, Charo IF, Hebert CA, Horuk R, Matsushima K, Miller LH, Oppenheim J, Power CA. International Union of Pharmacology, XXII. Nomenclature for chemokine receptors. Pharm Rev 2000;52:145–176.
51. Kennedy J, Kleyensteuber S, Schall TJ, Weiss MC, Yssel H, Schneider PV, Cocks GS, Bacon KB, Zlotnik A. Molecular cloning and function of human lymphotactin. J Immunol 1995;155:203–209.
52. Pan Y, Lloyd C, Zhou H, Dolich S, Deeds J, Gonzalo JA, Vath J, Gosselin M, Ma J, Dussault B, Woolf E, Alperin G, Culpepper J, Guttierez-Ramos JC, Gearing D. Neurotactin, a membrane-anchored chemokine upregulated in brain inflammation. Nature 1997;387:611–617.
53. Bazan JF, Bacon KB, Hardiman G, Wang W, Soo K, Rossi D, Greaves DR, Zlotnik A, Schall TJ. A new class of membrane-bound chemokine with a CX3C motif. Nature 1997;385:640–644.
54. Kuno K, Sukegawa K, Ishikawa Y, Orii T, Matsushima K. Acid sphingomyelinase is not essential for the IL-1 and tumor necrosis factor receptor signaling pathway leading to NFkB activation. Int Immunol 1994;8:1269–1272.
55. Barna BP, Pettay J, Barnett GH, Zhou P, Iwasaki K, Estes ML. Regulation of adult human non-neoplastic astrocytes is sensitive to tumor necrosis factor (TNF) or antibody to the 55 kDa TNF receptor. J Neuroimmunol 1994;50: 101–107.
56. Hayashi M, Luo Y, Laning J, Strieter RM, Dorf ME. Production and function of monocyte chemotactic protein-1 and other β-chemokines in murine glial cells. J Neuroimmunol 1995;60:143–150.
57. Hurwitz AA, Lyman WD, Berman JW. Tumor necrosis factor α and transforming growth factor β upregulate astrocyte expression of monocyte chemoattractant protein-1. J Neuroimmunol 1995;57:193–198.
58. Larsen CG, Zachariae COC, Oppenheim JJ, Matsushima K. Production of monocyte chemotactic and activating factor (MCAF) by human dermal fibroblast in response to interleukin 1 or tumor necrosis factor. Biochem Biophys Res Commun 1989;160:1403–1408.
59. Vanguri P, Farber JM. IFN and virus-inducible expression of an immediate early gene, crg-2/IP-10, and a delayed gene, I-Aα, in astrocytes and microglia. J Immunol 1994; 152:1411–1418.
60. Yamagami S, Tamura M, Hayashi M, Endo N, Tanabe H, Katsuura Y, Komoriya K. Differential production of MCP-1 and cytokine-induced neutrophil chemoattractant in the ischemic brain after transient focal ischemia in rats. J Leukoc Biol 1999;65:744–749.
61. Liu T, Young PR, McDonnell PC, White RF, Barone FC, Feuerstein GZ. Cytokine-induced neutrophil chemoattractant mRNA expressed in cerebral ischemia. Neurosci Lett 1993;164:125–128.
62. Ivacko JA, Sun R, Silverstein FS. Hypoxic–ischemic brain injury induces an acute microglial reaction in perinatal rats. Pedriatr Res 1996;39:39–47.
63. Wang X, Li X, Schmidt DB, Foley JJ, Barone FC, Ames RS, Sarau HM. Identification and molecular characterization of rat CXCR3: receptor expression and interferon-inducible protein-10 binding are increased in focal stroke. Mol Pharmacol 2000;57:1190–1198.
64. Harada A, Sekido N, Akahoshi T, Wada T, Mukaida N, Matsushima K. Essential involvement of interleukin-8 (IL-8) in acute inflammation. J Leukoc Biol 1994;56:559–564.
65. Peveri P, Walz A, Dewald B, Baggiolini M. A novel neutrophil-activating factor produced by human mononuclear phagocytes. J Exp Med 1988;167:1547–1559.
66. Masure S, Proost P, Van Damme J, Opdenakker G. Purification and identification of 91-kDa neutrophil gelatinase. Release by the activating peptide interleukin-8. Eur J Biochem 1991;198:391–398.
67. Baggiolini M, Imboden P, Detmers A. Neutrophil activation and the effects of interleukin-8/neutrophil-activating peptide-1 (IL-8/NAP-1). Cytokines 1992;4:1–17.
68. Detmers PA, Lo SK, Olsen-Egbert E, Walz A, Baggiolini M, Cohn ZA. Neutrophil-activating protein 1/interleukin 8 stimulates the binding activity of the leukocyte adhesion receptor CD11b/CD18 on human neutrophils. J Exp Med 1990;171:1155–1162.
69. Kostulas N, Kivisäkk P, Huang Y, Matusevicius D, Kostulas V, Link H. Ischemic stroke is associated with a systemic increase of blood mononuclear cells expressing interleukin-8 mRNA. Stroke 1998;29:462–466.
70. Zhang W, Smith C, Shapiro A, Monette R, Hutchison J, Stanimirovic D. Increased expression of bioactive chemokines in human cerebromicrovascular endothelial cells and astrocytes subjected to simulated ischemia in vitro. J Neuroimmunol 1999;101:148–160.
71. Cushing SD, Fogelman AM. Monocytes may amplify their recruitment into inflammatory lesions by inducing monocyte chemotactic protein. Arterioscler Thromb 1992; 12:78–82.
72. Ivacko J, Szaflarski J, Malinak C, Flory C, Warren JS, Silverstein FS. Hypoxic–ischemic injury induces monocyte chemoattractant protein-1 expression in neonatal rat brain. J Cereb Blood Flow Metab 1997;17:759–770.
73. Boddeke EW, Meigel I, Frentzel S, Gourmala NG, Harrison JK, Buttini M, Spleiss O, Gebicke-Harter P. Cultured rat microglia express functional beta-chemokine receptors. J Neuroimmunol 1999;98:176–184.
74. McManus CM, Weidenheim K, Woodman SE, Nunez J, Hesselgesser J, Nath A, Berman JW. Chemokine and chemokine-receptor expression in human glial elements: induction by the HIV protein, Tat, and chemokine autoregulation. Am J Pathol 2000;156:1441–1453.
75. Heesen M, Tanabe S, Berman MA, Yoshizawa I, Luo Y, Kim RJ, Post TW, Gerard C, Dorf ME. Mouse astrocytes respond to the chemokines MCP-1 and KC, but reverse transcriptase–polymerase chain reaction does not detect mRNA for the KC or new MCP-1 receptor. J Neurosci Res 1996;45:382–391.
76. Jiang Y, Salafranca MN, Adhikari S, Xia Y, Feng L, Sontag MK, deFiebre CM, Pennell NA, Streit WJ, Harrison JK. Chemokine receptor expression in cultured glia and rat experimental allergic encephalomyelitis. J Neuroimmunol 1998;86:1–12.
77. Fahey TJ, Tracey KJ, Tekamp-Olson P, Cousens LS, Stones WG, Shires GT, Cerami A, Sherry B. Macrophage inflammatory protein-1 modulates immune function. J Immunol 1992;148:2764–2769.
78. Taub DD, Conlon K, Lloyd AR, Oppenheim JJ, Kelvin DJ. Preferential migration of activated $CD4^+$ and $CD8^+$

cells in response to MIP-1α and MIP-1β. Science 1993; 260:355–358.
79. Biber K, Sauter A, Brouwer N, Copray JCVM, Boddeke HWGM. Ischemia induced neuronal expression of the microglia attracting chemoline Secondary Lymphoid-tissue Chemokine (SLC). Glia 2001;34:121–133.
80. Wang X, Ellison JA, Siren A-L, Lysko PG, Yue T-L, Barone FC, Shatzman A, Feuerstein GZ. Prolonged expression of interferon-inducible protein-10 in ischemic cortex after permanent occlusion of the middle cerebral artery in rat. J Neurochem 1998;71:1194–1204.
81. Soto H, Wang W, Strieter RM, Copeland NG, Gilbert DJ, Jenkins NA, Hedrick J, Zlotnik A. The CC chemokine 6ckine binds the CXC chemokine receptor CXCR3. Proc Natl Acad Sci USA 1998;95:8205–8210.
82. Spleiss O, Gourmala N, Boddeke HWGM, Sauter A, Fiebich BL, Berger M, Gebicke-Haerter PJ. Cloning of rat HIV-1-chemokine coreceptor CKR5 from microglia and upregulation of its mRNA in ischemic and endotoxinemic rat brain. J Neurosci Res 1998;53:16–28.
83. Utans-Schneitz U, Lorez H, Klinkert WE, da Silva J, Lesslauer W. A novel rat CC chemokine, identified by targeted differential display, is upregulated in brain inflammation. J Neuroimmunol 1998;92:179–190.
84. Liao F, Rabin RL, Smith CS, Sharma G, Nutman TB, Farber JM. CC-chemokine receptor 6 is expressed on diverse memory subsets of T cells and determines responsiveness to macrophage inflammatory protein 3 alpha. J Immunol 1999;162:186–194.

Universes in Delicate Balance: Chemokines and the Nervous System
R.M. Ransohoff, K. Suzuki, A.E.I. Proudfoot, W.F. Hickey and J.K. Harrison, editors

Chapter 3.5

Chemokine Responses in Virus-Induced Neurologic Disease: Balancing Host Defense and Neuropathology

Thomas E. Lane [1] and Michael J. Buchmeier [2]

[1] Department of Molecular Biology and Biochemistry, University of California, Irvine, CA 92697, USA; [2] Division of Virology, Department of Neuropharmacology (CVN-8), The Scripps Research Institute, 10550 N. Torrey Pines Road, La Jolla, CA 92037, USA

1. INTRODUCTION

Viral infection of the central nervous system (CNS) presents unique challenges to the immune system's ability to, recognize, and eliminate foreign antigen. Factors that increase the complexity of CNS immune responses include (1) the presence of the blood–brain barrier (BBB) which provides a physical and physiological obstruction that is difficult for cells and macromolecules to cross, (2) the relative absence of MHC class I and II expression on crucial CNS cells like astrocytes and neurons, and (3) lack of abundant antigen presenting cells (APC) which are required for the generation of an adaptive immune response. However, in spite of these obstacles, activated leukocytes do enter the CNS and participate in elimination of virus through a variety of highly efficient mechanisms. In addition, the presence of activated immune cells, e.g. T lymphocytes and macrophages, has been proven to contribute to the development of neuropathology in several viral model systems.

The underlying molecular and cellular mechanisms governing leukocyte migration and entry into the CNS following viral infection are just now being understood. Over the past several years, there has been a surge of interest reflected in the literature demonstrating the importance of chemokines in orchestrating events surrounding inflammation following microbial infection or injury. The chemokines represent an ever-growing family of small, molecular weight secreted proteins that have been shown to target the migration of specific populations of leukocytes during periods of inflammation. The chemokines are currently divided into four subfamilies based upon the arrangement of conserved cysteine residues in the amino-terminus of the protein (structural criteria) and what leukocytes are targeted, e.g. monocytes, macrophage, granulocytes, neutrophil, and T lymphocytes (functional criteria). Chemokines bind to specific receptors expressed on the surface of a wide variety of cells. The majority of chemokine receptors identified to date belong to the seven-transmembrane G protein-coupled receptor superfamily [1]. Following chemokine binding to cognate receptor, signaling events are initiated that result in various cellular processes including increases in intracellular calcium, production of cytokines and chemokines, adhesion to the endothelial matrix, and chemotaxis [2]. In addition, there is growing evidence that chemokine receptors influence the polarization of T cell immune response (i.e. Th1 vs. Th2) in tissue [3].

Viral infection of the CNS can result in a temporal expression of both chemokines and chemokine receptors by resident cells of the CNS as well as by inflammatory cells. Numerous studies have demonstrated a relationship between viral infection of the CNS and the expression of chemokine and/or chemokine receptors with leukocyte entry and disease development. There is an increasing amount of evidence indicating that chemokine expression represents a double-edged sword in the sense that expression is beneficial to the host by attracting T lymphocytes and macrophages which participate in host defense through elimination of foreign antigen, e.g. virus. Conversely, chemokine expression in response to CNS viral infection also clearly contributes to neuropathology by attracting activated T lymphocytes and macrophages into the CNS which then release cytotoxic factors. This chapter will focus on the chemokine response to viral infection of the CNS with an emphasis on the functional significance of chemokine expression as it relates to both host defense and neuropathology.

2. VIRAL INFECTION OF GLIAL CELLS AND CHEMOKINE EXPRESSION

At first glance the CNS does not appear equipped to effectively cope with viral infection due to the aforementioned barriers. However, experimental evidence indicates otherwise. Activated glial cells can express MHC class I and/or class II molecules suggesting that these cells may function in an APC capacity and contribute to the development of an adaptive response and viral clearance [4]. Moreover, both activated microglia and astrocytes release reactive oxygen and nitrogen intermediates that may interfere with viral replication [5,6].

In addition, cultured glial cells release cytokines in response to viral infection that may work in both an autocrine and/or paracrine manner in the regulation of the local immune response as well as enhance recruitment of protective T lymphocytes required for efficient clearance of viral antigen or conversely, disease development. Viral infection of glial cells results in robust expression of numerous chemokines. In vitro studies have demonstrated that human microglia and astrocytes are able to express chemokine genes such as MCP-1/CCL2, RANTES/CCL5, and MIP-1α/CCL3 and β/CCL4 [7] following infection with various viral pathogens including measles virus [8] and human coronavirus [9]. Infection of rat astrocytes and microglia with the Paramyxovirus, Newcastle disease virus (NDV), results in rapid expression of mRNA transcripts for CCL5 and non-ELR (glutamic acid–leucine–arginine) CXC chemokine IP-10/CXCL10 [10,11]. Unlike TNF-α, which requires live virus to trigger expression, UV-inactivated NDV was sufficient to trigger CCL5 and CXCL10 suggesting that signals initiated following viral binding to the surface receptor was sufficient to trigger chemokine gene expression [11]. Further analysis found that CCL5 expression was dependent upon tyrosine kinases and protein kinases C and A whereas CXCL10 expression was dependent on PKC activity but neither on PKA nor tyrosine kinase activity. Thus, the signals required for chemokine expression differ depending on the gene and can be controlled at the level of transcription. Infection of mouse astrocytes with the coronavirus, mouse hepatitis virus (MHV) results in the rapid induction of numerous chemokine genes including MIP-2/CXCL1, CXCL10, CCL2, CCL4, and MCP-3/CCL7 [12]. Similar to NDV, UV-inactivated MHV was sufficient to trigger low level chemokine gene expression. Interestingly, chemokine transcripts are readably detectable even following infection with a low dose of virus suggesting that a soluble factor(s), e.g. cytokines, may be released which triggers chemokine expression by neighboring cells that are uninfected. In support of this are studies demonstrating that interferon treatment of glial cells can result in robust chemokine gene expression [13,14]. Moreover, a recent report has indicated that treatment of primary cultures of mouse astrocytes with nanomolar concentrations of KC/CXCL2 and CXCL1 results in chemokine gene expression [15]. Therefore, cytokines and chemokines produced by virally infected cells are sufficient to function in a paracrine manner to induce expression of these soluble molecules in uninfected cells. In addition, it is important not to underestimate the impact of glial cell production of chemokines as it relates to acute and/or chronic disease development following viral infection of the CNS. Normally, chemokine expression is associated with inflammation suggesting that stimulatory factors provided by leukocytes are required for expression. However, it is clear that cultured glial cells may also exhibit prolonged expression of chemokine genes in the apparent absence of exogenous stimuli such as cytokines suggesting that these molecules can either function in promoting tissue damage or possibly in a repair/protective mechanism within the CNS.

3. IN VIVO MODELS OF VIRAL-INDUCED NEURODEGENERATIVE DISEASE: CORRELATION OF DISEASE SEVERITY WITH CHEMOKINE GENE EXPRESSION

From the studies discussed above, it is apparent that glial cells respond to viral infection by expressing numerous chemokine genes. Expression may be in response to a primary event or represent a secondary response to cytokine(s) released from infected cells as discussed above. Numerous laboratories have extended in vitro investigations involving viral triggering of chemokine expression in glial cells and have turned their attention to determining whether viral infection of the CNS of humans and/or animals results in chemokine gene expression.

Chemokines are expressed within the CNS of virus-infected humans and current evidence suggests that these molecules contribute to the pathogenesis of disease. For example, during herpes simplex type 1 (HSV-1) encephalitis, increased levels of CCL2 detected within the cerebral spinal fluid (CSF) of patients correlated with clinical disease severity [16]. Similarly, CSF levels of IL-8/CXCL8 correlated with increased neutrophil accumulation and increased disease severity in patients infected with Japanese encephalitis virus [17].

The demonstration of chemokine gene expression within the CNS of humans infected with neurotropic viruses supports the hypothesis that these molecules are important during the disease process. This is further evidenced by recent studies using various animal models of viral pathogenesis. CNS infection with different classes of viruses results in robust inflammatory responses characterized by recruitment of large numbers of leukocytes into the CNS suggesting that chemokines may contribute to this process. In support of this are numerous studies demonstrating expression of multiple chemokines within the CNS following infection with a wide variety of viruses (see Table 1). The following offers brief descriptions of recent in vivo studies analyzing the chemokine response to CNS infection with several different types of viruses.

Table 1. Chemokine gene expression within the CNS following infection with different viruses

Host	Virus [a]	Type	Chemokines	Ref.
Human	HSV-1	DNA	CCL2	[16]
Human	JEV	RNA	CXCL8	[17]
Mouse	LCMV	RNA	CXCL10, CCL2, CCL4–7	[20]
Mouse	MAV	DNA	CXCL1, CXCL10, CCL3–5	[24]
Mouse	BDV	RNA	CXCL10, CCL5	[26]
Mouse	TMEV	RNA	CXCL2, CXCL10, CCL2–5	[29,30]
Mouse	MHV	RNA	CXCL10, CCL2–5, CCL7	[12]

[a] HSV-1, herpes simplex virus-1; JEV, Japanese encephalitis virus; LCMV, lymphocytic choriomeningitis virus; MAV, mouse adenovirus; BDV, Borna disease virus; TMEV, Theiler's murine encephalomyelitis virus; MHV, mouse hepatitis virus.

3.1. Lymphocytic Choriomeningitis Virus (LCMV)

LCMV is an ambisense RNA virus and a member of the Arenaviridae. Infection of the murine CNS with LCMV results in a well-established model of viral meningitis [18,19]. LCMV infection of adult, immunocompetent mice results in a monophasic disease characterized by leukocyte infiltration into distinct anatomic regions of the brain and ultimately death between 6–8 days post-infection (p.i.) [18]. Others have determined that multiple chemokine genes are expressed following LCMV infection of the CNS during the acute stage of disease which correlates with onset of choriomeningitis. Chemokines expressed include CCL2, CCL4, CCL5, C10/CCL6, CCL7, and CXCL10 [20,21]. Expression of the CXC chemokine CXCL10 was analyzed in more detail as this chemokine was prominently expressed throughout the infection. In situ hybridization for CXCL10 mRNA and viral RNA revealed overlap in expression suggesting that CXCL10 expression is sensitive to areas of viral infection and replication. Interestingly, CNS infection of athymic nude mice (lacking conventionally educated lymphocytes) resulted in an identical chemokine profile when compared to LCMV-infected euthymic mice indicating that resident glial cells are responsible for local production of chemokines and do not require secondary signals provided by inflammatory leukocytes. This is supported by dual-labeling analysis which revealed that astrocytes were the major cellular source of CXCL10 following LCMV infection [20].

3.2. Adenovirus

Adenoviruses are icosahedral, double-stranded DNA viruses that are capable of infecting a wide variety of vertebrate hosts. Infection of certain strains of adult mice with mouse adenovirus (MAV) can result in fatal encephalitis within 4–6 days p.i. CNS disease is characterized by viral infection of cerebral vascular endothelial cells and resulting vascular pathology. C57BL/6 (B6) mice are susceptible to MAV-induced CNS disease whereas BALB/c mice are resistant [22,23]. Experiments designed to determine the underlying difference in disease susceptibility between B6 and BALB/c mice found no alterations in innate and/or adaptive immune responses or differences in cytokine expression profiles following viral infection [24]. MAV infection of B6 mice results in increased expression of the chemokines CXCL10 and CCL3–5 within the CNS [24]. CXCL10 was the prominent chemokine expressed within the CNS of infected B6 mice. In situ hybridization analysis determined that expression was associated with meningeal vessels as well as perivascular glial cells. MAV infection of resistant BALB/c mice resulted in restricted chemokine gene expression with only CXCL1 being detected. Interestingly, both strains of mice exhibited equivalent levels of expression of chemokine receptors CCR1–5. The authors conclude that productive infection in the CNS with MAV in mice susceptible to disease, i.e. B6 mice, results in a regulated pattern of chemokine gene expression and may point to a contribution by these molecules in disease pathogenesis.

3.3. Borna Disease Virus (BDV)

BDV is a negative-strand RNA virus that can cause a lethal meningoencephalitis in susceptible hosts [25]. Experimental infection of mice with BDV results in mice and rats results in a $CD8^{+}$ T lymphocyte-mediated neurologic disease [26]. The underlying mechanisms contributing to leukocyte entry into the CNS of BDV-infected rodents are just now being understood. Numerous studies have demonstrated expression of proinflammatory cytokines within the CNS of BDV-infected mice and rats. In addition, these investigators have identified a dramatic increase in chemokine gene expression following BDV infection. CXCL10 and CCL5 transcripts are strongly expressed within the CNS of infected mice and rats. Similar to LCMV infection of the CNS, BDV infection of immunodeficient mice, e.g. lacking T lymphocytes or receptors for types I and II interferons, resulted in CXCL10 and CCL5 expression suggesting that resident glial cells are capable of expressing these chemokines in the absence of stimulation by factors released by inflammatory lymphocytes. Astrocytes were determined to be the primary source of CXCL10 within the brains of infected rodents. The authors speculate that localized expression of chemokine genes by resident glial cells is important in the early immune response to viral infection and is required for the initiation of virus-induced neuroinflammation. Furthermore, the authors consider the possibility that sustained production of chemokines may contribute to behavioral and anatomic changes in BDV-infected rodents. For example, BDV infection of newborn rats ultimately results in the selective loss of cerebellar Purkinje cells and chronic expression of chemokines may contribute to this loss. Additional studies are required to further sort out the functions/roles chemokines may have in the complex model of BDV infection of the CNS.

3.4. Theiler's Murine Encephalomyelitis Virus (TMEV)

Intracranial infection of mice with TMEV, a member of the Picornaviridae, results in an acute polioencephalitis followed by a chronic demyelinating disease in susceptible strains of mice [27]. TMEV infection of the CNS results in dramatic increase in cytokine expression accompanied by lymphocyte infiltration [28]. Recent studies have indicated that chemokines are expressed within the CNS of TMEV-infected mice during both acute and chronic stages of disease [29,30]. Chemokines expressed during acute disease included CXCL2, CXCL10, and CCL2-5 and it is thought that these molecules cooperate in the initiation of events involved in leukocyte recruitment into the CNS. Following the acute stage of disease, the level of inflammation is reduced which coincides with a reduction in chemokine gene expression [30]. However, the onset of chronic demyelination within these mice coincides with at resurgence in CXCL10, CCL2, and CCL5 expression suggesting these molecules may contribute to the maintenance of chronic inflammation and demyelination accompanied by persistent viral replication. At this point, no functional studies have indicated whether expression of CXCL10, CCL2, or CCL5 is required for driving demyelination through exerting a chemotactic effect upon leukocytes. Collectively, these studies imply that chemokine expression is strictly regulated and expression is dictated by the phase of disease (e.g. acute versus chronic).

3.5. Mouse Hepatitis Virus (MHV)

Infection of susceptible mice with neurotropic strains of MHV results in an acute encephalomyelitis characterized by widespread viral infection of neurons and glial cells. Depending upon the viral strain, cellular infiltration usually peaks around 5–7 days and consists primarily of T lymphocytes and macrophages. We have recently determined that a complex, well-orchestrated expression of chemokine genes occurs following MHV infection of the CNS which appears to be dictated, in part, by viral burden [12]. As with the LCMV model, CXCL10 mRNA transcripts are detected soon after CNS infection and co-localize with areas of MHV replication (Fig. 1). By day 6 p.i., virus has spread throughout the brain parenchyma and a robust inflammatory response, characterized primarily by $CD4^+$ and $CD8^+$ T cells and macrophages, is established within the brain. Chemokines expressed at this time include CXCL10, CCL2–5, and CCL7. Double-labeling revealed that astrocytes and microglia expressed mRNA transcripts for CXCL10 during the acute stage of disease. Furthermore, analysis of the chemokine receptor mRNA profile indicates increased expression of CCR1, CCR5, and CXCR3 (Fig. 2). Confocal microscopy revealed that macrophages (determined by F4/80 antigen expression) express CCR5 (Fig. 3). In addition, studies have shown that both $CD4^+$ and $CD8^+$ T cells infiltrating into the brain express CXCR3 [31].

By day 12 p.i., MHV-infected mice that have survived the acute stage of disease develop an immune-mediated demyelinating disease with clinical and histologic features similar to the human demyelinating disease multiple sclerosis (MS). A recent study demonstrating that infiltration of $CD4^+$ T lymphocytes and macrophages perhaps driven by CCL5 expression is a key event linked to demyelination in mice persistently infected with MHV [32]. Mice have cleared infectious virus (as determined by plaque assay) by 12 days, yet viral RNA persists within white matter tracts for months after infection. Similar to the scenario with TMEV, the level of CNS infiltration subsides and this correlates with reduced expression of chemokine transcripts. Analysis of chemokine message expression within the brains and spinal cords of MHV-infected mice during the chronic demyelinating phase of disease (day 35) indicated that CXCL10 and CCL5 are the two prominent chemokines expressed [12]. In situ hybridization for chemokine transcripts indicated that expression was limited primarily to areas of viral persistence undergoing demyelination (Fig. 4). Astrocytes were determined to be the cellular source

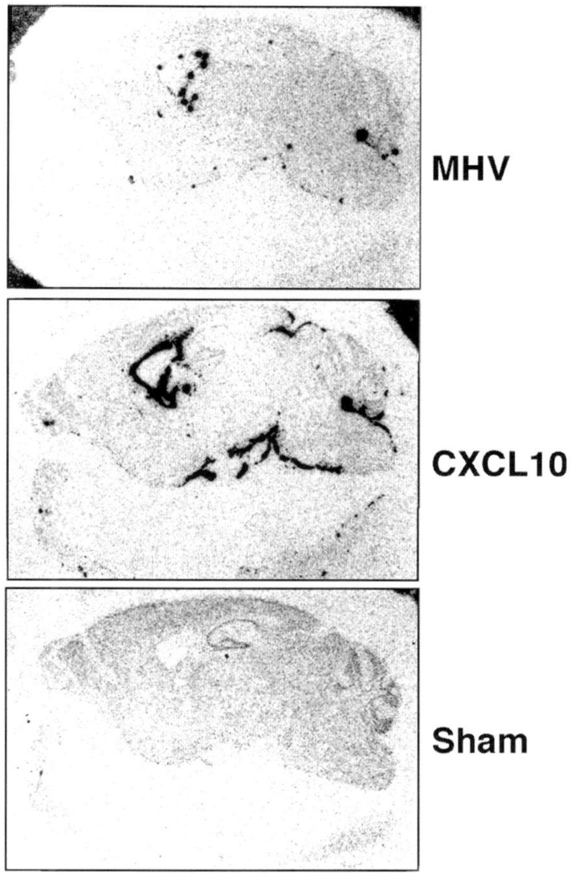

Figure 1. Co-localization of MHV RNA and CXCL10 mRNA transcripts. In situ hybridization was performed on sequential brain sections representing day 2 p.i. with MHV. Sections were probed with ^{35}S-labeled antisense riboprobes specific for MHV (top panel), CXCL10 (middle panel), or a labeled CXCL10 sense probe (bottom panel). Film autoradiography was performed to localize signal within the brain sections. Note the co-localization of MHV signal with CXCL10 specifically around the lateral ventricle and meninges.

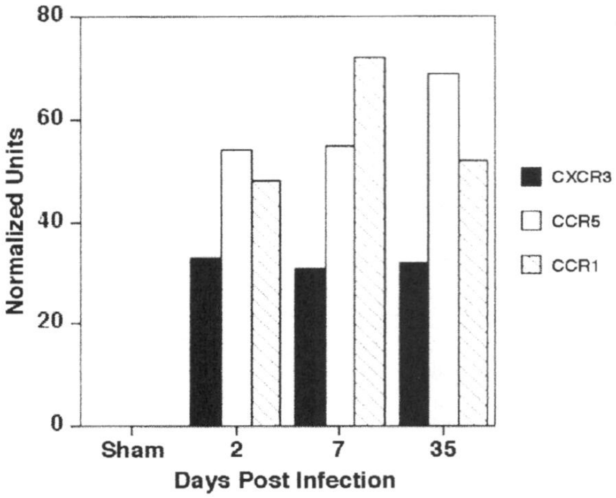

Figure 2. Chemokine receptor mRNA expression within the CNS of MHV-infected mice. Mice were infected intracranially with MHV and the chemokine receptor mRNA profile within the brain was analyzed by ribonuclease protection assay at days 2, 7, and 35 p.i. Levels of receptor mRNAs are presented as normalized units determined by performing densitometric analysis of receptor signal intensity. Prominent chemokine receptors expressed following viral infection included CXCR3, CCR5, and CCR1. Data presented represent average signal intensities determined from a minimum of three mice per time point.

of CXCL10 at this stage of disease whereas inflammatory cells, presumably CD4$^+$ T lymphocytes, expressed CCL5. Chemokine receptors expressed late in the disease process included CXCR3 and CCR5. Previous studies have demonstrated that CCL5 is capable of attracting T lymphocytes and macrophages following binding to CCR5 (as well as other chemokine receptors) expressed on the surface of these target cells [1]. CXCL-10 exerts a potent chemotactic effect on T lymphocytes through binding to CXCR3 [33].

4. FUNCTIONAL SIGNIFICANCE OF CHEMOKINE EXPRESSION DURING ACUTE VIRAL ENCEPHALOMYELITIS

The demonstration of chemokine and chemokine receptor expression within the CNS of virally infected mice is provocative and suggests that these molecules are important in contributing to events culminating in leukocyte entry into the CNS. However, despite these compelling observations, experimental evidence supporting a functional role for chemokines in events surrounding neuroinflammation following viral infection of the CNS is lacking. Studies by Karpus and colleagues have convincingly demonstrated that antibody-mediated neutralization of CCL2 or CCL3 is effective in reducing disease severity in mice with experimental autoimmune encephalomyelitis (EAE) by blocking monocyte entry into the CNS [3,34]. Therefore, it is not unreasonable to suggest that targeted inhibition by antibody of selected chemokines expressed within the CNS of virally infected mice may modulate the severity of inflammation. By extension, these experiments may illuminate specific targets or pathways for chemotherapeutic targeting in chronic neuroinflammatory disease.

In support of this hypothesis are a series of recent studies examining the contributions of Mig/CXCL9,

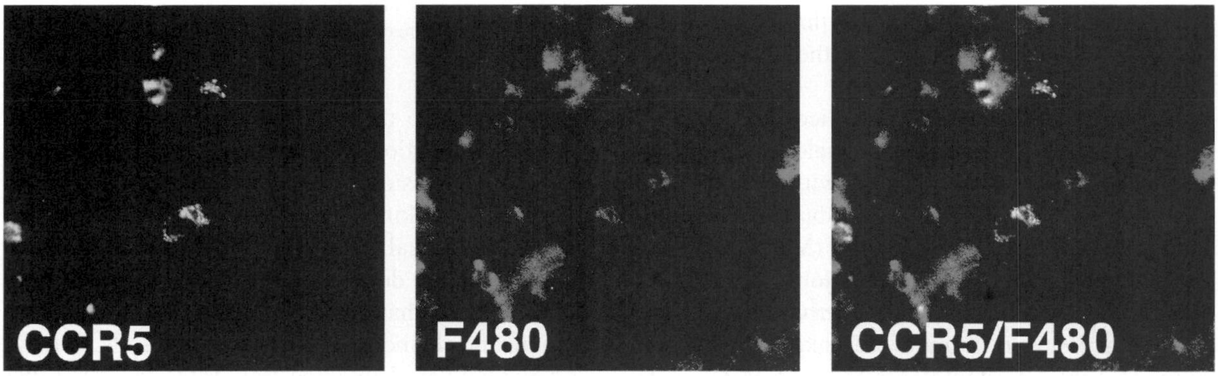

Figure 3. Macrophage/microglia express CCR5 within the CNS of MHV-infected mice. Confocal microscopy was performed to co-localize CCR5 expression with macrophage/microglial cells (determined by F4/80 antigen) expression at day 6 p.i. with MHV. Representative staining showed individual cells staining for CCR5 (green) and F4/80 (red) while dual-labeling indicated double-positive cells (yellow).

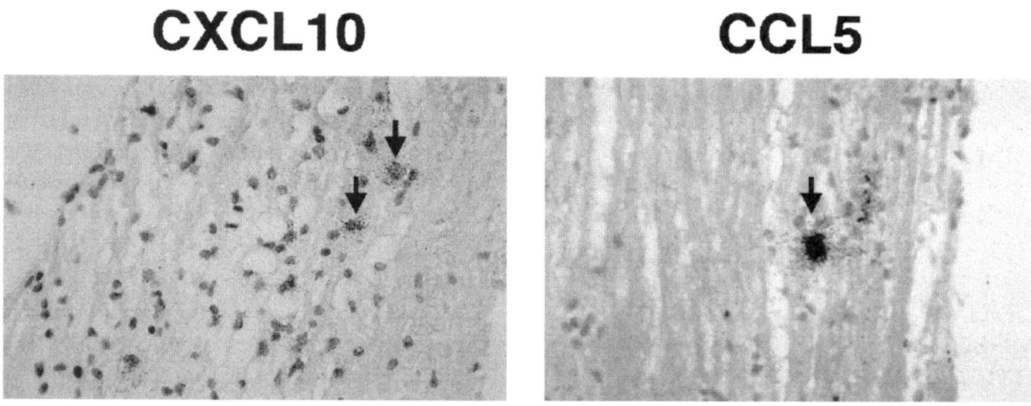

Figure 4. CXCL10 and CCL5 are expressed within areas of demyelination. In situ hybridization was performed to localize CXCL10 and CCL5 mRNA transcripts within spinal cords of MHV-infected mice during chronic demyelinating disease. Spinal cords from mice at day 35 p.i. were probed with either ^{35}S-labelled antisense riboprobes specific for CXCL10 or CCL5. The results indicate that these chemokines are prominently expressed within white matter tracts undergoing myelin stripping. Positive cells are indicated by overlaying silver grains (arrows).

CXCL10, and CCL5 to neuroinflammation, host defense, and neurologic disease following MHV infection of mice. Both CD4$^+$ and CD8$^+$ T lymphocytes are required for efficient clearance of MHV from the CNS and one mechanism by which activated T lymphocytes eliminate MHV is through the release of IFN-γ [31,32,35]. CXCL9 and CXCL10 are non-ELR CXC chemokines that attract activated T lymphocytes following binding to CXCR3. CXCL10 expression is inducible by both IFN-α/β and IFN-γ while CXCL9 expression is strictly dependent upon IFN-γ [36–38]. Analysis of CXCL9 and CXCL10 mRNA expression within the CNS of MHV-infected mice revealed that CXCL10 was clearly detectable by day 2 p.i. and was prominently expressed at days 7, 12, and 35 p.i. (Fig. 5). In contrast, CXCL9 transcripts were only detected at days 7 and 12 p.i. and this correlated with the presence of IFN-γ mRNA transcripts (not shown). Based upon these data, it is interesting to speculate that early expression of CXCL10 may reflect local production of IFN-α/β in response to MHV infection which, in turn, activates both infected and non-infected cells to produce CXCL10. CXCL9 expression is delayed until T lymphocytes enter the CNS and produce IFN-γ. Furthermore, these data suggested that both CXCL9 and CXCL10 may be important in host defense by attracting anti-viral T lymphocytes into the CNS. In support of this is the observation that administration of rabbit antisera specific for either CXCL9 or CXCL10 to MHV-infected mice during the acute stage

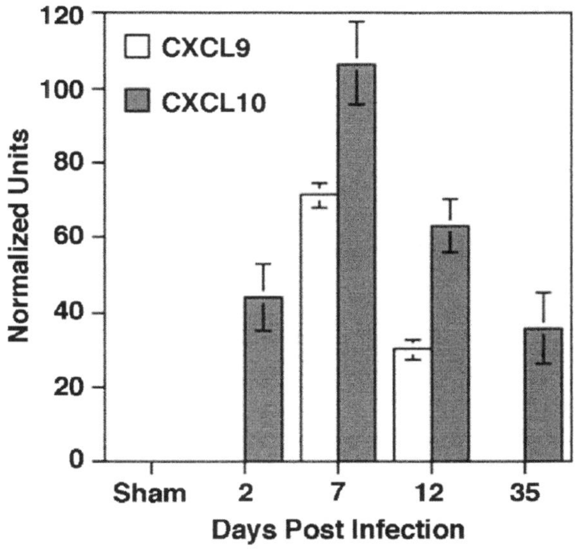

Figure 5. CXCL10 and CXCL9 are expressed following MHV infection of the CNS. CXCL10 and CXCL9 mRNA transcripts within the brains of MHV-infected mice were detected by ribonuclease protection assay at days 2, 7, 12, and 35 p.i. mRNA levels are presented as normalized units determined by performing densitometric analysis of signal intensity from film autoradiography. CXCL10 is prominently expressed at all stages of disease examined whereas CXCL9 expression is limited to days 7 and 12 p.i. Data presented represent average ± SEM and were determined from a minimum of three mice per time point.

of disease results in a dramatic increase in mortality. Treatment of MHV-infected mice with either anti-CXCL9 or anti-CXCL10 resulted in a significant decrease in numbers of CD4$^+$ and CD8$^+$ T lymphocyte infiltrating into the CNS which correlated with decreased expression of IFN-γ mRNA and protein and increased levels of virus [31,39]. Therefore, the collective evidence points to roles for both CXCL9 and CXCL10 as important sentinel molecules in promoting a protective Th1 response against MHV infection of the CNS.

5. FUNCTIONAL SIGNIFICANCE OF CHEMOKINE EXPRESSION DURING CHRONIC VIRAL-INDUCED DISEASE

CXCL10 and CCL5 are the predominant chemokines expressed in the CNS of persistently infected mice undergoing demyelination [12]. Moreover, CXCR3 and CCR5 expression (receptors for CXCL10 and CCL5, respectively) is also detected at this time. These findings have parallels in human neurodegenerative diseases. Glial cell expression of CCL5 and CXCL10 as well as CXCR3- and CCR5-positive mononuclear cells have been reported within demyelinating lesions present in MS patient [40]. Moreover, increased levels of CXCL9, CXCL10 and CCL5 are present within the cerebral spinal fluid of MS patients during periods of clinical attack and their presence correlates with increased numbers of inflammatory cells [40]. Collectively, these data illustrate the immunopathological consequences of expression of chemokines in human demyelinating diseases.

Experimental evidence supporting this hypothesis comes from studies using the MHV model of viral-induced demyelination. Intracranial infection of $CD4^{-/-}$ and $CD8^{-/-}$ mice with MHV resulted in increased mortality and delayed clearance of virus from the CNS which is consistent with previous studies demonstrating an important role for T lymphocytes in host defense against MHV-induced CNS disease [32]. Interestingly, infected $CD4^{-/-}$ mice displayed significantly less severe inflammation and demyelination as compared to $CD8^{-/-}$ and wild-type C57BL/6 (B6) mice. FACS analysis of the cellular infiltrate present within the CNS of infected mice revealed that $CD4^{-/-}$ mice contain fewer activated macrophages and significantly lower levels of CCL5 mRNA transcripts and protein when compared to $CD8^{-/-}$ and B6 mice. These data suggested that $CD4^{+}$ T cells are the predominant source of CCL5 following MHV infection of the CNS, although it is also possible that $CD4^{+}$ T cells influence expression of CCL5 by other cell populations through the release of cytokines and/or chemokines. Additional cellular sources such as $CD8^{+}$ T cells, macrophage and microglia must be considered due to the fact that CCL5 mRNA transcripts and protein were detected, albeit at lower levels, within the CNS of $CD4^{-/-}$ mice. In light of the fact that CCL5 exerts a potent chemotactic effect on both T cells and monocytes, these data suggest that the reduction in macrophage infiltration and the severity of demyelination into the CNS of $CD4^{-/-}$ mice is causally related to the reduced CCL5 levels observed.

To provide a direct test of CCL5 importance in contributing to MHV-induced CNS inflammation and demyelination, MHV-infected C57BL/6 mice were treated with anti-CCL5 antisera and the severity of disease evaluated. Treatment led to a disease in B6 mice similar to the phenotype observed in $CD4^{-/-}$ mice with delayed viral clearance from the brain and decreased cellular infiltration as well as a significant reduction in the severity of demyelination. The decreased capacity to clear virus from the brains is explained by the limited infiltration of $CD4^{+}$ and $CD8^{+}$ T cells into the brain during the acute stage of disease. Furthermore, the decrease in macrophage infiltration correlated with the reduced severity of demyelination supporting the observations with MHV-infected $CD4^{-/-}$ mice. These observations reinforce the functional significance of CCL5 expression during virus-induced CNS disease indicating that this chemokine has an important role in recruitment of both $CD4^{+}$ and $CD8^{+}$ T cells as well as macrophages into the CNS following MHV infection.

The functional significance of CXCL10 in contributing to chronic demyelinating disease in mice persistently infected with MHV has also been examined. Administration of neutralizing antibodies specific to CXCL10, but not CXCL9, to MHV-infected mice with established disease, e.g. hind-limb paralysis and demyelination, resulted in a dramatic reduction in clinical disease severity accompanied by an inhibition in the progression of demyelination that correlated with a significant decrease in $CD4^{+}$ T cell and macrophage infiltration into the CNS [41]. Importantly, the reduction in neurologic and histologic disease severity in anti-CXCL10-treated mice correlated with a dramatic increase in the number of remyelinated axons. Further, CCL5 expression (mRNA and protein) was significantly reduced in mice treated with anti-CXCL10. Based on previous studies examining the contribution of CCL5 to MHV-induced CNS disease [32], these observations suggest that impaired CD4$^+$ T cell infiltration into the CNS of anti-CXCL10-treated mice results in reduced expression of the macrophage chemoattractant CCL5 and this ultimately results in a dramatic reduction in demyelination. Molecular analysis of CXCR3 expression (RT-PCR) on infiltrating CD4$^+$ and CD8$^+$ T cells in persistently infected mice undergoing demyelination revealed that CD4$^+$ T cells express markedly higher levels of CXCR3 mRNA when compared to CD8$^+$ T cells. These data indicate that during chronic disease CD4$^+$ T cells are the major responding T cell population to chronic CXCL10 expression. Collectively, these data indicate that elimination of effector cells, e.g. CD4$^+$ T cells and macrophages, following establishment of disease allows the CNS time for repair of previously damaged axons. In addition, these studies support previous work in other models of neuroinflammatory disease that antibody-mediated targeting of chemokines may offer a viable treatment strategy for treatment of human neuroinflammatory diseases.

6. CHEMOKINE RECEPTORS AND VIRAL-INDUCED DISEASE

MHV infection of the CNS of susceptible mice results in a dramatic increase in both chemokines as well as chemokine receptors. Among the chemokine receptors that are prominently expressed during acute disease are CCR2 (major signaling receptors for CCL2) and CCR5 (receptor for various chemokines including CCL5) [42]. In order to evaluate the contributions that expression of these receptors may have in contributing to either host defense or disease following MHV infection of the CNS, mice lacking CCR2 (CCR2$^{-/-}$) were intracranially infected with MHV and disease severity evaluated. Such an infection resulted in a dramatic increase in mortality in CCR2$^{-/-}$ mice as compared to CCR2$^{+/+}$ mice such that by day 10 p.i. 100% of CCR2$^{-/-}$ mice had died [43]. Correlating with the increased mortality in infected CCR2$^{-/-}$ mice was an increase in viral burden within the brain accompanied by a significant reduction in T cell and macrophage migration into the CNS. In order to further evaluate the contributions of CCR2 to leukocyte trafficking, splenocytes obtained from immunized CCR2$^{-/-}$ mice were adoptively transferred into MHV-infected RAG1$^{-/-}$ mice (recombination activating gene 1 – lack functional T and B lymphocytes). These experiments revealed an overall lack of migration of both T cells and macrophages into the CNS of recipient mice which correlated with increased viral titers within the brains and a dramatic reduction in the severity of demyelination as compared to RAG1$^{-/-}$ mice that received splenocytes from immunized CCR2$^{+/+}$ mice. These data indicate that CCR2 is important in contributing to T cell and macrophage entry into the CNS following MHV infection. Interestingly, adoptive transfer of immune splenocytes obtained from CCR2$^{+/+}$ mice into MHV-infected CCR2$^{-/-}$ mice did not protect recipient mice from viral-induced neurologic disease and all mice died by day 7 following transfer. These results suggest that endogenous expression of CCR2, e.g. on blood–brain barrier endothelial cells, is also important in allowing leukocytes to enter the CNS. In addition, peripheral CD4$^+$ and CD8$^+$ T cells from immunized CCR2$^{-/-}$ mice displayed a marked reduction in IFN-γ production in response to defined MHV antigens indicating that CCR2 is required for the generation of virus-specific T cells.

Having established that CCL5, a major signaling ligand of CCR5, is important in contributing to demyelination in MHV-infected mice by attracting macrophages into the CNS, studies were performed to further characterize the contributions of CCR5 to neuroinflammation and demyelination in MHV-infected mice. Analysis of CCR5 expression within the CNS of MHV-infected mice revealed that the majority of cells expressing this receptor were macrophage/microglia [42]. MHV infection of CCR5$^{-/-}$ mice did not result in an increase in either morbidity or mortality as compared to infected CCR5$^{+/+}$ mice. In addition, clearance of virus from the CNS of infected CCR5$^{-/-}$ mice was not impaired and this correlated with equivalent levels of infiltrating T cells into the CNS when compared to MHV-infected CCR5$^{+/+}$ mice. However, demyelination was significantly reduced in MHV-infected CCR5$^{-/-}$ mice and this correlated with reduced levels of infiltrating macrophages as compared to $CCR5^{+/+}$ mice. These data suggest that CCL5 signaling through CCR5 is a primary mechanism whereby macrophages are able to enter the CNS and contribute to demyelination.

Taken together, the studies on both CCR2 and CCR5 illustrated the complexity of chemokine receptor expression within the CNS as it relates to both host defense and disease development in a model of viral-induced CNS disease.

7. SUMMARY AND CONCLUSIONS

Available evidence demonstrates clearly that viral infection of the CNS results in a dramatic increase in chemokine gene expression. Moreover, production of chemokines in response to infection is highly focussed within areas of viral replication early in the disease process and areas of viral RNA persistence during the chronic stages of disease. Resident glial cells, e.g. microglia and astrocytes, are capable of generating a robust chemokine response following viral infection in the absence of inflammatory cells suggesting that this response may reflect an innate CNS response against viral infection analogous to the response of phagocytic cells in the periphery. Differences in virus and cellular tropism are likely explanations for the slight differences in chemokine profiles and duration of expression observed in the different models. However, what is striking in the studies discussed are the similarities with regards to the types of chemokines expressed. CC chemokines are the prominent chemokines expressed. In addition, the non-ELR CXC chemokine CXCL10 is often the predominant chemokine expressed early following viral infection suggesting an important role as a sentinel molecule in initiating neuroinflammation. Interestingly, the presence of T lymphocytes and macrophages within the CNS of virally infected mice is often associated with either host defense, aiding in clearance of virus, or the development of immunopathologic demyelinating disease. Studies on MHV-infected mice clearly illustrate the delicate balance that exists between the sentinel role of chemokine expression in host defense and the development of disease. Early production of chemokines such as CXCL9, CXCL10 and CCL5 are beneficial as they serve to attract T lymphocytes into the CNS which contribute to viral clearance [31,32,39]. However, chronic expression of CCL5 is detrimental as illustrated by experiments demonstrating a significant decrease in macrophage infiltration and demyelination following CCL5 depletion by antibody. The use of chemokine blocking agents, either receptor antagonists or antibodies, or chemokine/chemokine receptor knock-out mice will further our understanding of how these molecules drive defense and disease within the CNS following viral infection and may help to identify important targets for the development of selective new anti-inflammatory agents to attack MS. Based on studies presented in this brief overview, it is clear that targeting chemokines during either acute or chronic viral-induced CNS disease may offer exciting new insights into potentially novel interventional mechanisms in treating human neuroinflammatory diseases.

ACKNOWLEDGEMENTS

The authors gratefully acknowledge support from the following sources: T. Lane is supported by grants NS37336 from the National Institutes of Health and RG30393A1/T from the National Multiple Sclerosis Society. M. Buchmeier is supported by grants AI43103 and AI 25913 from the National Institutes of Health.

REFERENCES

1. Premack BA, Schall TJ. Chemokine receptors: gateways to inflammation and infection. Nat Med 1996;2(211): 1174–1178.
2. Ward SG, Bacon K, Westwick J. Chemokines and T lymphocytes: more than an attraction. Immunity 1998;9: 1–11.
3. Kennedy KJ, Karpus WJ. Role of chemokines in the regulation of Th1/Th2 and autoimmune encephalomyelitis. J Clin Immunol 1999;19(5):273–279.
4. Cornet A, Bettelli E, Oukka M et al. Role of astrocytes in antigen presentation and naive T-cell activation. J Neuroimmunol 2000;106(1–2):69–77.
5. Murphy S. Production of nitric oxide by glial cells: regulation and potential roles in the CNS. Glia 2000;29(1): 1–13.
6. Persichini T, Colasanti M, Fraziano M, Colizzi V, Ascenzi P, Lauro GM. Nitric oxide inhibits HIV-1 replication in human astrocytoma cells. Biochem Biophys Res Commun 1999;254(1):200–202.
7. Zlotnik A, Yoshie O. Chemokines: A new classification

system and their role in immunity. Immunity 2000;12(2):121–127.

8. Xiao BG, Mousa A, Kivisakk P, Seiger A, Bakhiet M, Link H. Induction of beta-family chemokines mRNA in human embryonic astrocytes by inflammatory cytokines and measles virus protein. J Neurocytol 1998;27(8):575–580.
9. Edwards JA, Denis F, Talbot PJ. Activation of glial cells by human coronavirus OC43 infection. J Neuroimmunol 2000;108(1–2):73–81.
10. Vanguri P, Farber JM. IFN and virus-inducible expression of an immediate early gene, crg-2/IP-10, and a delayed gene, I-A alpha in astrocytes and microglia. J Immunol 1994;152(3):1411–1418.
11. Fisher SN, Vanguri P, Shin HS, Shin ML. Regulatory mechanisms of MuRANTES and CRG-2 chemokine gene induction in central nervous system glial cells by virus. Brain Behav Immun 1995;9(4):331–344.
12. Lane TE, Asensio VC, Yu N, Paoletti AD, Campbell IL, Buchmeier MJ. Dynamic regulation of alpha and beta chemokine expression in the central nervous system during mouse hepatitis virus-induced demyelinating disease. J Immunol 1998;160:970–978.
13. Hua LI, Lee SC. Distinct patterns of stimulus-inducible chemokine mRNA accumulation in human fetal astrocytes and microglia. Glia 2000;30:74–81.
14. McManus CM, Liu JS, Hahn MT, Hua LL, Brosnan CF, Berman JW, Lee SC. Differential induction of chemokines in human microglia by type I and II interferons. Glia 2000;29(3):273–280.
15. Luo Y, Fischer FR, Hancock WW, Dorf ME. Macrophage inflammatory protein-2 and KC induce chemokine production by mouse astrocytes. J Immunol 2000;165(7):4015–4023.
16. Rosler A, Pohl M, Braune HJ, Oertel WH, Gemsa D, Sprenger H. Time course of chemokines in the cerebrospinal fluid and serum during herpes simplex type 1 encephalitis. J Neurol Sci 1998;157(1):82–89.
17. Singh A, Kulshreshtha R, Mathur A. Secretion of the chemokine interleukin-8 during Japanese encephalitis virus infection. J Med Microbiol 2000;49(7):607–612.
18. Buchmeier MJ, Welsh RM, Dutko FJ, Oldstone MB. The virology and immunobiology of lymphocytic choriomeningitis virus infection. Adv Immunol 1980;30:275–331.
19. Buchmeier MJ, Zajac AJ. Lymphocytic choriomeningitis virus. In: Ahmed R, Chen I, eds. Persistent Virus Infections. Chichester: Wiley, 1999.
20. Asensio VC, Campbell IL. Chemokine gene expression in the brains of mice with lymphocytic choriomeningitis. J Virol 1997;71:7832–7840.
21. Asensio VC, Kincaid C, Campbell IL. Chemokines and the inflammatory response to viral infection in the central nervous system with a focus on lymphocytic choriomeningitis virus. J Neurovirol 1999;5(1):65–75.
22. Guida JD, Brosnan CF, Fejer G, Pirofski L, Horwitz MS. Mouse adenovirus type 1 causes a fatal hemorrhagic encephalomyelitis in adult C57BL/6 mice but not BALB/c mice. J Virol 1995;69:7674–7681.
23. Kring SC, King CS, Spindler KR. Susceptibility and signs associated with mouse adenovirus type 1 infection of adult outbred mice. J Virol 1995;69:8084–8088.
24. Charles PC, Chen X, Horwitz MS, Brosnan CF. Differential chemokine induction by the mouse adenovirus type-1 in the central nervous system of susceptible and resistant strains of mice. J Neurovirol 1999;5(1):55–64.
25. Rott R, Becht H. Natural and experimental Borna disease in animals. Curr Top Microbiol Immunol 1995;190:17–30.
26. Sauder C, Hallensleben W, Pagenstecher A, Schneckenburger S, Biro L, Pertlik D, Hausmann J, Suter M, Staeheli P. Chemokine gene expression in astrocytes of Borna disease virus-infected rats and mice in the absence of inflammation. J Virol 2000;74(19):9267–9280.
27. Rodriguez M, David CS. Demyelination induced by Theiler's virus: influence of the H-2 haplotype. J Immunol 1985;135:2145–2148.
28. Tsunoda I, Fujinami RS. Two models of multiple sclerosis: experimental allergic encephalomyelitis and Theiler's murine encephalomyelitis virus. J Neuropathol Exp Neurol 1996;55:673–686.
29. Theil DJ, Tsunoda I, Libbey JE, Derfuss TJ, Fujinami RS. Alterations in cytokine but not chemokine mRNA expression during three distinct Theiler's virus infections. J Neuroimmunol 2000;104:22–30.
30. Murray PD, Krivacic K, Chernosky A, Wei T, Ransohoff RM, Rodriguez M. Biphasic and regionally-restricted chemokine expression in the central nervous system in the Theiler's virus model of multiple sclerosis. J Neurovirol 2000;6:S44–52.
31. Liu MT, Chen BP, Oertel P, Buchmeier MJ, Armstrong D, Hamilton TA, Lane TE. The T cell chemoattractant IFN-inducible protein IP-10 is essential in host defense against viral-induced neurologic disease. J Immunol 2000; 165:2327–2330.
32. Lane TE, Liu MT, Chen BP, Asensio VC, Samawi RM, Paoletti AD, Campbell IL, Kunkel SL, Fox HS, Buchmeier MJ. A central role for $CD4^+$ T cells and RANTES in virus-induced central nervous system inflammation and demyelination. J Virol 2000;74(3):1415–1424.
33. Piali L, Weber C, LaRosa G, Mackay CR, Springer TA, Clark-Lewis I, Moser B. The chemokine receptor CXCR3 mediates rapid and shear-resistant adhesion-induction of effector T lymphocytes by the chemokines IP10 and Mig. Eur J Immunol 1998;28(3):961–972.
34. Karpus WJ, Lukacs NW, McRae BL, Strieter RM, Kunkel SL, Miller SD. An important role for the chemokine macrophage inflammatory protein-1 alpha in the pathogenesis of the T cell-mediated autoimmune disease, experimental autoimmune encephalomyelitis. J Immunol 1995; 155(10):5003–5010.
35. Parra B, Hinton DR, Marten NW et al. IFNgamma is required for viral clearance from central nervous system oligodendroglia. J Immunol 1999;162:1641–1647.
36. Vanguri P, Farber JM. An interferon-inducible mRNA predicted to encode a murine monokine. Proc Natl Acad Sci 1990;265(25):15049–15057.
37. Farber JM. HuMig: a new human member of the chemokine family of cytokines. Biochem Biophys Res Commun 1993;192(1):223–230.
38. Farber JM. Mig and IP-10: CXC chemokines that target lymphocytes. J Leukoc Biol 1997;61(3):246–257.

39. Liu MT, Armstrong D, Hamilton TA, Lane TE. Expression of Mig (monokine induced by interferon-γ) is important in T lymphocyte recruitment and host defense following viral infection of the central nervous system. J Immunol 2001;166:1790–1795.

40. Sorensen T, Tani M, Jensen J, Pierce V, Lucchinetti C, Folcik VA, Qin S, Rottman J, Sellebjerg F, Strieter RM, Frederikson JL, Ransohoff RM. Expression of specific chemokines and chemokine receptors in the central nervous system of multiple sclerosis patients. J Clin Invest 1999;103(6):807–815.

41. Liu MT, Keirstead HS, Lane TE. Neutralization of the chemokine CXCL10 reduces inflammatory cell invasion and demyelination and improves neurological function in a viral model of multiple sclerosis. J Immunol 2001; 167(7):4091–4097.

42. Glass WG, Liu MT, Kuziel WA, Lane TE. Reduced macrophage infiltration and demyelination in mice lacking the chemokine receptor CCR5 following infection with a neurotropic coronavirus. Virology 2001;288:8–17.

43. Chen BP, Kuziel WA, Lane TE. Lack of chemokine receptor CCR2 results in increased mortality and impaired leukocyte activation and trafficking following infection of the central nervous system with a neurotropic coronavirus. J Immunol 2001;167:4585–4592.

Universes in Delicate Balance: Chemokines and the Nervous System
R.M. Ransohoff, K. Suzuki, A.E.I. Proudfoot, W.F. Hickey and J.K. Harrison, editors

Chapter 3.6

Cell Recruitment in the Axotomized Facial Nucleus: Role of Cytokines, Chemokines and Cell Adhesion Molecules

Gennadij Raivich

Perinatal Brain Repair Group, Department of Obstetrics and Gynaecology, Department of Anatomy, University College London, Chenies Mews 86–96, London WC1E 6HX, UK

1. INTRODUCTION

Injury to the nervous system initiates a set of well-characterized morphologic and metabolic changes which appear to play a key role in two crucial physiological processes: protection against infectious agents and repair of the damaged tissue. These changes are accompanied by a rapid recruitment of different cell types to the site of injury, with a profound influence on the further course of events. Basically, there are two fundamentally different sets of recruitment: the entry of blood-borne leukocytes, and the migration to the damaged site by the tissue-resident cells such as astrocytes and the microglial cells. Both forms of recruitment are observed in the damaged brain. Compared with other tissues, however, the entry of circulating leukocytes is less extensive, which may contribute to the immune privilege of the damaged central nervous system.

This extravasation of hematopoietic cells is particularly restricted in mild or indirect forms of brain injury as exemplified in the axotomized facial motor nucleus. On the whole, this recruitment pattern is very similar in different injured motor nuclei or in other forms of retrograde reaction [1,2]. However, there are a number of important constraints and technical advantages associated with this cranial motor nucleus. The facial nucleus is the largest brainstem motor nucleus in the mouse, containing approximately 2000 motoneurons innervating the facial, platysmal and auricular muscles. It is a highly compact nucleus of 0.3 mm^3, which provides enough tissue not only for histology, but also for biochemical analysis. With the exception of a tiny nerve branch innervating the stapedius muscle, with an exit in the middle ear, all other facial nerve branches stay together during their passage through internal acoustic meatus and tympanic bulla and radiate away to their target muscles only after exiting the stylomastoid foramen. A complete transection of the facial nerve close to the stylomastoid foramen, including the main and the auricular branch, will thus lead to a uniform axotomy of almost all facial motoneurons in the affected nucleus. This situation is very different from trauma or ischemia, where there is a steep gradient of brain pathology between the completely necrotic and the more or less healthy tissue. With the advent of molecular biology, the axotomized facial motor nucleus has also turned into one of the key in vivo models to dissect the molecular mechanisms involved in neuronal survival, axonal regeneration and the activation of microglia and their role in the overall network of posttraumatic response. This chapter will describe the different forms of cellular recruitment in this injury model, summarize the current data on the associated molecular mechanisms, and briefly discuss the biological function subserved by this highly conserved cellular response.

2. MICROGLIAL RESPONSE TO FACIAL AXOTOMY

The microglial cells constitute a substantial part of the total glial population that is functionally related to peripheral tissue macrophages and other cells of the monocyte lineage. Like macrophages in other tissues, the normal, resting microglia appears to participate in the immune surveillance of the nervous system. They express receptors for key immune molecules immunoglobulin (FcγR) and complement fragments (αMβ2 integrin, complement receptor type 3) and their surface is covered with moderate levels of endogenous immunoglobulin and thrombospondin, which may enable microglia to monitor their local environment for foreign antigens, opsonized particles and apoptotic cells [3–8]. The highly ramified form of resting microglia, extending over territories 50 μm in diameter, probably supports their immune surveillance function.

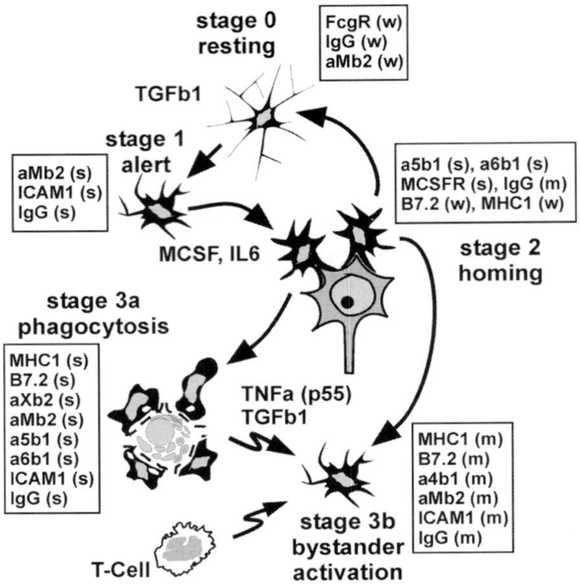

Figure 1. Microglial activation, a schematic summary. Cellular changes in activated microglia proceed through a series of steps, which differ in their morphology and molecular profile. Neural injury leads to a rapid transformation of the highly ramified resting microglia (stage 0) to more stout and deramified form (stage 1: state of alert), which home on damaged structures, like injured neurons, (stage 2: homing), but without additional damage, gradually return to the normal, resting state (stage 0). Cell death leads to a further transformation of microglia into phagocytic cells (stage 3a: phagocytosis). These foci of phagocytosis also lead to an activation of the surrounding, non-phagocytic microglia (stage 3b: bystander activation). The transformation of microglia into phagocytic, amoeboid macrophages in the presence of cell debris appears to be cytokine-independent. The early stages of activation are affected by MCSF, IL6 and TGFβ1, the late bystander activation by TGFβ1, by TNF and its p55 receptor (TNFR1). The boxes show the microglial activation markers specific for the activation stages in the mouse. Their staining intensity is shown in brackets: w, weak; m, moderate; s, strong. All integrins are encoded with standard nomenclature (a, alpha; b, beta). Modified from [2], Figure 2.

After injury, the ramified microglia transform rapidly from a resting to an activated state [2,9], with an increase in cell body size, a thickening of proximal processes and decreased ramification of the distal branches. On the molecular level, this activation proceeds through a series of steps — the very early state of alert, followed by homing and adhesion, phagocytosis, and, finally, the bystander activation of the neighboring, non-phagocytic cells, summarized in Fig. 1. These activated microglia are strongly attracted by damaged neurons, but in two different ways: by adhering to injured but surviving motoneurons, peaking at days 4–7 after facial axotomy, and by the phagocytosis of the neuronal cell debris, with a maximum at day 14. In addition to temporal and functional differences, recent molecular studies also show that these microglial responses are controlled by quite different signaling pathways.

2.1. Homing to Living Neurons

The adhesion of microglia to injured, living neurons is a multi-step process. Habitual contacts between neuronal cell bodies and the long, ramified processes of resting microglia are already present in the normal brain. In vitro, tenascin-R and galactose-binding lectins interfere with the adhesion of microglia to tissue sections of normal CNS gray matter [10,11] and could be involved in regulating normal neuron–microglia attachment. Following neuronal injury, these contacts are intensified, and the microglial cell body is moved to the neuronal cell surface, coming to lie face to face with the injured neuron. This is followed by a close enwrapping of the neuron with microglial processes, and finally, the engorgement of the apposed microglial cell body.

Studies using cytokine-deficient animal models support this concept of a gradual, multi-step process of microglial adhesion to injured neurons. Thus, microglia in mice lacking the macrophage colony stimulating factor/MCSF will freeze during the first step (Fig. 2): their processes will still touch axotomized neurons, but will not show the normally ensuing apposition, enwrapment or cell body swelling [12]. Absence of transforming growth factor beta-1/TGFβ1 will interfere with enwrapment [13,14],

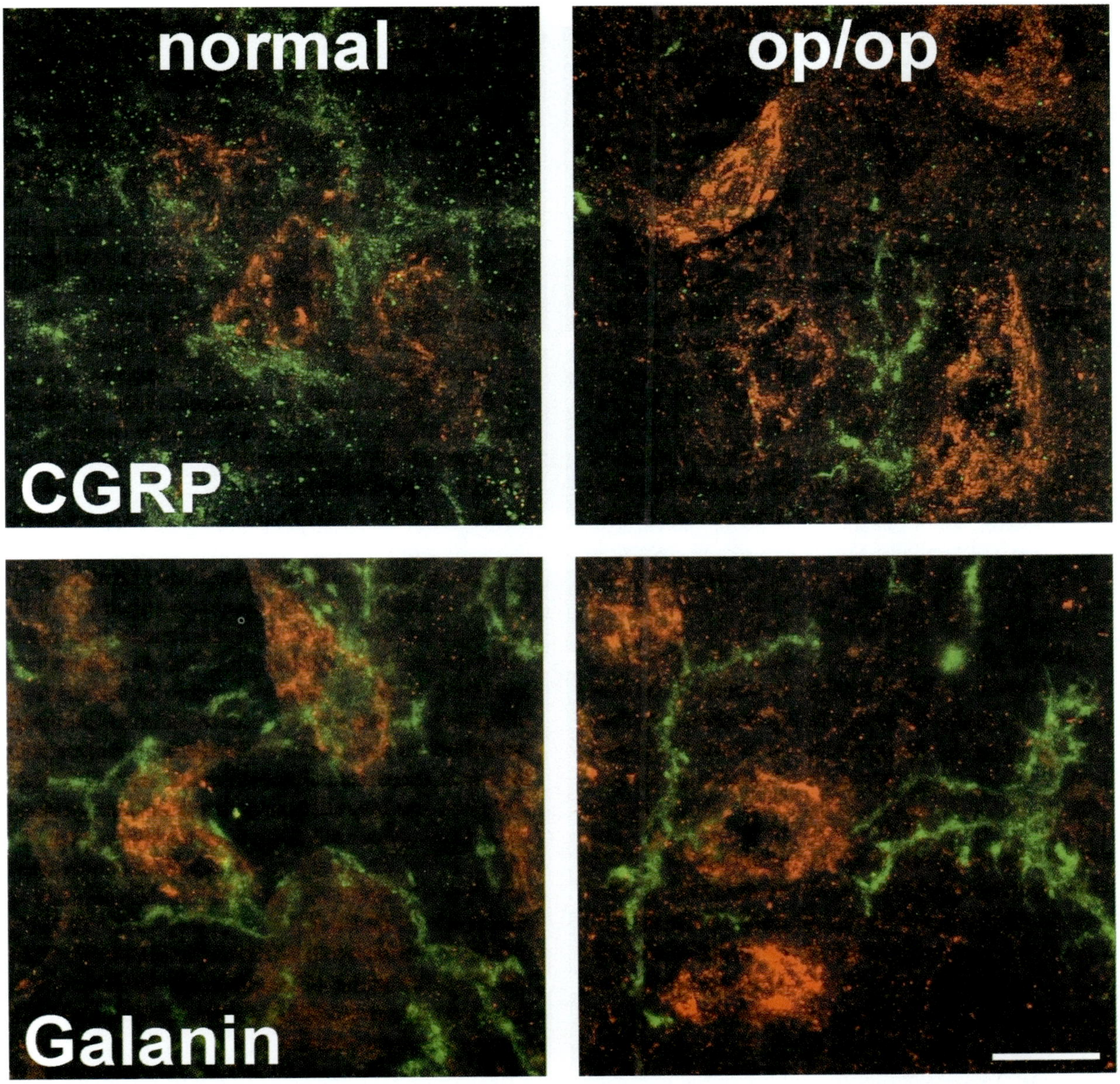

Figure 2. Absence of MCSF in the homozygous, op/op mice interferes with the attachment of microglia to axotomized facial motoneurons 4 days after injury. The microglial cells are labelled with the immunoreactivity for αMβ2 (green), axotomized neurons for CGRP or galanin (red). In the op/op animals lacking MCSF, the microglia touch the neurons, but are not observed to spread on the neuronal surface; in the normal mice, numerous activated microglia spread avidly on the surface of the axotomized neurons. This normal behavior could due to a downstream, stimulating effect of MCSF and its receptor on microglial receptors (CCR2? CX3CR1?) for a neuronal chemoattractive signal such as CCL2 (MCP1) or CX3CL1 (fractalkine). Scale bar: 10 μm. Reproduced from [12].

and that of interleukin-6/IL6 will show normal attachment, apposition and enwrapment, but no cell body swelling [15].

Interestingly, facial motoneurons also show clear expression for a list of different chemokines, including CCL2 (macrophage chemoattractant protein-1/MCP1), CCL5 (RANTES), CXCL10 (IP10) and CX3CL1 (fractalkine), summarized in Table 1. CCL5 and CXCL10 are primarily lymphocyte chemokines, and their expression does not change after axotomy (Fig. 3). In contrast, the expression of CCL2 and CX3CL1 on facial motoneurons is only

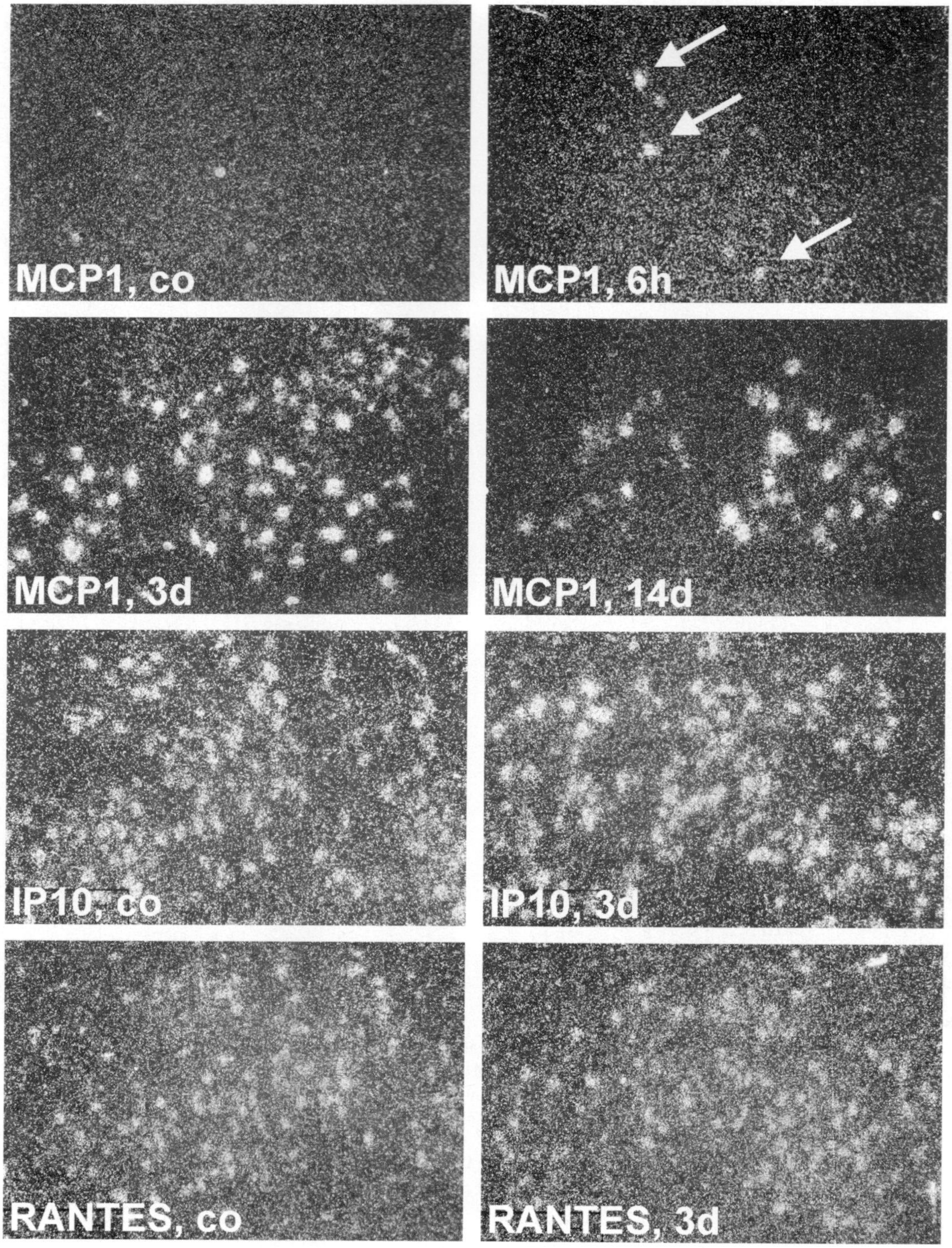

Figure 3. Chemokine expression in the facial motor nucleus. Axotomy of the facial nerve leads to a rapid induction of MCP1 (CCL2) mRNA on the injured motoneurons within the first 6 h (arrows) and peak expression 3 days (3 d) after injury, coinciding with the adhesion of microglia to the surface of injured motoneurons. IP10 (CXCL3) and RANTES (CCL5) mRNA is already expressed in normal control (co) motoneurons. This expression pattern does not change after axotomy. Reproduced from [18].

Table 1. Chemokines and chemokine receptors in the axotomized facial motor nucleus

Chemokine	+ve CT [a]	Change AA	Receptor	+ve CT [a]	Change AA	Function
CCL2 (MCP1)	Neuron	Increase	CCR2	Microglia	No change	RN? [b]
CCL5 (RANTES)	Neuron	No change	CCR1/3/5	ND	ND	ND
CXCL10 (IP10)	Neuron	No change	CXCR3	ND	ND	ND
CX3CRL1 (fractalkine)	Neuron	Increase	CX3CR1	Microglia	No change	RN? [b]

Source: [16–18]. Abbreviations: +ve, positive; CT, cell type; AA, after axotomy; RN, recruitment to neurons; ND, not determined.
[a] Positive cell type in the facial axotomy model.
[b] Deletions of the MCP1 receptor CCR2 or the fractalkine receptor CX3R1 do not inhibit recruitment of microglia to neurons. However, it is possible that several chemotactic signals work in concert and each alone is partially redundant.

apparent after facial axotomy. The de novo presence of fractalkine and macrophage chemoattractant protein-1/MCP1 in axotomized motoneurons [16,18] and their cognate receptors CX3CR1 and CCR2, respectively, on activated microglia has suggested chemokines as a source of chemical attraction between the two cell types. Notably, this attraction is not inhibited in single CX3CR1 or CCR2 knockouts ([17]; von Maltzan and Raivich, unpublished). This could point to a high degree of redundancy, at least on the level of chemokines; it is particularly striking, since both chemokines and their appropriate receptors at the right place and at the right time.

Although hypothetical, this sensitivity or lack of sensitivity to a single deletion could be due to a serial or parallel arrangement for a group of related signals, as suggested in Fig. 4. Serial arrangements would be quite sensitive to a deletion of one member. Parallel arrangements would show a high degree of redundancy. Why are some signals redundant and others critical for biological function in vivo? Theoretically, this could be due to the use of the same (redundant) or different (non-redundant) intracellular signaling pathways. Thus, the receptor for TGFβ1 is a serine–threonine kinase (STK), that for MCSF a tyrosine kinase (RTK) and for IL6, a non-kinase cytokine receptor (CYTR). In contrast, the chemokine receptors are all seven transmembrane domain glycoproteins coupled to G-proteins (7mGPR). In the case of neurons expressing several chemokines, and microglia all the appropriate receptors, a deletion of just one component could go unnoticed if all the receptors use the same downstream cascade in the same cell type.

As the activated microglia move into direct contact with injured neuronal cell bodies, they increase their expression of receptors for the microglial mi-

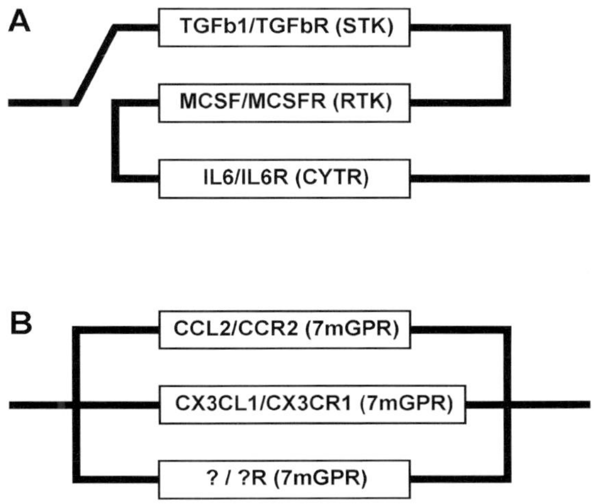

Figure 4. A schematic hypothesis. Redundancy or lack of redundancy for biological effects of a specific gene deficiency may depend on a serial or parallel arrangement. Signals arranged in series (A) will lead to a non-redundant, signals in parallel (B) to a highly redundant arrangement. Thus, single deletions of TGFβ1, MCSF or IL6, cited in A, all lead to a strong reduction in microglial proliferation in the axotomized facial motor nucleus and may serve as an example for a serial arrangement. The sequence shown in A (TGFβ1 → MCSF → IL6) is hypothetical, but it is clear that all 3 cytokines are needed for optimal microglial response. In contrast, single deletions of CCR2, the receptor for MCP1 (CCL2), or CX3CR1, the receptor for fractalkine (CX3CL1), cited in B, do not interfere with the adhesion of microglia to neurons or the subsequent microglial proliferation, even though they are at the right place and at the right time. This could suggest a parallel mode of action. Additional chemokines (?) and their receptors (?R) may also be involved. Also note that signals cited in A use different types of receptors: the receptor for TGFβ1 is a serine–threonine kinase (STK), that for MCSF a tyrosine kinase (RTK) and for IL6, a non-kinase carrying cytokine receptor (CYTR). In B, the chemokine receptors are all 7 transmembrane domain glycoproteins coupled to G-proteins (7mGPR).

togens MCSF and granulocyte–macrophage colony stimulating factor/GMCSF [19,20] and begin to proliferate [21–23]. Mitotic activity in these cells becomes maximal 2–3 days after axotomy [7], leading to an approximately 4–6-fold increase in the local microglial cell number [24]. Deletion of TGFβ1 [13,14], MCSF [7,12] or IL6 [15,24] strongly inhibits microglial mitogenesis. Proliferative activity subsides after day 4 [7,25,26]. This is accompanied by the IL6-dependent onset of immunoreactivity for major histocompatibility complex (MHC) class I molecules [15,27,28] and immunoaccessory glycoproteins like B7.2 [27]. Damaged cells are frequently leaky and the tight adhesion of microglia could allow for an efficient uptake of diffusible molecules and their presentation to infiltrating lymphocytes [12,15].

2.2. Phagocytosis

In the adult mouse, transection of the facial nerve will lead to a loss of approximately 30% of the neurons, peaking at 2 weeks after injury [29]. As a rule, this is normally followed by their phagocytosis by the neighboring activated microglial cells. Moreover, the removal of large cellular structures such as the degenerating motoneurons frequently leads to the formation of microglial nodules, consisting of 3–20 microglial phagocytes [26,29,30]. This state of phagocytosis causes a massive exacerbation in the expression of most activation markers already present during the state of alert and homing phases (Fig. 1). It includes a long list of cell adhesion molecules: the α5β1, α6β1 and αMβ2 integrins, thrombospondin and ICAM1 [8,29,31]. All of these molecules could participate in the adhesion and internalization to the many different and diverse components of the cellular debris. Most but not all microglial molecules with an immune function are augmented. Thus, there is a strong increase in Fcγ receptors, B7 and MHC class I [27,32,33]. There is also an upregulation of costimulatory molecules like B7.2, ICAM1 and αXβ2 integrin [8,27], which may serve as potent stimuli of T-cell activation in the presence of specific antigen [34,35]. Of particular interest is the phagocyte-specific induction of αXβ2 integrin [31]. In peripheral tissues, this integrin is a highly selective marker of dendritic cells [36]. It is a coreceptor for the bacterial lipopolysaccharide (LPS), which mediates activation of macrophages and dendritic cells [37] and may strengthen the binding between lymphocytes and the antigen-presenting cells [38]. Its expression on phagocytic microglia points to a similar function when they are contacted by infiltrating lymphocytes [30].

Unlike the early microglial response, the deletion of MCSF, IL6 or TGFβ1 will not interfere with microglial phagocytosis or their transformation into amoeboid phagocytes [12,14,15]. This also applies to interferon-gamma/IFNγ [27], which does play an important role in the induction of MHC2 and iNOS in infectious and autoimmune disease [39,40] and to CCR2, the receptor for MCP1 (Maltzan and Raivich, unpublished). Combined deletion of type 1 and type 2 receptors for tumor necrosis factor/TNF, TNFR1 and TNFR2, blocks almost completely the appearance of MHC1, B7.2 and IBA1-positive microglial nodules 14 days after facial nerve injury [27], the time point of maximal neuronal cell death in this experimental model of neural injury and microglial activation [29,30,41]. However, a later recount of the surviving motoneurons at day 30 showed a reduction of neuronal cell death by at least 75%, suggesting that the effects on phagocytic microglia are indirect, due to a primary action on neuronal survival [42]. Thus, none of the cytokines studied seem to interfere with the microglial transformation into phagocytes after the appearance of neural cell debris. This apparent lack of effect for cytokines in vivo is in line with in vitro studies in the microglia–astrocyte coculture, which did not show a morphological effect for a long list of microglial mitogens MCSF, GM-CSF or IL3, the inflammation-associated cytokines such as IL1, IL2, IL4, IL6, IL10, TNFα or IFNγ [43] and other proinflammatory molecules such as LPS [44]. A similar lack of morphological effect was also observed following neutralization of the TGFβ isoforms 1–3, although this neutralization strongly enhanced microglial proliferation [45].

Interestingly, the only physiological stimulus leading to deramification in this model is the addition of cell debris [46]. This signal is specific, only phagocytic microglia revert to the deramified, amoeboid phenotype; while the neighboring non-phagocytic cells remain ramified [46]. Several microglial adhesion molecules such as thrombospondin, sugar-binding lectins, integrins and immunoglobulin binding sites have been implicated in the phagocytosis in vitro [47,48]. These molecules are strongly upregulated on phagocytic microglia in vivo and are good candidates for the signaling pathways mediating microglial involvement with neural cell debris.

3. ASTROCYTES AND REACTIVE ASTROGLIOSIS

The astrocytes are the predominant neuroglial cell throughout the central nervous system, an essential component of the neural parenchyma involved in tissue homeostasis, blood–brain barrier and the maintenance of normal function. As with microglia, the astrocytes readily respond even to slight or indirect injury and this response proceeds through several stages. Early molecular activation, with the induction of the cytoskeletal protein glial fibrillary acid protein/GFAP and other activation markers [49,50] is followed by a redistribution of GFAP to the cell body and the main astrocyte processes [2,24], causing the appearance of stellar, GFAP-positive astrocytes. On the ultrastructural level, this is followed by the formation of thin, filament-poor astrocyte plates or lamellae that intercede between injured brain regions or even individual cells, and the surrounding, undamaged tissue. This long-term function appears aimed at creating a physical barrier between damaged and healthy cells or, on a macroscopic scale, between damaged and healthy tissue.

The overall extent of scarring depends on the severity of brain damage, but a rudimentary response around affected neurons is already observed in the mild, indirect trauma in the facial axotomy model. Beginning 4–7 days after axonal injury, the reactive astrocytes gradually replace microglia from the neuronal surface and enwrap injured neurons with thin and flat cytoplasmic processes, peaking at 4–8 weeks after trauma [51,52]. These processes also adhere to each other, forming an ordered, multilayered stack of astrocytic lamellae surrounding the axotomized neuron, which could act as a small glial scar. Successful regeneration leads to a partial retraction of these astrocyte processes and a gradual repopulation of the neuronal surface with some synaptic terminals. Thus, synaptic stripping is a reversible process, at least partially.

This transformation of resting into reactive astrocytes is tightly controlled by cytokines in the neural parenchyma, which regulate different steps in this activation repertoire. Absence of the neurokines leukemia inhibitory factor/LIF, ciliary neurotrophic factor/CNTF and to lesser extent, IL6, interferes with the early molecular activation ([24,53]; Raivich and Cullinan, unpublished). Transgenic deletion of IL6 strongly inhibits the redistribution of GFAP to the cell body and main astrocytes branches, causing the appearance of 30–60 μm large, GFAP-positive astrocytes with round, non-stellate morphology [2]. TGFβ1 has a reverse, but more generalized effect, both in normal and damaged brain. Thus, absence of TGFβ1 leads to the increase in the number of GFAP-positive, slender stellate astrocytes, but also of CD44-positive cells that belong to the reactive, protoplasmic phenotype with velate morphology [13]. At present, the chemical basis for the adhesion of astrocyte lamellae to injured neurons is still unknown, but the production of thin astrocyte processes depends on the presence of vimentin [54].

4. LYMPHOCYTES: A BIPHASIC FORM OF LEUKOCYTE RECRUITMENT

The massive recruitment of leukocytes is a key tenet of inflammation, in the brain as well as in any other tissue. Nevertheless, low numbers of hematogenous cells already enter even the normal central nervous system. Most of these cells fall into one of two categories: (a) cells of the monocyte lineage that give rise to the perivascular brain macrophages [55]; and (b) the activated T-lymphocytes [56,57]. The lymphocyte extravasation is not dependent for a specific antigen, but it is activation-dependent and inactive T-cells do not enter the normal brain.

Facial axotomy leads to a rapid increase in leukocyte extravasation. However, this enhanced entry is normally restricted to T-lymphocytes and natural killer cells, it is selective for the affected brain region and it proceeds in two phases. The first phase, a 10-fold increase over the normal number of T-cells in the facial nucleus, already reaches a plateau at 1 day after injury [30]. This rapid recruitment of T-cells coincides with the earliest signs of activation in the neighboring neurons and microglia. This recruitment is strongly dependent on the presence of two inflammation-associated cytokines IL6 and MCSF and their receptors in the injured facial motor nucleus [20,24,58]. Transgenic deletion of IL6 leads to an almost complete, approx. 10-fold decrease in lymphocyte entry [15], the biological absence of MCSF in the homozygous, osteopetrotic mice leads to a milder, 2.5-fold reduction in the number of recruited T-cells [12]. IL6 is strongly expressed by axotomized motoneurons [59]. In the case of MCSF, the cellular source of the cytokine is unknown, but the receptor is localized selectively on the activated microglial cells [20]. This selective neuronal synthesis of IL6, and the microglial expression of MCSF receptors, implicates both cell types, the axotomized neurons and the activated microglial cells, in mediating early lymphocyte recruitment into the injured nervous system.

Posttraumatic cell death, peaking at day 14 in the axotomized mouse facial motor nucleus, is associated with a further increase, an approximately 100-fold elevation in the number of CD3-positive T-cells over the control, unaffected tissue [30]. CD4-positive and CD8-positive T-cells, as well as the $CD3^+CD4^-CD8^-$ cells, the NK T-cells, contribute to this influx (Zhou, Paweletz and Raivich, unpublished). They also express high amounts of CD44, ICAM1 and $\alpha L\beta 2$, characteristic of activated T-cells [1,3,8,31]. These lymphocytes aggregate around phagocytic, MHC1, B7.2 and $\alpha X\beta 2$-positive microglia, providing a structural basis for efficient antigen presentation. Neuronal cell death also coincides with a massive increase of late pro-inflammatory cytokines, interleukin 1-beta/IL1β, tumor necrosis factor-alpha/TNFα and interferon-gamma/IFNγ, in the injured tissue and the induction of cell adhesion molecules, particularly $\alpha 5\beta 1$ and $\alpha 6\beta 1$ integrins on the blood vessel endothelia, providing a rich source of cytokines and cell adhesion molecules that could be involved in lymphocyte recruitment. Experiments with transgenic deletion animals have now confirmed this role for IL1 and TNF. Thus, deletion of IL1 receptor type 1/IL1R1 or TNF receptor type 2/TNFR2 led to a 2–3-fold reduction in late lymphocyte recruitment at day 14 [60]. Deletion of IL6 also had a statistically significant effect (2.5-fold), but this effect was weaker than at day 1 (Galiano et al., submitted), suggesting that it is superseded by the advent of additional, late pro-inflammatory stimuli, such as IL1β and TNFα. Deletion of IFNγ receptor type 1/IFNγR1 had no effect [60]. The same also applies to CCR2-knockouts, which lack one of the receptors for MCP1 (von Maltzan and Raivich, unpublished).

Unlike the early phase of lymphocyte recruitment there is currently little evidence for the involvement of neurons or microglia. Deletion of TNFR1 leads to a strong reduction in microglial activation but does not affect lymphocyte recruitment. Combined deletion of TNFR1 and TNFR2 leads to an almost complete disappearance of neuronal cell death, and of microglial nodules at day 14, but little further reduction in lymphocyte numbers, beyond that observed in simple TNFR2 homozygously deficient animals [60]. The astrocyte synthesis of many different chemokines, such as MCP1, GRO1 and IP10 shown in other forms of inflammatory brain pathology [61–63], could also point to this cell type as a source of chemoattractant molecules that lead to the late lymphocyte influx into the axotomized facial motor nucleus.

5. GRANULOCYTES: ENDOTOXIN-DEPENDENT ENTRY

A key feature of the cellular response in the normal adult facial motor nucleus after a peripheral axotomy is the complete absence of any extravasation by cells of myelomonocytic lineage such as monocytes or granulocytes. Due to the distance between the site of the peripheral axotomy and the facial motor nucleus in the central nervous system, the cellular response in the affected facial motor nucleus is restricted to a few lymphocytes and the local resident cells: the axotomized motoneurons, astrocytes and microglia. Studies using systemic injection of horseradish peroxidase (HRP) or human serum albumin also show an intact blood–brain barrier to soluble molecules [30,64]. Thus, the cellular response is probably mediated by the locally synthesized signals, and not by the circulating molecules, reducing the number of potential activating pathways. Under normal conditions, there is also no infiltration by circulating granulocytes or by macrophages based on specific granulocyte markers, the endogenous peroxidase staining or on a prelabeling of both cell types with Indian ink or rhodamine isothiocyanate [1,30,64,65]. This absence of hematopoietic influx is particularly important in the case of macrophages that express the same markers as the activated microglial cells. If infiltration does occur, it becomes very difficult to differentiate between resident microglia and the invading monocytes on the basis of the immunohistochemical markers. The absence of monocyte influx in the facial motor nucleus clearly simplifies the interpretation of the acquired data and is one of the main advantages of the facial axotomy model.

This normal and apparently very stable pattern changes following a systemic application of LPS. Injection of very high LPS doses will lead to a slow and gradual appearance of peroxidase-positive cells in the neural parenchyma (Fig. 5A,B), with a plateau 1–4 days after injection [64]. These peroxidase-positive cells are immunoreactive for neutrophil antigen 7/4 and αMβ2 integrin, negative for IBA1 (monocytes) and CD3 (T-cells) and could be prelabeled by subcutaneous injection with rhodamine B isothiocyanate (RITC), confirming their origin as blood-borne granulocytes. All RITC-positive cells are IBA1-negative, reconfirming the absence of monocyte influx. This influx of granulocytes is accompanied by a disruption of the blood–brain barrier to albumin and induction of the cell adhesion molecule ICAM-1 on affected blood vessels.

Axotomy leads to a further 2–4-fold increase in the number of granulocytes recruited to the facial nucleus (Fig. 5A,B,D), 14 days after peripheral transection [64]. Thus, cellular changes elicited by facial axotomy do enhance granulocyte extravasation, even though they are not sufficient to allow entry in the absence of LPS. A key role in mediating this process, its selectivity profile and changes following injury and systemic endotoxemia, is apparently played by the adhesion molecules on the endothelium lining the vessels of the damaged neural parenchyma. Thus, transgenic deletion of ICAM-1 led to a more than 50% reduction in the number of infiltrating granulocytes compared to litter-matched wild-type controls, in normal brain as well as in axotomized facial motor nucleus (Fig. 5D). Endothelial ICAM1, shown in Fig. 5C, is strongly induced following a systemic injection of LPS [64], but it is not affected by facial axotomy [8]. Nerve injury does lead, however, to a strong upregulation of the endothelial α5β1 and, to a lesser extent, α6β1 integrins, peaking at day 14 [31], and this could assist this form of leukocyte recruitment. This peak endothelial integrin expression coincides with neuronal cell death and synthesis of late pro-inflammatory cytokines IL1β, TNFα and IFNγ [30] providing a rich source of potential molecular signals that can influence leukocyte recruitment.

6. SUMMARY AND OVERVIEW

As described in this chapter, facial axotomy leads to a complex, highly elaborate and precisely timed recruitment of different cell types to the injured motoneurons and their environment. The resident microglia will rapidly become activated, adhere to neuronal surfaces with their cell bodies, enwrap the neurons with thick processes and interact with an early wave of T-lymphocytes which enter the injured brain and appear to survey the damaged parenchyma for cues to possible infection. The neighboring astrocytes will gradually surround injured neurons with several layers of thin sheets or lamellae, insulating them from the surrounding environment. Neuronal cell death, timed during the late stages of cellular response after axotomy, will cause an even stronger response, with microglia transforming into phagocytes that break down cell debris and interact with droves of second phase lymphocytes that are attracted to sites of cell death and phagocytosis. The entry of granulocytes is not a regular feature of the normal facial axotomy response, but it is strongly enhanced by high concentrations of systemic endotoxin, pointing to the presence of preformed recruitment pathways that could be opened in times of severe bacterial infection. All of these changes appear well coordinated and aimed at mounting defence mechanisms against infectious agents and the ensuing repair of the damaged tissue.

Recent studies on the inflammation-associated molecules have started to uncover the mechanisms that underlie this posttraumatic cellular activation. Thus, early stages of microglial activation and lymphocyte recruitment are dependent on MCSF and IL6, and the late phase on IL1, TNF and their receptors. The astrocyte response is regulated by different neurokines such as LIF, CNTF and IL6, but also held in check by anti-inflammatory molecules such as TGFβ1. Adhesion molecules such as ICAM1 are important for granulocyte recruitment. A number of chemokines such as fractalkine (CX3CL1) and MCP1 (CCL2) are also strongly upregulated in the

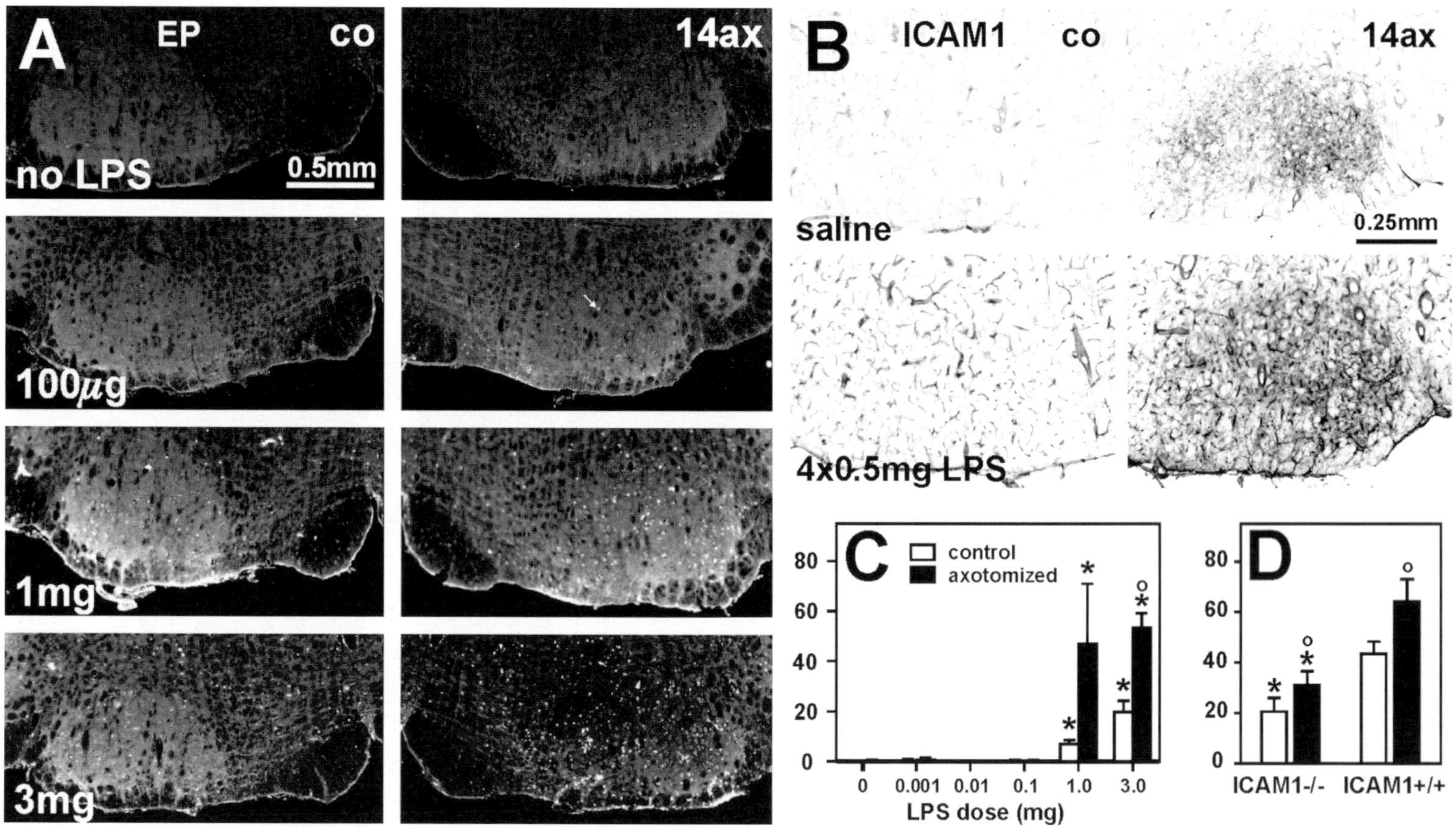

Figure 5. (A) Systemic injection of LPS leads to the influx of peroxidase-positive granulocytes throughout the brain, 48 h after injection. However, it is particularly prominent in the axotomized facial motor nucleus 14 days after axotomy (14ax), compared with the contralateral (co), non-operated side. These peroxidase-positive cells are very rare in the normal brain and up to the dose of 0.1 mg LPS (arrow), but injection of 1 and 3 mg LPS leads to a strong granulocyte recruitment. (B) Repeated injection of LPS (every 12 h) leads to a strong increase in ICAM-1 immunoreactivity in the facial nucleus 48 h after begin of LPS treatment (4×0.5 mg) and 14 days after axotomy. Control animals (injected with saline) show weak to moderate immunoreactivity on medium and large size blood vessels, respectively (co, ax), and moderate parenchymal staining on the axotomized side (ax). Note the strong vascular staining throughout the brainstem in the LPS-treated animals. (C) Dose response of granulocyte recruitment (mean + SEM) in the axotomized (14 d, filled bars) and contralateral (empty bars) facial nucleus to systemic injection of LPS. The onset of granulocyte recruitment is observed at the dose of 1 mg. Asterisks: statistically significant difference to untreated controls (A: 0 mg LPS, B: 0 h) using one-way ANOVA ($P < 0.002$), followed by Duncan's multiple range test ($P < 0.05$). ○: statistically significant difference (paired t-test, $P < 0.05$) between axotomized and control sides. (D) affect of ICAM-1 deficiency on LPS-mediated granulocyte recruitment. All animals received 4 doses of 0.5 mg LPS, with a 12-hour interval, beginning 12 days after axotomy. Both groups of animals were the F1 progeny of ICAM-1 heterozygous mice (ICAM1+/−). ICAM-1 deficiency leads to more than 50% reduction in granulocyte recruitment, both on the axotomized, and on the contralateral side. Asterisks: statistically significant difference (unpaired t-test, $P < 0.01$) between ICAM1+/+ and −/− animals. ○: statistically significant difference (paired t-test, $P < 0.003$) between axotomized and control sides. Reproduced from [64], in press.

aftermath of nerve injury and may play a pivotal role in recruiting different cell types, although some degree of redundancy could be expected to guarantee stable execution of the inflammatory response. A more complete elucidation will greatly enhance our insight into the cellular and molecular activation cascade in the injured neural tissue. They could also provide us with possible targets for pharmacological intervention, which could improve the quality of neural repair in the damaged central nervous system.

REFERENCES

1. Jones LL, Liu ZQ, Shen J, Werner A, Kreutzberg GW, Raivich G. Regulation of the cell adhesion molecule CD44 after nerve transection and direct trauma to the mouse brain. J Comp Neurol 2000;426:468–492.
2. Raivich G, Bohatschek M, Kloss CUA, Werner A, Jones LL, Kreutzberg GW. Neuroglial activation repertoire in the injured brain: Graded response, molecular mechanisms and cues to physiological function. Brain Res Rev 1999;30:77–105.
3. Engelhardt B, Martin–Simonet MT, Rott LS, Butcher EC, Michie SA. Adhesion molecule phenotype of T lymphocytes in inflamed CNS. J Neuroimmunol 1998;84 92–104.
4. Fishman PS, Savitt JM. Selective localization by neuroglia of immunoglobulin G in normal mice. J Neuropathol Exp Neurol 1989;48 212–220.
5. Liu L, Tornqvist E, Mattsson P, Eriksson NP, Persson JK, Morgan BP, Aldskogius H, Svensson M. Complement and clustering in the spinal cord dorsal horn and gracile nucleus following sciatic nerve injury in the adult rat. Neuroscience 1995;68:167–179.
6. Perry VH, Hume DA, Gordon S. Immunohistochemical localization of macrophages and microglia in the adult and developing mouse brain. Neuroscience 1985;15:313–326.
7. Raivich G, Moreno-Flores MT, Möller JC, Kreutzberg GW. Inhibition of posttraumatic microglial proliferation in a genetic model of macrophage colony-stimulating factor deficiency in the mouse. Eur J Neurosci 1994;6:1615–1618.
8. Werner A, Kloss CUA, Walter J, Kreutzberg GW, Raivich G. Intercellular adhesion molecule-1 (ICAM1) in the regenerating mouse facial motor nucleus. J Neurocytol 1998;27:219–232.
9. Raivich G, Kreutzberg GW. Inflammatory response following nerve injury. In: Ingoglia N, Murray M, eds. CNS Regeneration. New York: Marcel Decker, 2000:287–314.
10. Angelov DN, Walther M, Streppel M, Guntinas-Lichius O, Neiss WF, Probstmeier R, Pesheva P. Tenascin-R is antiadhesive for activated microglia that induce downregulation of the protein after peripheral nerve injury: a new role in neuronal protection. J Neurosci 1998;18:6218–6229.
11. Brown H, Townsend M, Fearn S, Perry VH. The monoclonal antibody HB1 recognizes an adhesion molecule for macrophages in the brain. J Neurocytol 1998;27:867–876.
12. Kalla R, Liu Z, Xu S, Koppius A, Imai Y, Kloss CU, Kohsaka S, Gschwendtner A, Moller JC, Werner A, Raivich G. Microglia and the early phase of immune surveillance in the axotomized facial motor nucleus: impaired microglial activation and lymphocyte recruitment but no effect on neuronal survival or axonal regeneration in macrophage-colony stimulating factor-deficient mice. J Comp Neurol 2001;436:182–201.
13. Jones LL, Doetschman T, Kreutzberg GW, Raivich G. Neuroglial activation in the injured central nervous system: role of transforming growth factor-β1. Soc Neurosci Abstr 1998;24:710.2.
14. Rong LY, Jones LL, Bueringer D, Koppius A, Doetschman T, Raivich G. Axon/myelin pathology, astrogliosis and perturbed microglial response in TGFb1-deficient mice. Soc Neurosci Abstr 2001;28:880.
15. Galiano M, Liu ZQ, Kalla R, Bohatschek M, Koppius A, Gschwendtner A, Xu SL, Werner A, Kloss C, Bluethmann H, Raivich G. Interleukin-6 (IL6) and the cellular response following facial nerve injury: Effects on lymphocyte recruitment, early microglial activation and axonal outgrowth in IL6-deficient mice. Eur J Neurosci 2001; 14:327–341.
16. Harrison JK, Jiang Y, Chen S, Xia Y, Maciejewski D, McNamara RK, Streit WJ, Salafranca MN, Adhikari S, Thompson DA, Botti P, Bacon KB, Feng L. Role for neuronally derived fractalkine in mediating interactions between neurons and CX3CR1-expressing microglia. Proc Natl Acad Sci USA 1998;95:10896–10901.
17. Jung S, Aliberti J, Graemmel P, Sunshine MJ, Kreutzberg GW, Sher A, Littman DR. Analysis of fractalkine receptor CX(3)CR1 function by targeted deletion and green fluorescent protein reporter gene insertion. Mol Cell Biol 2000;20:4106–4114.
18. Flugel A, Hager G, Horvat A, Spitzer C, Singer GM, Graeber MB, Kreutzberg GW, Schwaiger FW. Neuronal MCP-1 expression in response to remote nerve injury. J Cereb Blood Flow Metab 2001;21:69–76.
19. Raivich G, Gehrmann J, Kreutzberg GW. Increase of macrophage colony-stimulating factor and granulocyte-macrophage colony stimulating factor receptors in the regenerating rat facial nucleus. J Neurosci Res 1991;30: 682–686.
20. Raivich G, Haas S, Werner A, Klein MA, Kloss CUA, Kreutzberg GW. Regulation of MCSF receptors on microglia in the normal and injured mouse central nervous system: a quantitative immunofluorescence study using confocal laser microscopy. J Comp Neurol 1998;395:342–358.
21. Graeber MB, Tetzlaff W, Streit WJ, Kreutzberg GW. Microglial cells but not astrocytes undergo mitosis following rat facial nerve axotomy. Neurosci Lett 1988;85:317–321.
22. Kreutzberg GW. Autoradiographische Untersuchungen über die Beteiligung von Gliazellen an der axonalen Reaktion im Fazialiskern der Ratte. Acta Neuropathol 1966; 7:149–161.
23. Sjöstrand J. Studies on glial cells in the hypoglossal nucleus of the rabbit during nerve regeneration. Acta Physiol Scand 1966(Suppl. 67);270:1–17.

24. Klein MA, Möller JC, Jones LL, Bluethmann H, Kreutzberg GW, Raivich G. Impaired neuroglial activation in Interleukin-6 deficient mice. Glia 1997;19:227–233.
25. Horvat A, Schwaiger F, Hager G, Brocker F, Streif R, Knyazev P, Ullrich A, Kreutzberg GW. A novel role for protein tyrosine phosphatase shp1 in controlling glial activation in the normal and injured nervous system. J Neurosci 2001;21:865–874.
26. Streit WJ, Kreutzberg GW. Response of endogenous glial cells to motor neuron degeneration induced by toxic ricin. J Comp Neurol 1988;268:248–263.
27. Bohatschek M, Gschwendtner A, von Maltzan X, Kloss CUA, Pfeffer K, Labow M, Bluthmann H, Kreutzberg GW, Raivich G. Cytokine-mediated regulation of MHC1, MHC2 and B7-2, in the axotomized mouse facial motor nucleus. Soc Neurosci Abstr 1999;25:1535.
28. Raivich G, Gehrmann J, Graeber MB, Kreutzberg GW. Quantitative immunohistochemistry in the rat facial nucleus with iodine-125 iodinated secondary antibodies and in-situ autoradiography non-linear binding characteristics of primary monoclonal and polyclonal antibodies. J Histochem Cytochem 1993;41:579–592.
29. Möller JC, Klein MA, Haas S, Jones LL, Kreutzberg GW, Raivich G. Regulation of thrombospondin in the regenerating mouse facial nucleus. Glia 1996;17:121–132.
30. Raivich G, Jones LL, Kloss CUA, Werner A, Neumann H, Kreutzberg GW. Immune surveillance in the injured nervous system: T-lymphocytes invade the axotomized mouse facial motor nucleus and aggregate around sites of neuronal degeneration. J Neurosci 1998;18:5804–5816.
31. Kloss CUA, Werner A, Shen J, Klein MA, Kreutzberg GW, Raivich G. The integrin family of cell adhesion molecules in the injured brain: regulation and cellular localization in the normal and regenerating mouse facial nucleus. J Comp Neurol 1999;441:162–178.
32. De Simone R, Giampaolo A, Giometto B, Gallo P, Levi G, Peschle C, Aloisi F. The costimulatory molecule B7 is expressed on human microglia in culture and in multiple sclerosis acute lesions. J Neuropathol Exp Neurol 1995;54:175–187.
33. McGeer PL, Kawamata T, Walker DG, Akiyama H, Tooyama I, McGeer EG. Microglia in degenerative neurological disease. Glia 1993;7:84–92.
34. Bretscher P, Cohn M. A theory of self–nonself discrimination. Science 1970;169:1042–1049.
35. Robey E, Allison JP. T-cell activation: integration of signals from the antigen receptor and costimulatory molecules. Immunol Today 1995;16:306–310.
36. Brocker T, Riedinger M, Karjalainen K. Targeted expression of major histocompatibility complex (MHC) class II molecules demonstrates that dendritic cells can induce negative but not positive selection of thymocytes in vivo. J Exp Med 1997;185:541–550.
37. Ingalls RR, Golenbock DT. CD11c/CD18, a transmembrane signaling receptor for lipopolysaccharide. J Exp Med 1995;181 1473–1479.
38. Hynes RO. Integrins: versatility, modulation, and signaling in cell adhesion. Cell 1992;69:11–25.
39. Deckert-Schluter M, Bluethmann H, Rang A, Hof H, Schluter D. Crucial role of TNF receptor type 1 (p55), but not of TNF receptor type 2 (p75), in murine toxoplasmosis. J Immunol 1998;160:3427–3436.
40. Willenborg DO, Fordham S, Bernard CC, Cowden WB, Ramshaw IA. IFN-gamma plays a critical down-regulatory role in the induction and effector phase of myelin oligodendrocyte glycoprotein-induced autoimmune encephalomyelitis. J Immunol 1996;157:3223–3232.
41. Torvik A, Skjorten F. Electron microscopic observations on nerve cell regeneration and degeneration after axon lesions. II. Changes in the glial cells. Acta Neuropathol 1971;17:265–282.
42. Raivich G, Liu ZQ, Bohatschek M, Kloss CUA, Labow M, Bluethmann H. Cytotoxic potential of proinflammatory cytokines: combined deletion of TNFR1 and TNFR2 prevents motoneuron cell death after facial axotomy. Soc Neurosci Abstr 2001;28:874.
43. Kloss CUA, Kreutzberg GW, Raivich G. Proliferation of ramified microglia on an astrocyte monolayer: characterization of stimulatory and inhibitory cytokines. J Neurosci Res 1997;49:248–254.
44. Kloss CUA, Bohatschek M, Kreutzberg GW, Raivich G. Effect of lipopolysaccharide on the morphology and integrin immunoreactivity of ramified microglia in the mouse brain and in cell culture. Exp Neurol 2001;168:32–46.
45. Jones LL, Kreutzberg GW, Raivich G. Transforming growth factor β's 1, 2 and 3 inhibit proliferation of ramified microglia on an astrocyte monolayer. Brain Res 1998; 795:301–306.
46. Bohatschek M, Kloss CU, Kalla R, Raivich G. In vitro model of microglial deramification: Ramified microglia transform into amoeboid phagocytes following addition of brain cell membranes to microglia–astrocyte cocultures. J Neurosci Res 2001;64:508–522.
47. Stangel M, Joly E, Scolding NJ, Compston DA. Normal polyclonal immunoglobulins ('IVIg') inhibit microglial phagocytosis in vitro. J Neuroimmunol 2000;106:137–144.
48. Witting A, Muller P, Herrmann A, Kettenmann H, Nolte C. Phagocytic clearance of apoptotic neurons by microglia/brain macrophages in vitro: involvement of lectin-, integrin-, and phosphatidylserine-mediated recognition. J Neurochem 2000;75:1060–1070.
49. Eddleston M, Mucke L. Molecular profile of reactive astrocytes. Neuroscience 1993;54:15–36.
50. Norenberg MD. Astrocyte responses to CNS injury. J Neuropathol Exp Neurol 1994;53:213–220.
51. Graeber MB, Kreutzberg GW. Delayed astrocyte reaction following facial nerve axotomy. J Neurocytol 1988;17:209–220.
52. Guntinas-Lichius O, Martinez-Portillo F, Lebek J, Angelov DN, Stennert E, Neiss WF. Nimodipine maintains in vivo the increase in GFAP and enhances the astroglial ensheathment of surviving motoneurons in the rat following permanent target deprivation. J Neurocytol 1997;26:241–248.
53. Penkowa M, Moos T, Carrasco J, Hadberg H, Molinero A, Bluethmann H, Hidalgo J. Strongly compromised inflammatory response to brain injury in interleukin-6-deficient mice. Glia 1999;25:343–357.
54. Colucci-Guyon E, Gimenez y Ribotta M, Maurice T, Babinet C, Privat A. Cerebellar defect and impaired motor

coordination in mice lacking vimentin. Glia 1999;25:33–43.

55. Hickey WF, Kimura H. Perivascular microglial cells of the CNS are bone marrow-derived and present antigen in vivo. Science 1988;239:290–292.

56. Hickey WF, Hsu BL, Kimura H. T-lymphocyte entry into the central nervous system. J Neurosci Res 1991;28:254–260.

57. Wekerle H, Linington C, Lassmann H, Meyermann R. Cellular immune reactivity within the CNS. Trends Neurosci 1986;9:271–277.

58. Kiefer R, Lindholm D, Kreutzberg GW. Interleukin-6 and transforming growth factor-beta-1 mRNAs are induced in rat facial nucleus following motoneuron axotomy. Eur J Neurosci 1993;5:775–781.

59. Murphy PG, Borthwick LS, Johnston RS, Kuchel G, Richardson PM. Nature of the retrograde signal from injured nerves that induces interleukin-6 mRNA in neurons. J Neurosci 1999;19:3791–3800.

60. Raivich G, Galiano M, Jones LL, Kloss CUA, Werner A, Bluethmann H, Kreutzberg GW. Lymphocyte infiltration into the injured brain: Role of proinflammatory cytokines IL1, IL6 and TNFa. Ann Anat 1999;181:S43–S44.

61. Hua LL, Lee SC. Distinct patterns of stimulus-inducible chemokine mRNA accumulation in human fetal astrocytes and microglia. Glia 2001;30:74–81.

62. Wu Q, Miller RH, Ransohoff RM, Robinson S, Bu J, Nishiyama A. Elevated levels of the chemokine GRO-1 correlate with elevated oligodendrocyte progenitor proliferation in the jimpy mutant. J Neurosci 2000;20:2609–2617.

63. Xia MQ, Bacskai BJ, Knowles RB, Qin SX, Hyman BT. Expression of the chemokine receptor CXCR3 on neurons and the elevated expression of its ligand IP-10 in reactive astrocytes: in vitro ERK1/2 activation and role in Alzheimer's disease. J Neuroimmunol 2000;108:227–235.

64. Bohatschek M, Werner A, Raivich G. Systemic lipopolysaccharide injection leads to granulocyte influx into normal and injured brain: Effects of ICAM-1 deficiency. Exp Neurol 2001;172:137–52.

65. Rieske E, Graeber MB, Tetzlaff W, Czlonkowska A, Streit WJ, Kreutzberg GW. Microglia and microglia-derived brain macrophages in culture: generation from axotomized rat facial nuclei, identification and characterization in vitro. Brain Res 1989;492:1–14.

Universes in Delicate Balance: Chemokines and the Nervous System
R.M. Ransohoff, K. Suzuki, A.E.I. Proudfoot, W.F. Hickey and J.K. Harrison, editors

Chapter 3.7

Chemokines and Neural Inflammation in Experimental Brain Abscesses

Tammy Kielian [1] and William F. Hickey [2]

[1] *Department of Anatomy and Neurobiology, University of Arkansas for Medical Sciences, Little Rock, AR 72205;* [2] *Department of Pathology, Dartmouth Medical School, Dartmouth-Hitchcock Medical Center, Lebanon, NH 03756, USA*

1. INTRODUCTION

Unlike many diseases, it currently appears that the incidence of brain abscesses in the human population will remain constant, in part, due to the emergence of antibiotic resistant organisms. Despite advances made in detection and therapy, our knowledge of the pathogenesis of this serious malady remains largely unknown. Little information is available regarding the array of chemokine mediators induced by *Staphylococcus aureus* and their role in brain abscess pathogenesis. In this chapter, we will present the current knowledge regarding chemokine expression in experimental brain abscesses and the functional significance of some of these mediators using chemokine receptor knockout mice. Although we are beginning to unravel the innate response to *S. aureus* in the brain, many aspects of the complex interactions between host immunity and *S. aureus* in the CNS remain to be elucidated.

2. THE DISEASE

Brain abscesses develop in response to pyogenic bacteria, beginning as a localized area of cerebritis and evolving into a suppurative lesion surrounded by a well-vascularized fibrotic capsule. Brain abscesses account for approximately 1 in 10,000 hospital admissions in the U.S. [1]. The most common sources of brain abscess are direct or indirect cranial infection arising from the paranasal sinuses, middle ear, and teeth. Other routes include seeding of the brain from distant sites of infection in the body (i.e. endocarditis) or penetrating trauma to the head. The leading etiologic agents of brain abscesses are the *Streptococcal* strains and *Staphylococcus aureus* [1,2]. Even with the advances made in imaging techniques, the clinical diagnosis of brain abscess is often delayed since patients often present with non-specific clinical signs. If not detected early, an abscess has the potential to rupture into the ventricular space, a serious complication with an 80% mortality rate. In addition to sequestering the lesion, the immune response that is an essential part of abscess formation participates in the destruction of surrounding normal brain tissue, which can have detrimental effects. Indeed, some long-term effects following brain abscess resolution include seizures, mental retardation, and focal neurological defects [2]. The current treatment for brain abscesses includes surgical drainage to relieve pressure and long-term (6–8 weeks) systemic antimicrobial therapy.

3. THE LESION

Brain abscesses are one of the very few CNS pathological processes resulting in fibrosis rather than gliosis. The mouse and rat brain abscess models we have developed closely mimic human disease, in that the abscess progresses through a series of well-defined stages. In the rodent, each stage is defined on the basis of predominant cell types and the histological appearance of the abscess [3]. In the human, these stages have been identified based upon the use of CT scanning techniques [4]. The early stage, or early cerebritis, occurs from days 1 to 3 and is typified by neutrophil accumulation, tissue necrosis, and edema. The intermediate, or late

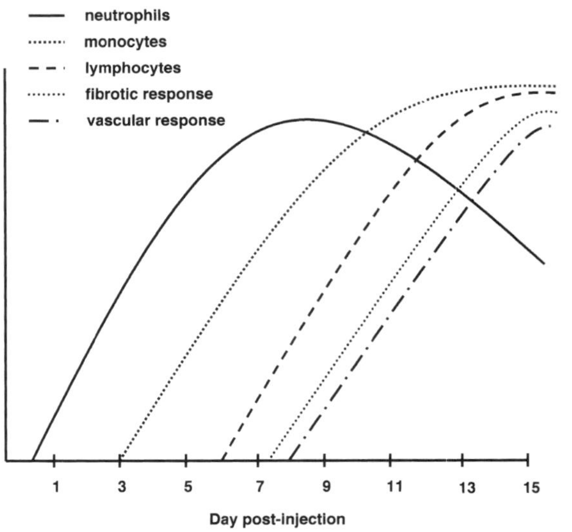

Figure 1. Graphical representation of the temporal influx of hematogenous cells into brain abscesses.

cerebritis stage, occurs from days 4 to 9 and is associated with a predominant macrophage and lymphocyte infiltrate. The final, or capsule stage, occurs from day 10 onward and is associated with the formation of a well-vascularized abscess wall, in effect sequestering the lesion and protecting the surrounding normal brain parenchyma. Myofibroblasts are found associated with the developing abscess wall at this stage and plasma cells predominate within lesions. A graphic representation of the time course of hematogenous cell influx and vascular/fibrotic responses in brain abscess is depicted in Fig. 1. Currently, it is not clear which cells are responsible for mediating the immunopathological response to *S. aureus* in the brain.

4. THE MODEL

Experimental brain abscesses are induced in the rodent by an intracerebral inoculation of live *Staphylococcus aureus* encapsulated in agarose beads [3,5]. The use of agarose beads prevents bacterial dissemination or rapid wound sterilization by the host. As previously mentioned, the rodent abscess model accurately reflects the course of disease progression in humans, providing an excellent system to study immunological pathways influencing abscess pathogenesis [3].

5. RATIONALE FOR EXAMINING CHEMOKINE EXPRESSION IN EXPERIMENTAL BRAIN ABSCESSES

It has been well established that pyogenic bacteria, such as *S. aureus*, can lead to abscess formation in peripheral tissues. In the CNS, we have demonstrated that *S. aureus* leads to the formation of brain abscesses in both the rat [3,5] and mouse (Kielian, manuscript submitted). As early as 24 h, neutrophils begin to accumulate within the area of cerebritis which is associated with significant edema and necrosis of the affected brain parenchyma. The influx of macrophages and lymphocytes soon follows at days 5 and 10 following *S. aureus* exposure, respectively. Throughout the progression of an evolving brain abscess, neutrophils continue to represent the majority of the inflammatory infiltrate. Based upon this sequential influx of hematogenous cells into brain abscesses, it appeared likely that chemokines were elicited in response to *S. aureus* in the brain. Recently, we have begun to characterize the chemokines expressed during the evolution of experimental brain abscesses and the cell type(s) responsible for their synthesis.

6. ADHESION MOLECULE EXPRESSION IN EXPERIMENTAL BRAIN ABSCESSES

In order to effectively recruit target cells into areas of active inflammation, the correct adhesion molecule repertoire must be expressed on the vascular endothelium of the target organ. Extravasation of hematogenous cells into tissues is a complex phenomenon which is thought to require three or more steps [6]. Central to this is the expression of the appropriate adhesion molecule repertoire on the target cells and vascular endothelium. The initial step is leukocyte rolling along the vascular bed which is mediated by the family of selectins. The next step is activation of the leukocyte to induce actin cytoskeletal rearrangement such that the cell can conform to the narrow space between endothelial cells. An example is the presentation of chemokines on the endothelial cell surface which activates specific chemokine receptors on the target cell. Indeed, chemokine signaling has been shown to induce actin polymerization in neutrophils. This step

Table 1. Adhesion receptor–counterreceptor pairs

Leukocyte [a]	Endothelium
L-selectin (L, N, M)	Lymph node addressin
sialyl Lewis x (N, M)	E-selectin
LFA-1 (L, N, M)	ICAM-1, ICAM-2
Mac-1 (CD11b) (N, M)	ICAM-1
VLA-4 (M, mT, nT)	VCAM-1

[a] L, lymphocytes; N, neutrophils; M, monocytes; mT, memory T cells; nT, naive T cells.

is followed by tight adhesion of the leukocyte to the endothelial cell which is integrin-mediated. The final step in leukocyte emigration is the extravasation of cells into tissues where they exert their actions. In the case of brain abscesses, this translates into the expression of adhesion molecules specific for ligands on neutrophils, monocytes, and lymphocytes. Some of these adhesion molecules and their respective ligands are listed in Table 1. Some endothelial adhesion molecules, such as ICAM-1, are constitutively expressed at low levels on cerebral microvessels. However, their expression and the levels of activation-induced adhesion molecules are upregulated by proinflammatory cytokines such as IL-1, TNF-α, and IL-6. We have recently demonstrated the strong induction of IL-1, TNF-α, and IL-6 in the rat brain abscess model which correlates with the expression of ICAM-1 and PECAM on vessels associated with brain abscesses [5]. It has been well documented in both in vivo and in vitro systems that these proinflammatory cytokines markedly alter CNS endothelial cells [7–15]. This transformation affects multiple types of adhesion molecules and in sites of evolving inflammation these cytokines prompt the endothelia to become HEV-like [16,17]. The functional significance of these adhesion molecules in the recruitment of hematogenous cells into brain abscesses is currently not known. However, the proinflammatory cytokines and chemokines detected in brain abscesses probably provide the signals necessary to decorate the endothelium with adhesion molecules critical for cell extravasation.

7. CHEMOKINE EXPRESSION IN EXPERIMENTAL BRAIN ABSCESSES

The influx of neutrophils and macrophages into the brain occurs very rapidly following *S. aureus* exposure. We have used RT-PCR and RNAse protection assays (RPA) to examine the expression of chemokines which may be responsible for the recruitment of these cells into brain abscesses in vivo. In the rat brain abscess model, KC/CXCL1, MCP-1/CCL2, MIP-1α/CCL3, and IP-10/CXCL10 were induced within 24 h following *S. aureus* exposure in the brain and could serve as the signals for neutrophil, monocyte, and lymphocyte recruitment [5]. In the mouse, *S. aureus* led to the rapid induction of MIP-1α, MIP-1β/CCL4, MIP-2/CXCL1, MCP-1, and TCA-3/CCL1 within 1 to 6 h following *S. aureus* exposure (Fig. 2). Furthermore, protein levels of the CXC chemokines, MIP-2 and KC, were significantly elevated

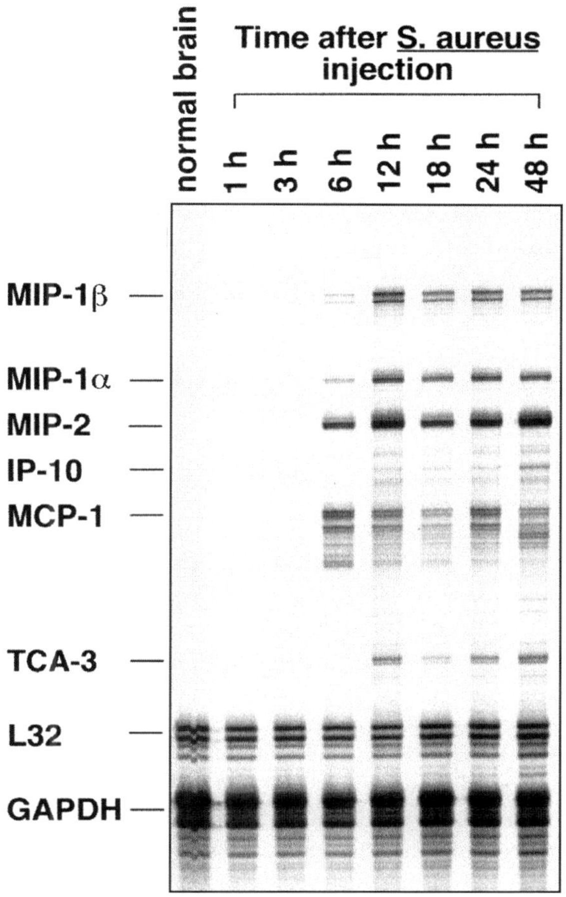

Figure 2. *S. aureus* induces rapid chemokine expression in the brain. Mice were implanted with *S. aureus*-encapsulated agarose beads at time 0. Total RNA was isolated from brain abscesses at the indicated time points and chemokine expression evaluated by RPA. RNA isolated from normal brain was included as a control. The identity of experimental mRNAs is denoted at the left.

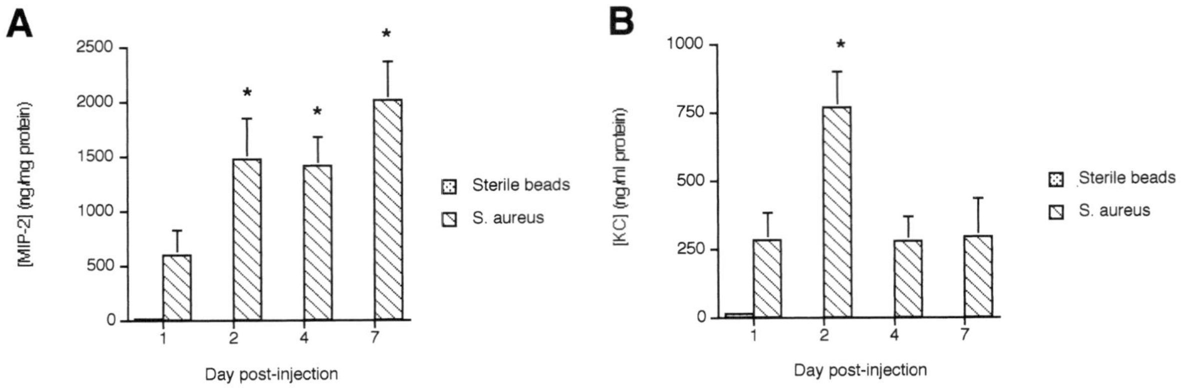

Figure 3. *S. aureus* induces the rapid and sustained expression of MIP-2 and KC. Mice were implanted with *S. aureus*-encapsulated or sterile beads on day 0. Brain abscess homogenates were prepared at the indicated time points and analyzed for MIP-2 (A) and KC (B) protein expression by ELISA. Chemokine concentrations are reported as ng per mg total protein to normalize results between individual samples.

in *S. aureus*-treated but not control animals (Fig. 3). The kinetics of MIP-2 and KC expression were rapid and sustained, with significant levels of both chemokines still detected at day 7 following bacterial exposure. The continued expression of MIP-2 and KC provides a signal for the continuous recruitment of neutrophils into evolving brain abscesses which is observed in vivo.

The rapidity at which chemokine production is induced in brain abscesses suggested that resident glial cells, such as astrocytes and/or microglia were responsible for their synthesis as opposed to recruited hematogenous cells. As the brain abscess evolves, infiltrating leukocytes such as neutrophils and monocytes/macrophages may also contribute to chemokine expression. To determine whether glial cells express chemokines in response to *S. aureus*, the response of primary astrocytes and microglia to heat-killed *S. aureus* was assessed in vitro using RPA. Primary microglia were found to be the major source of MIP-1α, whereas astrocytes were the major producers of RANTES/CCL5. Both cell types produced equivalent amounts of MIP-1β, MCP-1, and TCA-3. These results indicate that microglia and astrocytes can express numerous chemokines in response to *S. aureus* and most likely serve as the source of these mediators during the early phase of experimental brain abscess pathogenesis prior to inflammatory cell recruitment. However, it remains possible that other CNS resident cells such as cerebral endothelial cells and perivascular microglia also contribute to the initial synthesis of chemokines immediately following *S. aureus* exposure. Perivascular cells are known to be rapid responders to LPS as evidenced by their elaboration of COX II for eicosinoid synthesis within a few hours of LPS exposure [18]. However, the ability of these cells to express chemokines in the context of early brain abscesses remains unknown.

An interesting question to consider is what effect does local chemokine expression have on resident microglia and astrocytes in the context of brain abscesses? Indeed, microglia and astrocytes have been shown to express numerous chemokine receptors including CCR1, CCR2, and CCR5. Chemokines have recently been shown to directly stimulate chemokine expression in astrocytes [19]. The activation of resident glial cells by secreted chemokines would add another level of complexity to the responses induced by *S. aureus* in the brain.

8. REQUIREMENT FOR THE CXCR2 LIGANDS, MIP-2 AND KC, IN NEUTROPHIL-MEDIATED HOST DEFENSE IN EXPERIMENTAL BRAIN ABSCESSES

Neutrophils constitute the majority of infiltrating leukocytes during an evolving brain abscess. In addition to bacterial neutralization, neutrophils have the ability to mediate tissue injury through the release of proteolytic enzymes and production of reactive oxygen intermediates. Therefore, we were interested in defining the importance of neutrophils during the first 48 h following introduction of *S. aureus* into the brain. To perform these studies, mice were depleted of neutrophils using the mAb RB6-C85 which has been used extensively for in vivo neutrophil depletion studies [20,21]. RB6-C85-treated mice had significantly fewer circulating neutrophils throughout the duration of the experiment, whereas neutrophil levels in animals receiving an isotype-matched control Ab remained unaffected. Brain abscesses were collected from isotype control- or RB6-C85-treated animals 48 h following *S. aureus* implantation and compared for differences in morphology. Abscess lesions were significantly more severe in neutrophil-depleted mice compared to isotype control-treated animals. Immunohistochemical analysis revealed that abscesses in RB6-C85-treated mice had fewer infiltrating neutrophils and were more necrotic as demonstrated by the total loss of normal brain architecture compared to isotype control-treated animals. In addition, numerous extracellular bacterial foci were detected in abscesses from RB6-C85-treated animals, a phenomenon which is never observed in immunocompetent animals. These results reveal the importance of neutrophils in the early host immune response to *S. aureus* in the brain. The finding that tissue necrosis occurs in spite of a paucity of neutrophils, suggests that tissue damage in this model is not solely neutrophil-mediated. During the acute phase of infection, neutrophils appear critical for containing bacterial levels and preventing widespread necrosis of surrounding normal brain parenchyma.

We have established that two neutrophil-attracting CXC chemokines, MIP-2 and KC, are strongly induced in brain abscesses. To demonstrate the importance of MIP-2 and KC in neutrophil recruitment into abscesses and bacterial neutralization, we examined the pathogenesis of *S. aureus* in CXCR2 knockout mice. CXCR2 is the sole neutrophil chemokine receptor in the mouse and binds both MIP-2 and KC with high affinity [22]. Interestingly, we found higher mortality rates and elevated bacterial burdens in CXCR2 knockout mice compared to wild type littermates, suggesting that the ability to control bacterial replication is compromised in CXCR2 knockout mice. To determine whether this defect was related to the inability to recruit neutrophils into brain abscesses, an immunohistochemical analysis was performed on brain abscesses from CXCR2 knockout and control mice. Neutrophils infiltrating the brain parenchyma were rare in CXCR2 knockout mice, with significantly smaller numbers than detected in wild type mice (Fig. 4). Strikingly, the majority of neutrophils in CXCR2 knockout mice were sequestered within small blood vessels surrounding the developing abscess. This is most likely due to their inability to respond to MIP-2 and KC. Such an observation indicates that there is a defect in the mechanisms of neutrophil extravasation into the brain parenchyma but not in their adhesion to vascular endothelium in CXCR2 knockout mice. The relative paucity of neutrophils in the brain parenchyma of CXCR2 knockout mice indicates that CXCR2 ligands are the major chemotactic signals for neutrophil influx into brain abscesses and their activity cannot be substituted by any redundant chemotactic signals such as complement split products (C3a, C5a) or other chemokines. However, these studies have not yet established which ligand of CXCR2 is critical for neutrophil extravasation into brain abscesses in vivo, a question we are currently investigating. The nature of the transmembrane signals needed to activate the machinery for neutrophil migration across the blood–brain barrier is complex; however, signaling through CXCR2 seems to be critical based on this data.

Another intriguing observation was that in spite of the relative inability to recruit neutrophils into the infected brain parenchyma, abscesses still developed. This finding suggests that neutrophils per se are not absolutely required for abscess progression but do play a pivotal role in controlling bacterial burdens. This finding also argues that cells in addition to neutrophils or the bacteria themselves are responsible for initiating widespread necrosis of the brain parenchyma, which is supported by the findings obtained from neutrophil depletion studies. However, at this time, the origin of this tissue damage remains unknown.

9. IMPLICATIONS OF CHEMOKINE EXPRESSION IN EXPERIMENTAL BRAIN ABSCESSES

If not tightly regulated, the sustained expression of the very chemokines detected in evolving brain

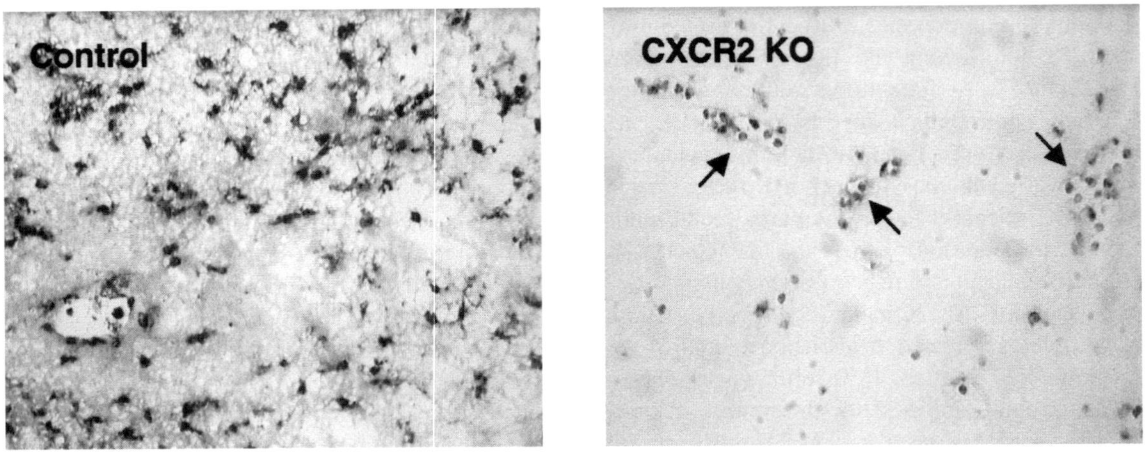

Figure 4. Impairment in neutrophil extravasation into the brain parenchyma in CXCR2 knockout mice. CXCR2 knockout and homozygous wild type littermates were implanted with *S. aureus*-encapsulated agarose beads on day 0. Animals were sacrificed 4 days following bacterial exposure and perfusion-fixed to collect brain abscesses for immunohistochemical analysis. Brain abscesses from control (CXCR2 wild type) and CCXR2 knockout mice were stained with the neutrophil-specific antibody GR-1. Arrows denote small vessels congested with neutrophils in CXCR2 knockout mice.

abscesses have the potential to cause excessive tissue damage by prompting the chronic influx and activation of newly recruited hematogenous cells. This could have detrimental effects on surrounding normal brain parenchyma leading to exaggerated tissue destruction and the manifestation of clinical neurological deficits. Therefore, understanding the chemokine expression profile in the context of *S. aureus*-induced brain abscesses might provide insights on how to best control the immune response to bacteria in the CNS. The optimal response would be to recruit sufficient numbers of effector cells to efficiently neutralize bacteria without causing overt damage to surrounding brain parenchyma. This result could be approached at one of two different levels required for the trafficking of hematogenous cells into the parenchyma adjacent to brain abscesses. One is through regulating cell entry into the CNS by blocking key adhesion molecules required for cell extravasation. This approach as been used to elucidate the critical role of VCAM-1/VLA-4 [23,24], P-selectin [25,26], and even ICAM-1 [27–30] in the recruitment of cells into the CNS in experimental autoimmune encephalomyelitis (EAE). An alternative mechanism would be to interfere with chemokine function using neutralizing antibodies or receptor antagonists. Again, this has been used effectively in the EAE system [31,32]. However, in practice these methods may prove difficult to achieve. And, it must be noted, that a response to a pyogenic bacterial infection may be far more fulminant than an autoimmune T cell-mediated event.

An evolving brain abscess involves the trafficking of leukocytes as well as the induction of vascular and fibrotic responses (Fig. 1). We have established the importance of MIP-2 and KC in neutrophil recruitment during the acute phase of brain abscess development. However, it appears likely that a different array of chemokines will be responsible for the accumulation of monocytes into brain abscesses which occurs at day 5 following bacterial exposure. Likewise, the influx of lymphocytes into abscesses may require a distinct repertoire of chemokine signals, or their accumulation may be chemokine-independent. The chemokine-independent trafficking of lymphocytes into brain abscesses may be a consequence of *Staphylococcal* antigen-specific T and B lymphocytes seeking their cognate antigen in the CNS. Bacterial antigens are available to the periphery and classical antigen presentation occurs in the cervical lymph nodes draining the brain, in addition to the migration of perivascular macrophages harboring bacterial antigens out of the CNS into the spleen or lymph nodes. Finally, the appearance of myofibroblasts associated with the developing abscess wall and the ensuing vascular response may also be affected by chemokines. Indeed, numerous ELR-containing CXC chemokines have been shown to be angiogenic [33–35]. The role of chemokines in regulating these changes in the developing abscess remain to be defined, but represent areas that we are actively pursuing.

By comparing the chemokine expression profiles elicited in other disease models, it appears as if the host chemokine response to various CNS insults may be conserved at some level. Several studies have examined the expression of chemokines in response to CNS insults such as stab wound injury in the brain [36] and experimental autoimmune encephalomyelitis [31,37,38]. In these models, KC, IP-10, MCP-1, and MIP-1α are upregulated following injury. These are among the same chemokines whose expression is induced following *S. aureus* exposure in the experimental brain abscess model. This raises an interesting possibility; can an effective interventive therapy aimed at attenuating chemokine expression have beneficial effects in disparate CNS diseases?

In summary, the rodent experimental brain abscess model provides an excellent model system to dissect the functional significance of chemokines in the pathogenesis of disease and may lead to therapeutic interventions for abscesses and other bacterial brain infections.

REFERENCES

1. Townsend GC, Scheld WM. Infections of the central nervous system. Adv Intl Med 1998;43:403–447.
2. Mathisen GE, Johnson JP. Brain abscess. Clinical Infect Dis 1997;25:763–781.
3. Flaris N, Hickey WF. Development and characterization of an experimental model of brain abscess in the rat. Am J Pathol 1992;141:1299–1307.
4. Wispelwey B, Dacey RG, Scheld WM. Brain abscess. In: Scheld WM, Whitely RJ, Durack DT, eds. Infections of the Central Nervous System. New York, NY: Raven Press, 1991:457–486.
5. Kielian T, Hickey WF. Proinflammatory cytokine, chemokine, and cellular adhesion molecule expression during the acute phase of experimental brain abscess development. Am J Pathol 2000;157:647–658.
6. Butcher EC. Leukocyte-endothelial cell recognition: Three (or more) steps to specificity and diversity. Cell 1991;67: 1033–1036.
7. Wong D, Dorovini-Zis K. Upregulation of intercellular adhesion molecule-1 (ICAM-1) expression in primary cultures of human brain microvessel endothelial cells by cytokines and lipopolysaccharide. J Neuroimmunol 1992; 39:11–21.
8. Kuchroo VK, Martin CA, Greer JM, Ju ST, Sobel RA, Dorf ME. Cytokines and adhesion molecules contribute to the ability of myelin proteolipid protein specific T cell clones to meditate experimental allergic encephalomyelitis. J Immunol 1993;151:4371–4379.
9. Male D, Rahman G, Price G, Tamatani T, Miyasaka M. Lymphocyte migration into the CNS modeled in vitro: role of LFA-1, ICAM-1 and VLA-4. Immunol 1994;81: 366–374.
10. De Vries HE, Moor AC, Blom-Roosmalen MC, de Boer AG, Breimer DD, van Berkel TJ, Kuiper J. Lymphocyte adhesion to brain capillary endothelial cells in vitro. J Neuroimmunol 1994;52:1–8.
11. Dore-Duffy P, Washington RA, Balabanov R. Cytokine-mediated activation of cultured CNS microvessels: a system for examining antigenic modulation of CNS endothelial cells, and evidence for long-term expression of the adhesion protein E-selectin. J Cereb Blood Flow Metabol 1994;14:837–844.
12. Watkins LR, Maier SF, Goehler LE. Cytokine-to-brain communication: a review and analysis of alternative mechanisms. Life Sci 1995;57:1011–1026.
13. Eng LF, Ghirnikar RS, Lee YL. Inflammation in EAE: role of chemokine/cytokine expression by resident and infiltrating cells. Neurochem Res 1996;21:511–525.
14. Joo F. Endothelial cells of the brain and other organ systems: some similarities and differences. Prog Neurobiol 1996;48:255–273.
15. Rubin LL, Staddon JM. The cell biology of the blood–brain barrier. Annu Rev Neurosci 1999;22:11–28.
16. Cannella B, Cross AH, Raine CS. Adhesion related molecules in the central nervous system: upregulation correlates with inflammatory cell influx. Lab Invest 1991; 65:23–31.
17. Raine CS. The Dale McFarlin Memorial Lecture: the immunology of multiple sclerosis. Ann Neurol 1994;36: S61–S72.
18. Elmquist JK, Breder CD, Sherin JE, Scamell TE, Hickey WF, Dewitt D, Saper CB. Intravenous lipopolysaccharide induced cyclooxygenase 2-like immunoreactivity in rat brain perivascular microglia and meningeal macrophages. J Comp Neurol 1997;381:119–129.
19. Luo Y, Fischer FR, Hancock WW, Dorf ME. Macrophage inflammatory protein-2 and KC induce chemokine production by mouse astrocytes. J Immunol 2000;165:4015–4023.
20. Mehrad B, Strieter RM, Moore TA, Tsai WC, Lira SA, Standiford TJ. CXC chemokine receptor-2 ligands are necessary components of neutrophil-mediated host defense in invasive pulmonary aspergillosis. J Immunol 1999;163: 6086–6094.
21. McColl SR, Staykova MA, Wozniak A, Fordham S, Bruce J, Willenborg DO. Treatment with anti-granulocyte antibodies inhibits the effector phase of experimental autoimmune encephalomyelitis. J Immunol 1998;161:6421–6426.
22. Lee J, Cacalano G, Camerato T, Toy K, Moore MW, Wood WI. Chemokine binding and activities mediated by the mouse IL-8 receptor. J Immunol 1995;155:2158–2164.
23. Yednock TA, Cannon C, Fritz LC, Sanchez-Madri F, Steinman L, Karin N. Prevention of experimental autoimmune encephalomyelitis by antibodies against α4β1 integrin. Nature 1992;356:63–66.
24. Tanaka M, Sato A, Makino M, Tabira T. Binding of an SJL T cell clone specific for myelin basic protein to SJL

brain microvessel endothelial cells is inhibited by anti-VLA-4 or its ligand, anti-vascular cell adhesion molecule 1 antibody. J Neuroimmunol 1993;46:253–257.

25. Hickey WF. P-selectin, pioneer cells and the path to inflammation [editorial]. Brain 2000;123:1073–1074.

26. Carrithers MD, Visintin I, Kang SJ, Janeway CA Jr. Differential adhesion molecule requirements for immune surveillance and inflammatory recruitment. Brain 2000; 123:1092–101.

27. Archelos JJ, Jung S, Maurer M, Schmied M, Lassmann H, Tamatani T, Miyasaka M, Toyka KV, Hartung HP. Inhibition of experimental autoimmune encephalomyelitis by an antibody to intercellular adhesion molecule ICAM-1. Ann Neurol 1993;34:145–154.

28. McCarron RM, Lang L, Racke MK, McFarlin DE, Spatz M. Effect of cytokines on ICAM expression and T cell adhesion to cerebrovascular endothelial cells. Adv Exp Med Biol 1993;331:237–242.

29. Steffen BJ, Butcher EC, Engelhardt B. Evidence for involvement of ICAM-1 and VCAM-1 in lymphocyte interaction with endothelium in experimental autoimmune encephalomyelitis in the central nervous system in the SJL/J mouse. Am J Pathol 1994;145:189–201.

30. Rieckmann P, Michel U, Albrecht M, Bruck W, Wockel L, Felgenhauer K. Soluble forms of intercellular adhesion molecule-1 (ICAM-1) block lymphocyte attachment to cerebral endothelial cells. J Neuroimmunol 1995;60:9–15.

31. Karpus WJ, Lukacs NW, McRae BL, Strieter RM, Kunkel SL, Miller SD. An important role for the chemokine macrophage inflammatory protein-1α in the pathogenesis of the T cell-mediated autoimmune disease, experimental autoimmune encephalomyelitis. J Immunol 1995;155: 5003–5010.

32. Kennedy KJ, Strieter RM, Kunkel SL, Lukacs NW, Karpus WJ. Acute and relapsing experimental autoimmune encephalomyelitis are regulated by differential expression of the CC chemokines macrophage inflammatory protein-1 alpha and monocyte chemotactic protein-1. J Neuroimmunol 1998;92:98–108.

33. Koch AE, Polverini PJ, Kunkel SL, Harlow LA, DiPietro LA, Elner VM, Elner SG, Strieter RM. Interleukin-8 (IL-8) as a macrophage-derived mediator of angiogenesis. Science 1992;258:1798–1801.

34. Strieter RM, Polverini PJ, Arenberg DA, Walz A, Opdenakker G, Van Damme J, Kunkel SL. Role of C-X-C chemokines as regulators of angiogenesis in lung cancer. J Leukoc Biol 1995;57:752–762.

35. Luster AD. Chemokines–Chemotactic cytokines that mediate inflammation. New Eng J Med 1998;338:436–445.

36. Ghirnikar RS, Lee YL, He TR, Eng LF. Chemokine expression in rat stab wound brain injury. J Neurosci Res 1996;46:727–733.

37. Glabinski AR, Tani M, Strieter RM, Tuohy VK, Ransohoff RM. Synchronous synthesis of α- and β-chemokines by cells of diverse lineage in the central nervous system of mice with relapses of chronic experimental autoimmune encephalomyelitis. Am J Pathol 1997;150:617–630.

38. Ransohoff RM, Glabinski A, Tani M. Chemokines in immune-mediated inflammation of the central nervous system. Cytokine Growth Factor Rev 1996;7:35–46.

Universes in Delicate Balance: Chemokines and the Nervous System
R.M. Ransohoff, K. Suzuki, A.E.I. Proudfoot, W.F. Hickey and J.K. Harrison, editors

Chapter 3.8

Insights from Transgenic and Knockout Mice

Iain L. Campbell and Valerie C. Asensio

The Scripps Research Institute, La Jolla, CA, USA

1. INTRODUCTION

Although the ability to induce leukocyte chemotaxis is a prototypic biological action of chemokines, an important emerging view of these small proteins is that they are multifunctional mediators of cellular communication in the central nervous system (CNS) in both physiologic and pathophysiologic states [1,2]. Cells intrinsic to the nervous system, including neurons, the macroglia and microglia, all have the ability to produce chemokines. Moreover, the surfaces of these same cells are decorated with a variety of different chemokine receptors. In the nervous system, as in the periphery, the chemokine gene superfamily can be generally divided into two broad groups based on their expression properties. The first group are constitutively expressed chemokines of which the non-ELR CXC family chemokine SDF-1 (CXCL12) and the CX_3C family member fractalkine (CX_3CL1) are the best characterized. In contrast to the constitutively expressed chemokines, a second larger group of chemokines consisting of many members of both the CXC and CC families, are not found in the CNS under normal conditions. However, abundant CNS production of chemokines belonging to this second group may be induced following a variety of insults including trauma, ischemia, infection and inflammation. Of course, the distinct expression characteristics between these two different groups of chemokines implies that their members may have fundamentally different biological functions. Some important questions are therefore raised concerning the biology of chemokines in the CNS. What is the function of these different chemokines in different states? What are the cellular targets of chemokine action? How do chemokines exert their actions? Can pivotal molecular targets be identified that might allow manipulation of the actions of chemokines?

In view of the breadth of potential chemokine interactions and the complexity of the milieu in which they are produced, to begin to address these questions in a satisfactory and relevant fashion, non-invasive, organism-based approaches are required. Advances over the past 15 years in the fields of molecular biology, molecular genetics, cellular biology and embryology have given us the technology to achieve the targeted delivery and germline transmission of foreign genes (transgenes) in the mouse. To date, the most commonly used techniques for generating transgenic mice fall into two categories which in simple terms result in mutant mice with gene expression either added or lost, respectively. In the first, a fusion gene construct with the gene of interest placed under the transcriptional control of a cell-specific promoter, is microinjected directly into the pronucleus of fertilized eggs which are then implanted in the oviduct of pseudopregnant recipient mice. In the second approach, modified homologous DNA constructs are introduced into embryonal stem cells (ES cells) in vitro and recombination of the introduced DNA with the target sequence results in disruption of the endogenous gene. Positive ES cells then contribute to the germline when injected into host blastocysts that are then allowed to develop following oviduct implantation in pseudopregnant recipients. In either case, once made, the stable mutant lines provide an unrestricted source of animals identical for the introduced genetic and molecular alteration, hence permitting systematic analysis of the function of the mutant gene in the intact organism (for more detailed reviews of these techniques see [3,4]).

As in other areas, the application of genetically manipulated animal models to the study of the CNS biology of the chemokines and their receptors has emerged as an important experimental approach. In this chapter, we will discuss findings from studies based on the use of CNS–chemokine transgenic and knockout models. In particular, we will focus on the contribution of these models to our understanding of basic functions and mechanisms of actions of chemokines in the CNS. For the sake of clarity our discussion will focus first on the constitutively expressed chemokines followed by those that are induced in the CNS during pathological states.

2. THE CONSTITUTIVELY EXPRESSED CHEMOKINES

Currently, two chemokines, SDF-1 and fractalkine, are known to be constitutively produced in the CNS as well as in many other organs. Transgenic approaches to determine the function of these chemokines have focused exclusively on the use of homologous recombination to generate mutant mice with targeted deletion of the chemokine receptors and/or their corresponding ligand.

2.1. CXCL12/SDF-1 and the CXCR4 Receptor

SDF-1 RNA is found normally in a wide variety of tissues including the brain [5]. SDF-1 is present predominantly in astrocytes and neurons in the brain and in Schwann cells in the periphery. SDF-1 exists in three isoforms, SDF-1α, SDF-1β and SDF-1γ that are produced by alternative splicing from a single gene transcript [5,6]. SDF-1α RNA is present at high levels in cultured rat astrocytes, at low levels in neurons and is not detectable in microglia, while SDF-1β RNA is found at low levels in all these cell types [7]. Although astrocytes are a predominant source for SDF-1 synthesis in vitro, little is known about the location or temporal production of this chemokine in vivo. However, the receptor for SDF-1, CXCR4, is found during early development (embryonic-day 9) both in the immune system and in proliferative areas of the rat brain [8]. In the rat, CXCR4 receptor homologue, LCR1, is localized to the germinal neuroepithelium of the brain during early stages of development, but then declines to undetectable levels in adult brain. Similarly, in the embryonic mouse brain, CXCR4 is found in many regions of the brain but is particularly high in proliferative regions of the cerebellum, including neuroepithelium, rhombic lip and external granular cell layer (EGL) [9]. These observations are consistent with a role for CXCR4 in neurogenesis and neuronal differentiation. Finally, SDF-1 is highly conserved across species (human and mouse SDF-1 proteins differ only by one amino-acid residue). In all, these observations imply that SDF-1 has a vital function perhaps during CNS development.

In recent studies the genes for both SDF-1 [10,11] and its receptor [9,11,12] have been disrupted in mutant mice. These animals die soon after birth with major defects in their vascular, hematopoietic and nervous systems. In particular, in the brain, the aberrant and premature migration of the granule neurons causes severe abnormalities in the development of the cerebellum. This alteration appears to be relatively specific since other aspects of cerebellar development appear to be normal in these mice. These findings indicate that SDF-1/CXCR4 signaling prevents the premature migration of proliferating granule cells from the external granule layer. However, the precise mechanism involved in this process is not known. The clear conclusions from these observations are that (1) SDF-1/CXCR4 interaction is essential for normal neuronal-cell migration and the patterning that occurs during development, and (2) there is a lack of redundancy in SDF-1/CXCR4 signaling which cannot be compensated for by other chemokines or their receptors. A number of questions are also raised by these findings. What is the temporal and spatial pattern of SDF-1 expression during CNS development? How does SDF-1 influence neuronal migration? There is a wide distribution of CXCR4 receptors in the developing brain, raising the question as to whether regions other than the cerebellum might be influenced by SDF-1? No doubt we will begin to see the answers to these issues in the years to come.

2.2. CX_3CL1/Fractalkine

The CX_3C chemokine CX_3CL1/fractalkine is also produced constitutively in the brain [13,14]. Fractalkine mRNA is expressed at high levels by neurons in the olfactory bulb, cerebral cortex, hippocampus, caudate, putamen and nucleus accumbens [15–17]. In vitro studies have confirmed that the fractalkine gene is constitutively expressed by neurons but can also be induced in astrocytes by TNF-α and IL-1β [18]. What could be the function of fractalkine in the nervous system? Fractalkine is unique amongst the chemokines and is tethered to the membrane by a mucin stalk, which is well suited for communication with adjacent cells expressing the fractalkine receptor CX_3CR1. One major candidate for the action of fractalkine is the microglial cell which expresses high levels of the fractalkine receptor, CX_3CR1. Indeed, in a model of peripheral nerve injury, where the facial motor nerve was transected, Harrison et al. showed that fractalkine and its receptor were highly upregulated at affected sites in the brain with motor neurons being the primary source of the chemokine while microglia expressed the receptor [15]. Moreover, fractalkine treatment of primary cultured rat microglial cells induces vigorous increases in intracellular Ca^{2+} levels and chemotaxis, which are inhibited by an antibody against the fractalkine receptor, CX_3CR1 [15].

A mouse strain in which CX_3CR1 gene was replaced by a green fluorescent protein (GFP) reporter gene was generated to study the physiological function of fractalkine/CX_3CR1 signaling [19]. These mice do not exhibit any developmental abnormalities, nor do they exhibit leukocyte recruitment defects. Interestingly, no defect in microglial migration, proliferation or differentiation nor in neuron damage or recovery occurs in the brain following peripheral nerve injury in CX_3CR1 deficient mice [19]. Also, neuronal–glial communication is not impaired in the absence of the fractalkine receptor in this model suggesting interaction between neurons and microglia through CX_3CR1 and its ligand is not pivotal in nerve injury. In addition, the apparent absence of other nervous system defects in the CX_3CR1-deficient mouse also suggests that fractalkine, unlike SDF-1, is dispensable for normal CNS development and neuronal survival. Therefore, we presently have to say that the function of fractalkine in the CNS is uncertain. One explanation for the surprising lack of any phenotype in the CXC_3R1 mice is that there is redundancy or compensation for the loss of fractalkine signaling. For example, fractalkine might bind to and activate another as yet unidentified receptor. Alternatively, other chemokines in the CNS might substitute for the actions of fractalkine.

2.3. Other

Neurons possess a wide variety of chemokine receptors including CXCR2, CXCR4, CCR1, CCR4, CCR5, CCR9, CX_3CR1 and the Duffy antigen related chemokine (DARC). Expression of the genes for these receptors in the CNS exhibits some regional variance, implying that individual chemokines could have differential effects on different neurons. For example DARC is found almost exclusively on Purkinje cells in the cerebellum [20], while the CXCR2 receptor is most prominent in the dentate gyrus of the human hippocampus [21]. In contrast with other receptors, the binding of ligands to DARC does not induce signal transduction [22] which suggested that DARC may serve as a sink for chemokines. Mutant mice generated to be deficient in the DARC receptor do not show any obvious developmental abnormalities including the brain [23,24] nor any neurological impairment [24]. However, after LPS challenge DARC KO mice have an exaggerated response to intraperitoneal administration of LPS confirming that DARC may serve as a chemokine sink and that normal expression of DARC regulates leukocyte trafficking during inflammation. Whether a similar function can be ascribed to DARC in the brain remains unclear and awaits testing. It should be noted that no adverse neurological consequences appear to afflict humans who carry a promoter region mutation that prevents expression of the DARC gene [25].

CXCR2 is expressed at high levels by neurons [21]. As mentioned above, CXCR2 was found on specific subpopulations of neurons, which raises the possibility that GRO-α (CXCL1) and other CXCR2 ligands might also regulate the development of these cells. However, mice that have been generated to be deficient in CXCR2 exhibit no abnormalities in hematopoiesis and outwardly appear normal, which suggests the absence of gross neurological perturbations [26]. Nevertheless, it remains possible that these animals have more subtle alterations that affect the white and possibly the gray matter.

3. THE INDUCIBLE CHEMOKINES

A great many chemokines belonging in particular to the CXC and CC families are known to be induced in the brain during a variety of different pathologic states [1,2,27,28]. The consequences of chemokine expression in the unmanipulated CNS have been examined in transgenic mice with CNS-targeted expression of specific chemokines. More recently, with the availability of chemokine receptor knock-out mice, the absence of specific chemokine/receptor interactions has been studied in the context of experimentally induced disease such as in experimental autoimmune encephalomyelitis (EAE).

3.1. CNS-Targeted Chemokine Expression Models

Leukocyte infiltration of the CNS is a feature of many inflammatory neurological disorders, ranging from bacterial and viral meningoencephalitis to multiple sclerosis, HIV-1-associated dementia, cerebral malaria and cerebral ischemia. In these disorders there is overlapping induction of the CNS expression of a number of chemokines belonging to the CXC, CC, C and CX_3C subfamilies (reviewed by [1,2,27]). Since individual chemokines show a chemotactic 'signature' then the pattern of chemokine expression likely determines the composition of leukocytes in the CNS inflammatory cell infiltrates. This assumption has been decisively confirmed by transgenic studies in which the expression of the specific chemokine genes, GRO-α/KC/N51 (CXCL1) or MCP-1 (CCL2) was targeted to the brain.

KC is an ELR-containing chemokine that belongs to the α-chemokine family and is a ligand for the CXCR2 receptor. The CNS function of this chemokine was examined by expressing CXCL1 under the control of the myelin basic protein (MBP) promoter in transgenic mice (MBP-KC) [29]. During the period of maximal transgene expression at 3–4 weeks of age, the brain of MBP-KC mice was heavily infiltrated with neutrophils. At this time, despite loss of blood–brain barrier (BBB) integrity and focal gliosis, there was no evidence of brain tissue destruction nor did affected animals display abnormal neurological signs. These results underscore the ability of KC to promote neutrophil migration to and extravasation in the CNS. However, the absence of any tissue destruction suggests that this chemokine does not activate the neutrophil. Curiously, in one high transgene expressing line, from >40 days of age mice exhibited a syndrome of progressive neurological dysfunction consisting of slowing of the righting reflex, clumsiness, increase of rigidity of the hindlimbs and tail and profound truncal instability. Despite the severity of their neurological disorder which resulted in premature death, there was no histological evidence of damage to neurons, myelin or axons. Although the basis for the neurological dysfunction developing in the MBP-KC mice is unknown, Tani et al. [29] speculate that this might be due to the glial and BBB perturbations. It is also possible as the authors suggest that chronic exposure of neurons to the chemokine may directly compromise their function. In support of such a mechanism, KC can enhance both elicited and spontaneous postsynaptic currents [30], and increase neurotransmitter release and reduce long-term depression [31], in Purkinje neurons.

A similar experimental approach was used to target the expression of MCP-1 to oligodendrocytes under the control of the MBP promoter in transgenic mice [32]. However, in contrast to KC, expression of MCP-1 in the brain produced a discrete monocytic cell infiltration that largely remained perivascular. The intensity of this infiltrate could be increased as well as some parenchymal infiltration stimulated, following systemic LPS administration to the MBP-MCP-1 transgenic mice. In all, there was no overt evidence of CNS injury in these transgenic mice indicating that while MCP-1 can provide a signal for the recruitment of monocytic cells to the CNS it is unable to activate these cells and does not promote an active inflammatory process.

Despite the limited number of chemokines examined, these transgenic studies raise a number of important points pertinent to our understanding of chemokine actions in the CNS. First, they show that chemokine production within the CNS alone is sufficient to drive the recruitment of leukocytes. Second, they demonstrate very clearly the fine specificity of chemokine action in targeting the recruitment of distinct leukocyte populations. Thus, KC targeted the recruitment of neutrophils while MCP-1 attracted monocytes. Finally, they underscore the inability of these chemokines to bring about functional activation of the leukocytes that they attracted to the brain.

3.2. Chemokine/Chemokine Receptor Gene Disruption Models

For many chemokine receptors expressed in the CNS the corresponding ligand is not present under physiological conditions in the adult brain but is induced during pathologic states. This suggests that these receptors and their ligands are primarily involved in regulating inflammatory processes and other pathophysiological responses. To date, through the use of homologous recombination techniques, the genes for a number of key chemokine receptors have been disrupted including CCR1 [33,34], CCR2 [35–37] and CCR5 [38]. Perhaps not surprisingly, there is an apparent absence of any abnormal phenotype including the CNS, observed in any of these chemokine receptor knockout mice. However, distinct defects in immune responses were noted under conditions of immune challenge in the absence of a specific chemokine receptor function. For example, CCR1 deficient mice, have abnormalities in neutrophil function that results in inefficient clearance of the fungus *Aspergillus fumigatus* [33]. In addition, compared with wild type controls, these mice form smaller lung granulomas after SEA injection suggesting a defect in the recruitment of T-cells, macrophages and eosinophils. In contrast to CCR1, disruption of CCR2 function leads to marked and largely selective abnormality in macrophage recruitment [35–37]. Thus, despite the well known redundancy of the CC chemokine receptors, ablation in the function of an individual chemokine receptor such as CCR2 can still result in very specific alterations in immune function. A number of recent reports document the use of these chemokine receptor knockout mice to study the role of these receptors in experimentally induced immune responses in the CNS.

Various chemokines are expressed in the brain during EAE, including C10 (CCL6), IP-10 (CXCL10), MIP-1α (CCL3), MCP-1, (CCL2) and RANTES (CCL5) [39–42]. There is a tight correlation between the expression of some of these chemokines, e.g. C10 [42] or MCP-1 and MIP-1α [43–45] and the distribution of inflammatory cells in the CNS and the development of clinical disease. The importance of MIP-1α in the development of EAE was inferred by the ability of anti-MIP-1α antibody to block PLP-specific T-cell transfer of EAE in SJL/J mice [46]. Moreover, DNA vaccination with naked DNA encoding MIP-1α protects Lewis rats from MBP-induced EAE [47]. The role of MIP-1α in MOG-induced EAE was further examined in C57BL/6 mice deficient either for the MIP-1α receptor CCR5 or for CCL3/MIP-1α [48]. In either case mutant mice were fully susceptible to EAE showing no differences in either incidence or severity compared with wild type animals. Therefore, these findings indicate that MIP-1α is not necessary for the initiation or progression of MOG-induced EAE. In parallel with these mouse studies, it has been reported that MS can occur in patients who are naturally deficient for CCR5 [49]. These findings clearly contradict those of the earlier studies with anti-MIP-1α antibody or naked DNA encoding MIP-1α, which suggested a key role of MIP-1α in the development of EAE. The reason for the discrepancies between these studies remains to be resolved. However, it should be noted that these separate studies each employed different EAE paradigms, different strains of animals and applied different strategies for the intervention in MIP-1α function.

In contrast to CCR5, two studies show a pivotal role of CCR2 in MOG-induced EAE. CCR2 is expressed predominantly by cells of the monocytic lineage and by activated T-cells and serves as the receptor for MCP-1–5 [50]. In addition to EAE, members of the MCP family and CCR2 are associated with inflammatory demyelinating lesions in MS [51–54]. The role of CCR2 in EAE was investigated in mice with a genetic disruption of this chemokine receptor gene. In two independent studies and in contrast to wild type control animals, CCR2 deficient mice fail to develop EAE following MOG-immunization [55,56]. Lack of clinical EAE in these mice was associated with an absence of mononuclear cell infiltrates in the brain and a complete diminution in the expression of a number of chemokine and chemokine receptor genes [56].

The findings document an essential role of CCR2 in the development of EAE, but what is the basis for this result? Although one study found that resistance to EAE in the CCR2-deficient animals is associated with a decreased antigen-specific T_h1 effector [56], this was not confirmed by the other study [55]. However, a fundamental defect in T-cell antigen activation and responsiveness seems an unlikely reason for the lack of EAE in the CCR2 mice since it was possible to transfer EAE to C57Bl/6 recipients with MOG-reactive T-cells from CCR2-deficient [55] mice. However, transfer of T-cells from the CCR2-deficient animals or wild type C57Bl/6 mice to CCR2-deficient recipients did not transfer EAE. These results suggest that the presence of CCR2 is not critical for the function of the transferred MOG-reactive T-cells. However, the failure to manifest disease in CCR2-deficient hosts firmly points to an essential role of CCR2 in the recipient mononuclear cell compartment. Therefore, it is likely that in the absence of CCR2, monocytes are unable to respond to chemotactic signaling by MCP-1 or possibly other members of the MCP-1 family — a process that obviously cannot be substituted for by another CCR. The resultant functional crippling of these cells translates to an inability to traffic and accumulate within the CNS where normally they would be involved in causing tissue damage and demyelination leading to clinical disease. This mechanism is consistent with other disease models where loss of CCR2 was found to abrogate macrophage tissue accumulation [35–37]. Recent studies show that MCP-1/CCL2 is the relevant functional ligand for CCR2 in EAE [57]. Since, as mentioned above, MCP-1 and CCR2 are implicated in the pathogenesis of MS, then antagonism of CCR2 might therefore prove to be an effective therapeutic approach.

CCRI is expressed by a number of different leukocytes including macrophages, neutrophils and lymphocytes where it binds a variety of chemokines, e.g. RANTES, MCP-2 and -3 and MIP-1α. The role of CCR1 in the progression of MOG-induced EAE has been studied in CCR1 knockout mice [58]. Compared with wild type controls, CCR1 KO mice had a lower incidence and less severe disease associated with a reduction in the number of infiltrating leukocytes and chemokines expressed in the brain. Both groups of animals exhibited similar in vitro lymphocyte responses to MOG challenge, suggesting that CCR1 KO mice were not systemically immunosuppressed. Thus, similar to CCR2, the loss of CCR1 appears to be associated with a fundamental defect in the trafficking of leukocytes to the CNS. However, in contrast with CCR2, and despite the wide number of ligands that will bind to CCR1, loss of this receptor provided only partial protection against the development of EAE.

These studies utilizing various CCR KO mice highlight necessary and non-redundant roles for CCR2 and partly CCR1 but not CCR5 in the development of EAE. On balance, the findings indicate that it is primarily the trafficking of mononuclear cells to the CNS, rather than a defect in T-cell activation and reactivity, that is disturbed in the absence of CCR1 or CCR2. Finally, based on these studies it is apparent that targeting, in particular CCR2, may be a powerful therapeutic approach in the clinic.

4. CONCLUDING PERSPECTIVES

The advent of molecular genetic approaches that permit the targeted manipulation of gene expression in an intact organism have provided powerful ap-

Table 1. Some functional and structural changes in the CNS of chemokine transgenic mice

Chemokine	Promoter/targeted cell	CNS disease	Infiltrating leukocytes	References
CCL2/MCP-1	MBP/oligodendrocytes	None	Monocytes	[32]
CXCL1/KC	MBP/oligodendrocytes	Late onset, ataxia, wasting	Neutrophils	[29]

Table 2. Some functional and structural changes in the CNS of chemokine/receptor knockout mice

Gene KO	Phenotype	EAE induction	References
CCL3/MIP-1α	No abnormalities	Fully susceptible to MOG-induced EAE	[48]
CCR1	Abnormalities in myeloid progenitor cells	Partially susceptible to MOG-induced EAE	[33,58]
CCR2	Abnormalities in macrophage recruitment	Fully resistant to MOG-induced EAE	[35–37,55,56]
CCR5	No abnormalities	Fully susceptible to MOG-induced EAE	[48]
CXCR4	Lethal–abnormal neuronal migration	nd	[9,11,12]

nd = not determined.

proaches towards understanding the actions of chemokines in the CNS in physiologic and pathologic states (see Tables 1 and 2). Promoter-driven expression of different biologically relevant chemokines in the CNS of transgenic mice highlight the potent action of these small molecules to stimulate the recruitment but not the functional activation of specific subsets of leukocytes to the CNS. While studies in knockout mice have established the essential role of SDF-1 in brain ontogenesis and the non-essential role of fractalkine in neuron–microglial cell communication. Finally, the availability of various CCR knockout mice has revealed the necessary and non-redundant roles for MCP/CCR2 signaling in EAE. What is certain is that studies in genetically manipulated mice as exemplified here, will continue to further advance our understanding of the functions of these important mediators and their receptors in the CNS and may pave the way to the identification of new targets for therapeutic intervention.

ACKNOWLEDGEMENTS

The authors studies referred to in this article were funded by USPHS grants MH 47680, MH 50426 and NS36979. This is paper number 13722-NP from the Scripps Research Institute.

REFERENCES

1. Asensio VC, Campbell IL. Chemokines and their receptors in the CNS: plurifunctional mediators in diverse states. Trends Neurosci 1999;22:504–512.
2. Mennicken F, Maki R, de Souza EB, Quirion R. Chemokines and chemokine receptors in the CNS: a possible role in neuroinflammation and patterning. Trends Pharmacol Sci 1999;20:73–77.
3. Campbell IL, Gold LH. Transgenic modeling of neuropsychiatric disorders. Mol Psychiatry 1996;1:105–120.
4. Galli-Taliadoros LA, Sedgwick JD, Wood SA, Korner H. Gene knock-out technology: a methodological overview for the interested novice. J Immunol Methods 1995;181: 1–15.
5. Tashiro K, Tada H, Heilker R, Shirozu M, Nakano T, Honjo T. Signal sequence trap: a cloning strategy for secreted proteins and type I membrane proteins. Science 1993;261:600–603.
6. Gleichmann M, Gillen C, Czardybon M et al. Cloning and characterization of SDF-1gamma, a novel SDF-1 chemokine transcript with developmentally regulated expression in the nervous system. Eur J Neurosci 2000;12:1857–1866.
7. Ohtani Y, Minami M, Kawaguchi N et al. Expression of stromal cell-derived factor-1 and CXCR4 chemokine receptor mRNAs in cultured rat glial and neuronal cells. Neurosci Lett 1998;249:163–166.
8. Jazin EE, Soderstrom S, Ebendal T, Larhammar D. Embryonic expression of the mRNA for the rat homologue of the fusin/CXCR-4 HIV-1 co-receptor. J Neuroimmunol 1997;79:148–154.
9. Zou YR, Kottmann AH, Kuroda M, Taniuchi I, Littman DR. Function of the chemokine receptor CXCR4 in haematopoiesis and in cerebellar development. Nature 1998;393:595–599.
10. Nagasawa T, Hirota S, Tachibana K et al. Defects of B-cell lymphopoiesis and bone-marrow myelopoiesis in mice lacking the CXC chemokine PBSF/SDF-1. Nature 1996;382:635–638.
11. Ma Q, Jones D, Borghesani PR et al. Impaired B-lymphopoiesis, myelopoiesis, and derailed cerebellar neuron migration in CXCR4- and SDF-1-deficient mice. Proc Natl Acad Sci USA 1998;95:9448–9453.
12. Tachibana K, Hirota S, Iizasa H et al. The chemokine receptor CXCR4 is essential for vascularization of the gastrointestinal tract. Nature 1998;393:591–594.
13. Bazan JF, Bacon KB, Hardiman G et al. A new class of membrane-bound chemokine with a CX3C motif. Nature 1997;385:640–644.
14. Pan Y, Lloyd C, Zhou H et al. Neurotactin, a membrane-anchored chemokine upregulated in brain inflammation. Nature 1997;387:611–617.
15. Harrison JK, Jiang Y, Chen S et al. Role for neuronally derived fractalkine in mediating interactions between neurons and CX3CR1-expressing microglia. Proc Natl Acad Sci USA 1998;95:10896–10901.
16. Nishiyori A, Minami M, Ohtani Y et al. Localization of fractalkine and CX3CR1 mRNAs in rat brain: does fractalkine play a role in signaling from neuron to microglia? FEBS Lett 1998;429:167–172.
17. Schwaeble WJ, Stover CM, Schall TJ et al. Neuronal expression of fractalkine in the presence and absence of inflammation. FEBS Lett 1998;439:203–207.
18. Maciejewski-Lenoir D, Chen S, Feng L, Maki R, Bacon KB. Characterization of fractalkine in rat brain cells: Migratory and activation signals for CX_3CR-1-expressing microglia. J Immunol 1999;163:1628–1635.
19. Jung S, Aliberti J, Graemmel P et al. Analysis of fractalkine receptor CX(3)CR1 function by targeted deletion and green fluorescent protein reporter gene insertion. Mol Cell Biol 2000;20:4106–4114.
20. Horuk R, Martin A, Hesselgesser J et al. The Duffy antigen receptor for chemokines: structural analysis and expression in the brain. J Leukoc Biol 1996;59:29–38.
21. Horuk R, Martin AW, Wang Z-x et al. Expression of chemokine receptors by subsets of neurons in the central nervous system. J Immunol 1997;158:2882–2890.
22. Neote K, Mak JY, Kolakowski LF, Jr, Schall TJ. Functional and biochemical analysis of the cloned Duffy antigen: identity with the red blood cell chemokine receptor. Blood 1994;84:44–52.
23. Dawson TC, Lentsch AB, Wang Z et al. Exaggerated response to endotoxin in mice lacking the Duffy

antigen/receptor for chemokines (DARC). Blood 2000; 96:1681–1684.
24. Luo H, Chaudhuri A, Zbrzezna V, He Y, Pogo AO. Deletion of the murine Duffy gene (Dfy) reveals that the Duffy receptor is functionally redundant. Mol Cell Biol 2000;20: 3097–3101.
25. Tournamille C, Colin Y, Cartron JP, Le Van Kim C. Disruption of a GATA motif in the Duffy gene promoter abolishes erythroid gene expression in Duffy-negative individuals. Nat Genet 1995;10:224–228.
26. Cacalano G, Lee J, Kikly K et al. Neutrophil and B cell expansion in mice that lack the murine IL-8 receptor homolog. Science 1994;265:682–684.
27. Ransohoff RM. Chemokines in neurological disease models: correlation between chemokine expression patterns and inflammatory pathology. J Leukoc Biol 1997;62:645–652.
28. Miller RJ, Meucci O. AIDS and the brain: is there a chemokine connection? Trends Neurosci 1999;22:471–479.
29. Tani M, Fuentes ME, Peterson JW et al. Neutrophil infiltration, glial reaction, and neurological disease in transgenic mice expressing the chemokine N51/KC in oligodendrocytes. J Clin Invest 1996;98:529–539.
30. Ragozzino D, Giovannelli A, Mileo AM, Limatola C, Santoni A, Eusebi F. Modulation of the neurotransmitter release in rat cerebellar neurons by GROBeta. Neuroreport 1998;9:3601–3606.
31. Giovannelli A, Limatola C, Ragozzino D et al. CXC chemokines interleukin-8 (IL-8) and growth-related gene product alpha (GROα) modulate Purkinje neuron activity in mouse cerebellum. J Neuroimmunol 1998;92:122–132.
32. Fuentes ME, Durham SK, Swerdel MR et al. Controlled recruitment of monocytes and macrophages to specific organs through transgenic expression of monocyte chemoattractant protein-1. J Immunol 1995;155:5769–5776.
33. Gao J-L, Wynn TA, Chang Y et al. Impaired host defense, hematopoiesis, granulomatous inflammation and type 1–type 2 cytokine balance in mice lacking CC chemokine receptor 1. J Exp Med 1997;185:1959–1968.
34. Gerard C, Frossard J-L, Bhatia M et al. Targeted disruption of the β-chemokine receptor CCR1 protects against pancreatitis-associated lung injury. J Clin Invest 1997;100: 2022–2027.
35. Kuziel WA, Morgan SJ, Dawson TC et al. Severe reduction in leukocyte adhesion and monocyte extravasation in mice deficient in CC chemokine receptor 2. Proc Natl Acad Sci USA 1997;94:12053–12058.
36. Boring L, Gosling J, Chensue SW et al. Impaired monocyte migration and reduced type 1 (Th1) cytokine responses in C-C chemokine receptor 2 knockout mice. J Clin Invest 1997;100:2552–2561.
37. Kurihara T, Warr G, Loy J, Bravo R. Defects in macrophage recruitment and host defense in mice lacking the CCR2 chemokine receptor. J Exp Med 1997;186: 1757–1762.
38. Huffnagle GB, McNeil LK, McDonald RA et al. Cutting edge: Role of C-C chemokine receptor 5 in organ specific and innate immunity to Cryptococcus neoformans. J Immunol 1999;163:4642–4646.
39. Berman JW, Guida MP, Warren J, Amat J, Brosnan CF. Localization of monocyte chemoattractant peptide-1 expression in the central nervous system in experimental autoimmune encephalomyelitis and trauma in the rat. J Immunol 1996;156:3017–3023.
40. Glabinski AR, Tani M, Strieter RM, Tuohy VK, Ransohoff RM. Synchronous synthesis of α- and β-chemokines by cells of diverse lineage in the central nervous system of mice with relapses of chronic experimental autoimmune encephalomyelitis. Am J Pathol 1997;150:617–630.
41. Godiska R, Chantry D, Dietsch GN, Gray PW. Chemokine expression in murine experimental allergic encephalomyelitis. J Neuroimmunol 1995;58:167–176.
42. Asensio VC, Lassmann S, Pagenstecher A, Steffensen SC, Henriksen SJ, Campbell IL. C10 is a novel chemokine expressed in experimental inflammatory demyelinating disorders that promotes the recruitment of macrophages to the central nervous system. Am J Pathol 1999;154:1181–1191.
43. Ransohoff RM, Hamilton TA, Tani M et al. Astrocyte expression of mRNA encoding cytokines IP-10 and JE/MCP-1 in experimental autoimmune encephalomyelitis. FASEB J 1993;7:592–600.
44. Hulkower K, Brosnan CF, Aquino DA et al. Expression of CSF-1, c-fms, and MCP-1 in the central nervous system of rats with experimental allergic encephalomyelitis. J Immunol 1993;150:2525–2533.
45. Karpus WJ, Ransohoff RM. Chemokine regulation of experimental autoimmune encephalomyelitis:temporal and spatial expression patterns govern disease pathogenesis. J Immunol 1998;161:2667–2671.
46. Karpus WJ, Lukacs NW, McRae BL, Strieter RM, Kunkel SL, Miller SD. An important role for the chemokine macrophage inflammatory protein-1α in the pathogenesis of the T cell-mediated autoimmune disease, experimental autoimmune encephalomyelitis. J Immunol 1995;155: 5003–5010.
47. Youssef S, Wildbaum G, Maor G et al. Long-lasting protective immunity to experimental autoimmune encephalomyelitis following vaccination with naked DNA encoding C-C chemokines. J Immunol 1998;161:3870–3879.
48. Tran EH, Kuziel WA, Owens T. Induction of experimental autoimmune encephalomyelitis in C57BL/6 mice deficient in either the chemokine macrophage inflammatory protein-1α or its CCR5 receptor. Eur J Immunol 2000;30: 1410–1415.
49. Bennetts BH, Teutsch SM, Buhler TM, Heard RNS, Stewart GJ. The CCR5 deletion mutation fails to protect against multiple sclerosis. Hum Immunol 1997;58:52–59.
50. Charo IF. CCR2: from cloning to the creation of knockout mice. Chem Immunol 1999;72:30–41.
51. McManus C, Berman JW, Brett FM, Staunton H, Farrell M, Brosnan CF. MCP-1, MCP-2 and MCP-3 expression in multiple sclerosis lesions: an immunohistochemical and in situ hybridization study. J Neuroimmunol 1998;86:20–29.
52. Van Der Voorn P, Tekstra J, Beelen RHJ, Tensen CP, Van Der Valk P, De Groot CJA. Expression of MCP-1 by reactive astrocytes in demyelinating multiple sclerosis lesions. Am J Pathol 1999;154:45–51.

53. Sorensen TL, Tani M, Jensen J et al. Expression of specific chemokines and chemokine receptors in the central nervous system of multiple sclerosis patients. J Clin Invest 1999;103:807–815.
54. Simpson JE, Newcombe J, Cuzner ML, Woodroofe MN. Expression of monocyte chemoattractant protein-1 and other β-chemokines by resident glia and inflammatory cells in multiple sclerosis lesions. J Neuroimmunol 1998; 84:238–249.
55. Fife BT, Huffnagle GB, Kuziel WA, Karpus WJ. CC chemokine receptor 2 is critical for induction of experimental autoimmune encephalomyelitis. J Exp Med 2000; 192:899–905.
56. Izikson L, Klein RS, Charo IF, Weiner HL, Luster AD. Resistance to experimental autoimmune encephalomyelitis in mice lacking the CC chemokine receptor (CCR)2. J Exp Med 2000;192:1075–1080.
57. Huang D, Wang J-T, Kivisakk P, Rollins BJ, Ransohoff RM. Absence of monocyte chemoattractant protein CMCP-1 leads to decreased local macrophage recruitment and antigen-specific Th1 immune response in experimental autoimmune encephalomyelitis. J Exp Med 2001;193: 713–726.
58. Rottman JB, Slavin AJ, Silva R, Weiner HL, Gerard CG, Hancock HL. Leukocyte recruitment during onset of experimental allergic encephalomyelitis is CCR1 dependent. Eur J Immunol 2000;30:2372–2377.

Section 4

CHEMOKINES AND EFFECTS ON OTHER CNS PROCESSES AND RESIDENT CELLS

Section Editor: J.K. Harrison

Universes in Delicate Balance: Chemokines and the Nervous System
R.M. Ransohoff, K. Suzuki, A.E.I. Proudfoot, W.F. Hickey and J.K. Harrison, editors

Chapter 4.1

Constitutive Roles for SDF-1/CXCR4 and Fractalkine/CX3CR1 in the CNS

Jeffrey K. Harrison

Department of Pharmacology and Therapeutics, University of Florida College of Medicine, Gainesville, FL 32610-0267, USA

1. OVERVIEW

The biological complexity of chemokines is as diverse as the gene family is numerous. Amongst the many functions for chemokines are potential roles in neuronal physiology, concepts that are well supported by the current literature. Two chemokine systems, namely SDF-1 (CXCL12) and fractalkine (CX3CL1) and their respective receptors CXCR4 and CX3CR1, are prominently and constitutively expressed in the CNS. This expression pattern suggests that these chemokine ligand–receptor pairs exhibit fundamental roles in neurophysiology. Functional insights into the specific capacities of these chemokine receptor systems in neurophysiology have also been suggested from in vitro studies, gene knockout mice, and at least for CXCR4, from analysis of the effects of HIV-1 and its gene products on the cellular function of CXCR4. Despite some progress, much more is needed to determine the precise role(s) of these chemokines in various aspects of CNS physiology.

2. HISTORICAL: IDENTIFICATION OF THE GENES ENCODING SDF-1, FRACTALKINE, CXCR4, AND CX3CR1

The identification of the genes encoding the SDF-1/CXCR4 and fractalkine/CX3CR1 ligand receptor pairs followed similar paths. In each case, cDNAs encoding the receptors were described first and reported as orphan G-protein-coupled receptors. Many GPCRs have been identified based on homology screening approaches or searches of EST databases for novel receptor sequences. The isolation of cDNAs ultimately determined to be CXCR4 and CX3CR1 were similarly explored via hybridization and/or reverse transcriptase-PCR (using degenerate oligonucleotides) techniques.

The formal identification of CXCR4 as an SDF-1 receptor was preceded by controversy as the gene product, initially isolated from a bovine brain cDNA library was incorrectly proposed to be a receptor for neuropeptide (NPY) [1]. The cDNA, termed LCR-1, was isolated via a reverse transcriptase-PCR using degenerate oligonucleotides based upon transmembranes 3 and 6 from a number of GPCRs. cDNAs encoding the human ortholog were subsequently reported by several groups and an appreciation that this receptor is expressed on peripheral leukocytes was made [2–7]. The receptor displays a widespread distribution pattern, and in fact, many cell types have been shown to express this gene. Nonetheless, the realization that cells expressing the human cDNA (fusin) supported entry of HIV-1 into cells in a CD4-dependent manner [8], increased the significance of the leukocyte expression and catapulted this receptor into the spotlight.

In recent years, the chemokine gene family has seen considerable expansion, and with this, the realization that these proteins have additional activities beyond their roles in inflammation. SDF-1/CXCL12 was initially identified from a signal sequence trap cloning strategy [9]. The SDF-1 gene has at least two distinct transcripts encoding proteins that differ in their C-termini. An additional form, SDF-1 gamma, has been characterized in the rat [10], although its relationship to α and β forms is not entirely understood. SDF-1 was initially appreciated as a bone marrow stromal cell derived chemoattractant for hematopoietic progenitor cells [11]. In 1996, two groups independently reported that SDF-1 is a ligand for the fusin/LESTR/HUMSTR receptor. The receptor was formally identified as an SDF-1 binding protein and SDF-1 dependent signaling through this receptor was established [12,13]. In keeping with established rules for the nomenclature of chemokine receptors, the receptor was given the name CXCR4.

A cDNA that was eventually characterized to be the rat fractalkine receptor (ratCX3CR1) was initially isolated from a rat brainstem cDNA library using the DNA sequences encoding the rat angiotensin receptor (rat AT1R) [14]. Ligand binding analysis of cells transfected with this cDNA (RBS11) failed to detect specific binding to a number of ligands including angiotensin, opioid, galanin, PAF, etc. At the time this receptor cDNA was isolated, the only known chemokine receptors were CXCR1 and CXCR2, formally known as IL-8RA and IL-8RB. The amino acid similarity of RBS11 to these two chemokine receptors is modest (~37%) and no more similar to these receptors than a number of other GPCRs for peptide ligands (e.g. angiotensin, opioid and somatostatin). However, the identification of CCR1 and CCR2 as receptors for CC chemokines revealed sequences of the GPCR superfamily reasonably similar to the RBS11 receptor sequence that suggested it could encode a novel chemokine receptor. Subsequent to the identification of RBS11, the human ortholog was reported independently by two groups [15,16]. The widespread tissue distribution of RBS11 (rat ortholog) and V28/CMKBRL1 (human ortholog) and the notable expression of this gene in CNS tissues provided initial clues to the potential role of chemokines in fundamental physiological processes of the CNS.

In early 1997, a unique chemokine was reported from an EST database search for novel chemokines [17]. This chemokine, termed fractalkine/CX3CL1, contains a unique three amino acid spacing between the first two highly conserved cysteine residues and defines the only known member of the CX3C subfamily of chemokines. The murine ortholog of fractalkine, termed neurotactin, was reported in the same year [18]. More notable in the structure of fractalkine, is the presence of extended structural elements, including a single transmembrane (TM) spanning domain and an approximate 240 amino acid mucin-like domain that tethers a structurally conserved chemokine domain to the TM region and a short intracellular C-terminus. The dramatic similarity in the tissue distributions of this chemokine and the RBS11 or V28/CMKBRL1 receptors suggested that these latter gene products could encode functional fractalkine receptors. Formal confirmation of this hypothesis soon followed [19]. Both receptors were identified to bind the cognate species fractalkines [19–21]; RBS11 binds human and rat fractalkine with comparable affinities. This unique chemokine also possesses the property of promoting the adhesion of CX3CR1 expressing cells [19], and subsequent reports demonstrated that fractalkine can mediate firm capture of CX3CR1 expressing cells under conditions mimicking physiological blood flow [22,23]. This phenomenon is likely a very important functional property of this chemokine in peripheral organ systems. However, the relevance of this specific property to the function of this ligand in the CNS is uncertain.

3. EXPRESSION PATTERNS IN THE CNS

Fractalkine, SDF-1, CX3CR1, and CXCR4 are prominently expressed in CNS tissue. Some controversy exists regarding the expression of fractalkine in the brain. Murine microglia were initially identified as the major cell type in the CNS expressing fractalkine [18]. However, subsequent to this first report, neurons were shown to be the predominant fractalkine expressing cell type. The in vivo neuronal expression pattern of fractalkine has been demonstrated using techniques of in situ hybridization and immunohistochemistry and has been determined in several species including humans, monkeys, and rats [21,24,25]. Possible species differences in the expression of fractalkine may exist although it is also possible that false-positive staining of the murine microglia was observed. Fractalkine receptor, CX3CR1, is expressed by microglia in vivo, an observation documented by two methods. In situ hybridization analysis has indicated microglia as the principal cell type expressing CX3CR1 [21]. In CX3CR1 null mice, replacement of the CX3CR1 coding sequences with GFP, yields GFP-positive microglial cells [26]. These latter data confirm indirectly that microglia are the principal fractalkine receptor expressing cells in the intact CNS. While microglia have been the only CNS cell type to be identified to express CX3CR1 in vivo, cultured neurons, astrocytes and microglia express CX3CR1 [27–29]. CX3CR1 protein and mRNA as well as functional responses to fractalkine have been reported in primary cultures of each of these cell types. The relevance of astrocytic and neuronal expressed CX3CR1 remains to be determined given that these cell types have not been shown to express the receptor in vivo. However, it is possible that these cells express the receptor at early stages in the development of the CNS or in response to specific neuropathological challenges.

While the expression of SDF-1 and CXCR4 are widespread and relatively abundant in the CNS, it is less clear if specific cell–cell communication networks are operating with this ligand receptor pair. Cultured neurons and glia have been identified as both SDF-1 and CXCR4 expressing cells [30–38]. Immunohistochemical and in situ hybridization analysis have confirmed both neuronal and Schwann cell expression of SDF-1 [10] while neuron and glial cell elements have been determined to express CXCR4 [39–42]. Thus there is potential for this ligand receptor system to operate in both an autocrine and paracrine manner, and contrasts to the apparently more selective role that fractalkine and CX3CR1 may play in mediating specific neuronal : microglial cell interactions. Specific cellular communication links, utilizing SDF-1 and CXCR4, may be dependent on temporal and spatial expression patterns of these genes in both normal and pathological states of the CNS. For instance, autocrine mechanisms may predominate under certain situations while specific paracrine interactions could prevail under other physiological or pathological states.

Discussion of the localization of SDF and fractalkine to elements of the CNS extends beyond the exclusive expression of these molecules to neuronal and/or glial cells. The presence of CXCR4 and CX3CR1 on leukocytic populations would give rise to increased CNS levels of the receptors under states in which leukocytic infiltration is evident, for instance injury and disease (neurodegenerative, ischemia, etc). Furthermore, cerebral vasculature endothelial cells are worthy of mention. These chemokine expressing cells are potentially very relevant in participating in physiological functions of these genes that are initiated at the blood–brain barrier. Evidence for the ability of SDF-1 to induce transmigration of CXCR4 expressing cells into the brain is supported by in vitro studies using a model of the blood–brain barrier [43]. Thus it is entirely conceivable that these chemokines could orchestrate receptor expressing cells to traffic through the CNS under normal physiological states, as part of intrinsic immune surveillance mechanisms.

4. FUNCTIONAL INSIGHTS FROM GENE-DISRUPTED MICE AND IN VITRO CELL SYSTEMS

The function of either chemokine ligand : receptor pair in the CNS is far from a clear understanding. The expression analysis summarized above provides clues into specific cellular communication networks that involve SDF-1/CXCR4 or fractalkine/CX3CR1 signaling mechanisms. However, the role of these chemokines in global functional neurophysiology requires a greater level of knowledge, revealed from further experimentation. Toward this goal, additional insights have come largely through analysis of gene-disrupted mice and studies from in vitro cell and in situ slice preparations.

SDF-1 and CXCR4 gene-disrupted mice display similar phenotypes, and die perinatally, presumably due to specific effects on organogenesis [44–47]. The mutant mice exhibit defects in B-cell lymphopoiesis, bone marrow myelopoiesis, and cardiac septal formation. Also, at this early stage of development, SDF-1- and CXCR4-deficient animals display an altered cerebellar anatomical architecture, primarily characterized by an absence of foliation that typically distinguishes normal cerebellum [46,47]. An irregular external granule cell layer and ectopically located Purkinje cells are notable features of the cerebellar cytoarchitecture in the mutant mice. This resultant phenotype suggests a role for SDF-1 and CXCR4 in normal development of the cerebellum; embryonic expression of SDF-1 and CXCR4 has been formally demonstrated [48,49]. These effects were specific for cerebellar development, at this embryonic stage, as normal neuronal architecture was observed in other areas of the CNS. Having stated this, given the widespread distribution of SDF-1 and

CXCR4 in the CNS, other critical functions of these genes in the more developed brain are likely. Additional phenotypes in the adult brain may be revealed from studies using conditional knockout animals or via pharmacological antagonism.

Phenotypes distinguishing mice deficient in CX3CR1, from their wild type pairs, has not been as forthcoming. Similar monocyte, dendritic, and microglial cell responses are seen in CX3CR1-deficient mice and wild-type animals [26]. While a role for fractalkine and CX3CR1 in neuronal : microglial cell interactions in response to nerve injury and regeneration is suggested based upon the dynamic regulation of these genes in the rat model of facial motor nerve axotomy [21], no significant differences were seen in this response in the CX3CR1 gene-disrupted mice. The characteristic increase in microglial cell numbers subsequent to peripheral nerve injury, due primarily to a local proliferative response, was seen in the fractalkine receptor-deficient mice. Both the kinetics and extent of the biphasic response were indistinguishable in the two sets of animals. Thus, a more subtle role for fractalkine in the response of the facial motor nucleus to injury is quite likely. Preliminary studies also showed that CX3CR1-null mice developed EAE that was at least as severe as that observed in wild-type animals [50]. Additional neurological challenges to fractalkine- or CX3CR1-gene-disrupted mice may also reveal other unique functional properties of this chemokine system.

Functional insights into SDF-1 and fractalkine neurobiology are also suggested from studies of the effects of these ligands on signal transduction pathways in isolated neuronal and glial cells. Specific signaling pathways downstream from CXCR4 have been identified in cultured nervous system cells and include increased calcium mobilization, a response typically seen with all chemokine receptors in a variety of cell types. Purkinje neurons in cerebellar slice preparations respond to application of SDF-1, via an increase in a slow inward current in these cells [51]. This response to SDF-1 is blocked by CNQX, suggesting an involvement of ionotropic glutaminergic receptors. The regulation of neuronal excitability indicates a potential role for SDF-1 in neurotransmission.

Effects of SDF-1 on neuronal viability have been documented. There is some controversy in that some groups report that SDF-1 can promote neuronal apoptosis [31,32,52], while other observations suggest that SDF-1 inhibits programmed cell death [27]. SDF-1 has been shown to prevent neurotoxicity induced by HIV-1 gp120. In some cases, gp120 and SDF-1 produce similar effects, although differential signaling by SDF-1 and HIV gp120 are noted in astrocytes and neurons [37]. It is conceivable that disruption of normal CXCR4 signaling by HIV-1 or its products may contribute to the dementia associated with AIDS.

Characterization of in vitro responses of CX3CR1 expressing cells to fractalkine has indicated that a number of signal transduction processes can be modulated. Fractalkine stimulates a number of downstream effects in cultured microglia, including mobilization of intracellular calcium, actin rearrangement, and activation of the kinases, MEK/ERK and Akt/PKB [21,28,53]. Intracellular calcium mobilization and effects on the cytoskeleton are likely to account for the efficacy of fractalkine in stimulating microglia chemotaxis. On the other hand, activation of cellular Akt is associated with a number of critical cellular responses including those involved in cell survival and proliferation. A role for fractalkine in cell survival is supported by data indicating that fractalkine inhibits Fas-mediated microglial cell death [53]. Overall, the specific contribution of these various signaling pathways to the function of microglia in vivo remains to be determined. Akt and ERK activation, by fractalkine, has also been observed in cultured neurons [54]. Once again, the relevance of this finding to neuronal survival in vivo is questionable. Fractalkine has also been shown to inhibit LPS induced TNF-α release from microglia [55] and therefore may be acting as an anti-inflammatory chemokine. Reciprocal effects of LPS, on microglial cell expression of CX3CR1 have also been reported. LPS treatment of microglia results in diminished microglia responses to fractalkine [56]. The concept of fractalkine and CX3CR1 acting to suppress inflammatory mechanisms is an attractive hypothesis and is supported by the relative abundance of these gene products in normal CNS tissue. On the other hand, roles for fractalkine as a proinflammatory chemokine are also supported by data in the literature. Fractalkine has been shown to stimulate matrix metaloproteinase activity and production of tissue inhibitors of MMPs (TIMPs) by cultured microglia [57]. A similar response to other proinflammatory chemokines (MCP-1/CCL2, MIP-1β/CCL4, RANTES/CCL5, and IL-8/CXCL8) was observed and this effect may contribute to the breakdown of the blood–brain barrier in diseases such as MS. A role for fractalkine in promoting migration of CX3CR1 expressing leukocytes into the injured CNS is also supported by in vitro studies of the effects of excitotoxic stimuli on the cleavage of fractalkine from the neuronal cell surface [58]. In response to excitotoxic doses of glutamate, cultured neurons release fractalkine in a matrix metalloproteinase dependent manner. An alternative, and equally attractive, hypothesis would consider these responses to constitute homeostatic mechanisms, involving fractalkine and CX3CR1, in order to limit the inflammatory response induced by other factors.

5. REGULATION IN DISEASE AND GENETIC POLYMORPHISMS

Often, insights into the physiological function of a given gene, are derived from studies of the genes in disease. Chemokines and chemokine receptors have been analyzed in a variety of neurological disorders including multiple sclerosis (and its recognized animal model, EAE) Alzheimer's disease, trauma, and ischemia (in this volume). In rat EAE, elevated levels of both CXCR4 and CX3CR1 mRNA are found in diseased spinal cords [29]. In a peripheral nerve injury model, elevated levels of CX3CR1 are found associated with microglia [21]. The expression of CXCR4 and CX3CR1 in human MS and AD has not yet been examined, and this may be due, in part, to a lack of quality reagents suitable for immunohistochemical detection. Regardless, dissection of fundamental processes regulated by chemokines, in neuropathologies where a significant peripheral inflammatory cell infiltrate is present, will not be easy and thus the interpretation of any expression data is limited. Discerning resident vs. inflammatory cell receptor expression, as well as the effect of inflammatory cell products on resident cell expression, will contribute to the difficulty in understanding physiological roles of these systems.

Polymorphisms in the SDF-1 and CX3CR1 genes have been characterized, and have been shown to affect disease outcomes in HIV-1 infected patients. A variant in the SDF-1 gene has been identified in the 3′-non-coding sequence of the gene, a region highly identical in the human and murine genes [59]. The biological consequences of this polymorphism are not known, outside of the fact that individuals with this variant display slower onset of AIDS. The basis of this phenomena is proposed to be due to increased expression of the protein product and antagonism of CXCR4, i.e. inhibition of HIV-1 entry into CXCR4 expressing T-cells. Given that this is the mechanism for the protective effect of this polymorphism, other more fundamental issues, unrelated to HIV-1 infection, are worth considering. If SDF-1 is 'overexpressed' in the CNS, what then is the effect on CXCR4 signaling in the brains of these individuals, and specifically the neurons or glial cells expressing the receptor? Perhaps a more careful neurophysiological or behavior analysis of these individuals would reveal specific neuronal functions of this ligand receptor system.

Structural variants of CX3CR1 have been identified and shown to be associated with more rapid progression to AIDS in at least one French cohort of HIV-infected patients [60]; there was no such association in a North American cohort [61]. The biochemical consequences of the altered amino acid sequence on receptor function are currently not known. While this polymorphism appears to affect AIDS progression in the French cohort, a similar question, as proposed above for the SDF-1 variant, arises as to whether or not there are functional consequences associated with this receptor variant. Once again, a neurophysiological analysis of these individuals, or animal models made to express these variants, could reveal insights into the function of fractalkine in the CNS.

6. CONCLUDING REMARKS

Roles for chemokines in fundamental neurophysiological processes are clearly indicated from data derived from studies of the two highlighted chemokine systems. Both SDF-1 and fractalkine, and their respective receptors, are prominently expressed in nervous system tissues. A specific intercellular communication network, between neurons and microglia, is suggested by the predominant expression of fractalkine on neurons and CX3CR1 on microglia. However, the function of this system in the CNS is far from a clear understanding. Disruption of these genes in the mouse has not revealed any specific neurophysiological phenotype. Thus, we are left with translating information derived from in vitro studies of the effect of fractalkine on receptor expressing cells in culture, to what is happening in vivo. Many signal transduction pathways are activated by fractalkine in cultured nervous system cells, including microglia, astrocytes, and neurons, and these mechanisms are shared by other chemokines acting on other cell types, most notably peripheral leukocytes. Given the activation of survival signals in CX3CR1 expressing cells, the anti-inflammatory actions of fractalkine on cultured microglia, and the prominent, constitutive expression of fractalkine and CX3CR1, it is attractive to suggest that this ligand receptor pair regulates homeostatic processes of the CNS. Similarly, the relative abundance of SDF-1 and CXCR4 in the CNS suggests fundamental roles for this chemokine system in neurophysiology. These clearly include roles for SDF-1 and CXCR4 in the developing nervous system as a very notable alteration in cerebellar development was revealed from analysis of SDF-1 and CXCR4 gene-disrupted mice. However, it also seems certain that additional functions of this chemokine system will be revealed from additional experimentation. Pleiotropic roles for SDF-1 in the CNS seem likely given the expression pattern of this chemokine and its receptor, CXCR4, on cells of the CNS, i.e. the multiple cell types expressing each of these genes. A specific role for SDF-1 in neurotransmission is suggested from in vitro studies, as are roles in neuronal viability. This field of chemokine research is at an exciting stage and investigators are poised to make valuable contributions toward understanding other, non-traditional roles for chemokines, as well as enhancing our knowledge of fundamental physiological processes of the brain.

REFERENCES

1. Rimland J, Xin W, Sweetnam P, Saijoh K, Nestler EJ, Duman RS. Sequence and expression of a neuropeptide Y receptor cDNA. Mol Pharmacol 1991;40:869–875.
2. Jazin EE, Yoo H, Blomqvist AG, Yee F, Weng G, Walker MW, Salon J, Larhammar D, Wahlestedt C. A proposed bovine neuropeptide Y (NPY) receptor cDNA clone, or its human homologue, confers neither NPY binding sites nor NPY responsiveness on transfected cells. Regul Pept 1993;47:247–258.
3. Herzog H, Hort YJ, Shine J, Selbie LA. Molecular cloning, characterization, and localization of the human homolog to the reported bovine NPY Y3 receptor: lack of NPY binding and activation. DNA Cell Biol 1993;12: 465–471.
4. Federsppiel B, Melhado IG, Duncan AM, Delaney A, Schappert K, Clark-Lewis I, Jirik FR. Molecular cloning of the cDNA and chromosomal localization of the gene for a putative seven-transmembrane segment (7-TMS) receptor isolated from human spleen. Genomics 1993;16: 707–712.
5. Nomura H, Nielsen BW, Matsushima K. Molecular cloning of cDNAs encoding a LD78 receptor and putative leukocyte chemotactic peptide receptors. Int Immunol 1993;5:1239–1249.
6. Loetscher M, Geiser T, O'Reilly T, Zwahlen R, Baggiolini M, Moser B. Cloning of a human seven-transmembrane domain receptor, LESTR, that is highly expressed in leukocytes. J Biol Chem 1994;269:232–237.
7. Raport CJ, Schweickart VL, Chantry D, Eddy RL Jr, Shows TB, Godiska R, Gray PW. New members of the chemokine receptor gene family. J Leukocyte Biol 1996; 59:18–23.
8. Feng Y, Broder CC, Kennedy PE, Berger EA. HIV-1 entry cofactor: functional cDNA cloning of a seven-transmembrane, G protein-coupled receptor. Science 1996; 272:872–877.
9. Tashiro K, Tada H, Heilker R, Shirozu M, Nakano T, Honjo T. Signal sequence trap: a cloning strategy for secreted proteins and type I membrane proteins. Science 1993;261:600–603.
10. Gleichmann M, Gillen C, Czardybon M, Bosse F, Greiner-Petter R, Auer J, Muller HW. Cloning and characterization of SDF-1gamma, a novel SDF-1 chemokine transcript with developmentally regulated expression in the nervous system. Eur J Neurosci 2000;12:1857–1866.
11. Aiuti A, Webb IJ, Bluel C, Springer T, Gutierrez-Ramos JC. The chemokine SDF-1 is a chemoattractant for human $CD34^+$ hematopoietic progenitor cells and provides a new mechanism to explain the mobilization of $CD34^+$ progenitors to peripheral blood. J Exp Med 1997;185: 111–120.
12. Oberlin E, Amara A, Bachelerie F, Bessia C, Virelizier JL, Arenzana-Seisdedos F, Schwartz O, Heard JM, Clark-Lewis I, Legler DF, Loetscher M, Baggiolini M, Moser B. The CXC chemokine SDF-1 is the ligand for LESTR/fusin and prevents infection by T-cell-line-adapted HIV-1. Nature 1996;382:833–835.
13. Bleul CC, Farzan M, Choe H, Parolin C, Clark-Lewis I, Sodroski J, Springer TA. The lymphocyte chemoattractant SDF-1 is a ligand for LESTR/fusin and blocks HIV-1 entry. Nature 1996;382:829–833.
14. Harrison JK, Barber CM, Lynch KR. cDNA cloning of a G-protein coupled receptor expressed in rat spinal cord and brain related to chemokine receptors. Neurosci Lett 1994;169:85–89.
15. Raport CJ, Schweickart VL, Eddy RL Jr, Shows TB, Gray PW. The orphan G-protein-coupled receptor-encoding gene *V28* is closely related to genes for chemokine re-

ceptors and is expressed in lymphoid and neural tissues. Gene 1995;163:295–299.

16. Combadiere C, Ahuja SK, Murphy PM. Cloning, chromosomal localization, and RNA expression of a human β chemokine receptor-like gene. DNA Cell Biol 1995;14: 673–680.
17. Bazan JF, Bacon KB, Hardiman G, Wang W, Soo K, Rossi D, Greaves DR, Zlotnik A, Schall TJ. A new class of membrane-bound chemokine with a CX3C motif. Nature 1997;385:640–644.
18. Pan Y, Lloyd C, Zhou H, Dolich S, Deeds J, Gonzalo JA, Vath J, Gosselin M, Ma J, Dussault B, Woolf E, Alperin G, Culpepper J, Gutierrez-Ramos JC, Gearing D. Neurotactin, a membrane-anchored chemokine upregulated in brain inflammation. Nature 1997;387:611–617.
19. Imai T, Hieshima K, Haskell C, Baba M, Nagira M, Kakizaki M, Takagi S, Nomiyama H, Schall TJ, Yoshie O. Identification and molecular characterization of fractalkine receptor CX3CR1, which mediates both leukocyte migration and adhesion. Cell 1997;91:521–530.
20. Combadiere C, Salzwedel K, Smith ED, Tiffany HL, Berger EA, Murphy PM. J Biol Chem 1998;273:23799–23804.
21. Harrison JK, Jiang Y, Chen S, Xia Y, Maciejewski D, McNamara RK, Streit WJ, Salafranca MN, Adhikari S, Thompson DA, Botti P, Bacon KB, Feng L. Role for neuronally-derived fractalkine in mediating interactions between neurons and CX3CR1-expressing microglia. Proc Natl Acad Sci USA 1998;95:10896–10901.
22. Fong AM, Robinson LA, Steeber DA, Tedder TF, Yoshie O, Imai T, Patel DD. Fractalkine and CX3CR1 mediate a novel mechanism of leukocyte capture, firm adhesion, and activation under physiologic flow. J Exp Med 1998;188: 1413–1419.
23. Haskell CA, Cleary MD, Charo IF. Molecular uncoupling of fractalkine-mediated cell adhesion and signal transduction; rapid flow arrest of CX3CR1-expressing cells is independent of G-protein activation. J Biol Chem 1999; 274:10053–10058.
24. Nishiyori A, Masbumi M, Ohtani Y, Takami S, Yamamoto J, Kawaguchi N, Kume T, Akaike A, Satoh M. Localization of fractalkine and CX3CR1 mRNAs in rat brain: does fractalkine paly a role in signaling from neuron to microglia. FEBS Lett 1998;429:167–172.
25. Schwaeble WJ, Stover CM, Schall TJ, Dairaghi DJ, Trinder PKE, Linington C, Iglesias A, Schubart A, Lynch NJ, Weihe E, Schafer MKH. Neuronal expression of fractalkine in the presence and absence of inflammation. FEBS Lett 1998;439:203–207.
26. Jung S, Aliberti J, Graemmel P, Sunshine MJ, Kreutzberg GW, Sher A, Littman DR. Analysis of fractalkine receptor CX(3)CR1 function by targeted deletion and green fluorescent protein reporter gene insertion. Mol Cell Biol 2000;20:4106–4114.
27. Meucci O, Fatatis A, Simen AA, Bushell TJ, Gray PW, Miller RJ. Chemokines regulate hippocampal neuronal signaling and gp120 neurotoxicity. Proc Natl Acad Sci USA 1998;95:14500–14505.
28. Maciejewski-Lenoir D, Chen S, Feng L, Maki R, Bacon KB. Characterization of fractalkine in rat brain cells: migratory and activation signals for CX3CR-1-expressing microglia. J Immunol 1999;163:1628–1635.
29. Jiang Y, Salafranca MN, Adhikari S, Xia Y, Feng L, Sontag MK, deFiebre CM, Pennell NA, Streit WJ, Harrison JK. Chemokine receptor expression in cultured glia and rat experimental allergic encephalomyelitis. J Neuroimmunol 1998;86:1–12.
30. Hesselgesser J, Halks-Miller M, DelVecchio V, Peiper SC, Hoxie J, Kolson DL, Taub D, Horuk R. CD4-independent association between HIV-1 gp120 and CXCR4: functional chemokine receptors are expressed in human neurons. Curr Biol 1997;7:112–121.
31. Kaul M, Lipton SA. Chemokines and activated macrophages in HIV gp120-induced neuronal apoptosis. Proc Natl Acad Sci USA 1999;96:8212–8216.
32. Zheng J, Thylin MR, Ghorpade A, Xiong H, Persidsky Y, Cotter R, Niemann D, Che M, Zeng YC, Gelbard HA, Shepard RB, Swartz JM, Gendelman HE. Intracellular CXCR4 signaling, neuronal apoptosis and neuropathogenic mechanisms of HIV-1-associated dementia. J Neuroimmunol 1999;98:185–200.
33. Klein RS, Williams KC, Alvarez-Hernandez X, Westmoreland S, Force T, Lackner AA, Luster AD. Chemokine receptor expression and signaling in macaque and human fetal neurons and astrocytes: implications for the neuropathogenesis of AIDS. J Immunol 1999;163:1636–1646.
34. Bajetto A, Bonavia R, Barbero S, Piccioli P, Costa A, Florio T, Schettini G. Glial and neuronal cells express functional chemokine receptor CXCR4 and its natural ligand stromal cell-derived factor 1. J Neurochem 1999; 73:2348–2357.
35. McManus CM, Weidenheim K, Woodman SE, Nunez J, Hesselgesser J, Nath A, Berman JW. Chemokine and chemokine-receptor expression in human glial elements: induction by the HIV protein, Tat, and chemokine autoregulation. Am J Pathol 2000;156:1441–1453.
36. Pandey V, Bolsover SR. Immediate and neurotoxic effects of HIV protein gp120 act through CXCR4 receptor. Biochem Biophys Res Commun 2000;274:212–215.
37. Lazarini F, Casanova P, Tham TN, De Clercq E, Arenzana-Seisdedos F, Baleux F, Dubois-Dalcq M. Differential signalling of the chemokine receptor CXCR4 by stromal cell-derived factor 1 and the HIV glycoprotein in rat neurons and astrocytes. Eur J Neurosci 2000;12:117–125.
38. Coughlan CM, McManus CM, Sharron M, Gao Z, Murphy D, Jaffer S, Choe W, Chen W, Hesselgesser J, Gaylord H, Kalyuzhny A, Lee VM, Wolf B, Doms RW, Kolson DL. Expression of multiple functional chemokine receptors and monocyte chemoattractant protein-1 in human neurons. Neuroscience 2000;97:591–600.
39. Wong ML, Xin WW, Duman RS. Rat LCR1: cloning and cellular distribution of a putative chemokine receptor in brain. Mol Psychiatry 1996;1:133–140.
40. Lavi E, Strizki JM, Ulrich AM, Zhang W, Fu L, Wang Q, O'Connor M, Hoxie JA, Gonzalez-Scarano F. CXCR-4 (Fusin), a co-receptor for the type 1 human immunodeficiency virus (HIV-1), is expressed in the human brain in a variety of cell types, including microglia and neurons. Am J Pathol 1997;151:1035–1042.

41. Vallat AV, De Girolami U, He J, Mhashilkar A, Marasco W, Shi B, Gray F, Bell J, Keohane C, Smith TW, Gabuzda D. Localization of HIV-1 co-receptors CCR5 and CXCR4 in the brain of children with AIDS. Am J Pathol 1998; 152:167–178.
42. Westmoreland SV, Rottman JB, Williams KC, Lackner AA, Sasseville VG. Chemokine receptor expression on resident and inflammatory cells in the brain of macaques with simian immunodeficiency virus encephalitis. Am J Pathol 1998;152:659–665.
43. Wu DT, Woodman SE, Weiss JM, McManus CM, D'Aversa TG, Hesselgesser J, Major EO, Nath A, Berman JW. Mechanisms of leukocyte trafficking into the CNS. J Neurovirol 2000;6:S82–5.
44. Nagasawa T, Hirota S, Tachibana K, Takakura N, Nishikawa S, Kitamura Y, Yoshida N, Kikutani H, Kishimoto T. Defects of B-cell lymphopoiesis and bone-marrow myelopoiesis in mice lacking the CXC chemokine PBSF/SDF-1. Nature 1996;382:635–638.
45. Tachibana K, Hirota S, Iizasa H, Yoshida H, Kawabata K, Kataoka Y, Kitamura Y, Matsushima K, Yoshida N, Nishikawa S, Kishimoto T, Nagasawa T. The chemokine receptor CXCR4 is essential for vascularization of the gastrointestinal tract. Nature 1998;393:591–594.
46. Zou Y-R, Kottmann AH, Kuroda M, Taniuchi I, Littman DR. Function of the chemokine receptor CXCR4 in haematopoiesis and in cerebellar development. Nature 1998;393:595–599.
47. Ma Q, Jones D, Borghesani PR, Segal RA, Nagasawa T, Kishimoto T, Bronson RT, Springer TA. Impaired B-lymphopoiesis, myelopoiesis, and derailed cerebellar neuron migration in CXCR4- and SDF-1-deficient mice. Proc Natl Acad Sci USA 1998;95:9448–9453.
48. Jazin EE, Soderstrom S, Ebendal T, Larhammar D. Embryonic expression of the mRNA for the rat homologue of the fusin/CXCR-4 HIV-1 co-receptor. J Neuroimmunol 1997;79:148–154.
49. McGrath KE, Koniski AD, Maltby KM, McGann JK, Palis J. Embryonic expression and function of the chemokine SDF-1 and its receptor, CXCR4. Dev Biol 1999;213:442–456.
50. Haskell CA, Hancock WW, Salant DJ, Gao W, Csizmadia V, Peters W, Faia K, Fituri O, Rottman JB, Charo IF. Targeted deletion of CX3CR1 reveals a role for fractalkine in cardiac allograft rejection. J Clin Invest 2001;108:679–688.
51. Limatola C, Giovannelli A, Maggi L, Ragozzino D, Castellani L, Ciotti MT, Vacca F, Mercanti D, Santoni A, Eusebi F. SDF-1alpha-mediated modulation of synaptic transmission in rat cerebellum. Eur J Neurosci 2000;12: 2497–2504.
52. Hesselgesser J, Taub D, Baskar P, Greenberg M, Hoxie J, Kolson DL, Horuk R. Neuronal apoptosis induced by HIV-1 gp120 and the chemokine SDF-1 alpha is mediated by the chemokine receptor CXCR4. Curr Biol 1998;8: 595–598.
53. Boehme SA, Lio FM, Maciejewski-Lenoir D, Bacon KB, Conlon PJ. The chemokine fractalkine inhibits Fas-mediated cell death of brain microglia. J Immunol 2000; 165:397–403.
54. Meucci O, Fatatis A, Simen AA, Miller RJ. Expression of CX3CR1 chemokine receptors on neurons and their role in neuronal survival. Proc Natl Acad Sci USA 2000;97: 8075–8080.
55. Zujovic V, Benavides J, Vige X, Carter C, Taupin V. Fractalkine modulates TNF-alpha secretion and neurotoxicity induced by microglial activation. Glia 2000;29:305–315.
56. Boddeke EW, Meigel I, Frentzel S, Biber K, Renn LQ, Gebicke-Harter P. Functional expression of the fractalkine (CX3C) receptor and its regulation by lipopolysaccharide in rat microglia. Eur J Pharmacol 1999;374:309–313.
57. Cross AK, Woodroofe MN. Chemokine modulation of matrix metalloproteinase and TIMP production in adult rat brain microglia and a human microglial cell line in vitro. Glia 1999;28:183–189.
58. Chapman GA, Moores K, Harrison D, Campbell CA, Stewart BR, Strijbos PJ. Fractalkine cleavage from neuronal membranes represents an acute event in the inflammatory response to excitotoxic brain damage. J Neurosci 2000;20:RC87.
59. Winkler C, Modi W, Smith MW, Nelson GW, Wu X, Carrington M, Dean M, Honjo T, Tashiro K, Yabe D, Buchbinder S, Vittinghoff E, Goedert JJ, O'Brien TR, Jacobson LP, Detels R, Donfield S, Willoughby A, Gomperts E, Vlahov D, Phair J, O'Brien SJ. Genetic restriction of AIDS pathogenesis by an SDF-1 chemokine gene variant. Science 1998;279:389–393.
60. Faure S, Meyer L, Costagliola D, Vaneensberghe C, Genin E, Autran B, Delfraissy JF, McDermott DH, Murphy PM, Debre P, Theodorou I, Combadiere C. Rapid progression to AIDS in HIV+ individuals with a structural variant of the chemokine receptor CX3CR1. Science 2000;287: 2274–2277.
61. McDermott DH, Colla JS, Kleeberger CA, Plankey M, Rosenberg PS, Smith ED, Zimmerman PA, Combadiere C, Leitman SF, Kaslow RA, Goedert JJ, Berger EA, O'Brien TR, Murphy PM. Genetic polymorphism in CX3CR1 and risk of HIV disease. Science 2000;290:2031a.

Universes in Delicate Balance: Chemokines and the Nervous System
R.M. Ransohoff, K. Suzuki, A.E.I. Proudfoot, W.F. Hickey and J.K. Harrison, editors

Chapter 4.2

The Role of the Chemokine GRO-1 in the Development of Vertebrate CNS Glial Cells

Robert H. Miller [1] and Shenandoah Robinson [2]

[1] *Department of Neurosciences, Case Western Reserve University School of Medicine, University Hospitals of Cleveland, Cleveland, OH 44106, USA;* [2] *Department of Neurosurgery, University Hospitals of Cleveland, Cleveland, OH 44106, USA*

1. INTRODUCTION

Normal functioning of the vertebrate central nervous system (CNS) is critically dependent upon the generation of matching populations of neurons and glial cells. One of the clearest demonstrations of balanced numbers of neurons and glia occurs with the formation of myelin, the fatty insulation of axons required for the rapid saltatory conduction of action potentials along their length. Within the CNS myelin is produced by oligodendrocytes and how the number of oligodendrocytes is regulated in particular regions of the emerging vertebrate central nervous system is not well understood. Here we discuss recent data suggesting that the chemokine GRO-1 CXCL1 plays critical roles in regulating local oligodendrogenesis in both development and disease.

2. DEVELOPMENT OF OLIGODENDROCYTE PRECURSORS

During development all the major classes of cells in the vertebrate CNS arise from neuroepithelial cells of the neural tube. In any specific CNS region the first cell type to develop are neurons followed by astrocytes and the last cell type to develop is oligodendrocytes. In caudal regions of the CNS such as the spinal cord, spatially restricted domains within the neuroepithelium give rise to distinct cell types [1]. For example, cells in ventral midline regions of the neural tube give rise to motor neurons [2] and subsequently oligodendrocyte precursors [3–6] while cells in dorsal regions of the neural tube give rise to interneurons and astrocytes [3]. Much of molecular signaling underlying the spatial patterning of cell type specification in the CNS is beginning to be understood [7,8]. By contrast, temporal patterning of neural cell emergence is poorly understood.

It is perhaps not surprising that cells of the oligodendrocyte lineage would be generated later than neurons, since the primary function of this lineage is to produce myelin that wraps the axons of a variable number of long projection neurons [9] and enhances conduction velocity. Likewise, a similar localized source of neurons and oligodendrocytes would ensure that both cells had their appropriate partners with which to interact during subsequent development. Indeed, several lines of evidence suggest that individual progenitor cells in the early neural tube may give rise to both neurons and oligodendrocyte precursors [10,11].

In vitro studies have defined a series of distinct stages in the development of the rodent oligodendrocyte lineage [12]. Immature oligodendrocyte precursors can be identified by labeling with the monoclonal antibody (mAb) A2B5 [13], more mature precursors can be identified by the binding of mAb O4 [14] and differentiated cells can be identified by labeling with antibodies to galactocerebroside, one of the major glycolipids in myelin [15]. These antibodies can also be used to isolate essentially pure populations of the relevant cell types by cell sorting or immuno-panning approaches [16,17].

The spatially restricted ventricular origin of oligodendrocyte precursors mandates that successful myelination of white matter depends on the subsequent migration of progenitor cells from their source [18]. In vitro and transplant analyses suggest that the majority of long distance migration of oligodendrocyte lineage cells is accomplished by immature precursors that have a characteristic bipolar morphology and express cell surface antigens recognized by the mAbA2B5 [19,20]. How this migration is regulated is unclear although several candidate molecules including integrins [21], growth factors [22] and the chemokine GRO-1 (now termed CXCL1) have been proposed.

The majority of proliferation in the oligodendrocyte lineage occurs after cells have migrated away from the ventricular midline and entered the presumptive white matter [18,23]. Immature oligodendrocyte precursors are the most active proliferative phase of the oligodendrocyte lineage and a significant subpopulation utilize platelet derived growth factor (PDGF-AA) as their major mitogen [24,25]. Platelet derived growth factor binds to PDGFαR on the surface of immature oligodendrocyte precursors which may be the predominant cell population expressing the PDGFαR in the postnatal CNS [4,26]. Not only is PDGF a major mitogen for immature oligodendrocyte precursors, but it is also an important survival factor and provides a chemoattractant stimulus [22]. More mature precursors lose their response to PDGF, proliferate primarily in response to basic fibroblast growth factor (bFGF) [14,27,28] and are not particularly migratory [20] suggesting that the sequential action of multiple factors regulate oligodendrocyte development.

The proliferation of oligodendrocyte precursors in developing white matter is spatially and temporally regulated [23,29] to ensure that all projection axons are myelinated effectively. In ventral regions of the rat spinal cord most oligodendrocyte precursor proliferation is complete and myelination begins in the first postnatal week. By contrast, in dorsal regions such as the cortical spinal tract, maximal oligodendrocyte proliferation and myelination does not occur until the second or third postnatal week [29,30] implying an influence of the local environment on the generation of oligodendrocytes. Elevated local proliferation of oligodendrocyte precursors also occurs in a number of pathological conditions in the adult CNS [31–33] suggesting that environmental factors contribute to spatially localized oligodendrocyte precursor proliferation even in the adult CNS. This localization of oligodendrocyte proliferation is not simply a reflection of the differential distribution of PDGF. Platelet derived growth factor is synthesized by neurons [34] and astrocytes [35] and is ubiquitously distributed in the developing spinal cord [36] indicating that additional signals regulate the local proliferation of oligodendrocyte precursors [37]. One of these signals is the chemokine GRO-1.

3. CHEMOKINES AND DEVELOPMENT

The chemokine growth regulated oncogene-1, previously known as GRO-α was initially identified and termed KC, a PDGF inducible gene expressed in quiescent murine NIH 3T3 fibroblasts [38]. Subsequently, human, rat and hamster orthologs were identified and their characterization has provided insights into the functions of GRO peptides. For example, a gene overexpressed in human diploid fibroblasts, rendered tumorigenic by cDNAs from Chinese hamster ovary cells [39], but expressed in lower levels in non-tumorigenic sibling cell lines was designated ‘gro’. Independently, a peptide, designated melanocyte growth stimulatory activity (MGSA), secreted from human melanoma cells that possessed autocrine growth-promoting properties was identified [40]. The molecular identity of GRO-1 and MGSA and their homology to KC were recognized early [41] and a family of closely related human genes was characterized [42]. An avian homolog of GRO-1, 9E3/CEF-4 was isolated from chick embryo fibroblasts during exponential growth and shown to be expressed in developing wings and in tissues undergoing neovascularization [43]. It was suggested that CEF-4 might be involved in regulating angiogenesis in the avian system; an extension of this hypothesis to the mammalian CXC chemokines IL-8 and GRO-1 has gained substantial support [44]. Growth regulation by GRO peptides has now been described for cells of diverse lineages including fibroblasts, melanocytes and hemopoietic progenitors [45]. In parallel, a substantial literature describing the roles of these molecules in inflammatory responses has developed [46,47].

The roles of CXC chemokines in the CNS are only beginning to be examined. Elevated levels of CXC chemokine expression in the CNS are found in infection [48], tumors [49] ischemia [50] and demyelination [51]. Overexpression of GRO-1 in oligodendrocytes induced neutrophil invasion and astrogliosis [52]. Similarly, leukocyte infiltration and blood–brain barrier breakdown occurred following overexpression of a CXC-chemokine by astrocytes [53]. Recently, it has become clear that GRO-1 also acts directly on neural cell lineages in the developing CNS [17,54].

4. GRO-1 SYNERGIZES WITH PDGF TO PROMOTE THE PROLIFERATION OF OLIGODENDROCYTE PRECURSORS

The ability to isolate purified populations of oligodendrocyte precursors facilitated detailed characterization of their proliferative response to distinct mitogenic stimuli. The proliferative response of purified immature oligodendrocyte precursors to PDGF was enhanced four fold by addition of conditioned medium from neonatal rat spinal cord, even though conditioned medium alone had little inherent mitogenic capacity [17]. Fractionation of conditioned medium indicated that the active fraction had a mass of less than 10 kDa consistent with the size of a typical chemokine [17].

The chemokine GRO-1 has established mitogenic activity [39,40,55], and a number of independent observations provide compelling evidence of a role for GRO-1 in control of $A2B5^+$ oligodendrocyte precursor proliferation [17,54]. In cultures of purified $A2B5^+$ cells addition of 0.5 ng/ml GRO-1 resulted in a 4-fold increase in proliferation over that seen with PDGF alone as determined using a BrdU incorporation assay (Fig. 1). The synergistic effect between GRO-1 and PDGF was totally abolished by the addition of 2 μg/ml neutralizing anti-GRO antibodies (Fig. 1) and the neutralizing effects of the anti-GRO-1 antibodies were reversed by addition of 100 ng/ml GRO-1. More importantly, addition of neutralizing anti-GRO-1 antibodies to biologically active spinal cord conditioned medium essentially abolished the synergistic effect on PDGF-mediated proliferation of oligodendrocyte precursors (Fig. 1).

In the absence of PDGF, GRO-1 had no effect on precursor cell proliferation indicating that PDGF is essential for $A2B5^+$ cell proliferation. In the presence of 10 ng/ml PDGF, GRO-1 elicited a characteristic concentration-dependent proliferative response with a peak at 0.5 ng/ml GRO-1 (Fig. 2). This relatively narrow concentration dependence suggests that GRO-1 enhancement of PDGF-driven oligodendrocyte precursor proliferation is precisely regulated by

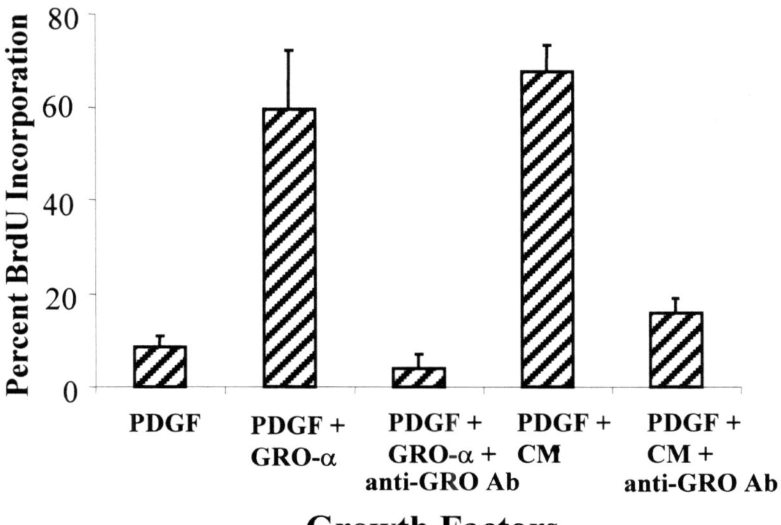

Figure 1. Quantification of the effect of GRO-1 on the proliferation of immature spinal cord oligodendrocyte precursors. Addition of 0.5 ng/ml GRO-1 (GRO-α) enhances PDGF-driven proliferation to the same extent as spinal cord conditioned medium (CM). The amplification of proliferation by both GRO-1 and CM is blocked by addition of anti-GRO antibody, suggesting that the active factor in CM is GRO-1. Data represent the mean $\pm$ standard deviation.

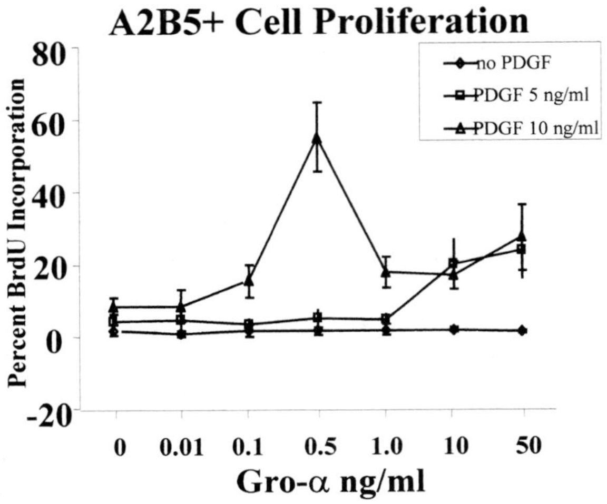

Figure 2. The GRO-1 enhancement of oligodendrocyte precursor proliferation requires a threshold of PDGF and is optimal at a concentration of 0.5 ng/ml GRO-1. At 10 ng/ml PDGF there is a greater than 4-fold increase in BrdU incorporation seen with 0.5 ng/ml GRO-1.

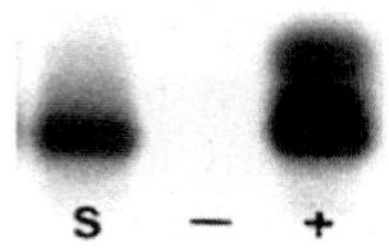

Figure 3. Neonatal rat spinal cord astrocytes contain GRO-1 mRNA. Cultures of purified spinal cord astrocytes were subjected to RT-PCR analysis with GRO-1 specific primers. The sample (s) had a strong band of the appropriate size that co-migrated with the major band detected in extracts of LPS-treated murine macrophages (+). No band was detected in control (−) samples from unstimulated macrophages.

Table 1. Spinal cord and astrocyte conditioned media contain biologically active concentrations of GRO-α

Culture	GRO-α (ng/ml)	GRO-β (ng/ml)	IL-8 (ng/ml)
Astrocytes	0.463	0.022	0
Astrocytes	0.370	0	0
Whole spinal cord	0.424	0.026	0

The concentrations of three different chemokines in the conditioned media from whole spinal cord were determined by ELISA assays. The conditioned media contained 0.5 ng/ml GRO-1 (GRO-α), low levels of GRO-β and no detectable IL-8.

local availability of chemokine. Indeed, cell intrinsic properties as well as ligand concentrations regulate glial precursor responses. For example, more mature ($O4^+$) oligodendrocyte precursors are refractory to the effects of PDGF and GRO-1, consistent with previous studies indicating that PDGF is not a major mitogen for these cells [14,27].

5. A SUBSET OF SPINAL CORD ASTROCYTES IS THE CELLULAR SOURCE OF GRO-1

The major GRO-1-producing cell type in spinal cord cultures is type-1 astrocytes [17]. Analyses of pure type-1 astrocyte cultures by RT/PCR using GRO-1-specific primers demonstrated a single band at the expected size by Southern blotting (Fig. 3). Double labeling of astrocyte cultures with antibodies to GFAP and GRO-1 indicated that a subpopulation (approx. 25%) of the $GFAP^+$ cells were GRO-1 immunoreactive. The $GRO\text{-}1^+$ cells had no obvious distinctive morphological or other phenotypic characteristics (Fig. 4).

Not only do cultured type-1 astrocytes express GRO-1 but they also release biologically relevant quantities into the culture medium. Chemokine-specific ELISA assays [56] indicated that conditioned medium from astrocyte or whole spinal cord cultures contained concentrations of GRO-1 (0.5 ng/ml), similar to that which induced maximal increases in BrdU incorporation in the in vitro dose–response assay (Table 1). By contrast, conditioned media from both cultures contained only low levels of GRO-β and the negative control human chemokine, IL-8, was not detectable.

6. THE DISTRIBUTION OF $GRO\text{-}1^+$ SPINAL CORD ASTROCYTES IS DEVELOPMENTALLY REGULATED

During development, the proliferation of oligodendrocyte precursors occurs in ventral white matter considerably earlier than in dorsal white matter [18] and is temporally correlated with the local expression of GRO-1. In the first postnatal week, the majority of $GRO\text{-}1^+$ astrocytes were located in ventral spinal cord white matter while fewer $GRO\text{-}1^+$ cells were located in dorsal white matter [17]. In gray matter, $GRO\text{-}1^+$ cells were present both ventrally and dorsally, some of which appeared to be neurons. During

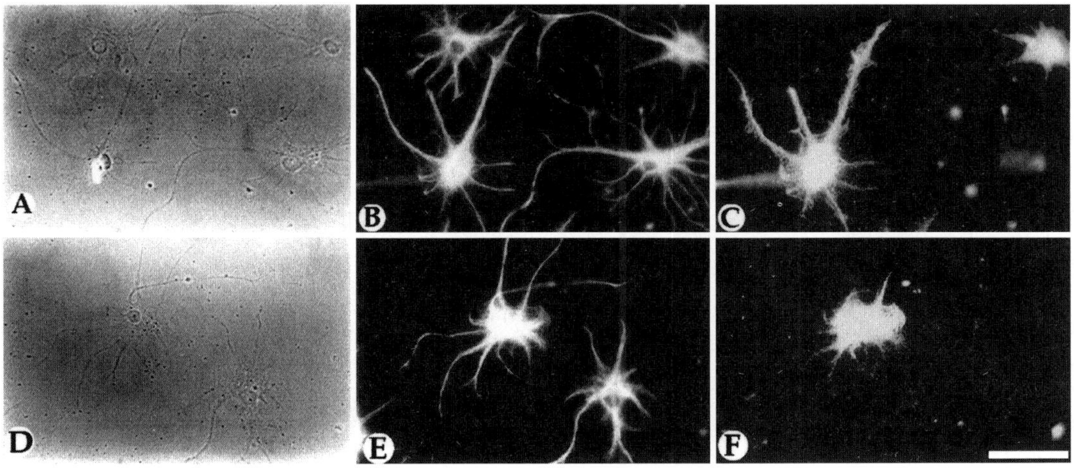

Figure 4. A subpopulation of spinal cord astrocytes contains GRO-1 protein. Cultures of neonatal rat spinal cord astrocytes were double labeled with anti-GFAP antibodies (B,E) and anti-GRO-1 antibodies (C,F). Approximately 25% of the cells in the cultures labeled strongly with the anti-GRO-1 antibodies. A,D are corresponding phase contrast micrographs. Bar = 30 μm.

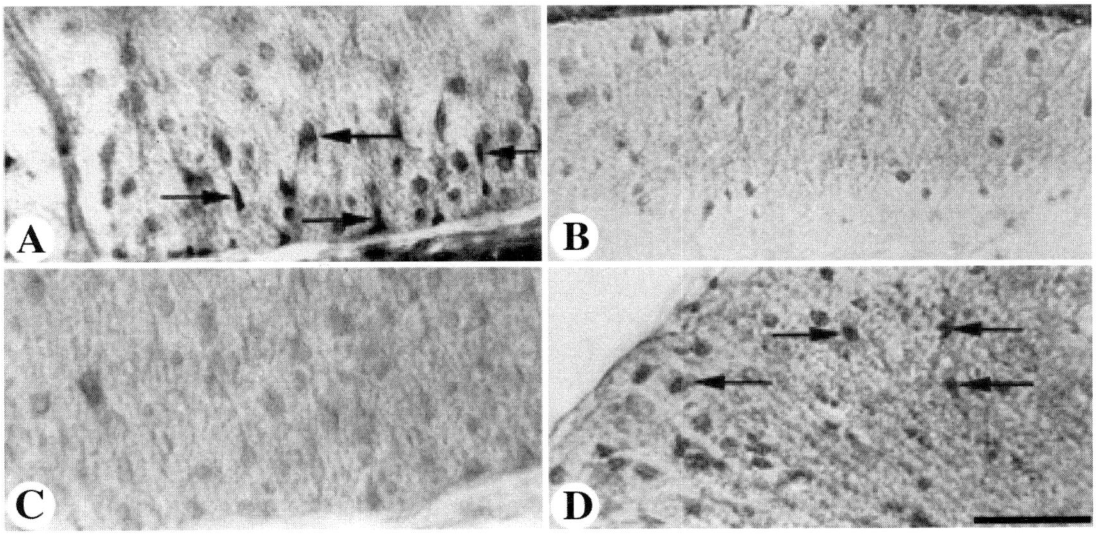

Figure 5. The expression of GRO-1 in spinal cord white matter astrocytes is developmentally regulated. In ventral regions (A,C), astrocytes express relatively high levels of GRO-1 at PI (A) (arrows) but low levels at P8 (C). By contrast, in dorsal spinal cord (B,D) there are relatively low levels of GRO-1 at P1 (B) but higher levels at P8 (D) (arrows). Bar = 30 μm.

the second and third postnatal weeks, however, the pattern changed significantly. Far fewer GRO-1^+ cells were present throughout the cord and most were localized in dorsal white matter [17] (Fig. 5). This pattern of GRO-1 expression in the developing spinal cord is coincident with the pattern of emergence of oligodendrocytes [23,29] suggesting that during development local control of proliferation of spinal cord oligodendrocyte precursors is regulated by the local expression of GRO-1.

7. GRO-1 REGULATES OLIGODENDROCYTE PRECURSOR PROLIFERATION IN PATHOLOGICAL CONDITIONS

The proliferation of oligodendrocyte precursors is not restricted to developmental epochs. In response to demyelinating insults and other injuries, the local proliferation of oligodendrocytes is upregulated [31–33]. Increases in precursor proliferation also occur in dysmyelinating mutants such as *jimpy* mice (jp)

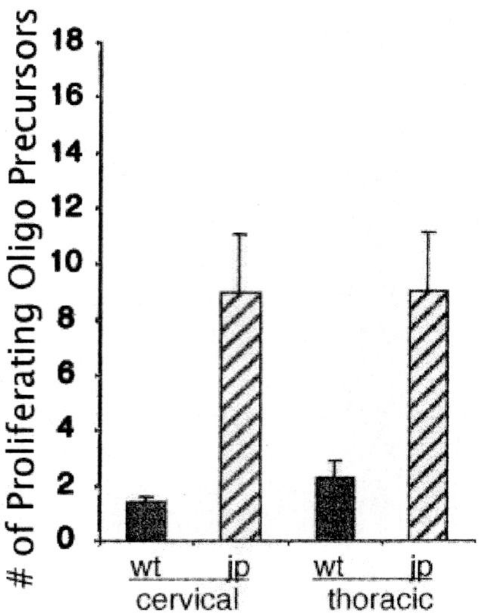

Figure 6. The number of proliferating oligodendrocyte precursors is increased in the jimpy spinal cord. The numbers represent the number of BrdU$^+$ oligodendrocyte precursors per unit area of white matter in the cervical and thoracic spinal cord. There is a three- to eight-fold increase in the number of BrdU$^+$ oligodendrocyte precursors in the jp spinal cord white matter. Data represent the mean ± standard deviation.

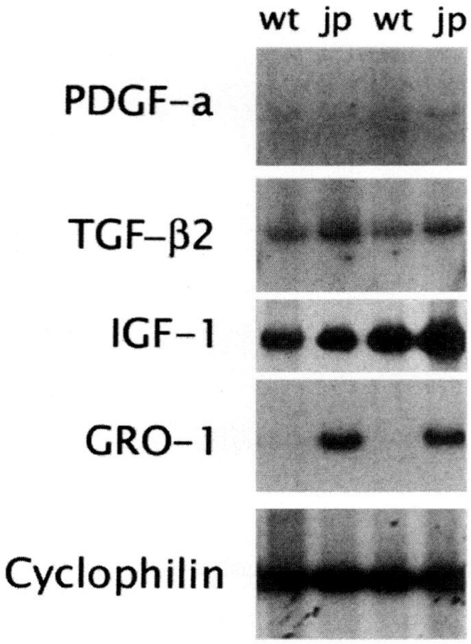

Figure 7. Semi-quantitative RT-PCR demonstrates a significant increase in the expression of GRO-1 mRNA in jimpy compared to wild type spinal cord. The levels of several other growth factors that regulate oligodendrocyte development including IGF-1 PDGF-A and TGF beta 2 are not significantly altered in jimpy spinal cord.

[54,57]. The jp mutation is an X-linked recessive mutation in the proteolipid protein gene (PLP) [58] that results in dysmyelination in selected white matter tracts in affected males [57]. In affected regions, the reduction in myelin was accompanied with an increase in the proliferation of oligodendrocyte precursors [54,57]. For example, throughout the length of the jp spinal cord, the number of proliferating oligodendrocyte precursors in white matter was 3–6 fold higher than in control animals (Fig. 6). This increase in cell proliferation reflected local signaling since there was virtually no change in the number of proliferating precursors in adjacent spinal cord gray matter. Furthermore, the increased levels of precursor cell proliferation resulted in increased numbers of oligodendrocyte precursors in the mutant spinal cord.

Comparison of the expression levels of several oligodendrocyte precursor growth factors in jp and normal spinal cord by PT-PCR demonstrated a specific and significant increase in expression of GRO-1 mRNA. For example, relatively low levels of PDGF-A mRNA were expressed in both *jp* and *wt* spinal cord (Fig. 7). Similarly, there was no significant difference in the level of mRNA encoding insulin-like growth factor-I (IGF-I) (Fig. 7), another putative oligodendrocyte progenitor cell mitogen [59]. Similarly, expression levels of proliferative inhibitors such as transforming growth factor (TGF) family members, known to be present in the CNS [60,61] and inhibit oligodendrocyte progenitor proliferation [61] were not significantly different between jp and wild type spinal cords. By contrast, semi-quantitative RT-PCR revealed a 17-fold higher level of GRO-1 mRNA in *jp* spinal cord compared with that in *wt* spinal cord (Fig. 7) [54]. Thus, the elevated levels of oligodendrocyte precursor proliferation in *jp* spinal cord correlate with expression of elevated levels of mRNA for the chemokine KC/GRO-1. As during development, immunohistochemical analyses demonstrate the major cellular source of the GRO-1 in the jp spinal cord to be a subset of astrocytes in the dorsal columns.

The elevated levels of GRO-1 in jp spinal cord mediate elevated levels of oligodendrocyte precursor proliferation [54]. Extracts of jp spinal cord induced elevated levels of proliferation of purified oligodendrocyte precursors compared to wild type extracts

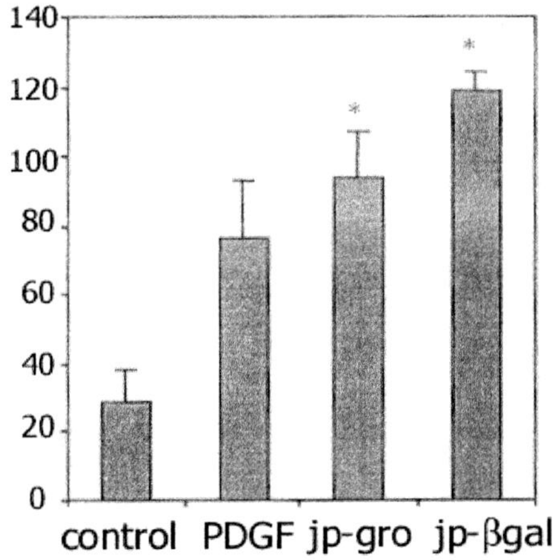

Figure 8. Jimpy spinal cord extracts enhance the proliferation of oligodendrocyte precursors to a greater extent than wild type spinal cord extracts. The jp induced enhanced proliferation is inhibited by depletion of GRO-α from the extract. The results represent the relative proliferation levels in control cultures, as well as in the presence of PDGF, and GRO-depleted jp (jp-gro) and control jp (jp-βgal) spinal cord extracts.

(Fig. 8). Immunodepletion of the jp extract with anti-GRO-1 antibodies resulted in a 2–3-fold reduction in the induced proliferative response [54], demonstrating that the enhanced proliferation induced by jp extracts was the result of increased expression of GRO-1.

8. GRO-1 REGULATES OLIGODENDROCYTE PRECURSOR MIGRATION IN VITRO

A primary function of chemokines is to act as chemotactic agents for a wide variety of cell types. Immature oligodendrocyte precursors are highly motile cells and successful myelination in the vertebrate CNS is dependent on their long distance migration [18]. Recent studies suggest that GRO-1 not only regulates oligodendrocyte precursor proliferation but also acts as an effective inhibitor of cell translocation [62]. In a number of in vitro assays PDGF stimulates oligodendrocyte precursor migration, and acts as a chemotactic agent [22]. By contrast, GRO-1 acts as a potent stop signal and inhibits PDGF stimulated migration in a rapid and reversible manner [62]. The GRO-1 concentration dependence of this inhibition is similar to that seen for proliferative control suggesting that the enhanced cell proliferation and inhibition of cell migration may be functionally linked.

9. A MODEL OF SPINAL CORD OLIGODENDROGENESIS

The notion that combinations of GRO-1 with PDGF results in a cessation of cell translocation and enhanced cell proliferation provides the framework for a model of oligodendrocyte precursor proliferation during development of the spinal cord (Fig. 9). In this model oligodendrocyte precursors arise in the ventral ventricular zone of developing spinal cord and disperse to the peripheral presumptive white matter possibly under the influence of ubiquitously expressed PDGF [22]. During early postnatal periods, oligodendrocyte precursors entering ventral white matter tracts experience locally elevated levels of the chemokine GRO-1 [17] which inhibits further migration [62] and stimulates proliferation [17]. As a result the numbers of oligodendrocyte

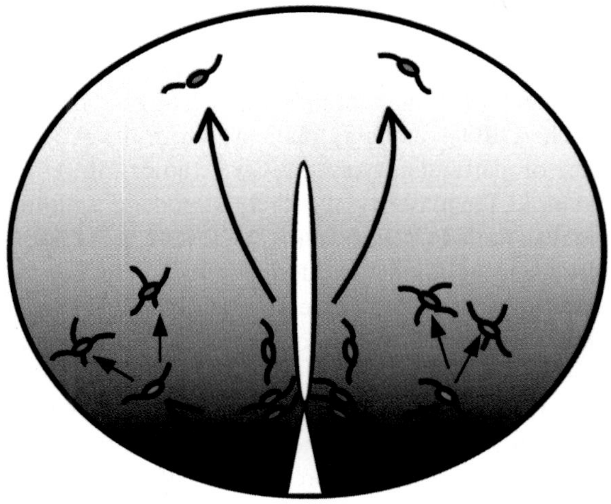

Figure 9. A model for the spatial regulation of oligodendrocyte development in the spinal cord. Immature oligodendrocyte precursors (red) arise in the ventral ventricular zone of the embryonic spinal cord and migrate to lateral and dorsal presumptive white matter. In the first postnatal week those cells that migrate ventrally encounter locally elevated levels of the chemokine GRO-1 which inhibits their further migration and promotes their proliferation in combination with ubiquitously expressed PDGF-A. The progeny of these cells then differentiates to oligodendrocytes that myelinate adjacent axons. One to two weeks later in development the expression of GRO-1 decreases in ventral and increases in dorsal spinal cord resulting in myelination of dorsal axonal tracts.

precursors increase and these cells then differentiate and myelinate all local axons. Oligodendrocyte precursors that migrate into dorsal white matter where there is no elevated expression of GRO-1 at this time continue to migrate and proliferate at a relatively low level. Later in development the expression of GRO-1 decreases in ventral and increases in dorsal spinal cord regions [17]. At this stage, oligodendrocyte precursors that migrate into dorsal regions will encounter high levels of GRO-1. Consequently, their migration will be inhibited [62] and their proliferation stimulated [17]. Subsequently, oligodendrocytes that differentiate from these precursors will myelinate adjacent axons. This model has a number of attractive features. It provides a functional rationale for the mechanistic association of elevated proliferation and cessation of migration. It illuminates the requirement for multi-factorial regulation of oligodendrocyte development and it provides insights into potential dysregulation of oligodendrogenesis in neoplasia and demyelinating diseases.

10. GRO-1 SIGNALING AND GLIAL TUMORS

Analyses of expression of chemokines and their receptors in human brain tumors suggest that the PDGF/GRO-1 pathway plays a role in the progression of particular types of CNS tumors [63]. For example, the majority of cells in oligodendrogliomas express both PDGFαR and CXCR2. If, as is likely, oligodendrogliomas are derived from oligodendrocyte precursors, this expression is not surprising. In other types of CNS tumors such as astrocytomas, there was lower expression of these receptors. Not only were the receptors of the PDGF/GRO-1 pathway expressed in oligodendrogliomas, but the appropriate ligands were also present [63]. Platelet derived growth factor A and GRO-1 were detectable in samples of oligodendrogliomas suggesting that these tumors have the capacity for autocrine stimulation of cell proliferation [63]. In vitro perturbation studies with antibodies against GRO-1 or CXCR2 demonstrated that this pathway is of functional importance in the proliferation of these tumors [63]. It seems likely, that the expansion of oligodendroglioma is driven in part by the synergy between PDGF and GRO-1 and that a major contributor to the neoplastic phenotype is the ectopic expression of the growth factors leading to autocrine stimulation of proliferation. Whether other types of CNS tumors utilize similar mechanisms and distinct chemokine pathways is not clear.

11. CONCLUSIONS AND FUTURE DIRECTIONS

The realization that chemokines may play important roles in the genesis of neuroepithelial-derived neural cells opens a new direction in the analyses of chemokine function. While it has been clear for several years that chemokines and their receptors play critical roles in inflammatory responses in the CNS [48,50,51] it is only recently that direct influences of these molecules on the cells of the CNS have been described [17,64–66]. Not surprisingly, the majority of these studies have addressed issues of cell migration in the CNS. For example, murine astrocytes express CCR1 a MIP-1α receptor and demonstrate directed migration in response to gradients of MIP-1α [65]. Likewise rat type-1 astrocytes have been reported to express the chemokines stromal cell derived factor 1 (SDF1) and its receptor CXCR4 [66]. Analyses of CXCR4-deficient animals indicate that this receptor and its ligand play an important role in regulating the migration of granule cells in the developing cerebellum [64].

It seems likely that as our understanding of the roles of chemokines in the development and functioning of the vertebrate CNS becomes clearer, the participation of these pleiotropic molecules in a variety of normal and pathological activities will be identified. Such identification will open new avenues for potential therapeutic intervention during CNS trauma and disease.

ACKNOWLEDGEMENTS

Much of this work has been supported by grants NIH NS36674 and funding from the MS society, and by an American Cancer Society Institutional Research Grant to the Cancer Center at Case Western Reserve University and University Hospitals of Cleveland.

REFERENCES

1. Jessell TM, Dodd J. Floor plate-derived signals and the control of neural cell pattern in vertebrates. Harvey Lect 1990;86(87):87–128.
2. Jessell TM, Bovolenta P, Placzek M, Tessier LM, Dodd J. Polarity and patterning in the neural tube: the origin and function of the floor plate. Ciba Found Symp 1989; 144(255):255–276.
3. Warf BC, Fok-Seang J, Miller RH. Evidence for the ventral origin of oligodendrocyte precursors in the rat spinal cord. J Neurosci 1991;11(8):2477–2488.
4. Pringle NP, Richardson WD. A singularity of PDGF alpha-receptor expression in the dorsoventral axis of the neural tube may define the origin of the oligodendrocyte lineage. Development 1993;117(2):525–533.
5. Ono K, Bansal R, Payne J, Rutishauser U, Miller RH. Early development and dispersal of oligodendrocyte precursors in the embryonic chick spinal cord. Development 1995;121:1743–1754.
6. Timsit S, Martinez S, Allinquant B, Peyron F, Puelles L, Zalc B. Oligodendrocytes originate in a restricted zone of the embryonic ventral neural tube defined by DM-20 mRNA expression. J Neurosci 1995;15:1012–1024.
7. Roelink H, Porter JA, Chiang DT, Beachy PA, Jessell TM. Floor plate and motor neuron induction by different concentrations of the amino-terminal cleavage product of sonic hedgehog autoproteolysis. Cell 1995;81:445–455.
8. Echelard Y, Epstein DJ, St JB, Shen L, Mohler J, McMahon JA. Sonic hedgehog, a member of a family of putative signaling molecules is implicated in the regulation of CNS polarity. Cell 1993;75:1417–1430.
9. Bunge RP. Glial cells and the central myelin sheath. Physiol Rev 1968;48:197–251.
10. Richardson WD, Pringle NP, Yu WP, Hall AC. Origin of spinal cord oligodendrocytes:possible developmental and evolutionary relationships with motor neurons. Dev Neurosci 1997;19:58–68.
11. Orentas DM, Hayes J, Dyer K, Miller RH. Sonic hedgehog signaling is required during the appearance of spinal cord oligodendrocyte precursors. Development 1999;126: 2419–2429.
12. Pfeiffer SE, Warrington AE, Bansal R. The oligodendrocyte and its many cellular processes. Trends Cell Biol 1993;3:191–197.
13. Raff MC, Miller RH, Noble M. A glial progenitor cell that develops in vitro into an astrocyte or an oligodendrocyte depending on culture medium. Nature 1983;303(5916): 390–396.
14. Fok-Seang J, Miller RH. Distribution and differentiation of A2B5$^+$ glial precursors in the developing rat spinal cord. J Neurosci Res 1994;37:219–235.
15. Raff MC, Mirsky R, Fields KL, Lisak RP, Dorfman SH, Silberberg DH et al. Galactocerebroside is a specific cell surface antigenic marker for oligodendrocytes in culture. Nature 1978;274:813–816.
16. Barres BA, Hart IK, Coles HS, Burne JF, Voyvodic JT, Richardson WD et al. Cell death and control of cell survival in the oligodendrocyte lineage. Cell 1992;70(1): 31–46.
17. Robinson S, Tani M, Streiter RM, Ransohoff RM, Miller RH. The chemokine growth related oncogene-a promotes spinal cord precursor proliferation. J Neurosci 1998;18: 10457–10463.
18. Miller RH, Payne J, Milner L, Zhang H, Orentas D. Spinal cord oligodendrocytes develop from a limited number of migratory, highly proliferative precursors. J Neurosci Res 1997;50:157–168.
19. Noble M, Murray K. Purified astrocytes promote the in vitro division of a bipotential glial progenitor cell. EMBO J 1984;3(10):2243–2247.
20. Warrington AE, Barbarese E, Pfeiffer SE. Differential myelinogenic capacity of specific stages of the oligodendrocyte upon transplantation into hypomyelinating hosts. J Neurosci Res 1993;34:1–13.
21. Milner R, ffrench-Constant C. A developmental analysis of oligodendroglial integrins in primary cells: changes in αv-associated B subunits during differentiation. Development 1994;120:3497–3506.
22. Armstrong R, Harvath L, Dubois-Dalcq M. Astrocytes and O-2A progenitors migrate toward distinct molecules in a microchemotaxis chamber. Ann NY Acad Sci 1991; 633(520):520–522.
23. Gilmore SA. Neuroglial populations in the spinal white matter of neonatal and early postnatal rats: an autoradiographic study of numbers of neuroglia and changes in their proliferative activity. Anat Rec 1971;171:283–292.
24. Noble M, Murray K, Stroobant P, Waterfield MD, Riddle P. Platelet-derived growth factor promotes division and motility and inhibits premature differentiation of the oligodendrocyte/type-2 astrocyte progenitor cell. Nature 1988;333(6173):560–562.
25. Richardson WD, Pringle N, Mosley MJ, Westermark B, Dubois-Dalcq M. A role for platelet-derived growth factor in normal gliogenesis in the central nervous system. Cell 1988;53(2):309–319.
26. Pringle NP, Mudhar HS, Collarini EJ, Richardson WD. PDGF receptors in the rat CNS: during late neurogenesis, PDGF alpha-receptor expression appears to be restricted to glial cells of the oligodendrocyte lineage. Development 1992;115(2):535–551.
27. Gard AL, Pfeiffer SE. Two proliferative stages of the oligodendrocyte lineage (A2B5$^+$O4$^-$ and O4$^+$GalC$^-$) under different mitogenic control. Neuron 1990;5(5):615–625.
28. Gard AL, Pfeiffer SE. Glial cell mitogens bFGF and PDGF differentially regulate development of O4$^+$ GalC$^-$ oligodendrocyte progenitors. Dev Biol 1993;159:618–630.
29. Schwab ME, Schnell L. Region-specific appearance of myelin constituents in the developing rat spinal cord. J Neurocytol 1989;18(2):161–169.
30. Schwab ME, Schnell L. Channeling of developing rat corticospinal tract axons by myelin-associated neurite growth inhibitors. J Neurosci 1991;11(3):709–721.
31. Amat JA, Fields KL, Schubart UK. Distribution of Phosphoprotein p19 in Rat Brain During Ontogeny — Stage-Specific Expression in Neurons and Glia. Dev Brain Res 1991;60(2):205–218.
32. Prineas JW, Barnard RO, Kwon EE, Sharer LR, Cho ES. Multiple Sclerosis: Remyelination of nascent lesions. Ann Neurol 1993;33:137–151.

33. Raine CS, Scheinberg L, Waltz JM. Multiple Sclerosis: Oligodendrocyte survival and proliferation in an active established lesion. Lab Invest 1981;45:534–538.
34. Yeh HJ, Ruit KG, Wang YX, Parks WC, Snider WD, Deuel TF. PDGF A-chain gene is expressed by mammalian neurons during development and in maturity. Cell 1991;64(1):209–216.
35. Pringle N, Collarini EJ, Mosley MJ, Heldin CH, Westermark B, Richardson WD. PDGF A chain homodimers drive proliferation of bipotential (O-2A) glial progenitor cells in the developing rat optic nerve. EMBO J 1989; 8(4):1049–1056.
36. Calver AR, Hall AC, Yu W-P, Walsh FS, Heath JK, Betsholtz C et al. Oligodendrocyte population dynamics and the role of PDGF in vivo. Neuron 1998;20:869–882.
37. Robinson S, Miller RH. Environmental enhancement of growth factor mediated oligodendrocyte precursor proliferation. Mol Cell Neurosci 1996;8:38–52.
38. Cochran BJ, Reffel AC, Stiles CD. Molecular cloning of gene sequences regulated by platelet derived growth factor. Cell 1983;33:939–947.
39. Anisowicz A, Bardwell L, Sager R. Constitutive over expression of a growth-related gene in transformed Chinese hamster and human cells. Proc Natl Acad Sci USA 1987; 84:7188–7192.
40. Richmond A, Thomas HG. Melanoma growth stimulatory activity: isolation from human melanoma tumors and characterization of tissue distribution. J Cell Biochem 1988; 36:185–198.
41. Oquendo P, Alberta J, Wen D, Graycar JL, Derynck R, Stiles CD. The platelet-derived growth factor-inducible KC gene encodes a secretory protein related to platelet α-granule proteins. J Biol Chem 1989;264(7):4133–4137.
42. Haskill S, Peace A, Morris J, Sporn SA, Anisowicz A, Lee SW et al. Identification of three related human GRO genes encoding cytokine functions. Proc Natl Acad Sci USA 1990;87:7732–7736.
43. Martins-Green M, Bissell MJ. Localization of 9E3/CEF-4 in avian tissues: expression is absent in Rous sarcoma virus induced tumors but is stimulated by injury. J Cell Biol 1990;110:581–595.
44. Strieter RM, Polverini PJ, Kunkel SL, Arenberg DA, Burdick MD, Kasper J et al. The functional role of the ELR motif in CXC chemokine-mediated angiogenesis. J Biol Chem 1995;270:27348–27357.
45. Broxmeyer HE, Sherry B, Cooper S, Lu L, Maze R, Beckmann MP et al. Comparative analysis of the human macrophage inflammatory protein family of cytokines (chemokines) on proliferation of human myeloid progenitor cells. Interacting effects involving suppression, synergistic suppression and blocking of suppression. J Immunol 1993;150:3448–3458.
46. Rollins B. Chemokines. Blood 1997;90:909–928.
47. Baggiolini M. Chemokines and leukocyte traffic. Nature 1998;392:565–568.
48. Spanaus KS. C-X-C and C-C chemokines are expressed in the cerebrospinal fluid in bacterial meningitis and mediate chemotactic activity on peripheral blood-derived polymorphonuclear and mononuclear cells in vitro. J Immunol 1997;158:1956–1964.
49. Van Meir E, Ceska M, Effenberger F, Walz A, Grouzmann I, Desbaillets I et al. Interleukin-8 is produced by neoplastic and infectious diseases of the human central nervous system. Cancer Res 1992;52:4297–4305.
50. Liu T. Cytokine-induced neutrophil chemoattractant mRNA expressed in cerebral ischemia. Neurosci Lett 1993;164:125–128.
51. Glabinski A, Tani M, Streiter R, Tuohy V, Ransohoff R. Synchronous synthesis of a and b chemokines by cells of diverse lineage in the central nervous system of mice with relapses of experimental autoimmune encephalomyelitis. Am J Pathol 1997;150:617–630.
52. Tani M. Neutrophil infiltration, glial reaction and neurological disease in transgenic mice expressing the chemokine N51/KC in oligodendrocytes. J Clin Invest 1996;98: 529–539.
53. Bell MD. Recombinant human adenovirus with rat MIP-2 gene insertion causes prolonged PMN recruitment to the murine brain. Eur J Neurosci 1996;8:1803–1811.
54. Wu Q, Miller RH, Ransohoff RM, Robinson S, Nishiyama A. Elevated levels of the chemokine GRO-1 correlate with elevated oligodendrocyte progenitor proliferation in the jimpy mutant. J Neurosci 2000;20:2609–2617.
55. Richmond A, Thomas HG, Roy RG. Separation of melanoma growth stimulatory activity and human type-alpha transforming growth factor. Methods Enzymol 1987; 146:112–126.
56. Keane MP. The CXC chemokines. IL8 and IP10 regulate angiogenic activity in idiopathic pulmonary fibrosis. J Immunol 1997;159:1437–1443.
57. Skoff RP. Increased proliferation of oligodendrocytes in the hypomyelinated mouse mutant-jimpy. Brain Res 1982; 248(1):19–31.
58. Nave KA, Bloom FE, Milner RJ. A single nucleotide difference in the gene for myelin proteolipid protein defines the jimpy mutation in mouse. J Neurochem 1987;49(6): 1873–1877.
59. McMorris FA, Dubois-Dalcq M. Insulin-like growth factor I promotes cell proliferation and oligodendroglial commitment in rat glial progenitor cells developing in vitro. J Neurosci Res 1988;21(2–4):199–209.
60. Unsicker K, Flanders KC, Cissel DS, Lafyatis R, Sporn MB. Transforming growth factor beta isoforms in the adult rat central and peripheral nervous system. Neuroscience 1991;44:613–625.
61. McKinnon RD, Piras G, Ida JA, Dubois-Dalcq M. A Role for TGF-B in Oligodendrocyte Differentiation. J Cell Biol 1993;121(6):1397–1407.
62. Tsai H, Frost E, Robinson S, Ransohoff R, Sussman C, ffrench-Constant C et al. The inhibitory effect on oligodendrocyte precursor migration by growth regulated oncogene-a (GRO-a) is mediated by CXCR2. J Neurosci Abs 2000;26 part 2:506.16.
63. Robinson S, Cohen M, Prayson R, Ransohoff RM, Tabrizi N, Miller RH. Constitutive expression of Growth Regulated Oncogene and its receptor in oligodendrogliomas. Neurosurgery 2001;48:864–873.
64. Zou RY, Kottman AH, Kuroda M, Taniuchi I, Littman DR. Function of the chemokine receptor CXCR4 in hematopoiesis and cerebellar development. Nature 1998; 393:595–599.

65. Tanabe S. Murine astrocytes express a functional chemokine receptor. J Neurosci 1997;17:6522–6528.
66. Bajetto A, Bonavia R, Barbero S, Piccioli P, Costa A, Florio T et al. Glial and neuronal cells express functional chemokine receptor CXCR4 and its natural ligand stromal cell-derived factor 1. J Neurochem 1999;73:2348–2357.

Universes in Delicate Balance: Chemokines and the Nervous System
R.M. Ransohoff, K. Suzuki, A.E.I. Proudfoot, W.F. Hickey and J.K. Harrison, editors

Chapter 4.3

Chemokine Interactions with Astrocytes

Martin E. Dorf, Falko R. Fischer, Michael A. Berman and Yi Luo

Department of Pathology, Harvard Medical School, Boston, MA 02115, USA

1. INTRODUCTION

The chemokines KC, MIP-2, SDF-1α/CXCL12, RANTES/CXCL15, eotaxin/CCL11, MCP-1/CCL2, MIP-1α/CCL3, MIP-1β/CCL4, and fractalkine/CX3CL1 stimulate biologic responses in astrocytes. These responses include chemotaxis, cytokine/chemokine synthesis, intracellular calcium mobilization, receptor modulation, and phosphorylation of signaling molecules. Multiple lines of evidence are also presented documenting the expression of chemokine receptors CXCR4, CCR1, CCR5, and CX3CR1. Accumulating data suggest several additional chemokine and orphan receptors are also present on astrocytes. Most chemokine receptors are probably expressed at low levels and are modulated following stimulation with chemokines or other inflammatory mediators. Thus, chemokine receptors are only detected on a subset of astrocytes during disease or following induction. The presence of multiple chemokine receptors implies a wider role for chemokines in the regulation of central nervous system functions.

Astrocytes are the most abundant glial cell type, comprising about one-third of the cells in the central nervous system (CNS). Astrocytes are cells of neuroectodermal origin; they project foot processes that envelop the basement membrane of neurons, synapses, and capillaries. Astrocytes are a heterogeneous class of cells with diverse functions. The regional heterogeneity and functional specialization of astrocytes implies that these cells adapted to perform functions specific to their respective anatomical locales [1,2]. It is now clear that astrocytes actively participate in bi-directional communications with neurons. Neurons signal astrocytes through synaptically released neurotransmitters [3] and in turn astrocytes signal neurons by releasing glutamate which increases neuronal calcium levels and modulates synaptic transmission [4–6]. Astrocytes participate in metabolism of serotonin and have evolved mechanisms of neurotransmitter uptake that protect neurons from neurotoxic injury during glutamate toxicity [7]. Astrocytes also participate in neuronal repair by producing oligodendrocyte growth factors that promote remyelination [8]. In addition, to these interactions with neurons and other glia, astrocytes participate in maintenance of the blood–brain barrier. Astrocytic endfeet surround the abluminal surfaces of brain microvascular endothelial cells to help form the blood–brain barrier [9]. This barrier consists of tight junctions of nonfenestrated endothelial cells. The blood–brain barrier is a critical shield protecting the brain from exchanges of macromolecules and cells from the circulation. Ablation of reactive astrocytes disrupts the barrier and inhibits its repair [10]. Because of their essential role in maintenance of the blood–brain barrier, astrocytes help control the traffic of leukocytes into injured CNS parenchyma [10]. Thus astrocytes through their plexus of fine processes form commensal relationships with the neuronal, glial, and endothelial elements of the CNS.

Beyond these physiological roles, astrocytes are implicated in the pathophysiology of inflammatory and neurodegenerative diseases including microbial infections [11,12], Alzheimer's disease [13] and multiple sclerosis [14,15]. These diseases are characterized by scarring lesions containing reactive hypertrophic astrocytes with an increased content of glial fibrillary acidic protein. Reactive astrocytes are major sources of the cytokines that orchestrate migration and activation of leukocytes into neuronal lesions. Many of these inflammatory cytokines belong to a family of genetically and structurally related molecules termed chemokines. Chemokines are highly basic, low molecular weight peptides and glycopeptides that mediate leukocyte chemotaxis and increased cell adhesion. These essential features are required to direct cells to sites of infection or inflammation. Over 40 chemokines have been molecularly cloned [16]. These molecules are classified into four distinct families according to the configuration of the first two of four-conserved cysteine residues. Chemokines are classified into four groups known as CXC, CC, XC, and CX3C chemokines [16]. The CXC subfamily members have a single amino acid residue intervening between the first two-conserved cysteine residues. Based on the presence or absence of a glutamic acid–leucine–arginine motif immediately preceding the first cysteine, the CXC chemokines are further subdivided into two subclasses (ELR versus non-ELR). The ELR tripeptide sequence in the N-terminal domain is involved in interactions with neutrophil chemokine receptors [17]. All ELR CXC chemokines (e.g. KC, MIP-2, and IL-8) are neutrophil chemoattractants, while the non-ELR CXC chemokines (e.g. SDF-1α) generally attract lymphocytes. The largest collection of chemokines belong to the CC subfamily (e.g. MCP-1, MIP-1α, MIP-1β, RANTES, eotaxin) in which the first two cysteine residues are adjacent. Fractalkine (also known as neurotactin) is the only CX3C chemokine identified to date; it has three amino acid residues between the first two cysteines. Disulfide bonds form between the first and third and between the second and fourth cysteines to establish the tertiary structure of CXC, CC, and CX3C chemokines. In contrast, members of the XC chemokine family lack one of the first two cysteine residues. The sources and biological function of chemokines along with their cognate receptors are well described on leukocytes [18,19]. Only recently have astrocytes been implicated as additional targets of these molecules. This review will summarize the effects of chemokines on astrocytes and the potential receptors responsible for these interactions.

Chemokines mediate their effects on cells by activating members of the G-protein coupled seven-transmembrane spanning receptor family. Five receptors for CXC and 11 for CC chemokines have been cloned [16,20]. In addition, one receptor for CX3C and one for XC chemokines have also been identified. Generally each chemokine receptor binds multiple chemokines and most chemokines bind to more than one receptor [21].

2. CXC CHEMOKINES STIMULATE ASTROCYTES

The ability to attract cells toward a chemical gradient is a common characteristic of chemokines. Directed migration was also used to demonstrate chemokine activity in early reports of chemokine–astrocyte interaction [22]. KC and MIP-2, murine ELR CXC chemokines, attract cultured astrocytes in vitro at concentrations ranging from 0.1 to 10 nM (Fig. 1). The physiological relevance of such movements is a matter of speculation as astrocytes are generally regarded as relatively non-motile parenchymal cells. We speculate that chemokines released by developing endothelial cells may potentially help guide astrocytes to form the endothelial tight junctions associated with the blood–brain barrier [9]; similarly chemokines released by neurons may promote associations with glutamatergic synapses [23].

In studies conducted a decade ago, Rus et al. [24] reported that incubation of rat neonatal astrocyte cultures with 0.1–2.5 nM IL-8 for as little as 2 h resulted in production of mRNA for the complement component C3. IL-8 treatment also induced accumulation of C3 proteins in culture supernatants. Recently two related murine ELR CXC chemokines, KC and MIP-2, were shown to stimulate chemokine and cytokine synthesis in astrocytes. The latter murine chemokines are related to the human growth-regulated oncogene proteins (known as GRO-α, GRO-β, GRO-γ), but exact homologies remain uncertain [25–27]. One to 10 nM KC or MIP-2 was sufficient to induce mRNA for proinflammatory cytokines (TNF-α, IL-6, and IFN-γ) and chemokines (KC, MIP-2, IP-10, MCP-1, and RANTES). In addition, KC and other chemokine proteins were detected by ELISA [28]. This chemokine amplification cascade was noted in astrocytes and mesangial cells while microglia and renal epithelial cells failed to produce similar responses [28,29]. Interestingly, KC and MIP-2 dis-

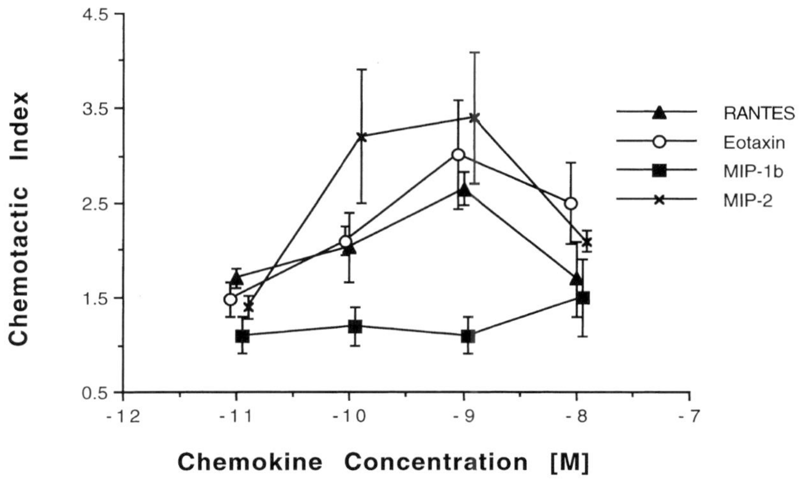

Figure 1. Astrocyte chemotaxis. Comparisons of astrocyte migration across a membrane with 14-μm pores in response to various concentrations of the murine chemokines RANTES (▲), eotaxin (○), MIP-2 (X), and MIP-1β (■). The average numbers of migrating cells per high-power field were divided by the background migration in medium alone for presentation as the chemotactic index. Procedural details have been described previously [22,69].

played autocrine activity, stimulating production of additional KC and MIP-2 [28,29]. Similar autocrine properties were previously reported using GRO-α on human umbilical vein endothelial cells [30]. Measurements of the levels of KC protein produced indicate that the autocrine response is self-limiting, but sufficient to prolong inflammatory responses within the CNS [28]. Other inflammatory proteins and products of the clotting cascade induce similar patterns of chemokine synthesis [31] potentially accounting for the overlapping chemokine profiles in a variety of pathologies including stroke, trauma, and multiple sclerosis.

In addition to up-regulation of chemokine and cytokine transcripts, treatment with KC, MIP-2, RANTES, TNF-α, thrombin, or LPS up-regulated the level of message for an orphan chemokine receptor, termed Eo1, by at least 4-fold (Fig. 2). The mouse Eo1 gene (GenBank accession numbers AF030185 and AF316576) encodes a seven transmembrane spanning receptor with sequence similarity to CCR1, CCR3, and CCR5 [32]. Down-modulation of CX3CR1 and CXCR4 was also noted following chemokine stimulation, especially with RANTES (Fig. 2). These effects are specific, as treatment with chemokines TCA4, MCP-1, and SDF-1α failed to significantly modulate expression of these receptors (Fig. 2). The receptor modulation that follows chemokine stimulation further emphasizes the dynamic nature of astrocytes and their ability to react to environmental factors.

In attempts to explore pathways of astrocyte signaling, changes in phosphorylation patterns after chemokine stimulation were examined. Astrocytes were treated with 10 nM chemokine for the indicated times, cell lysates were prepared subsequently and immunoblotted with antibodies specific for phosphotyrosine or phosphorylated MAP kinase (Fig. 3). This approach established chemokines mediated astrocyte activation via the MAP kinase pathway [33]. As shown in Fig. 3, KC treatment triggered phosphorylation of astrocyte proteins within 10 min. Specifically, the increased levels of the phosphorylated 44- and 42-kDa extracellular signal-regulated kinases (Erk1 and Erk2) were most apparent, indicating participation of the MAP kinase cascade following chemokine activation. Similar results were noted following treatment with 100 ng/ml MCP-1 or RANTES (Fig. 3). Inhibition of Erk1 and Erk2 phosphorylation with the MEK-1 inhibitor UO126 and the protein tyrosine kinase inhibitor genistein (Fig. 3) support involvement of the MAP kinase pathway in chemokine-mediated astrocyte signaling. Addition of these inhibitors also reduced chemokine RNA induction (40–70%) [32]. It should be noted that dexamethasone and IFN-β failed to inhibit astrocyte chemokine synthesis (data not shown) suggesting that these standard anti-inflammatory therapies

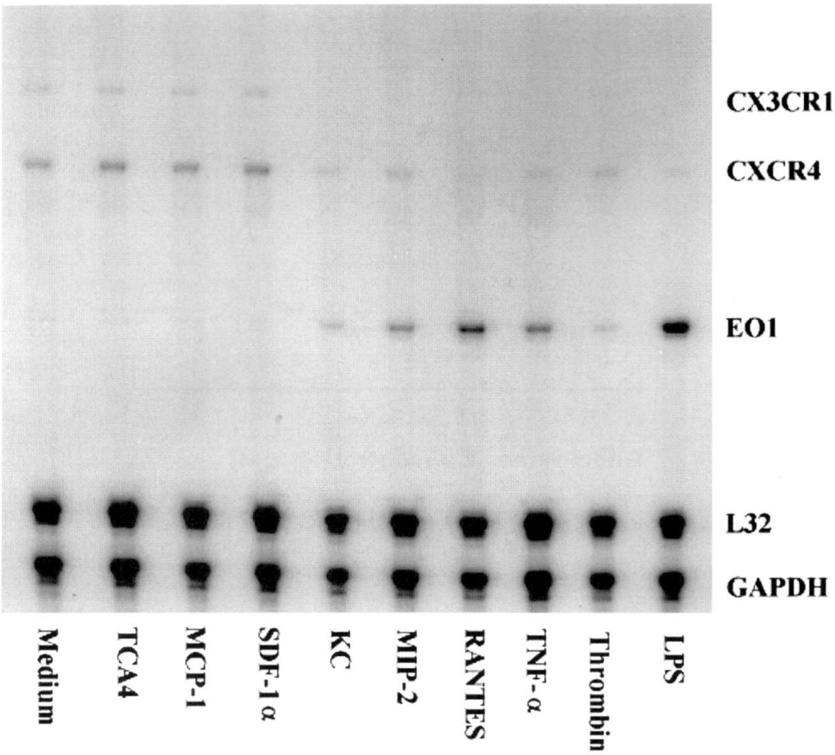

Figure 2. Chemokines modulate receptor expression. RNA was prepared from astrocytes after 16 h culture with 100 ng/ml TCA4/CCL21, MCP-1, SDF-1α, KC, MIP-2, RANTES, 20 ng/ml TNF-α, 0.5 U/ml thrombin, or medium. RNase protection assays were used to examine expression of CXCR4, CX3CR1, and Eo1. To quantify the data, gels were examined on a phosphoimager and the data were normalized based on the levels of the L32 housekeeping control.

for multiple sclerosis may be ineffective at blocking this source of proinflammatory mediators.

Bell et al. [34] examined the effects of injecting the MIP-2 cDNA (driven by a CMV promoter and inserted into a replication-deficient adenovirus) into the brain. Mice that received the MIP-2-adenovirus construct displayed transient expression of MIP-2 primarily localized to astrocytes. The introduction of MIP-2 was accompanied by a dramatic increase in neutrophil recruitment to the brain parenchyma and a breakdown of the blood–brain barrier. The leukocyte infiltrate was sustained for several days. Interestingly, however, the mice never developed signs of neurologic disease. Transgenic mice expressing the KC gene under the control of a myelin basic protein promoter, thereby restricting expression to oligodendrocytes, have also been generated [35]. At 3 weeks of age, the KC transgenic animals displayed extensive perivascular, meningeal, and parenchymal infiltration primarily with neutrophils. Subsequently many transgenic mice developed a neurological syndrome of postural instability and rigidity. Histologic analysis revealed that the neurological impairment was associated with microglial activation and disruption of the blood–brain barrier. Specifically, brain microvessels demonstrated a disrupted astrocytic barrier with displacement of foot processes and surrounding astrocyte activation near sites of leukocyte infiltration in the brain parenchyma [35]. The combined results demonstrate that expression of the ELR^+ CXC chemokines KC or MIP-2 in the brain is associated with disruption of the astrocyte–endothelial barrier and prolonged recruitment of inflammatory cells into the parenchyma. Thus, the combination of in vivo and in vitro data support the hypothesis that KC and MIP-2 induce movement of astrocytic endfeet connections with endothelium resulting in disruption of the blood–brain barrier and amplification of chemokine production by astrocytes prolonging inflammatory responses in the parenchyma [28,34].

The known receptors for the KC and MIP-2

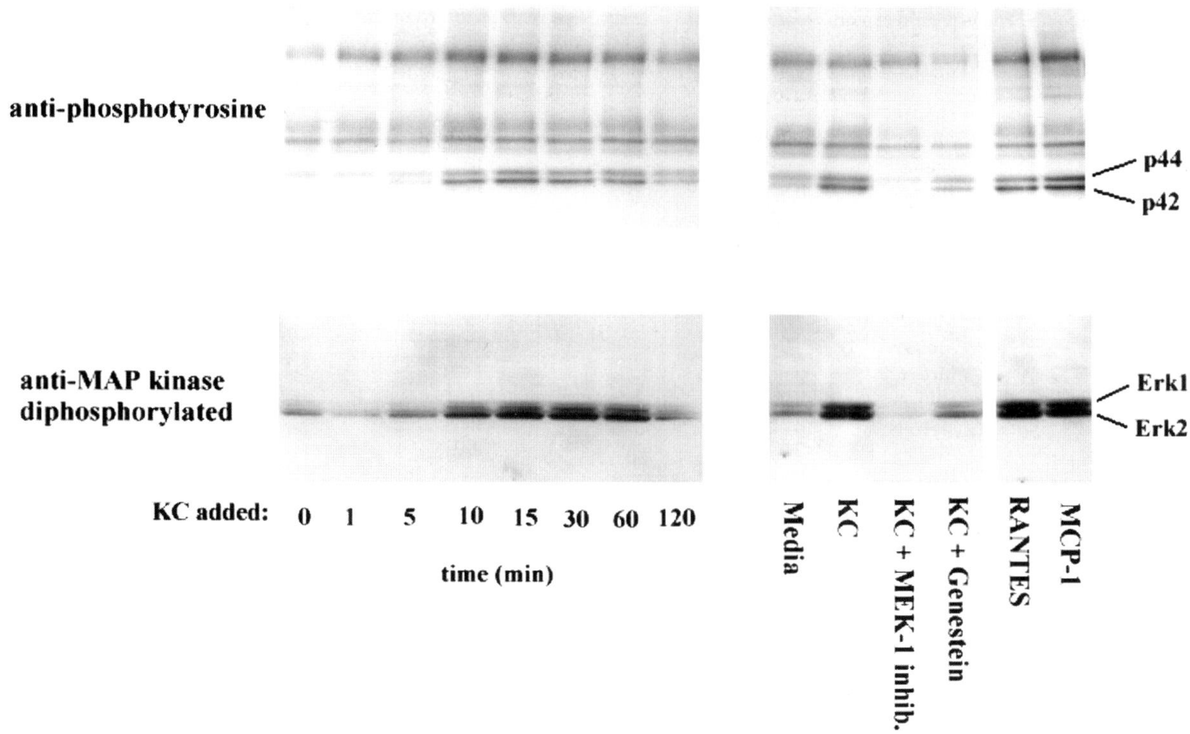

Figure 3. Chemokines activate MAP kinase signal cascade. Left panels: immunoblots of astrocyte lysates after 0–120 min stimulation with 10 nM KC. Right panels: cultured astrocytes were treated for 10 min with media, 10 nM KC, 10 nM KC plus 20 μM MEK-1 inhibitor UO126, 10 nM KC plus 100 μM genestein, 10 nM RANTES or 10 nM MCP-1. Blots were stained with anti-phosphotyrosine (top panels) and antibodies specific for phosphorylated MAP kinase (bottom panels).

chemokines include CXCR2 and DARC. Although DARC was expressed in Purkinje cells of the cerebellum it was not detected on astrocytes [36,37]. CXCR2 was also identified in the CNS primarily localized to projection neurons in the hippocampus, dentate nucleus of the cerebellum, and locus coeruleus [37]. Of greater relevance to this review, cultured human astrocytes and an astrocytic cell line were shown to express CXCR2 by flow cytometry and RT-PCR [38–40]. These results were confirmed by immunohistochemistry on human brain tissues [39]. The ligands of CXCR2 include IL-8, KC, MIP-2, and GRO-α that could well account for the ability of these molecules to stimulate functional responses in astrocytes. However, attempts to identify CXCR2 mRNA in cultured mouse astrocytes by RT-PCR have been unsuccessful [22]. Direct verification of a functional role for CXCR2 on murine astrocytes was recently evaluated by analysis of astrocytes from CXCR2-deficient mice. MIP-2 and KC induced chemokine expression in astrocytes that lack CXCR2, indicating that in the mouse alternative receptors must be sufficient for responses to these ligands [28]. The ability of KC and MIP-2 to stimulate chemokine production in CXCR2-deficient cells was confirmed using genotyped mesangial cells to exclude the presence of CXCR2 products (Luo and Dorf, unpublished observations).

Although species specific differences could account for the disparities in CXCR2 expression, this explanation can not account for all discrepancies noted in astrocyte chemokine receptor expression. Some differences are probably attributable to the use of different techniques and cell sources [13,37,41–46]. False-negative results stemming from the reduced sensitivity of some techniques may partially account for the confusion. Serological approaches can detect cell surface protein expression but the methods are generally less sensitive than amplification assays for detecting the presence of messenger RNA. However, the presence of RNA transcripts does not guarantee expression of the corresponding

Table 1. Chemokine and orphan receptors on astrocytes

Chemokine receptor	Species	Astrocytes in vivo or ex vivo	Cultured astrocytes	References
CCR1	Human		IF	[68]
	Mouse		PCR, RPA, FACS, ISH	[69,70]
CCR2	Human		IF	[40,68]
CCR3	Human	IHC	PCR	[41,50]
	Macaque	IHC		[78]
CCR5	Human	IHC	PCR, IF, FACS	[44,45,48,50,68,77,79]
	Macaque	IHC, IF	FACS	[48,77,78]
	Rat		PCR	[76]
	Mouse		ISH	[31]
CCR10	Mouse		PCR	[31]
CCR11	Mouse		PCR	[31]
CXCR2	Human	IHC	IF, FACS, PCR	[38,39,61,79]
CXCR4	Human	IHC	PCR, IF, FACS	[43,44,48,53,59]
	Macaque	IHC, IF	FACS	[48,78]
	Rat		North, PCR, IF	[42,52,58]
	Mouse		North, PCR, FACS, RPA	[51,55,57] Fig. 2
CX3CR1	Rat		North, RPA, FACS	[42,102] Fig. 2
	Mouse		PCR, RPA	[31]
Apj	Human		PCR	[44,108]
	Mouse		PCR	[31]
Bonzo/CXCR6	Human		PCR	[44]
	Mouse		PCR	Fig. 6
RDC1	Mouse		North	[111]
GPR30	Mouse		PCR	Fig. 6
ChemR23	Mouse		PCR	Fig. 6
CRTH2	Mouse		PCR	Fig. 6
Eo1	Mouse		RPA, PCR	Figs. 2 and 6

FACS, flow cytometry; IF, immunofluorescence; IHC, immunohistochemistry; ISH, in situ hybridization; North, Northern blot; PCR, reverse transcription-polymerase chain reaction and RPA, RNase protection assay.

protein [44,45]. False-positive results attributable to cross-reactivity of serological and molecular probes must also be carefully controlled. Since most researchers examined primary cultured fetal or neonatal cortical type-1 astrocytes, we should caution that these populations may contain small percentages (<5%) of contaminating cells including fibroblasts, microglia, and bipotential glial progenitor cells which can differentiate into oligodendrocytes or type-2 astrocytes complicating interpretations [47]. Furthermore the levels of astrocyte chemokine receptors were reported to decrease with time in culture [48] and modulation of chemokine receptor expression has been noted [39,49,50]. It has proven much more difficult to directly identify chemokine receptors on astrocytes in situ. The data compiled in Table 1 summarize the strategies used to identify these receptors on either fresh or cultured astrocytes.

Human and mouse astrocytes respond to SDF-1α by increases in the concentration of intracellular free calcium [48,51,52]. These responses are at least partially sensitive to pertussis toxin treatment indicating that signal transduction is associated with $G\alpha_i$ proteins [51–53]. SDF-1α also inhibits forskolin-stimulated cAMP production in astrocytes in a pertussis toxin-sensitive fashion [52,53]. These observations confirm previous reports demonstrating involvement of $G\alpha_i$ proteins in leukocyte chemokine signaling [18,19]. In addition, SDF-1α induces an increase in tyrosine phosphorylation of various astrocyte proteins including Erk1 and Erk2; this response is sensitive to treatment with pertussis toxin [33].

There are important contrasts between astrocyte responses to SDF-1α versus other CXC chemokines (Table 2). While KC and MIP-2 stimulated astrocyte chemotaxis, astrocytes do not migrate toward SDF-1α [33,51]. In fact, SDF-1α pretreatment prevented astrocyte migration toward other chemokines [31]. Furthermore, while KC and related ELR^+ CXC ligands stimulated extensive chemokine synthesis in astrocytes, SDF-1α treated astrocytes produce little

Table 2. Astrocyte responses to chemokines

Chemokine family	Chemokines	Astrocyte responses	References
CXC (ELR)	KC	Chemotaxis	[22]
		Cytokine synthesis	[28]
		Chemokine synthesis	[28]
		Receptor modulation	Fig. 2
		Erk1/2 phosphorylation	Fig. 3
CXC (ELR)	MIP-2	Chemotaxis	Fig. 1
		Cytokine synthesis	[28]
		Chemokine synthesis	[28]
		Receptor modulation	Fig. 2
CXC (ELR)	IL-8/CXCL8	C3 synthesis	[24]
CXC (non-ELR)	SDF-1α/CXCL12	Ca^{2+} mobilization	[48,51–53]
		Inhibit forskolin induced cAMP	[52,53]
		Erk1/2 phosphorylation	[33]
		Inhibition of chemotaxis	[31]
CC	MIP-1α/CCL3	Chemotaxis	[69,70]
		Chemokine synthesis	[45] Fig. 4
CC	MIP-1β/CCL4	Ca^{2+} mobilization	[48]
		Chemokine synthesis	[45] Fig. 4
		Inhibition of chemotaxis	[31]
CC	RANTES/CCL5	Chemotaxis	Fig. 1
		Ca^{2+} mobilization	[48]
		Chemokine synthesis	Fig. 4 and 5
		Cytokine synthesis	[32]
		Receptor modulation	Fig. 2
		Erk1/2 phosphorylation	Fig. 3
CC	Eotaxin/CCL11	Chemotaxis	[31]
		Chemokine synthesis	Fig. 5
		Cytokine synthesis	[32]
		Receptor modulation	[32]
		Inhibit forskolin induced cAMP	[53]
		IP_3 accumulation	[89]
CC	MCP-1/CCL2	Chemotaxis	[22]
		Erk1/2 phosphorylation	Fig. 3
CX3C	Fractalkine/CX3CL1	Ca^{2+} mobilization	[102]
		Growth factor synthesis	[102]

or no chemokine message following SDF-1α stimulation [31]. Reciprocally, SDF-1α triggers a vigorous calcium flux but MIP-2 failed to induce calcium mobilization in astrocytes [51]. This suggests a complex pattern of astrocyte signaling and effector mechanisms that can be initiated by chemokines. In addition to chemokine/cytokine synthesis, KC or MIP-2 stimulation can also modulate chemokine receptor expression. Thus, as shown in Fig. 2, astrocytes responded to stimulation with KC and MIP-2 by increased expression of a seven transmembrane spanning receptor, termed Eo1, and decrease of CX3CR1 mRNA levels. In contrast, SDF-1α treatment failed to modulate transcripts for either of these receptors (Fig. 2). Thus, the chemokine balance existing in the local CNS environment may dramatically alter astrocyte functional capacities. The significance of astrocytes responses to SDF-1 was underlined by the observation that this pathway may contribute to glutamate-mediated excitotoxicity [54].

CXCR4 and its alternate splice variant CXCR4B [55,56] are the only identified receptors for SDF-1α. Murine CXCR4 and its splice variant differ by the insertion of two amino acid residues. Both receptor isoforms respond to ligand with a rise in intracellular calcium and message for both forms is present in murine astrocytes [55]. Numerous reports describe expression of CXCR4 message [22,42,57,58] and protein [51,52,59] on cultured astrocytes, and on astrocytes examined immediately ex vivo [48] and in situ [59,60].

CXCR4 is a major co-receptor for some strains

of HIV-1 that arise during progression to immunodeficiency and AIDS dementia complex. Expression of CXCR4 may be up-regulated in vitro upon exposure to IL-1β [61] and on reactive astrocytes during HIV or SIV infection [60,62], but was modestly down-regulated following treatment with KC or MIP-2 (Fig. 2) as well as by inflammatory cytokines [63]. Viral envelope glycoprotein (gp120) released from virions may bind to astrocytes and neurons via CXCR4 causing direct neuronal, but not astrocyte, injury [64]. There are additional distinctions in the responses of astrocytes and neurons to gp120 signaling through CXCR4 [60]. Neurons undergo apoptosis following treatment with gp120 and kinase activation is not detected [33]. In contrast, Erk1 and Erk2 are activated by gp120 in astrocytes [33]. These dramatic tissue-specific differences underscore the diversity of responses and tissue specificity of chemokine receptors. The significance of these observations to the progression of AIDS dementia and HIV encephalitis remains to be defined.

The critical roles of the CXCR4/SDF-1α receptor-ligand pair to neuronal development have been best documented using mice genetically deficient for either SDF-1α or CXCR4. Specifically, mice lacking these genes display aberrant migration of external granule layer cells into the internal granule layer, indicating a role for CXCR4 and its ligand in proper neuronal patterning during cerebellar development [65–67].

3. CC CHEMOKINES STIMULATE ASTROCYTES

The extensive redundancy of CC chemokines and the promiscuity of their receptors makes systematic analysis more difficult. The CC chemokine MIP-1α binds receptors on astrocytes with a high affinity ($K_d = 0.3$ nM) [68]. Two groups have demonstrated that nanomolar concentrations of ligand stimulated astrocyte chemotaxis [69,70]. In addition MIP-1α triggered synthesis of MCP-1, MIP-1β, and autoinduction of MIP-1α in human astrocyte cultures [45]. The activities of individual chemokines on astrocytes are summarized in Table 2.

At least three chemokine receptors are reported to bind MIP-1α: CCR1, CCR4, and CCR5 [71–74]. Although RT-PCR failed to demonstrate CCR4 expression on mouse [69] or human [45] astrocytes, evidence exists for expression of both CCR1 and CCR5 (Table 1). Three groups independently detected CCR1 expression on cultured fetal/neonatal astrocytes using both molecular [69] and serological [68,70] techniques. The level of CCR1 expression is low, but can be increased by treatment with the cytokine TGFβ1. The modulation of CCR1 is functionally relevant because TGFβ1-stimulated astrocytes displayed augmented chemotactic responses to MIP-1α [70]. CCR5 is another receptor known to bind MIP-1α [75]. Again accumulating evidence from studies on mouse [31], rat [76], monkey [48] and human astrocytes [44,45,48,68] detected CCR5 message using RT-PCR [44,76] and in situ hybridization [31] and CCR5 protein was demonstrated by immunofluorescence [45,68,77], flow cytometry [48], and immunocytochemistry [77,78]. The levels of astrocyte CCR5 were up-regulated following interaction with the HIV-1 protein gp41 [79] and in some patients with multiple sclerosis [50]. The redundancy of these chemokine receptors may also account for the failure to note neurological defects in either CCR1 or CCR5-deficient mice [80–83].

MIP-1β induced calcium mobilization [48] and chemokine synthesis [45] in human cultured astrocytes at concentrations of 1–10 nM. At similar concentrations, MIP-1β inhibited mouse astrocyte migration to known chemoattractants [31]. The ability to prevent astrocyte migration may prove to be a protective response intended to maintain CNS homeostasis. MIP-1β binds at least two chemokine receptors [71,84] expressed on astrocytes, CCR1 and CCR5. Thus, one or more chemokine receptors may also control astrocyte responses to this ligand.

RANTES, like KC, MIP-2, TNF-α, and thrombin, is a very potent inducer of chemokine and cytokine synthesis (Fig. 4). In contrast to the effects of MIP-1α and MIP-1β, RANTES induced >10-fold more RANTES, MIP-2, IP-10, and MCP-1 message (Fig. 4). RANTES treatment also modulated chemokine receptors (Fig. 2) and calcium flux in cultured astrocytes [48]. RANTES is a ligand for CCR1, CCR3, and CCR5 [73,84–86]. Astrocytes express at least two RANTES receptors, CCR1 and CCR5 [71,84]. Thus, more than one chemokine receptor may control astrocyte responses to this ligand. CCR3 was also detected on cultured monkey [78] and human astrocytes [41] by immunohistochemi-

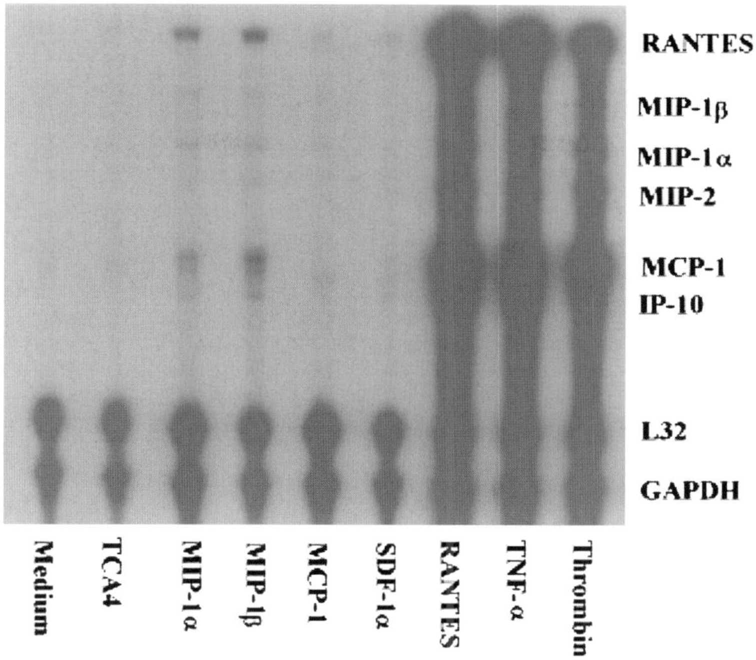

Figure 4. Induction of chemokine transcripts. RNA samples from cultured BALB/c astrocytes incubated for 16 h with medium or 100 ng/ml TCA4, MIP-1α, MIP-1β, MCP-1, RANTES, TNF-α, or 0.5 U/ml thrombin were assayed in a RNase protection assay for expression of lymphotactin, RANTES, eotaxin, MIP-1β, MIP-1α, MIP-2, IP-10/CXCL10, MCP-1, TCA3/CCL1, and the housekeeping genes L32 and GAPDH. The protected bands are identified. The gel was deliberately overexposed to reveal relative chemokine production following stimulation with MIP-1α or MIP-1β versus RANTES.

cal techniques. Some patients with multiple sclerosis display CCR3-bearing astrocytes in areas around vessels and at the glia limitans [50]. In contrast, cultured mouse astrocytes express CCR1 and CCR5, but messenger RNA for CCR3 was not detected [32]. Whether the apparent discrepancy of CCR3 expression is a species difference between mouse and man or reflects astrocyte heterogeneity is unknown. RANTES also supports survival of human fetal cortical astrocytes, through a mechanism that requires interferon-gamma [87].

Eotaxin is a highly selective ligand that primarily binds CCR3 to transduce signals [88]. Eotaxin induces astrocyte chemotaxis [31], synthesis of chemokines and cytokines (Fig. 5), and receptor modulation [32]. Eotaxin stimulates astrocytes through a Gαi protein-associated receptor as demonstrated by pertussis toxin-sensitive inhibition of forskolin induced cAMP responses [53] and accumulation of inositol 1,4,5-triphosphate [89]. However, eotaxin and RANTES stimulation of chemokine synthesis may involve signaling mechanisms that are insensitive to pertussis toxin [32]. Dissecting the complexity of this collection of chemokine receptors and ligands has proven to be challenging as the redundancy of the chemokines, promiscuity of their receptors, and the entanglement of signaling mechanisms continue to confuse the interpretation of results. This difficulty is highlighted by the finding that astrocytes from CCR1-, CCR3-, and CCR5-deficient mice respond to RANTES and eotaxin indicating that none of these receptors is solely responsible for controlling responsiveness to RANTES or eotaxin (Fig. 5).

Among the first reports suggesting the importance of chemokines in CNS inflammatory diseases was the finding in mice with experimental allergic encephalomyelitis that astrocytes located within the inflammatory cuffs contained MCP-1 [90,91]. Anti-MCP-1 antibodies reduced relapses of this experimental disease, confirming a role for MCP-1 in pathogenesis [92]. Transgenic expression of MCP-1 in oligodendrocytes induces perivascular accumulation of monocytes/macrophages without parenchymal infiltration or evidence of neurological damage [93]. Even though MCP-1 was produced in the parenchyma it was localized near astrocytic endfeet

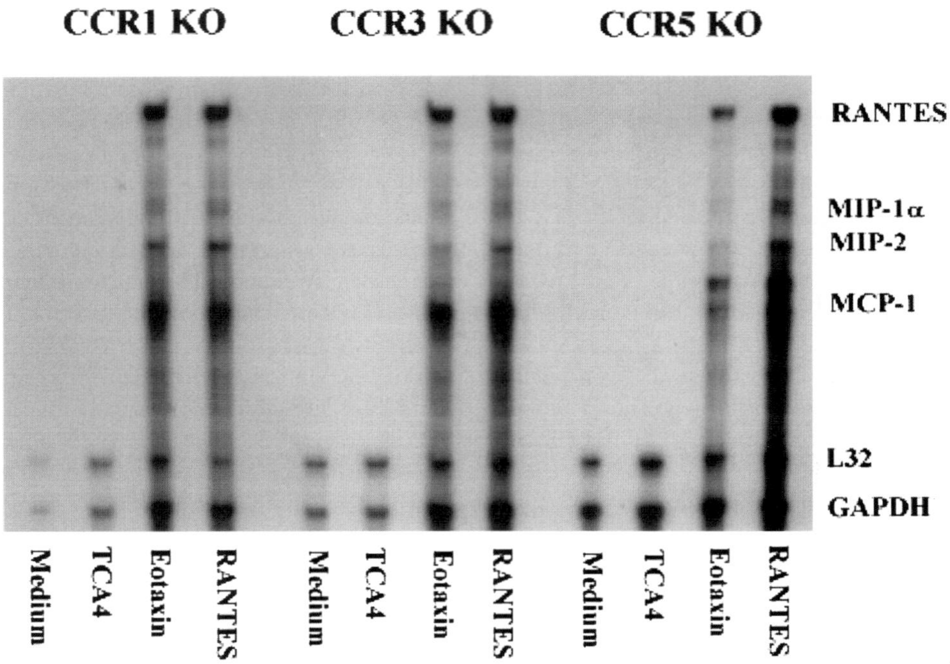

Figure 5. RANTES and eotaxin stimulate astrocytes from CCR1-, CCR3-, and CCR5-deficient mice. Cultured astrocytes derived from mice genetically deficient for CCR1, CCR3, or CCR5 were stimulated with either medium, 100 ng/ml eotaxin, 100 ng/ml RANTES, or 20 ng/ml TNF-α for 16 h. RNA was prepared and examined by RNase protection assay for expression of chemokine transcripts. Protected bands for RANTES, MIP-1α, MIP-2, MCP-1, L32, and GAPDH are identified.

at the abluminal surface of the endothelium [93]. Evidence for the transport of selected chemokines to specific capillary locations has been reported [94,95] but the mechanisms remain unknown. We conclude that production of MCP-1 within the CNS causes disruption of the blood–brain barrier and recruitment of monocytes into the perivascular space. This model of chemokine function has been affirmed with in vitro studies [96], but the nature of the chemokine–target cell interactions was not dissected.

By varying the concentrations of MCP-1 on either side of the Boyden chamber's membrane it was demonstrated that astrocyte movements were only directed toward the ligand (chemotaxis) rather than triggering random migration (chemokinesis) [22]. The potent chemotactic activity of this ligand was noted as concentrations of $\geq$0.1 nM stimulated chemotaxis [22]. These observations are consistent with the high affinity of MCP-1 binding ($K_d = 1$ nM) to cultured human astrocytes [68]. Competitive inhibition studies using MCP3 and MIP-1α suggested the specificity of the human MCP-1 receptor was compatible with that for CCR2 [68]. Furthermore, treatment with cytokines regulated expression of astrocyte MCP-1 binding sites [68]. One group using immunofluorescence with quantitative image analysis reported the direct demonstration of CCR2 on cultured astrocytes [68]. Another group noted up-regulation of CCR2 expression following HIV infection of fetal astrocyte cultures or an astrocytic cell line [40]. However, we failed to detect CCR2 mRNA in mouse astrocytes even though the cells were capable of directed migratory responses to the MCP-1 ligand [22]. The nature of the receptor(s) that binds MCP-1 [97] to mouse astrocytes remains unknown.

4. EXPRESSION OF FRACTALKINE AND ORPHAN CHEMOKINE RECEPTORS ON ASTROCYTES

Fractalkine/CX3CL1 is the only known member of the CX3C class of chemokines. Fractalkine consists of a chemokine domain attached to a membrane-spanning domain via a mucin-like stalk [98,99]. Fractalkine is expressed on neurons [46,100] and endothelial cells [101], especially those in brain. Fractalkine is up-regulated in rodents with experimental allergic encephalomyelitis [99] or following treatment of astrocyte cultures with cytokines [102]. However, fractalkine can be proteolytically cleaved from its membrane anchor to release a soluble ligand. Both the full-length and soluble forms of fractalkine demonstrated high affinity binding on astrocytes with a disassociation constant of 0.1–0.2 nM [102]. The presence of CX3CR1 on rodent astrocytes and microglia was demonstrated by Northern blot [42], PCR [31], and flow cytometry [102]. Fractalkine-mediated calcium mobilization was used to establish CX3CR1 function in astrocytes [102]. The involvement of Gαi proteins in fractalkine-mediated signaling was suggested by inhibition of the calcium flux response with pertussis toxin and anti-CX3CR1 antibody inhibited ligand-induced calcium mobilization [102]. Although fractalkine failed to induce chemotaxis [102] or chemokine synthesis (Luo, unpublished), one of the intriguing effects of fractalkine on astrocytes was secretion of a factor that promoted microglial proliferation, suggesting an additional mechanism for astrocyte–glial communications [102].

The ability of some chemokines (e.g. KC, MIP-2, RANTES, eotaxin) to affect astrocyte movements may be sufficient to loosen astrocyte endfeet interactions with the endothelium thereby disturbing the integrity of the blood–brain barrier. Furthermore, astrocyte stimulation with the same chemokines down-regulates CX3CR1 expression (Fig. 2). The down-modulation of CX3CR1 may further contribute to disassociation of astrocytes from fractalkine-bearing endothelial cells promoting disruption of the blood–brain barrier.

There are several seven transmembrane spanning G-protein-coupled receptors that share sequence homology with the conventional series of chemoattractant receptors, but for which endogenous ligands have not yet been identified. Some of these orphan molecules function as co-receptors for infection with human or simian immunodeficiency viruses [103–105]. RT-PCR, Northern blotting, and RNase protection assays have been used to detect transcripts of orphan chemokine receptors in astrocytes. Additional data presented in Fig. 6 extend the list of orphan receptors expressed by mouse astrocytes. The astrocyte orphan receptors include the murine homologues of Bonzo [104,106], ChemR23 [107], and Apj [103,108] which can potentially function as co-receptors for either HIV-1 or SIV. Astrocytes also express CCR5, CXCR4 and CX3CR1, all of which can also serve as co-receptors for selected HIV-1 strains. HIV-1 infection of astrocytes has been reported for both pediatric and adult AIDS [109,110]. The mechanisms involved in HIV-1 entry into primary astrocytes are not well understood, but the potentially synergistic expression of so many HIV-permissive chemoattractant receptors provides a theoretical mechanism for direct interactions between astrocytes and virus. In addition, astrocytes express message for RDC1 [111] and a murine GPR30 homologue [112], which display homologies to CXC receptors, and to Eo1 which shares homology with members of the CC receptor family [32]. Finally mouse astrocytes also express CRTH2 which is related to the *N*-formyl peptide, C5a, and chemokine receptors [113]. Definition of the role these orphan receptors play in CNS physiology and/or pathology awaits characterization of their endogenous ligands.

5. CONCLUSIONS

The scope of chemokine actions has been extended beyond their traditional roles in leukocyte biology. Astrocytes, the most abundant cell type in the CNS, mount autocrine or paracrine responses to nanomolar concentrations of several chemokines including KC, MIP-2, RANTES and eotaxin. Chemokines stimulate a variety of astrocyte responses in vitro including cell migration, cytokine and chemokine production, and modulation of heterologous receptors. In vivo evidence supporting each type of response was reported using genetically engineered mouse models in which chemokine expression was targeted to the CNS or in inflammatory diseases affecting the CNS.

Although there is ample evidence at the RNA and protein level that chemokine receptors are expressed on astrocytes, the functional significance of these astrocyte receptors remains an item of conjecture. Although most astrocyte chemokine receptors are expressed in both primates and rodents, species differences may regulate receptor distribution for some chemokine receptors. The signaling mechanisms responsible for triggering astrocyte responses are under investigation, with preliminary data hinting that distinct signaling pathways may differentially control astrocyte responsiveness. The receptor modula-

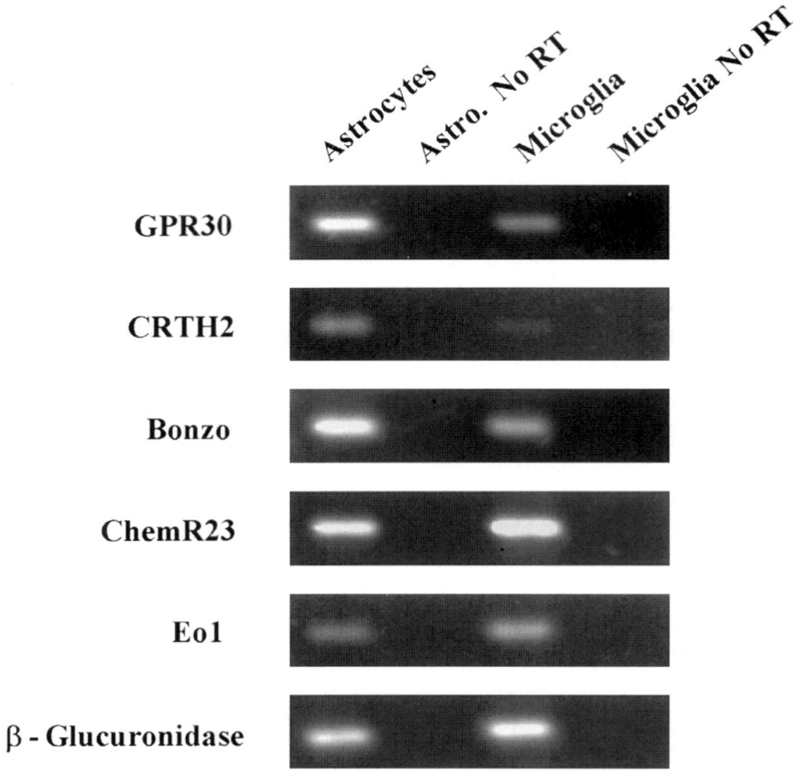

Figure 6. Mouse astrocytes express orphan chemokine receptors. Representative reverse transcription (RT)-PCR results using primers for mouse GPR30, CHTH2, Bonzo, ChemR23, Eo1, and the housekeeping gene β-glucuronidase. Astrocyte or microglial RNA treated with or without RT enzyme followed by 28–32 cycles of PCR. Positive controls for each primer pair revealed bands at the predicted sizes. Procedural details are described elsewhere [22,69].

tion that accompanies chemokine stimulation further contributes to the dynamic nature of astrocytes.

ACKNOWLEDGEMENTS

Portions of this review have been presented elsewhere [31]. This study was supported by National Institute of Health Grants CA67416 and NS37284 and by the Multiple Sclerosis Society Grant 2989-A-2. The authors would like to thank Drs. C. Gerard and W. Kuziel for the generous gifts of chemokine receptor-deficient mice.

REFERENCES

1. Aschner M. Astrocytic functions and physiological reactions to injury: the potential to induce and/or exacerbate neuronal dysfunction — a forum position paper. Neurotoxicology 1998;19(1):7–17.
2. Ojeda SR, Ma YJ. Glial–neuronal interactions in the neuroendocrine control of mammalian puberty: facilitatory effects of gonadal steroids. J Neurobiol 1999;40(4): 528–540.
3. Porter JT, McCarthy KD. Hippocampal astrocytes in situ respond to glutamate released from synaptic terminals. J Neurosci 1996;16(16):5073–5081.
4. Pasti L, Volterra A, Pozzan T, Carmignoto G. Intracellular calcium oscillations in astrocytes: a highly plastic, bidirectional form of communication between neurons and astrocytes in situ. J Neurosci 1997;17(20):7817–7830.
5. Bezzi P, Carmignoto G, Pasti L, Vesce S, Rossi D, Rizzini BL et al. Prostaglandins stimulate calcium-dependent glutamate release in astrocytes. Nature 1998; 391(6664):281–285.
6. Hertz L, Dringen R, Schousboe A, Robinson SR. Astrocytes: glutamate producers for neurons. J Neurosci Res 1999;57(4):417–428.
7. Rothstein JD, Dykes-Hoberg M, Pardo CA, Bristol LA, Jin L, Kuncl RW et al. Knockout of glutamate transporters reveals a major role for astroglial transport in excitotoxicity and clearance of glutamate. Neuron 1996;

16(3):675–686.
8. Robinson S, Tani M, Strieter RM, Ransohoff RM, Miller RH. The chemokine growth-regulated oncogene-alpha promotes spinal cord oligodendrocyte precursor proliferation. J Neurosci 1998;18(24):10457–10463.
9. Janzer RC, Raff MC. Astrocytes induce blood–brain barrier properties in endothelial cells. Nature 1987; 325(6101):253–257.
10. Bush TG, Puvanachandra N, Horner CH, Polito A, Ostenfeld T, Svendsen CN et al. Leukocyte infiltration, neuronal degeneration, and neurite outgrowth after ablation of scar-forming, reactive astrocytes in adult transgenic mice. Neuron 1999;23(2):297–308.
11. Lane TE, Asensio VC, Yu N, Paoletti AD, Campbell IL, Buchmeier MJ. Dynamic regulation of alpha- and beta-chemokine expression in the central nervous system during mouse hepatitis virus-induced demyelinating disease. J Immunol 1998;160(2):970–978.
12. Sauder C, Hallensleben W, Pagenstecher A, Schneckenburger S, Biro L, Pertlik D et al. Chemokine gene expression in astrocytes of Borna disease virus-infected rats and mice in the absence of inflammation. J Virol 2000;74(19):9267–9280.
13. Xia MQ, Hyman BT. Chemokines/chemokine receptors in the central nervous system and Alzheimer's disease. J Neurovirol 1999;5(1):32–41.
14. Tani M, Glabinski AR, Tuohy VK, Stoler MH, Estes ML, Ransohoff RM. In situ hybridization analysis of glial fibrillary acidic protein mRNA reveals evidence of biphasic astrocyte activation during acute experimental autoimmune encephalomyelitis. Am J Pathol 1996; 148(3):889–896.
15. Van Der Voorn P, Tekstra J, Beelen RH, Tensen CP, Van Der Valk P, De Groot CJ. Expression of MCP-1 by reactive astrocytes in demyelinating multiple sclerosis lesions. Am J Pathol 1999;154(1):45–51.
16. Zlotnik A, Yoshie O. Chemokines: a new classification system and their role in immunity. Immunity 2000;12(2): 121–127.
17. Hebert CA, Vitangcol RV, Baker JB. Scanning mutagenesis of interleukin-8 identifies a cluster of residues required for receptor binding. J Biol Chem 1991;266(28): 18989–18994.
18. Baggiolini M, Dewald B, Moser B. Human chemokines: an update. Annu Rev Immunol 1997;15:675–705.
19. Luster AD. Chemokines — chemotactic cytokines that mediate inflammation. New Engl J Med 1998;338(7): 436–445.
20. Gosling J, Dairaghi DJ, Wang Y, Hanley M, Talbot D, Miao Z et al. Cutting edge: identification of a novel chemokine receptor that binds dendritic cell- and T cell-active chemokines including ELC, SLC, and TECK. J Immunol 2000;164(6):2851–2856.
21. Baggiolini M. Chemokines and leukocyte traffic. Nature 1998;392(6676):565–568.
22. Heesen M, Tanabe S, Berman MA, Yoshizawa I, Luo Y, Kim RJ et al. Mouse astrocytes respond to the chemokines MCP-1 and KC, but reverse transcriptase-polymerase chain reaction does not detect mRNA for the KC or new MCP-1 receptor. J Neurosci Res 1996;45(4): 382–391.
23. Araque A, Sanzgiri RP, Parpura V, Haydon PG. Astrocyte-induced modulation of synaptic transmission. Can J Physiol Pharmacol 1999;77(9):699–706.
24. Rus HG, Kim LM, Niculescu FI, Shin ML. Induction of C3 expression in astrocytes is regulated by cytokines and Newcastle disease virus. J Immunol 1992;148(3): 928–933.
25. Haskill S, Peace A, Morris J, Sporn SA, Anisowicz A, Lee SW et al. Identification of three related human GRO genes encoding cytokine functions. Proc Natl Acad Sci USA 1990;87(19):7732–7736.
26. Tekamp-Olson P, Gallegos C, Bauer D, McClain J, Sherry B, Fabre M et al. Cloning and characterization of cDNAs for murine macrophage inflammatory protein 2 and its human homologues. J Exp Med 1990;172(3): 911–919.
27. Bozic CR, Kolakowski LF Jr, Gerard NP, Garcia-Rodriguez C, von Uexkull-Guldenband C, Conklyn MJ et al. Expression and biologic characterization of the murine chemokine KC. J Immunol 1995;154(11):6048–6057.
28. Luo Y, Fischer FR, Hancock WW, Dorf ME. Macrophage inflammatory protein-2 and KC induce chemokine production by mouse astrocytes. J Immunol 2000;165(7):4015–4023.
29. Luo Y, Lloyd C, Gutierrez–Ramos JC, Dorf ME. Chemokine amplification in mesangial cells. J Immunol 1999; 163(7):3985–3992.
30. Wen DZ, Rowland A, Derynck R. Expression and secretion of gro/MGSA by stimulated human endothelial cells. EMBO J 1989;8(6):1761–1766.
31. Dorf ME, Berman MA, Tanabe S, Heesen M, Luo Y. Astrocytes express functional chemokine receptors. J Neuroimmunol 2000;111(1–2):109–121.
32. Luo Y, Berman MA, Fischer FR, Abromson-Leeman SR, Zhang Y, Kuziel WA et al. RANTES stimulates inflammatory cascades and receptor modulation in murine astrocytes, submitted.
33. Lazarini F, Casanova P, Tham TN, De Clercq E, Arenzana-Seisdedos F, Baleux F et al. Differential signalling of the chemokine receptor CXCR4 by stromal cell-derived factor 1 and the HIV glycoprotein in rat neurons and astrocytes. Eur J Neurosci 2000;12(1):117–125.
34. Bell MD, Taub DD, Kunkel SJ, Strieter RM, Foley R, Gauldie J et al. Recombinant human adenovirus with rat MIP-2 gene insertion causes prolonged PMN recruitment to the murine brain. Eur J Neurosci 1996;8(9):1803–1811.
35. Tani M, Fuentes ME, Peterson JW, Trapp BD, Durham SK, Loy JK et al. Neutrophil infiltration, glial reaction, and neurological disease in transgenic mice expressing the chemokine N51/KC in oligodendrocytes. J Clin Invest 1996;98(2):529–539.
36. Giovannelli A, Limatola C, Ragozzino D, Mileo AM, Ruggieri A, Ciotti MT et al. CXC chemokines interleukin-8 (IL–8) and growth-related gene product alpha (GROalpha) modulate Purkinje neuron activity in mouse cerebellum. J Neuroimmunol 1998;92(1–2):122–132.

37. Horuk R, Martin AW, Wang Z, Schweitzer L, Gerassimides A, Guo H et al. Expression of chemokine receptors by subsets of neurons in the central nervous system. J Immunol 1997;158(6):2882–2890.
38. Lacy M, Jones J, Whittemore SR, Haviland DL, Wetsel RA, Barnum SR. Expression of the receptors for the C5a anaphylatoxin, interleukin-8 and FMLP by human astrocytes and microglia. J Neuroimmunol 1995;61(1): 71–78.
39. Muller-Ladner U, Jones JL, Wetsel RA, Gay S, Raine CS, Barnum SR. Enhanced expression of chemotactic receptors in multiple sclerosis lesions. J Neurol Sci 1996; 144(1–2):135–141.
40. Cota M, Kleinschmidt A, Ceccherini-Silberstein F, Aloisi F, Mengozzi M, Mantovani A et al. Upregulated expression of interleukin-8, RANTES and chemokine receptors in human astrocytic cells infected with HIV-1. J Neurovirol 2000;6(1):75–83.
41. Ghorpade A, Xia MQ, Hyman BT, Persidsky Y, Nukuna A, Bock P et al. Role of the beta-chemokine receptors CCR3 and CCR5 in human immunodeficiency virus type 1 infection of monocytes and microglia. J Virol 1998; 72(4):3351–3361.
42. Jiang Y, Salafranca MN, Adhikari S, Xia Y, Feng L, Sonntag MK et al. Chemokine receptor expression in cultured glia and rat experimental allergic encephalomyelitis. J Neuroimmunol 1998;86(1):1–12.
43. Reeves JD, Hibbitts S, Simmons G, McKnight A, Azevedo-Pereira JM, Moniz-Pereira J et al. Primary human immunodeficiency virus type 2 (HIV-2) isolates infect CD4-negative cells via CCR5 and CXCR4: comparison with HIV-1 and simian immunodeficiency virus and relevance to cell tropism in vivo. J Virol 1999;73(9): 7795–7804.
44. Sabri F, Tresoldi E, Di Stefano M, Polo S, Monaco MC, Verani A et al. Nonproductive human immunodeficiency virus type 1 infection of human fetal astrocytes: independence from CD4 and major chemokine receptors. Virology 1999;264(2):370–384.
45. McManus CM, Weidenheim K, Woodman SE, Nunez J, Hesselgesser J, Nath A et al. Chemokine and chemokine-receptor expression in human glial elements: induction by the HIV protein, tat, and chemokine autoregulation. Am J Pathol 2000;156(4):1441–1453.
46. Bacon KB, Harrison JK. Chemokines and their receptors in neurobiology: perspectives in physiology and homeostasis. J Neuroimmunol 2000;104(1):92–97.
47. Zelenaia O, Schlag BD, Gochenauer GE, Ganel R, Song W, Beesley JS et al. Epidermal growth factor receptor agonists increase expression of glutamate transporter GLT-1 in astrocytes through pathways dependent on phosphatidylinositol 3-kinase and transcription factor NF-kappaB. Mol Pharmacol 2000;57(4):667–678.
48. Klein RS, Williams KC, Alvarez-Hernandez X, Westmoreland S, Force T, Lackner AA et al. Chemokine receptor expression and signaling in macaque and human fetal neurons and astrocytes: implications for the neuropathogenesis of AIDS. J Immunol 1999;163(3):1636–1646.
49. Galasso JM, Harrison JK, Silverstein FS. Excitotoxic brain injury stimulates expression of the chemokine receptor CCR5 in neonatal rats. Am J Pathol 1998;153(5): 1631–1640.
50. Simpson J, Rezaie P, Newcombe J, Cuzner ML, Male D, Woodroofe MN. Expression of the beta-chemokine receptors CCR2, CCR3 and CCR5 in multiple sclerosis central nervous system tissue. J Neuroimmunol 2000; 108(1–2):192–200.
51. Tanabe S, Heesen M, Yoshizawa I, Berman MA, Luo Y, Bleul CC et al. Functional expression of the CXC-chemokine receptor-4/fusin on mouse microglial cells and astrocytes. J Immunol 1997;159(2):905–911.
52. Bajetto A, Bonavia R, Barbero S, Piccioli P, Costa A, Florio T et al. Glial and neuronal cells express functional chemokine receptor CXCR4 and its natural ligand stromal cell-derived factor 1. J Neurochem 1999;73(6): 2348–2357.
53. Zheng J, Ghorpade A, Niemann D, Cotter RL, Thylin MR, Epstein L et al. Lymphotropic virions affect chemokine receptor-mediated neural signaling and apoptosis: implications for human immunodeficiency virus type 1-associated dementia. J Virol 1999;73(10):8256–8267.
54. Bezzi P, Domercq M, Brambilla L, Galli R, Schols D, De Clercq E, Vescovi A, Bagetta G, Kollias G, Medolesi J, Volterra A. CXCR4-activated astrocyte glutamate release via TNFalpha: amplification by microglia triggers neurotoxicity. Nat Neurosci 2001;4:702–710.
55. Heesen M, Berman MA, Hopken UE, Gerard NP, Dorf ME. Alternate splicing of mouse fusin/CXC chemokine receptor-4: stromal cell-derived factor-1alpha is a ligand for both CXC chemokine receptor-4 isoforms. J Immunol 1997;158(8):3561–3564.
56. Gupta SK, Pillarisetti K. Cutting edge: CXCR4-Lo: molecular cloning and functional expression of a novel human CXCR4 splice variant. J Immunol 1999;163(5): 2368–2372.
57. Heesen M, Berman MA, Benson JD, Gerard C, Dorf ME. Cloning of the mouse fusin gene, homologue to a human HIV-1 co-factor. J Immunol 1996;157(12):5455–5460.
58. Ohtani Y, Minami M, Kawaguchi N, Nishiyori A, Yamamoto J, Takami S et al. Expression of stromal cell-derived factor-1 and CXCR4 chemokine receptor mRNAs in cultured rat glial and neuronal cells. Neurosci Lett 1998;249(2–3):163–166.
59. Lavi E, Strizki JM, Ulrich AM, Zhang W, Fu L, Wang Q et al. CXCR-4 (Fusin), a co-receptor for the type 1 human immunodeficiency virus (HIV-1), is expressed in the human brain in a variety of cell types, including microglia and neurons. Am J Pathol 1997;151(4):1035–1042.
60. Sanders VJ, Everall IP, Johnson RW, Masliah E. Fibroblast growth factor modulates HIV coreceptor CXCR4 expression by neural cells. HNRC Group. J Neurosci Res 2000;59(5):671–679.
61. Hesselgesser J, Horuk R. Chemokine and chemokine receptor expression in the central nervous system. J Neurovirol 1999;5(1):13–26.
62. Sanders VJ, Pittman CA, White MG, Wang G, Wiley CA, Achim CL. Chemokines and receptors in HIV encephalitis. AIDS 1998;12(9):1021–1026.

63. Han Y, Wang J-T, Ransohoff RM. TNF-α downregulates CXCR4 in primary murine astrocytes. Brain Res 2001; 888:1–10.
64. Lipton SA. Neuronal injury associated with HIV-1: approaches to treatment. Annu Rev Pharmacol Toxicol 1998;38:159–177.
65. Nagasawa T, Hirota S, Tachibana K, Takakura N, Nishikawa S, Kitamura Y et al. Defects of B-cell lymphopoiesis and bone-marrow myelopoiesis in mice lacking the CXC chemokine PBSF/SDF-1. Nature 1996; 382(6592):635–638.
66. Zou YR, Kottmann AH, Kuroda M, Taniuchi I, Littman DR. Function of the chemokine receptor CXCR4 in haematopoiesis and in cerebellar development. Nature 1998;393(6685):595–599.
67. Ma Q, Jones D, Borghesani PR, Segal RA, Nagasawa T, Kishimoto T et al. Impaired B-lymphopoiesis, myelopoiesis, and derailed cerebellar neuron migration in CXCR4- and SDF-1-deficient mice. Proc Natl Acad Sci USA 1998;95(16):9448–9453.
68. Andjelkovic AV, Kerkovich D, Shanley J, Pulliam L, Pachter JS. Expression of binding sites for beta chemokines on human astrocytes. Glia 1999;28(3):225–235.
69. Tanabe S, Heesen M, Berman MA, Fischer MB, Yoshizawa I, Luo Y et al. Murine astrocytes express a functional chemokine receptor. J Neurosci 1997;17(17): 6522–6528.
70. Han Y, Wang J, Zhou Z, Ransohoff RM. TGFbeta1 selectively up-regulates CCR1 expression in primary murine astrocytes. Glia 2000;30(1):1–10.
71. Post TW, Bozic CR, Rothenberg ME, Luster AD, Gerard N, Gerard C. Molecular characterization of two murine eosinophil beta chemokine receptors. J Immunol 1995; 155(11):5299–5305.
72. Gao JL, Murphy PM. Cloning and differential tissue-specific expression of three mouse beta chemokine receptor-like genes, including the gene for a functional macrophage inflammatory protein-1 alpha receptor. J Biol Chem 1995;270(29):17494–17501.
73. Hoogewerf A, Black D, Proudfoot AE, Wells TN, Power CA. Molecular cloning of murine CC CKR-4 and high affinity binding of chemokines to murine and human CC CKR-4. Biochem Biophys Res Commun 1996;218(1): 337–343.
74. Power CA, Meyer A, Nemeth K, Bacon KB, Hoogewerf AJ, Proudfoot AE et al. Molecular cloning and functional expression of a novel CC chemokine receptor cDNA from a human basophilic cell line. J Biol Chem 1995; 270(33):19495–19500.
75. Meyer A, Coyle AJ, Proudfoot AE, Wells TN, Power CA. Cloning and characterization of a novel murine macrophage inflammatory protein-1 alpha receptor. J Biol Chem 1996;271(24):14445–14451.
76. Bajetto A, Bonavia R, Barbero S, Florio T, Costa A, Schettini G. Expression of chemokine receptors in the rat brain. Ann New York Acad Sci 1999;876:201–209.
77. Rottman JB, Ganley KP, Williams K, Wu L, Mackay CR, Ringler DJ. Cellular localization of the chemokine receptor CCR5. Correlation to cellular targets of HIV-1 infection. Am J Pathol 1997;151(5):1341–1351.
78. Westmoreland SV, Rottman JB, Williams KC, Lackner AA, Sasseville VG. Chemokine receptor expression on resident and inflammatory cells in the brain of macaques with simian immunodeficiency virus encephalitis. Am J Pathol 1998;152(3):659–665.
79. Speth C, Joebstl B, Barcova M, Dierich MP. HIV-1 envelope protein gp41 modulates expression of interleukin-10 and chemokine receptors on monocytes, astrocytes and neurones. AIDS 2000;14(6):629–636.
80. Gao JL, Wynn TA, Chang Y, Lee EJ, Broxmeyer HE, Cooper S et al. Impaired host defense, hematopoiesis, granulomatous inflammation and type 1–type 2 cytokine balance in mice lacking CC chemokine receptor 1. J Exp Med 1997;185(11):1959–1968.
81. Gerard C, Frossard JL, Bhatia M, Saluja A, Gerard NP, Lu B et al. Targeted disruption of the beta-chemokine receptor CCR1 protects against pancreatitis-associated lung injury. J Clin Invest 1997;100(8):2022–2027.
82. Zhou Y, Kurihara T, Ryseck RP, Yang Y, Ryan C, Loy J et al. Impaired macrophage function and enhanced T cell-dependent immune response in mice lacking CCR5, the mouse homologue of the major HIV-1 coreceptor. J Immunol 1998;160(8):4018–4025.
83. Sato N, Kuziel WA, Melby PC, Reddick RL, Kostecki V, Zhao W et al. Defects in the generation of IFN-gamma are overcome to control infection with Leishmania donovani in CC chemokine receptor (CCR) 5-, macrophage inflammatory protein-1 alpha-, or CCR2-deficient mice. J Immunol 1999;163(10):5519–5525.
84. Samson M, Labbe O, Mollereau C, Vassart G, Parmentier M. Molecular cloning and functional expression of a new human CC-chemokine receptor gene. Biochemistry 1996;35(11):3362–3367.
85. Neote K, DiGregorio D, Mak JY, Horuk R, Schall TJ. Molecular cloning, functional expression, and signaling characteristics of a C–C chemokine receptor. Cell 1993; 72(3):415–425.
86. Ponath PD, Qin S, Post TW, Wang J, Wu L, Gerard NP et al. Molecular cloning and characterization of a human eotaxin receptor expressed selectively on eosinophils. J Exp Med 1996;183(6):2437–2448.
87. Bakhiet M, Tjernlund A, Mousa A, Gad A, Stromblad S, Kuziel WA, Seiger A, Anderson J. RANTES promotes growth and survival of human first-trimester forebrain astrocytes. Nat Cell Biol 2001;3:150–157.
88. Kitaura M, Nakajima T, Imai T, Harada S, Combadiere C, Tiffany HL et al. Molecular cloning of human eotaxin, an eosinophil-selective CC chemokine, and identification of a specific eosinophil eotaxin receptor, CC chemokine receptor 3. J Biol Chem 1996;271(13):7725–7730.
89. Zheng J, Thylin MR, Ghorpade A, Xiong H, Persidsky Y, Cotter R et al. Intracellular CXCR4 signaling, neuronal apoptosis and neuropathogenic mechanisms of HIV-1-associated dementia. J Neuroimmunol 1999;98(2):185–200.
90. Ransohoff RM, Hamilton TA, Tani M, Stoler MH, Shick HE, Major JA et al. Astrocyte expression of mRNA encoding cytokines IP-10 and JE/MCP-1 in experimental autoimmune encephalomyelitis. FASEB J 1993;7(6): 592–600.

91. Hulkower K, Brosnan CF, Aquino DA, Cammer W, Kulshrestha S, Guida MP et al. Expression of CSF-1, c-fms, and MCP-1 in the central nervous system of rats with experimental allergic encephalomyelitis. J Immunol 1993;150(6):2525–2533.
92. Karpus WJ, Kennedy KJ. MIP-1alpha and MCP-1 differentially regulate acute and relapsing autoimmune encephalomyelitis as well as Th1/Th2 lymphocyte differentiation. J Leukocyte Biol 1997;62(5):681–7.
93. Fuentes ME, Durham SK, Swerdel MR, Lewin AC, Barton DS, Megill JR et al. Controlled recruitment of monocytes and macrophages to specific organs through transgenic expression of monocyte chemoattractant protein-1. J Immunol 1995;155(12):5769–5776.
94. Middleton J, Neil S, Wintle J, Clark-Lewis I, Moore H, Lam C et al. Transcytosis and surface presentation of IL-8 by venular endothelial cells. Cell 1997;91(3):385–395.
95. Stein JV, Rot A, Luo Y, Narasimhaswamy M, Nakano H, Gunn MD et al. The CC chemokine thymus-derived chemotactic agent 4 (TCA-4, secondary lymphoid tissue chemokine, 6Ckine, exodus-2) triggers lymphocyte function-associated antigen 1-mediated arrest of rolling T lymphocytes in peripheral lymph node high endothelial venules. J Exp Med 2000;191(1):61–76.
96. Weiss JM, Nath A, Major EO, Berman JW. HIV-1 Tat induces monocyte chemoattractant protein-1-mediated monocyte transmigration across a model of the human blood–brain barrier and up-regulates CCR5 expression on human monocytes. J Immunol 1999;163(5):2953–2959.
97. Chaudhuri A, Zbrzezna V, Polyakova J, Pogo AO, Hesselgesser J, Horuk R. Expression of the Duffy antigen in K562 cells. Evidence that it is the human erythrocyte chemokine receptor. J Biol Chem 1994;269(11):7835–7838.
98. Bazan JF, Bacon KB, Hardiman G, Wang W, Soo K, Rossi D et al. A new class of membrane-bound chemokine with a CX3C motif. Nature 1997;385(6617):640–644.
99. Pan Y, Lloyd C, Zhou H, Dolich S, Deeds J, Gonzalo JA et al. Neurotactin, a membrane-anchored chemokine upregulated in brain inflammation. Nature 1997;387(6633):611–617.
100. Harrison JK, Jiang Y, Chen S, Xia Y, Maciejewski D, McNamara RK et al. Role for neuronally derived fractalkine in mediating interactions between neurons and CX3CR1-expressing microglia. Proc Natl Acad Sci USA 1998;95(18):10896–10901.
101. Fischer FR, Santambrogio L, Luo Y, Berman MA, Hancock WW, Dorf ME. Modulation of experimental autoimmune encephalomyelitis: effect of altered peptide ligand on chemokine and chemokine receptor expression. J Neuroimmunol 2000;110(1–2):195–208.
102. Maciejewski-Lenoir D, Chen S, Feng L, Maki R, Bacon KB. Characterization of fractalkine in rat brain cells: migratory and activation signals for CX3CR-1-expressing microglia. J Immunol 1999;163(3):1628–1635.
103. Choe H, Farzan M, Konkel M, Martin K, Sun Y, Marcon L et al. The orphan seven-transmembrane receptor apj supports the entry of primary T-cell-line-tropic and dualtropic human immunodeficiency virus type 1. J Virol 1998;72(7):6113–6118.
104. Deng HK, Unutmaz D, KewalRamani VN, Littman DR. Expression cloning of new receptors used by simian and human immunodeficiency viruses. Nature 1997;388(6639):296–300.
105. Shimizu N, Soda Y, Kanbe K, Liu HY, Mukai R, Kitamura T et al. A putative G protein-coupled receptor, RDC1, is a novel coreceptor for human and simian immunodeficiency viruses. J Virol 2000;74(2):619–626.
106. Liao F, Alkhatib G, Peden KW, Sharma G, Berger EA, Farber JM. STRL33, A novel chemokine receptor-like protein, functions as a fusion cofactor for both macrophage-tropic and T cell line-tropic HIV-1. J Exp Med 1997;185(11):2015–2023.
107. Samson M, Edinger AL, Stordeur P, Rucker J, Verhasselt V, Sharron M et al. ChemR23, a putative chemoattractant receptor, is expressed in monocyte-derived dendritic cells and macrophages and is a coreceptor for SIV and some primary HIV-1 strains. Eur J Immunol 1998;28(5):1689–1700.
108. Choe W, Albright A, Sulcove J, Jaffer S, Hesselgesser J, Lavi E et al. Functional expression of the seven-transmembrane HIV-1 co-receptor APJ in neural cells. J Neurovirol 2000;6(Suppl 1):S61–69.
109. Bagasra O, Lavi E, Bobroski L, Khalili K, Pestaner JP, Tawadros R et al. Cellular reservoirs of HIV-1 in the central nervous system of infected individuals: identification by the combination of in situ polymerase chain reaction and immunohistochemistry. AIDS 1996;10(6):573–585.
110. Lavi E, Kolson DL, Ulrich AM, Fu L, Gonzalez-Scarano F. Chemokine receptors in the human brain and their relationship to HIV infection. J Neurovirol 1998;4(3):301–311.
111. Heesen M, Berman MA, Charest A, Housman D, Gerard C, Dorf ME. Cloning and chromosomal mapping of an orphan chemokine receptor: mouse RDC1. Immunogenetics 1998;47(5):364–370.
112. O'Dowd BF, Nguyen T, Marchese A, Cheng R, Lynch KR, Heng HH et al. Discovery of three novel G-protein-coupled receptor genes. Genomics 1998;47(2):310–313.
113. Abe H, Takeshita T, Nagata K, Arita T, Endo Y, Fujita T et al. Molecular cloning, chromosome mapping and characterization of the mouse CRTH2 gene, a putative member of the leukocyte chemoattractant receptor family. Gene 1999;227(1):71–77.

Universes in Delicate Balance: Chemokines and the Nervous System
R.M. Ransohoff, K. Suzuki, A.E.I. Proudfoot, W.F. Hickey and J.K. Harrison, editors

Chapter 4.4

Why Do Neurons Express Chemokine Receptors?

Richard J. Miller and Seog Bae Oh

Northwestern University Medical School, Department of Molecular Pharmacology and Biology Chemistry, 303 E. Chicago Ave., Chicago, IL 60611, USA

1. INTRODUCTION

Chemokines and their receptors constitute a large family of 'CHEMOtactic cytoKINES' that play a fundamental role in the control of leukocyte chemotaxis. Chemokines are small proteins that fall into four related subfamilies as determined by specific structural parameters. All of these molecules produce their effects by activating a family of G-protein-coupled receptors (GPCRs). These chemokine receptors also fall into related subfamilies, each receptor potentially transducing signals from one or more chemokines. The precise structures and nomenclature for the various chemokines and their receptors have been discussed in detail in several excellent reviews [1–6].

In addition to their normal physiological roles, the fact that chemokines control leukocyte chemotaxis has also focused considerable attention on their involvement in inflammation [7,8]. Furthermore, the discovery that some chemokine receptors can function as cellular receptors for the HIV-1 virus, has highlighted the critical role of chemokines and their receptors in the pathogenesis of AIDS [9–15]. Indeed, the fact that the major HIV-1 envelope protein gp120 can not only bind to certain chemokine receptors, but sometimes also activate them [16–21], adds a further degree of complexity to the potential cellular consequences of chemokine receptor activation.

It has now also been demonstrated that chemokines and their receptors are expressed throughout the central and peripheral nervous systems [22–24]. It is clear that all of the major cell types found in the nervous system, i.e. microglia, glia and neurons, express chemokine receptors and can also synthesize and release chemokines. Moreover, the expression levels of chemokines and their receptors in the nervous system have been shown to be subject to all kinds of intriguing fluctuations — the significance of which are not completely understood (e.g. [25–27]). Indeed, although the existence of a chemokine system that is endogenous to the brain is now well established, its raison d'etre is not. There have been several suggestions as to what chemokines might be doing in the brain, mostly by analogy with their functions in the immune system. For example, it has been suggested that chemokines play a role as mediators of the inflammatory component associated with many brain disorders [25–29], or as chemotactic cues for neuronal migration during development [30–33]. In support of these suggestions, it is clear that chemokines can be synthesized in the brain by activated microglia and astrocytes in response to a variety of insults, and there is considerable evidence that they play a key role as local mediators of inflammation [34]. However, the finding that neurons can produce and respond to chemokines is much more surprising. What exactly is the reason for the chemokine sensitivity of neurons? How are chemokines involved in the behavior of neurons under both normal and pathophysiological circumstances? Attempts to answer such questions are in their infancy. We really know very little about what controls the release of chemokines from cells in the brain and under what circumstances they actually produce functional effects on neurons in particular. This chapter describes what is currently known about the ways in which neurons can respond to chemokines and how these effects may be important for the function of the nervous system under normal and pathophysiological circumstances.

2. EXPRESSION OF CHEMOKINES AND THEIR RECEPTORS BY NEURONS

At this time there are officially 11 CC(β), 6 CXC(α), 1 CX3C and 1C types of chemokine receptors, as well as a number of related 'orphan' receptors [1–6,24]. Many of these receptors have been detected in the nervous system [22–24], and some are expressed by neurons, both in the developing and mature brain. Studies employing in situ hybridization or immunohistochemistry have identified neuronal chemokine receptors in the brains of several species, including humans (e.g. [35]), as well as in cultured neurons from these species (e.g. [19]). Furthermore, many of these receptors have been shown to be expressed by a variety of neuronally derived cell lines, which have been used as models for studying the consequences of chemokine receptor activation in neurons. The NT2 cell line, which is derived from the human nervous system, has been particularly widely studied in this regard [31,36–39].

2.1. α-Chemokine Receptors

Of the α-chemokine receptors, the CXCR4 receptor has been the most widely studied in the brain. There are several reasons for this. First of all this receptor is extensively expressed in the nervous system throughout development [33,40,41]. Gene targeting of either the CXCR4 receptor or of its one known agonist, the chemokine SDF-1α (CXCL12), results in a spectacular phenotype involving disruption of brain development [33,42]. Furthermore, the CXCR4 receptor is known to be a major receptor for T-tropic versions of the HIV-1 virus [10], and so its role in the pathogenesis of HIV-1 associated dementia (HAD) has been of obvious interest and has been intensively studied [22,43]. The CXCR4 receptor has been shown to be expressed by microglia, Astrocytes and neurons in various parts of the brain [44–46], and so probably has a large number of different functions in the nervous system.

Although the CXCR4 receptor is expressed by neurons in the mature nervous system [47–49], it is particularly highly expressed in the developing embryo [33,40,41]. Indeed, McGrath et al. [41] showed that of 11 chemokine receptors examined, CXCR4 was by far the most extensively expressed. CXCR4 receptors could be clearly demonstrated in the earliest embryos examined (E8.5–9.5) where they appear in the neural epithelium and subsequently in the developing spinal cord/hind brain, forebrain and cerebellum. Interestingly, SDF-1α, the selective ligand for the CXCR4 receptor, is not highly expressed in the nervous system during development, but in the Mesenchyme surrounding the developing neural tissue [50] (Fig. 1). It is therefore quite possible that SDF-1α might act as a chemotactic guide for the movement of developing neurons. Indeed, in some in vitro studies chemokines, including SDF-1α, have been shown to produce chemotactic effects on neurons, such as cerebellar granule cells, dorsal root ganglion neurons (DRG) and neuronal progenitor cells [37,51,52]. Thus, gradients of SDF-1α, secreted by mesenchymal cells, may act as guides for developing neurons. In keeping with this hypothesis CXCR4 receptor KO mice exhibit severe abnormalities in cerebellar development, involving a deficit in granule cell migration [33,42]. In these mice neuronal precursors migrate prematurely from the external granule layer (EGL) resulting in severe malformation of the cerebellum. Similarly, in recent studies, we have observed abnormalities in the development of the Dentate Gyrus of the hippocampus in CXCR4 KO mice (Fig. 1). This is also the result of deficits in granule cell migration, as well as a reduction in the number of dividing neuronal precursors, resulting in a smaller than normal dentate gyrus. Thus, CXCR4-mediated signaling seems to be important for the development of both the hippocampus and cerebellum — two of the areas of the brain where CXCR4 receptors are highly expressed by neurons [41]. SDF-1α can elicit a variety of receptor-mediated intracellular biochemical signals in cultured granule cells from the hippocampus and cerebellum, some of which could certainly be responsible for its chemotactic effects on these cells (see below). Of particular interest, it has recently been demonstrated [53] that CXCR4-mediated leukocyte chemotaxis can be opposed by the action of the protein Slit acting through its receptor Roundabout (Robo). The Slit/robo System has

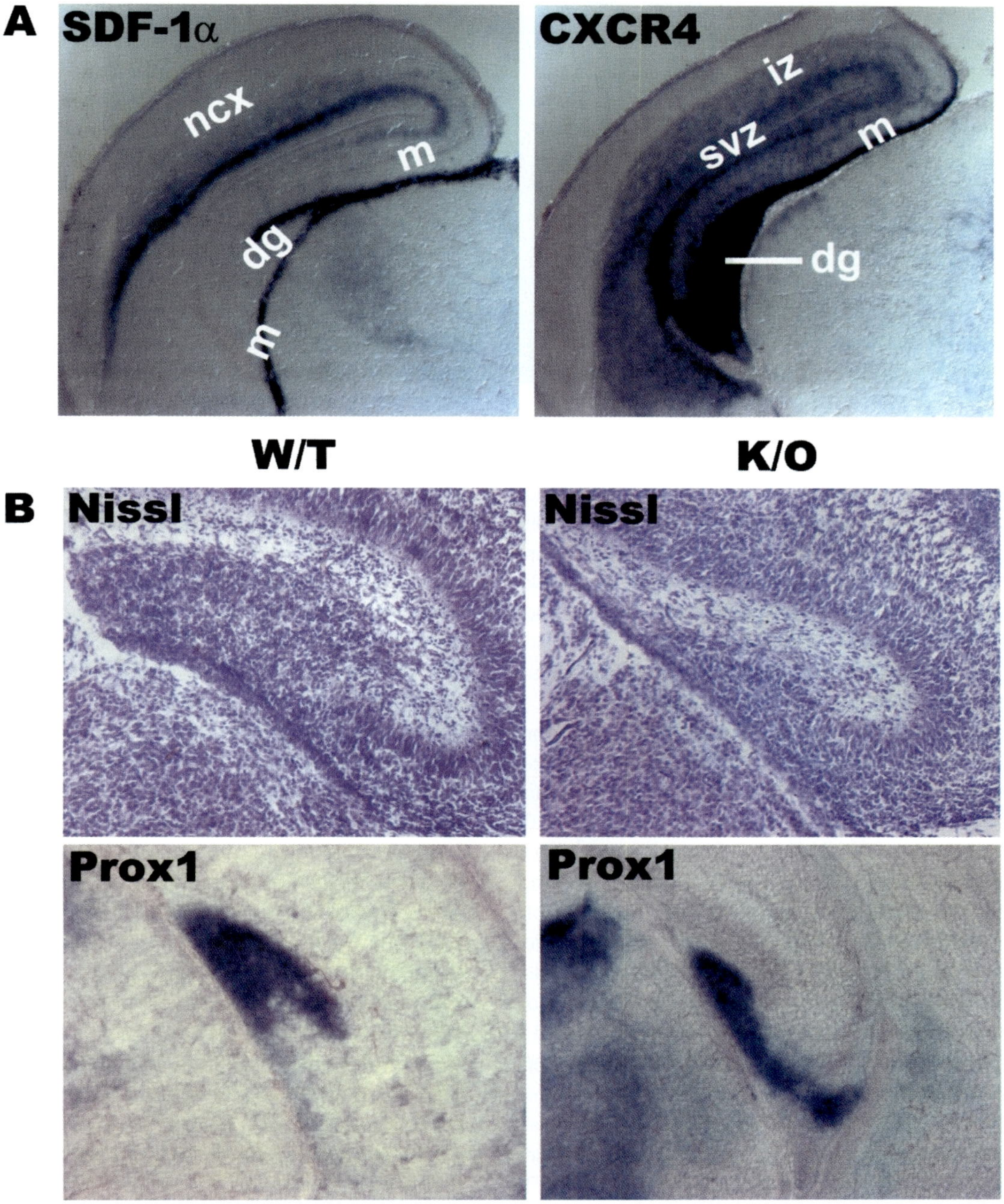

Figure 1. (A) SDF-1α and CXCR4 expression in the forebrain of embryonic mice. Coronal sections (40 μm) through the forebrain of E15.5 mice were processed for in situ hybridization. ncx, neocortex; svz, subventricular zone; m, meninges; dg, dentate gyrus; iz, intermediate zone. Both SDF-1α and CXCR4 message are expressed strongly in the neocortex where neurons are migrating. However, SDF-1α expression is mostly seen in meninges overlying the hippocampus, whereas CXCR4 is strongly expressed at the dentate gyrus and CA3 end of the hippocampus. (B) The development of the hippocampal dentate gyrus is disrupted in E18.5 CXCR4 knockout mice. Top panel: Cresyl violet staining of coronal sections (20 μm) through the forebrain of mice using cryostat. Bottom panel: Prox1 expression in the hippocampal dentate gyrus. Coronal sections (40 μm) through the forebrain of mice were processed for in situ hybridization. (Data, courtesy of Dr. Meiling Lu.)

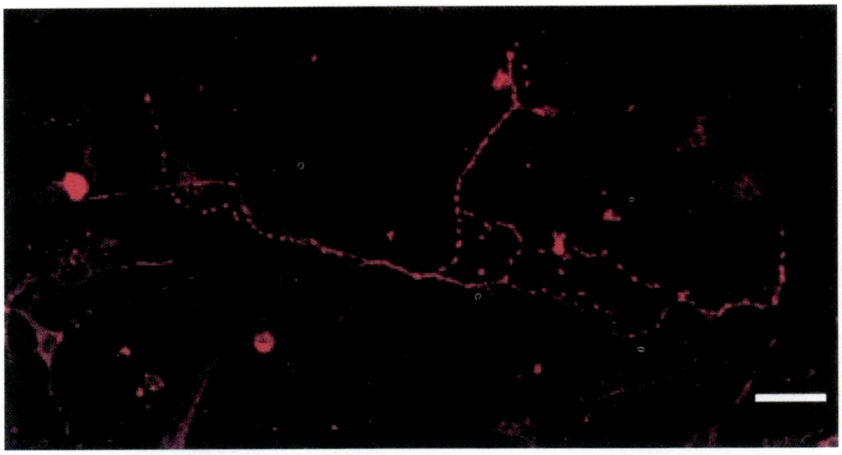

Figure 2. Immunohistochemical localization of CXCR4 receptors expressed by cultured rat dorsal root ganglion neurons. Scale bar: 30 μm.

been previously shown to be very important in the developing nervous system, and Slit/Robo function has been defined mainly in terms of its effects on axonal guidance. Thus, by analogy with the work on leukocytes, it is an intriguing possibility that the SDF-1α/CXCR4 and Slit/Robo systems may also play interacting roles in controlling aspects of neuronal development.

Although there is now good evidence for a role for CXCR4 receptors in the regulation of neuronal development, these receptors are also expressed in the brains of mature animals (as demonstrated in several species) [47–49]. This includes expression by different types of neurons in the central nervous system, and also the DRG [54] (Fig. 2). The functions of these neuronally expressed chemokine receptors are unclear. However, it has been shown that the levels of expression of CXCR4 receptors and of SDF-1α may change under different circumstances, indicating that they may be important in the responses of the nervous system in a variety of pathological states [34,48].

In addition to CXCR4 receptors, there is also evidence for the neuronal expression of CXCR2 receptors [19,55], and possibly CXCR1 receptors [56]. Originally it was reported that CXCR1 receptors were not detectable, although CXCR2 receptors were found to be widely expressed in several parts of the brain [55]. However, recently, it has been shown that cholinergic septal neurons express both CXCR1 and CXCR2 receptors and are capable of mediating IL-8 signaling through these receptors [56]. CXCR3 receptors have been shown to be expressed by forebrain neurons and stimulation of these receptors by IP-10 (CXCL10) or Mig (CXCL9) produces activation of MAP kinase in cultures of these neurons [38,57]. Finally, the CXCR5 receptor has been shown to be expressed in the central nervous system, in the cerebellum in particular [58,59].

2.2. β-Chemokine Receptors

As in the case of α-chemokine receptors, the expression of several types of β-chemokine receptors has been demonstrated in the brain, including expression by neurons [3,24,60]. Here again several types of receptors including CCR1, CCR2, CCR3, CCR4, and CCR5 have been shown to be expressed by a variety of cultured neuronal cell lines [31] and by primary neuronal cultures [19]. Studies on brain tissues from various species under normal and pathophysiological circumstances have demonstrated the expression of CCR3 and CCR5 by neurons in addition to microglia [60–63]. The populations of neurons that express these receptors have not been very well defined at this point, although large pyramidal neurons in the cortex and hippocampus seem to be included in this population. In addition to β-chemokine receptors, chemokine ligands for these receptors have also been shown to be synthesized in the brain by different types of cells including neurons [60]. Upregulation of many of these chemokines by reactive astrocytes and microglia in connection with different neuropathological states has been one type of observation that has encouraged the belief that they might be important mediators of the inflammatory components of brain disease [27,34,64–67].

2.3. Fractalkine Receptors (CX3CR1)

The unique CX3CR1 receptor and its ligand fractalkine (CX3CL1) represent an unusual case in the canon of chemokine signaling. Fractalkine is synthesized as a membrane-bound molecule, attached to a mucin-like transmembrane stalk [68,69]. Soluble fractalkine can be released into the extracellular milieu by proteolytic cleavage in the stalk region [70]. Both membrane-bound and free fractalkine are bioactive [71]. Membrane-tethered fractalkine may be important in cell/cell adhesion interactions both in the nervous system and elsewhere. On the other hand, bioactive 'soluble' fractalkine might be capable of signaling over considerable distances. The regulation of fractalkine cleavage and release could therefore constitute an important control point in fractalkine signaling and an important therapeutic target. It is certainly clear that fractalkine can be synthesized and released from neurons [70–72]. The CX3CR1/fractalkine receptor has also been reported to be expressed by cultured hippocampal pyramidal and DRG neurons [19,54]. Cultured neurons of both types have been shown to respond to fractalkine in a variety of ways [19,54,73]. It should be noted, however, that mice in which GFP has been used to replace the CX3CR1 gene exhibit no expression in neurons (microglia being the major site of expression) [74], and other studies on normal brain appears to confirm this lack of neuronal expression [75]. It might therefore be supposed that something associated with culturing neurons — perhaps this might be 'traumatic' in some way, causes them to express CX3CR1 receptors. In this regard, it is interesting to note that Tong et al. [76] reported CX3CR1 expression by neurons in the brains of children infected with HIV-1, but not in uninfected brains. Curiously, in spite of the widespread expression of fractalkine and of CX3CR1 receptors, both in the nervous system and elsewhere [77,78], and in spite of the several reasonable suggestions concerning physiological roles for this chemokine/receptor pair, mice lacking either the CX3CR1 gene or the fractalkine gene are without a discernable phenotype [72,74].

3. EFFECTS OF CHEMOKINES ON NEURONAL BEHAVIOR

The expression of chemokine receptors by neurons under non-pathological circumstances suggests that they also play a role in the regulation of normal neuronal behavior. As a guide as to what to expect, we should consider the consequences of activating GPCRs other than chemokine receptors on neuronal excitability and synaptic communication — something that has been studied in considerable detail [79]. Upon activation of a GPCR, the α or β/γ subunits of heterotrimeric G-proteins can mediate an enormous number of signal transduction options. Some of these signals are important for relatively slow phenomena related to cell growth, development and survival. However, the G-protein-mediated regulation of ion channels and components of the transmitter vesicle exocytosis machinery affords rapid control of neuronal excitability and synaptic communication [79]. Regulation of the state of ion channel phosphorylation by intracellular second messengers is another mechanism for the rapid receptor-mediated control of cellular excitability [79]. Experiments using heterologous expression systems and with leukocytes have shown that activation of chemokine receptors can produce effects on a number of ion channels, including voltage-sensitive Ca and K channels [16,80–82]. In general, these effects of chemokines have been reported to be inhibited following treatment with PTX, indicating that chemokine receptor signaling is transduced by G_0 or G_i [81]. On the other hand, it is not clear that this always holds true for chemokine signaling in neurons, where PTX resistant effects have also been reported [83]. Several studies have indicated that chemokines can produce rapid effects on neuronal excitability and communication — presumably resulting from effects on ion channels and transmitter release [19,56,83–85]. In the DRG, different chemokine receptors are expressed by separate subpopulations of neurons and many neurons appear to express multiple types of chemokine receptors [51,54]. Activation of chemokine receptors on small DRG nociceptors produces rapid excitation, reminiscent of that produced by bradykinin which also activates a GPCR expressed by these cells (Fig. 3). As might therefore be expected, by analogy with bradykinin, chemokines produce potent allodynia

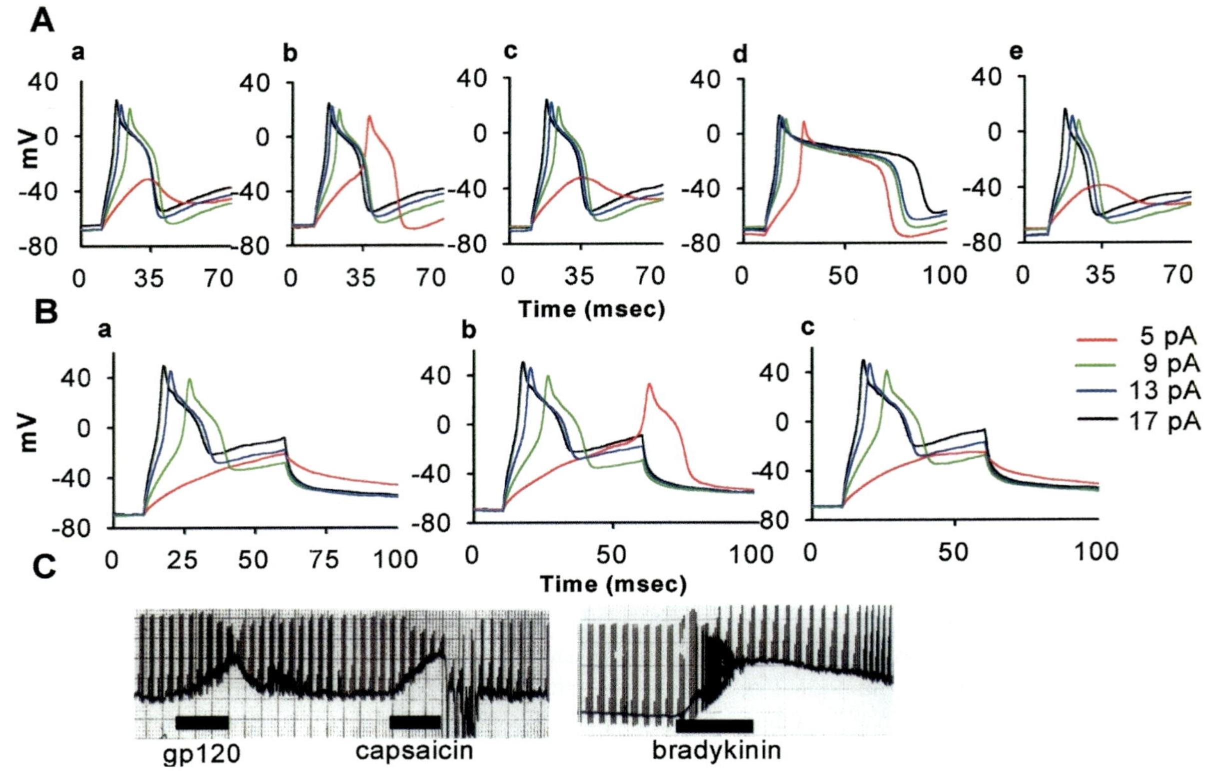

Figure 3. Chemokines and gp120 produce excitatory effects on cultured rat dorsal root ganglion neurons. (A) The effects of 50 nM MDC (b) and 1 μM capsaicin (d). c, e, washout of the effects of MDC and capsaicin, respectively. (B) The effects of 200 pM gp120 (b) with washout (c). (C) Depolarization of a DRG neuron by 200 pM gp120. The effects of 1 μM bradykinin and 1 μM capsaicin are shown as a comparison. Scale bar: 3 min.

when injected into the rat paw [54]. Thus, it is possible that chemokines released from leukocytes at sites of inflammation might directly excite DRG pain responses. Interestingly, DRG chemokine receptors can also be activated by the HIV-1 coat protein gp120, and injection of both T and M tropic gp120s into the rat paw also produces potent allodynia [54]. It is therefore possible that the chronic pain syndrome often associated with AIDS patients may result from direct excitation of the DRG by HIV-1 envelope proteins [86]. It is interesting to note that chronic exposure of DRG neurons to gp120 also causes apoptotic death of the cells, as it does with neurons in the CNS (unpublished observations) — a phenomenon that could well contribute to the sensory neuropathies frequently observed in HIV-1 infected patients [86].

There are also indications that chemokines can rapidly regulate the properties of central neurons as well. For example, both the CXCR4 agonist SDF-1α, and agonists at CXCR2 receptors such as GRO-α (CXCL1) and IL-8 (CXCL8), evoke Ca transients in granule and Purkinje cells in cerebellar slices, and increase the spontaneous release of neurotransmitters, such as GABA and glutamate — both synaptically and from surrounding glial cells [83–85]. Such effects appear to be mediated through the activation of MAP kinases in presynaptic neurons and possibly glia. Furthermore, GRO-α was shown to reduce the degree of long-term synaptic depression (LTD) that can be obtained at the parallel fiber synapses between granule cells and Purkinje cells [84]. Such an observation is consistent with the known role of Ca mobilization from intracellular stores in Purkinje cells in the induction of LTD in the cerebellum. These results indicate that chemokines may play a role in shaping long-term synaptic plasticity in the cerebellum and elsewhere. Perhaps locally produced chemokines might normally be important in regulating the strength of synaptic connections. Several studies have also demonstrated the expression of CXCR4 and CXCR2 receptors by cerebellar neurons both in vivo [83,87] and in primary neuronal cultures [88].

Cultured hippocampal pyramidal neurons have been shown to express a variety of chemokine receptors including the CX3CR1/fractalkine receptor [19]. Excitatory transmission between pyramidal neurons in culture was blocked by fractalkine [19] and this may be associated with inhibition of voltage dependent Ca currents [19,82]. Similarly, it has recently been demonstrated that cholinergic septal neurons express CXCR1 and CXCR2 receptors, and that activation of these receptors inhibited Ca currents in these cells [56]. Thus, evidence is accumulating that activation of neuronal chemokine receptors can produce rapid changes in neuronal excitability and synaptic communication.

In addition to these rapid effects, there are also several reports illustrating long-term effects of chemokines on neuronal survival (e.g. [19,73]) — a phenomenon that could have considerable importance in the context of neuropathology. Thus, it is certainly possible that upregulation of chemokine production in the brain, as a result of injury or some other insult, might represent an attempt by the nervous system to limit damage by employing the survival-promoting effects of chemokines. On the other hand, it has also been suggested that under some circumstances chemokines might actively participate in neuronal death and contribute to neurodegeneration [89]. These different effects could result from the diverse second messenger signaling pathways that chemokines have been shown to activate in neurons and other cell types (see below), and so their pro- or anti-survival effects could depend on the cellular context.

The first report of the survival-promoting effects of chemokines was that of Araujo et al. [90] who demonstrated that IL-8 increased the survival of cultured hippocampal pyramidal neurons. Furthermore, activation of IL-8 receptors on cerebellar granule cells also increased their survival [91]. These observations were extended by Meucci et al. [19] who showed that cultured hippocampal pyramidal neurons expressed a wide variety of chemokine receptors and that activation of many of these receptors increased survival to varying extents. Fractalkine, for example, seems to increase survival by directly stimulating neuronal CX3CR1 receptors expressed by hippocampal pyramidal cells and cerebellar granule cells in culture [19,73,76]. In contrast, however, Kaul and Lipton [89] demonstrated that, although several chemokines promoted the survival of cortical neurons in mixed neuronal/glial cultures, SDF-1α was exceptional in that it actually promoted apoptosis — perhaps through a direct effect on the neurons. In support of this contention, SDF-1α also increased apoptosis in CXCR4 expressing NT2 cells [37].

A related issue concerns the effects of the HIV-1 envelope protein gp120, some versions of which are also potential ligands for the CXCR4 receptor. These proteins have been shown to produce apoptosis when added to neuronal cultures in vitro [19,37,89,92] and to produce a HAD-like neuropathology when expressed in vivo [93]. Indeed, the proapoptotic effects of gp120 have been reported in a variety of cell types in addition to neurons. Thus, there is an increasing literature suggesting that binding of envelope proteins to chemokine receptors expressed by CD4 positive T-lymphocytes actually plays a proactive role in promoting the apoptosis of these cells in addition to mediating viral fusion [94–97]. Recent reports also suggest that it is the binding of the gp120 to CXCR4 receptors, rather than CD4 molecules, that is important in this regard [97]. Nevertheless, exactly how the interaction of gp120 with CXCR4 induces apoptotic signaling in neurons and other types of cells is not clear. In the case of neurons, SDF-1α and gp120 appear to produce diametrically opposite effects on neuronal survival in the same paradigm according to some reports [19], although in others both molecules are proapoptotic [89,98]. This might be due to differential effects of the two ligands when it comes to signaling via the CXCR4 receptor — as we shall now discuss.

4. EFFECTS OF CHEMOKINES ON NEURONAL SIGNAL TRANSDUCTION

Although activation of GPCRs was originally thought to influence a limited number of second messenger systems, we now realize that receptor activation is potentially a very information-rich event. Thus, not only are heterotrimeric G-proteins activated, but the further recruitment of scaffold proteins, such as β-arrestin, affords a multitude of additional signaling options as well. Furthermore, some GPCRs, including chemokine receptors, have been shown to influence downstream pathways by interacting with them directly through a variety of protein/protein interaction motifs [99]. As is the case with GPCRs in general, activation of chemokine receptors elicits a wide spectrum of signal transduction events in neurons and other types of cells [100]. Typically activation of chemokine receptors produces an increase in Ca, as has been shown in many types of cells [17,19,52,54,83,101]. However, in addition, activation of chemokine receptors initiates a host of other responses including the regulation of ion channels, the MAP kinase pathway (including ERKs, p38 and JNK), the cAMP pathway, the JAK/STAT pathway, the PI3 kinase/Akt pathway and the transcription factors CREB and NF-kB [19,73,89,91,102–106]. Chemokine receptor-mediated transactivation of tyrosine kinase receptors has also been observed [107]. From the perspective of the present discussion it is important to understand which of these pathways are activated in neurons, under what circumstances and what their consequences might be.

The functional consequences of chemokine induced Ca increases in neurons are not clear, but they may certainly contribute to the stimulatory effects of chemokines on the release of neurotransmitters or chemokine induced effects on synaptic plasticity such as those in the cerebellum discussed above [83–85]. Similarly chemokine induced Ca increases in astrocytes may underlie chemokine induced release of glutamate, a possible contributing effect to HAD [98]. The activation of neuronal chemokine receptors may have consequences for neuronal excitability through the release of G-protein β/γ subunits and their effects on ion channels and other downstream targets [79]. Effects of chemokines on other neuronal signaling pathways may be of importance when considering their effects on neuronal survival. Thus, activation of the ERK, NF-kB and Akt pathways might all contribute to the pro-survival effects of chemokines [19,73], whereas activation of p38, JNK and ceramide production might do the opposite [89,105].

There is also evidence that gp120 can bind to, and sometimes also activate, neuronal chemokine receptors. T-tropic gp120s bind to the CXCR4 molecule, presumably in conjunction with a co-receptor such as the human CD4 molecule. Diverse reports in the literature have indicated that the consequences of gp120 binding to CXCR4 receptors are variable. Gp120 can act as an antagonist of the effects of SDF-1α in some reports [81,108] or as a receptor agonist [16–21], mimicking one or more aspects of SDF-1α signaling. One reason for the variable results obtained with gp120 might be that these studies have used different versions of gp120 as well as different types of target cells. Subtle differences in the way in which these gp120s interact with the CXCR4 receptor may affect its efficacy as an agonist. If we consider the literature covering a variety of cell types, including neurons, it is clear that gp120 can recapitulate at least some of the agonist effects of SDF-1α, including effects on receptor trafficking [109], Ca mobilization [19,54], regulation of ion channels [16] and activation of enzymatic pathways, such as the MAP kinase pathway [16–21]. This is true for a number of different types of gp120, including versions that are specific for both the CCR5 and CXCR4 receptors. In neurons, it is clear that gp120IIIB for example (a T-tropic HIV-1 envelope protein) produces Ca transients in a number of neuronal types [17,19,54,110–112]. These Ca transients may result from a combination of effects, including mobilization from intracellular stores, synaptic release of glutamate or combinations of these effects [92,112]. Interestingly, it has been suggested that both Ca mobilization and the activation of glutamate receptors are important in the neurotoxic effects of gp120 [112–115]. That such effects of gp120 are truly mediated through the activation of chemokine receptors, as is generally assumed, remains to be conclusively demonstrated in most instances. However, in a few cases [17,52], the effects of gp120 on neurons have been shown to be blocked by CXCR4 receptor antagonists, such as the bicyclam AMD3100, which completely blocked the ability of gp120 to activate ERKs in cortical neuron cultures [52]. Although it seems that gp120 can elicit signaling through the CXCR4 receptor in neurons, the way in which gp120 and SDF-1α interact with CXCR4 receptors is different and so there may be differences in the signals they produce. These differences might be significant for understanding the diverse effects of these two molecules on neuronal survival. For example, although chemokines can activate Akt in cultured hippocampal pyramidal neurons — and this is clearly important in their survival-promoting effects — gp120 kills these same neurons and does not activate this pathway [19,73]. In neuronal progenitor cells, SDF-1α activated ERK1/2, but here again this effect was not reproduced by gp120 [52]. In summary, it is possible that binding of gp120 to neuronal chemokine receptors might initiate apoptosis through normal or abnormal signaling events or else, in its role as an antagonist, by blocking pro-survival signals provided by chemokine activation of receptors. Clearly, it is important for us to understand exactly what consequences result from the interaction of gp120 with neuronal chemokine receptors and how this signaling influences neuronal survival. As discussed above, an identical debate is currently underway when considering the way gp120 interacts with chemokine receptors on T-lymphocytes and how this influences the death of these cells.

5. CHEMOKINES AND NEUROPATHOLOGY

Upregulation of chemokines and their receptors have been observed in the CNS in association with a variety of neuropathological states that have an associated inflammatory component [26,29,34,49]. The production of chemokines under these circumstances can occur in all the major cell types in the brain, including neurons, although little is known about how these events are regulated [27,34]. There is good reason to believe that chemokines produced in the brain might be important mediators of inflammatory responses by virtue of their effects on microglia, macrophages and other immune cells. Thus, a key role for locally produced chemokines in conditions such as experimental autoimmune encephalomyelitis, multiple sclerosis (MS), stroke, trauma, HIV-1 encephalitis and Alzheimer's disease (AD) is strongly suggested by analogy with their chemotactic effects on leukocytes in the periphery [34]. A related question, however, concerns whether activation of chemokine receptors expressed directly by neurons themselves is significant in any of these disorders.

Fractalkine, for example, is normally synthesized by neurons where it is attached to a transmembrane mucin-like stalk (see above). Membrane-tethered fractalkine can be released by the action of enzymes, such as metalloproteinases [70]. Soluble fractalkine could certainly participate in brain inflammatory responses by acting on CX3CR1 receptors, which are most abundantly expressed on microglia [74]. Indeed, it has been suggested that fractalkine may have both pro- and anti-inflammatory effects by this route [72,76,116,117]. As discussed above, expression of CX3CR1 receptors by neurons in culture has been demonstrated [19,54], and the receptor was also reported to be upregulated by neurons in the brains of HIV-1-infected patients [76]. Under such circumstances, the effects of fractalkine on neurons might be important. In all cases to date, fractalkine has been reported to have a survival-promoting (anti-apoptotic) effect on neurons (e.g. in hippocampal pyramidal cells and cerebellar granule cells) [19,73,76]. In studies on cultured hippocampal pyramidal cells, activation of CX3CR1 receptors resulted in a variety of signals that could promote survival — including inhibition of Ca currents and excitatory neurotransmission, as well as activation of the Akt kinase, ERK1/2 and NF-kB pathways [19,73]. One scenario suggests that diverse insults to the brain might trigger the release of fractalkine from neurons, perhaps through the activation of Ca-sensitive metalloproteinases [70]. This seems quite likely, as many insults are associated with increases in neuronal Ca. This is certainly the case, for example, in instances where the insult has an excitotoxic component involving the release of glutamate and activation of glutamate receptors. The soluble fractalkine released might be neuroprotective through its paracrine and autocrine survival-promoting effects on local neurons.

Similarly, Xia et al. [57] have demonstrated the expression of CXCR3 receptors by cortical and subcortical neurons in both normal and AD brains. IP-10, a chemokine ligand for this receptor, was significantly upregulated in AD. CXCR2 receptors were also found to be associated with dystrophic neurites in AD plaques. These authors found that stimulation of CXCR3 or CXCR2 receptors in cortical neurons by the appropriate chemokines produced strong activation of ERK1/2, which could play a survival-promoting role in AD. On the other hand, as the authors suggest, it is possible that long-term activation of ERKs or similar pathways could actually play a detrimental role in AD by encouraging the abnormal phosphorylation of tau or some similar mechanism.

The role of chemokine receptors in the brain has obviously been of particular interest in attempts to understand the cellular/molecular basis for HAD-

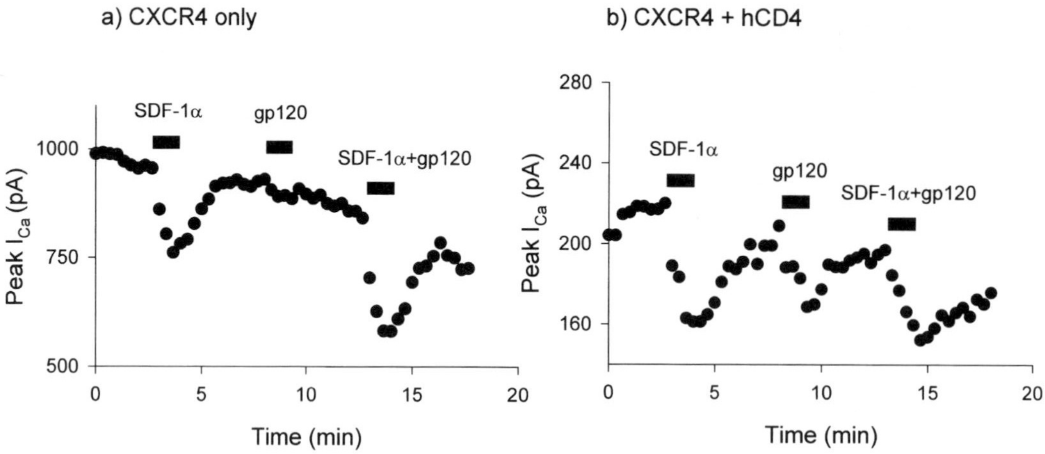

Figure 4. Inhibitory effects of SDF-1α and gp120 on voltage-dependent Ca currents (α_{1B}, β_{1-3}, α_2/δ) expressed in HEK293 cells. In a, only the rat CXCR4 receptor was expressed. Note the effects of SDF-1α, but lack of effect of gp120. In b, the human CD4 molecule was expressed in addition to the rat CXCR4 receptor. Now agonist effects of gp120 are also apparent.

related neuropathology [43,118]. Of special interest here is the relative interplay between the effects of chemokines and those of gp120s, which are also potential chemokine receptor ligands (see above). Although the neurotoxic factors that are responsible for HAD have not been precisely identified, there is a long list of possibilities and gp120 is certainly near the very top of this list [22,43]. Thus, it has been speculated that gp120 released from viruses replicating in cells, such as microglia, might be the initial stimulus for a neurotoxic cascade. This view has been encouraged by several observations, such as the fact that mice overexpressing gp120 in the brain develop HAD-like neuropathology [93]. Furthermore, gp120s of various types have been shown to produce neuronal apoptosis in diverse paradigms both in vivo [119–123] and in vitro [19,37,89,124]. It should be stressed that in order for gp120 to produce its deleterious effects on the nervous system, there is no necessity for it to act directly upon neurons. Several reasonable scenarios have been suggested in which the primary locus of action of gp120 is either the microglia or astrocytes from which it stimulates the release of further potential neurotoxins, such as glutamate [43,98,125]. Nevertheless, the observation that neurons do express potential HIV-1 co-receptors such as CXCR4 and CCR5 receptors certainly makes a direct neuronal effect a possibility [22].

The evolutionary constraints upon HIV-1 development in the CNS are distinct from those in the periphery and it is instructive to ask which kinds of viruses and gp120s may actually contribute to HAD. According to Gabuzda and colleagues [126,127] (Chapter 5.3) these seem most likely to be viruses with some selectivity for the CXCR4 receptor (although other tropisms cannot be ruled out as well). Indeed, most of the literature on this topic has utilized CXCR4 selective gp120s, such as gp120IIIB. However, the hypothesis that gp120 (and by implication HIV-1) produces neuropathology through its effects on the CXCR4 receptor is far from secure. For example, it is generally believed that the interaction of gp120 with cells requires it to interact not only with a chemokine receptor such as CXCR4, but also with the CD4 molecule. Indeed, gp120 does not really show effective interactions with rat CXCR4 receptors in heterologous expression systems without co-expression of the human CD4 molecule (e.g. Fig. 4). Considering that many studies investigating the toxic effects of gp120 employ mouse or rat tissues, it is not clear how the gp120 might interact with such cells. One hypothesis might be that gp120 is producing its effects through a receptor other than the CXCR4 receptor. However, even in the case of potential alternative receptors such as the APJ receptor [128,129], the requirement for human CD4 is maintained. Furthermore, where this question has been directly examined the effects of gp120 on neurons do seem to involve CXCR4 receptors. For example, SDF-1α and gp120 interact with each other directly in binding assays using brain membranes from non-human species [130]. In addition, a number of effects of gp120 have been shown to be blocked by CXCR4 receptor antagonists (see above). One is therefore left with the hypothesis that molecules other than CD4 can assist with the binding of gp120 to neuronal CXCR4 receptors. A number of possibilities have been suggested. Furthermore, the CD4 'independent' effects of gp120s are not restricted to the nervous system [131–140]. However, the identity of these mysterious co-receptors has not been clearly elucidated.

6. CONCLUSION

The fact that chemokine receptors are expressed by all the major cell types in the brain indicates that chemokines probably have many different important roles to play in nervous system. At this point in time there is good evidence for chemokines being responsible for some of the things we already expect them to do — e.g. brain inflammation and chemotaxis during development. However, as indicated in the preceding discussion, the great diversity of effects that chemokines produce on neurons makes it likely that this is only the tip of the iceberg. There is a strong likelihood that chemokines and their receptors are concerned in diverse aspects of neuropathology as well as aspects of the normal neuronal physiology.

ACKNOWLEDGEMENTS

We are indebted to Dr. Meiling Lu for Fig 1. This work was supported by NIH Grants DA13141 and MH40165 to R.J.M.

REFERENCES

1. Murphy PM, Baggiolini M, Charo IF, Hebert CA, Horuk R, Matsushima K, Miller LH, Oppenheim JJ, Power CA. International union of pharmacology. XXII. Nomenclature for chemokine receptors. Pharmacol Rev 2000;52: 145–176.
2. Rojo D, Suetomi K, Navarro J. Structural biology of chemokine receptors. Biol Res 1999;32:263–272.
3. Rossi D, Zlotnik A. The biology of chemokines and their receptors. Annu Rev Immunol 2000;18:217–242.
4. Berson JF, Doms RW. Structure–function studies of the HIV-1 coreceptors. Semin Immunol 1998;10:237–248.
5. Baggiolini M, Dewald B, Moser B. Human chemokines: an update. Annu Rev Immunol 1997;15:675–705.
6. Bacon KB, Harrison JK. Chemokines and their receptors in neurobiology: perspectives in physiology and homeostasis. J Neuroimmunol 2000;104:92–97.
7. Baggiolini M. Chemokines and leukocyte traffic. Nature 1998;392:565–568.
8. Luster AD. Chemokines — chemotactic cytokines that mediate inflammation. New Engl J Med 1998;338:436–445.
9. Alkhatib G, Combadiere C, Broder CC, Feng Y, Kennedy PE, Murphy PM, Berger EA. CC CKR5: a RANTES, MIP-1alpha, MIP-1beta receptor as a fusion cofactor for macrophage-tropic HIV-1. Science 1996;272:1955–1958.
10. Feng Y, Broder CC, Kennedy PE, Berger EA. HIV-1 entry cofactor: functional cDNA cloning of a seven-transmembrane, G protein-coupled receptor. Science 1996;272:872–877.
11. Doranz BJ, Rucker J, Yi Y, Smyth RJ, Samson M, Peiper SC, Parmentier M, Collman RG, Doms RW. A dual-tropic primary HIV-1 isolate that uses fusin and the beta-chemokine receptors CKR-5, CKR-3, and CKR-2b as fusion cofactors. Cell 1996;85:1149–1158.
12. Berger EA, Murphy PM, Farber JM. Chemokine receptors as HIV-1 coreceptors: roles in viral entry, tropism, and disease. Annu Rev Immunol 1999;17:657–700.
13. Choe H, Farzan M, Sun Y, Sullivan N, Rollins B, Ponath PD, Wu L, Mackay CR, LaRosa G, Newman W, Gerard N, Gerard C, Sodroski J. The beta-chemokine receptors CCR3 and CCR5 facilitate infection by primary HIV-1 isolates. Cell 1996;85:1135–1148.
14. Combadiere C, Salzwedel K, Smith ED, Tiffany HL, Berger EA, Murphy PM. Identification of CX3CR1. A chemotactic receptor for the human CX3C chemokine fractalkine and a fusion coreceptor for HIV-1. J Biol Chem 1998;273:23799–23804.
15. Deng HK, Unutmaz D, KewalRamani VN, Littman DR. Expression cloning of new receptors used by simian and human immunodeficiency viruses. Nature 1997;388:296–300.
16. Liu QH, Williams DA, McManus C, Baribaud F, Doms RW, Schols D, De Clercq E, Kotlikoff MI, Collman RG, Freedman BD. HIV-1 gp120 and chemokines activate ion channels in primary macrophages through CCR5 and CXCR4 stimulation. Proc Natl Acad Sci USA 2000;97: 4832–4837.
17. Pandey V, Bolsover SR. Immediate and neurotoxic effects of HIV protein gp120 act through CXCR4 receptor. Biochem Biophys Res Commun 2000;274:212–215.
18. Lannuzel A, Barnier JV, Hery C, Huynh VT, Guibert B, Gray F, Vincent JD, Tardieu M. Human immunodeficiency virus type 1 and its coat protein gp120 induce apoptosis and activate JNK and ERK mitogen-activated protein kinases in human neurons. Ann Neurol 1997;42: 847–856.
19. Meucci O, Fatatis A, Simen AA, Bushell TJ, Gray PW, Miller RJ. Chemokines regulate hippocampal neuronal signaling and gp120 neurotoxicity. Proc Natl Acad Sci USA 1998;95:14500–14505.
20. Davis CB, Dikic I, Unutmaz D, Hill CM, Arthos J, Siani MA, Thompson DA, Schlessinger J, Littman DR. Signal transduction due to HIV-1 envelope interactions

with chemokine receptors CXCR4 or CCR5. J Exp Med 1997;186:1793–1798.

21. Weissman D, Rabin RL, Arthos J, Rubbert A, Dybul M, Swofford R, Venkatesan S, Farber JM, Fauci AS. Macrophage-tropic HIV and SIV envelope proteins induce a signal through the CCR5 chemokine receptor. Nature 1997;389:981–985.
22. Miller RJ, Meucci O. AIDS and the brain: is there a chemokine connection? Trends Neurosci 1999;22:471–479.
23. Hesselgesser J, Horuk R. Chemokine and chemokine receptor expression in the central nervous system. J Neurovirol 1999;5:13–26.
24. Bajetto A, Bonavia R, Barbero S, Florio T, Schettini G. Chemokines and their receptors in the central nervous system. Front Neuroendocrinol 2001;22:147–184.
25. Xia MQ, Hyman BT. Chemokines/chemokine receptors in the central nervous system and Alzheimer's disease. J Neurovirol 1999;5:32–41.
26. Glabinski AR, Ransohoff RM. Chemokines and chemokine receptors in CNS pathology. J Neurovirol 1999;5: 3–12.
27. Panenka W, Jijon H, Herx LM, Armstrong JN, Feighan D, Wei T, Yong VW, Ransohoff RM, MacVicar BA. P2X7-like receptor activation in astrocytes increases chemokine monocyte chemoattractant protein-1 expression via mitogen-activated protein kinase. J Neurosci 2001; 21:7135–7142.
28. Huang D, Han Y, Rani MR, Glabinski A, Trebst C, Sorensen T, Tani M, Wang J, Chien P, O'Bryan S, Bielecki B, Zhou ZL, Majumder S, Ransohoff RM. Chemokines and chemokine receptors in inflammation of the nervous system: manifold roles and exquisite regulation. Immunol Rev 2000;177:52–67.
29. Mennicken F, Maki R, de Souza EB, Quirion R. Chemokines and chemokine receptors in the CNS: a possible role in neuroinflammation and patterning. Trends Pharmacol Sci 1999;20:73–78.
30. Asensio VC, Campbell IL. Chemokines in the CNS: plurifunctional mediators in diverse states. Trends Neurosci 1999;22:504–512.
31. Halks-Miller M, Hesselgesser J, Miko IJ, Horuk R. Chemokine receptors in developing human brain. Methods Enzymol 1997;288:27–38.
32. Horuk R. Chemokines beyond inflammation. Nature 1998;393:524–525.
33. Zou YR, Kottmann AH, Kuroda M, Taniuchi I, Littman DR. Function of the chemokine receptor CXCR4 in haematopoiesis and in cerebellar development. Nature 1998;393:595–599.
34. Ransohoff RM, Tani M. Do chemokines mediate leukocyte recruitment in post-traumatic CNS inflammation? Trends Neurosci 1998;21:154–159.
35. Lavi E, Kolson DL, Ulrich AM, Fu L, Gonzalez-Scarano F. Chemokine receptors in the human brain and their relationship to HIV infection. J Neurovirol 1998;4:301–311.
36. Hesselgesser J, Halks-Miller M, DelVecchio V, Peiper SC, Hoxie J, Kolson DL, Taub D, Horuk R. CD4-independent association between HIV-1 gp120 and CXCR4: functional chemokine receptors are expressed in human neurons. Curr Biol 1997;7:112–121.
37. Hesselgesser J, Taub D, Baskar P, Greenberg M, Hoxie J, Kolson DL, Horuk R. Neuronal apoptosis induced by HIV-1 gp120 and the chemokine SDF-1 alpha is mediated by the chemokine receptor CXCR4. Curr Biol 1998;8:595–598.
38. Coughlan CM, McManus CM, Sharron M, Gao Z, Murphy D, Jaffer S, Choe W, Chen W, Hesselgesser J, Gaylord H, Kalyuzhny A, Lee VM, Wolf B, Doms RW, Kolson DL. Expression of multiple functional chemokine receptors and monocyte chemoattractant protein-1 in human neurons. Neuroscience 2000;97:591–600.
39. Catani MV, Corasaniti MT, Navarra M, Nistico G, Finazzi-Agro A, Melino G. gp120 induces cell death in human neuroblastoma cells through the CXCR4 and CCR5 chemokine receptors. J Neurochem 2000; 74:2373–2379.
40. Jazin EE, Soderstrom S, Ebendal T, Larhammar D. Embryonic expression of the mRNA for the rat homologue of the fusin/CXCR-4 HIV-1 co-receptor. J Neuroimmunol 1997;79:148–154.
41. McGrath KE, Koniski AD, Maltby KM, McGann JK, Palis J. Embryonic expression and function of the chemokine SDF-1 and its receptor, CXCR4. Dev Biol 1999; 213:442–456.
42. Ma Q, Jones D, Borghesani PR, Segal RA, Nagasawa T, Kishimoto T, Bronson RT, Springer TA. Impaired B-lymphopoiesis, myelopoiesis, and derailed cerebellar neuron migration in CXCR4- and SDF-1-deficient mice. Proc Natl Acad Sci USA 1998;95:9448–9453.
43. Kaul M, Garden GA, Lipton SA. Pathways to neuronal injury and apoptosis in HIV-associated dementia. Nature 2001;410:988–994.
44. Lavi E, Strizki JM, Ulrich AM, Zhang W, Fu L, Wang Q, O'Connor M, Hoxie JA, Gonzalez-Scarano F. CXCR-4 (Fusin), a co-receptor for the type 1 human immunodeficiency virus (HIV-1), is expressed in the human brain in a variety of cell types, including microglia and neurons. Am J Pathol 1997;151:1035–1042.
45. Tanabe S, Heesen M, Yoshizawa I, Berman MA, Luo Y, Bleul CC, Springer TA, Okuda K, Gerard N, Dorf ME. Functional expression of the CXC-chemokine receptor-4/fusin on mouse microglial cells and astrocytes. J Immunol 1997;159:905–911.
46. Jiang Y, Salafranca MN, Adhikari S, Xia Y, Feng L, Sonntag MK, deFiebre CM, Pennell NA, Streit WJ, Harrison JK. Chemokine receptor expression in cultured glia and rat experimental allergic encephalomyelitis. J Neuroimmunol 1998;86:1–12.
47. Ohtani Y, Minami M, Kawaguchi N, Nishiyori A, Yamamoto J, Takami S, Satoh M. Expression of stromal cell-derived factor-1 and CXCR4 chemokine receptor mRNAs in cultured rat glial and neuronal cells. Neurosci Lett 1998;249:163–166.
48. Bajetto A, Bonavia R, Barbero S, Florio T, Costa A, Schettini G. Expression of chemokine receptors in the rat brain. Ann New York Acad Sci 1999;876:201–209.
49. Sanders VJ, Pittman CA, White MG, Wang G, Wiley CA, Achim CL. Chemokines and receptors in HIV encephalitis. AIDS 1998;12:1021–1026.

50. Tham TN, Lazarini F, Franceschini IA, Lachapelle F, Amara A, Dubois-Dalcq M. Developmental pattern of expression of the alpha chemokine stromal cell-derived factor 1 in the rat central nervous system. Eur J Neurosci 2001;13:845–856.
51. Bolin LM, Murray R, Lukacs NW, Strieter RM, Kunkel SL, Schall TJ, Bacon KB. Primary sensory neurons migrate in response to the chemokine RANTES. J Neuroimmunol 1998;81:49–57.
52. Lazarini F, Casanova P, Tham TN, De Clercq E, Arenzana-Seisdedos F, Baleux F, Dubois-Dalcq M. Differential signalling of the chemokine receptor CXCR4 by stromal cell-derived factor 1 and the HIV glycoprotein in rat neurons and astrocytes. Eur J Neurosci 2000;12: 117–125.
53. Wu JY, Feng L, Park HT, Havlioglu N, Wen L, Tang H, Bacon KB, Jiang Z, Zhang X, Rao Y. The neuronal repellent Slit inhibits leukocyte chemotaxis induced by chemotactic factors. Nature 2001;410:948–952.
54. Oh SB, Tran PB, Gillard SE, Hurley RW, Hammond DL, Miller RJ. Chemokines and glycoprotein120 produce pain hypersensitivity by directly exciting primary nociceptive neurons. J Neurosci 2001;21:5027–5035.
55. Horuk R, Martin AW, Wang Z, Schweitzer L, Gerassimides A, Guo H, Lu Z, Hesselgesser J, Perez HD, Kim J, Parker J, Hadley TJ, Peiper SC. Expression of chemokine receptors by subsets of neurons in the central nervous system. J Immunol 1997;158:2882–2890.
56. Puma C, Danick M, Quirion R, Ramon F, Williams S. The Chemokine interleukine-8 acutely reduces Ca^{2+} currents ion identified cholinergic septal neurons expressing CXCR1 and CXCR2 receptor mRNAs. J Neurochem 2001;78:960–971.
57. Xia MQ, Bacskai BJ, Knowles RB, Qin SX, Hyman BT. Expression of the chemokine receptor CXCR3 on neurons and the elevated expression of its ligand IP-10 in reactive astrocytes: in vitro ERK1/2 activation and role in Alzheimer's disease. J Neuroimmunol 2000;108: 227–235.
58. Kaiser E, Forster R, Wolf I, Ebensperger C, Kuehl WM, Lipp M. The G protein-coupled receptor BLR1 is involved in murine B cell differentiation and is also expressed in neuronal tissues. Eur J Immunol 1993;23: 2532–2539.
59. Kouba M, Vanetti M, Wang X, Schafer M, Hollt V. Cloning of a novel putative G-protein-coupled receptor (NLR) which is expressed in neuronal and lymphatic tissue. FEBS Lett 1993;321:173–178.
60. Xia MQ, Qin SX, Wu LJ, Mackay CR, Hyman BT. Immunohistochemical study of the beta-chemokine receptors CCR3 and CCR5 and their ligands in normal and Alzheimer's disease brains. Am J Pathol 1998;153: 31–37.
61. Westmoreland SV, Rottman JB, Williams KC, Lackner AA, Sasseville VG. Chemokine receptor expression on resident and inflammatory cells in the brain of macaques with simian immunodeficiency virus encephalitis. Am J Pathol 1998;152:659–665.
62. Zhang L, He T, Talal A, Wang G, Frankel SS, Ho DD. In vivo distribution of the human immunodeficiency virus/simian immunodeficiency virus coreceptors: CXCR4, CCR3, and CCR5. J Virol 1998;72:5035–5045.
63. Rottman JB, Ganley KP, Williams K, Wu L, Mackay CR, Ringler DJ. Cellular localization of the chemokine receptor CCR5. Correlation to cellular targets of HIV-1 infection. Am J Pathol 1997;151:1341–1351.
64. Galasso JM, Harrison JK, Silverstein FS. Excitotoxic brain injury stimulates expression of the chemokine receptor CCR5 in neonatal rats. Am J Pathol 1998;153: 1631–1640.
65. Galasso JM, Liu Y, Szaflarski J, Warren JS, Silverstein FS. Monocyte chemoattractant protein-1 is a mediator of acute excitotoxic injury in neonatal rat brain. Neuroscience 2000;101:737–744.
66. Cota M, Kleinschmidt A, Ceccherini-Silberstein F, Aloisi F, Mengozzi M, Mantovani A, Brack-Werner R, Poli G. Upregulated expression of interleukin-8, RANTES and chemokine receptors in human astrocytic cells infected with HIV-1. J Neurovirol 2000;6:75–83.
67. Barnes DA, Huston M, Holmes R, Benveniste EN, Yong VW, Scholz P, Perez HD. Induction of RANTES expression by astrocytes and astrocytoma cell lines. J Neuroimmunol 1996;71:207–214.
68. Pan Y, Lloyd C, Zhou H, Dolich S, Deeds J, Gonzalo JA, Vath J, Gosselin M, Ma J, Dussault B, Woolf E, Alperin G, Culpepper J, Gutierrez-Ramos JC, Gearing D. Neurotactin, a membrane-anchored chemokine upregulated in brain inflammation. Nature 1997;387:611–617.
69. Bazan JF, Bacon KB, Hardiman G, Wang W, Soo K, Rossi D, Greaves DR, Zlotnik A, Schall TJ. A new class of membrane-bound chemokine with a CX3C motif. Nature 1997;385:640–644.
70. Chapman GA, Moores K, Harrison D, Campbell CA, Stewart BR, Strijbos PJ. Fractalkine cleavage from neuronal membranes represents an acute event in the inflammatory response to excitotoxic brain damage. J Neurosci (Online) 2000;20:RC87.
71. Harrison JK, Jiang Y, Chen S, Xia Y, Maciejewski D, McNamara RK, Streit WJ, Salafranca MN, Adhikari S, Thompson DA, Botti P, Bacon KB, Feng L. Role for neuronally derived fractalkine in mediating interactions between neurons and CX3CR1-expressing microglia. Proc Natl Acad Sci USA 1998;95:10896–10901.
72. Cook DN, Chen SC, Sullivan LM, Manfra DJ, Wiekowski MT, Prosser DM, Vassileva G, Lira SA. Generation and analysis of mice lacking the chemokine fractalkine. Mol Cell Biol 2001;21:3159–3165.
73. Meucci O, Fatatis A, Simen AA, Miller RJ. Expression of CX3CR1 chemokine receptors on neurons and their role in neuronal survival. Proc Natl Acad Sci USA 2000; 97:8075–8080.
74. Jung S, Aliberti J, Graemmel P, Sunshine MJ, Kreutzberg GW, Sher A, Littman DR. Analysis of fractalkine receptor CX(3)CR1 function by targeted deletion and green fluorescent protein reporter gene insertion. Mol Cell Biol 2000;20:4106–4114.
75. Maciejewski-Lenoir D, Chen S, Feng L, Maki R, Bacon KB. Characterization of fractalkine in rat brain cells: migratory and activation signals for CX3CR-1-expressing microglia. J Immunol 1999;163:1628–1635.

76. Tong N, Perry SW, Zhang Q, James HJ, Guo H, Brooks A, Bal H, Kinnear SA, Fine S, Epstein LG, Dairaghi D, Schall TJ, Gendelman HE, Dewhurst S, Sharer LR, Gelbard HA. Neuronal fractalkine expression in HIV-1 encephalitis: roles for macrophage recruitment and neuroprotection in the central nervous system. J Immunol 2000;164:1333–1339.
77. Raport CJ, Schweickart VL, Eddy RL Jr, Shows TB, Gray PW. The orphan G-protein-coupled receptor-encoding gene V28 is closely related to genes for chemokine receptors and is expressed in lymphoid and neural tissues. Gene 1995;163:295–299.
78. Harrison JK, Barber CM, Lynch KR. cDNA cloning of a G-protein-coupled receptor expressed in rat spinal cord and brain related to chemokine receptors. Neurosci Lett 1994;169:85–89.
79. Miller RJ. Presynaptic receptors. Annu Rev Pharmacol Toxicol 1998;38:201–227.
80. Saito M, Sato R, Hisatome I, Narahashi T. RANTES and platelet-activating factor open Ca^{2+}-activated K^+ channels in eosinophils. FASEB J 1996;10:792–798.
81. Madani N, Kozak SL, Kavanaugh MP, Kabat D. gp120 envelope glycoproteins of human immunodeficiency viruses competitively antagonize signaling by coreceptors CXCR4 and CCR5. Proc Natl Acad Sci USA 1998; 95:8005–8010.
82. Oh SB, Endoh T, Simen AA, Ren D, Miller RJ. Regulation of calcium currents by chemokines and their receptors. 2002; in press.
83. Limatola C, Giovannelli A, Maggi L, Ragozzino D, Castellani L, Ciotti MT, Vacca F, Mercanti D, Santoni A, Eusebi F. SDF-1alpha-mediated modulation of synaptic transmission in rat cerebellum. Eur J Neurosci 2000;12: 2497–2504.
84. Giovannelli A, Limatola C, Ragozzino D, Mileo AM, Ruggieri A, Ciotti MT, Mercanti D, Santoni A, Eusebi F. CXC chemokines interleukin-8 (IL-8) and growth-related gene product alpha (GROalpha) modulate Purkinje neuron activity in mouse cerebellum. J Neuroimmunol 1998; 92:122–132.
85. Ragozzino D, Giovannelli A, Mileo AM, Limatola C, Santoni A, Eusebi F. Modulation of the neurotransmitter release in rat cerebellar neurons by GRO beta. NeuroReport 1998;9:3601–3606.
86. Griffin JW, Crawford TO, McArthur JC, Peripheral neuropathies associated with HIV infection. In: Gendelman HE, Lipton SA, Epstein L, Swindells S, eds. The Neurology of AIDS. New York: Chapman and Hall, 1998: 275–291.
87. Wong ML, Xin WW, Duman RS. Rat LCR1: cloning and cellular distribution of a putative chemokine receptor in brain. Mol Psychiatry 1996:1;133–140.
88. Gillard SE, Mastracci RM, Miller RJ. Expressions of functional chemokine receptors by rat cerebellar neurons. 2002; in press.
89. Kaul M, Lipton SA. Chemokines and activated macrophages in HIV gp120-induced neuronal apoptosis. Proc Natl Acad Sci USA 1999:96;8212–8216.
90. Araujo DM, Cotman CW. Trophic effects of interleukin-4, -7 and -8 on hippocampal neuronal cultures: potential involvement of glial-derived factors. Brain Res 1993: 600;49–55.
91. Limatola C, Ciotti MT, Mercanti D, Vacca F, Ragozzino D, Giovannelli A, Santoni A, Eusebi F, Miledi R. The chemokine growth-related gene product beta protects rat cerebellar granule cells from apoptotic cell death through alpha-amino-3-hydroxy-5-methyl-4-isoxazolepropionate receptors. Proc Natl Acad Sci USA 2000; 97:6197–6201.
92. Meucci O, Miller RJ. gp120-induced neurotoxicity in hippocampal pyramidal neuron cultures: protective action of TGF-beta1. J Neurosci 1996;16:4080–4088.
93. Toggas SM, Masliah E, Rockenstein EM, Rall GF, Abraham CR, Mucke L. Central nervous system damage produced by expression of the HIV-1 coat protein gp120 in transgenic mice. Nature 1994;367:188–193.
94. Blanco J, Jacotot E, Cabrera C, Cardona A, Clotet B, De Clercq E, Este JA. The implication of the chemokine receptor CXCR4 in HIV-1 envelope protein-induced apoptosis is independent of the G protein-mediated signalling. AIDS 1999;13:909–917.
95. Berndt C, Mopps B, Angermuller S, Gierschik P, Krammer PH. CXCR4 and CD4 mediate a rapid CD95-independent cell death in CD4($^+$) T cells. Proc Natl Acad Sci USA 1998;95:12556–12561.
96. Herbein G, Mahlknecht U, Batliwalla F, Gregersen P, Pappas T, Butler J, O'Brien WA, Verdin E. Apoptosis of $CD8^+$ T cells is mediated by macrophages through interaction of HIV gp120 with chemokine receptor CXCR4. Nature 1998;395:189–194.
97. Roggero R, Robert-Hebmann V, Harrington S, Roland J, Vergne L, Jaleco S, Devaux C, Biard-Piechaczyk M. Binding of human immunodeficiency virus type 1 gp120 to CXCR4 induces mitochondrial transmembrane depolarization and cytochrome c-mediated apoptosis independently of Fas signaling. J Virol 2001;75:7637–7650.
98. Bezzi P, Domercq M, Brambilla L, Galli R, Schols D, De Clercq E, Vescovi A, Bagetta G, Kollias G, Meldolesi J, Volterra A. CXCR4-activated astrocyte glutamate release via TNFalpha: amplification by microglia triggers neurotoxicity. Nat Neurosci 2001;4:702–710.
99. Snow BE, Hall RA, Krumins AM, Brothers GM, Bouchard D, Brothers CA, Chung S, Mangion J, Gilman AG, Lefkowitz RJ, Siderovski DP. GTPase activating specificity of RGS12 and binding specificity of an alternatively spliced PDZ (PSD-95/Dlg/ZO-1) domain. J Biol Chem 1998;273:17749–17755.
100. Thelen M. Dancing to the tune of chemokines. Nat Immunol 2001;2:129–134.
101. Klein RS, Williams KC, Alvarez-Hernandez X, Westmoreland S, Force T, Lackner AA, Luster AD. Chemokine receptor expression and signaling in macaque and human fetal neurons and astrocytes: implications for the neuropathogenesis of AIDS. J Immunol 1999;163:1636–1646.
102. Mellado M, Rodriguez-Frade JM, Manes S, Martinez AC. Chemokine signaling and functional responses: the role of receptor dimerization and TK pathway activation. Annu Rev Immunol 2001;19:397–421.
103. Vila-Coro AJ, Rodriguez-Frade JM, Martin De Ana A, Moreno-Ortiz MC, Martinez AC, Mellado M. The che-

mokine SDF-1alpha triggers CXCR4 receptor dimerization and activates the JAK/STAT pathway. FASEB J 1999;13:1699–1710.

104. Bajetto A, Bonavia R, Barbero S, Piccioli P, Costa A, Florio T, Schettini G. Glial and neuronal cells express functional chemokine receptor CXCR4 and its natural ligand stromal cell-derived factor 1. J Neurochem 1999; 73:2348–2357.

105. Limatola C, Mileo AM, Giovannelli A, Vacca F, Ciotti MT, Mercanti D, Santoni A, Eusebi F. The growth-related gene product beta induces sphingomyelin hydrolysis and activation of c-Jun N-terminal kinase in rat cerebellar granule neurones. J Biol Chem 1999;274: 36537–36543.

106. Ganju RK, Brubaker SA, Meyer J, Dutt P, Yang Y, Qin S, Newman W, Groopman JE. The alpha-chemokine, stromal cell-derived factor-1alpha, binds to the transmembrane G-protein-coupled CXCR-4 receptor and activates multiple signal transduction pathways. J Biol Chem 1998;273:23169–23175.

107. Venkatakrishnan G, Salgia R, Groopman JE. Chemokine receptors CXCR-1/2 activate mitogen-activated protein kinase via the epidermal growth factor receptor in ovarian cancer cells. J Biol Chem 2000;275:6868–6875.

108. Staudinger R, Wang X, Bandres JC. HIV-1 envelope is a neutral antagonist to CXCR4 in T-cells. Biochem Biophys Res Commun 2001;280:1003–1007.

109. Tarasova NI, Stauber RH, Michejda CJ. Spontaneous and ligand-induced trafficking of CXC-chemokine receptor 4. J Biol Chem 1998;273:15883–15886.

110. Lo TM, Fallert CJ, Piser TM, Thayer SA. HIV-1 envelope protein evokes intracellular calcium oscillations in rat hippocampal neurons. Brain Res 1992;594:189–196.

111. Holden CP, Haughey NJ, Nath A, Geiger JD. Role of Na^+/H^+ exchangers, excitatory amino acid receptors and voltage-operated Ca^{2+} channels in human immunodeficiency virus type 1 gp120-mediated increases in intracellular Ca^{2+} in human neurons and astrocytes. Neuroscience 1999;91:1369–1378.

112. Medina I, Ghose S, Ben-Ari Y. Mobilization of intracellular calcium stores participates in the rise of $[Ca^{2+}]i$ and the toxic actions of the HIV coat protein GP120. Eur J Neurosci 1999;11:1167–1178.

113. Lipton SA, Sucher NJ, Kaiser PK, Dreyer EB. Synergistic effects of HIV coat protein and NMDA receptor-mediated neurotoxicity. Neuron 1991;7:111–118.

114. Lipton SA. Models of neuronal injury in AIDS: another role for the NMDA receptor? Trends Neurosci 1992;15: 75–79.

115. Lannuzel A, Lledo PM, Lamghitnia HO, Vincent JD, Tardieu M. HIV-1 envelope proteins gp120 and gp160 potentiate NMDA-induced $[Ca^{2+}]i$ increase, alter $[Ca^{2+}]i$ homeostasis and induce neurotoxicity in human embryonic neurons. Eur J Neurosci 1995;7:2285–2293.

116. Schwaeble WJ, Stover CM, Schall TJ, Dairaghi DJ, Trinder PK, Linington C, Iglesias A, Schubart A, Lynch NJ, Weihe E, Schafer MK. Neuronal expression of fractalkine in the presence and absence of inflammation. FEBS Lett 1998;439:203–207.

117. Fong AM, Robinson LA, Steeber DA, Tedder TF, Yoshie O, Imai T, Patel DD. Fractalkine and CX3CR1 mediate a novel mechanism of leukocyte capture, firm adhesion, and activation under physiologic flow. J Exp Med 1998; 188:1413–1419.

118. Wesselingh SL, Thompson KA. Immunopathogenesis of HIV-associated dementia. Curr Opin Neurol 2001;14: 375–379.

119. Gray F, Adle-Biassette H, Brion F, Ereau T, le Maner I, Levy V, Corcket G. Neuronal apoptosis in human immunodeficiency virus infection. J Neurovirol 2000; 6(Suppl 1):S38–43.

120. Adle-Biassette H, Bell JE, Creange A, Sazdovitch V, Authier FJ, Gray F, Hauw JJ, Gherardi R. DNA breaks detected by in situ end-labelling in dorsal root ganglia of patients with AIDS. Neuropathol Appl Neurobiol 1998; 24:373–380.

121. Shi B, De Girolami U, He J, Wang S, Lorenzo A, Busciglio J, Gabuzda D. Apoptosis induced by HIV-1 infection of the central nervous system. J Clin Invest 1996;98:1979–1990.

122. Gelbard HA, James HJ, Sharer LR, Perry SW, Saito Y, Kazee AM, Blumberg BM, Epstein LG. Apoptotic neurons in brains from paediatric patients with HIV-1 encephalitis and progressive encephalopathy. Neuropathol Appl Neurobiol 1995;21:208–217.

123. Petito CK, Kerza-Kwiatecki AP, Gendelman HE, McCarthy M, Nath A, Podack ER, Shapshak P, Wiley CA. Review: neuronal injury in HIV infection. J Neurovirol 1999;5:327–341.

124. Hery C, Wingertsmann L, Levy Y, Gray F, Tardieu M. HIV-1 induces apoptosis in cultured human neurons and microglial cells but not in astrocytes. Neuropathol Appl Neurobiol 1997;23:352–353.

125. Allen NJ, Attwell D. A chemokine–glutamate connection. Nat Neurosci 2001;4:676–678.

126. Gabuzda D, Wang J. Chemokine receptors and mechanisms of cell death in HIV neuropathogenesis. J Neurovirol 2000;6(Suppl 1):S24–32.

127. Gabuzda D, Wang J, Gorry PD. HIV-1-associated dementia. In: Ransohoff RM, Harrison JK, Hickey WF, Suzuki K, Proudfoot AEI, eds. Universes in Delicate Balance: Chemokines and the Nervous System. Amsterdam: Elsevier:000–000.

128. Choe W, Albright A, Sulcove J, Jaffer S, Hesselgesser J, Lavi E, Crino P, Kolson DL. Functional expression of the seven-transmembrane HIV-1 co-receptor APJ in neural cells. J Neurovirol 2000;6(Suppl 1):S61–69.

129. Cayabyab M, Hinuma S, Farzan M, Choe H, Fukusumi S, Kitada C, Nishizawa N, Hosoya M, Nishimura O, Messele T, Pollakis G, Goudsmit J, Fujino M, Sodroski J. Apelin, the natural ligand of the orphan seven-transmembrane receptor APJ, inhibits human immunodeficiency virus type 1 entry. J Virol 2000;74:11972–11976.

130. Banisadr G, Dicou E, Berbar T, Rostene W, Lombet A, Haour F. Characterization and visualization of [125I] stromal cell-derived factor-1alpha binding to CXCR4 receptors in rat brain and human neuroblastoma cells. J Neuroimmunol 2000;110:151–160.

131. Reeves JD, Hibbitts S, Simmons G, McKnight A, Azevedo-Pereira JM, Moniz-Pereira J, Clapham PR. Primary human immunodeficiency virus type 2 (HIV-2) iso-

lates infect CD4-negative cells via CCR5 and CXCR4: comparison with HIV-1 and simian immunodeficiency virus and relevance to cell tropism in vivo. J Virol 1999; 73:7795–7804.
132. Reeves JD, Heveker N, Brelot A, Alizon M, Clapham PR, Picard L. The second extracellular loop of CXCR4 is involved in CD4-independent entry of human immunodeficiency virus type 2. J Gen Virol 1998;79:1793–1799.
133. Iyengar S, Schwartz DH, Hildreth JE. T cell-tropic HIV gp120 mediates CD4 and CD8 cell chemotaxis through CXCR4 independent of CD4: implications for HIV pathogenesis. J Immunol 1999;162:6263–6267.
134. Hoxie JA, LaBranche CC, Endres MJ, Turner JD, Berson JF, Doms RW, Matthews TJ. CD4-independent utilization of the CXCR4 chemokine receptor by HIV-1 and HIV-2. J Reprod Immunol 1998;41:197–211.
135. Misse D, Cerutti M, Noraz N, Jourdan P, Favero J, Devauchelle G, Yssel H, Taylor N, Veas F. A CD4-independent interaction of human immunodeficiency virus-1 gp120 with CXCR4 induces their cointernalization, cell signaling, and T-cell chemotaxis. Blood 1999; 93:2454–2462.
136. Edinger AL, Mankowski JL, Doranz BJ, Margulies BJ, Lee B, Rucker J, Sharron M, Hoffman TL, Berson JF, Zink MC, Hirsch VM, Clements JE, Doms RW. CD4-independent, CCR5-dependent infection of brain capillary endothelial cells by a neurovirulent simian immunodeficiency virus strain. Proc Natl Acad Sci USA 1997; 94:14742–14747.
137. Endres MJ, Clapham PR, Marsh M, Ahuja M, Turner JD, McKnight A, Thomas JF, Stoebenau-Haggarty B, Choe S, Vance PJ, Wells TN, Power CA, Sutterwala SS, Doms RW, Landau NR, Hoxie JA. CD4-independent infection by HIV-2 is mediated by fusin/CXCR4. Cell 1996;87: 745–756.
138. Reeves JD, McKnight A, Potempa S, Simmons G, Gray PW, Power CA, Wells T, Weiss RA, Talbot SJ. CD4-independent infection by HIV-2 (ROD/B): use of the 7-transmembrane receptors CXCR-4, CCR-3, and V28 for entry. Virology 1997;231:130–134.
139. Potempa S, Picard L, Reeves JD, Wilkinson D, Weiss RA, Talbot SJ. CD4-independent infection by human immunodeficiency virus type 2 strain ROD/B: the role of the N-terminal domain of CXCR-4 in fusion and entry. J Virol 1997;71:4419–4424.
140. Martin KA, Wyatt R, Farzan M, Choe H, Marcon L, Desjardins E, Robinson J, Sodroski J, Gerard C, Gerard NP. CD4-independent binding of SIV gp120 to rhesus CCR5. Science 1997;278:1470–1473.

Universes in Delicate Balance: Chemokines and the Nervous System
R.M. Ransohoff, K. Suzuki, A.E.I. Proudfoot, W.F. Hickey and J.K. Harrison, editors

Chapter 4.5

Microglial Chemokines and Chemokine Receptors

Knut Biber

Department of Medical Physiology, State University Groningen, A. Deusinglaan 1, 9700 AD Groningen, The Netherlands

1. INTRODUCTION

The microglial cell, originally described by Del Rio-Hortega [1], is the third major type of brain glial cells. Despite a long and still ongoing debate on their developmental origin [2] they are today generally considered the resident macrophages of the brain. It has been shown that bone marrow derived monocytes invade the CNS during development and give rise to microglia after the blood–brain barrier (BBB) has been established [2,3]. Microglia are involved in phagocytosis and removal of degenerating cells during the development of the CNS. After the first few weeks of postnatal development microglia undergo morphological changes, developing from a typical macrophage like cell to a ramified morphology. Ramified microglia are characterized by small, long cell bodies, which are highly branched with tiny, extended processes. Although ramified microglia constitute approximately 10–20% of the total glial population [3,4] in the adult CNS, the function of these cells remains still largely unknown, due in part to the lack of techniques recording the activity of these cells in vivo. For example, although it is known that microglia are more associated with gray matter than with white matter a physiological role in synaptic transmission has not been defined yet. Obviously, microglia exert their major functions during pathological events in the brain and it is widely accepted now that microglia are the major endogenous immune-competent cells of the brain [5–7]. The ramified or resting microglia show a down-regulated immune phenotype adapted to the immune-suppressive environment of the CNS. In this state microglia express several pattern recognition receptors like the LPS receptor [8], which enables them to screen their microenvironment and make them the sentinels of the CNS. Microglia generally are the first cells within the CNS to become activated in response to a variety of pathological events. Irrespective of the stimulus, activation of microglia proceeds via a stereotypic multi-step process. The first morphological changes are characterized by a retraction of their fine processes followed by a step by step development into a macrophage-like, phagocytic cell. This process, which occurs fast (i.e. within a few hours), has been recently shown in fascinating 3D pictures by Daily and Waite [9]. Activated microglia are characterized by expression and release of a large variety of immunological mediators such as major histocompatibility complex, complement proteins, prostanoids, nitric oxide and cytokines, but also neurotrophins and other beneficial molecules [5,6,8,10–12]. It is now widely accepted that microglia initiate and maintain the inflammatory response of the CNS during pathological conditions. This makes microglia key players orchestrating a multicell cascade that follows neuronal damage. Considerable evidence has accumulated in the last few years that microglia activation not only is detrimental to brain tissue but also may have significant protective functions [5,7,8,13]. In order to develop more sophisticated therapies for neurodegenerative disorders, it is therefore of particular interest to gain more knowledge of the intercellular signaling pathways which control microglial activation. In this chapter I will discuss the role of microglial chemokines and chemokine receptors in CNS inflammation.

2. CHEMOKINES EXPRESSED BY MICROGLIA

2.1. Cultured Microglia

Due to mouse and human genome projects and EST databases, a large variety of chemokines and chemokine receptors have been found in the last few years [14]. Currently, more than 40 human chemokines have been identified [15]. One of the first reports showing the expression of chemokines in cultured microglia was published by Hayashi et al. [16], showing microglial expression of MIP-1α. So far the following chemokines have been found in cultured microglia: I-309 (CCL1), MCP-1 (CCL2), MIP-1α (CCL3), MIP-1β (CCL4), RANTES (CCL5), IL-8 (CXCL8), IP-10 (CXCL10), GROα (CXCL1), and SDF-1β CXCL12) (see Table 1). Most microglial chemokines are not expressed under control culture conditions, although it is well known that cultured microglia are always in a partially activated state. One chemokine, which was reproducibly shown to be expressed by unstimulated microglia is MCP-1 [17–24]. Investigating the mRNA expression of chemokines in human microglia cultured from adult post-mortem brain we also detected MCP-1 mRNA expression under unstimulated culture conditions (K. Biber, C. de Groot, unpublished observations, manuscript in preparation). This baseline expression of MCP-1 in culture may be a feature of human or rat microglia since this was not observed in cultured mouse microglia [16,25]. Moreover, constitutive expression of SDF-1β was found in cultured rat microglia [26]. The microglial expression of SDF-1β differed from other chemokines, since it was insensitive to stimulation with lipopolysaccharide (LPS) [26].

Generally, activation of microglia in culture by pro-inflammatory stimuli is required to induce expression of chemokines. One of the most thoroughly investigated and most pronounced pro-inflammatory stimuli for glial cells is LPS, a cell wall component of gram negative bacteria. LPS is a strong inducer of almost all chemokines which have been found in microglia (see Table 1): MIP-1α and β, MCP-1 [19,24,25], RANTES [27], IL-8 [28,29], I-309 and IP-10 [17,30]. Another bacterial toxin, capsular polysaccharide glucuronoxylomannan (GXM) derived from cryptococci, induces the expression of IL-8 in human fetal microglia [88].

Other strong inductors of a variety of chemokines expressed in cultured microglia include pro-inflammatory cytokines such as IL-1β and TNFα [18,19, 27,31,32] or interferon (IFN) β and γ [20,25,31] (Table 1).

Chemokine expression in cultured microglia is also induced by proteins that are involved in CNS pathology such as amyloid β-protein (Aβ). Aβ stimulated the expression of MCP-1 [25], MIP-1α, MIP-1β and IL-8 [33]. The effect of Aβ on microglia seems not to be due to the amino acid sequence of Aβ. Amylin which forms similar fibrils as Aβ induced

Table 1. Induction of chemokine expression in cultured microglia

Chemokine	Species	Stimulus
CC chemokines		
I-309 (CCL1)	Human	LPS [17], IFNβ [17], IFNγ [17]
MCP-1 (CCL2)	Human, mouse, rat	LPS [17–19,22,23,25,89], IFNβ [20], IFNγ [17,20,25], TNFα + IL-1β [19,23], Aβ [25], TNFα [23], IL-6 [23], HIV Tat [21]
MIP-1α (CCL3)	Human, rat, mouse	LPS [16–19,24,63,89,91], IFNβ [17,20], IL-1β [17–19,31,32], TNFα [17–19,31,32], IFNγ [19,31], Aβ [33], HIV Tat [21,34], gp41 [34], TNFα + IL-1β [19,32]
MIP-1β (CCL4)	Human, rat	LPS [17–19,22,24,63], TNFα [17–19], IL-1β [17–19], IFNβ [17,20], Aβ [33], HIV Tat [21], TNFα + IL-1β [19,32]
RANTES (CCL5)	Human, rat, mouse	LPS [17,22,24,27,63,89], IL-1β [27,31], TNFα [27,31], IFNβ [20], NDV [90], IFNγ [31], HIV Tat [34], gp41 [34], TNFα + IFNγ [32]
CXC chemokines		
GROα (CXCL1)	Human	TNFα + IL-1β [32]
IL-8 (CXCL8)	Human	LPS [17,22,28,29], IL1β [17,28,29,32], TNFα [17,28,32], Aβ [33], GXM [88], TNFα + IL1β [32]
IP-10 (CXCL10)	Human, rat, mouse	LPS [17,30,89], IFNβ [17], IFNγ [17,89], IL-1β [17], NDV [90], TNFα [17,89]

Abbreviation: NDV, Newcastle disease virus.

comparable effects on microglial chemokine synthesis, indicating that microglia most likely respond to the fibrillar structure of the peptide [33]. HIV Tat, the HIV transactivator protein which is released from HIV-infected cells induced the synthesis of MCP-1, MIP-1α, MIP-1β and RANTES in cultured human fetal microglia [21,34]. These results indicate that cultured microglia respond to pro-inflammatory stimuli by synthesizing and releasing chemokines. Chemokines, once released, may contribute to additional chemokine synthesis in the CNS. For example, it has been shown by McManus and colleagues that MIP-1α and MIP-1β induce the release of MCP-1, MIP-1α and MIP-1β, indicating an autocrine role of chemokines in microglial chemokine synthesis [21].

Whereas pro-inflammatory cytokines induce microglial chemokine synthesis, anti-inflammatory cytokines down-regulate microglial chemokine expression. In human fetal microglia the expression of both RANTES and IL-8 was induced by LPS, IL-1β or TNFα. The induction of both chemokines was dependent on the activation of nuclear factor (NF)-κB [27,29], a transcription factor which is involved in the expression of a great variety of genes, among them several cytokines. Co-stimulation with IL-4, IL-10 or TGFβ inhibited the expression of microglial RANTES and IL-8 [27–29]. The inhibitory effect of anti-inflammatory cytokines was most likely due to an inhibition of the transcription of chemokine genes, since activation of NFκB was inhibited by IL-4, IL-10 and TGFβ [27,29]. Inhibition of NFκB activity might also be responsible for the inhibition of LPS-induced release of IL-8 by microglia by thalidomide, a psychoactive drug that is known to down-regulate microglial cytokine synthesis [22]. Surprisingly, the synthesis of RANTES, MIP-1β and MCP-1 was not affected by thalidomide treatment, indicating that different microglial chemokines are regulated by different sets of transcription factors [22].

Differences in the regulation of microglial CC and CXC chemokines have been published by Janabi et al. [32]. They showed that prostaglandins inhibited the release of RANTES, MIP-1α and MIP-1β, but had no effect on the secretion of IL-8 and GROα. On the other hand the CXC chemokines IL-8 and GROα had inhibitory effects on the expression of PGE_2, whereas the CC chemokines RANTES, MIP-1α and MIP-1β did not affect microglial prostaglandin synthesis [32].

Moreover, it has been shown that IFNβ and IFNγ differentially regulate microglial chemokines of CXC or CC families [17,20]. IFNβ induced the release of RANTES, MIP-1α and MIP-1β, whereas IFNγ only stimulated the synthesis of microglial MCP-1. However, I-309 and IP-10 were induced by both IFNβ and IFNγ [17,20]. Taken together these results indicate that microglia react specifically to a given stimulus with the release of distinct chemokines.

2.2. Chemokines Expressed by Microglia in Vivo

Apart from SDF and fractalkine that is found in neurons, there is little chemokine expression in healthy brain ([35–37]; Chapter 4.1). Induction of chemokine expression in the CNS has been observed in a number of neurodegenerative diseases and animal models [38–40]. In theses instances astrocytes have been identified as the cellular source of chemokines in damaged brain tissue. Unfortunately an identification of chemokine producing microglia in the damaged CNS is more complicated. Microglia are invariably activated during CNS pathology and express macrophage-specific surface markers. Furthermore, many pathological conditions in the brain are accompanied by an infiltration of the CNS by hematogenous leukocytes. Considering there are yet no available methods to readily distinguish between blood-derived monocytes and activated microglia in brain parenchyma, a clear demonstration of chemokine expression by endogenous microglia is difficult. The identification of microglia as a cellular source for chemokines can, however, be made by more indirect evidence, i.e. morphological features and time point of chemokine expression. If chemokine positive macrophage-like cells show ramification, or are expressed at early time points before the onset of monocyte infiltration, it is likely that microglia are the cellular source.

Using different detection techniques expression of MCP-1, MIP-1α, MIP-1β, RANTES, MIP-2 and IP-10 in mononuclear macrophage-like cells within the CNS has been shown (see Table 2). Using morphological criteria and considering the time point of expression, most of these chemokines have been found in microglia. Based on the morphology the following chemokines were detected in microglia-like cells. MIP-1α and MIP-1β in AIDS patients with HIV neu-

Table 2. Expression of chemokines in human disease and animal models in mononuclear cells

Disease or model	Species	Cellular source	Chemokine	Reference
HIV	Human	MI	MIP-1α, MIP-1β	[41]
MS	Human	PC	RANTES	[42]
		MM	MIP-1α, MCP-1	
		MM, MI	MIP-1β	
Alzheimer's disease	Human	MI, MM	MCP-1, MIP-1α	[64,92]
Glioblastoma	Human	MM, MI	MCP-1	[43]
Ischemia (MCAO)	Rat	MI	MIP-1α, MIP-1β	[44–46]
		MM	MCP-1	
Experimental autoimmune encephalomyelitis	Rat	MM	RANTES, MIP-1α, MIP-β	[93,96]
Experimental bacterial meningitis	Rat	MM	RANTES, MCP-1, MIP-2, MIP-1α	[82]
Experimental african trypanosomiasis	Rat	MI	MIP-1α, MIP-2, RANTES	[47]
LPS injection	Rat	MI	MIP-1α, MIP-1β	[45]
MHV infection	Mouse	MM	IP-10	[94]
SIV	Monkey	PC	MIP-1α, MIP-1β	[49]
IL-1β injection	Rabbit	PC	MCP-1	[50]

Abbreviations: MI, microglia; MM, macrophages/microglia; PC, perivascular cells; MHV, mouse hepatitis virus.

rodegeneration [41] and MIP-1β in normally appearing white matter in the vicinity of lesions in postmortem brains of MS patients [42]. Furthermore, MCP-1 was found in microglia-like cells adjacent to glioblastomas [43]. MIP-α and MIP-1β expression were also found in microglia like cells at early time points after induction of ischemia. Also after LPS injection chemokine expression in microglia-like cells was detected [44–46]. In addition expression of MIP-1α, MIP-2 and RANTES was observed after experimental exposure to African trypanosomiasis pathogens [47]. These data show that brain microglia in vivo may react on a variety of CNS pathologies with the expression of specific chemokines. This is in line with the results observed in vitro.

Chemokines are also expressed by another mononuclear macrophage-like cell in the CNS, the perivascular macrophage or perivascular cells. These cells, often termed 'perivascular microglia', are different from CNS resident microglia. Perivascular cells are elongated, flattened cells which are located between the neuronal parenchyma and the vascular endothelium (the Virchow–Robin space) [8,48]. Presumably because these specialized cells are localized between the blood–brain barrier and the CNS parenchyma they may display a very specific chemokine expression pattern. For example MIP-1α and MIP-1β expression in perivascular cells was detected in simian immunodeficiency virus-induced encephalitis in macaque monkeys [49]. Furthermore, MCP-1 expression in perivascular cells was found in rabbit retina 2 h after the injection of IL-1β [50].

3. MICROGLIAL CHEMOKINE RECEPTORS

3.1. Expression of Chemokine Receptors in Cultured Microglia

Investigating the expression of chemotactic receptors in human glial cells Lacy et al. found expression of IL-8 receptor, now known as CXCR1, in cultured human microglia [51]. This was one of the first reports indicating that cultured microglia express chemokine receptors. Recently, the expression of chemokine receptors in cultured microglia has been studied by various groups (see Table 3). So far the expression of CCR1, 2, 3 and 5 has been found in fetal cultured microglia from rat and human sources [20,21,52–55]. Furthermore, cultured microglia express CXCR1, CXCR4 [26,54–57] and the fractalkine receptor CX3CR1 [55,58–60].

Contradictory reports on the expression of CCR2 and CCR3 mRNA in cultured microglia have been published. Low microglial CCR2 mRNA expression was found by Boddeke et al. [52] and McManus et al. [21], whereas no CCR2 mRNA in cultured microglia was found by others [55]. CCR3 mRNA expression in cultured microglia was found by He et al. [60] and McManus et al. [61] but not by Jiang et al. [55] and

Table 3. Expression of chemokine receptors in cultured microglia

Chemokine receptor	Species	Cell type	Reference
CCR1	Rat, human	Primary microglia	[21,52]
CCR2	Rat, human	Primary microglia	[21,52]
CCR3	Human	Primary microglia	[21,53,54]
CCR5	Human, rat	Primary microglia	[21,52–55,63]
CXCR1	Human	Primary microglia	[51]
CXCR4	Mouse, rat, human	Cell line N9, primary microglia	[26,54–57]
CX3CR	Rat	Primary microglia	[55,58–60]

Boddeke et al. [52]. Our own studies on the expression of CC chemokine receptors in cultured mouse microglia showed expression of CCR1, 3 and 5, but not of CCR2, 4, 6, 7 and 8. Recently, we detected expression of an orphan chemokine receptor in cultured mouse microglia (Biber et al., unpublished observations, manuscript in preparation).

In contrast to the expression of chemokines, most chemokine receptors are expressed constitutively in microglia. Chemokine receptor numbers per cell have been determined using antibody binding in human fetal microglia [54]. The rank order of expression was CCR5 (4.5×10^4 binding sites (bs)) > CXCR4 (2×10^4 bs) > CCR3 (5×10^3 bs). Subsequent experiments investigating the signaling of these receptors indicated that stimulation of CCR5 and CXCR4 induce intracellular calcium signaling. Stimulation of intracellular calcium signaling via CCR3 has not been demonstrated, possibly due to limitations in the sensitivity of the detection technique [54]. In rat microglia it was found that CX3CR1 is expressed in high levels, approx. $3–5 \times 10^4$ bs per cell [62]. Although the level of chemokine receptor expression may vary depending on culture conditions, these results implicate that cultured microglia express high numbers of chemokine receptors.

Little information is available on the regulation of chemokine receptor expression in cultured microglia. Most experiments on regulation of chemokine receptor expression have been performed using LPS stimulation of microglia cultivated from newborn rats. An upregulation of CCR1, 2 and 5 mRNA by LPS treatment was found for rat microglia [52,63]. LPS treatment, however, did not affect expression of CXCR4 mRNA [26]. The expression of CX3CR1, however, was markedly decreased in response to LPS treatment [59,60]. Jiang and coworkers reported upregulation of CCR5 mRNA by stimulation with IFNγ [55]. These data indicate that pro-inflammatory conditions, which often induce the synthesis of microglial chemokines, may have differential effects on microglial chemokine receptor expression.

3.2. Microglial Chemokine Receptors in Vivo

Several papers concerning chemokine receptor expression in ramified microglia in healthy brain have been published (Table 4). Expression of CCR3 and CCR5 was found in resting microglia and upregulation of both receptors was found in neuronal tissue affected by HIV-induced neurodegeneration [64,65]. Additional CXCR4 expression in microglia was found in control and HIV-infected brains. The level of CXCR4 expression in microglia, however, did not correlate with disease severity [57,65]. Similar findings have been published in macaque monkeys, where expression of CCR3, CCR5 and CXCR4 was found in microglia but only CCR5 expression was upregulated in simian immunodeficiency virus (SIV)-infected animals [66]. Several papers also indicate the expression of the fractalkine receptor CX3CR1 in ramified microglia in vivo [35–37]. While CCR1 expression on cultured microglial cells has been documented, this receptor has not been determined to be expressed by microglia in vivo (Table 4).

A comparison of human post-mortem control brain tissues with post-mortem tissue from MS patients, showed that microglial expression of CCR5 but not CCR1 or 2 was found in non-diseased controls, whereas in MS tissue macrophages/microglia were positive for CCR1, CCR5 and occasionally CCR2 [67]. Expression of CCR5 in macrophages/ microglia in MS lesions was also detected by Sorensen and colleagues [68]. Also expression of CCR2 and CCR5 was found in the damaged brain

Table 4. Expression of microglial chemokine receptors in vivo

Chemokine receptor	Cellular source	Disease	Reference
CCR2	MM	MS brain, excitotoxicity	[67,70,95]
CCR3	MI, MM	Alzheimer's brain, SIV monkey, MS brain, HIV	[61,64,66,67,95]
CCR5	MI, MM	Alzheimer's brain, SIV monkey, MS brain, excitotoxicity, HIV	[64–69,95]
CXCR4	MI, MM	SIV monkey, HIV	[57,65,66]
CX3CR	MI, MM	HIV, rat and mouse brain	[35–37]

region after NMDA-induced neurotoxicity [69,70]. These data show that microglia express distinct chemokine receptors not only under pathological conditions but also in their resting state. This series of observations parallels the situation observed in vitro, where unstimulated microglia expressed specific chemokine receptors.

4. EFFECTS OF MICROGLIAL CHEMOKINE RECEPTOR STIMULATION

Microglial chemokine receptors are functional. This has been shown in a variety of papers where the effects of chemokine stimulation on microglial calcium signaling and chemotaxis were investigated. Intracellular calcium transients and/or chemotaxis of cultured microglia were induced by stimulation with SDF-1α [54,56], fractalkine [58,59], MIP-α [71], MIP-1β [54,71], MCP-1 [52,71], RANTES [52,71,72], eotaxin [72], IL-8 and IP-10 [71]. In an interesting study the effect of several chemokines on microglial production of matrix metalloproteinases and their natural inhibitors was investigated [73]. It was found that certain chemokines (MCP-1, MIP-1β, RANTES, IL-8 and fractalkine) stimulated the production of both matrix metalloproteinases and their natural inhibitors. It was therefore suggested that the functional balance of microglial matrix metalloproteinases was affected by chemokines [73]. Two recent publications suggested a role for fractalkine in modulating microglial activity and in promoting microglial survival. The secretion of TNFα and neurotoxic properties of microglia were inhibited by fractalkine treatment [60]. Boehme and colleagues reported that treatment with fractalkine inhibited Fas-mediated microglial death, by up-regulating the anti-apoptotic protein Bcl-x_L and down-regulating the pro-apoptotic members Bad and BID [62]. Interestingly, no direct effect of fractalkine on microglial proliferation was observed by Maciejewski-Lenoir and co-workers. However, these workers showed that fractalkine stimulated the release of a factor from astrocytes which promoted microglial proliferation [58]. It is not yet known whether chemokines may have similar effects on microglia in vivo, but if so, this would suggest an important role for chemokines in the regulation of the physiology of microglia.

5. POTENTIAL FUNCTIONS OF MICROGLIAL CHEMOKINES AND CHEMOKINE RECEPTORS

The results presented in this chapter revealed that microglia contribute to the synthesis of chemokines in the CNS during pathological conditions. Today it is generally assumed that the orchestration of infiltration of the brain by blood leukocytes is mediated by locally induced chemokines [74–76]. In various models of the blood–brain barrier chemokines stimulate the migration of different types of blood leukocytes [24,77–79]. Moreover it has been shown in chemokine receptor knockout mice that leukocyte infiltration of the CNS is impaired [76,80] or can be inhibited by blocking the bio-activity of chemokines [81–83].

Since most of the endogenous cells of the CNS express receptors for chemokines, a broad spectrum of chemokines in CNS physiology and pathophysiology has been proposed. Chemokines are most likely involved in development and patterning of the brain. Furthermore, chemokines may contribute to CNS synaptic transmission and plasticity. Chemokines may have neuroprotective effects and possibly contribute to tumorigenicity and angiogenesis [17,39,75,76,84]. Studies to elucidate the contribution of microglially derived chemokines using genetically modified animals (knockout of chemokines in microglia or overexpression of chemokines in microglia) have not been conducted yet. Due to this lack of experimental evidence it remains a matter of speculation whether chemokines derived from microglia contribute to physiological or pathological processes in the CNS.

It has been shown that cultured microglia under basal conditions and ramified microglia in healthy CNS tissue express a variety of chemokine receptors (see above). This raises the question on the function of these microglial chemokine receptors. As indicated in the introduction of this chapter microglia are the endogenous sensors of the CNS, which continuously screen their microenvironment and react rapidly to any form of brain pathology. To fulfill this role microglia express a variety of different receptors and ion-channels [7,8]. It is therefore tempting to speculate that chemokine receptors might belong to the microglial receptor set enabling the detection of pathological events within the CNS. Both astrocytes and neurons express and release chemokines in culture and in vivo. Due to their chemokine receptors microglia have the capacity to detect these chemokines and respond in an appropriate way. The sensing of neuronal damage is one of the most important tasks of microglia. Thus, it is attractive to suggest that chemokines might be acting as the critical signaling molecules in relaying information from damaged neuronal structures to the microglia cell. Several chemokines are expressed by neurons after neuronal damage, including MIP-1α [64], IP-10 [85] and SLC [97]. Expression of IP-10 and SLC has been found at an early stage (3–12 h) in neurons after ischemic brain damage. Furthermore, our group observed that microglia respond to both IP-10 and SLC with an increase in intracellular calcium and chemotaxis [97], indicating that microglia express the corresponding chemokine receptors. It is therefore tempting to speculate that both IP-10 and SLC might be involved in a neuron–microglia signaling system related to ischemic pathology. A similar function has already been suggested for CX3CR1 and its ligand fractalkine. The membrane-bound chemokine fractalkine is constitutively expressed in neurons. It has been found that neurons damaged by lesions of the nervus fascialis may release fractalkine [35]. Since microglia constitutively express the corresponding receptor CX3CR1 and activation of microglia is the first detectable cellular event after fascialis lesion, a role of fractalkine in neuron–microglia signaling has been suggested [35]. Recent findings, however, did not corroborate this assumption. In a CX3CR1 knockout model the microglial activation after fascialis lesion was unchanged compared to wild-type controls [36], which implies that fractalkine does not significantly contribute to the neuron–microglia signaling system which activates microglia after lesion of the fascialis nerve [36]. On the other hand it was shown in vitro that fractalkine might modulate the activity of microglia [60]. These results clearly show that more experimental evidence is necessary to clarify the function of chemokines in neuron–microglia signaling.

Several papers suggest that microglia and microglial chemokine receptors play an important role in the infection of the brain by HIV and in the pathogenesis of AIDS-related dementia [86,87]. Microglia express all major chemokine receptors (CCR3, CCR5 and CXCR4) which are used by HIV as co-receptors to enter cells. Microglia, therefore, are the major CNS endogenous cells infected by HIV and the predominant viral reservoir within the CNS. Conflicting results are published on the receptor subtype, responsible for the infection of microglia with HIV. He and colleagues suggested CCR3 and CCR5, since they could show that antibodies for both receptors inhibited microglial HIV infection [61]. Based on a high expression level of CCR5 in microglia, which was more pronounced than that of CXCR4 or CCR3, Albright and coworkers suggested a major involvement of CCR5 in HIV infection of microglia [54]. In a study by Ghorpade et al. [53] antibodies against CCR3 and CCR5 failed to prevent microglial HIV infection, suggesting the existence of another chemokine receptor involved in HIV infection. More experiments are therefore needed to elucidate the role of microglial chemokine receptors in HIV infection of the brain.

6. CONCLUSIONS AND KEY ISSUES FOR FUTURE CONSIDERATION

Based on the current literature, it is likely that chemokine signaling in the CNS is involved in physiological as well as pathological processes. Microglia contribute to the synthesis of chemokines within the CNS and respond to chemokines via chemokine receptors. Microglia, therefore, are a substantial part of the *chemokinergic* network within the CNS. Since the scope of functions of chemokines and their receptors has broadened considerably to extend beyond leukocyte-chemoattraction, microglial chemokines and chemokine receptors might be important factors in controlling the activity of microglia. The identification of the precise relationship of chemokines and microglial activity and the determination whether microglial activation induced by chemokines will have protective or pathogenic consequences, will be the future challenge.

REFERENCES

1. Del-Rio Hortega P. Microglia. In: Penfield W, ed. Cytology and Cellular Pathology of the Nervous System. Oxford: Oxford University Press, 1932:481–534.
2. Ling EA, Wong WC. The origin and nature of ramified and amoeboid microglia: a historical review and current concepts. Glia 1993;7:9–18.
3. Perry VH, Andersson PB, Gordon S. Macrophages and inflammation in the central nervous system. Trends Neurosci 1993;16:268–273.
4. Lawson LJ, Perry VH, Dri P, Gordon S. Heterogeneity in the distribution and morphology of microglia in the normal, adult mouse brain. Neuroscience 1991;39:151–170.
5. Bruce Keller AJ. Microglial–Neuronal interactions in synaptic damage and recovery. J Neurosci Res 1999;58: 191–201.
6. Minghetti L, Levi G. Microglia as effector cells in brain damage and repair: focus on prostanoids and nitric oxide. Prog Neurobiol 1998;54:99–125.
7. Kreutzberg GW. Microglia: A sensor for pathological event is the CNS. Trends Neurosci 1996;19:312–318.
8. Raivich G, Bohatschek M, Kloss CUA, Werner A, Jones LL, Kreutzberg GW. Neuroglial activation repertoire in the injured brain: graded response molecular mechanisms and cues to physiological function. Brain Res Rev 1999; 30:77–105.
9. Daily ME, Waite M. Confocal imaging of microglial cell dynamics in hippocampal slice cultures. Methods 1999; 18:222–230.
10. Frade JM, Barde YA. Microglia-derived nerve growth factor causes cell death in the developing retina. Neuron 1999;20:35–41.
11. Moore S, Thanos S. The concept of microglia in relation to central nervous system disease and regeneration. Prog Neurobiol 1996;35:21–30.
12. Streit WJ, Walter SA, Pennell NA. Reactive microgliosis. Prog Neurobiol 1999;57:563–581.
13. Batchelor PE, Liberatore GT, Wong JYF, Porritt MJ, Frerichs F, Donnan GA, Howelss DW. Activated macrophages and microglia induce dopaminergic sprouting in the injured striatum and express brain-derived neurotrophic factor and glial cell line-derived neurotrophic factor. J Neurosci 1999;19:1708–1716.
14. Rossi D, Zlotnik A. The biology of chemokines and their receptors. Annu Rev Immunol 2000;18:217–242.
15. Murphy PM, Baggiolini M, Charo IF, Hebert CA, Horuk R, Matsushima K, Miller LH, Oppenheim JJ, Power CA. Nomenclature for chemokine receptors. Pharmacol Rev 2000;52:146–168.
16. Hayashi M, Luo Y, Laning J, Strieter RM, Dorf ME. Production and function of monocyte chemoattractant protein-1 and other β-chemokines in murine glial cells. J Neuroimmunol 1995;60:143–150.
17. Hua LL, Lee SC. Distinct patterns of stimulus-inducible chemokine mRNA accumulation in human fetal astrocytes and microglia. Glia 2000;30:74–81.
18. Peterson PK, Hu S, Salak-Johnson J, Molitor TW, Chao CC. Differential production of and migratory response to β chemokines by human microglia and astrocytes. J Infect Dis 1997;175:478–481.
19. McManus CM, Brosnan CF, Berman JW. Cytokine induction of MIP-1α and MIP-1β in human fetal microglia. J Immunol 1998;160:1449–1455.
20. McManus CM, Liu JSH, Hahn MT, Hua LL, Brosnan CF, Berman JW, Lee SC. Differential induction of chemokines in human microglia type I and II interferons. Glia 2000; 29:273–280.
21. McManus CM, Weidenheim K, Woodman SE, Nunez J, Hesselgesser J, Nath A, Berman JW. Chemokine and chemokine-receptor expression in human glial elements. Am J Pathol 2000;156:1441–1453.
22. Lokensgard JR, Hu S, van Fenema EM, Sheng WS, Peterson PK. Effect of thalidomide on chemokine production by human microglia. J Infect Dis 2000;182:983–987.
23. Calvo CF, Yoshimura T, Gelman M, Mallta M. Production of monocyte chemotactic protein-1 by rat brain macrophages. Eur J Neurosci 1996;8:1725–1734.
24. Persidsky Y, Ghorpade A, Rasmussen J, Limoges J, Liu XJ, Stins M, Fiala M, Way D, Kim KS, Witte MH, Weinand M, Carhart LR, Gendelman HE. Microglial and astrocyte chemokines regulate monocyte migration through the blood–brain barrier in human immunodeficiency Virus-1 encephalitis. Am J Pathol 1999;155:1599–1611.
25. Meda L, Bernasconi S, Bonaiuto C, Sozzani S, Zhou D, Otvos L Jr, Mantovani A, Rossi F, Cassatella A. β-amyloid (25–35) peptide and IFN-γ synergistically induce the production of the chemotactic cytokine MCP-1/JE in monocytes and microglial cells. J Immunol 1996;157: 1213–1218.
26. Ohtani Y, Minami M, Kawaguchi N, Nishiyori A, Yamamoto J, Takami S, Satoh M. Expression of stromal cell-derived factor-1 and CXCR4 chemokine receptor mRNAs in cultured rat glial and neuronal cells. Neurosci Lett 1998;249:163–166.
27. Hu S, Chao CC, Ehrlich LC, Sheng WS, Sutton RL, Rockswold GL, Peterson PK. Inhibition of microglial cell RANTES production by IL-10 and TGF-β. J Leukoc Biol 1999;65:815–821.
28. Ehrlich LC, Hu S, Sheng WS, Sutton RL, Rockswold GL, Peterson PK, Chao CC. Cytokine regulation of human

microglial cell IL-8 production. J Immunol 1998;160: 1944–1948.
29. Ehrlich LC, Hu S, Peterson PK, Chao CC. IL-10 down-regulates human microglial IL-8 inhibition of NF-kappaB activation. Neuroreport 1998;9:1723–1726.
30. Ren LQ, Gourmala N, Boddeke HW, Gebicke-Haerter PJ. Lipopolysaccharide-induced expression of IP-10mRNA in rat brain and in cultured rat astrocytes and microglia. Brain Res Mol Brain Res 1998;59:256–263.
31. Janabi N, di Stefano M, Wallon C, Hery C, Chiodi F, Tardieu M. Induction of human immunodeficiency virus type 1 replication in human glial cells after proinflammatory cytokines stimulation: effect of IFNγ, Il1β, and TNFα on differentiation and chemokine production in glial cells. Glia 1998;23:304–315.
32. Janabi N, Hau I, Tardieu M. Negative feedback between prostaglandin and α- and β-chemokine synthesis in human microglial cells and astrocytes. J Immunol 1999;162: 1701–1706.
33. Yates SL, Burgess LH, Kocsis-Angle J, Antal JM, Dority MD, Embury PB, Piotrkowski AM, Brunden KR. Amyloid β and amylin fibrils induce increases in proinflammatory cytokine and chemokine production by THP-1 cells and murine microglia. J Neurochem 2000;74:1017–1025.
34. Sheng WS, Hu S, Hegg CC, Thayer SA, Peterson PK. Activation of human microglial cells by HIV-1 gp41 and Tat proteins. Clin Immunol 2000;96:243–251.
35. Harrison JK, Jiang Y, Chen S, Xia Y, Maciejewski D, McNamara RK, Streit WJ, Salafranca MN, Adhikari S, Thompson DA, Botti P, Bacon KB, Feng L. Role for neuronally derived fractalkine in mediating interactions between neurons and CX3CR1-expressing microglia. Proc Natl Acad Sci USA 1998;95:10896–10901.
36. Jung S, Aliberti J, Graemmel P, Sunshine MJ, Kreutzberg GW, Sher A, Littman DR. Analysis of fractalkine receptor CX_3CR1 function by targeted deletion and green fluorescent protein reporter gene insertion. Mol Cell Biol 2000; 20:1–13.
37. Tong N, Perry SW, Zhang Q, James HJ, Guo H, Brooks A, Bal H, Kinnear SA, Fine S, Epstein LG, Dairaghi D, Schall TJ, Gendelman HE, Dewhurst S, Sharer LR, Gelbard HA. Neuronal fractalkine expression in HIV-1 encephalitis: roles for macrophage recruitment and neuroprotection in the central nervous system. J Immunol 2000; 164:1333–1339.
38. Glabinski AR, Ransohoff RM. Chemokines and chemokine receptors in CNS pathology. J NeuroVirol 1999;5: 3–12.
39. Hesselgesser J, Horuk R. Chemokine and chemokine receptor expression in the central nervous system. J NeuroVirol 1999;5:13–26.
40. Asensio VC, Kincaid C, Campbell IL. Chemokines and the inflammatory response to viral infection in the central nervous system with a focus on lymphocytic choriomeningitis virus. J NeuroVirol 1999;5:65–75.
41. Schmidtmayerova H, Nottet HSL, Nuovo G, Raabe T, Flanagan CR, Dubrovsky L, Gendelman HE, Cerami A, Bukrinsky M, Sherry B. Human immunodeficiency virus type 1 infection alters chemokine β peptide expression in human monocytes: implications for recruitment of leukocytes into brain and lymph nodes. Proc Natl Acad Sci USA 1996;93:700–704.
42. Simpson JE, Newcombe J, Cuzner ML, Woodroofe MN. Expression of monocyte chemoattractant protein-1 and other β-chemokines by resident glia and inflammatory cells in multiple sclerosis lesions. J Neuroimmunol 1998; 84:238–249.
43. Leung SY, Wong MP, Chung LP, Chan ASYC, Yuen ST. Monocyte chemoattractant protein-1 expression and macrophage infiltration in gliomas. Acta Neuropathol 1997;93:518–527.
44. Takami S, Nishikawa H, Minami M, Nishiyori A, Sato M, Akaike A, Satoh M. Induction of macrophage inflammatory protein MIP-1alpha mRNA on glial cells after focal cerebral ischemia in the rat. Neurosci Lett 1997;227:173–176.
45. Gourmala NG, Buttini M, Limonta S, Sauter A, Boddeke HWGM. Differential and time-dependent expression of monocyte chemoattractant protein-1 mRNA by astrocytes and macrophages in rat brain: effects of ischemia and peripheral lipopolysaccharide administration. J Neuroimmunol 1997;74:35–44.
46. Gourmala NG, Limonta S, Bochelen D, Sauter A, Boddeke HWGM. Localization of macrophage inflammatory protein: Macrophage inflammatory protein-1 expression in rat brain after peripheral administration of lipopolysaccharide and focal cerebral ischemia. Neuroscience 1999;88: 1255–1266.
47. Sharafeldin A, Eltayeb R, Pashenkov M, Bakhiet M. Chemokines are produced in the brain early during the course of experimental African trypanosomiasis. J Neuroimmunol 2000;103:165–170.
48. Thomas WE. Brain macrophages: on the role of pericytes and perivascular cells. Brain Res Rev 1999;31:42–57.
49. Sasseville VG, Smith MM, Mackay CR, Pauley DR, Mansfield KG, Ringler DJ, Lackner AA. Chemokine expression in simian immunodeficiency virus-induced AIDS encephalitis. Am J Pathol 1996;149:1459–1467.
50. Cuff CA, Berman JW, Brosnan CF. Interleukin-1β-induced expression of monocyte chemotactic protein-1 in the rabbit retina: an in situ and immunohistochemical study. Glia 2000;30:279–289.
51. Lacy M, Jones J, Whittemore SR, Haviland DL, Wetsel RA, Barnum SR. Expression of the receptors for the C5a anaphylatoxin, interleukin-8 and FMLP by human astrocytes and microglia. J Neuroimmunol 1995;61:71–78.
52. Boddeke EWGM, Meigel I, Frentzel S, Gourmala NG, Harrison JK, Buttini M, Spleiss O, Gebicke-Härter P. Cultured rat microglia express functional β-chemokine receptors. J Neuroimmunol 1999;98;176–184.
53. Ghorpade A, Xia MQ, Hyman BT, Persidsky Y, Nukuna A, Bock P, Che M, Limoges J, Gendelman HE, Mackay CR. Role of the β-chemokine receptors CCR3 and CCR5 in human immunodeficiency virus type 1 infection of monocytes and microglia. J Virol 1998:72;3351–3361.
54. Albright AV, Shieh JT, Itoh T, Lee B, Pleasure D, O'Connor MJ, Doms RW, González-Scarano F. Microglia express CCR5, CXCR4, CCR3, but of these, CCR5 is the principal coreceptor for human immunodeficiency virus type 1 dementia isolates. J Virol 1999:73;205–213.

55. Jiang Y, Salafranca MN, Adhikari S, Xia Y, Feng L, Sonntag MK, deFiebre CM, Pennel NA, Streit WJ, Harrison JK. Chemokine receptor expression in cultured glia and rat experimental allergic encephalomyelitis. J Neuroimmunol 1998;86:1–12.
56. Tanabe S, Heesen T, Yoshizawa I, Berman MA, Luo Y, Bleul CC, Springer TA, Okuda K, Gerard N, Dorf ME. Functional expression of the CXC-chemokine receptor-4/fusin on mouse microglial cells and astrocytes. J Immunol 1997;159:905–911.
57. Lavi E, Strizki JM, Ulrich AM, Zhang W, Fu L, Wang Q, O'Connor M, Hoxie JA, González-Scarano F. CXCR-4 (Fusin), a co-receptor for the type 1 human immunodeficiency virus (HIV-1) is expressed in the human brain in a variety of cell types, including microglia and neurons. Am Soc Invest Pathol 1997;151:1035–1042.
58. Maciejewski-Lenoir D, Chen S, Feng L, Maki R, Bacon KB. Characterization of fractalkine in rat brain cells: migratory and activation signals for CX_3CR-1-expressing microglia. J Immunol 1999;163:1628–1635.
59. Boddeke HWGM, Meigel I, Frentzel S, Biber K, Ren L, Gebicke-Härter PJ. Functional expression of the fractalkine (CX_3C) receptor and its regulation by lipopolysaccharide in rat microglia. Eur J Pharmacol 1999;374:309–313.
60. Zujovic V, Benavides J, Vigé X, Carter C, Taupin V. Fractalkine modulates TNF-α secretion and neurotoxicity induced by microglial activation. Glia 2000;29:305–315.
61. He J, Chen Y, Farzan M, Choe H, Ohagen A, Gartner S, Busciglio J, Yang X, Hofmann W, Newman W, Mackay CR, Sodroski J, Gabuzda D. CCR3 and CCR5 are coreceptors for HIV-1 infection of microglia. Nature 1997; 385:645–649.
62. Boehme SA, Lio FM, Maciejewski-Lenoir D, Bacon KB, Conlon PJ. The chemokine fractalkine inhibits Fas-mediated cell death of brain microglia. J Immunol 2000; 165:397–403.
63. Spleiss O, Gourmala N, Boddeke HWGM, Sauter A, Fiebich BL, Berger M, Gebicke-Haerter PJ. Cloning of rat HIV-1-chemokine coreceptor CKR5 from microglia and upregulation of its mRNA in ischemic and endotoxinemic rat brain. J Neurosci Res 1998;53:16–28.
64. Xia M, Qin S, Wu L, Mackay CR, Hyman BT. Immunohistochemical study of the β-chemokine receptors CCR3 and CCR5 and their ligands in normal and Alzheimer's disease brains. Am J Pathol 1998;153:31–37.
65. Vallat A-V, de Giromali U, He J, Mhashilkar A, Marasco W, Shi B, Gray F, Bell J, Keohane C, Smith TW, Gabuzda D. Localization of HIV-1 co-receptors CCR5 and CXCR4 in the brain of children with aids. Am J Pathol 1998;152: 167–178.
66. Westmoreland SV, Rottman JB, Williams KC, Lackner AA, Sasseville VG. Chemokine receptor expression on resident and inflammatory cells in the brain of macaques with simian immunodeficiency virus encephalitis. Am J Pathol 1998;152:659–665.
67. Balashov KE, Rottman JB, Weiner HL, Hancock WW. $CCR5^+$ and $CXCR3^+$ T cells are increased in multiple sclerosis and their ligands MIP-1α and IP-10 are expressed in demyelinating brain lesions. Proc Natl Acad Sci USA 1999;96:6873–6878.
68. Sorensen TL, Tani M, Jensen J, Pierce V, Lucchinetti C, Folcik VA, Qin S, Rottman J, Sellebjerg F, Strieter RM, Frederiksen JL, Ransohoff RM. Expression of specific chemokines and chemokine receptors in the central nervous system of multiple sclerosis patients. J Clin Invest 1999;103:807–815.
69. Galasso JM, Harrison JK, Silverstein FS. Excitotoxic brain injury stimulates expression of the chemokine receptor CCR5 in neonatal rats. Am J Pathol 1998;153: 1631–1640.
70. Galasso JM, Miller MJ, Cowell RM, Harrison JK, Warren JS, Silverstein FS. Acute excitotoxic injury induces expression of monocyte chemoattractant protein-1 and its receptor, CCR2, in neonatal rat brain. Exp Neurol 2000; 165:295–305.
71. Cross AK, Woodroofe MN. Chemokines induce migration and changes in actin polymerization in adult rat brain microglia and a human fetal microglial cell line in vitro. J Neurosci Res 1999;55:17–23.
72. Hegg CC, Hu S, Peterson PK, Thayer SA. β-Chemokines and human immunodeficiency virus type-1 proteins evoke intracellular calcium increases in human microglia. Neuroscience 2000;98:191–199.
73. Cross AK, Woodroofe MN. Chemokine modulation of matrix metalloproteinase and TIMP production in adult rat brain microglia and a human microglial cell line in vitro. Glia 1999;28:183–189.
74. Ransohoff RM, Tani M. Do chemokines mediate leukocyte recruitment in post-traumatic CNS inflammation? Trends Neurosci. 1998;21:154–159.
75. Asensio VC, Campbell IL. Chemokines in the CNS: plurifunctional mediators in diverse states. Trends Neurosci 1999;22:504–512.
76. Mennicken F, Maki R, de Souza EB, Quirion R. Chemokines and chemokine receptors in the CNS: a possible role in neuroinflammation and patterning. Trends Pharmacol Sci 1999;20:73–77.
77. Weiss JM, Downie SA, Lyman WD, Berman JW. Astrocyte-derived monocyte-chemoattractant protein-1 directs the transmigration of leukocytes across a model of the human blood–brain barrier. J Immunol 1998;161: 6896–6903.
78. Wu DT, Woodman SE, Weiss JM, McManus CM, Dáversa TG, Hesselgesser J, Major EO, Nath A, Berman JW. Mechanisms of leukocyte trafficking into the CNS. J Neurovirol 2000;6:82–85.
79. Fiala M, Zhang L, Gan X, Sherry B, Taub D, Graves MC, Hama S, Way D, Weinand M, Witte M, Lorton D, Kuo YM, Roher AE. Amyloid-beta induces chemokine secretion and monocyte migration across a human blood–brain barrier model. Mol Med 1998;4;480–489.
80. Rottman JB, Slavin AJ, Silva R, Weiner HL, Gerard CG, Hancock WW. Leukocyte recruitment during onset of experimental allergic encephalomyelitis is CCR1 dependent. Eur J Immunol 2000;30:2372–2377.
81. Ghirnikar RS, Lee YL, Eng LF. Chemokine antagonist infusion attenuates cellular infiltration following spinal cord contusion injury in rat. J Neurosci Res 2000;59:63–73.
82. Diab A, Abdalla H, Li HL, Shi FD, Zhu J, Höjberg A, Lindquist L, Wretlind B, Bakhiet M, Link H. Neutraliza-

tion of macrophage inflammatory protein 2 (MIP-2) and MIP-1α attenuates neutrophil recruitment in the central nervous system during experimental bacterial meningitis. Infect Immun 1999;67:2590–2601.
83. Karpus WJ, Lukacs NW, McRae BL, Strieter RM, Kunkel SL, Miller SD. An important role for the chemokine macrophage inflammatory protein-1 alpha in the pathogenesis of the T cell-mediated autoimmune disease, experimental autoimmune encephalomyelitis. J Immunol 1995; 155:5003–5010.
84. Bacon KB, Harrison JK. Chemokines and their receptors in neurobiology: perspectives in physiology and homeostasis. J Neuroimmunol 2000;104:92–97.
85. Wang X, Ellison JA, Siren A-L, Lysko PG, Yue T-L, Barone FC, Shatzman A, Feuerstein GZ. Prolonged expression of interferon-inducible protein-10 in ischemic cortex after permanent occlusion of the middle cerebral artery in rat. J Neurochem 1998;71:1194–1204.
86. Miller RJ, Meucci O. AIDS and the brain: is there a chemokine connection? Trends Neurosci 1999;22:471–479.
87. Zink WE, Zheng J, Persidsky Y, Poluektova L, Gendelman HE. The neuropathogenesis of HIV-1 infection. FEMS Immunol Med Microbiol 1999;26:233–241.
88. Lipovsky MM, Gekker G, Hu S, Ehrlich LC, Hoepelman AI, Peterson PK. Cryptococcal glucuronoxylomannan induces interleukin (IL)-8 production by human microglia but inhibits neutrophil migration toward IL-8. J Infect Dis 1998;177:260–263.
89. Sun D, Hu X, Whitaker JN, Walker WS. Expression of chemokine genes in rat glial cells: the effect of myelin basic protein-reactive encephalitogenic T cells. J Neurosci Res 1997;47:192–200.
90. Fisher SN, Vanguri P, Shin HS, Shin ML. Regulatory mechanisms of MuRantes and CRG-2 chemokine gene induction in central nervous system glial cells by virus. Brain Behav Immun 1995;9:331–344.
91. Murphy GM, Jia XC, Song Y, Ong E, Shrivastava R, Bocchini V, Lee YL, Eng LF. Macrophage inflammatory protein 1-alpha mRNA expression in an immortalized microglial cell line and cortical astrocyte cultures. J Neurosci Res 1995;40:755–763.
92. Ishizuka K, Kumura T, Igata-yi R, Katsuragi S, Takamatsu J, Miyakawa T. Identification of monocyte chemoattractant protein-1 in senile plaques and reactive microglia of Alzheimer's disease. Psychiatry Clin Neurosci 1997;3: 135–138.
93. Miyagishi R, Kikuchi S, Takayama C, Inoue Y, Tashiro K. Identification of cell types producing RANTES, MIP-1α and MIP-1β in rat experimental autoimmune encephalomyelitis by in situ hybridization. J Neuroimmunol 1997;77:17–26.
94. Lane TE, Asensio VC, Yu N, Paoletti AD, Campbell IL, Buchmeier MJ. Dynamic regulation of α- and β-chemokine expression in the central nervous system during mouse hepatitis virus-induced demyelinating disease. J Immunol 1998;160:970–978.
95. Simpson J, Rezaie P, Newcombe J, Cuzner ML, Male D, Woodroofe MN. Expression of the β-chemokine receptors CCR2, CCR3 and CCR5 in multiple sclerosis central nervous system tissue. J Neuroimmunol 2000;108:192–200.
96. Berman JW, Guida MP, Warren J, Amat J, Brosnan CF. Localization of monocyte chemoattractant peptide-1 expression in the central nervous system in experimental autoimmune encephalomyelitis and trauma in the rat. J Immunol 1996;156:3017–3023.
97. Biber K, Sauter A, Brouwer N, Copray SCVM, Boddeke HWGM. Ischemia-induced neuronal expression of the microglia attracting chemokine secondary lymphoid-tissue chemokine (SLC). Glia 2001;34:121–133.

Universes in Delicate Balance: Chemokines and the Nervous System
R.M. Ransohoff, K. Suzuki, A.E.I. Proudfoot, W.F. Hickey and J.K. Harrison, editors

Chapter 4.6

Chemokines and Chemokine Receptors along the Brain Microvasculature

Anuska V. Andjelkovic and Joel S. Pachter

Blood–Brain Barrier Laboratory, Department of Pharmacology, University of Connecticut Health Center, 263 Farmington Avenue, Farmington, CT 06030, USA

1. INTRODUCTION

Increasing evidence highlights the prominence of chemokines in a variety of physiological and pathophysiological processes in the central nervous system (CNS). In particular, chemokines have been shown to be critical determinants in the development of CNS inflammatory and infectious conditions, wherein these specialized cytokines are thought to drive the influx of leukocytes into the neural parenchyma. Recent results from this laboratory and others further indicate that receptors for chemokines exist along the microvascular endothelial cells that comprises the blood–brain barrier (BBB), and suggest that these receptors may play a significant role in relaying chemokine signals between the CNS and circulation. Such microvascular receptors might prove effective targets for therapeutics aimed at mitigating both CNS inflammatory episodes and pathogen dissemination to and from the CNS.

Ever since the earliest characterizations of chemokines, appreciation of the diverse roles played by this superfamily of proteins has continued to evolve. There are currently four recognized chemokine classes, known as α, β, γ and δ, or CXC, CC, C and CXXXC, respectively, where the C terminology refers to the presence of conserved cysteine residues, and X the number of amino acids between cysteines [1,2]. Notwithstanding their initial designation as *chemo*tactic cyto*kines*, reflecting the ability to stimulate chemotaxis of specific subclasses of leukocytes, chemotactic activity is not the sole, nor perhaps in some cases even the principle, functional domain of chemokines. Furthermore, it is widely acknowledged that leukocytes are not the exclusive cellular target of chemokine action. On the contrary, the current spectrum of chemokine activities encompasses such various and sundry themes as inflammation, angiogenesis, hematopoiesis, B and T cell differentiation, tumor metastasis and organ development. Indeed, with the list of chemokines growing to include some fifty members, perception of the functional responsibilities of chemokines has expanded nearly as drastically as the chemokine number itself.

Nowhere is this perception of chemokine pleiotropy more resonant than in the central nervous system (CNS). Chemokines gained early recognition as possible effectors of CNS inflammation associated with demyelinating disease [3–6], chemical and traumatic injury [7–9], ischemia [10–14] and infection [14–18]. Though the actions of chemokines in these examples were considered only in the context of stimulating leukocyte entry into the CNS, the repertoire of chemokine targets in the brain has widened to include glia [19–27], neurons [28–30] and vascular elements [31–35]. Counterintuitively, the relatively recent awareness that select chemokine receptors also function as co-receptors for human immunodeficiency virus-1 (HIV-1) has stimulated novel considerations of chemokine action in the CNS [36–38], and spawned a plethora of theories regarding the pathogenesis of HIV-1-associated neurologic disease [39–41].

Prompted by heightened awareness of chemokines and their receptors in the CNS, the objective of this article is to specifically draw attention to an issue that has only been modestly addressed: *the expression of chemokine receptors along the brain microvasculature, and their potential role in neuroinflammatory and neuroinfectious processes.* As we hope to convey, the unique properties of the cerebral microvasculature might significantly dictate the course and extent of chemokine action within the CNS, and critically modulate both physiologic and pathophysiologic processes. It is our expectation that in focusing on this topic, the already expanded range of chemokine activities may be broadened still.

2. INFILTRATION OF LEUKOCYTES INTO THE CNS: ROLE OF CHEMOKINES

Evidence that infiltration of leukocytes into the CNS is linked to the expression of chemokines has been most profoundly and consistently demonstrated in varied paradigms of the demyelinating condition experimental autoimmune encephalomyelitis (EAE), a generally recognized rodent model of the human neurologic disease multiple sclerosis (MS). Those chemokines that have garnered the most attention in this regard include β chemokines monocyte chemoattractant protein-1 (MCP-1/CCL2), macrophage inflammatory protein-1 alpha (MIP-1α/CCL3), regulated upon activation normal T cell expressed and secreted (RANTES/CCL5), and α chemokines interferon-gamma inducible protein-10 (IP-10) and growth-related oncogene alpha (GRO-α/CXCL1) [3,42–45]. With respect to MCP-1 and MIP-1α, evidence that their expression is a causal factor in leukocyte extravasation in EAE, and not merely an epiphenomenon, has been demonstrated by in vivo studies employing neutralizing antibodies to each of these chemokines. Specifically, anti-MIP-1α antibody was found to attenuate the clinical severity of acute EAE disease, while anti-MCP-1 antibody mitigated the chronic-relapsing form and corresponding leukocyte infiltrates [43,46]. The specificity of these antibody effects, and with it a further indictment of MCP-1 and MIP-1α in instigating leukocyte extravasation, has most recently been confirmed utilizing MCP-1 and MIP-1α naked DNA vaccines generated from RT-PCR products [47].

Even outside the context of EAE, there is strong support for the argument that the presence of chemokines in the CNS is a critical driving force for leukocyte extravasation. Strikingly, intracerebral injections of recombinant forms of MCP-1, or IL-8/CXCL8 and MIP-2 (macrophage inflammatory protein-2, a neutrophil-selective chemokine) were found to overcome the brain's intrinsic resistance to stimulation of leukocyte recruitment, resulting in parenchymal accumulations of monocytes or neutrophils, respectively [48]. Overexpression of MCP-1 [49], N51/KC [50] (also a neutrophil-selective chemokine) and MIP-2 [51] transgenes within cells of the brain parenchyma have yielded similar results, implying that chemokines hold the keys to unlocking the doors that intern leukocytes within the lumen of microvessels in the brain.

Interestingly, in numerous inflammatory, traumatic and infectious conditions of the CNS, astrocytes and infiltrating leukocytes have been determined to be major sources of chemokine production [6,9,25,45,52–58] [59]. Inasmuch as astrocytes lie juxtaposed to the perivascular space immediately surrounding brain microvessels, and recently emigrated leukocytes either occupy this space or a proximate locale, both cell types are in optimal position to exert a profound influence on the extravasation process. However, if chemokines released by these cells are to stimulate leukocyte extravasation solely or largely via their recognized chemotactic actions, as is presently perceived, then we are obliged to ask the following: *How can leukocytes respond to chemokines from which they are physically separated by an intact vascular barrier, especially one so restrictive in its permeability as the blood–brain barrier (BBB)?* It is one thing to demonstrate leukocyte chemotaxis across an inert, semipermeable filter through which chemokines can readily diffuse, yet quite another to assume that an analogous situation exists across a continuous, and relatively impervious, biological membrane. Perhaps clues to this apparent conundrum lie within the sphere of as yet unappreciated chemokine activities at the BBB?

Before proceeding with this argument, however, it is best to briefly reconnoiter the BBB's unique properties and cellular composition. The BBB has as its anatomical substrate the highly specialized endothelial cells that form the walls of the brain microvasculature. Interconnecting these brain microvascular endothelial cells (BMEC) are a series of continuous tight junctions (TJ) that form physical anastomoses between apposing endothelial membranes, effectively occluding the aqueous space separating individual cells (for recent reviews on the cell biology of the BBB see [60,61]). By so doing, TJ significantly limit the paracellular passage of both soluble and cellular elements between the CNS and the circulation. Reflecting this quality is the high electrical resistance across brain capillaries (approximately 1000 ohms cm^2), which is nearly three orders of magnitude higher than that across the endothelium of peripheral vessels (1–5 ohms cm^2). Were it not for such relative vascular impermeance, the susceptibility of the circulation to routine fluctuations in ionic and nutrient composition, and exposure to pathogens, would wreak havoc on CNS homeostasis. However, with the BBB able to limit even the transendothelial flux of small metal ions, it is only natural to question how chemokines, of average molecular weight in the 10 kDa range, can cross this substantial anatomic divide and interact with leukocytes in the lumen of brain microvessels.

A priori, a few possibilities regarding the mechanism(s) of chemokine action within the CNS may be considered when bearing in mind the constraints of the BBB. In the simplest case, chemokines deposited in the perivascular space might 'leak' through interendothelial junctions, in a diffusion-mediated manner, to engage leukocytes in the vascular lumen. Given the limitations imposed on paracellular transport in nonpathological states, this pathway is most conceivable only under conditions where the integrity of the BBB has been compromised, as, for example, during traumatic brain injury or sustained inflammation. Alternatively, chemokines could be relayed across brain endothelial cells, rather than between them. Such transcellular communication might occur by transcytosis, as has been described to effect the abluminal to luminal movement of IL-8 across dermal microvessels [62], or by facilitated diffusion/carrier-mediated transport. In this regard, several different receptor-mediated transporter systems are recognized to exist in BMEC, which function in delivering a variety of specific ligands across the BBB in a vectorial manner — both abluminal-to-luminal and luminal-to-abluminal directions being exemplified [63–65]. It is of further interest to note that caveolae, nonclathrin-coated vesicles involved in endocytosis and transcytosis in several cell types [66,67], have been described in brain [68], and shown to be operative in the vectorial movement of low density lipoprotein across an in vitro model of the BBB [69]. Adding a final element of complexity, intravenous CINC/CXCL1 was shown to enter brain, probably by sample diffusion [70]. Thus, maintenance of the transvascular chemokine gradient must be multi-vectorial.

3. CHEMOKINE BINDING SITES AND RECEPTORS ON BRAIN MICROVESSELS

The relaying of chemokine signals across the BBB, either by transport or signal transduction events, would arguably require the presence of chemokine receptors or chemokine binding sites on the endothelium. For the purpose of clarification, chemokine receptors and chemokine binding sites are functionally distinguished in the following manner. The former are defined by their ability to signal following binding one or more members of the chemokine superfamily, while the latter do not manifest, or have yet to evidence, a detectable biologic response to chemokine engagement [1]. Chemokine receptors are designated by the abbreviations CCR, CXCR, CXXXCR and XCR, based on the specific class of chemokines they bind, followed by a number. They are considered part of the rhodopsin family of seven-transmembrane domain, G protein-coupled receptors. To date, a total of eighteen receptors — CCR1 through 11, CXCR1 through 5, CXXXR1 and XCR1 — are recognized, each of which has a signature chemokine binding profile by which it is functionally distinguished.

While the focus of attention of chemokine receptor expression has been directed at leukocytes and other cells of hemopoietic lineage, endothelial cells have emerged as bona fide targets of chemokine action (reviewed in [71]). Indeed, chemokine receptors have been reported to be expressed by endothelial cells from a wide variety of tissues, in both tissue culture and in situ settings (Table 1), and implicated in mediating several aspects of angiogenesis and angiostasis. Such expression by endothelial cells prompts consideration that specialized interactions of chemokines with the brain microvascular endothelium might, among other functions, constitute the incipient steps in directing leukocyte traffic into the CNS.

In support of this hypothesis, recent evidence from this laboratory has indicated that separate, high-

Table 1. Chemokine receptor expression in endothelial cells

Chemokine receptor	Species	Endothelial type	Detection method	Ref.
CXCR1	human	HUVEC and ECV304 cells (in culture)	flow cytometry, RT-PCR	[72]
CXCR2		HUVEC and astrocyte co- culture	RNAse protection assay	[73]
		DMEC (in culture)	immunofluorescence, flow cytometry	[74]
		HUVEC (in culture)	immunofluorescence, flow cytometry	[74]
CXCR3	human	DMEC (in culture)	immunofluorescence, flow cytometry	[74]
		HUVEC (in culture)	immunofluorescence, flow cytometry	[74]
CXCR4	macaque	BMEC (in culture)	RT-PCR	[34] [a]
	human	HUVEC (in culture)	RT-PCR, Northern and Western blot	[75]
			RT-PCR, Northern blot	[76]
			Northern blot, in situ hybridization, immunofluorescence	[77]
			immunofluorescence, flow cytometry	[74]
		HUVEC, BMEC and CAEC (in culture)	flow cytometry, immunofluorescence	[33] [a]
		HUVEC and ECV304 cells (in culture)	flow cytometry, RT-PCR	[72]
		HUVEC and astrocyte co- culture	RNAse protection assay	[73]
		HUVEC and HIAEC (in culture)	flow cytometry, immunofluorescence	[78]
		HDMEC (in culture)	immunofluorescence, flow cytometry	[74]
		CEAC (tissue sections)	immunocytochemistry, immunofluorescence	[33,78]
		DMEC (in culture)	RT-PCR	[75]
	bovine	BAEC (in culture)	Northern and Western blot, flow cytometry, immunofluorescence	[75]
			Northern and Western blot	[77]
		BAEC (tissue sections)	in situ hybridization, immunofluorescence	[77]
CCR1	human	BMEC (brain sections)	immunocytochemistry, immunofluorescence	[79] [a]
		BMEC (brain microvessels)	chemokine binding, immunofluorescence	[31,32] [a]
		HUVEC and astrocyte co- culture	RNAse protection assay	[73]
CCR2	human	BMEC (brain microvessels)	chemokine binding, immunofluorescence	[31,32] [a]
		BMEC and CAEC (in culture)	flow cytometry, immunofluorescence	[33] [a]
		CAEC (tissue sections)	immunofluorescence	[33]
		arterial, venous, venular endothelium (tissue sections)	immunocytochemistry	[80]
		HUVEC (in culture)	RT-PCR, RNAse protection assay, Western blot, flow cytometry	[80]
CCR3	macaque	BMEC (in culture)	RT-PCR	[34] [a]
	human	HUVEC, BMEC and CAEC (in culture)	flow cytometry, immunofluorescence	[33] [a]
		CEAC (tissue sections)	immunofluorescence	[33]
		HUVEC and astrocyte co- culture	RNAse protection assay	[73]
CCR5	macaque	BMEC (in culture)	RT-PCR	[34] [a]
	human	endothelium of brain and other tissues (tissue sections)	immunocytochemistry	[35] [a]
		BMEC (brain microvessels)	chemokine binding, immunofluorescence	[31,32] [a]
		HUVEC, BMEC and CAEC (in culture)	flow cytometry, immunofluorescence	[33] [a]

Above, are listed examples of the chemokine receptors expressed in different types of endothelial cells, as detected by a variety of methods.

BAEC, bovine aortic endothelial cells; BMEC, brain microvessel endothelial cells; CAEC, coronary artery endothelial cells; DMEC, dermal microvessel endothelial cells; HUVEC, human umbilical vein endothelial cells.

[a] Denote examples of expression in brain.

affinity binding sites for MCP-1 and MIP-1α reside on the abluminal surface of human brain microvessels [31,32]. These sites were revealed utilizing biotinylated derivatives of these chemokines along with confocal microscopy and three-dimensional image reconstruction, and shown to possess numerous characteristics of the major recognized receptors for MCP-1 and MIP-1α: CCR2, and CCR1 and CCR5,

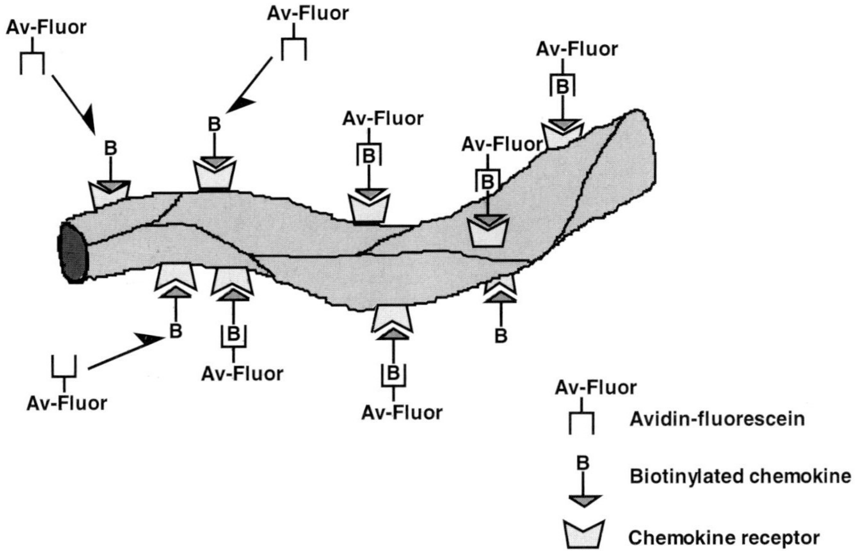

Figure 1. Paradigm for detecting chemokine binding sites along brain microvessels. Freshly isolated, unfixed microvessels are first incubated with biotinylated chemokine. To visualize sites of chemokine binding, microvessels are subsequently reacted with fluorescein-conjugated avidin, and viewed by fluorescence microscopy. This application has been used by the authors to determine both the distribution of abluminal chemokine binding sites as well as their pharmacologic characteristics [31,32].

respectively (Figs. 1 and 2). Holistic distributions of the two labeled chemokines around the microvascular surface are strikingly distinguishable from each other, strongly implying the existence of at least two independent chemokine binding sites. Specifically, the MCP-1 binding pattern is relatively continuous, nearly completely ensheathing the abluminal microvascular surface. MIP-1α binding, in contrast, displays a more punctate arrangement along the abluminal face. Binding patterns of both chemokines can be additionally resolved from perivascular macrophages and astrocytes, the main perivascular cell types in situ, and lie more closely in register with the endothelial surface. MCP-1 and MIP-1α binding patterns also differ from major basement membrane components such as heparan sulfate, collagen and laminin, and are resistant to the enzymatic action of heparinase. Such lack of dependence of MCP-1 and MIP-1α binding on basement membrane heparan sulfate is particularly noteworthy, as this and other glycosaminoglycans (GAGs) have a demonstrated affinity, albeit several orders of magnitude lower than that of cognate receptors, for many chemokines [81]. Hence, while GAGs present along the luminal surface of the endothelium and in the extracellular matrix may aid in the presentation of chemokines to leukocytes and other cell types [82–84], they are not the predominant vehicle responsible for the observed MCP-1 or MIP-1α binding to the abluminal microvascular surface. These features collectively implicate BMEC as the main source of the chemokine binding sites detected along isolated brain microvessels.

Competitor chemokine binding assays (Fig. 3) and quantitative fluorescence microscopy has further indicated that binding of MCP-1 and MIP-1α occurs in a saturable manner with respective K_d values of approximately 2.0 nM and 0.5 nM [32]. Remarkably, these affinity constants closely approximate those determined for MCP-1 and MIP-1α binding to their cognate receptors expressed endogenously in leukocytes or by transfection in other cell types (reviewed in [81,85]). They additionally closely compare to the concentration of 1 nM noted for MCP-1 stimulation of HUVEC chemotaxis [80], implying that chemokine binding to brain microvessels occurs within the range of physiologic action on endothelial cells. Emphasizing the limited number of highly selective sites available, labeled MCP-1 binding to brain microvessels can be antagonized by unlabeled MCP-1, as well as by another CCR2-binding chemokine MCP-3 [86], but not by MIP-1α, which does not recognize CCR2. Binding of labeled MIP-1α is likewise inhibitable by its unlabeled homologue, but

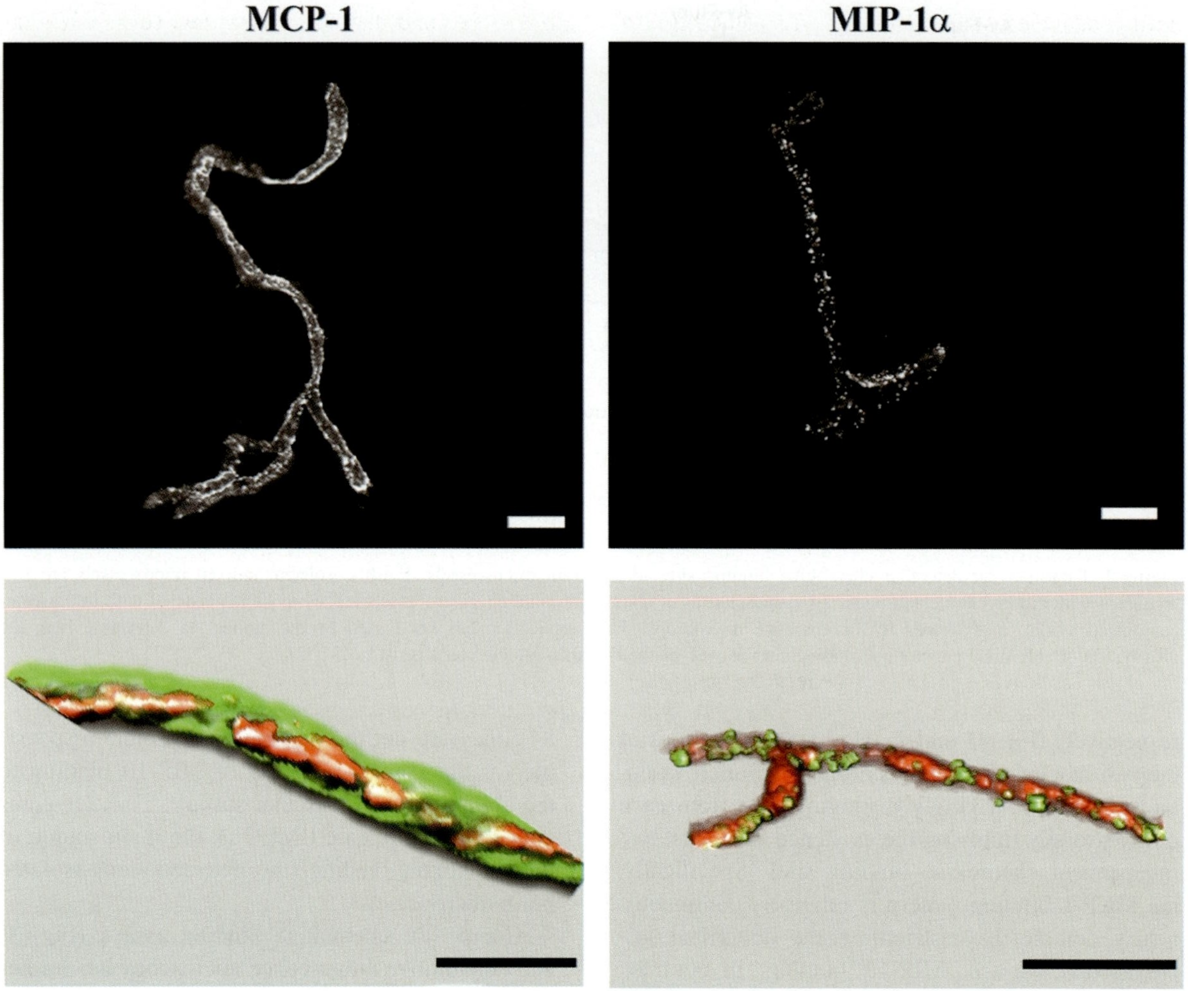

Figure 2. Distribution of chemokine binding sites along brain microvessels. (Top row) Human brain microvessels reacted with biotinylated MCP-1 and biotinylated MIP-1α as described in Fig. 1. (Bottom row) Three-dimensional reconstructions of microvessels that have been additionally stained with anti-factor VIII antibody and rhodamine-conjugated secondary antibody (to reveal endothelial cells) [32], showing the holistic distribution of sites of chemokine binding (green) around the endothelial cells (red). Binding patterns for biotinylated MCP-1 and biotinylated MIP-1α are vastly different, with the former appearing near continuous, and the latter decidedly punctate. Such distinct distributions suggest different binding sites for the two chemokines. Scale bar represents 80 microns.

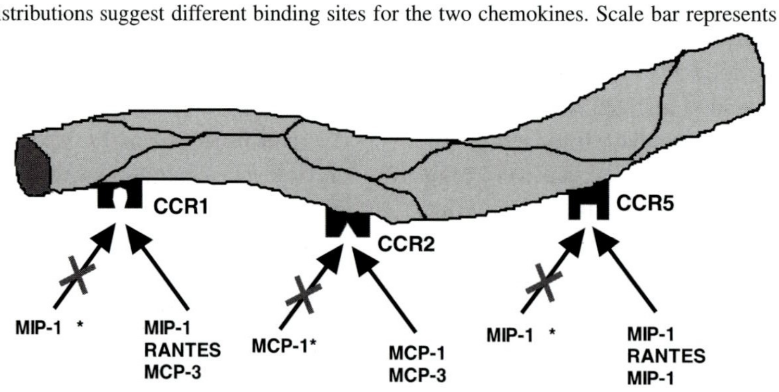

Figure 3. Competitive chemokine binding along brain microvessels. To characterize the binding sites recognizing biotinylated MCP-1 (MCP-1*) and MIP-1α (MIP-1*), microvessels are incubated with each biotinylated chemokine in the presence of unlabeled chemokines known to bind to specific chemokine receptors. The relative degree of MCP-1* and MIP-1α* binding is assessed by quantitative fluorescence microscopy [31,32].

not by unlabeled MCP-1. Moreover, labeled MIP-1α binding is antagonized by RANTES, which recognizes CCR1, CCR3 and CCR5 [87,88], as well as by MCP-3, which also recognizes CCR-1 [89,90], and MIP-1β, which binds to CCR5 [91]. Such pharmacodynamic data significantly challenges the idea that MCP-1 and MIP-1α binding to isolated brain microvessels is nonspecific, and, instead, advocates that highly selective, receptor-mediated interactions are responsible. Immunocytochemical detection of several chemokine receptors, including CCR1, CCR2, CCR3 and CCR5, along isolated human brain microvessels [32], human and macaque brain microvessels in situ [35], and in cultured human BMEC [33], further supports this conclusion. And in what may be an analogous situation, the retinal pigment epithelium, which forms part of the blood–retina barrier, has recently been shown to demonstrate CXCR4 expression [92]. Taken together, these findings advance the argument that the binding sites on the abluminal brain microvascular surface reflect the expression of cognate chemokine receptors by BMEC. Interestingly, chemokine binding may not be restricted to the abluminal surface, as Berger et al. [33] reported anti-CCR3 and anti-CCR5 immunoreactivity along the luminal surface of coronary vessels in situ and on the apical surface (analogous to the luminal surface in situ) of cultured BMEC and coronary artery endothelial cells (CAEC). Bilateral expression of chemokine binding sites along both abluminal and luminal surfaces of brain in vivo, may carry important implications for both chemokine and pathogen transport across the BBB.

Additional studies have attested to the functional competence of these chemokine binding sites. This laboratory has reported that both MCP-1 and MIP-1α undergo endothelial internalization following their binding to the surface of brain microvessels at 37°C, a hallmark response to the interactions of these chemokines with their respective cognate receptors [93–95]. Given the competitive nature by which these chemokines bind to brain microvessels, internalization likely proceeds via receptor-mediated endocytosis, and not fluid phase or adsorptive endocytosis, the latter two processes being nonspecific or dictated by electrostatic interactions, respectively [96]. MCP-1 and MIP-1α binding sites also display a transient downmodulation, with prior exposure of brain microvessels to each of these chemokines causing an initial loss and then subsequent recovery of binding of the homologous, labeled chemokine [31]. Interestingly, MCP-1 and MIP-1α endocytosed at the abluminal surface appear to relocate to the periluminal region [31], thus possibly reflecting receptor-mediated transcytosis across BMEC. These observations, at the very least, suggest that brain microvascular chemokine binding sites are physiologically responsive in a ligand-selective manner.

Despite the accumulating data evidencing both their expression and functional potential, the precise role(s) of these binding sites in executing chemokine actions in the CNS awaits further experimentation and investigative scrutiny. However, speculation that such sites along the brain microvasculature may be directly involved in leukocyte extravasation is encouraged by recent findings regarding chemokine receptors at other endothelial and epithelial barriers. In what may be a similar scenario for MCP-1 and MIP-1α binding sites at the BBB, Godaly et al. [97] reported that transepithelial migration of neutrophils during urinary tract infection (UTI) is dependent upon expression of CXCR1, a receptor for IL-8, by epithelial cells of the urinary tract mucosa. The crucial role of uroepithelial CXCR1 in this process was demonstrated in an in vitro UTI model by showing that exposure of confluent cultures of infected human kidney epithelial cells to CXCR1-specific antibody blocked the transepithelial migration of neutrophils by approximately 60%. This phenomenon was corroborated in an in vivo murine UTI paradigm, which further established that, in mice rendered deficient in the murine IL-8 receptor homologue by homologous recombination, neutrophils were unable to cross the uroepithelial barrier into the lumen. IL-8 has also been shown to directly increase endothelial permeability [33], a feature that MCP-1 and MIP-1α may duplicate at the level of brain microvessels, as intracerebral injection of CC chemokines has been shown to result in BBB breakdown [48]. It is of interest to mention here that such disruption in endothelial integrity might actually be a considered as a temperate angiogenic response, as both heightened permeability and angiogenesis require dismantling of junctional complexes and restructuring of the cytoskeleton [98]. Lastly, binding of chemokines to the abluminal surface of brain microvessels might trigger chemokine release and presentation at the luminal surface, as chemokine-stimulated secretion of chemokines by endothelial cells [99] and polarized epithelial cells has been reported [92,100].

4. MICROVASCULAR CHEMOKINE RECEPTORS AND HIV-1 ENTRY INTO THE BRAIN

With the crossing of chemokine receptor functions at the nexus of HIV-1 co-receptor and vascular effector, a potential role for chemokine receptors in dissemination of virus to the brain cannot be overlooked. The relative celerity with which HIV-1 productively infects the CNS appears to lie in marked contrast to the impervious nature of the BBB. Nevertheless, HIV-1 can be detected in the cerebrospinal fluid as early as one week after infection [101]. Such accessibility to the CNS could imply that a mechanism(s) exists for facilitating the entry of virus across the normally restrictive BBB endothelium. While transport within infected leukocytes has been most consistently offered as the means by which HIV-1 initially crosses the BBB (reviewed in [102]), other, less conventional, routes of entry are also feasible. As one intriguing example, some free viral particles in the bloodstream might penetrate the BBB by co-opting chemokine receptors located along the luminal microvascular surface, and enter endothelial cells via receptor-mediated endocytosis. Virus entering in this manner might then transit through the endothelial cytoplasm and emerge from the abluminal microvascular surface in viable form. Once deposited on the parenchymal side of the BBB, virus would then be free to interact with perivascular glial elements (both astrocytes and perivascular microglia) that also express chemokine receptors, support virus replication and are considered potential reservoirs for HIV-1 in the brain [103,104]. In support of such an interpretation are a collection of several independent findings. Recently, Devito et al. [105], described the transcytosis of infectious HIV-1 across human epithelial cells, providing a precedent for transcellular passage of viable virus. Transcellular passage of HIV-1 across cultured BMEC has further been noted in a preliminary account [106]. As for the link to chemokine receptors, Banks et al. [107] reported that endocytosis of the HIV-1 coat protein gp120 by BMEC is mediated by an undefined cofactor bearing biochemical features of a chemokine receptor. Additionally, binding of gp120 to cultured human embryonic kidney cells has been observed to result in internalization of CCR5 and CXCR4 [108], both of which receptors are expressed by BMEC [31–35]. Thus, the relatively high expression of select chemokine receptors by BMEC, as compared to other endothelial cells [33], might contribute to the viral predilection to invade the CNS.

5. CONCLUSION

To be sure, the realm of vascular chemokine receptors remains controversial. Such skepticism is born, in part, of inconsistencies among different laboratories in the detection of endothelial chemokine receptor expression, and also possibly to a cautious disinclination to prematurely extend the physiological boundaries of chemokine activities. As regards the former argument, differences in detecting a given chemokine receptor's expression in a particular tissue may be due to methodological issues. For example, Berger et al. [33] stressed that cell passage number and fixation conditions are critical variables affecting chemokine receptor detection by immunocytochemistry in cultured HUVEC and CAEC, and this laboratory has encountered similar experiences in revealing chemokine binding sites on cultured astrocytes (unpublished data). Demonstration of differential expression and responsiveness of chemokine receptors by endothelial cells from different tissues [33,74] further argues that what holds for one endothelial type may not necessarily hold for another. Moreover, as their particular function along vascular elements may not require chemokine receptors to be expressed at the relatively high levels displayed by hemopoietic cells, lower receptor expression might, at times, compromise receptor identification on endothelial and other cell types. Keeping in mind these considerations regarding detection of vascular chemokine receptors, it may be that, at least in certain circumstances, absence of proof is not proof of absence.

With respect to the reservation of broadening the scope of chemokine targets to encompass the brain microvasculature, additional experiments are clearly necessary to mitigate this hesitation. Studies with knockout mice rendered deficient in the expression of chemokine receptors in the vasculature, selectively, as well as in vitro paradigms employing purified endothelial cells from chemokine receptor knockout mice, will ultimately serve to resolve the significance of endothelial chemokine receptors along brain microvessels. Confirmation of a role for chemokine receptors at the BBB will undoubtedly expand the endothelium as a target for therapeutics in the treatment of inflammatory, infectious and vascular diseases of the brain.

Acknowledgements

Work from the authors' laboratory was supported in parts from grants from the National Institutes of Health and the National Multiple Sclerosis Society to J.S.P., and from the American Heart Association to A.V.A. A debt of gratitude is owed to Dr. Diane Biegel and Mr. Kirk Dzenko for their assistance in the preparation of the manuscript.

REFERENCES

1. Murphy PM, Baggliolini M, Charo IF, Hebert CA, Horuk R, Mutsuhima J, Miller LH, Oppenheim JJ, Power CA. International Union of Pharmacology. XXII. Nomenclature for chemokine receptors. Pharmacol Rev 2000;52: 1–176.
2. Rossi D, Zlotnick A. The biology of chemokines and chemokine receptors. Ann Rev Immunol 2000;18:217–242.
3. Godiska R, Chantry D, Dietsch GN, Gray PW. Chemokine expression in murine experimental allergic encephalomyelitis. J Neuroimmunol 1995;58:167–176.
4. Hulkower K, Brosnan CF, Aquino DA, Cammer W, Kulshrestha S, Guida MP, Rapoport DA, Berman JW. Expression of CSF-1, c-fms, and MCP-1 in the central nervous system of rats with experimental allergic encephalomyelitis. J Immunol 1993;150:2525–1233.
5. Karpus WJ, Lukacs NW, MacRae BL, Streiter RM, Kunkel SL, Miller SD. An important role for the chemokine macrophage inflammatory protein-1α in the pathogenesis of the T cell-mediated autoimmune disease, experimental autoimmune encephalomyelitis. J Immunol 1995;155:5003–5010.
6. Ransohoff RM, Hamilton TA, Tani M, Stoler M, Shick HE, Major JA, Estes ML, Thomas DM, Tuohy VK. Astrocyte expression of mRNA encoding cytokines IP-10 and JE/MCP-1 in experimental autoimmune encephalomyelitis. FASEB J 1993;7:592–600.
7. Calvo CF, Yoshimura T, Gelman M, Mallat M. Production of monocyte chemotactic protein-1 by rat brain macrophages. Eur J Neurosci 1996;8:1725–1734.
8. Ghirnikar RS, Lee YL, He TR, Eng LF. Chemokine expression in rat stab wound injury. J Neurosci Res 1996;46:727–733.
9. Glabinski AR, Balasingame V, Tani M, Kunkel SL, Streiter RM, Yong VW, Ransohoff RM. Chemokine monocyte chemoattractant protein-1 is expressed by astrocytes after mechanical injury to the brain. J Immunol 1996; 156:4363–4368.
10. Liu T, Young PR, McDonnell PC, White RF, Barone FC, Feuerstein GZ. Cytokine-induced neutrophil chemoattractant mRNA expressed in cerebral ischemia. Neurosci Lett 1993;163:125–128.
11. Morita M, Kasahara T, Mukaida N, Matsishima K, Nagahsima T, Nishizawa M, Yoshida M. Induction and regulation of IL-8 and MCAF production in human brain tumor cell lines and brain tumor tissues. Eur Cytokine Netw 1993;4:351–358.
12. Nitta T, Allegretta M, Okumura K, Sato K, Steinman L. Neoplastic and reactive human astrocytes express interleukin-8 gene. Neurosurg Rev 1992;15:203–207.
13. Takeshima H, Kuratsu J, Takeya M, Yoshimura T, Ushio Y. Expression and localization of messenger RNA and protein for monocyte chemoattractant protein-1 in human malignant glioma. J Neurosurg 1994;80:1956–1962.
14. Van Meir E, Ceska M, Effenberger F, Walz A, Groutzmann E, Desbaillets I, Frei K, Fontana A, de Tribolet N. Interleukin-8 is produced in neoplastic and infectious diseases of the human central nervous system. Cancer Res 1992;52:4297–4305.
15. Morimoto K, Hooper DC, Bornhorst A, Corisdeo S, Bette M, Fu ZF, Schafer MK, Koprowski H, Weihe E, Dietzschold B. Intrinsic response to Borna disease virus infection of the central nervous system. Proc Natl Acad Sci USA 1996;93:13345–13350.
16. Sasseville VG, Smith MM, Mackay CR, Pauly DR, Mansfield KG, Ringler DJ, Lackner AA. Chemokine expression in simian immunodeficiency virus-induced AIDS encephalitis. Am J Pathol 1996;149:1459–1467.
17. Schmidtmayerova H, Nottet HS, Nuovo G, Raabe T, Flanagan CR, Dubrovsky L, Gendelman HE, Cerami A, Bukrinsky M, Sherry B. Human immunodeficiency virus type 1 infection alters chemokine beta peptide expression in human monocytes: implications for the recruitment of leukocytes in the brain and spinal cord. Proc Natl Acad Sci USA 1996;93:700–704.
18. Seebach J, Bartholdi D, Frei K, Spanaus KS, Ferraro E, Widner U, Isenmann S, Streiter RM, Schwab M, Pfister H et al. Experimental Listeria meningoencephalitis. Macrophage inflammatory protein-1 alpha and -2 are produced intrathecally and mediate chemotactic activity in cerebrospinal fluid of infected mice. J Immunol 1995; 155:4367–4375.
19. Albright AV, Shieh JT, Itoh T, Lee B, Pleasure D, O'Conner MJ, Doms RW, Gonzalez-Scarano F. Microglia express CCR5, CXCR4 and CCR3, but of these, CCR5 is the principle coreceptor for human immunodeficiency virus type 1 dementia isolates. J Virol 1999;73: 205–213.
20. Andjelkovic AV, Kerkovich D, Pulliam L, Shanley JD, Pachter, JS. Expression of binding sites for β chemokines on human astrocytes. Glia 1999;28:225–235.

21. Bajetto A, Bonavia R, Barbero S, Florio T, Costa A, Schettini G. Expression of chemokine receptors in rat brain. Ann NY Acad Sci 1999;876:201–209.
22. Boddeke EW, Meigel I, Frentzel S, Gourmala NG, Harrison JK, Buttini M, Spleiss O, Gebiicke-Harter P. Cultured rat microglia express functional beta-chemokine receptors. J Neuroimmunol 1999;98:176–184.
23. Han Y, Wang J, Zhou Z, Ransohoff RM. TGFβ1 selectively up-regulates CCR1 expression in primary murine astrocytes. Glia 2000;30:1–10.
24. McManus CM, Weidenheim J, Woodman SE, Nunez J, Hesselgesser J, Nath A, Berman JW. Chemokine and chemokine–receptor expression in human glial elements: induction by the HIV protein TAT and chemokine autoregulation. Am J Pathol 2000;156:1441–1453.
25. Simpson J, Rezaie P, Newcombe J, Cuzner ML, Male D, Woodroofe MN. Expression of the beta-chemokine receptors CCR2, CCR3 and CCR5 in multiple sclerosis central nervous system tissue. J Neuroimmunol 2000; 108:192–200.
26. Tanabe S, Heesen M, Berman MA, Fisher MB, Yoshizawa I, Luo Y, Dorf ME. Murine astrocytes express a functional chemokine receptor. J Neurosci 1997; 17:6522–6628.
27. Xia MQ, Qin SX, Wu LJ, Mackay CR, Hyman BT. Immunohistochemical study of the beta-chemokine receptors CCR3 and CCR5 and their ligands in normal and Alzheimer's disease brains. Am J Pathol 1998;153: 31–37.
28. Coughlan CM, McManus CM, Sharron M, Gao Z, Murphy D, Jaffer S, Choe W, Chen W, Hesselgesser J, Gaylord H, Kalyuzhny A, Less VM, Wolf B, Doms RW, Kolson DL. Expression of multiple functional chemokine receptors and monocyte chemoattractant protein-1 in human neurons. Neuroscience 2000;97:591–600.
29. Horuk R, Martin AW, Wang Z-X, Schweitzer L, Gerassimides A, Guo H, Lu Z-H, Hesselgesser J, Perez HD, Kim J, Parker J, Hadley TJ, Peiper SC. Expression of chemokine receptors on subsets of neurons in the central nervous system. J Immunol 1997;158:2882–2890.
30. Speth C, Joebstl B, Barcova M, Dietrich MP. HIV-1 envelope protein gp41 modulates expression of interleukin-10 and chemokine receptors on monocytes, astrocytes and neurons. AIDS 2000;14:629–636.
31. Andjelkovic AV, Pachter JS. Characterization of MCP-1 and MIP-1α binding sites along human brain microvessels. J Neurochem 2000;75:1898–1906.
32. Andjelkovic AV, Spencer DD, Pachter JS. Visualization of chemokine binding sites along human brain microvessels. J Cell Biol 1999;145:403–412.
33. Berger O, Gan X, Gujuluva C, Burns AR, Suur G, Stins M, Way D, Witte M, Weinand M, Said J, Kim K-S, Taub D, Graves MC, Fiala M. CXC and CC chemokine receptors on coronary and brain endothelia. Mol Med 1999;5:795–805.
34. Edinger AL, Mankowski JL, Doranz BJ, Margulies BJ, Lee B, Rucker J, Sharron M, Hoffman TL, Berson JF, Zink MC, Hirsch VM, Clements JE, Doms RW. CD4-independent, CCR5-dependent infection of brain capillary endothelial cells by a neurovirulent simian immunodeficiency virus strain. Proc Natl Acad Sci USA 1997; 94:14742–14747.
35. Rottman JB, Ganley KP, Williams K, Wu L, Mackay CR, Ringler DJ. Cellular localization of the chemokine receptor CCR5. Am J Pathol 1997;151:1341–1351.
36. Asencio VC, Campbell IL. Chemokines in the CNS: plurifunctional mediators in diverse states. Trends Neurosci 2000;22:504–512.
37. Bacon KB, Harrison JK. Chemokines and their receptors in neurobiology: perspectives in physiology and homeostasis. J Neuroimmunol 2000;104:92–97.
38. Hesselgesser J, Horuk R. Chemokines and chemokine receptors in the brain. In: Hebert CA, ed. Chemokines in Disease: Biology and Clinical Research. Totowa, NJ: Humana Press, 1999:295–322.
39. Gabuzda D, Wang J. Chemokine receptors and virus entry into the central nervous system. J Neurovirol 1999; 5:643–658.
40. Gabuzda D, Wang J. Chemokine receptors and mechanisms of cell death in HIV neuropathogenesis. J Neurovirol 2000;6(Suppl 1):S24–32.
41. Miller RJ, Meucci O. AIDS and the brain: is there a chemokine connection? Trends Neurosci 1999;22:471–479.
42. Glabinski AR, Tuohy VK, Ransohoff RM. Expression of chemokines RANTES, MIP-1alpha and GRO-alpha correlates with inflammation in acute experimental autoimmune encephalomyelitis. Neuroimmunomodulation 1998;5:166–171.
43. Kennedy KJ, Streiter RM, Kunkel SL, Lukacs NW, Karpus WJ. Acute and relapsing experimental autoimmune encephalomyelitis are regulated by differential expression of the CC chemokines macrophage inflammatory protein-1alpha and monocyte chemotactic protein-1. J Neuroimmunol 1998;92:98–108.
44. Matejuk A, Vandenbark AA, Burrows GG, Bebo BF Jr, Offner H. Reduced chemokine and chemokine receptor expression in spinal cords of TCR BV8S2 transgenic mice protected against experimental autoimmune encephalomyelitis with BV8S2 protein. J Immunol 2000; 164:3924–3931.
45. Miyagishi R, Kikuchi S, Takayama C, Inoue Y, Tashiro K. Identification of cell types producing RANTES, MIP-1 alpha and MIP-1 beta in rat experimental autoimmune encephalomyelitis by in situ hybridization. J Neuroimmunol 1997;77:17–26.
46. Karpus WJ, Kennedy KJ. MIP-1alpha and MCP-1 differentially regulate acute and relapsing autoimmune encephalomyelitis as well as TH1/Th2 lymphocyte differentiation. J Leukoc Biol 1997;62:681–687.
47. Youssef S, Wildbaum D, Nathan K. Prevention of experimental autoimmune encephalomyelitis by MIP-1α and MCP-1 naked DNA vaccines. J Autoimmun 1999;13:21–29.
48. Bell MD, Taub DD, Perry VH. Overriding the brain's intrinsic resistance to leukocyte recruitment with intraparenchymal injection of recombinant chemokines. Neurosciences 1996;74:283–292.
49. Fuentes ME, Durham SK, Swerdel MR, Lewin AC, Barton DS, Megill JR, Bravo R, Lira SA. Controlled recruitment of monocytes and macrophages to specific organs

through transgenic expression of monocyte chemoattractant protein-1. J Immunol 1995;155:5769–4776.
50. Tani M, Fuentes ME, Peterson JW, Trapp BD, Durham SK, Loy JK, Bravo R, Ransohoff RM, Lira SA. Neutrophil infiltration, glial reaction, and neurological disease in transgenic mice expressing the chemokine N51/KC in oligodendrocytes. J Clin Invest 1996;98: 529–539.
51. Bell MD, Taub DD, Kunkel SJ, Streiter RM, Foley R, Gauldie J, Perry VH. Recombinant human adenovirus with rat MIP-2 gene insertion causes prolonged PMN recruitment to the murine brain. Eur J Neurosci 1996;8: 1803–1811.
52. Balashov KE, Rottman JB, Weiner HL, Hancock MW. CCR5(+) and CXCR3(+) T cells are increased in multiple sclerosis and their ligands MIP-1alpha and IP-10 are expressed in demyelinating brain lesions. Proc Natl Acad Sci USA 1999;96:6873–6888.
53. Berman JW, Guida MP, Warren J, Amat J, Brosnan CF. Localization of monocyte chemoattractant peptide-1 expression in the central nervous system in experimental autoimmune encephalomyelitis and trauma in the rat. J Immunol 1996;156:3017–3023.
54. Diab A, Abdalla H, Li HL, Shi FD, Zhu J, Hojberg B, Linquist L, Wretland B, Bakhiet M, Link. Neutralization of macrophage inflammatory protein 2 (MIP-2) and MIP-1 alpha attenuates neutrophil recruitment in the central nervous system during experimental bacterial meningitis. Infect Immun 1999;67:2590–2601.
55. McManus C, Berman JW, Brett FM, Staunton H, Farrell M, Brosnan CF. MCP-1, MCP-2 and MCP-3 expression in multiple sclerosis lesions: an immunohistochemical and in situ hybridization study. J Neuroimmunol 1998; 86:20–29.
56. Nygardas PY, Maatta JA, Hinkkanen AE. Chemokine expression by central nervous system resident cells and infiltrating neutrophils during experimental autoimmune encephalomyelitis in the BALB/C mouse. Eur J Immunol 2000;30:1911–1918.
57. Sauder C, Hallensleben W, Pagenstecher A, Schneckenburger S, Biro L, Perlick D, Hausmann Stuter M, Staeheli P. Chemokine gene expression in astrocytes of Borna disease virus-infected rats and mice in the absence of inflammation. J Virol 2000;74:9267–9280.
58. Sharafeldin A, Eltayeb R, Pashenkov M, Bakhiet M. Chemokines are produced in the brain early during the course of experimental African trypanosomiasis. J Neuroimmunol 2000;103:165–170.
59. Van Der Voorn P, Tekstra J, Beelen RH, Tensen CP, Van Der Walk P, De Groot CJ. Expression of MCP-1 by reactive astrocytes in demyelinating multiple sclerosis lesions. Am J Pathol 1999;154:45–51.
60. Kniesel U, Wolburg H. Tight junctions of the blood–brain barrier. Cell Mol Neurobiol 2000;20:57–76.
61. Rubin LL, Staddon JM. The cell biology of the blood–brain barrier. Annu Rev Neurosci 1999;22:11–28.
62. Middleton J, Neil S, Wintle J, Clark-Lewis I, Moore H, Lam C, Auer M, Hub E, Rot A. Transcytosis and surface expression of IL-8 by venular endothelial cells. Cell 1997;91:385–395.
63. Abbott NJ, Romero IA. Transporting therapeutics across the blood–brain barrier. Mol Med Today 1996;2:106–113.
64. Makic J, Stins M, McComb JG, Calero M, Ghiso J, Kim KS, Yan DS, Stern D, Schmidt AM, Fangione B, Zlokovic BV. Human blood–brain barrier receptors for Alzheimer's amyloid 1–40. J Clin Invest 1998;102:734–743.
65. Rose JM, Audus KL. Receptor-mediated angiotensin by brain microvessel endothelial cells. Peptides 1998;19: 1023–1030.
66. Montesano R, Roth J, Robert A, Orci L. Non-coated membrane invaginations are involved in binding and internalization of cholera and tetanus toxins. Nature 1992; 296:651–653.
67. Parton RG. Caveolae and caveolins. Curr Opin Cell Biol 1996;8:542–548.
68. Ikezu T, Ueda H, Trapp B, Nishiyama K, Sha JF, Volonte D, Galbiati F, Byrd AL, Bassel G, Serizawa H, Lane WS, Lisanti MP, Okamoto T. Affinity purification and characterization of caveolins from brain: differential expression of caveolin-1, -2, and -3 in brain endothelial and astroglia cell types. Brain Res 1998;804:177–192.
69. Dehouck B, Fenart L, Dehouck MP, Pierce A, Torpier G, Cecchelli RA. new function for the LDL receptor: transcytosis of LDL across the blood–brain barrier. J Cell Biol 1997;138:877–889.
70. Pan W, Kastin AJ. Changing the chemokine gradient: CINC1 across the blood–brain barrier. J Neuroimmunol 2001;115(1–2):64–70.
71. Murdoch C, Finn A. Chemokine receptors and their role in vascular biology. J Vasc Biol 1999;37:1–7.
72. Murdoch C, Monk PN, Finn A. CXC chemokine receptor expression on human endothelial cells. Cytokine 1998; 11:704–712.
73. Shaw KTY, Greig NH. Chemokine receptor mRNA expression at the in vitro blood–brain barrier during HIV infection. NeuroReport 1999;10:53–56.
74. Salcedo R, Resau JH, Halverson D, Hudson EA, Dambach M, Powell D, Wasserman K, Oppenheim JJ. Differential expression and responsiveness of chemokine receptors (CXCR1–3) by human microvascular endothelial cells and umbilical vein endothelial cells. FASEB J 2000;14:2055–2064.
75. Feil C, Augustin HG. Endothelial cells differentially express functional CXC-chemokine receptor-4 (CXCR4/Fusin) under the control of autocrine activity and exogenous cytokines. Biochem Biophys Res Commun 1998;247:38–45.
76. Gupta SK, Lysko PG, Pillarisetti K, Ohlstein E, Stadel JM. Chemokine receptors in human endothelial cells. J Biol Chem 1998;273:4282–4287.
77. Volin MV, Joseph L, Shockley MS, Davis PF. Chemokine receptor CXCR4 expression in endothelium. Biochem Biophys Res Commun 1998;242:46–53.
78. Molino M, Woolkalis MJ, Prevost N, Pratico D, Barnathan ES, Taraboletti G, Haggarty BC, Hesselgesser J, Horuk R, Hoxie JA, Brass LF. CXCR4 on human endothelial cells can serve as both a mediator of biological responses and as a receptor for HIV-2. Biochem Biophys Acta 2000;1500:227–240.

79. Sanders VJ, Pittman CA, Whit MG, Wang W, Wiley CA, Achim CL. Chemokines and receptors in HIV encephalitis. AIDS 1998;12:1021–1026.
80. Weber KSC, Nelson PJ, Grone H-J, Weber C. Expression of CCR2 by endothelial cells. Implications for MCP-1 mediated wound injury repair and in vivo inflammatory activation of endothelium. Arterioscler Thromb Vasc Biol 1999;19:2085–2093.
81. McFadden G, Kelvin D. New strategies for chemokine inhibition and modulation. Biochem Pharmacol 1997;54: 1271–1280.
82. Ali S, Palmer ACV, Banerjee B, Fritchley SJ, Kirby JA. Examination of the function of RANTES, MIP-1α, and MIP-1β following interaction with heparin-like glycosaminoglycans. J Biol Chem 2000; 275:11721–11727.
83. Hoogewerf A, Kuschert GSV, Proudfoot AEI, Borlat F, Clark-Lewis I, Power CA, Wells TNC. Glycosaminoglycans mediate cell surface oligomerization of chemokines. Biochemistry 1987;36:13570–13578.
84. Kuschert GSV, Coulin F, Power CA, Proudfoot AEI, Hubbard RE, Hoogewerf AJ, Wells TNC. Glycosaminoglycans interact selectively with chemokines and modulate receptor binding and cellular responses. Biochemistry 1999;38:12959–12968.
85. Sozznani S, Locati M, Zhou M, Rieppi M, Luini W, Lamorte G, Bianchi G, Polentarutti N, Allavena P, Mantovani A. Receptors, signal transduction, and spectrum of action of monocyte chemotactic protein-1 and related chemokines. J Leukoc Biol 1995;59:39–44.
86. Charo IF. CCR2: from cloning to the creation of knockout mice. In: A. Mantovani, ed. Chemokines. Basel: Karger, 1999:30–41.
87. Daugherty BL, Siciliano SJ, DeMartino JA, Malkowitz L, Sirotina A, Springer MS. Cloning, expression and characterization of the human eosinophil exotaxin receptor. J Exp Med 1996;183:2349–2354.
88. Proudfoot AEI, Power CA, Hoogewerf A, Montjovent M-O, Borlat F, Wells A. Characterization of the RANTES/MIP-1α receptor (CC CKR-1) stably transfected in KEK 293 cells and the recombinant ligands. FEBS Lett 1995;376:19–23.
89. Combadiere C, Ahuja SK, Van Damme J, Tiffany HL, Gao JL, Murphy PM. Monocyte chemoattractant protein-3 is a functional ligand for CC chemokine receptors 1 and 2B. J Biol Chem 1995;270:29671–29675.
90. Sarau HM, Rush JA, Foley JJ, Brawner ME, Schmidt D, White JR, Barnette MS. Characterization of functional chemokine receptors (CCR1 and CCR2) on EoL-3 cells: a model system to examine the role of chemokines in cell function. J Pharmacol Exp Ther 1997;283:411–418.
91. Samson M, Labbe O, Mollereau C, Vassart G, Parmentier M. Molecular cloning and functional expression of a new human CC-chemokine receptor gene. Biochemistry 1996;35:3362–3367.
92. Crane IJ, Wallace CA, McKillop-Smith S, Forrester, JV. CXCR4 receptor expression on human retinal pigment epithelial cells from the blood–retina barrier leads to chemokine secretion and migration in response to stromal cell-derived factor 1α. J Immunol 2000;165:4372–4378.
93. Blanpain C, Migeotte I, Lee B, Vakilil J, Doranz BJ, Govaertz C, Vassart G, Doms RW, Parmentier M. CCR5 binds multiple CC-chemokines: MCP-3 acts as a natural antagonist. Blood 1999;94:1899–1905.
94. Signoret N, Rosenkilde MM, Klasse PJ, Schwartz TW, Malim MH, Hoxie JA, Marsh M. Differential regulation of CXCR4 and CCR5 endocytosis. J Cell Sci 1998;111: 2819–2830.
95. Solari R, Offord RE, Remy S, Aubry JP, Wells TN, Whitehorn E, Oung T. Proudfoot AE. Receptor-mediated endocytosis of CC-chemokines. J Biol Chem 1997;272: 9617–9620.
96. Broadwell RD, Banks WA. Cell biological perspective for the transcytosis of peptides and proteins through the mammalian blood–brain fluid barriers. In: Pardridge WM, ed. The Blood–Brain Barrier. New York, NY: Raven Press, Ltd., 1993:165–199.
97. Godaly G, Hang L, Frendeus B, Svanburg C. Transepithelial neutrophil migration is CXCR1 dependent in vitro and is defective in IL-8 receptor knockout mice. J Immunol 2000;165:5287–5294.
98. Vestweber D. Molecular mechanisms that control endothelial contacts. J Pathol 2000;190:281–291.
99. Wen DZ, Rowland A, Derynck R. Expression and secretion of gro/MGSA by stimulated human endothelial cells. EMBO J 1989;8:1761–1766.
100. Dwinell MB, Eckmann L, Leopard JD, Varki NM, Kagnoff MF. Chemokine receptor expression by human intestinal epithelial cells. Gastroenterology 1999;117:359–367.
101. Davis LE, Hjelle BL, Miller VE, Palmer DL, Llewellyn AL, Merlin TL, Young SA, Mills RG, Wachsman W, Wiley CA. Early viral invasion in iatrogenic human immunodeficiency virus infection. Neurology 1992;42: 1736–1739.
102. Nottet HSM, Dhawan S. HIV-1 entry into the brain: Mechanisms for the infiltration of HIV-1 infected macrophages across the blood–brain barrier. In: Gendelman H, Lipton SA, Epstein L, Strindells S, eds. The Neurology of AIDS. New York, NY: International Thompson Publishing, 1999:49–60.
103. Albright AV, Shieh JT, O'Conner MJ, Gonzalez-Scarano F. Characterization of cultured microglia that can be infected by HIV-1. J Neurovirol 2000;6(Suppl 1):S53–60.
104. Messam CA, Major EO. Stages of restricted astrocyte infection in astrocyte cultures derived from human fetal brain tissue. J Neurovirol 2000;6(Suppl 1):S90–94.
105. Devito C, Broliden K, Kaul R, Svensson L, Johansen K, Kiama P, Kimani P, Kimani J, Lopalco J, Piconi S, Bwayo JJ, Plummer F, Clerici M, Hinkula J. Mucosal and plasma IgA from HIV-1-exposed uninfected individuals inhibit HIV-1 transcytosis across human epithelial cells. J Immunol 2000;165:5170–5176.
106. Fiala M, Gujuluva C, Berger O, Popok W, Pusharsky T, Bukrinsky M, Burns A. Transcellular and paracellular pathways of HIV neuroinvasion and heart invasion. FASEB J 2000;14:A937.
107. Banks WA, Akerstrom V, Kastin AJ. Adsorptive endocytosis mediates the passage of HIV-1 across the blood–

brain barrier: evidence for a post-internalization coreceptor. J Cell Sci 1998;111:533–540.

108. Wang JM, Ueda H, Howard OMZ, Grimm MC, Chertov O, Gong X, Gong W, Reseau JH, Broder CC, Evans G, Arthur LO, Ruscetti FW, Oppenheim JJ. HIV-1 envelope gp120 inhibits monocyte response to chemokines through CD4 signal-dependent chemokine receptor down-regulation. J Immunol 1998;161:4309–4317.

Section 5

CHEMOKINES AND NEUROLOGICAL DISEASES

Section Editor: R. Ransohoff

Universes in Delicate Balance: Chemokines and the Nervous System
R.M. Ransohoff, K. Suzuki, A.E.I. Proudfoot, W.F. Hickey and J.K. Harrison, editors

Chapter 5.1

Chemokines and Chemokine Receptors in Multiple Sclerosis: A Few Answers and Many More Questions

Richard M. Ransohoff [1] and Corinna Trebst [2]

[1] *Department of Neurology (Mellen Center for MS Treatment and Research) and Department of Neurosciences of the Lerner Research Institute, Cleveland Clinic Foundation, NC30, 9500 Euclid Avenue, Cleveland, OH 44195, USA;* [2] *Department of Neurosciences of the Lerner Research Institute, Cleveland Clinic Foundation, NC30, 9500 Euclid Avenue, Cleveland, OH 44195, USA*

1. MULTIPLE SCLEROSIS

1.1. Clinical Characteristics

Multiple sclerosis (MS) is a common acquired disease of the human central nervous system (CNS). Typically, onset occurs between ages of 20 and 40 with a sharp peak at the age of 30. The disease does not shorten life so that it remains chronic for approximately 40 years in the majority of affected individuals. MS affects 1 per 1000 individuals in the United States and Western Europe, therefore involving a patient population of approximately one million individuals worldwide. MS is regarded as the most significant primary neurological cause of disability among adults of Western European origin, both in Europe and in the United States [1].

MS presents in about 85% of patients with intermittent episodes of neurological dysfunction [1]. This pattern of symptoms, termed *relapsing–remitting MS*, is followed after 10–20 years in the majority of cases by a stage of continuous functional decline, designated as *secondary-progressive MS*. Approximately 10–15% of patients experience slow symptomatic worsening from onset. This pattern of disease is called *primary-progressive MS*. It is considered essential to examine comparable patient groups in clinical treatment trials and natural history studies, so that these differing symptom patterns are typically used to stratify patients for entry into clinical protocols. Recent data showed that patients of either subtype who have reached a stage of irreversible neurological disability characterized by moderate difficulty in walking will thereafter experience progression with a predictable time course [2].

For the majority of patients, MS will produce a significant impact on daily function. Thus, after 15–20 years of disease, 50% of patients require aids to walk, which translates into a population of individuals aged 45–50 that are dependent on canes, walkers or wheelchairs for routine activities. Although prior natural-history studies suggest that as many as 30% of patients have 'benign' non-disabling disease, contemporary follow-up studies demonstrate that 2/3 of patients classified as benign MS after 10 years of disease are disabled after 25 years [3].

The clinical spectrum of MS comprises several less common variants, including a fulminant, fatal disease presentation (Marburg's disease) [4]. Other eponymic subtypes of MS such as Devic's syndrome (selective affection of the optic nerves and spinal cord) are uncommon but important to understand, given the likelihood that a diversity of underlying processes (see below) can probably culminate in the clinical phenomena that characterize MS [5–7].

1.2. Pathology

The pathology of MS has been studied continuously since its identification as a nosological entity in the mid-nineteenth century [8]. In almost all cases, MS specifically and exclusively affects the CNS, which is composed of the brain and spinal cord. Aside from complications of immobility, there are no pathological findings in extra-neural tissues. Furthermore, MS typically spares peripheral nervous system (PNS) myelin, a fact that is not well understood, because many of the myelin protein targets of the putative autoimmune process are shared between the CNS and the PNS [9].

Histopathological hallmarks of MS include primary demyelination, inflammation, gliosis, and axonal damage [8]. Demyelination is the most important single pathologic feature of the disease [10], and studies of very acute lesions indicate clearly that the initial and primary injured component in the CNS of MS patients is myelin [4]. Myelin may be repaired, sometimes extensively, giving rise to lightly myelinated structures termed 'shadow plaques'. Persistently demyelinated fiber tracts become encased in the fibrous processes of reactive astrocytes, creating a cirrhotic end-stage plaque.

1.3. Beyond Demyelination: Axonal Pathology in MS Lesions

In addition to promoting rapid conduction of nerve impulses, myelin provides axons with both trophic support and a protective barrier. Therefore it is not surprising that demyelinated nerve fibers in MS lesions undergo transection, an irreversible event. This fact was appreciated and exquisitely illustrated in the earliest pathological descriptions of MS [11]. Recently, with newer quantitative immunohistochemical techniques, it has become apparent how frequently this event occurs [11–17]. In particular, as many as 1% of nerve fibers within the lesional microenvironment may be destroyed with each demyelinating event. Molecular mechanism(s) that underlie nerve fiber transection in MS lesions are unknown [11]. Exposure to components of the inflammatory microenvironment has been proposed but specific toxic elements remain to be identified [1]. Importantly, however, remyelinated axons appear substantially protected from transection [13]. In active demyelinating lesions, there is a tight, direct and consistent relationship between numbers of transected axons and activated macrophages [4,14].

The axon may also suffer from the consequences of chronic demyelination. It has been known for some years that diseases of the PNS in which myelin is either damaged or is aberrantly formed (dysmyelination) can culminate in axonal pathology because of the deprivation of the trophic support of myelin [18]. Loss of myelin's trophic influence is also deleterious to CNS fibers: mice that produce abnormal CNS myelin due to deletion of specific myelin protein genes suffer axonal pathology during adulthood [19,20]. By analogy, chronically demyelinated, non-inflammatory lesions are now known to be sites of chronic, smoldering axon loss and probable substrates for neurological progression [13,16,17,21].

1.4. Hematogenous Inflammatory Cells in the CNS of MS Patients

The hematogenous inflammatory component of MS can be examined and characterized both in the cerebrospinal fluid (CSF) and in tissue lesions. The CSF of MS patients frequently contains an excess of leukocytes (a condition termed pleiocytosis). Interestingly, these cells do not reflect the relative abundance of leukocyte populations in the blood, with approximately 100% of CSF cells in MS being mononuclear (90% T-cells and 10% monocytes). Neutrophils are rarely detected, except in hyperacute syndromes or in the highly destructive variants such as Devic's disease. The CD4/CD8 ratio of T-cells in MS CSF is 2 : 1, as seen in blood. In CNS tissue lesions, mononuclear cells dominate, with granulocytes observed infrequently. However, the cellular sub-populations in CSF and tissue differ: about 10% of lesional marrow-derived cells are T-cells, with predominance of CD8+ T-cells. Approximately 90% of lesional cells of bone-marrow origin are macrophages, derived both from resident microglia and hematogenous monocytes.

In summary, the pathological cascade in multiple sclerosis begins with invasion of the nervous system by leukocytes and culminates with irreversible injury to myelin and axons [22]. A significant recent report indicated that MS is pathologically heterogeneous among different patients, with lesions in a single brain all exhibiting a similar character [23]. The implications of this observation are that different patients may be afflicted with a variety of fundamentally differing pathological processes. This pathological heterogeneity was captured largely by examining the patterns of demyelination and fate of oligodendroglia [23–25]. Three general patterns of tissue injury seemed apparent. One, in which myelin was destroyed in an inflammatory environment including T-cells, macrophages and (in some cases) immunoglobulin and activated complement; this pattern has been duplicated in all essential aspects by various animal models of autoimmune demyelination [26]. A second major pattern of MS pathology was dominated by selective early disappearance of the myelin-associated glycoprotein (MAG), with varying degrees of attendant inflammation and oligodendrocyte cell death. This second pattern has not been replicated in animal models of myelin-specific autoimmunity, and is of uncertain pathogenesis. A third pattern, observed in a small minority of cases, involved death of oligodendrocytes in the periplaque white matter, with demyelination being a presumptive consequence of oligodendrocyte loss. The cumulative message of this work is that the clinical and pathological phenomena of MS can be generated by diverse processes, entering a final common pathway to demyelination, axonal loss and disability [1,23].

These recent observations introduce a wholesome note of caution to dogmatic, sweeping statements about MS pathogenesis. However, it remains likely that, in the majority of cases, the intensity of inflammation is directly related to the extent of tissue injury. This conclusion rests on multiple lines of evidence:

(1) Studies of autopsy and biopsy MS brain sections indicate that destruction of myelin [4] and axons [14,15] as well as death of oligodendrocytes [24] are all directly related to numbers of activated inflammatory hematogenous cells.

(2) Magnetic resonance imaging (MRI) studies show that clinical deterioration is predicted by the number of inflammatory events, typified by disruption of the blood–brain barrier (BBB), as revealed by foci of gadolinium leakage into brain parenchymal lesions [27,28].

1.5. Etiology and Pathogenesis

The cause of MS is unknown [29]. Based on tissue pathology, genetic and epidemiological data, it has been proposed that MS may frequently represent an autoimmune T-cell-dependent inflammatory disorder that occurs in genetically susceptible individuals after an exposure to environmental triggers [30–35]. According to this concept, it is widely accepted that virus infection can trigger such a process [36–38]. These notions are consistent with the geographical distribution of MS, which occurs in susceptible populations in temperate climates at a much higher frequency than in such populations in equatorial zones [39,40]. As defined by migration studies, disease expression typically follows at a considerable latency after exposure to putative viral or other environmental influences. It remains controversial whether a single pathogen is etiologically related to MS in any subgroup of patients [29]. Therefore, investigators favor a 'hit and run' mechanism of disease causation in the majority of patients [41–43]. These debates will likely continue until it is demonstrated that MS is caused by one or more pathogens, or until definitive studies put such suspicions to rest.

2. CHEMOKINES AND LEUKOCYTE TRAFFICKING IN THE CNS

2.1. Chemokines and Leukocyte Trafficking

Considerable attention has recently been devoted to understanding leukocyte migration and retention during inflammatory processes affecting the CNS [44]. Fundamental mechanisms underlying leukocyte trafficking during inflammation and development have received much attention during the past decade, as summarized elsewhere in this volume [45–52]. The roles of chemokines in these events have been elucidated in some detail. After about a decade of confusion, a simplifying nomenclature for both chemokines and chemokine receptors has been proposed; these systematic names are juxtaposed below with the traditional chemokine designations [53,54].

Cells respond to chemokines only by virtue of the expression of specific high-affinity receptors [55–60]. An overall scheme of leukocyte extravasation has been applied with great success to events as diverse as inflammation, developmental, and recirculatory leukocyte trafficking, and colonization of gut and skin with resident populations of leukocytes [61–63]. Therefore, it appears that variants of this general scheme underlie all leukocyte trafficking. Combinatorial diversity is adduced to account for the enormous variety of leukocyte extravasation events explained by the paradigm.

Ligand–receptor relationships among chemokines and their receptors are complex. For example, the prototype monocyte-selective chemokine, monocyte chemoattractant protein (MCP)-1/CCL2 acts on circulating monocytes through the receptor CCR2. The presence of four (possibly five) MCPs in the murine genome produces an apparent redundancy that would suggest that individual MCPs should be nonessential. However, gene-targeting studies in mice have shown repeatedly that redundancy in the chemokine/chemokine receptor system is more apparent than real.

2.2. Unusual Character of CNS Inflammation

Clearly, the current paradigm of leukocyte trafficking during inflammation has great appeal in attempts to account for processes that affect the CNS. The CNS is highly intolerant of acute inflammation, which can produce edema. Why is inflammatory swelling so dangerous to the CNS? First, the cerebral contents consist of brain parenchyma, CSF volume and blood volume, rigidly confined within the skull and dural meninges. Additional fluid volume, produced by inflammation, must therefore be compensated by reduced blood volume, which rapidly produces secondary ischemic changes. Second, swelling must be accommodated by shift of the cerebral contents, typically with compression of vital brainstem regulatory structures that govern cardiac and respiratory functions, often producing fatal outcomes. Apparently, the CNS has evolved mechanisms to avoid such catastrophes. Inflammation in the CNS appears to be severely restricted and controlled by multiple mechanisms including the BBB that excludes most macromolecules and inflammatory cells [64–71]. The BBB is composed of the brain microvascular cells that exhibit two unusual characteristics: non-fenestrated plasma membranes and intercellular tight junctions. These attributes are induced in cerebrovascular endothelial cells by astrocytes; each CNS capillary is invested by a network of astrocyte and microglial processes termed the glial limitans. Furthermore, cerebrovascular endothelial cells express lower levels of adhesion molecules (compared to other vascular beds) under resting conditions. However, in an activated state, cerebrovascular endothelial cells express high levels of cell adhesion molecules (CAMs) that provide substrates for leukocyte recruitment [72].

Neuronal injury, mediated by intracerebral injection of an excitotoxin, kainic acid, provides an instructive example of the unusual character of CNS inflammation. Kainic acid produces a necrotizing neuronal injury [73]. In peripheral tissues, necrosis of parenchymal cells would be expected to elicit a brisk neutrophilic infiltrate with attendant edema [72]. However, kainic acid challenge of the CNS induces an inflammatory response without neutrophils, characterized by gradual recruitment of macrophages and activation of the resident microglia [72].

As further example of atypical response to physical injury, percussion injury to the rodent spinal cord produces a sparse, transitory neutrophilic reaction, followed by robust recruitment of macrophages with a minority population of T-cells [74]. This pattern is mirrored by a brief low-level expression of neutrophil-specific chemokines followed by vigorous expression of MCP-1/CCL2 and MCP-5/CCL12 [75].

Therefore, selective expression of monocyte-targeted chemokines may constitute a major driving force for CNS accumulation of monocytes in preference to neutrophils. Supporting this concept, when neutrophil-specific chemokines are expressed in unusual circumstances (such as brain abscess), neutrophils do accumulate at high levels within the CNS [76].

As a unifying statement, it appears plausible that precisely regulated production of chemokines provides a mechanism for the tendency of the mammalian CNS to develop features of chronic rather than acute inflammation. This concept, that chemokines are intimately implicated in leukocyte recruitment to the CNS, has been subjected to numerous 'proof of principal' experiments using animal models. Much of this work has been performed in gene-targeted and transgenic mice, often using a model of multiple sclerosis, experimental autoimmune encephalomyelitis (EAE). These studies are described in detail elsewhere in this volume. Two topics that have been examined in these experimental models are particularly pertinent to our understanding of the functions of chemokines in MS and will be discussed below.

2.3. Sources of Chemokines in the CNS in EAE

Given that the CNS precisely regulates chemokine production, it has been considered important to identify cellular sources of these products. Gratifyingly, sources of chemokines in the CNS in EAE have been addressed by several groups, and generally concordant findings have been reported. Several chemokines (MIP-1α/CCL3 and MIP-1β/CCL4 and RANTES/CCL5) are made predominantly by infiltrating leukocytes [77–79]. Other chemokines (IP-10/CXCL10, MCP-1/CCL2) are produced mainly by the resident population of astrocytes [77,78,80].

2.4. Regulation of Chemokine Expression in EAE

What drives the expression of chemokines in EAE? Based on simple descriptive considerations, it appears likely that inflammatory cytokines produced by activated T-cells and macrophages are responsible. From tissue-culture studies it is known that such cytokines (interferon-γ (IFN-γ), tumor necrosis factor-α (TNF-α), IL-1) can stimulate cells to produce huge quantities of chemokines [81–83]. Furthermore, the distribution of chemokine-expressing cells, around aggregates of inflammatory cells, suggests that products of the inflammatory cells stimulate chemokine expression [80]. Finally, a recent study in gamma interferon knockout (GKO) mice provided data that support this hypothesis [84]. Surprisingly, GKO mice are more susceptible to EAE than their wild-type counterparts, and exhibit typical inflammatory pathology. Further, there is high-level intrathecal production of several chemokines including MCP-1/CCL2 in the CNS of GKO mice with EAE. IP-10/CXCL10 was strikingly absent from the array of chemokines produced during GKO-EAE [84]. IP-10/CXCL10 was initially cloned in a differential hybridization experiment using IFN-γ to stimulate U937 cells [85], and most disease models confirm that IFN-γ is essential for the production of IP-10/CXCL10 in vivo. Therefore, the absence of IP-10/CXCL10 in the CNS of GKO-EAE mice suggests that inflammatory cytokines direct the expression of specific chemokines during EAE. This concept has also been strongly supported by interventional studies in which inflammatory cytokine production was modified by treatments with altered peptide ligands, producing both cytokine and chemokine modulation.

Time course studies of CNS chemokine expression indicated that chemokine products were elaborated in an explosive fashion immediately after the earliest T-cell infiltrates appeared in the affected CNS [86,87]. In summary, it appears likely that initial invasion of the CNS by activated antigen-specific T-cells occurs independently of production of chemokines within the CNS tissue [88,89]. However, immediately after T-cells become resident within the nervous system and produce inflammatory cytokines, large-scale local expression of multiple chemokines ensues and contributes to the dramatic amplification of the inflammatory response in EAE [79].

3. CHEMOKINES AND THEIR RECEPTORS IN MULTIPLE SCLEROSIS

Studies of chemokines and their receptors in MS have been, to date, descriptive [90–93]. Much of the information that has been reported concerning chemokines and chemokine receptors in MS can be readily correlated to findings in animal models, including EAE.

3.1. Genetic Studies

There are clues that chemokine and chemokine receptor genes are related to either the likelihood of getting MS (susceptibility) or the severity of the disease. Fiten and colleagues reported that the risk to develop MS was decreased in Swedish patients that possessed specific microsatellite polymorphisms in the upstream region of the MCP-3/CCL7 gene [94]. Different polymorphisms conferred protection for patients with high-risk HLA haplotypes as compared with those with a low-risk HLA haplotypes. To date, this is the only study that reported a susceptibility-associated chemokine- or chemokine receptor-gene polymorphism.

These results have correlates in experimental model systems. Teuscher and colleagues identified a locus termed eae7, one of several that control the pattern of EAE in congenic inbred mouse strains [95]. This quantitative trait locus (QTL) proved to be coincident with the β-chemokine gene cluster. Sequence polymorphisms for several chemokines distinguished the two eae7 alleles and predicted significant structural alterations in the chemokine proteins involved. Of interest, mice deficient for β-chemokine receptors are in some cases resistant to induction of EAE, or experience atypical, mild disease [96,97]. Therefore, in both spontaneous human disease and in the mouse model, inflammatory demyelination is linked, both in susceptibility and in severity, to polymorphic genes in the chemokine system.

Additional work has focused on the well-characterized Δ32 polymorphism of CCR5. Homozygotes for this allele (about 1% of the Western European population) are functional 'knockouts' for CCR5. Bennets and colleagues showed that the homozygous Δ32 CCR5 genotype did not protect from multiple sclerosis in a study of 120 unrelated patients in Australia. Specifically, two patients in this small sample were CCR5 Δ32 homozygotes, a frequency that would be expected by chance given the background frequency of the CCR5 Δ32 allele and prevalence of MS [98]. Sellebjerg and colleagues found an intriguing effect of the CCR5 Δ32 heterozygote state on MS severity. They identified patients at the time of first attack, and performed genotyping studies for CCR5. In this cohort, Δ32 heterozygotes progressed to clinically definite MS (defined by the occurrence of a second attack) significantly more slowly than did patients with wild-type CCR5 genes [99]. Oksenberg and coworkers showed in a well-documented cohort of familial MS cases, that Δ32 heterozygotes exhibited a mean three-year delay in the onset of their MS, as compared with patients possessing wild-type CCR5 alleles [100]. However, this observation was not confirmed when the simplex MS families in this database were compared [100]. Further, the disease course after onset in the familial CCR5 Δ32 heterozygotes was not milder than that observed in patients with wild-type CCR5 alleles (J. Oksenberg, pers. commun.).

3.2. Chemokine and Chemokine Receptor Expression in MS: Overview

Chemokine and chemokine receptor expression in MS have also been reported. For the most part, studies have focused on either body fluids (blood, CSF) or cells therein, or on autopsy brain tissue.

Interestingly, the first study of a chemokine-related product in MS was reported in 1987 by Cananzi et al. before the chemokine family was defined [101]. In this study, platelet factor 4 (PF4/CXCL4) was significantly increased in plasma of MS patients as compared to healthy controls. Aspirin treatment did not affect PF4 production by cells isolated from patients, suggesting that the product originated from mast-cells. Because the function and receptor specificity of PF4/CXCL4 remain uncertain, this observation has not been extensively followed-up.

3.3. Chemokines in the CSF

There have been limited reports of chemokine levels in the CSF of MS patients (Table 1). Our group investigated chemokine levels by enzyme-linked immunosorbent assay (ELISA) in the CSF of 38 patients with active symptomatic MS, in comparison to 21 healthy individuals without neurological disease [102]. Several chemokines (IP-10/CXCL10, mig/CXCL11 and RANTES/CCL5) were significantly elevated while levels of MCP-1/CCL2 were significantly decreased in the CSF of patients with MS, as compared to CSF of controls. Other groups have independently confirmed both the increased levels of IP-10/CXCL10 in CSF of MS patients and the selective decrease of MCP-1/CCL2 [103–106]. As reported previously, MCP-1/CCL2 was dramatically increased in CSF from patients with either bacterial or viral meningitis, demonstrating the relative specificity of this finding for MS [104,107,108]. In several studies, MIP-1α/CCL3 was found either at low-picogram levels or was undetectable [102,109].

The sources of chemokines remain to be further elucidated. In a study by Kivisäkk and coworkers, the production of MCP-1/CCL2 and RANTES/CCL5 by CSF mononuclear cells was similar in patients with MS and in other inflammatory controls, consistent with the production of chemokines by CNS parenchymal cells during inflammation, rather than hematogenous infiltrating cells [110].

Furthermore, very little is known about the effects of standard treatment (methylprednisolone, interferon-β-1a or -1b, glatiramer acetate) on the expression of chemokines. In a recent study by our laboratory, we serially investigated levels of IP-10/CXCL10 and MCP-1/CCL2 in the CSF of patients with active MS before and after treatment with methylprednisolone or placebo and in a control population. Higher levels of IP-10/CXCL10 in the CSF were found before and after treatment in patients with MS as compared to controls. Treatment with methylprednisolone had no influence on

Table 1. Chemokines in MS clinical material: overview

Chemokines	Circulation	CSF	CNS tissue
MCP-1/CCL2	CCL2 serum levels in RR-MS unaffected by IFN-β-1a or methylprednisolone treatment [106].	CCL2 CSF levels decreased in patients with active MS compared to controls (consistently in multiple studies, [102–104,106]). In patients with bacterial or viral meningitis dramatically increased as compared with controls [104,107,108]. The production of CCL2 by CSF mononuclear cells is similar in MS and controls [110]. CSF CCL2 levels affected by treatment with methylprednisolone [111].	In active MS lesion CCL2 is expressed by astrocytes and inflammatory cells [92,125–127].
IP-10/CXCL10	CXCL10 serum levels in RR-MS unaffected by IFN-β-1a or methylprednisolone treatment [106].	CXCL10 CSF levels elevated in patients with active MS as compared with controls (consistently in multiple studies, [102,105,106]). CSF CXCL10 levels unaffected by methylprednisolone treatment [111].	In active MS lesions CXCL10 is expressed by astrocytes [102,113,123].
RANTES/CCL5	CCL5 serum levels elevated in MS patients compared to controls, reduced by treatment with IFN-β-1b [112].	CCL5 CSF levels elevated in patients with active MS as compared with controls (in some, but not all studies, [102]). The production of CCL5 by CSF mononuclear cells is similar in MS and controls [110].	In active MS lesion CCL5 is expressed by perivascular cells and, to a lesser extent, by astrocytes [92,127–130].
MIP-1α/CCL3		Either at low-picogram levels or undetectable. No differences between MS and controls [102,109].	In active MS lesion CCL3 is expressed by macrophages and microglial cells [92,127,128].
MIP-1β/CCL4			In active MS lesion CCL4 is expressed by inflammatory cells and activated neuroglial cells [92,127,128].
MCP-3/CCL7			In active MS lesion CCL7 is expressed by astrocytes and inflammatory cells [92,126,127].
MCP-2/CCL8			In active MS lesion CCL8 is expressed by astrocytes and inflammatory cells [92,126,127].
Mig/CXCL9		CXCL9 CSF levels elevated in patients with active MS compared with controls, in some but not all studies [102].	
PF4/CXCL4	CXCL4 levels increased in plasma from MS patients as compared with controls [101].		

IP-10/CXCL10 CSF levels, indicating a role for IP-10/CXCL10 in maintaining inflammation. On the other hand, MCP-1/CCL2 CSF levels were altered by methylprednisolone treatment. A significant increase in the CSF concentration of MCP-1/CCL2 after treatment with methylprednisolone was observed, in contrast to a non-significant increase in the placebo-group. These data suggested differential involvement in intrathecal inflammation for the two studied chemokines in MS [111].

Interpreting measurements of circulating chemokines are fraught with complexity, in part due to the presence of a putative chemokine 'sink', the Duffy antigen for chemokines (DARC) on erythrocytes. One study reported that serum concentrations of RANTES/CCL5 were elevated in MS patients, but reduced in those receiving IFN-β-1b; further, exposure of monocytic cells in vitro to IFN-β-1b reduced RANTES/CCL5 expression at both mRNA and protein levels [112].

3.4. Chemokine Receptors on CSF and Circulating Cells

Comparisons of chemokine receptor expression on mononuclear cells in the CSF and in the circulation are of special interest for understanding cell trafficking into the CNS (Table 2). Our group focused on receptors corresponding to IP-10/CXCL10 and RANTES/CCL5, the chemokines that we found most-significantly elevated in the CSF of patients with attacks [102]. We observed that T-cells in the CSF expressed the IP-10/CXCL10/mig/CXCL9 receptor CXCR3 on approximately 90% of cells. This frequency was significantly increased over that observed in blood, where less than 40% of T-cells were CXCR3+ [102]. This enrichment was striking and consistent among all patients studied; however, it was also found that controls without neurological disease exhibited an identical level of CXCR3 expression on CSF T-cells. Therefore, it can be concluded that cells within the CSF compartment express in the majority CXCR3, but this cell population is not specific for MS.

In the case of one RANTES/CCL5 receptor, CCR5, we observed that a small fraction (less than 10%) of circulating T-cells were CCR5+. While the percentage of CCR5+ T-cells in MS CSF was significantly larger than that found in circulating cells,

Table 2. Chemokine receptors in MS clinical material: overview

Chemokine receptors	Circulation	CSF	CNS tissue
CCR1 (Ligands: CCL3, 5, 7, 14, 15, 16 and 23)	Low expression of CCR1 detectable on circulating T-cells [114]. Expressed on the majority of monocytes.	CCR1 expressed by the majority of CSF monocytes, independent of CNS inflammation [115].	In active MS lesion CCR1 expressed by newly infiltrating monocytes [115].
CCR2 (Ligands: CCL2, 7 and 13)	Low expression of CCR2 detectable on circulating T-cells, but highly expressed on monocytes [114].		In active MS lesion low levels of CCR2 expressed on macrophages [124].
CCR3 (Ligands: CCL5, 7, 8, 11, 13, 15, 24 and 26)	Low expression of CCR3 detectable on circulating T-cells [113,114].		In active MS lesion low levels of CCR3 expressed on macrophages [124].
CCR5 (Ligands: CCL3, 4 and 5)	CCR5 expression elevated on circulating T-cells in patients with MS compared with controls [102,113,114,117,118]. Associated with higher secretion of IFN-γ and TNF-α [113,117]. Reduced CCR5 expression on circulating T-cells in remission compared to relapse in RR-MS patients [114].	CCR5 expression enriched on CSF T-cells, compared with blood [102,114]. Expressed by the majority of CSF monocytes in MS patients and in controls [99,115]. Reduced on CSF T-cells during remission compared to relapse in RR-MS patients [114].	In active MS lesion CCR5 expressed on perivascular lymphocytes [102,113,123], infiltrating monocytes and activated macrophages/microglial cells [102,113,115,124].
CXCR3 (Ligands: CXCL9, 10 and 11)	CXCR3 expression elevated on T-cells in patients with MS compared with controls in some, but not all, studies [102,113,116]. Associated with IFN-γ and TNF-α secretion [113].	CXCR3 expressed by the majority of CSF T-cells, independent of CNS inflammation [102]. Expression similar in remission and relapse in RR-MS patients [114].	In active MS lesion CXCR3 expressed by perivascular lymphocytes [102,113,123].

CCR5+ T-cells constituted a minority of total CSF T-cells [102].

Several groups concurrently performed studies that confirmed and extended these results [113,114]. Misu and coworkers showed additionally that during remission, CCR5 expression, but not CXCR3 expression, was reduced on CSF T-cells compared with the expression during relapse [114].

Less information is available about the expression of chemokine receptors on monocytes in the blood and CSF of patients with MS. In six patients with MS, Sellebjerg and coworkers found that the majority of CSF monocytes (76%) expressed CCR5, compared with only 4% in the circulation [99]. In a larger collaborative study including these workers together with our laboratory, and comprising patients with MS, with optic neuritis, and without neurological disease, the majority of CSF monocytes expressed CCR1 and CCR5, while CCR1+/CCR5+ monocytes constituted a minority in the circulating pool. The presence of CCR1+/CCR5+ monocytes in the CSF was independent of CNS inflammation [115].

Investigators have also addressed the question whether patients with MS show preferential expression of chemokine receptors on circulating cells. The expression of CCR5 on circulating T-cells was significantly increased in patients with MS compared with a control population [102,113,114,116–118]. In addition, CXCR3 expression on circulating T-cells was significantly higher in MS patients compared with controls in some, but not all, studies [113,116]. CCR5+ and CXCR3+ subpopulations of T-cells were associated with higher secretion of interferon-γ and tumor necrosis factor-α [113,117]. This result suggested that CCR5 and CXCR3 expression on Th1-committed T-cells (previously described in vitro) might be pertinent for the pathogenesis of MS. However, it is clear that the majority of CNS-infiltrating lymphocytes in MS are not CXCR3+/CCR5+, since only a minority of either CD4+ or CD8+ T-cells in MS CSF were identified as expressing CCR5 by Sørensen et al. [102]. The relative lack of CCR5 expression on CSF T-cells may not, however, exclude a pathogenic role for these cells. Correlating with this observation, Zang and colleagues reported a modest tendency for circulating mononuclear cells from MS patients to migrate in response to MIP-1α/CCL3 or RANTES/CCL5 than are cells from control individuals [116]. This report suggested that CCR5 on T-cells from MS patients could possibly be more responsive to chemokine signaling than were other receptors in their preparation. It may be pertinent that recent studies from Mantovani and colleagues [119] have shown that chemokine receptors can be rendered hypo-responsive to ligand-induced signaling by exposure to certain cytokines, such as IL-10. Misu and coworkers additionally compared the number of CCR5+ circulating CD4+ cells during an acute exacerbation and three weeks later during remission in six patients with relapsing–remitting MS. CCR5 expression was increased during relapse, as compared to control individuals. In remission, CCR5 values decreased to control levels, suggesting an association of CCR5+ T-cells with disease activity [114].

Little is known about expression of chemokine receptors and MS disease course. As a first pass at this difficult question, Sørensen and Sellebjerg compared chemokine receptor expression on circulating mononuclear cells in patients with relapsing–remitting MS and secondary-progressive MS in a cross-sectional study, showing slight differences in the expression patterns of chemokine receptors CCR2 and CCR5 between the two patient groups [120]. Further comparisons, using longitudinal study design, are awaited.

Of additional interest are the effects of standard treatments such as IFN-β-1a or -1b and glatiramer acetate on the expression of chemokine receptors. Our knowledge in this important field of research is quite restricted due to reports using small patient sets studied in cross-sectional fashion, and reporting directly conflicting results. Zang and coworkers showed that in vitro exposure of T-cells to IFN-β-1a selectively inhibited mRNA expression for MIP-1α/CCL3, RANTES/CCL5 and CCR5 [121]. In this study, T-cell expression of CCR5 was significantly reduced in MS patients treated with IFN-β-1a, compared to untreated MS patients and reduction of CCR5 surface expression was correlated with decreased T-cell migration toward MIP-1α/CCL3 and RANTES/CCL5. However, in a recent study by Wandinger and colleagues, using DNA microarrays, CCR5 gene expression was found to be upregulated in peripheral blood mononuclear cells cultured for 24 h in the presence of IFN-β-1b. Additionally, peripheral blood mononuclear cells of MS patients treated with IFN-β-1b for 6 months showed an upregulation of CCR5 gene expression, compared to baseline expression [122]. Therefore, our understanding of effects of interferons and glatiramer acetate on the chemokine system in MS patients remains quite immature and will undoubtedly be a fertile field for study.

3.5. Analysis of Chemokines and Chemokine Receptors in MS Autopsy Brain Sections

The results discussed above derived from body fluids were extended by immunohistochemical studies of autopsy brain sections from patients with MS (Tables 1 and 2). We found CXCR3+ cells in the perivascular space around more than 99% of vessels within active demyelinating MS lesions [102]. In serial sections, immunohistochemistry for CD3 and CXCR3 showed identical distribution and morphology of immunoreactive cells (Fig. 1), indicating that T-cells in MS brain lesions, as in MS CSF, were in the majority CXCR3+. Balashov and coworkers also detected large numbers of CXCR3+ perivascular cells in MS lesions [113]. Compatible findings were also reported by Simpson and colleagues [123]. Expression of one of the CXCR3 ligands, IP-10/CXCL10, was associated with astrocytes and perivascular astrocytic processes within MS lesions [102,113,123]. Taken together, these results suggested that interactions between CXCR3 and its ligand IP-10/CXCL10 might mediate trafficking, retention or activation of T-cells in MS lesions.

Immunohistochemistry of lesional material indicated that relatively few perivascular cells were CCR5+ but that a large population of activated phagocytic macrophages expressed CCR5 (Fig. 2) [102]. These results suggested that the interactions between RANTES/CCL5 and CCR5 on monocytes and macrophages might be implicated in recruitment and activation of effectors in MS. Simpson and colleagues also noted CCR5 expression by macrophages, but additionally reported CCR2 and CCR3 expression by these cells; immunoreactivity for β-chemokine receptors was also noted on astrocytes in these studies [124]. In a recent study from our laboratory, the expression of CCR1 and CCR5 on mononuclear phagocytes was examined in relation to demyelinating activity and spatial distribution in MS lesions. Within early-active lesions hematogenous monocytes were CCR1+/CCR5+. Interestingly, CCR1 and CCR5 expression on monocytes and microglial cells was activation-dependent, with no CCR1, but abundant CCR5 expression found on mononuclear phagocytes in completely demyelinated inactive lesions [115].

Several groups have focused on the production of β-chemokines within MS lesions, with generally consistent findings. Simpson et al., McManus

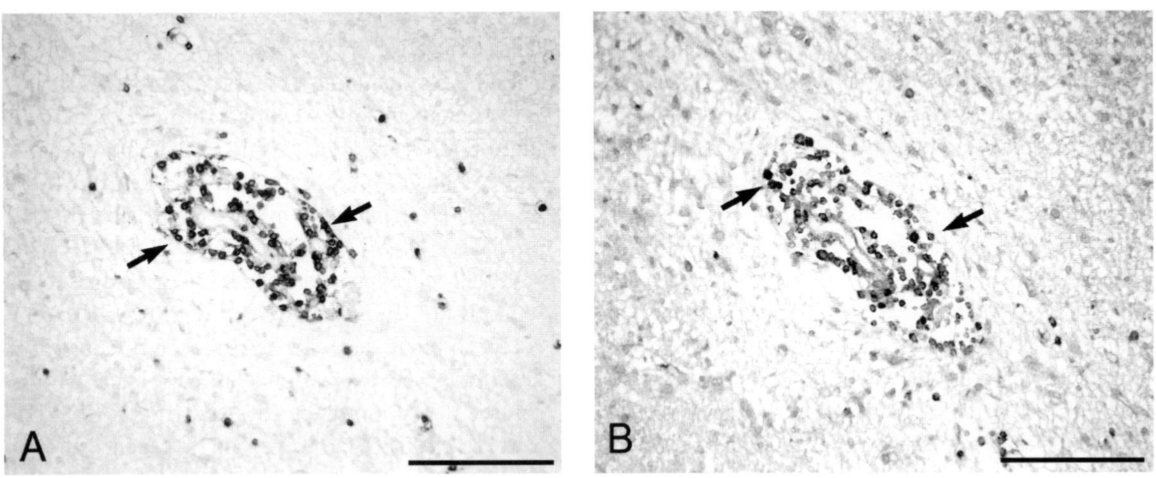

Figure 1. Similar morphology and distribution of CD3+ and CXCR3+ cells in MS lesions. Formalin-fixed, paraffin-embedded serial sections (7–10 μm) from an actively demyelinating MS lesion were analyzed by immunohistochemistry for CD3 (A) and CXCR3 (B), as described [102]. Immunoreactive cells (arrows) are abundant in perivascular aggregates, and are similar in morphology and distribution, indicating that the majority of CD3+ T-cells in MS lesions express CXCR3. Bar indicates 100 μm.

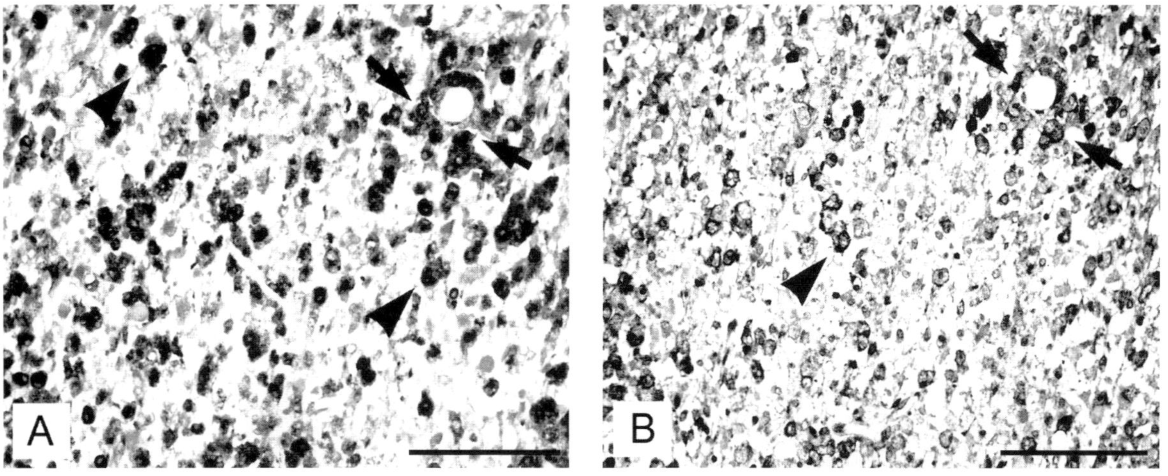

Figure 2. Monocytes and phagocytic macrophages in MS lesions express CCR5. Formalin-fixed, paraffin-embedded serial sections (7–10 μm) from an actively demyelinating MS lesion were analyzed by immunohistochemistry for CD68 (A) and CCR5 (B), as described [102]. Immunoreactive cells are detected in perivascular aggregates (arrows) and in the parenchyma (arrowheads). CD68-positive perivascular cells exhibit the appearance of activated monocytes, while the parenchymal cells display the morphology of phagocytic macrophages. Both of these cell populations are immunoreactive with CCR5 antibodies. Bar indicates 100 μm.

and co-workers and Van Der Voorn et al. have all described MCP-1/CCL2-immunoreactive astrocytes within MS lesions [125–127]. McManus et al. further described the production of MCP-2/CCL7 and MCP-3/CCL8 by glial cells within these lesions [126]. Simpson et al. and Boven et al. detected MIP-1α/CCL3 and MIP-1β/CCL4 immunoreactivity in inflammatory and activated glial cells [127,128]. Hvas et al. showed that RANTES/CCL5 mRNA could be localized to perivascular inflammatory cells in MS lesions [129,130]. How can the data concerning transient decline in CSF MCP-1/CCL2 concentration during MS attacks be reconciled with high-level expression of this chemokine in CNS inflammatory lesions by astrocytes? It would seem most likely that CSF MCP-1/CCL2 reflects primarily the production of this chemokine by leukocytes in the subarachnoid space, with the astrocyte-derived chemokine peptide being immobilized in parenchymal extracellular matrix. A plausible hypothesis would be that CSF leukocytes, exposed to a strongly polarized Th1 environment during MS attacks, would down-regulate MCP-1/CCL2 expression transiently, while production of the chemokine by parenchymal astrocytes would continue, unabated.

In the aggregate, these results appear to support in a general sense the cellular sources of chemokines as described in EAE. In that model, our group showed that IP-10/CXCL10 and MCP-1/CCL2 were produced by astrocytes [131] while Godiska et al. showed that RANTES/CCL5 was produced by perivascular inflammatory cells [132], a finding that we and others subsequently confirmed [77,78]. Berman and colleagues provided compatible results concerning the sources of other chemokines as well [133,134]. In summary, it appears likely that the conceptual scheme of leukocyte trafficking into inflammatory lesions, that was derived in studies using the EAE animal model, may be applicable as well to the human disease MS. This situation provides a favorable circumstance for the evaluation of experimental anti-inflammatory strategies that are targeted to either production of chemokines or to the function of chemokines (via blocking chemokine receptors).

3.6. 'Miles to go before we sleep . . . ' [1]

There is an enormous amount that we do not yet know about the roles of chemokines and their receptors in MS [135]. For the completion of a basic descriptive database, one awaits the performance of studies that document:

[1] Adapted from 'Stopping by woods on a snowy evening' by Robert Frost (http://www.ketzle.com/frost/snowyeve.htm)

(1) serial changes in blood and CSF chemokine levels;
(2) serial analysis of chemokine receptor expression on blood and CSF leukocytes;
(3) establishment of clear-cut relationships between expression of chemokine receptors and demyelinating activity and/or axonal damage in MS lesions;
(4) analysis of possible differential expression of chemokines or chemokine receptors in the variant MS pathologies such as Devic's disease;
(5) relationships of chemokines or chemokine receptor expression to disease activity as defined by MRI scanning.

There are daunting complexities in drawing firm conclusions from these types of data. Certainly, temporal specificities (including possible down-regulation of receptors during trafficking) add a further dimension in the consideration of the meaning of detecting specific chemokine receptor–ligand pairs in lesions. Somewhat surprisingly, there is little information about the effects of transmigration across an endothelial monolayer on chemokine receptor expression. This issue is more complex in the case of CNS and CSF leukocytes, which have crossed the BBB and the associated basement membrane during extravasation and attainment of the perivascular subarachnoid space. Also meriting consideration are the complex interactions of chemokines with selectins, adhesion molecules and extracellular matrix components (including heparin). Finally, it has recently become clear that naturally occurring truncated variants of individual chemokines can be generated by peptidases such as CD26 and matrix metalloproteases, that are present in MS tissues; the products of these reactions can function as potent inhibitors of chemokine receptors. Ultimately, in the complicated milieu of the MS lesion, actions of individual chemokines may promote destructive inflammation or recruitment of vital and protective regulatory cells or even tissue remodeling and repair processes. The final definitive test of theories about chemokine action in MS will be provided by the efficacy (or lack thereof) that emanates from blocking individual receptors.

REFERENCES

1. McDonald WI. Relapse, remission and progression in multiple sclerosis. N Engl J Med 2000;343:1486–1487.
2. Confavreux C, Vukusic S, Moreau T, Adeleine P. Relapses and progression of disability in multiple sclerosis. N Engl J Med 2000;343:1430–1438.
3. Hawkins SA, McDonnell GV. Benign multiple sclerosis? Clinical course, long term follow up, and assessment of prognostic factors. J Neurol Neurosurg Psychiatry 1999; 67:148–152.
4. Bitsch A, Wegener C, da Costa C, Bunkowski S, Reimers CD, Prange HW, Brück W. Lesion development in Marburg's type of acute multiple sclerosis: from inflammation to demyelination. Mult Scler 1999;5:138–146.
5. Wingerchuk DM, Hogancamp WF, O'Brien PC, Weinshenker BG. The clinical course of neuromyelitis optica (Devic's syndrome). Neurology 1999;53:1107–1114.
6. Mandler RN, Davis LE, Jeffery DR, Kornfeld M. Devic's neuromyelitis optica: a clinicopathological study of 8 patients. Ann Neurol 1993;34:162–168.
7. O'Riordan JI, Gallagher HL, Thompson AJ, Howard RS, Kingsley DP, Thompson EJ, McDonald WI, Miller DH. Clinical, CSF, and MRI findings in Devic's neuromyelitis optica. J Neurol Neurosurg Psychiatry 1996;60:382–387.
8. Hickey WF. The pathology of multiple sclerosis: a historical perspective. J Neuroimmunol 1999;98:37–44.
9. Lassmann H. Comparative Neuropathology of Chronic Experimental Allergic Encephalomyelitis and Multiple Sclerosis. Berlin: Springer-Verlag, 1983.
10. Lassmann H, Raine C, Antel J, Prineas J. Immunopathology of multiple sclerosis. Report on an international meeting held at the Institute of Neurology of the University of Vienna. J Neuroimmunol 1998;86:213–218.
11. Kornek B, Lassmann H. Axonal pathology in multiple sclerosis. A historical note. Brain Pathol 1999;9:651–656.
12. Silber E, Sharief MK. Axonal degeneration in the pathogenesis of multiple sclerosis. J Neurol Sci 1999;170:11–18.
13. Kornek B, Storch MK, Weissert R, Wallström E, Stefferl A, Olsson T, Linington C, Schmidbauer M, Lassmann H. Multiple sclerosis and chronic autoimmune encephalomyelitis: a comparative quantitative study of axonal injury in active, inactive, and remyelinated lesions. Am J Pathol 2000;157:267–276.
14. Ferguson B, Matyszak MK, Esiri MM, Perry VH. Axonal damage in acute multiple sclerosis lesions. Brain 1997;120:393–399.
15. Trapp BD, Peterson J, Ransohoff RM, Rudick R, Mörk S, Bö L. Axonal transection in the lesions of multiple sclerosis. N Engl J Med 1998;338:278–285.
16. Mews I, Bergmann M, Bunkowski S, Gullotta F, Brück W. Oligodendrocyte and axon pathology in clinically silent multiple sclerosis lesions. Mult Scler 1998;4:55–62.
17. Trapp BD, Ransohoff RM, Rudick R. Axonal pathology in multiple sclerosis: relationship to neurologic disability. Curr Opin Neurol 1999;12:295–302.
18. Griffin JW, George EB, Chaudhry V. Wallerian degener-

ation in peripheral nerve disease. Baillieres Clin Neurol 1996;5:65–75.
19. Griffiths I, Klugmann M, Anderson T, Yool D, Thomson C, Schwab MH, Schneider A, Zimmermann F, McCulloch M, Nadon N, Nave KA. Axonal swellings and degeneration in mice lacking the major proteolipid of myelin. Science 1998;280:1610–1613.
20. Werner H, Jung M, Klugmann M, Sereda M, Griffiths IR, Nave KA. Mouse models of myelin diseases. Brain Pathol 1998;8:771–793.
21. Waxman SG. Demyelinating diseases — new pathological insights, new therapeutic targets. N Engl J Med 1998; 338:323–325.
22. Sørensen TL, Ransohoff RM. Etiology and pathogenesis of multiple sclerosis. Semin Neurol 1998;18:287–294.
23. Lucchinetti C, Brück W, Parisi J, Scheithauer B, Rodriguez M, Lassmann H. Heterogeneity of multiple sclerosis lesions: implications for the pathogenesis of demyelination. Ann Neurol 2000;47:707–717.
24. Lucchinetti C, Brück W, Parisi J, Scheithauer B, Rodriguez M, Lassmann H. A quantitative analysis of oligodendrocytes in multiple sclerosis lesions. A study of 113 cases. Brain 1999;122:2279–2295.
25. Lucchinetti CF, Brück W, Rodriguez M, Lassmann H. Distinct patterns of multiple sclerosis pathology indicates heterogeneity in pathogenesis. Brain Pathol 1996;6:259–274.
26. Storch MK, Stefferl A, Brehm U, Weissert R, Wallström E, Kerschensteiner M, Olsson T, Linington C, Lassmann H. Autoimmunity to myelin oligodendrocyte glycoprotein in rats mimics the spectrum of multiple sclerosis pathology. Brain Pathol 1998;8:681–694.
27. Simon JH. From enhancing lesions to brain atrophy in relapsing MS. J Neuroimmunol 1999;98:7–15.
28. Coles AJ, Wing MG, Molyneux P, Paolillo A, Davie CM, Hale G, Miller D, Waldmann H, Compston A. Monoclonal antibody treatment exposes three mechanisms underlying the clinical course of multiple sclerosis. Ann Neurol 1999;46:296–304.
29. Noseworthy JH. Progress in determining the causes and treatment of multiple sclerosis. Nature 1999;399:A40–47.
30. Hogancamp WE, Rodriguez M, Weinshenker BG. Identification of multiple sclerosis-associated genes. Mayo Clin Proc 1997;72:965–976.
31. Oksenberg JR, Hauser SL. New insights into the immunogenetics of multiple sclerosis. Curr Opin Neurol 1997;10:181–185.
32. Sadovnick AD, Dyment D, Ebers GC. Genetic epidemiology of multiple sclerosis. Epidemiol Rev 1997;19:99–106.
33. Ebers GC, Dyment DA. Genetics of multiple sclerosis. Semin Neurol 1998;18:295–299.
34. Chataway J, Feakes R, Coraddu F, Gray J, Deans J, Fraser M, Robertson N, Broadley S, Jones H, Clayton D, Goodfellow P, Sawcer S, Compston A. The genetics of multiple sclerosis: principles, background and updated results of the United Kingdom systematic genome screen. Brain 1998;121:1869–1887.
35. Compston A. The genetic epidemiology of multiple sclerosis. Philos Trans R Soc Lond B Biol Sci 1999;354: 1623–1634.
36. Weiner HL. A 21 point unifying hypothesis on the etiology and treatment of multiple sclerosis. Can J Neurol Sci 1998;25:93–101.
37. Dal Canto M, Melvold R, Kim B, Miller S. Two models of multiple sclerosis: experimental allergic encephalomyelitis (EAE) and Theiler's murine encephalomyelitis virus (TMEV) infection. A pathological and immunological comparison. Microsc Res Tech 1995;32:215–229.
38. Tsunoda I, Fujinami RS. Two models for multiple sclerosis: experimental allergic encephalomyelitis and Theiler's murine encephalomyelitis virus. J Neuropathol Exp Neurol 1996;55:673–686.
39. Weinshenker BG. Epidemiology of multiple sclerosis. Neurol Clin 1996;14:291–308.
40. Ebers GC, Sadovnick AD. The role of genetic factors in multiple sclerosis susceptibility. J Neuroimmunol 1994; 54:1–17.
41. Martino G, Hartung HP. Immunopathogenesis of multiple sclerosis: the role of T cells. Curr Opin Neurol 1999; 12:309–321.
42. Waksman BH. Demyelinating disease: evolution of a paradigm. Neurochem Res 1999;24:491–495.
43. Steinman L. Multiple sclerosis: a coordinated immunological attack against myelin in the central nervous system. Cell 1996;85:299–302.
44. Asensio VC, Campbell IL. Chemokines in the CNS: plurifunctional mediators in diverse states. Trends Neurosci 1999;22:504–512.
45. Von Andrian UH, Chambers JD, McEnvoy LM, Bargatze RF, Arfors KE, Butcher EC. Two-step model of leukocyte–endothelial cell interaction in inflammation: distinct roles for LECAM-1 and the leukocyte beta 2 integrins in vivo. Proc Natl Acad Sci USA 1991;88:7538–7542.
46. Butcher E. Leukocyte–endothelial cell recognition: three (or more) steps to specificity and diversity. Cell 1991;67: 1033–1036.
47. Schall TJ, Bacon KB. Chemokines, leukocyte trafficking, and inflammation. Curr Opin Immunol 1994;6:865–873.
48. Springer TA. Traffic signals for lymphocyte recirculation and leukocyte emigration: the multistep paradigm. Cell 1994;76:301–314.
49. Butcher EC, Picker LJ. Lymphocyte homing and homeostasis. Science 1996;272:60–66.
50. Premack BA, Schall TJ. Chemokine receptors: gateways to inflammation and infection. Nat Med 1996;2:1174–1178.
51. Foxman EF, Campbell JJ, Butcher EC. Multistep navigation and the combinatorial control of leukocyte chemotaxis. J Cell Biol 1997;139:1349–1360.
52. Baggiolini M. Chemokines and leukocyte traffic. Nature 1998;392:565–568.
53. Zlotnick A, Yoshie O. Chemokines: a new classification system and their role in immunity. Immunity 2000;12: 121.
54. Murphy PM, Baggiolini M, Charo IF, Hebert CA, Horuk R, Matsushima K, Miller LH, Oppenheim JJ, Power CA. International Union of Pharmacology, XXII. Nomencla-

ture for chemokine receptors. Pharmacol Rev 2000;52: 145–176.
55. Murphy P. The molecular biology of leukocyte chemoattractant receptors. Annu Rev Immunol 1994;12:593–633.
56. Kelvin D, Michiel D, Johnston J, Lloyd A, Sprenger H, Oppenhein J, Wang J-M. Chemokines and serpentines: the molecular biology of chemokine receptors. J Leukoc Biol 1993;54:604–612.
57. Horuk R. Molecular properties of the chemokine receptor family. Trends Pharmacol Sci 1994;15:159–165.
58. Oppenheim JJ, Zachariae C, Mukaida N, Matsushima K. Properties of the novel proinflammatory supergene 'intercrine' cytokine family. Annu Rev Immunol 1991;9: 617–648.
59. Taub DD, Oppenheim JJ. Chemokines, inflammation and the immune system. Ther Immunol 1994;1:229–246.
60. Oppenheim JJ. Overview of chemokines. Adv Exp Med Biol 1993;351:183–186.
61. Cyster JG. Chemokines and cell migration in secondary lymphoid organs. Science 1999;286:2098–2102.
62. Sallusto F, Palermo B, Lenig D, Miettinen M, Matikainen S, Julkunen I, Forster R, Burgstahler R, Lipp M, Lanzavecchia A. Distinct patterns and kinetics of chemokine production regulate dendritic cell function. Eur J Immunol 1999;29:1617–1625.
63. Bacon KB, Oppenheim JJ. Chemokines in disease models and pathogenesis. Cytokine Growth Factor Rev 1998; 9:167–173.
64. Bolton SJ, Perry VH. Differential blood–brain barrier breakdown and leucocyte recruitment following excitotoxic lesions in juvenile and adult rats. Exp Neurol 1998; 154:231–240.
65. Perry VH, Bolton SJ, Anthony DC, Betmouni S. The contribution of inflammation to acute and chronic neurodegeneration. Res Immunol 1998;149:721–725.
66. Perry VH. A revised view of the central nervous system microenvironment and major histocompatibility complex class II antigen presentation. J Neuroimmunol 1998;90: 113–121.
67. Perry VH, Anthony DC, Bolton SJ, Brown HC. The blood–brain barrier and the inflammatory response. Mol Med Today 1997;3:335–341.
68. Castano A, Bell MD, Perry VH. Unusual aspects of inflammation in the nervous system: Wallerian degeneration. Neurobiol Aging 1996;17:745–751.
69. Perry VH, Bell MD, Brown HC, Matyszak MK. Inflammation in the nervous system. Curr Opin Neurobiol 1995;5:636–641.
70. Lawson LJ, Perry VH. The unique characteristics of inflammatory responses in mouse brain are acquired during postnatal development. Eur J Neurosci 1995;7:1584–1595.
71. Fabry Z, Raine C, Hart M. Nervous tissue as an immune compartment: the dialect of the immune response in the CNS. Immunol Today 1994;15:218–224.
72. Bell MD, Perry VH. Adhesion molecule expression on murine cerebral endothelium following the injection of a proinflammagen or during acute neuronal degeneration. J Neurocytol 1995;24:695–710.
73. Sperk G, Lassmann H, Baran H, Kish S, Seitelberger F. Kainic acid induced seizures: neurochemical and histopathological changes. Neuroscience 1983;10:1301–1315.
74. Popovich P, Wei P, Stokes B. Cellular inflammatory response after spinal cord injury in Sprague–Dawley and Lewis rats. J Comp Neurol 1997;377:443–464.
75. McTigue D, Tani M, Kravacic K, Chernosky A, Kelner G, Maciejewski D, Maki R, Ransohoff RM, Stokes B. Selective chemokine mRNA accumulation in the rat spinal cord after contusion injury. J Neurosci Res 1998; 53:368–376.
76. Kielian T, Hickey WF. Proinflammatory cytokine, chemokine, and cellular adhesion molecule expression during the acute phase of experimental brain abscess development. Am J Pathol 2000;157:647–658.
77. Miyagishi R, Kikuchi S, Takayama C, Inoue Y, Tashiro K. Identification of cell types producing RANTES, MIP-1α and MIP-1β in rat experimental autoimmune encephalomyelitis by in situ hybridization. J Neuroimmunol 1997;77:17–26.
78. Glabinski A, Tani M, Strieter R, Tuohy V, Ransohoff RM. Synchronous synthesis of α- and β-chemokines by cells of diverse lineage in the central nervous system of mice with relapses of experimental autoimmune encephalomyelitis. Am J Pathol 1997;150:617–630.
79. Ransohoff RM, Glabinski A, Tani M. Chemokines in immune-mediated inflammation of the central nervous system. Cytokine Growth Factor Rev 1996;7:35–46.
80. Tani M, Glabinski AR, Tuohy VK, Stoler MH, Estes ML, Ransohoff RM. In situ hybridization analysis of glial fibrillary acidic protein mRNA reveals evidence of biphasic astrocyte activation during acute experimental autoimmune encephalomyelitis. Am J Pathol 1996;148: 889–896.
81. Hayashi M, Luo Y, Laning J, Strieter RM, Dorf ME. Production and function of monocyte chemoattractant protein-1 and other beta-chemokines in murine glial cells. J Neuroimmunol 1995;60:143–150.
82. Majumder S, Zhou Z-HL, Ransohoff RM. Transcriptional regulation of chemokine gene expression in astrocytes. J Neurosci Res 1996;45:758–769.
83. Majumder S, Zhou LZ, Chaturvedi P, Babcock G, Aras S, Ransohoff RM. p48/STAT-1alpha-containing complexes play a predominant role in induction of IFN-gamma-inducible protein, 10 kDa (IP-10) by IFN-gamma alone or in synergy with TNF-alpha. J Immunol 1998; 161:4736–4744.
84. Glabinski AR, Krakowski M, Han Y, Owens T, Ransohoff RM. Chemokine expression in GKO mice (lacking interferon-gamma) with experimental autoimmune encephalomyelitis. J Neurovirol 1999;5:95–101.
85. Luster AD, Unkeless JC, Ravetch JV. Gamma-interferon transcriptionally regulates an early-response gene containing homology to platelet proteins. Nature 1985;315: 672–676.
86. Glabinski A, Tani M, Tuohy VK, Tuthill R, Ransohoff RM. Central nervous system chemokine gene expression follows leukocyte entry in acute murine experimental autoimmune encephalomyelitis. Brain Behav Immun 1996; 9:315–330.
87. Glabinski AR, Tuohy VK, Ransohoff RM. Expression of chemokines RANTES, MIP-1alpha and GRO-alpha

correlates with inflammation in acute experimental autoimmune encephalomyelitis. Neuroimmunomodulation 1998;5:166–171.
88. Hickey WF. Migration of hematogenous cells through the blood–brain barrier and the initiation of CNS inflammation. Brain Pathol 1991;1:97–105.
89. Hickey WF, Hsu BL, Kimura H. T-lymphocyte entry into the central nervous system. J Neurosci Res 1991;28:254–260.
90. Ransohoff RM, Bacon KB. Chemokine receptor antagonism as a new therapy in multiple sclerosis. Expert Opin Invest Drugs 2000;5:1079–1097.
91. Zhang GX, Baker CM, Kolson DL, Rostami AM. Chemokines and chemokine receptors in the pathogenesis of multiple sclerosis. Mult Scler 2000;6:3–13.
92. Woodroofe N, Cross AK, Harkness K, Simpson JE. The role of chemokines in the pathogenesis of multiple sclerosis. Adv Exp Med Biol 1999;468:135–150.
93. Ransohoff RM. Mechanisms of inflammation in MS tissue: adhesion molecules and chemokines. J Neuroimmunol 1999;98:57–68.
94. Fiten P, Vandenbroeck K, Dubois B, Van Coillie E, Nelissen I, Van Damme J, Ligers A, Hillert J, Andersson M, Olsson T, Opdenakker G. Microsatellite polymorphisms in the gene promoter of monocyte chemotactic protein-3 and analysis of the association between monocyte chemotactic protein-3 alleles and multiple sclerosis development. J Neuroimmunol 1999;95:195–201.
95. Teuscher C, Butterfield RJ, Ma RZ, Zachary JF, Doerge RW, Blankenhorn EP. Sequence polymorphisms in the chemokines Scya1 (TCA-3), Scya2 (monocyte chemoattractant protein (MCP)-1), and Scya12 (MCP-5) are candidates for eae7, a locus controlling susceptibility to monophasic remitting/nonrelapsing experimental allergic encephalomyelitis. J Immunol 1999;163:2262–2266.
96. Fife BT, Huffnagle GB, Kuziel WA, Karpus WJ. CC chemokine receptor 2 is critical for induction of experimental autoimmune encephalomyelitis. J Exp Med 2000; 192:899–906.
97. Rottman JB, Slavin AJ, Silva R, Weiner HL, Gerard CG, Hancock WW. Leukocyte recruitment during onset of experimental allergic encephalomyelitis is CCR1 dependent. Eur J Immunol 2000;30:2372–2377.
98. Bennetts BH, Teutsch SM, Buhler MM, Heard RN, Stewart GJ. The CCR5 deletion mutation fails to protect against multiple sclerosis. Hum Immunol 1997;58: 52–59.
99. Sellebjerg F, Madsen HO, Jensen CV, Jensen J, Garred P. CCR5 Δ32, matrix metalloproteinase-9 and disease activity in multiple sclerosis. J Neuroimmunol 2000;102: 98–106.
100. Barcellos LF, Schito AM, Rimmler JB, Vittinghoff E, Shih A, Lincoln R, Callier S, Elkins MK, Goodkin DE, Haines JL, Pericak-Vance MA, Hauser SL, Oksenberg JR. CC-chemokine receptor 5 polymorphism and age of onset in familial multiple sclerosis. Immunogenetics 2000;51:281–288.
101. Cananzi AR, Ferro-Milone F, Grigoletto F, Toldo M, Meneghini F, Bortolon F, D'Andrea G. Relevance of platelet factor four (PF4) plasma levels in multiple sclerosis. Acta Neurol Scand 1987;76:79–85.
102. Sørensen T, Tani M, Jensen J, Pierce V, Lucchinetti C, Folcik V, Qin S, Rottman J, Sellebjerg F, Strieter R, Frederiksen J, Ransohoff RM. Expression of specific chemokines and chemokine receptors in the central nervous system of multiple sclerosis patients. J Clin Invest 1999;106:807–815.
103. Mahad D, Howell S, Woodroofe MN. Chemokines in the cerebrospinal fluid of patients with multiple sclerosis [abstract]. Ann Neurol 2000;48:452
104. Sindern E, Niederkinkhaus Y, Henschel M, Ossege LM, Patzold T, Malin JP. Differential release of β-chemokines in serum and CSF of patients with relapsing–remitting multiple sclerosis. Acta Neurol Scand 2001;104:88–91.
105. Prat E, Baron PL, Meda L, Galimberti D, Livraghi S, Ardolino G, Clerici R, Scarpini E, Scarlato G. Interferon-γ-inducible protein-10 is increased in the cerebrospinal fluid of patients with multiple sclerosis [abstract]. Mult Scler 2000;5:83.
106. Franciotta D, Martino G, Zardini E, Furlan R, Bergamaschi R, Andreoni L, Cosi V. Serum and SCF levels of MCP-1 and IP-10 in multiple sclerosis patients with acute and stable disease and undergoing immunomodulatory therapies. J Neuroimmunol 2001;115:192–198
107. Lahrtz F, Piali L, Spanaus KS, Seebach J, Fontana A. Chemokines and chemotaxis of leukocytes in infectious meningitis. J Neuroimmunol 1998;85:33–43.
108. Sprenger H, Rosler A, Tonn P, Braune HJ, Huffmann G, Gemsa D. Chemokines in the cerebrospinal fluid of patients with meningitis. Clin Immunol Immunopathol 1996;80:155–161.
109. Miyagishi R, Kikuchi S, Fukazawa T, Tashiro K. Macrophage inflammatory protein-1α in the cerebrospinal fluid of patients with multiple sclerosis and other inflammatory neurological diseases. J Neurol Sci 1995;129:223–227.
110. Kivisäkk P, Teleshova N, Özenci V, Huang Y, Matusevicius D, Fredrikson S, Söderström M, Link H. No evidence for elevated numbers of mononuclear cells expressing MCP-1 and RANTES mRNA in blood and CSF in multiple sclerosis. J Neuroimmunol 1998;91:108–112.
111. Sørensen T, Sellebjerg F, Jensen C, Strieter R, Ransohoff RM. Interferon-γ inducible protein, 10 kDa (IP-10) and monocyte chemoattractant protein (MCP)-1: differential involvement in intrathecal inflammation in multiple sclerosis. Eur J Neurol 2001;8:1–8.
112. Iarlori C, Reale M, Lugaresi A, De Luca G, Bonanni L, Di Iorio A, Feliciani C, Conti P, Gambi D. RANTES production and expression is reduced in relapsing–remitting multiple sclerosis patients treated with interferon-beta-1b. J Neuroimmunol 2000;107:100–107.
113. Balashov KE, Rottman JB, Weiner HL, Hancock WW. CCR5(+) and CXCR3(+) T cells are increased in multiple sclerosis and their ligands MIP-1alpha and IP-10 are expressed in demyelinating brain lesions. Proc Natl Acad Sci USA 1999;96:6873–6878.
114. Misu T, Onodera H, Fujihara K, Matsushima K, Yoshie O, Okita N, Takase S, Itoyama Y. Chemokine receptor expression on T cells in blood and cerebrospinal fluid at relapse and remission of multiple sclerosis: imbalance of Th1/Th2-associated chemokine signaling. J Neuroimmunol 2001;114:207–212.

115. Trebst C, Sørensen T, Kivisäkk P, Cathcart M, Hesselgesser J, Horuk R, Sellebjerg F, Lassmann H, Ransohoff RM. CCR1+/CCR5+ mononuclear phagocytes accumulate in the central nervous system of patients with multiple sclerosis. Am J Pathol 2001;159:1701–1710.
116. Zang YC, Samanta AK, Halder JB, Hong J, Tejada-Simon MV, Rivera VM, Zhang JZ. Aberrant T cell migration toward RANTES and MIP-1 α in patients with multiple sclerosis. Overexpression of chemokine receptor CCR5. Brain 2000;123:1874–1882.
117. Strunk T, Bubel S, Mascher B, Schlenke P, Kirchner H, Wandinger K-P. Increased numbers of CCR5+ interferon-γ- and tumor necrosis factor-α-secreting T lymphocytes in multiple sclerosis patients. Ann Neurol 2000;47:269–273.
118. Calabresi PA, Martin R, Jacobson S. Chemokines in chronic progressive neurological diseases: HTLV-1 associated myelopathy and multiple sclerosis. J Neurovirol 1999;5:102–108.
119. D'Amico G, Frascaroli G, Bianchi G, Transidico P, Doni A, Vecchi A, Sozzani S, Allavena P, Mantovani A. Uncoupling for inflammatory chemokine receptors by IL-10: generation of functional decoys. Nat Immunol 2000; 1:387–391.
120. Sørensen T, Sellebjerg F. Distinct chemokine receptor and cytokine expression profile in secondary progressive MS. Neurology 2001;57:1371–1376.
121. Zang YC, Halder JB, Samanta AK, Hong J, Rivera VM, Zhang JZ. Regulation of chemokine receptor CCR5 and production of RANTES and MIP-1alpha by interferon-beta. J Neuroimmunol 2001;112:174–180.
122. Wandinger KP, Sturzebecher CS, Bielekova B, Detore G, Rosenwald A, Staudt LM, McFarland HF, Martin R. Complex immunomodulatory effects of interferon-beta in multiple sclerosis include the upregulation of T helper 1-associated marker genes. Ann Neurol 2001;50:349–357.
123. Simpson JE, Newcombe J, Cuzner ML, Woodroofe MN. Expression of the interferon-γ-inducible chemokines IP-10 and mig and their receptor, CXCR3, in multiple sclerosis lesions. Neuropathol Appl Neurobiol 2000;26: 133–142.
124. Simpson J, Rezaie P, Newcombe J, Cuzner ML, Male D, Woodroofe MN. Expression of the β-chemokine receptors CCR2, CCR3 and CCR5 in multiple sclerosis central nervous system tissue. J Neuroimmunol 2000;108:192–200.
125. Van Der Voorn P, Tekstra J, Beelen RH, Tensen CP, Van Der Valk P, De Groot CJ. Expression of MCP-1 by reactive astrocytes in demyelinating multiple sclerosis lesions. Am J Pathol 1999;154:45–51.
126. McManus C, Berman JW, Brett FM, Staunton H, Farrell M, Brosnan CF. MCP-1, MCP-2 and MCP-3 expression in multiple sclerosis lesions: an immunohistochemical and in situ hybridization study. J Neuroimmunol 1998; 86:20–29.
127. Simpson JE, Newcombe J, Cuzner ML, Woodroofe MN. Expression of monocyte chemoattractant protein-1 and other β-chemokines by resident glia and inflammatory cells in multiple sclerosis lesions. J Neuroimmunol 1998; 84:238–249.
128. Boven LA, Montagne L, Nottet HS, De Groot CJ. Macrophage inflammatory protein-1α (MIP-1α), MIP-1β, and RANTES mRNA semiquantification and protein expression in active demyelinating multiple sclerosis (MS) lesions. Clin Exp Immunol 2000;122:257–263.
129. Hvas J, Bernard CC. Molecular detection and quantitation of the chemokine RANTES mRNA in neurological brain. APMIS 1998;106:598–604.
130. Hvas J, McLean C, Justesen J, Kannourakis G, Steinman L, Oksenberg JR, Bernard CC. Perivascular T cells express the pro-inflammatory chemokine RANTES mRNA in multiple sclerosis lesions. Scand J Immunol 1997;46: 195–203.
131. Ransohoff RM, Hamilton TA, Tani M, Stoler MH, Shick HE, Major JA, Estes ML, Thomas DM, Tuohy VK. Astrocyte expression of mRNA encoding cytokines IP-10 and JE/MCP-1 in experimental autoimmune encephalomyelitis. FASEB J 1993;7:592–602.
132. Godiska R, Chantry D, Dietsch G, Gray P. Chemokine expression in murine experimental autoimmune encephalomyelitis. J Neuroimmunol 1995;58:167–176.
133. Hulkower K, Brosnan CF, Aquino DA, Cammer W, Kulshrestha S, Guida MP, Rapoport DA, Berman JW. Expression of CSF-1, c-fms, and MCP-1 in the central nervous system of rats with experimental allergic encephalomyelitis. J Immunol 1993;150:2525–2533.
134. Berman J, Guida M, Warren J, Amat J, Brosnan C. Localization of monocyte chemoattractant peptide-1 expression in the central nervous system in experimental autoimmune encephalomyelitis and trauma in the rat. J Immunol 1996;156:3017–3023.
135. Trebst C, Ransohoff RM. Investigating chemokines and chemokine receptors in multiple sclerosis: opportunities and challenges. Arch Neurol 2001;58:1975–1980.

Universes in Delicate Balance: Chemokines and the Nervous System
R.M. Ransohoff, K. Suzuki, A.E.I. Proudfoot, W.F. Hickey and J.K. Harrison, editors

Chapter 5.2

Chemokines in the Central Nervous System and Alzheimer's Disease

MengQi Xia [1] and Bradley T. Hyman [2]

[1] *Department of Neurosurgery, Children's Hospital, Harvard Medical School, Boston, MA 02115, USA;* [2] *Alzheimer's Research, Department of Neurology, Massachusetts General Hospital, Harvard Medical School, Boston, MA 02114, USA*

1. INTRODUCTION

Alzheimer's disease (AD) is a devastating illness affecting the elderly. It is the most common cause of dementia and the fourth leading cause of death in the United States. Its clinical features include progressive dementia with gradual loss of cognitive function. Its main neuropathological features include neurofibrillary tangles, senile plaques, and loss of neurons and synapses [1]. A marked astrocytosis and microglial activation occurs throughout the cortex. Although this is a disease that has been studied extensively by neuroscientists, the need for input from immunologists has been implicated by the involvement of inflammatory responses in AD pathogenesis including activated glia [2–5], and a promising effect of anti-inflammatory treatment in delaying disease progression [6,7]. Due to the lack of apparent blood–brain barrier breakdown, or the infiltration of leukocytes in AD, the inflammatory response is most likely mediated by resident CNS cells. The demonstration of the constitutive presence of many chemokine receptors and their ligands in resident CNS cells and their upregulation in AD suggests a role for these molecules as important inflammatory mediators in AD and perhaps other neurodegenerative diseases [8–14].

Here, we will firstly summarize the chemokines and receptors so far found in the CNS, and their changes in AD brains (Table 1, Figs. 1–4). Secondly we will describe our most recent findings on the direct effect of some chemokines on neuronal signaling pathways. Thirdly, we present a hypothesis regarding the pathophysiological role of chemokines in the CNS and particularly in AD, and how this delicate balance can be achieved or disrupted under physiological or pathological conditions. Finally, we present the implications of this hypothesis for future therapeutic intervention.

2. CHEMOKINES AND THEIR RECEPTORS SO FAR DETECTED IN THE CNS AND IN AD BRAIN

Chemokines and their receptors detectable in brain tissues or neuronal and glial cultures under pathophysiological conditions by immunohistochemistry, by in situ or by reverse transcriptase-polymerase chain reaction (RT-PCR) are summarized in Table 1. Immunoreactivity for a number of chemokine receptors has also been detected in CNS cells: Three (CXCR2, CXCR3, CXCR4) out of the five CXCRs have been detected on neurons with the expression of CXCR2 being the strongest (Figs. 1 and 2) [8,11,12,15,16]; Two (CCR3 and CCR5) out of eight CCRs have been detected mainly on microglia cells and weakly on neurons (Fig. 3) [9]. Chemokines detected by immunohistochemistry in brain tissue include IP-10 (Fig. 4), MIP-1β in a subpopulation of resting and reactive astrocytes [9,11]; GROα in a subpopulation of neurons [17]; MIP-1α in neurons and weakly on some microglia [9,18]; and MCP-1 in some reactive microglia [19].

AD neuropathology contains two major features: intraneuronal paired helical filament containing neurofibrillary tangles which occur within the neuronal cell body and processes, and complex extracellular

Table 1. Chemokines/receptors in the CNS and in AD

	Human	Rodent
CXCR1	Not detectable [8]	
CXCR2	Cortical neurons and some neuritic plaques in AD [8,12]	Detectable on rat OLP primary culture and OLP cell line [71]
IL-8	Not detectable in AD brains by immunohistochemistry [8]; induced in microglia culture by LPS, IL-1β, TNF-α or bacteria [72,73]; induced in astrocytes by TNF-α and IL-1β [74] or by reduced microenvironmental oxygen pressure [75]; mRNA detectable in neoplastic astrocytes [76]	
GROα	Detectable in some neurons in normal and in AD brains by immunohistochemistry [17]	
CXCR3	Cortical neurons, purkinje cells and motor neurons [11]	
IP-10	Elevated in some reactive astrocytes in AD [11]; present in 79% in the CSF of viral meningitis patients [77]; elevated in CSF during MS attack [80]	Inducible in both astrocytes and microglia culture by LPS, TNF-α and IFN-γ [78] or noninfectious NDV [79]; the most prominently induced chemokine mRNA induced in resident CNS cells in LCM [81,82]
Mig	Not detectable in tissue by histology [11]; elevated in CSF during MS attack [80]	mRNA inducible in rat astrocytes and microglia by IFN-γ [83]
CXCR4	Neurons and microglia [16]	Gene deficient mice showed abnormal cerebellum development [32]
SDF-1	Not detectable [17]	Gene deficient mice showed abnormal cerebellum development [32,33]
CCR1	Not detectable [17]	mRNA detected in mouse astrocyte by PCR [84]
CCR2	Not available	
MCP-1	Reactive microglia of mature senile plaques in AD [19]; microglia of HIV encephalitis [77]; present in 97% in the CSF of viral meningitis patients [77]; CSF levels markedly higher in CMV encephalitis [85]	Protein inducible in rat and mouse glial culture by LPS [78]; mRNA or protein induced in rat CNS cells after focal ischemia [86] and excitotoxic injury [87]; in mouse EAE, accumulated in a striking, transient burst within astrocytes near the inflamed sites [88]; sharply induced in astrocytes near mechanical injury [89]
CCR3	Upregulated in some reactive microglia in AD [9]	
Eotaxin	Not detectable in AD [17]	
CCR5	Upregulated in some reactive microglia in AD [9]	mRNA upregulated in rat focal ischemia model [90]
MIP-1α	Neurons and some microglia [9,18]; mRNA and protein inducible by LPS, TNF-α or IL-1β in human fetal microglia culture [92]	mRNA induced in mouse resident CNS cells after spinal cord injury [91] and in rat CNS cells after focal ischemia [86,93]
MIP-1β	Upregulated in reactive astrocytes of AD [9]; mRNA and protein inducible by LPS, TNF-α or IL-1β in human fetal microglia culture [92]	mRNA induced in mouse resident CNS cells after spinal cord injury [91]
Rantes	Not detectable in AD [9]; microglia of HIV encephalitis [77]	mRNA induced in rat astrocytes and microglia by NDV [79]
CX3CR1	Microglia	Rat microglia by in situ hybridization [94]
Fractalkine	Neurons and reactive astrocytes [17]	mRNA detected in rat neurons by in situ hybridization [94]
Duffy antigen	Purkinje cells in cerebellum [95]	

Unless otherwise specified, all the results of brain tissues were obtained by immunohistochemistry. LCM, lymphocytic choriomeningitis; NDV, the neurotropic paramyxovirus; Newcastle disease virus; MV, Measles virus. MS, multiple sclerosis.

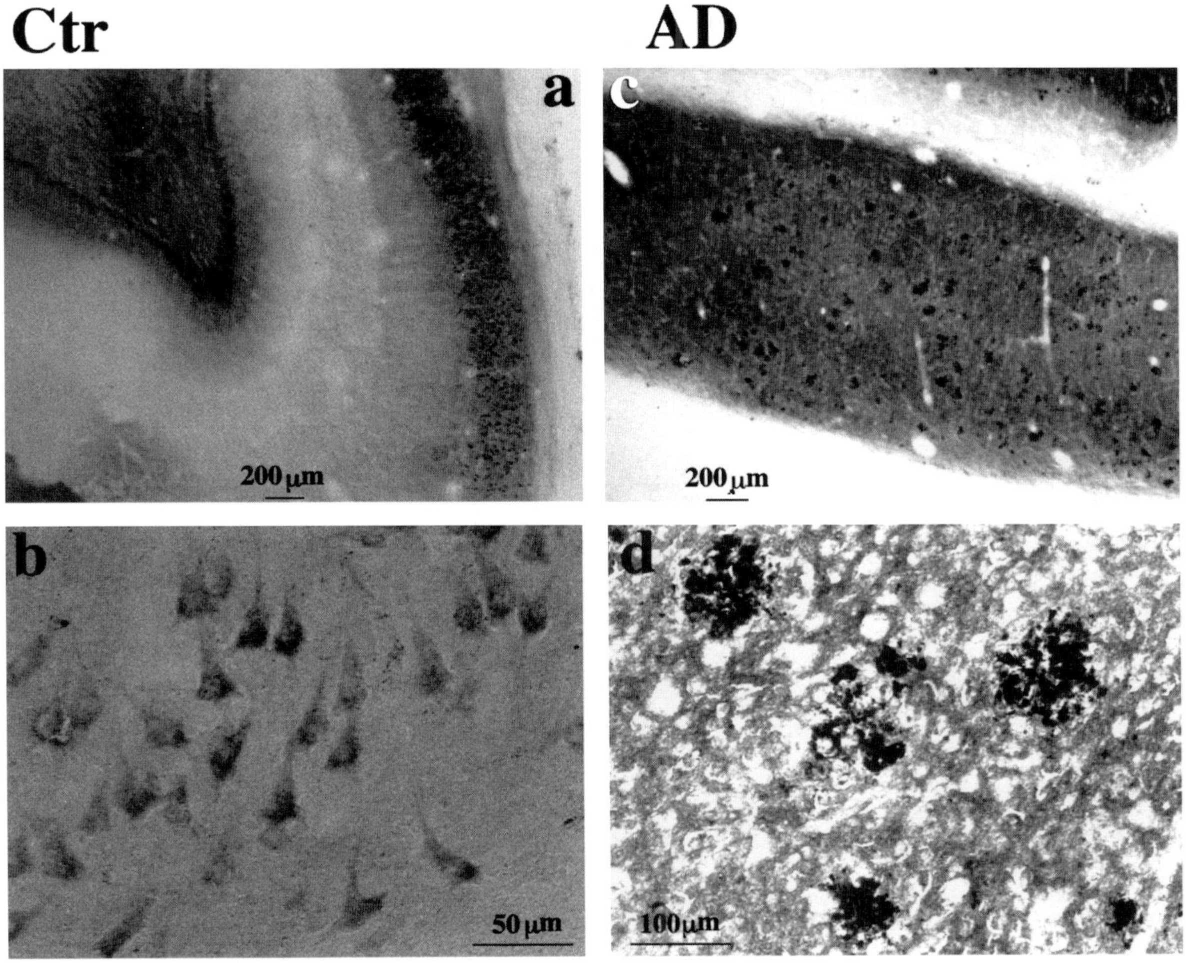

Figure 1. IL-8RB (CXCR2, Mab 6C6) immunoreactivity in (a,b) the hippocampal formation of a 66-year-old male control patient showing that (a) neurons and neuropil of the dentate gyrus and CA region are strongly immunoreactive; (b) pyramidal neurons in CA1 region are clearly stained; in c and d, the hippocampal formation of a 74-year-old male AD patient showing (c) a similar pattern of gray matter staining and an increased immunoreactivity on a subpopulation of plaques; and (d) in the CA1, an increased immunoreactivity was found on a subpopulation of neuritic plaques. These neuritic plaques were also shown by double immunofluorescent staining and confocal microscopic analysis to be positive for APP but not tangles. Modified with permission [8].

lesions called senile plaques. Senile plaques consist of a deposit of Amyloid β(Aβ), a 40–42 amino acid peptide derived from the amyloid precursor protein (APP). Frequently, plaques are surrounded by reactive microglia, reactive astrocytes and several types of dystrophic neurites.

In AD brain, upregulation of CXCR2 expression was observed on dystrophic neurites of senile plaques [8]. Simultaneous work by Horuk and colleagues [12] concurred with this finding. Despite the lack of obvious CXCR3 upregulation in neurons of AD brains (Fig. 2), its ligand IP-10 is found strongly upregulated in reactive astrocytes of these brains (Fig. 4) [11]; CC chemokine receptors CCR3 and CCR5 have been found on microglia of both normal and AD brains (Fig. 3), and their expression is increased on some reactive microglia in AD [9]. Among the CCR3 and CCR5 ligands tested (MIP-1β and MIP-1α, RANTES, eotaxin and MCP-3), only immunoreactivity for MIP-1β and MIP-1α were observed: MIP-1β was predominantly found in a subpopulation of reactive astrocytes which were more widespread and more strongly stained in AD than control brains. MIP-1α stained predominantly in neurons and weakly in some microglia particularly in the white matter of both AD and controls

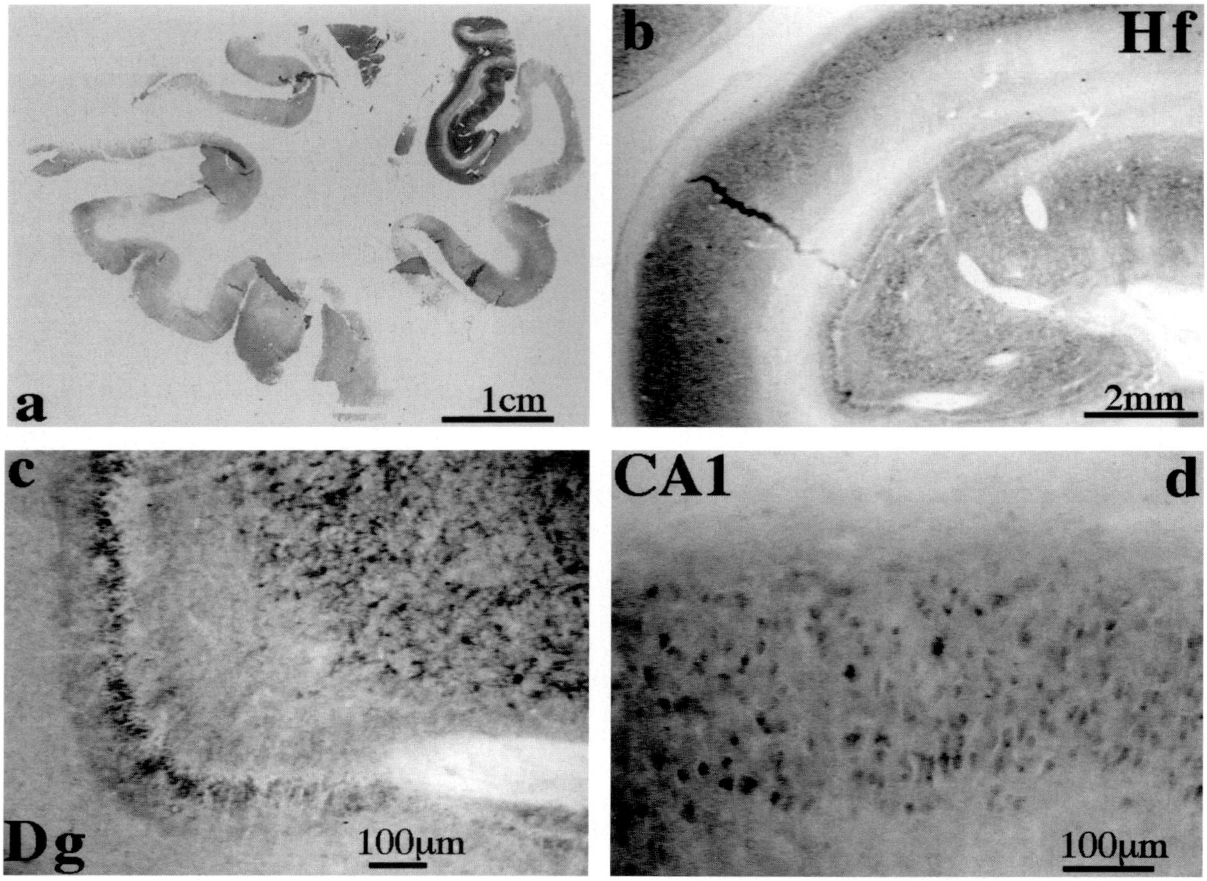

Figure 2. CXCR3 immunoreactivity in cortical and subcortical regions. CXCR3 immunoreactivity is shown at the temporal lobe of a 79-year-old female AD patient to be (a) globally confined to the gray matter of temporal neocortex and hippocampal formation; (b,c) at higher powers on granule cells and their dendrites at molecular layer of dentate gyrus, mossy fibers and CA region neurons; and (b,d) on pyramidal neurons and neuropils of CA1 region. Adopted with permission [11].

[9,18]. Immunoreactivity of MCP-1 (a CCR2 ligand) has also been reported in mature senile plaques and reactive microglia [19].

The present data support that chemokine expression in vivo is more restricted to particular cell types and is under fine regulation [20–23], despite the fact under in vitro conditions, many cell types have the capacity to express chemokines. As summarized in Table 1, various stimuli and stress conditions such as mitogens, free radicals, amyloid, hypoxia, viral infections, cytokines and even chemokines themselves, either alone or in combination, can stimulate different chemokines from different cell types.

3. PHYSIOLOGICAL ROLE OF CHEMOKINES IN THE CNS

3.1. Neurotrophic Factors or Neuronal Modulators

Recently, we have observed a remarkable triggering effect of GROα/KC for signaling pathways involving extracellular signal regulated kinase (ERK1/2) and PI-3 kinase in mouse cortical neurons [17]. The triggering efficacy of GROα/KC appeared to be only slightly weaker than that of BDNF, but stronger than that of bFGF or NGF. Moreover, the triggering effect by GROα/KC appeared to be concentration dependent in the tested range (1.25–100 nM) and more long lasting; whereas the triggering effect by BDNF

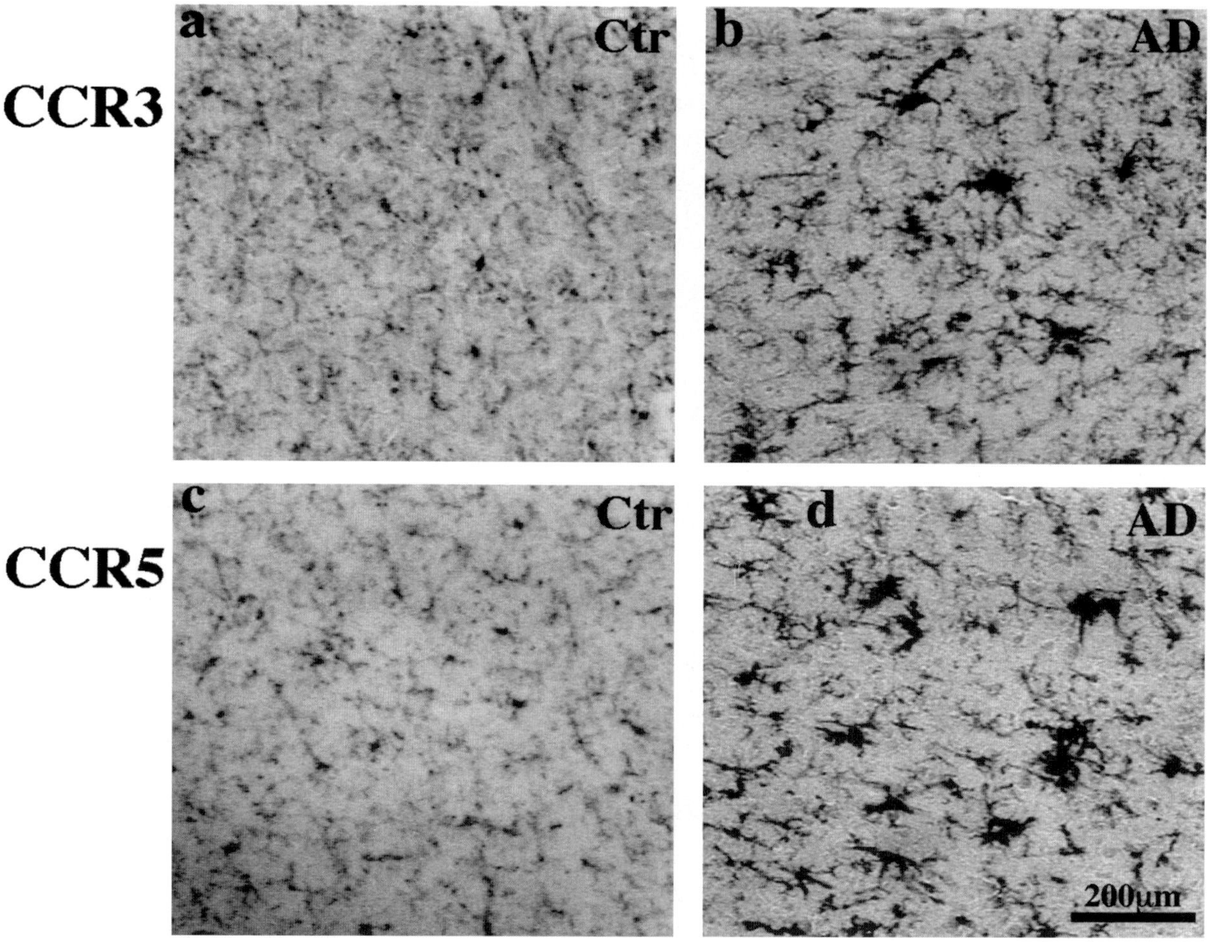

Figure 3. CCR3 (Mab 7B11) and CCR5 (Mab 3A9) immunoreactivity in the inferior temporal lobes of a 58-year-old control patient and an 81-year-old AD patient with duration of illness for 12 years. CCR3 (a,b) and CCR5 (c,d) immunoreactivities are clearly seen on microglia of both cases. In the control case, the majority of cells stained are resting microglia; while in the AD brain both resting and reactive microglia cells are clearly stained, some reactive microglia appears in clusters. All images have the same scale of magnification. Adapted with permission [9].

reached a plateau at a much lower concentration (0.7–7 nM) and did not appear to be as long lasting. Furthermore, in control human brains, low levels of GROα immunoreactivity have been detected in a subpopulation of neurons. Therefore there is a possibility that a physiologically relevant ligand such as GROα, or perhaps other CXCR2 ligands, are potent autocrine growth factors for neuronal survival and even regeneration. The growth factor/protective effects of CXCR2 ligands have been demonstrated in neurons and other cell types: IL-8 in rat hippocampal neurons [24]; and GROα and IL-8 in tumor growth [25,26]. In addition to GROα, we have also observed in mouse cortical neurons the activating effect of CXCR3 ligands IP-10 and Mig [11] and CXCR4 ligand SDF-1α [17] for ERK1/2 pathway. Interestingly, the cell culture conditions for the CXCR2, CXCR3 and CXCR4 ligands to reach their highest potency appeared to differ [17], suggesting that these chemokines may exert different pathophysiological effects. Neuroprotective effects of other chemokines, such as MDC, Rantes, SDF-1α and soluble fractalkine [27] have also been documented. Other neuronal modulatory effects reported include: the modulation of neurotransmitter release in the rat cerebellum by the CXCR2 ligand GROβ [28], and an increase of intracellular Ca^{2+} in rat hippocampal neurons by IL-8 (a CXCR2 ligand), SDF-1 (a CXCR4 ligand),

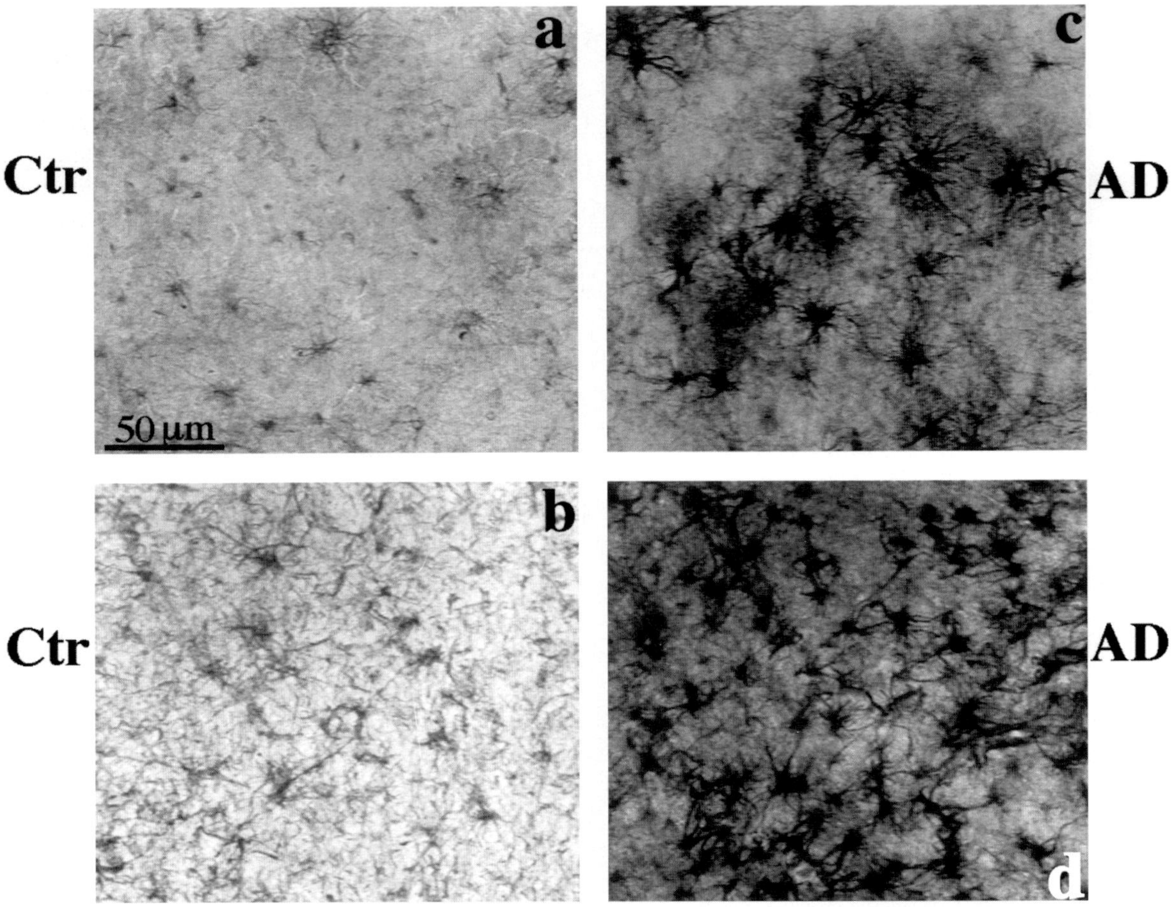

Figure 4. IP-10 (4G3) immunoreactivity in astrocytes of the temporal neocortices of both control and AD patients. a and b represent a 69-year-old patient with cerebral amyloid angiopathy (CAA) and a 94-year-old control patient, respectively; c and d represent an 81-year-old and a 82-year-old AD patients, respectively. In these non-AD control brains, IP-10 immunoreactivity was detected in a relatively small number of astrocytes, most of which appeared in a resting state, and the level of its expression appeared modest. However, in the AD cases, significantly larger numbers of astrocytes were strongly immunoreactive for IP-10. These strongly IP-10$^+$ cells were apparently in an activated state and were widespread at hippocampal formation and neocortex, and many of them were distributed in clusters and the overall background staining for IP-10 appeared to be higher in AD brains than that of controls. Images were taken under similar conditions. Scale of magnification is the same for all the images. Adapted with permission [11].

fractalkine (a CX3CR1 ligand) and MDC (a CCR4 ligand) [27].

3.2. Coordinators for Neuron–Neuron, Neuron–Glia and Glia–Glia Interactions

The detection of chemokine receptors on one cell type, and the ligand on others (i.e. both CXCR2 and GROα on neurons; CXCR3 on neurons, IP-10 on astrocytes; CCR5 on microglia, MIP-1β on astrocytes and MIP-1α on neurons; CX3CR1 on microglia, fractalkine on neurons and astrocytes) suggests a role for chemokines as mediators for neuron–neuron, neuron–glia and glia–glia interactions. This pattern of chemokine/receptor expression makes possible a mechanism for maintaining a state of homeostasis and for a coordinated response in events of various injuries. Indeed, the chemotactic ability of both microglia and astrocytes to certain chemokines has been demonstrated [29–31].

3.3. Neuronal Patterning

Other evidence for a physiological role of chemokines and chemokine receptors in the CNS comes from data that mice deficient in CXCR4 or its ligand SDF-1, showed an altered pattern of cerebellar development [32,33].

3.4. Regulator of Microvasculature

The angiogenic effect of ELR containing chemokines (such as IL-8, GROα, β and γ, ENA-78, etc.), and the anti-angiogenic activity of the ELR-negative chemokines, such as IP-10, Mig, SDF-1 have been reported [34–36]. Although not directly assayed in the CNS, it is very likely that chemokines also function as a regulator of microvasculature. Therefore, the change of chemokine expression in the CNS may ultimately affect the blood supply to CNS cells.

The above data suggest that chemokines may play many important roles in the CNS to maintain homeostasis and to participate in restorative events. It remains to be tested by in vivo studies how potent their effects can be in CNS physiology, and how their effects are modulated in relation to each other.

4. PATHOLOGICAL ROLE OF CHEMOKINES IN THE CNS

Given their classical role in the immune system, chemokines very likely constitute part of the local immune system in the brain, functioning as mediators and amplifiers of inflammatory responses during pathological events.

We have most recently observed in AD brains an apparently coordinated upregulation of IP-10 and MIP-1β in reactive astrocytes [11]. This coordinated upregulation of chemokines suggests an ongoing chemokine activation, therefore an ongoing chronic inflammation in AD brains. In addition, a strong upregulation of CXCR2 in a subpopulation of neuritic plaques in some AD brains may represent an aberrant regenerative phenomenon. Our data on the strong effect of GROα, IP-10, and Mig on ERK1/2 activation also suggests some detrimental neuronal effects if the stimulation is imbalanced. The followings are our predictions of the pathological roles of chemokines.

4.1. Role in ERK1/2 Activation, Tau Phosphorylation, and Tangle Formation

Many kinases (ERK, PKA, GSK3b, cdk2, cdk5, SAPK) have been suggested to be involved in tau phosphorylation and tangle formation in AD [37–46]. Activation of ERK1/2 [38,47–49] and PKA [50–52] have been considered particularly important for this process. Our recent demonstration of a direct triggering effect of CXCR2 ligand GROα for tau phosphorylation in vitro (detectable by the commonly used tangle detecting antibody AT8) suggests a role for imbalanced chemokine upregulation in tangle formation [17].

4.2. Role in Aberrant Neuronal Sprouting and in Neuronal Injuries

Abortive neuronal sprouting has been found in AD brain [53,54]. Upregulation of CXCR2 on dystrophic neurites has been found to correlate with that of APP expression [8]. Since APP-positive neuritic plaques are believed to represent aberrant regenerating neurites [55], dystrophic neurites with upregulated CXCR2 expression may represent an aberrant regenerating phenomenon, and CXCR2 ligands, other chemokines and growth promoting cytokines elevated in AD brain may all be responsible for this aborted compensatory response.

Despite the role of ERK1/2 activation in promoting neuronal growth, differentiation, survival, as well as long-term memory [56,57], sustained ERK1/2 activation has been associated with neuronal death in several experimental paradigms: i.e. death of hippocampal neurons in a cell-culture model of seizure [58], okadaic acid-induced neuronal injury of hippocampal slice cultures [59] as well as neuronal injury after focal ischemia [60]. The potent effect of certain chemokines on neuronal ERK1/2 pathways, and their upregulation in AD brains may potentially lead to imbalanced ERK activation and therefore neuronal injury.

In AD brains, the unexpected appearance of certain cell cycle markers has been found in neurons [61–63]. Some appear to predict sites of neuronal loss. It has been proposed that terminally differentiated neurons in AD brains may have been forced into the cell cycle which may eventually lead to cell death. Cytokines are known to play an important role in cell cycle control in many cell types [64,65]. However, data on chemokines in cell cycle control has been scarce. MIP-1α has been shown to inhibit the cell cycle progression of both primitive hemopoietic and epidermal progenitor cells [66]; effects on neuronal cell cycle have not yet been tested.

4.3. Effect on Cerebral Blood Circulation

In AD, a reduction of blood flow can be observed even at the early stage of the disease by PET scan. Due to the role of chemokines as a regulator of angiogenesis, the strong upregulation of the angiostatic chemokine IP-10 in reactive astrocytes of AD brain [11] implies a pathological effect of an imbalanced chemokine expression through the disturbance of microvasculature formation and blood flow, resulting in relative hypoxia. Hypoxia can then be another stress factor to further aggravate the pathogenesis.

5. THE DELICATE BALANCE OF CHEMOKINES IN THE CNS

Although the exact role of chemokines in the CNS is still not yet well defined, its multiple double edged roles in this region may be predicted by our knowledge of the dual role of Nuclear factor-kappaB (NF-κB). NF-κB is a ubiquitous transcription factor that governs the expression of genes encoding cytokines, chemokines, growth factors, cell adhesion molecules, and some acute phase proteins in health and in various disease states. NF-κB is activated by several agents, including cytokines, oxidant-free radicals, inhaled particles, ultraviolet irradiation, and bacterial or viral products. Inappropriate activation of NF-κB has been linked to many inflammatory events. However, activation of NF-κB is also necessary for neuronal growth and protection, and proper immune function [67,68]. Here we also want to emphasize the fact that chemokine expression is regulated by NF-κB activation, therefore the dual role of NF-κB also implies a dual role for chemokines.

6. IS INFLAMMATION A PRIMARY OR SECONDARY RESPONSE IN AD?

Whether inflammation is the primary or secondary cause in AD pathogenesis has been a central issue of debate in recent years. We favor the idea that inflammation is centrally involved in AD pathogenesis (even if it is initially a 'response' to Aβ deposition or other lesions) based on the following observations: (a) the distribution of IP-10^+ and MIP-$1\beta^+$ reactive astrocytes do not necessarily correlate with senile plaques, and these strongly chemokine positive reactive astrocytes have been found at a very early stage of the disease, even before AD symptoms became obvious [11]; (b) microglia activation does not always correlate with amyloid plaques [9]; (c) Our recent study showing that the secretion of APP from mouse cortical neurons can be modulated by chemokines such as GROα and IP-10 suggests a direct effect of chemokines on amyloid production (MengQi Xia et al., unpublished observations); (d) recent data implicate a genetic polymorphism in IL-1α as being a risk factor for AD [69,70]. Inflammation can also be interpreted as the response to a wide variety of local stresses and pathological states. We suggest that the proper control of inflammation (even if it is 'secondary') will have beneficial effect on AD and other neurodegenerative diseases.

7. PROSPECT OF THERAPEUTIC INTERVENTION THROUGH MODULATION OF CHEMOKINE SYSTEM

Many CXC and CC chemokines may potentially be important mediators of the inflammatory changes responsible for the CNS degeneration. Most significantly, they may play a critical role in the initiation stage of the pathological processes. At this stage, it is hard to predict which receptor and ligand pair would be the most important target for therapy. CXCR2 is an attractive target, given the high level of CXCR2 expression on neurons. The strong upregulation of CXCR3 ligand IP-10 in reactive astrocytes of AD brains also makes CXCR3 a likely candidate for intervention. Due to the complexity of the disturbance in AD, and the complexity of interactions among inflammatory mediators, intervention at multiple levels of inflammation rather than at a few selective receptors may turn out to be critical. Furthermore, modulation rather than complete inhibition of any target may prove to be an adequate choice for therapy because of the dual roles of many targets.

8. CONCLUSIONS AND FUTURE DIRECTIONS

The detection of many chemokines and their receptors on various resident cells of CNS suggest their physiological role in the CNS. Given the primary role of chemokines as inflammatory mediators in the body's defense system, the upregulation of many chemokines and their receptors in AD brain (even in the early stages of the disease) implies that localized inflammation is important in the disease pathogenesis. The direct neurotrophic or neuronal modulatory effects of some chemokines demonstrated by us and others provide strong supporting evidence for their physiological role in CNS. We hypothesize that there may be detrimental effects on neurons if chemokine expression is continuously triggered in a non-regulated fashion. Animal models (i.e. APP or PS-1 mutant transgenic mice, chemokine receptor CXCR2 knockout mice) are available to test these hypothesis. For potential therapeutic intervention through the chemokine system in AD, the double-edged role of chemokines suggests that pharmacological modulation of the system rather than complete blockade may help the body regain a state of homeostasis.

ACKNOWLEDGEMENTS

This research was supported by NIA P50AG05134. We thank colleagues at LeukoSite, Inc. (now part of Millennium Pharmaceutics, Inc.) for supplying many excellent reagents critical for these studies.

REFERENCES

1. Hyman BT. The neuropathological diagnosis of Alzheimer's disease: clinical-pathological studies. Neurobiol Aging 1997;18:S27–32.
2. Hull M, Strauss S, Berger M, Volk B, Bauer J. Inflammatory mechanisms in Alzheimer's disease. Euro Arch Psychiatry Clin Neurosci 1996;246:124–128.
3. Rogers J, Webster S, Lue LF, Brachova L, Civin WH, Emmerling M, Shivers B, Walker D, McGeer P. Inflammation and Alzheimer's disease pathogenesis. Neurobiol Aging 1996;17:681–686.
4. Sheng JG, Ito K, Skinner RD, Mrak RE, Rovnaghi CR, Vaneldik LJ, Griffin WS. In vivo and in vitro evidence supporting a role for the inflammatory cytokine interleukin-1 as a driving force in Alzheimer pathogenesis. Neurobiol Aging 1996;17:761–766.
5. Mrak RE, Sheng JG, Griffin WS. Glial cytokines in Alzheimer's disease: review and pathogenic implications. Human Pathol 1995;26:816–823.
6. McGeer PL, Rogers J. Anti-inflammatory agents as a therapeutic approach to Alzheimer's disease. Neurology 1992;42:447–449.
7. Rich JB, Rasmusson DX, Folstein MF, Carson KA, Kawas C, Brandt J. Nonsteroidal anti-inflammatory drugs in Alzheimer's disease. Neurology 1995;45:51–55.
8. Xia MQ, Qin SX, McNamara M, Mackay C, Hyman BT. Interleukin-8 receptor B immunoreactivity in brain and neuritic plaques of Alzheimer's disease. Am J Pathol 1997;150:1267–1274.
9. Xia MQ, Qin SQ, Wu LJ, Mackay CR, Hyman BT. Immunohistochemical study of CCR3 and CCR5 and their ligands in control and AD brains. Am J Pathol 1998;153: 31–37.
10. Xia MQ, Hyman BT. Chemokines/chemokine receptors in the central nervous system and Alzheimer's disease [Review]. J Neurovirol 1999;5:32–41.
11. Xia MQ, Bacskai BJ, Knowles RB, Qin SX, Hyman BT. Expression of the chemokine receptor CXCR3 on neurons and the elevated expression of its ligand IP-10 in reactive astrocytes: in vitro ERK1/2 activation and role in Alzheimer's disease. J Neuroimmunol 2000;108:227–235.
12. Horuk R, Martin AW, Wang ZX, Schweitzer L, Gerassimides A, Guo HH, Lu ZH, Hesselgesser J, Perez HD, Kim J, Parker J, Hadley TJ, Peiper SC. Expression of chemokine receptors by subsets of neurons in the central nervous system. J Immunol 1997;158:2882–2890.
13. He JL, Chen YZ, Farzan M, Choe HY, Ohagen A, Gartner S, Busciglio J, Yang XY, Hofmann W, Newman W, Mackay CR, Sodroski J, Gabuzda D. CCR3 and CCR5 are co-receptors for HIV-1 infection of microglia. Nature 1997;385:645–649.
14. Harrison JK, Jiang Y, Chen SZ, Xia YY, Maciejewski D, McNamara RK, Streit WJ, Salafranca MN, Adhikari S, Thompson DA, Botti P, Bacon KB, Feng LL. Role for neuronally derived fractalkine in mediating interactions between neurons and CX3CR1-expressing microglia. Proc Natl Acad Sci USA 1998;95:10896–10901.
15. Hesselgesser J, Halksmiller M, Delvecchio V, Peiper SC, Hoxie J, Kolson DL, Taub D, Horuk R. CD4-independent association between HIV-1 gp120 and CXCR4 — functional chemokine receptors are expressed in human neurons. Curr Biol 1997;7:112–121.
16. Lavi E, Strizki JM, Ulrich AM, Zhang W, Fu L, Wang Q, Oconnor M, Hoxie JA, Gonzalezscarano F. CXCR-4 (Fusin), a co-receptor for the type 1 human immunodeficiency virus (HIV-1), is expressed in the human brain in a variety of cell types, including microglia and neurons. Am J Pathol 1997;151:1035–1042.
17. Xia M and Hyman BT. GROα/KC, a chemokine receptor CXCR2 ligand, can be a potent trigger for ERK1/2 and PI-3 kinase pathways and for tau hyperphosphorylation – a role in Alzheimer's disease? J Neuroimmunol 2002;122: 55–64.
18. Ishizuka K, Igatayi R, Kimura T, Hieshima K, Kukita T, Kin Y, Misumi Y, Yamamoto M, Nomiyama H, Miura R, Takamatsu J, Katsuragi S, Miyakawa T. Expression and distribution of CC chemokine macrophage inflammatory

protein-1-alpha/LD78 in the human brain. NeuroReport 1997;8:1215–1218.

19. Ishizuka K, Kimura T, Igata-yi R, Katsuragi S, Takamatsu J, Miyakawa T. Identification of monocyte chemoattractant protein-1 in senile plaques and reactive microglia of Alzheimer's disease. Psychiatry Clin Neurosci 1997;51: 135–138.
20. Atta ur R, Harvey K, Siddiqui RA. Interleukin-8: An autocrine inflammatory mediator. Curr Pharmaceut Design 1999;5:241–253.
21. Dong G, Chen Z, Kato T, Van Waes C. The host environment promotes the constitutive activation of nuclear factor-kappaB and proinflammatory cytokine expression during metastatic tumor progression of murine squamous cell carcinoma. Cancer Res 1999;59:3495–3504.
22. Ondrey FG, Dong G, Sunwoo J, Chen Z, Wolf JS, Crowl-Bancroft CV, Mukaida N, Van Waes C. Constitutive activation of transcription factors NF-(kappa)B, AP-1, and NF-IL6 in human head and neck squamous cell carcinoma cell lines that express pro-inflammatory and pro-angiogenic cytokines. Mol Carcinogen 1999;26:119–129.
23. Duffey DC, Chen Z, Dong G, Ondrey FG, Wolf JS, Brown K, Siebenlist U, Van Waes C. Expression of a dominant-negative mutant inhibitor-kappaB alpha of nuclear factor-kappaB in human head and neck squamous cell carcinoma inhibits survival, proinflammatory cytokine expression, and tumor growth in vivo. Cancer Res 1999; 59:3468–3474.
24. Araujo DM, Cotman CW. Trophic effects of interleukin-4, -7 and -8 on hippocampal neuronal cultures: potential involvement of glial-derived factors. Brain Res 1993;600: 49–55.
25. Balentien E, Mufson BE, Shattuck RL, Derynck R, Richmond A. Effects of MGSA/GRO alpha on melanocyte transformation. Oncogene 1991;6:1115–1124.
26. Brew R, Erikson JS, West DC, Kinsella AR, Slavin J, Christmas SE. Interleukin-8 as an autocrine growth factor for human colon carcinoma cells in vitro. Cytokine 2000; 12:78–85.
27. Meucci O, Fatatis A, Simen AA, Bushell TJ, Gray PW, Miller RJ. Chemokines regulate hippocampal neuronal signaling and gp120 neurotoxicity. Proc Natl Acad Sci USA 1998;95:14500–14505.
28. Ragozzino D, Giovannelli A, Mileo An, Limatola C, Santoni A, Eusebi F. Modulation of the neurotransmitter release in rat cerebellar neurons by GRO beta. NeuroReport 1998:9;3601–3606.
29. Maeda K, Nakai M, Maeda S, Kawamata T, Yamaguchi T, Tanaka C. Possible different mechanism between amyloid-Beta (25–35)- and substance P-induced chemotaxis of murine microglia. Gerontology 1997;43:11–15.
30. Peterson PK, Hu SX, Salakjohnson J, Molitor TW, Chao CC. Differential production of and migratory response to beta chemokines by human microglia and astrocytes. J Infect Dis 1997;175:478–481.
31. Tanabe S, Heesen M, Berman MA, Fischer M, Yoshizawa I, Luo Y, Dorf ME. Murine astrocytes express a functional chemokine receptor. J Neurosci 1997;17:6522–6528.
32. Zou YR, Kottmann AH, Kuroda M, Taniuchi I, Littman DR. Function of the chemokine receptor CXCR4 in haematopoiesis and in cerebellar development. Nature 1998;393:595–599.
33. Nagasawa T, Hirota S, Tachibana K, Takakura N, Nishikawa S, Kitamura Y, Yoshida N, Kikutani H, Kishimoto T. Defects of B-cell lymphopoiesis and bone-marrow myelopoiesis in mice lacking the CXC chemokine PBSF/SDF-1. Nature 1996;382:635–638.
34. Belperio JA, Keane MP, Arenberg DA, Addison CL, Ehlert JE, Burdick MD, Strieter RM. CXC chemokines in angiogenesis. J Leukcyte Biol 2000;68:1–8.
35. Moore BB, Keane MP, Addison CL, Arenberg DA, Strieter RM. CXC chemokine modulation of angiogenesis: the importance of balance between angiogenic and angiostatic members of the family. J Invest Med 1998;46:113–120.
36. Strieter RM, Polverini PJ, Arenberg DA, Kunkel SL. The role of CXC chemokines as regulators of angiogenesis. Shock 1995;4:155–160.
37. Trojanowski JQ, Lee VM. Paired helical filament tau in Alzheimer's disease. The kinase connection. Am J Pathol 1994;144:449–453.
38. Trojanowski JQ, Mawal-Dewan M, Schmidt ML, Martin J, Lee VM. Localization of the mitogen activated protein kinase ERK2 in Alzheimer's disease neurofibrillary tangles and senile plaque neurites. Brain Res 1993;618: 333–337.
39. Trojanowski JQ, Lee VM. Phosphorylation of paired helical filament tau in Alzheimer's disease neurofibrillary lesions: focusing on phosphatases. FASEB J 1995;9:1570–1576.
40. Illenberger S, Zheng-Fischhofer Q, Preuss U, Stamer K, Baumann K, Trinczek B, Biernat J, Godemann R, Mandelkow EM, Mandelkow E. The endogenous and cell cycle-dependent phosphorylation of tau protein in living cells: implications for Alzheimer's disease. Mol Biol Cell 1998;9:1495–1512.
41. Leger J, Kempf M, Lee G, Brandt R. Conversion of serine to aspartate imitates phosphorylation-induced changes in the structure and function of microtubule-associated protein tau. J Biol Chem 1997;272:8441–8446.
42. Scott CW, Spreen RC, Herman JL, Chow FP, Davison MD, Young J, Caputo CB. Phosphorylation of recombinant tau by cAMP-dependent protein kinase. Identification of phosphorylation sites and effect on microtubule assembly. J Biol Chem 1993;268:1166–1173.
43. Baumann K, Mandelkow EM, Biernat J, Piwnica-Worms H, Mandelkow E. Abnormal Alzheimer-like phosphorylation of tau-protein by cyclin-dependent kinases cdk2 and cdk5. FEBS Lett 1993;336:417–424.
44. Goedert M, Hasegawa M, Jakes R, Lawler S, Cuenda A, Cohen P. Phosphorylation of microtubule-associated protein tau by stress-activated protein kinases. FEBS Lett 1997;409:57–62.
45. Reynolds CH, Nebreda AR, Gibb GM, Utton MA, Anderton BH. Reactivating kinase/p38 phosphorylates tau protein in vitro. J Neurochem 1997;69:191–198.
46. Shea TB, Cressman CM. The order of exposure of tau to signal transduction kinases alters the generation of 'AD-like' phosphoepitopes. Cell Mol Neurobiol 1999;19:223–233.
47. Hyman BT, Elvhage TE, Reiter J. Extracellular signal regulated kinases. Localization of protein and mRNA in

the human hippocampal formation in Alzheimer's disease. Am J Pathol 1994;144:565–572.
48. Roder HM, Eden PA, Ingram VM. Brain protein kinase PK40erk converts Tau into a PHF-like form as found in Alzheimer's disease. Biochem Biophys Res Commun 1993;193:639–647.
49. Blanchard BJ, devi Raghunandan R, Roder HM, Ingram VM. Hyperphosphorylation of human Tau by brain kinase PK40erk beyond phosphorylation by cAMP-dependent PKA: relation to Alzheimer's disease. Biochem Biophys Res Commun 1994;200:187–194.
50. Zheng-Fischhofer Q, Biernat J, Mandelkow EM, Illenberger S, Godemann R, Mandelkow E. Sequential phosphorylation of Tau by glycogen synthase kinase-3beta and protein kinase A at Thr212 and Ser214 generates the Alzheimer-specific epitope of antibody AT100 and requires a paired-helical-filament-like conformation. Eur J Biochem 1998;252:542–552.
51. Drewes G, Trinczek B, Illenberger S, Biernat J, Schmitt-Ulms G, Meyer HE, Mandelkow EM, Mandelkow E. Microtubule-associated protein/microtubule affinity-regulating kinase (p110mark). A novel protein kinase that regulates tau-microtubule interactions and dynamic instability by phosphorylation at the Alzheimer-specific site serine 262. J Biol Chem 1995;270:7679–7688.
52. Jicha GA, Weaver C, Lane E, Vianna C, Kress Y, Rockwood J, Davies P. cAMP-dependent protein kinase phosphorylations on tau in Alzheimer's disease. J Neurosci 1999;19:7486–7494.
53. Geddes JW, Anderson KJ, Cotman CW. Senile plaques as aberrant sprout-stimulating structures. Exp Neurol 1986; 94:767–776.
54. Masliah E, Mallory M, Hansen L, Alford M, Albright T, DeTeresa R, Terry R, Baudier J, Saitoh T. Patterns of aberrant sprouting in Alzheimer's disease. Neuron 1991;6: 729–739.
55. Masliah E, Mallory M, Hansen L, Alford M, DeTeresa R, Terry R, Baudier J, Saitoh T. Localization of amyloid precursor protein in GAP43-immunoreactive aberrant sprouting neurites in Alzheimer's disease. Brain Res 1992; 574:312–316.
56. Kornhauser JM, Greenberg ME. A kinase to remember: dual roles for MAP kinase in long-term memory [comment]. Neuron 1997;18:839–842.
57. Fukunaga K, Miyamoto E. Role of MAP kinase in neurons [Review]. Mol Neurobiol 1998;16:79–95.
58. Murray B, Alessandrini A, Cole AJ, Yee AG, Furshpan EJ. Inhibition of the p44/42 MAP kinase pathway protects hippocampal neurons in a cell-culture model of seizure activity. Proc Natl Acad Sci USA 1998;95:11975–11980.
59. Runden E, Seglen PO, Haug FM, Ottersen OP, Wieloch T, Shamloo M, Laake JH. Regional selective neuronal degeneration after protein phosphatase inhibition in hippocampal slice cultures: Evidence for a MAP kinase-dependent mechanism. J Neurosci 1998;18:7296–7305.
60. Alessandrini A, Namura S, Moskowitz MA, Bonventre JV. MEK1 protein kinase inhibition protects against damage resulting from focal cerebral ischemia. Proc Natl Acad Sci USA 1999;96:12866–12869.
61. Busser J, Geldmacher DS, Herrup K. Ectopic cell cycle proteins predict the sites of neuronal cell death in Alzheimers-disease brain. J Neurosci 1998;18:2801–2807.
62. Nagy Z, Esiri MM, Cato AM, Smith AD. Cell cycle markers in the hippocampus in Alzheimers-disease. Acta Neuropathologica 1997;94:6–15.
63. McShea A, Harris PL, Webster KR, Wahl AF, Smith MA. Abnormal expression of the cell cycle regulators P16 and CDK4 in Alzheimer's disease. Am J Pathol 1997;150: 1933–1939.
64. Danova M, Aglietta M. Cytokine receptors, growth factors and cell cycle in human bone marrow and peripheral blood hematopoietic progenitors. Haematologica 1997;82: 622–626.
65. Buckley CD, Simmons DL. Cell adhesion — a new target for therapy [Review]. Mol Med Today 1997;3:449–456.
66. Bernstein SH, Eaves CJ, Herzig R, Fay J, Lynch J, Phillips GL, Christiansen N, Reece D, Ericson S, Stephan M, Kovalsky M, Hawkins K, Rasmussen H, Devos A, Herzig GP. A randomized phase II study of bb-10010 — a variant of human macrophage inflammatory protein-1-alpha for patients receiving high-dose etoposide and cyclophosphamide for malignant lymphoma and breast cancer. Br J Haematol 1997;99:888–895.
67. Mattson MP, Goodman Y, Luo H, Fu W, Furukawa K. Activation of NF-kappaB protects hippocampal neurons against oxidative stress-induced apoptosis: evidence for induction of manganese superoxide dismutase and suppression of peroxynitrite production and protein tyrosine nitration. J Neurosci Res 1997;49:681–697.
68. Chen F, Castranova V, Shi X, Demers LM. New insights into the role of nuclear factor-kappaB, a ubiquitous transcription factor in the initiation of diseases. Clin Chem 1999;45:7–17.
69. Du Y, Dodel RC, Eastwood BJ, Bales KR, Gao F, Lohmuller F, Muller U, Kurz A, Zimmer R, Evans RM, Hake A, Gasser T, Oertel WH, Griffin WS, Paul SM, Farlow MR. Association of an interleukin 1 alpha polymorphism with Alzheimer's disease. Neurology 2000;55: 480–483.
70. Grimaldi LM, Casadei VM, Ferri C, Veglia F, Licastro F, Annoni G, Biunno I, De Bellis G, Sorbi S, Mariani C, Canal N, Griffin WS, Franceschi M. Association of early-onset Alzheimer's disease with an interleukin-1alpha gene polymorphism. Ann Neurol 2000;47:361–365.
71. Nguyen D, Stangel M. Expression of the chemokine receptors CXCR1 and CXCR2 in rat oligodendroglia cells. Brain Res. Dev. Brain Res. 2001;128:77–81.
72. Ehrlich LC, Hu S, Sheng WS, Sutton RL, Rockswold GL, Peterson PK, Chao CC. Cytokine regulation of human microglial cell IL-8 production. J Immunol 1998:160; 1944–1948.
73. Lipovsky MM, Gekker G, Hu S, Ehrlich LC, Hoepelman AI, Peterson PK. Cryptococcal glucuronoxylomannan induces interleukin (IL)-8 production by human microglia but inhibits neutrophil migration toward IL-8. J Infect Dis 1998;177:260–263.
74. Aloisi F, Borsellino G, Care A, Testa U, Gallo P, Russo G, Peschle C, Levi G. Cytokine regulation of astrocyte function: in-vitro studies using cells from the human brain. Int J Dev Neurosci 1995;13:265–274.

75. Desbaillets I, Diserens AC, Detribolet N, Hamou MF, Vanmeir EG. Upregulation of interleukin 8 by oxygen-deprived cells in glioblastoma suggests a role in leukocyte activation, chemotaxis, and angiogenesis. J Exp Med 1997;186:1201–1212.
76. Nitta T, Steinman L, Sato K. An analysis of lymphokine gene expression within astrocytoma. No Shinkei Geka. Neurol Surg 1992;20:763–768.
77. Lahrtz F, Piali L, Nadal D, Pfister HW, Spanaus KS, Baggiolini M, Fontana A. Chemotactic activity on mononuclear cells in the cerebrospinal fluid of patients with viral meningitis is mediated by interferon-gamma inducible protein-10 and monocyte chemotactic protein-1. Eur J Immunol 1997;27:2484–2489.
78. Sun DM, Hu XZ, Liu XH, Whitaker JN, Walker WS. Expression of chemokine genes in rat glial cells — the effect of myelin basic protein-reactive encephalitogenic T cells. J Neurosci Res 1997;48:192–200.
79. Fisher SN, Vanguri P, Shin HS, Shin ML. Regulatory mechanisms of MuRantes and CRG-2 chemokine gene induction in central nervous system glial cells by virus. Brain Behav Immun 1995;9:331–344.
80. Sorensen TL, Tani M, Jensen J, Pierce V, Lucchinetti C, Folcik VA, Qin SX, Rottman J, Sellebjerg F, Strieter RM, Frederiksen JL, Ransohoff RM. Expression of specific chemokines and chemokine receptors in the central nervous system of multiple sclerosis patients. J Clin Invest 1999;103:807–815.
81. Asensio VC, Campbell IL. Chemokine gene expression in the brains of mice with lymphocytic choriomeningitis. J Virol 1997;71:7832–7840.
82. Asensio VC, Kincaid C, Campbell IL. Chemokines and the inflammatory response to viral infection in the central nervous system with a focus on lymphocytic choriomeningitis virus. J Neurovirol 1999;5:65–75.
83. Vanguri P. Interferon-gamma-inducible genes in primary glial cells of the central nervous system: comparisons of astrocytes with microglia in Lewis brown Norway rats. J Neuroimmunol 1995;56:35–43.
84. Tanabe S, Heesen M, Berman MA, Fischer MB, Yoshizawa I, Luo Y, Dorf ME. Murine astrocytes express a functional chemokine receptor. J Neurosci 1997;17:6522–6528.
85. Bernasconi S, Cinque P, Peri G, Sozzani S, Crociati A, Torri W, Vicenzi E, Vago L, Lazzarin A, Poli G, Mantovani A. Selective elevation of monocyte chemotactic protein-1 in the cerebrospinal fluid of AIDS patients with cytomegalovirus encephalitis. J Infect Dis 1996;174:1098–1101.
86. Kim JS, Gautam SC, Chopp M, Zaloga C, Jones ML, Ward PA, Welch KM. Expression of monocyte chemoattractant protein-1 and macrophage inflammatory protein-1 after focal cerebral ischemia in the rat. J Neuroimmunol 1995;56:127–134.
87. Szaflarski J, Ivacko J, Liu XH, Warren JS, Silverstein FS. Excitotoxic injury induces monocyte chemoattractant protein-1 expression in neonatal rat brain. Mol Brain Res 1998;55:306–314.
88. Glabinski AR, Tani M, Tuohy VK, Tuthill RJ, Ransohoff RM. Central nervous system chemokine mRNA accumulation follows initial leukocyte entry at the onset of acute murine experimental autoimmune encephalomyelitis. Brain Behav Immun 1995;9:315–330.
89. Glabinski AR, Balasingam V, Tani M, Kunkel SL, Strieter RM, Yong VW, Ransohoff RM. Chemokine monocyte chemoattractant protein-1 is expressed by astrocytes after mechanical injury to the brain. J Immunol 1996;156:4363–4368.
90. Spleiss O, Gourmala N, Boddeke HWGM, Sauter A, Fiebich BL, Berger M, Gebickehaerter PJ. Cloning of rat HIV-1-chemokine coreceptor CKR5 from microglia and upregulation of its mRNA in ischemic and endotoxinemic rat brain. J Neurosci Res 1998;53:16–28.
91. Bartholdi D, Schwab ME. Expression of pro-inflammatory cytokine and chemokine mRNA upon experimental spinal cord injury in mouse: an in situ hybridization study. Eur J Neurosci 1997;9:1422–1438.
92. McManus CM, Brosnan CF, Berman JW. Cytokine induction of MIP-1-alpha and MIP-1-beta in human fetal microglial. J Immunol 1998;160:1449–1455.
93. Takami S, Nishikawa H, Minami M, Nishiyori A, Sato M, Akaike A, Satoh M. Induction of macrophage inflammatory protein MIP-1-alpha mRNA on glial cells after focal cerebral ischemia in the rat. Neurosci Lett 1997;227:173–176.
94. Nishiyori A, Minami M, Ohtani Y, Takami S, Yamamoto J, Kawaguchi N, Kume T, Akaike A, Satoh M. Localization of fractalkine and CX(3)CR1 mRNAs in rat brain: does fractalkine play a role in signaling from neuron to microglia? FEBS Lett 1998;429:167–172.
95. Horuk R, Martin A, Hesselgesser J, Hadley T, Lu ZH, Wang ZX, Peiper SC. The Duffy antigen receptor for chemokines: structural analysis and expression in the brain. J Leukocyte Biol 1996;59:29–38.

Universes in Delicate Balance: Chemokines and the Nervous System
R.M. Ransohoff, K. Suzuki, A.E.I. Proudfoot, W.F. Hickey and J.K. Harrison, editors

Chapter 5.3

HIV-1-Associated Dementia

Dana Gabuzda [1,2], Jianbin Wang [1,3] and Paul R. Gorry [1,3]

[1] *Department of Cancer Immunology and AIDS, Dana-Farber Cancer Institute, JF816, 44 Binney Street, Boston, MA 02115, USA;* [2] *Department of Neurology, Harvard Medical School, Boston, MA 02115, USA;* [3] *Department of Pathology, Harvard Medical School, Boston, MA 02115, USA*

1. INTRODUCTION

Human immunodeficiency virus type I (HIV-1) infects macrophages and microglia in the central nervous system (CNS) and frequently causes dementia and other neurological disorders in patients with acquired immunodeficiency syndrome (AIDS) [1,2]. Until a few years ago, HIV-1-associated dementia (HAD) and related cognitive and motor disorders (termed minor cognitive/motor disorder) occurred in up to 20–30% of AIDS patients. However, the incidence of these disorders has declined since the introduction of combination highly active antiretroviral therapies (HAART) in 1996. Most current antiviral drugs have relatively poor CNS penetration. Therefore, the CNS represents a reservoir for long-term viral persistence and HAD continues to be an important complication of HIV-1 infection. HIV-1-associated dementia results from complex interactions between viral and host factors. Four chemokine receptors and their ligands (CXCR4/SDF-1 [CXCL12], CCR5/MIP-1α [CCL3] and β [CCL4], RANTES [CCL5], CCR2/MCP-1 [CCL2], and CX_3CR1/fractalkine [CX_3CL1]) are of particular importance in HAD. This chapter will discuss the role of these and other chemokine receptors/chemokines in the disease mechanisms.

Neuroinvasion of HIV-1 occurs through the trafficking of infected monocytes and possibly $CD4^+$ lymphocytes, across the blood–brain barrier [2,3]. HIV-1 enters the CNS early in the course of infection. However, CNS infection is latent and typically does not cause HIV-1 encephalitis or HAD until more than 10 years later after clinical progression to AIDS. The neuropathology of HIV-1 encephalitis suggests a diffuse low grade viral infection of the CNS, with chronic inflammatory lesions and focal demyelination. The most commonly affected brain regions are the white matter and basal ganglia, followed by the cerebral cortex. Neuropathological abnormalities characteristic of HIV-1 encephalitis include reactive astrocytosis, myelin pallor, microglial nodules, perivascular inflammation, multinucleated giant cells, abnormal blood–brain barrier permeability, and neuronal loss [1,2]. Dendritic abnormalities and loss of synapses have also been documented.

Why does HAD typically occur only in the advanced stages of HIV-1 disease after progression to AIDS? Possible explanations include the loss of immune control associated with disease progression, increased immune activation, increased trafficking of activated and/or infected monocytes into the CNS, and the late emergence of viral variants that impact CNS disease progression [2,3]. In general, there is a positive correlation between viral load in the CSF and the severity of HAD. A positive correlation with viral load in brain is also generally observed, but is less consistent due to sampling error and regional variability in HIV-1 encephalitis. Despite relatively poor CNS penetration by most currently used drugs, antiretroviral therapy usually results in a decline in CSF viral load (in parallel with a decline in plasma viral load), and improvement in neurological function and neuroimaging studies. Thus, HAD and other neurological disorders associated with HIV-1 infection is at

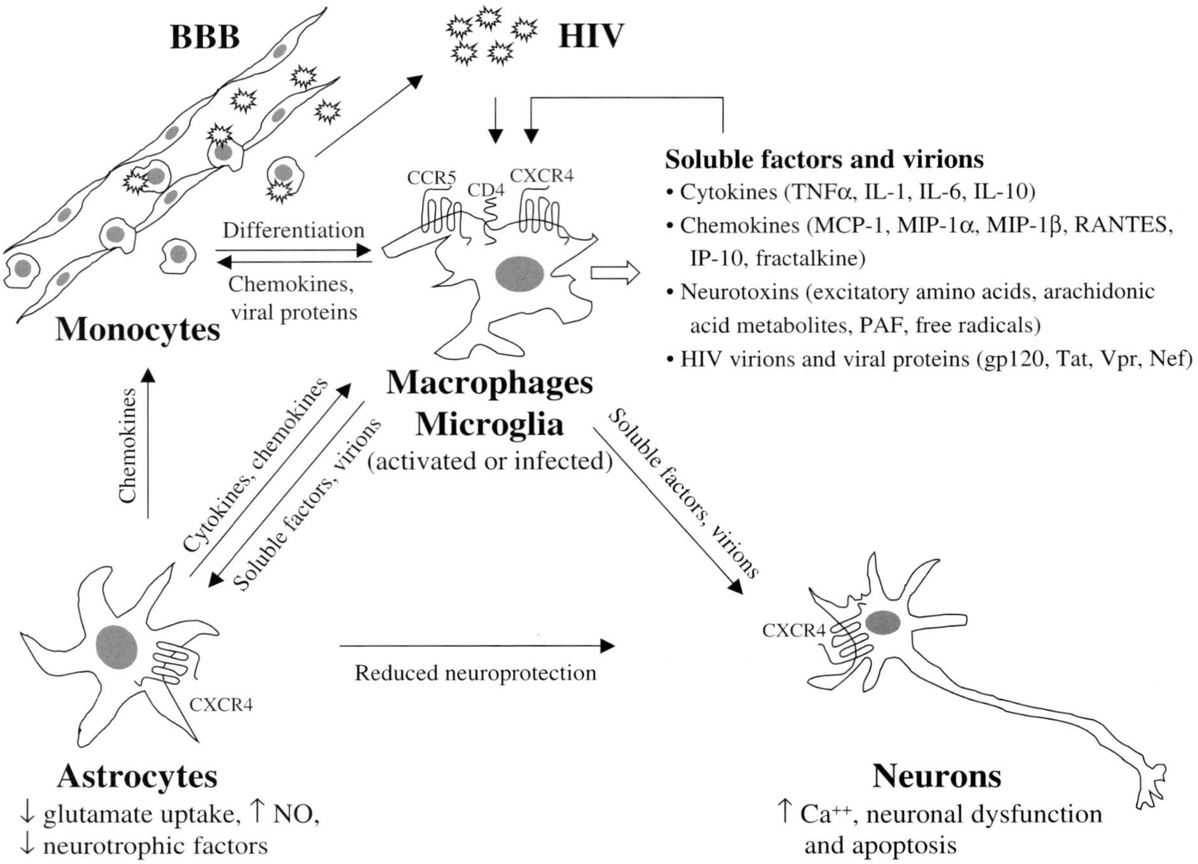

Figure 1. Model of pathways involved in HIV-1-associated dementia. BBB, blood–brain barrier.

least partially reversible if systemic infection can be controlled.

The mechanisms that lead to brain injury in HAD are poorly understood. Apoptosis of neurons and astrocytes is detected in autopsy brain tissues from AIDS patients with HAD [4–9]. The major pathway for neuronal injury is thought to involve the release of soluble neurotoxic factors from activated macrophages and microglia (Fig. 1) (reviewed in [1,10,11]). Relatively few macrophages and microglia in the CNS are infected with HIV-1 compared to the degree of neurological dysfunction in most AIDS patients. Furthermore, proliferation and activation of macrophages and microglia and the increased expression of macrophage activation markers (i.e. β_2-microglobulin, quinolinate, TNF-α) correlates with clinical dementia more closely than the numbers of infected cells [10,11]. Potential neurotoxins that contribute to HAD include viral proteins including soluble forms of the HIV-1 envelope, Tat, and possibly Vpr and Nef proteins, and factors released from activated or infected macrophages and microglia including excitatory amino acids (glutamate, quinolinate and L-cysteine), arachidonic acid metabolites, platelet activating factor, free radicals, and other yet unknown factors (Fig. 1) (reviewed in [1,10,11]). Increased expression of cytokines (e.g. TNF-α, IL-1β, IL-6, IL-10, IFN-γ, TGF-β), chemokines (e.g. MIP-1α, MIP-1β, RANTES, MCP-1, IP-10 [CXCL10], fractalkine), and matrix metalloproteinases by activated microglia and/or reactive astrocytes may also contribute to neurological dysfunction. In addition, astrocyte dysfunction may contribute to CNS injury in HAD. For example, the normal uptake of glutamate by astrocytes may be reduced, and astrocyte release of glutamate may contribute to neurotoxicity mediated by excitotoxic amino acids. Blood–brain barrier abnormalities may also contribute to CNS injury in HAD [12–14]. The in vivo role of specific factors that underlie the pathogenesis of HIV-1 dementia remains to be established.

The role of particular HIV-1 strains in causing HAD is unknown. The genetic evolution of HIV-1 within the brain is distinct from that in lymphoid tissues and other organs [15–21]. Specific sequences within the viral envelope glycoproteins (Env), particularly the V3 variable loop region, are associated with brain infection [15–21]. However, infection of the CNS is not sufficient to cause HAD [16,17,22,23], suggesting that neurovirulence is likely to be determined by distinct genetic and biological characteristics. Thus, HAD may be associated with particular strains of HIV-1 that replicate efficiently in macrophages and microglia, in addition to having the capacity to induce dysfunction or cell death of neurons and other cells in the CNS.

Table 1. HIV-1 coreceptors

Coreceptor	Ligand(s)	Expression in brain
Major coreceptors		
CCR5	MIP-1α, MIP-1β, RANTES	+
CXCR4	SDF-1	+
Minor coreceptors		
CCR2b	MCP-1-4	+
CCR3	Eotaxin, Eotaxin 2, RANTES, MCP-2-4	+
CCR8 (Chem R1/TER-1)	I-309	+
CX_3CR1 (V28)	Fractalkine	+
APJ	Apelin	+
STRL33/BONZO	CXCL16	+
GPR1	?	+
GPR15 (BOB)	?	–
Chem R23	?	?
RDC-1	?	+

The new chemokine nomenclature is given here in parentheses. MIP-1α (CCL3); MIP-1β (CCL4); RANTES (CCL5); SDF-1 (CXCL12); MCP-1 (CCL2); MCP-2 (CCL8); MCP-3 (CCL7); MCP-4 (CCL13); I-309 (CCL1); eotaxin (CCL11); eotaxin 2 (CCL24); fractalkine (CX_3CL1).

2. CCR5 AND CXCR4 ARE CORECEPTORS FOR HIV-1 ENTRY IN THE CNS

The cellular tropism of HIV-1 is determined by the interaction of the HIV-1 Env with CD4 and a particular chemokine coreceptor (reviewed in [24–29]). Coreceptors that can mediate HIV-1 infection are shown in Table 1. CCR5 and CXCR4 are the principal HIV-1 coreceptors [30–35]. HIV-1 viruses that use CCR5 (R5) can infect primary $CD4^+$ T-cells and macrophages. Viruses that use CXCR4 (X4) infect primary CD4 T-cells and T-cell lines, but in some cases can also infect macrophages. Dual-tropic viruses (R5X4) can use both coreceptors. A subset of viruses can also use alternative coreceptors for virus entry such as CCR3, CCR2b, CCR8, APJ, STRL33 (BONZO), Gpr1, Gpr15 (BOB), CX_3CR1, ChemR23, and RDC-1 ([33,34,36–48]; reviewed in [24,25,27]). However, most studies on these receptors have been performed using transfected cells. Therefore, the role of these alternative coreceptors in vivo is unknown. Furthermore, most of these coreceptors are used inefficiently for virus entry compared to CCR5 and CXCR4. R5 HIV-1 viruses are the predominant strains transmitted from person to person and are usually present during asymptomatic infection and throughout the course of disease. X4 or R5X4 viruses arise in the later stages of disease in 40–50% of infected adults [49,50] and their emergence is frequently associated with rapid $CD4^+$ T-cell decline and disease progression. In some individuals, disease progression is associated with general broadening of virus tropism by expansion of coreceptor usage [49,50]. It is late in the course of disease progression, when X4 and R5X4 isolates emerge, that neurological symptoms such as HAD typically arise.

CCR5 is the major coreceptor for HIV-1 infection of macrophages and microglia in the CNS ([51–54]; reviewed in [55]). Furthermore, previous studies suggest that CCR5 is the principal coreceptor used by viruses isolated from brain of individuals with AIDS [2,51,52,54,56–58]. CXCR4 is also expressed on macrophages and microglia and can support relatively efficient infection by a subset of primary X4 viruses [58–62]. X4 or R5X4 viruses are typically found in blood and lymphoid tissues, but are also present in brain, albeit at lower frequency than R5 viruses [16,21,58,63,64]. CCR3 is expressed on microglia and may facilitate infection by certain HIV-1 strains [21,51,53,54]. CCR3, CCR2b, APJ, CCR8, Gpr1, Gpr15, and STRL33 can be used at low efficiency as coreceptors for virus entry in transfected cells by some brain-derived viruses [21,52,58], but the role of these alternative coreceptors in mediating infection of macrophages and microglia in vivo is unknown. Among the alternative coreceptors, only CCR2b, CCR3, and possibly CCR8, are known to be expressed on macrophages and/or microglia.

Astrocytes and brain capillary endothelial cells can be infected at a very low level in vitro and in vivo (reviewed in [2,55]). However, most studies suggest that this low level of infection is probably not significant compared to infection of macrophages and microglia. In contrast to infection of other target cells, HIV-1 entry into astrocytes and endothelial cells is CD4-independent. Astrocytes and brain capillary endothelial cells express low levels of CXCR4 [65–67], and possibly other coreceptors [68–70]. There is some evidence that CXCR4 may mediate infection of endothelial cells. However, CXCR4 and other known chemokine coreceptors do not appear to mediate astrocyte infection by HIV-1 [71]. CD4-independent HIV-1 isolates have been generated in vitro [72–74], but naturally occurring HIV-1 variants that can use chemokine receptors for CD4-independent entry have not been identified. Further studies are required to identify the receptor(s) which mediate CD4-independent infection of astrocytes and brain capillary endothelial cells.

Genetic polymorphisms in CCR5 are associated with effects on susceptibility to HIV-1 infection or disease progression (reviewed in [24,75,76]). Individuals homozygous for CCR5 alleles with the Δ32 mutation (approximately 1% of Caucasians), a 32 amino acid deletion which results in a non-functional CCR5 receptor and lack of CCR5 expression on the cell surface, are highly resistant to HIV-1 infection. Heterozygosity for the CCR5 Δ32 mutation (approximately 10–15% of Caucasians) and genetic polymorphisms in the CCR5 promoter have also been shown to affect the rate of HIV-1 disease progression. Thus, the levels of CCR5 expression determine cellular susceptibility to HIV-1 infection [77] and can affect disease progression. CNS abnormalities have not been detected in individuals homozygous for the CCR5 Δ32 mutation, or in CCR5 knock-out mice, suggesting that CCR5 is not required for normal CNS development and function. CCR2 and SDF-1 genetic polymorphisms have also been associated with effects on HIV-1 disease progression (reviewed in [24,75]). Studies by Faure et al. [78] showed that allele M280 of CX_3CR1 is associated with increased risk of HIV-1-infection and accelerated disease progression. However, another study which analyzed several different cohorts did not confirm this finding [79]. Do these genetic polymorphisms affect susceptibility to HAD? A small preliminary study suggested that heterozygosity for the CCR5 Δ32 allele does not reduce the risk of developing HAD [80]. However, subsequent studies in children suggest that the risk of developing HAD may be reduced [81]. The effects of CCR2 and SDF-1 polymorphisms on the relative risk of CNS disease remain unknown.

The HIV-1 envelope glycoproteins are derived from a gp160 precursor, which is glycosylated and proteolytically cleaved to form the gp120 and gp41 subunits. The external subunit (gp120) and transmembrane subunit (gp41) remain noncovalently associated and oligomerize, mostly as trimers, on the surface of virions. The proposed model for virus entry is that high affinity binding of HIV-1 gp120 to CD4 induces a conformational change in gp120, which increases its affinity for CCR5 or CXCR4 by exposing the chemokine receptor binding site (reviewed in [24,26,82,83]). Subsequently, gp41 undergoes a conformational change that leads to fusion of the viral membrane with the target cell membrane by exposing the fusion domain of gp41. The efficiency of virus entry can be affected by multiple factors, including cell-specific post-translational modifications of CCR5 and CXCR4 (e.g. glycosylation or tyrosine sulfation) [26,82,84], the level of CD4 and coreceptor expression [85,86], and the efficiency of coreceptor–CD4 interactions at the cell surface [26,83,87,88]. CCR5, and to a lesser extent CXCR4, partially colocalizes with CD4 at the cell surface. HIV-1 gp120 binding increases coreceptor/CD4 colocalization and co-capping [26]. Thus, interactions between these cell surface receptors can be modified by extracellular stimuli. Env-coreceptor interactions can also occur in the absence of CD4 [73,74,82,89], but these CD4-independent interactions do not necessarily allow virus entry.

Signaling-defective mutants of CCR5 or CXCR4 can still support HIV-1 infection [27,90], suggesting that virus entry mediated by chemokine receptors does not appear to require signaling. However, a recent study suggested that signaling may be required for macrophage infection by a subset of R5 viruses [91]. Although signaling does not appear to be required for HIV-1 entry, binding of soluble or virion-associated gp120 to CCR5 or CXCR4 can activate cellular signaling pathways [91–95]. The gp120 interaction with CCR5 leads to calcium influx and activation of ERK-1 and -2 MAPK in naive T-cells [94]. In contrast, X4 viruses do not activate ERK-1 and -2 [94]. The gp120 interaction with CCR5 or CXCR4 has also been shown to activate ion channels in macrophages [96]. In addition, gp120 can activate focal adhesion kinase (FAK) and PYK2 (also known as RAFTK) via CCR5 or CXCR4 [92,93,95,97]. Not surprisingly, patterns of gp120–chemokine receptor signaling are highly cell-specific [94].

Chemokines can inhibit entry of HIV-1 viruses that use the same receptor(s) [24]. The CCR5 ligands MIP-1α, MIP-1β, and RANTES can inhibit HIV-1 infection by R5 viruses [30,34,98]. Similarly, the CXCR4 ligand SDF-1 inhibits entry by X4 or R5X4 HIV-1 viruses [99,100]. Infection by a subset of R5 or X4 viruses that also use CCR3 or APJ as coreceptors can be inhibited by a CCR3 ligand (eotaxin [CCL11]) and the APJ ligand (apelin) [34,101,102]. The mechanisms of chemokine inhibition of HIV-1 infection are complex. Furthermore, variable results can be obtained depending on the target cell type, viral isolate, and cell culture conditions [103,104]. Chemokines can inhibit HIV-1 infection by one or more of several different mechanisms: (1) direct competitive inhibition of the HIV-1 gp120–coreceptor interaction; (2) downregulation of the autologous chemokine receptor; (3) cross-regulation of expression or function of a heterologous chemokine receptor; or (4) effects on cellular signaling pathways that influence HIV-1 replication via effects on virus entry, post-entry events, viral gene expression, or other steps in the viral life cycle ([104,105]; reviewed in [24,26,28,83]).

3. CXCR4 MEDIATES NEURONAL DYSFUNCTION AND APOPTOSIS INDUCED BY THE HIV-1 ENV

Many studies over the last 10 years have shown that HIV-1 gp120 is neurotoxic to primary human or rat neurons at picomolar concentrations (reviewed in [1,11,106]). The induction of neurotoxicity occurs via interactions with CXCR4 [107]. Studies in CXCR4 knock-out mice have shown that CXCR4 plays a role in neuronal migration and patterning during brain development [108,109]. SDF-1/CXCR4 interactions can also mediate signaling in neurons, astrocytes and microglia, and direct migration of neuronal progenitors [11,54,89,110–112]. CXCR4, and possibly other chemokine receptors, is expressed on subpopulations of neurons in the cerebral cortex and other brain regions [7,65,66,70,113–115]. This may render these cells vulnerable to mechanisms of CNS injury associated with HIV-1 infection through interactions between soluble or virion-associated HIV-1 gp120 and CXCR4, and possibly other chemokine receptors.

Soluble or virion-associated HIV-1 gp120 can bind to CXCR4 on neurons in the absence of CD4 [89] and thereby induce neuronal signaling and apoptosis [107,116–118]. However, other studies suggest that neuronal apoptosis is induced through the release of soluble neurotoxic factors by activated or infected macrophages and microglia [110,117,118]; reviewed in [10,11]. A subset of HIV-1 isolates which use CXCR4 for virus entry can induce apoptosis of neurons and astrocytes in vitro [21,62,117]. Replacement of the *env* gene in an R5 non-apoptosis-inducing virus with the X4 Env of an apoptosis-inducing virus was sufficient to confer the apoptosis-inducing phenotype [21]. However, viral usage of CXCR4 is neither necessary nor sufficient to induce apoptosis since not all X4 strains induce apoptosis and a subset of R5 strains is apoptosis-inducing [21,58]. Together, these results provide evidence that the HIV-1 Env is a major determinant of neurodegenerative mechanisms associated with HIV-1 infection in vitro and suggest that X4 and R5X4 viruses are likely to play a role in the development of HAD. This model is consistent with the observation that HIV-1 gp120–CXCR4 interactions also contribute to viral cytopathicity and apoptosis of uninfected bystander cells in the immune system [97,119,120].

The possibility that HIV-1 strains which use CXCR4 may contribute to disease progression in the CNS is consistent with several previous observations: (1) X4 HIV-1 viruses emerge in the later stages of disease, which is the time when HAD occurs; (2) X4 HIV-1 viruses are generally more cytopathic than R5 HIV-1 viruses for uninfected bystander cells in the immune system, including $CD4^+$ T-cells, $CD8^+$ T-cells, and B-cells [119–121]; and (3) a neurovirulent SHIV (a chimeric simian-human immunodeficiency virus that contains the *tat*, *rev*, *vpu*, and *env* genes from the T-cell line-tropic laboratory-adapted HIV-1 IIIB isolate and adapted for growth in macrophages) that causes AIDS, neurological disease, and renal disease in rhesus macaques uses only CXCR4 for virus entry [122]. Furthermore, although X4 viruses are typically found in blood and lymphoid tissues, trafficking of virus from blood into brain occurs in a subset of AIDS patients [16,19,20,63,123]. Moreover, Env V3 region sequences typical of X4 or R5X4 HIV-1 strains are detected in the brain, and choroid plexus [16,63,123], albeit at much lower frequency than sequences characteristic of R5 HIV-1 strains. However, the in vivo role of CXCR4 and other chemokine receptors in HAD remains to be established. HIV-1 dementia or encephalitis can occur in individuals who progress to AIDS in the absence of X4 or R5X4 viruses [124], particularly in children [125]. These findings, together with the studies described above are consistent with a model in which viral determinants in addition to CXCR4 usage also contribute to viral cytopathicity in the CNS. These may include a subset of viruses that primarily use CCR5, or possibly an alternative coreceptor expressed in the CNS (e.g. CCR3, APJ), as well as other viral proteins such as Tat and Nef which have been proposed to have neurotoxic activities.

CXCR4-mediated mechanisms of neuronal injury may not necessarily require virus replication. For example, gp120 binding to CXCR4 on the surface of macrophages, microglia, or astrocytes could activate the production of a neurotoxin [110,117,118], or interfere with SDF-1-mediated CXCR4 signaling [111,126]. SDF-1 can induce signaling and chemotaxis in human neurons in vitro [89,118]. However, in some experimental systems, SDF-1 can also induce neuronal apoptosis [107,118]. This observation, together with the finding that expression of SDF-1 is increased in the brain of patients with HAD and upregulated in astrocytes exposed to HIV-1-infected or activated macrophage conditioned media [118] or in reactive astrocytes [111], raises the possibility that increased SDF-1 production by astrocytes is another mechanism that could potentially contribute to neuronal injury. Moreover, expression of CXCR4 is upregulated on reactive astrocytes and possibly other cells in the brain of AIDS patients [65,66,114].

Recent studies have begun to define the pathways involved in mediating neuronal apoptosis induced by the interaction of HIV-1 gp120 and CXCR4 [11]. gp120–CXCR4 interactions activate ERK-1 and -2 in rat or human neurons, but not astrocytes [112,127]. In contrast, SDF-1 activates ERK-1 and -2 in both cell types. Apoptosis induced by gp120- and SDF-1 in rat cortical neurons was inhibited by a p38 MAPK inhibitor, but not by an ERK-1 and-2 inhibitor [110]. TKP, a tripeptide that inhibits activation of macrophages and microglia, blocked gp120- but not SDF-1-induced neuronal apoptosis, suggesting that gp120-induced apoptosis may be dependent largely on the indirect pathways, such as activation of CXCR4 on mononuclear phagocytes. Other studies that examined neuronal apoptosis induced by HIV-1 virions representing diverse viral strains found evidence for involvement of the PKA pathway and an increase in inositol 1,4,5-triphosphate and $[Ca^{2+}]$ in the apoptotic signaling pathway [117,118]. These signals correlated with activation of caspase 3 and neuronal apoptosis. HA1004, a relatively broad kinase inhibitor which inhibits PKA, PKC, and calcium/calmodulin-dependent protein kinase II (but not drugs which selectively inhibit PKC or the ERK-1 and -2, or p38 MAPK) blocked CXCR4-mediated neuronal apoptosis. The different results obtained in these studies may reflect the different species used for neuronal cell cultures (i.e. rat vs. human), or other experimental variables such as cell culture conditions. The specific mechanisms which link intracellular signalling to neuronal apoptosis remain to be determined. Understanding these pathways may provide insights that could facilitate the development of new therapeutic approaches for the prevention or treatment of HAD.

4. CHEMOKINE INTERACTIONS IN HIV-1 ENCEPHALITIS

HIV-1 encephalitis is an inflammatory neurological disease associated with immune activation and leukocyte trafficking into the CNS. As in other inflammatory neurological diseases, CNS expression of a variety of chemokines and chemokine receptors is increased, particularly in areas of inflammation (reviewed in [113,128–130]). In addition to determining the composition of leukocytes in CNS inflammatory infiltrates, these changes in expression may impact HAD by promoting virus replication.

Table 2. Chemokine expression in brain and CSF of patients with HIV-1-associated dementia

Chemokine	Cellular source	Reference
CC		
MCP-1	Astrocytes, macrophages/microglia	[133–136,139]
MIP-1α, MIP-1β, RANTES	Macrophages/microglia, astrocytes	[66,131,132, 136,138,143]
IL-8, Gro-α	Astrocytes	[136]
IP-10, MIG	Astrocytes, macrophages/microglia	[137,139]
CXC		
SDF-1	Astrocytes	[118]
CX3C		
Fractalkine	Neurons (astrocytes)	[153,154]

The new chemokine nomenclature is given here in parentheses, or in the legend to Table 1. Gro-α (CXCL1); MIG (CXCL9).

Increased expression of chemokines and their receptors may also influence other processes mediated by interactions between the viral Env and chemokine receptors. Chemokines shown to be upregulated in the CNS of AIDS patients with HIV-1 encephalitis are shown in Table 2.

Expression of MIP-1α, MIP-1β, RANTES, MCP-1, SDF-1, IP-10, IL-8 (CXCL8) and fractalkine [66,118,131–139] is increased in the brain and/or CSF of AIDS patients with HAD. Elevated expression of MIP-1α, MIP-1β, RANTES, and IP-10 has also been shown in the brains of SIV-infected rhesus macaques [140]. Activated or infected macrophages and microglia produce increased levels of MIP-1α, MIP-1β and RANTES, and MCP-1 [141–144]. Astrocytes are also an important source of these chemokines, particularly MCP-1 [135]. These findings, together with the observation that increased expression of MCP-1 and IP-10 in brain or CSF shows a positive correlation with HAD [135,136,139] suggest that chemokines may increase the trafficking of monocytes and T-cells, respectively, in the CSF compartment and across the blood–brain barrier in HIV-1-infected individuals. Some studies suggested that MIP-1β or RANTES may protect neurons from apoptosis induced by gp120 [110,116]. Whether these chemokines are protective in vivo is an open question. vMIP-II, a chemokine homolog encoded by Kaposi's sarcoma-associated herpes virus (KSHV), was shown to inhibit HIV-1 infection via CCR3, CCR5, and other coreceptors [145]. This finding, together with the observation that AIDS patients with Kaposi's sarcoma have a lower incidence of HAD [145,146], led to speculation that increased expression of vMIP-II in these individuals may reduce CNS complications by inhibiting virus entry in macrophages and microglia. Whether increased expression of vMIP-II or other chemokines with potential antiviral activity such as MIP-1α, MIP-1β, and RANTES reduces viral entry and replication in the CNS in vivo remains to be determined.

The migration of monocytes across the blood–brain barrier plays a critical role in HAD (reviewed in [3]). The presence of perivascular infiltrates composed primarily of monocytes and macrophages is one of the pathological hallmarks of HIV-1 encephalitis. MCP-1 is a potent stimulus for monocyte chemotaxis and trafficking. Increased expression of MCP-1 in brain and CSF is associated with an increased risk of HAD [134–136]. Increased MCP-1 production was shown to occur primarily in reactive astrocytes [135]. Interestingly, the HIV-1 Tat protein is a potent stimulus for astrocyte production of MCP-1 [135]. These observations suggest that HIV-1 infection, possibly through expression of the HIV-1 Tat protein, may facilitate recruitment of monocytes into the CNS by inducing MCP-1 expression in astrocytes. Studies in SIV- infected rhesus macaques with encephalitis, as well as in vitro studies [147] provide further support for this model.

CCR5-positive perivascular monocytes and macrophages are detected at increased frequency in the brain of pediatric and adult patients with HIV-1 encephalitis and HAD compared to HIV-1-infected individuals without neurologic disease [7], suggesting that increased entry of CCR5-positive mononuclear cells into the brain may contribute to disease progression in the CNS. These cells could serve as carriers of virus into the brain or as targets for HIV-1 infection, in addition to contributing to disease pathogenesis by producing neurotoxic products [148]. Factors which may contribute to the increased trafficking of CCR5-positive mononuclear cells into the brain of AIDS patients include increased expression of monocyte chemoattractants, such as MIP-1α, MIP-1β, RANTES and MCP-1 [66,131,133,135,136,139], increased expression of adhesion molecules involved in monocyte trafficking through the blood–brain barrier [149], and abnormal blood–brain barrier permeability [12,13]. Furthermore, in vitro studies suggest that activated or HIV-1-infected monocytes are more likely to migrate through the blood–brain barrier [149,150]. Interestingly, an activated monocyte subset that expresses CD14, CD16, and CD69 is significantly increased in the peripheral blood of patients with HAD irrespective of other clinical parameters such as stage of disease, CD4 cell count, viral load, or opportunistic infections [148].

CCR5 expression is increased on macrophages and microglia within inflammatory lesions in the brain of pediatric and adult patients with HIV-1 encephalitis [7,66]. Monocyte/macrophage expression of CCR5 can be upregulated by IL-10 and HIV-1 Tat [147,151,152]. Thus, increased expression of these factors in the brain of AIDS patients may enhance CCR5 expression on brain macrophages and microglia. Previous studies demonstrated elevated levels of IL-10 in the brain and CSF of AIDS patients with HIV-1 encephalitis and HAD. In a recent study [152], IL-10 was shown to upregulate CCR5 expression on microglia, facilitate chemotactic migration in response to MIP-1β, and enhance HIV-1 infection and replication by R5 strains in microglia. These studies suggest that IL-10 may promote disease progression in the CNS by enhancing the entry and replication of R5 strains in microglia, as well as facilitating migration of mononuclear phagocytes.

Fractalkine and its receptor CX_3CR1 are upregulated in the brain of AIDS patients [153,154]. Fractalkine is constitutively expressed on neurons [155,156] and can be induced in astrocytes [157]. Its receptor, CX_3CR1, is expressed on neurons and microglia [116,156]. Studies by Tong et al. demonstrated increased expression of fractalkine in neurons in brain tissues from AIDS patients with HAD [153]. Furthermore, increased expression of CX_3CR1 was demonstrated on both neurons and microglia. Fractalkine was neuroprotective when coadministered with HIV-1 Tat or platelet activating factor in vitro. Other studies have also suggest that fractalkine may have a neuroprotective role [156,158]. However, fractalkine may also mediate recruitment of macrophages and/or microglia [155,157]. Whether fractalkine is neuroprotective in vivo remains to be determined.

The level of chemokine and chemokine receptor expression and the state of immune activation are likely to play important roles in the pathogenesis of HAD. The study of immunomodulators that regulate the expression or effects of chemokines and their receptors in the brain is an active area for drug development. In particular, therapies targeting CCR2 and/or CCR5 may be beneficial as therapeutic approaches to inhibit monocyte recruitment into the CNS, and drugs targeting CCR5 and/or CXCR4 may prevent virus entry into macrophages and microglia.

5. CHEMOKINE RECEPTORS AS THERAPEUTIC TARGETS IN HIV-1-ASSOCIATED DEMENTIA

HIV-1 coreceptors are promising targets for the prevention and treatment of HIV-1 infection and HAD. A variety of agents targeted at CCR5 and CXCR4 [159–166] can inhibit HIV-1 infection of T-cells, macrophages, and microglia [21,58,61]. CCR5 is a good therapeutic target since individuals homozygous for the CCR5Δ32 mutation do not exhibit any obvious clinical abnormalities. However, CXCR4 may not be practical as a therapeutic target since it is expressed on many cell types and blocking its function may therefore be associated with unacceptable side effects. Nevertheless, CXCR4-targeted inhibitors could be used in topical microbicides to prevent sexual transmission.

HIV-1 infection mediated by CCR5 can be inhibited by the small molecule inhibitors TAK-779 and Sch-C, and by AOP-RANTES, a modified RANTES peptide [160,162–166]. TAK779 and AOP-RANTES also inhibit HIV-1 entry mediated by CCR2 and CCR3, respectively. TAK-779, which inhibits binding of RANTES to CCR5, interacts with the region of CCR5 formed by the extracellular loops. Infection mediated by CXCR4 can be inhibited by AMD3100, a bicyclam compound which interacts with charged residues in the extracellular loops and inhibits binding of SDF-1 to CXCR4 [159,161]. CXCR4-mediated infection can also be inhibited by the small peptides ALX40-4C and T22 [159,165,166]. TAK-779 and AMD3100 are potent inhibitors of viral entry and infection in vitro. However, both small molecule inhibitors have poor bioavailability. Derivatives with better pharmacokinetic properties for clinical applications are currently being developed [166]. Sch-C, a potent inhibitor of CCR5-mediated HIV-1 entry, shows good bioavailability and is currently in early clinical trials. All of these compounds work by blocking the binding of HIV-1 gp120 to CCR5 or CXCR4. In addition, they downregulate cell surface expression of these receptors and in some cases inhibit recycling to the cell surface [167]. CCR5- and CXCR4-targeted inhibitors can be synergistic against HIV-1 when used in combination with other inhibitors of virus entry, such as T20, a peptide which inhibits virus–cell fusion [168,169]. Other new therapeutic strategies targeted at HIV-1 coreceptors, such as immunomodulatory therapies and gene therapies are also being developed [83,162,166].

6. CONCLUSIONS

HIV-1-associated dementia results from complex interactions between viral and host factors. Four chemokine receptors and their ligands (CXCR4/SDF-1, CCR5/MIP-1α and β, RANTES, CCR2/MCP-1, and CX_3CR1/fractalkine) are of particular importance in HAD. HIV-1 infection of macrophages and microglia in the CNS is primarily mediated by CCR5. A subset of strains can infect macrophages and microglia via CXCR4. Increased monocyte trafficking into the CNS of patients with HAD is likely to result from increased expression of MCP-1, in addition to CCR5 chemokines. Neurodegenerative mechanisms in HAD may involve interactions between the HIV-1 Env and CXCR4, or possibly other chemokine receptors on neurons, and possibly other CNS cell types. CCR5 and CXCR4 are promising targets for the prevention and treatment of HAD. Several compounds that selectively block virus entry without affecting physiological functions of these receptors are being developed, and some are in early clinical trials. Understanding the role of CCR5, CXCR4 and other chemokine receptors and their ligands in HIV-1 neuropathogenesis will be important for advancing the development of new therapeutic strategies for the prevention and treatment of HAD and other neurological complications of HIV-1 infection.

ACKNOWLEDGEMENTS

Supported by NS37277 and NS35734. Dana-Farber Cancer Institute is the recipient of a Center for AIDS Research grant AI28691 and a Cancer Center Grant. D.G. is an Elizabeth Glaser Scientist supported by the Pediatric AIDS Foundation.

REFERENCES

1. Lipton SA, Gendelman HE. Dementia associated with the acquired immunodeficiency syndrome. New Engl J Med 1995;332:934–940.
2. Price RW. Neurological complications of HIV infection. Lancet 1996;348:445–452.
3. Gartner S. HIV infection and dementia. Science 2000; 287:602–604.
4. Gelbard HA, James HJ, Sharer LR, Perry SW, Saito Y, Kazee AM, Blumberg BM, Epstein LM. Apoptotic neurons in brains from paediatric patients with HIV-1 encephalitis and progressive encephalopathy. Neuropathol Appl Neurobiol 1995;21:208–217.
5. Petito CK, Roberts B. Evidence of apoptotic cell death in HIV encephalitis. Am J Pathol 1995;146:1121–1131.
6. Shi B, De Girolami U, He J, Wang S, Lorenzo A, Busciglio J, Gabuzda D. Apoptosis induced by HIV-1 infection of the central nervous system. J Clin Invest 1996;98:1979–1990.
7. Vallat AV, De Girolami U, He J, Mhashilkar A, Marasco W, Shi B, Gray F, Bell J, Keohane C, Smith TW, Gabuzda D. Localization of HIV-1 coreceptors CCR5 and CXCR4 in the brain of children with AIDS. Am J Pathol 1998;152:167–178.
8. Gray F, Adie Biassette H, Brion F, Ereau T, Le Maner I, Levy V, Corcket G. Neuronal apoptosis in human immunodeficiency virus infection. J NeuroVirol 2000; 6(Suppl 1):538–543.
9. Gabuzda D, Wang J. Chemokine receptors and mechanisms of cell death in HIV neuropathogenesis. J NeuroVirol 2000;6(Suppl 1):S24–32.
10. Gonzalez-Scarano F, Baltuch G. Microglia as mediators of inflammatory and degenerative diseases. Annu Rev Neurosci 1999;22:219–240.
11. Kaul M, Garden G, Lipton SA. Pathways to neuronal injury and apoptosis in HIV-associated dementia. Nature 2001;410:988–904.
12. Petito CK, Cash KS. Blood–brain barrier abnormalities in the acquired immunodeficiency syndrome: immunohistochemical localization of serum proteins in postmortem brain. Ann Neurol 1992;32:658–666.
13. Power C, Kong P-A, Crawford TO, Wesselingh S, Glass JD, McArthur JC, Trapp BD. Cerebral white matter changes in acquired immunodeficiency syndrome dementia: alterations of the blood–brain barrier. Ann Neurol 1993;34:339–350.
14. Persidsky Y, Zheng J, Miller D, Gendelman HE. Mononuclear phagocytes mediate blood–brain barrier compromise and neuronal injury during HIV-1-associated dementia. J Leukocyte Biol 2000;68:413–422.
15. Donaldson YK, Bell JE, Holmes EC, Hughes ES, Brown HK, Simmonds, P. In vivo distribution and cytopathology of variants of human immunodeficiency virus type 1

showing restricted sequence variability in the V3 loop. J Virol 1994;68:5991–6005.
16. Korber BTM, Kunstman KJ, Patterson BK, Furtado M, McEvilly MM, Levy R, Wolinsky SM. Genetic differences between blood- and brain-derived viral sequences from human immunodeficiency virus type 1-infected patients: evidence of conserved elements in the V3 region of the envelope protein of brain-derived sequences. J Virol 1994;68:7467–7481.
17. Power C, McArthur JC, Johnson RT, Griffin DE, Glass JD, Perryman S, Chesebro B. Demented and nondemented patients with AIDS differ in brain derived human immunodeficiency virus type 1 envelope sequences. J Virol 1994;68:4643–4649.
18. Hughes ES, Bell JE, Simmonds P. Investigation of the dynamics of the spread of human immunodeficiency virus to brain and other tissues by evolutionary analysis of sequences from the $p17^{gag}$ and *env* genes. J Virol 1997;71:1272–1280.
19. Wong JK, Ignacio CC, Torriani F, Havlir D, Fitch NJS, Richman DD. In vivo compartmentalization of human immunodeficiency virus: evidence from the examination of *pol* sequences from autopsy tissues. J Virol 1997;71: 2059–2071.
20. Van'T Wout AB, Ran LJ, Kuiken CL, Kootstra NA, Pals ST, Schuitemaker H. Analysis of the temporal relationship between human immunodeficiency virus type 1 quasispecies in sequential blood samples and various organs obtained at autopsy. J Virol 1998;72:488–496.
21. Öhagen A, Devitt A, Kunstman KJ, Gorry P, Rose P, Korber B, Sodroski J, Taylor J, Levy R, Murphy R, Wolinsky S, Gabuzda D. Genetic and functional analysis of full-length HIV-1 *env* genes derived from brain and blood of AIDS patients, submitted.
22. Joag SV, Stephens EB, Galbreath D, Zhu W, Li Z, Foresman L, Zhao L-J, Pinson DM, Narayan O. Simian immunodeficiency virus SIV_{mac} chimeric virus whose *env* gene was derived from SIV-encephalitic brain is macrophage-tropic but not neurovirulent. J Virol 1995; 69:1367–1369.
23. Mankowski JL, Spelman JP, Ressetar HG, Strandberg JD, Laterra J, Carter DL, Clements JE, Zink MC. Neurovirulent simian immunodeficiency virus replicates productively in endothelial cells of the central nervous system in vivo and in vitro. J Virol 1994;68:8202–8208.
24. Berger EA, Murphy, PM, Farber JM. Chemokine receptors as HIV-1 coreceptors: roles in viral entry, tropism, and disease. Annu Rev Immunol 1999;17:657–700.
25. Doms RW, Peiper SC. Unwelcomed guests with master keys: how HIV uses chemokine receptors for cellular entry. Virology 1997;235:179–190.
26. Doms RW. Beyond receptor expression: The influence of receptor conformation, density, and affinity in HIV-1 infection. Virology 2000;276:229–237.
27. Littman DR. Chemokine receptors: keys to AIDS pathogenesis? Cell 1998;93:677–680.
28. Rucker J, Doms RW. Chemokine receptors as HIV coreceptors: implications and interactions. AIDS Res Hum Retroviruses 1998;14:(Suppl 3)S241–S246.
29. Edinger AL, Clements JE, Doms RW. Chemokine and orphan receptors in HIV-2 and SIV tropism and pathogenesis. Virology 1999;260:211–221.
30. Alkhatib G, Combadiere C, Broder CC, Feng Y, Kennedy PE, Murphy PM, Berger EA. CC CKR5: A RANTES, MIP-1α, MIP-1β receptor as a fusion cofactor for macrophage-tropic HIV-1. Science 1996;272:1955–1958.
31. Deng HK, Liu R, Ellmeier W, Choe S, Unutmaz D, Burkhart M, Di Marzio P, Marmon S, Sutton RE, Hill CM, Davis CB, Peiper SC, Schall TJ, Littman DR, Landau NR. Identification of a major co-receptor for primary isolates of HIV-1. Nature 1996;381:661–666.
32. Dragic T, Litwin V, Allaway GP, Martin SR, Huang Y, Nagashima KA, Cayanan C, Maddon PJ, Koup RA, Moore JP, Paxton WA. HIV-1 entry into $CD4^+$ cells is mediated by the chemokine receptor CC-CKR-5. Nature 1996;381:667–673.
33. Doranz BJ, Rucker J, Yi Y, Smyth RJ, Samson M, Peiper SC, Parmentier M, Collman RG, Doms RW. A dualtropic primary HIV-1 isolate that uses fusin and the β chemokine receptors CKR-5, CKR-3, CKR-2b as fusion cofactors. Cell 1996;85:1149–1158.
34. Choe H, Farzan M, Sun Y, Sullivan N, Rollins B, Ponath, PD, Wu L, MacKay CR, LaRosa G, Newman W, Gerard N, Gerard C, Sodroski J. The β-chemokine receptors CCR3 and CCR5 facilitate infection by primary HIV-1 isolates. Cell 1996;85:1135–1148.
35. Feng Y, Broder CC, Kennedy PE, Berger EA. HIV-1 entry cofactor: Functional cDNA cloning of a seven-transmembrane, G protein-coupled receptor. Science 1996;272:872–877.
36. Deng HK, Unutmaz D, KewalRamani VN, Littman DR. Expression cloning of new receptors used by simian and human immunodeficiency viruses. Nature 1997:388:296–300.
37. Farzan M, Choe H, Martin K, Marcon L, Hofmann W, Karlsson G, Sun Y, Barrett P, Marchand N, Sullivan N, Gerard N, Gerard C, Sodroski J. Two orphan seven-transmembrane segment receptors which are expressed in CD4-positive cells support simian immunodeficiency virus infection. J Exp Med 1997;186:405–411.
38. Liao F, Alkhatib G, Peden KWC, Sharma G, Berger EA, Farber JM. STRL33, A novel chemokine receptor-like protein, functions as a fusion cofactor for both macrophage-tropic and T cell line-tropic HIV-1. J Exp Med 1997;185:2015–2023.
39. Rucker J, Edinger AL, Sharron M, Samson M, Lee B, Berson JF, Yi Y, Margulies B, Collman RG, Doranz BJ, Parmentier M, Doms RW. Utilization of chemokine receptors, orphan receptors, and herpesvirus-encoded receptors by diverse human and simian immunodeficiency viruses. J Virol 1997;71:8999–9007.
40. Choe H, Farzan M, Konkel M, Martin K, Sun Y, Marcon L, Cayabyab M, Berman M, Dorf ME, Gerard N, Gerard C, Sodroski J. The orphan seven-transmembrane receptor apj supports the entry of primary T-cell-line-tropic and dualtropic human immunodeficiency virus type 1. J Virol 1998;72:6113–6118.
41. Edinger AL, Hoffman TL, Sharron M, Lee B, O'Dowd B, Doms RW. Use of GPR1, GPR15, and STRL33 as coreceptors by diverse human immunodeficiency virus

type 1 and simian immunodeficiency virus envelope proteins. Virology 1998;249:367–378.
42. Edinger AL, Hoffman TL, Sharron M, Lee B, Yi Y, Choe W, Kolson DL, Mitrovic B, Zhou Y, Faulds D, Collman RG, Hesselgesser J, Horuk R, Doms RW. An orphan seven-transmembrane domain receptor expressed widely in the brain functions as a coreceptor for human immunodeficiency virus type 1 and simian immunodeficiency virus. J Virol 1998;72: 7934–7940.
43. Horuk R, Hesselgesser J, Zhou Y, Faulds D, Halks-Millert M, Harvey S, Taub D, Samson M, Parmentier M, Rucker J, Doranz BJ, Doms RW. The CC chemokine I-309 inhibits CCR8-dependent infection by diverse HIV-1 strains. J Biol Chem 1998;273:386–391.
44. Combadiere C, Salzwedel K, Smith ED, Tiffany HL, Berger EA, Murphy PM. Identification of CX3CR1. A chemotactic receptor for the human CX3C chemokine fractalkine and a fusion coreceptor for HIV-1. J Biol Chem 1998;273:23799–23804.
45. Jinno A, Shimizu N, Soda Y, Haraguchi Y, Kitamura T, Hoshino H. Identification of the chemokine receptor TER1/CCR8 expressed in brain-derived cells and T cells as a new coreceptor for HIV-1 infection. Biochem Biophys Research Commun 1998;243:497–502.
46. Samson M, Edinger AL, Stordeur P, Rucker J, Verhasselt V, Sharron M, Govaerts C, Mollereau C, Vassart G, Doms RW, Parmentier M. ChemR23, a putative chemoattractant receptor, is expressed in monocyte-derived dendritic cells and macrophages and is a coreceptor for SIV and some primary HIV-1 strains. Eur J Immunol 1998; 28:1689–1700.
47. Shimizu N, Soda Y, Kanbe K, Liu H-Y, Jinno A, Kitamura T, Hoshino H. An orphan G protein-coupled receptor, GPR1, acts as a coreceptor to allow replication of human immunodeficiency virus Types 1 and 2 in brain-derived cells. J Virol 1999;73:5231–5239.
48. Shimizu N, Soda Y, Kanbe K, Liu H-Y, Mukai R, Kitamura T, Hoshino H. A putative G protein-coupled receptor, RDC1, is a novel coreceptor for human and simian immunodeficiency viruses. J Virol 2000;74:619–626.
49. Björndal Å, Deng H, Jansson M, Fiore JR, Colognesi C, Karlsson A, Albert J, Scarlatti G, Littman DR, Fenyö EM. Coreceptor usage of primary human immunodeficiency virus type 1 isolates varies according to biological phenotype. J Virol 1997;71:7478–7487.
50. Connor RI, Sheridan KE, Ceradini D, Choe S, Landau NR. Change in coreceptor use correlates with disease progression in HIV-1-infected individuals. J Exp Med 1997;185:621–628.
51. He J, Chen Y, Farzan M, Choe H, Öhagen A, Gartner S, Buscigilo J, Yang X, Hofmann W, Newman W, MacKay CR, Sodroski J, Gabuzda D. CCR3 and CCR5 are coreceptors for HIV-1 infection of microglia. Nature 1997; 385:645–649.
52. Shieh JTC, Albright AV, Sharron M, Gartner S, Strizki J, Doms RW and Gonzalez-Scarano F. Chemokine receptor utilization by human immunodeficiency virus type 1 isolates that replicate in microglia. J Virol 1998;72: 4243–4249.
53. Ghorpade A, Xia MQ, Hyman BT, Persidsky Y, Nukuna A, Bock P, Che M, Limoges J, Gendelman HE, MacKay CR. Role of the β-chemokine receptors CCR3 and CCR5 in human immunodeficiency virus type 1 infection of monocytes and microglia. J Virol 1998;72:3351–3361.
54. Albright AV, Shieh JTC, Itoh T, Lee B, Pleasure D, O'Connor MJ, Doms RW, Gonzalez-Scarano F. Microglia express CCR5, CXCR4, CCR3, but of these, CCR5 is the principal coreceptor for human immunodeficiency virus type 1 dementia isolates. J Virol 1999;73: 205–213.
55. Gabuzda D, Wang J. Chemokine receptors and virus entry in the central nervous system. J NeuroVirol 1999; 5:643–658.
56. Li S, Juarez J, Alali M, Dwyer D, Collman R, Cunningham A, Naif HM. Persistent CCR5 utilization and enhanced macrophage tropism by primary blood human immunodeficiency virus type 1 isolates from advanced stages of disease and comparison to tissue-derived isolates. J Virol 1999;73:9741–9755.
57. Smit TK, Wang B, Ng T, Osborne R, Brew B, Saksena NK. Varied tropism of HIV-1 isolates derived from different regions of adult brain cortex discriminate between patients with and without AIDS dementia complex (ADC): Evidence for neurotropic HIV variants. Virology 2001;279:509–526.
58. Gorry PR, Bristol G, Zack JA, Birch CJ, Bell JE, Bannert N, Crawford K, Schols D, De Clercq E, Kunstman K, Wolinsky SM, Gabuzda D. Macrophage tropism of human immunodeficiency virus Type 1 isolates from brain and lymphoid tissues predicts neurotropism independent of coreceptor specificity. J Virol 2001;75:10073–10089.
59. Simmons G, Reeves JD, McKnight A, Dejucq N, Hibbitts S, Power CA, Aarons E, Schols D, De Clercq E, Proudfoot AEI, Clapham PR. CXCR4 as a functional coreceptor for human immunodeficiency virus type 1 infection of primary macrophages. J Virol 1998;72:8453–8457.
60. Verani A, Pesenti E, Polo S, Tresoldi E, Scarlatti G, Lusso P, Siccardi AG, Vercelli D. CXCR4 is a functional coreceptor for infection of human macrophages by CXCR4-dependent primary HIV-1 isolates. J Immunol 1998;161:2084–2088.
61. Hibbitts S, Reeves JD, Simons G, Gray PW, Epstein LG, Schols D, De Clercq E, Wells TNC, Proudfoot AEI, Clapham PR. Coreceptor ligand inhibition of fetal brain cell infection by HIV Type 1. AIDS Res Hum Retroviruses 1999;15:989–1000.
62. Öhagen A, Ghosh S, He J, Huang K, Chen Y, Yuan M, Osathanondh R, Gartner S, Shi B, Shaw G, Gabuzda D. Apoptosis induced by infection of primary brain cultures with diverse human immunodeficiency virus type 1 isolates: Evidence for a role of the envelope. J Virol 1999; 73: 897–906..
63. Chang J, Jozwiak R, Wang B, Ng T, Ge YC, Bolton W, Dwyer DE, Randle C, Osborn R, Cunningham AC, Saksena ND. Unique HIV type 1 V3 region sequences derived from six different regions of brain: region-specific evolution within host-determined quasispecies. AIDS Res Hum Retroviruses 1998;14:25–30.
64. McGavin CH, Land SA, Sebire KL, Hooker DJ, Gurusinghe AD, Birch CJ. Syncytium-inducing phenotype and

zidovudine susceptibility of HIV-1 isolated from post-mortem tissue. AIDS 1996;10:47–53.

65. Lavi E, Strizki JM, Ulrich AM, Zhang W, Fu L, Wang Q, O'Connor M, Hoxie JA, González-Scarano F. CXCR-4 (Fusin), a co-receptor for the Type 1 human immunodeficiency virus (HIV-1), is expressed in the human brain in a variety of cell types, including microglia and neurons. Am J Pathol 1997;151:1035–1042.
66. Sanders VJ, Pittman CA, White MG, Wang G, Wiley CA, Achim CL. Chemokines and receptors in HIV encephalitis. AIDS 1998;12:1021–1026.
67. Klein RS, Williams KC, Alvarez-Hernandez X, Westmoreland S, Force T, Lackner AA, Luster AD. Chemokine receptor expression and signaling in macaque and human fetal neurons and astrocytes: Implications for the neuropathogenesis of AIDS. J Immunol 1999;163:1636–1646.
68. Rottman JB, Ganley KP, Williams K, Wu LJ, MacKay CR, Ringler DJ. Cellular localization of the chemokine receptor CCR5 — Correlation to cellular target of HIV-1 infection. Am J Pathol 1997;151:1341–1351.
69. Westmoreland SV, Rottman JB, Williams KC, Lackner AA, Sasseville VG. Chemokine receptor expression on resident and inflammatory cells in the brain of macaques with simian immunodeficiency virus encephalitis. Am J Pathol 1998;152:659–665.
70. Choe W, Albright A, Sulcove J, Jaffer S, Hesselgesser J, Lavi E, Crino P, Kolson DL. Functional expression of the seven-transmembrane HIV-1 co-receptor APJ in neural cells. J NeuroVirol 2000;Suppl 1:S61–69.
71. Sabri F, Tresoldi E, DiStefano M, Polo S, Monaco MC, Verani A, Fiore JR, Lusso P, Major E, Chiodi F, Scarlatti G. Nonproductive human immunodeficiency virus type 1 infection of human fetal astrocytes independence from CD4 and major chemokine receptors. Virology 1999; 264:370–384.
72. Dumonceaux J, Nisole S, Chanel C, Quivet L, Amara A, Baleux F, Briant P, Hazan U. Spontaneous mutations in the env gene of the human immunodeficiency virus type 1 NDK isolate are associated with a CD4-independent entry phenotype. J Virol 1998;72:512–519.
73. Hoffman TL, LaBranche CC, Zhang W, Canziani G, Robinson J, Chaiken I, Hoxie JA, Doms RW. Stable exposure of the coreceptor-binding site in a CD4-independent HIV-1 envelope protein. Proc Natl Acad Sci USA 1999;96:6359–6364.
74. Kolchinsky P, Mirzabekov T, Farzan M, Kiprilov E, Cayabyab M, Mooney LJ, Choe H, Sodroski J. Adaptation of a CCR5-using, primary human immunodeficiency virus Type 1 isolate for CD4-independent replication. J Virol 1999;72:8120–8126.
75. Michael NL. Host genetic influences on HIV-1 pathogenesis. Curr Opin Immunol 1999;11:466–474.
76. Carrington M, Dean M, Martin MP, O'Brien SJ. Genetics of HIV-1 infection: chemokine receptor CCR5 polymorphism and its consequences. Hum Mol Genet 1999;8: 1939–1945.
77. Wu L, Paxton WA, Kassam N, Ruffing N, Rottman JB, Sullivan N, Choe H, Sodroski J, Newman W, Koup RA, MacKay CR. CCR5 levels and expression pattern correlate with infectability by macrophage-tropic HIV-1, in vitro. J Exp Med 1997;1185:1681–1691.
78. Faure S, Meyer L, Costagliola D, Vaneensberghe C, Genin E, Autran B, Alt F, Immunoco Study Groups, Jean-Francois Delfraissy, Seroco Study Group, McDermott DH, Murphy PM, Debre P, Theiodorou I, Combadiere C. Rapid progression to AIDS in HIV^+ individuals with a structural variant of the chemokine receptor CX_3CR1. Science 2000;287:2274–2277.
79. McDermott DH, Colla JS, Kleeberger CA, Plankey M, Rosenberg PS, Smith ED, Zimmerman PA, Combadiere C, Leitman SF, Kaslow RA, Goedert JJ, Berger EA, O'Brien TR, Murphy PM. Genetic polymorphism in CX_3CR1 and risk of HIV disease. Science 2000;290: 2031a.
80. Barroga CF, Ellis R, Nelson J, Heaton RK, Atkinson JH, McCutchan JA, Grant I, Spector SA. HIV-1 neurocognitive disorders and chemokine receptors. AIDS 1997;11: 1651–1664.
81. Barroga CF, Raskino C, Fangon MC, Palumbo PE, Baker CJ, Englund JA, Spector SA. The *CCR5Δ32* allele slows disease progression of human immunodeficiency virus-1-infected children receiving antiretroviral treatment. J Infect Dis 2000;182:413–419.
82. Berson JF, Doms RW. Structure–function studies of the HIV-1 coreceptors. Semin Immunol 1998;10:237–248.
83. Dimitrov DS, Xiao X, Chabot DJ, Broder CC. HIV Coreceptors. J Membr Biol 1998;166:75–90.
84. Farzan M, Mirzabekov T, Kolchinsky P, Wyatt R, Cayabyab M, Gerard NP, Gerard C, Sodroski J, Choe H. Tyrosine sulfation of the amino terminus of CCR5 facilitates HIV-1 entry. Cell 1999;96:667–676.
85. Kozak SL, Platt EJ, Madani N, Ferro FE Jr, Peden K, Kabat D. CD4, CXCR-4, CCR-5 dependencies for infections by primary patient and laboratory-adapted isolates of human immunodeficiency virus type 1. J Virol 1997; 71:873–882.
86. Platt EJ, Madiani N, Kozak SL, Kabat D. Infectious properties of human immunodeficiency virus type 1 mutants with distinct affinities for the CD4 receptor. J Virol 1997;71:883–890.
87. Lapham CK, Zaitseva MB, Lee S, Romanstseva T, Golding H. Fusion of monocytes and macrophages with HIV-1 correlates with biochemical properties of CXCR4 and CCR5. Nat Med 1999;5:303–308.
88. Dimitrov DS, Norwood D, Stantchev TS, Feng Y, Xiao X, Broder CC. A mechanism of resistance to HIV-1 entry: Inefficient interactions of CXCR4 with CD4 and gp120 in macrophages. Virology 1999;259:1–6.
89. Hesselgesser J, Halks-Miller M, DelVecchio V, Peiper SC, Hoxie J, Kolson DL, Taub D, Horuk R. CD4-independent association between HIV-1 gp120 and CXCR4: functional chemokine receptors are expressed in human neurons. Curr Biol 1997;7:112–121.
90. Farzan M, Choe H, Martin KA, Sun Y, Sidiko M, Mackay CR, Gerard NP, Sodroski J, Gerard C. HIV-1 entry and macrophage inflammatory protein-1β-mediated signaling are independent functions of the chemokine receptor CCR5. J Biol Chem 1997;272:6854–6857.
91. Arthos J, Rubbert A, Rabin RL, Cicala C, Machado E, Wildt K, Hanbach M, Steenbeke TD, Swofford R, Farber

JM, Fauci AS. CCR5 signal transduction in macrophages by human immunodeficiency virus and simian immunodeficiency virus envelopes. J Virol 2000;74:6418–6424.
92. Davis CB, Dikie I, Unutmaz D, Hill CM, Arthos J, Siani MA, Thompson DA, Schlessinger J, Littman DR. Signal transduction due to HIV-1 envelope interactions with chemokine receptors CXCR4 or CCR5. J Exp Med 1997;186:1793–1798.
93. Weissman D, Rabin RL, Arthos J, Rubbert A, Dybul M, Swofford R, Venkatesan S, Farber JM, Fauci AS. Macrophage-tropic HIV and SIV envelope proteins induce a signal through the CCR5 chemokine receptor. Nature 1997;389:981–990.
94. Popik W, Pitha PM. Exploitation of cellular signaling by HIV-1: Unwelcome guests with master keys that signal their entry. Virology 2000;276:1–6.
95. Cicala C, Arthos J, Ruiz M, Vaccarezza M, Rubbert A, Riva A, Wildt K, Cohen O, Fauci AS. Induction of phosphorylation and intracellular association of CC chemokine receptor 5 and focal adhesion kinase in primary human $CD4^+$ T cells by macrophage-tropic HIV envelope. J Immunol 1999;163:420–426.
96. Liu QH, Williams DA, McManus C, Baribaud F, Doms RW, Schols D, De Clercq E, Kotlikoff MI, Collman RG, Freedman BD. HIV-1 gp120 and chemokines activate ion channels in primary macrophages through CCR5 and CXCR4 stimulation. Proc Natl Acad Sci 2000;97:4832–4837.
97. Cicala C, Arthos J, Rubbert A, Selig S, Wildt K, Cohen OJ, Fauci AS. HIV-1 envelope induces activation of caspase-3 and cleavage of focal adhesion kinase primary human $CD4^+$ T cells. Proc Natl Acad Sci USA 2000;97: 1178–1183.
98. Cocchi F, DeVico AL, Garzino-Demo A, Arya SK, Gallo RC, Lusso P. Identification of RANTES, MIP-1α, and MIP-1β as the major HIV-suppressive factors produced by CD8+ T cells. Science 1995;270:1811–1815.
99. Bleul CC, Farzan M, Choe H, Parolin C, Clark-Lewis I, Sodroski J, Springer TA. The lymphocyte chemoattractant SDF-1- is a ligand for LESTR/fusin and blocks HIV-1 entry. Nature 1996;382: 829–833.
100. Oberlin E, Amara A, Bachelerie F, Bessia C, Virelizier J-L, Arenzana-Seisdedos F, Schwartz O, Heard J-M, Clark-Lewis I, Legler DF, Loetscher M, Baggiolini M, Moser B. The CXC chemokine SDF-1 is the ligand for LESTR/fusin and prevents infection by T-cell-line-adapted HIV-1. Nature 1996;382:883–835.
101. Cayabyab M, Hinuma S, Farzan M, Choe H, Fukusumi S, Kitada C, Nishizawa N, Hosoya M, Nishimura O, Messele T, Pollakis G, Goudsmit J, Fujino M, Sodroski J. Apelin, the natural ligand of the orphan seven-transmembrane receptor APJ, inhibits human immunodeficiency virus Type 1 entry. J Virol 2000;74:11972–11976.
102. Puffer BA, Sharron M, Coughlan CM, Baribaud F, McManus CM, Lee B, David J, Price K, Horuk R, Tsang M, Doms RW. Expression and coreceptor function of APJ for primary immunodeficiency viruses. Virology 2000; 276:435–444.
103. Kelly MD, Naif HM, Adams SL, Cunningham AL, Lloyd AR. Dichotomous effects of β-chemokines on HIV replication in monocytes and monocyte-derived macrophages. J Immunol 1998;160:3091–3095.
104. Gordon CJ, Muesing MA, Proudfoot AEI, Power CA, Moore JP, Trkola A. Enhancement of human immunodeficiency virus type 1 infection by the CC-chemokine RANTES is independent of the mechanism of virus–cell fusion. J Virol 1999;73:684–694.
105. Schmidtmayerova H, Alfano M, Nuovo G, Bukrinsky M. Human immunodeficiency virus type 1 T-lymphotropic strains enter macrophages via a CD4- and CXCR4-mediated pathway: Replication is restricted at a postentry level. J Virol 1998;72:4633–4642.
106. Lipton SA. Neuronal injury associated with HIV-1: Approaches to treatment. Annu Rev Pharmacol Toxicol 1998;38:159–177.
107. Hesselgesser J, Taub D, Baskar P, Greenberg M, Hoxie J, Kolson DL, Horuk R. Neuronal apoptosis induced by HIV-1 gp120 and the chemokine SDF-1α is mediated by the chemokine receptor CXCR4. Curr Biol 1998;8: 595–598.
108. Ma Q, Jones D, Borghesani PR, Segal RA, Nagasawa T, Kishimoto T, Bronson RT, Springer TA. Impaired B-lymphopoiesis, myelopoiesis, derailed cerebellar neuron migration in CXCR4- and SDF-1-deficient mice. Proc Natl Acad Sci USA 1998;95:9448–9453.
109. Zou Y-R, Kottman AH, Koruda M, Taniuchi I, Littman DR. Function of the chemokine receptor CXCR4 in haematopoiesis and in cerebellar development. Nature 1998;393:595–599.
110. Kaul M, Lipton SA. Chemokines and activated macrophages in HIV gp120-induced neuronal apoptosis. Proc Natl Acad Sci USA 1999;96:8212–8216.
111. Bajetto A, Bonavia R, Barbero S, Piccioli P, Costa A, Florio T, Schettini G. Glial and neuronal cells express functional chemokine receptor CXCR4 and its natural ligand stromal cell-derived factor 1. J Neurochem 1999; 73:2348–2357.
112. Lazarini F, Casanova P, Tham TN, De Clercq E, Arenzana-Seisdedos F, Baleux F, Dubois-Dalcq M. Differential signalling of the chemokine receptor CXCR4 by stromal cell-derived factor 1 and the HIV glycoprotein in rat neurons and astrocytes. Eur J Neurosci 2000;12: 117–125.
113. Hesselgesser J, Horuk R. Chemokine and chemokine receptor expression in the central nervous system. J NeuroVirol 1999;5:13–26.
114. Lavi E, Kolson DL, Ulrich AM, Fu L, González-Scarano. Chemokine receptors in the human brain and their relationship to HIV infection. J NeuroVirol 1998;4:301–311.
115. Zhang L, He T, Talal A, Wang G, Frankel SS, Ho DD. In vivo distribution of the human immunodeficiency virus/simian immunodeficiency virus coreceptors: CXCR4, CCR3, and CCR5. J Virol 1998;72:5035–5045.
116. Meucci O, Fatatis A, Simen AA, Bushell TJ, Gray PW, Miller RJ. Chemokines regulate hippocampal neuronal signaling and gp120 neurotoxicity. Proc Natl Acad Sci USA 1998;95:14500–14505.
117. Zheng J, Ghorpade A, Niemann D, Cotter RL, Thylin MR, Epstein L, Swartz JM, Shepard RB, Liu X, Nukuna A, Gendelman HE. Lymphotropic virions affect chemokine receptor-mediated neural signaling and apoptosis:

implications for human immunodeficiency virus type 1-associated dementia. J Virol 1999;73:8256–8267.
118. Zheng J, Thylin MR, Ghorpade A, Xiong H, Persidsky Y, Cotter R, Niemann D, Che M, Zeng Y-C, Gelbard HA, Shepard RB, Swartz JM, Gendelman HE. Intracellular CXCR4 signaling, neuronal apoptosis and neuropathogenic mechanisms of HIV-1-associated dementia. J Neuroimmunol 1999;98:185–200.
119. Herbein G, Mahlknecht U, Batliwalla F, Gregersen P, Pappas T, Butler J, O'Brien WA, Verdin E. Apoptosis of CD8+ T cells is mediated by macrophages through interaction of HIV gp120 with chemokine receptor CXCR4. Nature 1998;395:189–194.
120. Penn ML, Grivel J-C, Schramm B, Goldsmith MA, Margolis L. CXCR4 utilization is sufficient to trigger $CD4^+$ T cell depletion in HIV-1-infected human lymphoid tissue. Proc Natl Acad Sci USA 1999;96:663–668.
121. Glushakova S, et al. Evidence for the HIV-1 phenotype switch as a causal factor in acquired immunodeficiency. Nat Med 1998;4:346–349.
122. Liu ZQ, Muhkerjee S, Sahni M, McCormick-Davis C, Leung K, Li Z, Gattone VH, Tian C, Doms RW, Hoffman TL, Raghavan R, Narayan O, Stephens EB. Derivation and biological characterization of a molecular clone of $SHIV_{KU-2}$ that causes AIDS, neurological disease, and renal disease in rhesus macaques. Virology 1999;260: 295–307.
123. Morris A, Marsden M, Halcrow K, Hughes ES, Brettle RP, Bell JE, Simmonds P. Mosaic structure of the human immunodeficiency virus type 1 genome infecting lymphoid cells and the brain: Evidence for frequent in vivo recombination events in the evolution of regional populations. J Virol 1999;73:8720–8731.
124. Brew BJ, Evans L, Byrne C, Pemberton L, Hurren L. The relationship between AIDS dementia complex and the presence of macrophage tropic and non-syncytium inducing isolates of human immunodeficiency virus type 1 in the cerebrospinal fluid. J NeuroVirol 1996;2:152–157.
125. Fitzgibbon JE, Gaur S, Gavai M, Gregory P, Frenkel LD, John JF Jr. Effect of the HIV-1 syncytium-inducing phenotype on disease stage in vertically infected children. J Med Virol 1998;55:56–63.
126. Madani N, Kozak SL, Kavanaugh MP, Kabat D. gp120 envelope glycoproteins of human immunodeficiency viruses competitively antagonize signaling by coreceptors CXCR4 and CCR5. Proc Natl Acad Sci USA 1998; 95:8005–8010.
127. Lannuzel A, Barnier JV, Hery C, Huynh VT, Guibert B, Gray F, Vincent JD, Tardieu M. Human immunodeficiency virus type 1 and its coat protein gp120 induce apoptosis and activate JNK and ERK mitogen-activated protein kinases in human neurons. Ann Neurol 1997;42: 847–856.
128. Asensio VC, Campbell IL. Chemokines in the CNS: plurifunctional mediators in diverse states. Trends Neurosci 1999;22:504–512.
129. Miller RJ, Meucci O. AIDS and the brain: is there a chemokine connection? Trends Neurosci 1999;22:471–479.
130. Glabinski AR, Ransohoff RM. Chemokines and chemokine receptors in CNS pathology. J NeuroVirol 1999;5: 3–12.
131. Schmidtmayerova H, Nottet HSLM, Nuovo G, Raabe T, Flanagan CR, Dubrovsky L, Gendelman HE, Cerami A, Bukrinsky M, Sherry B. Human immunodeficiency virus type 1 infection alters chemokine β peptide expression in human monocytes: amplifications for recruitment of leukocytes into brain and lymph nodes. Proc Natl Acad Sci USA 1996;93:700–704.
132. Nuovo GJ, Alfieri ML. AIDS dementia is associated with massive, activated HIV-1 infection and concomitant expression of several cytokines. Mol Med 1996;2:358–366.
133. Bernasconi S, Cinque P, Peri G, Sozzani S, Crociati A, Torri W, Vicenzi E, Vago L, Lazzarin A, Poli G, Mantovani A. Selective elevation of monocyte chemotactic protein-1 in the cerebrospinal fluid of AIDS patients with cytomegalovirus encephalitis. J Infect Dis 1996; 174:1098–1101.
134. Cinque P, Vago L, Mengozzi M, Torri V, Ceresa D, Vicenzi E, Transidico P, Vagani A, Sozani S, Mantovani A, Lazzarin A, Poli G. Elevated cerebrospinal fluid levels of monocyte chemotactic protein-1 correlate with HIV-1 encephalitis and local viral replication. AIDS 1998;12: 1327–1332.
135. Conant K, Garzino-Demo A, Nath A, McArthur JC, Halliday W, Power C, Gallo RC, Major EO. Induction of monocyte chemoattractant protein-1 in HIV-1 Tat-stimulated astrocytes and elevation in AIDS dementia. Proc Natl Acad Sci USA 1998;95:3117–3121.
136. Kelder W, McArthur JC, Nance-Sproson T, McClernon D, Griffin DE. β-chemokines MCP-1 and RANTES are selectively increased in cerebrospinal fluid of patients with human immunodeficiency virus-associated dementia. Ann Neurol 1998:44;831–835.
137. Kolb SA, Sporer B, Lahrtz F, Koedel U, Pfister HW, Fontana A. Identification of a T cell chemotactic factor in the cerebrospinal fluid of HIV-1-infected individuals as interferon-gamma inducible protein 10. J Neuroimmunol 1999:93;173–181.
138. Letendre SL, Lanier ER, McCutchan JA. Cerebrospinal fluid β chemokine concentrations in neurocognitively impaired individuals infected with human immunodeficiency virus Type 1. J Infect Dis 1999:180;310–319.
139. Franciotta D, Martino G, Zardini E, Furlan R, Bergamaschi R, Andreoni L, Cosi V. Serum and CSF levels of MCP-1 and IP-10 in multiple sclerosis patients with acute and stable disease and undergoing immunomodulatory therapies. J Neuroimmunol 2001:115;192–198.
140. Sasseville JG, Smith MM, MacKay CR, Pauley DR, Mensfield KG, Ringler DR, Lackner AA. Chemokine expression in simian immunodeficiency virus-induced AIDS encephalitis. Am J Path 1996:149;1459–1467.
141. Peterson PK, Hu S, Salak-Johnson J, Molitor TW, Chao CC. Differential production of and migratory response to beta chemokines by human microglia and astrocytes. J Infect Dis 1997:175;478–481.
142. McManus CM, Brosnan CF, Berman JW. Cytokine induction of MIP-1 alpha and MIP-1 beta in human fetal microglia. J Immunol 1998:160;1449–1455.

143. Persidsky Y, Ghorpade A, Rasmussen J, Limoges J, Liu XJ, Stins M, Flala M, Way D, Kim KS, Witte MH, Weinand M, Carhart LR, Gendelman HE. Microglial and astrocyte chemokines regulate monocyte migration through the blood–brain barrier in human immunodeficiency virus-1 encephalitis. Am J Pathol 1999;155:1599–1611.
144. Boven LA, Middel J, Breij ECW, Schotte D, Verhoef J, Soderland C, Nottet HSLM. Interactions between HIV-infected monocyte-derived macrophages and human brain microvascular endothelial cells result in increased expression of CC chemokines. J NeuroVirol 2000;6:382–389.
145. Baldeweg T, Catalan J, Gazzard BG, Weiss RA, Boshoff C. Kaposi's sarcoma and protection from HIV dementia. Science 1998;280:362.
146. Liestoel K, Goplen AK, Dunlop O, Bruun JN, Moehlen J. Kaposi's sarcoma and protection from HIV dementia. Science 1998;280:361–362.
147. Weiss JM, Nath A, Major EO, Berman JW. HIV-1 Tat induces monocyte chemoattractant protein-1-mediated monocyte transmigration across a model of the human blood–brain barrier and up-regulates CCR5 expression on human monocytes. J Immunol 1999;163:2953–2959.
148. Pulliam L, Gascon R, Stubblebine M, McGuire D, McGrath MS. Unique monocyte subset in patients with AIDS dementia. Lancet 1997;349:692–695.
149. Nottet HSLM, Persidsky Y, Sasseville VG, Nukuna AN, Bock P, Zhai Q-H, Sharer LR, McComb RD, Swindells S, Soderland C, Gendelman HE. Mechanisms for the trans-endothelial migration of HIV-1-infected monocytes into brain. J Immunol 1996;156:1284–1295.
150. Persidsky Y, Stins M, Way D, Witte MH, Weinand M, Kim KS, Bock P, Gendelman HE, Fiala M. A model for monocyte migration through the blood–brain barrier during HIV-1 encephalitis. J Immunol 1997;158:3499–3510.
151. Sozzani S, Ghezzi S, Iannolo G, Luini W, Borsatti A, Polentarutti N, Sica A, Locati M, Mackay C, Wells TNC, Biswas P, Vicenzi E, Poli G, Montani A. Interleukin 10 increases CCR5 expression and HIV infection in human monocytes. J Exp Med 1998;187:439–444.
152. Wang J, Crawford K, Yuan M, Wang H, Gorry PR, Gabuzda D. Th2 cytokines regulate CCR5 and CD4 expression and human immunodeficiency virus type 1 replication in human macrophages and microglia. J Infect Dis 2002; in press.
153. Tong N, Perry SW, Zhang Q, James HJ, Guo H, Brooks A, Bal H, Kinnear SA, Fine S, Epstein LG, Dairaghi D, Schall TJ, Gendelman HE, Dewhurst S, Sharer LR, Gelbard HA. Neuronal fractalkine expression in HIV-1 encephalitis: Roles for macrophage recruitment and neuroprotection in the central nervous system. J Immunol 2000;164:1333–1339.
154. Pereira CF, Middel J, Jansen G, Verhoef J, Nottet HSLM. Enhanced expression of fractalkine in HIV-1 associated dementia. J. Neuroimmunol 2001;115:168–175..
155. Pan Y, Lloyd C, Zhou H, Dolich S, Deeds J, Gonzalo JA, Vath J, Gosselin M, Ma J, Dussault B, Woolf E, Alperin G, Culpepper J, Gutierrez-Ramos JC, Gearing D. Neurotactin, a membrane-anchored chemokine upregulated in brain inflammation. Nature 1997;387:611–517.
156. Harrison JK, Jiang Y, Chen S, Xia Y, Maciejewski D, McNamara RK, Streit WJ, Salafranca MN, Adhikari S, Thompson DA, Botti P, Bacon KB, Feng L. Role for neuronally derived fractalkine in mediating interactions between neurons and CX3CR1-expressing microglia. Proc Natl Acad Sci USA 1998;95:10896–10901.
157. Maciejewski-Lenoir D, Chen S, Feng L, Maki R, Bacon KB. Characterization of fractalkine in rat brain cells: migratory and activation signals for CX_3CR-1-expressing microglia. J Immunol 1999;163:1628–1635.
158. Zujovic V, Benavides J, Vige X, Carter C, Taupin V. Fractalkine modulates TNF-α secretion and neurotoxicity induced by microglial activation. Glia 2000;29:305–315.
159. Schols D, Struyf S, Van Damme J, Este JA, Henson G, De Clercq E. Inhibition of T-tropic HIV strains by selective antagonization of the chemokine receptor CXCR4. J Exp Med 1997;186:1383–1388.
160. Simmons G, Clapham PR, Picard L, Offord RE, Rosenkilde MM, Schwartz TW, Buser R, Wells TNC, Proudfoot AE. Potent inhibition of HIV-1 infectivity in macrophages and lymphocytes by a novel CCR5 antagonist. Science 1997;276;276–279.
161. Donzella GA, Schols D, Lin SW, Este JA, Nagashima KA, Maddon PJ, Allaway GP, Sakmar TP, Henson G, De Clercq E, Moore JP. AMD3100, a small molecule inhibitor of HIV-1 entry via the CXCR4 co-receptor. Nat Med 1998:4;72–77.
162. Cairns JS, D'Souza MP. Chemokines and HIV-1 second receptors: The therapeutic connection. Nat Med 1998:4; 563–568.
163. Baba M, Nishimura O, Kanzaki N, Okamoto M, Sawada H, Iizawa Y, Shiraishi M, Aramaki Y, Okonogi K, Ogawa Y, Meguro K, Fujino M. A small-molecule, nonpeptide CCR5 antagonist with highly potent and selective anti-HIV-1 activity. Proc Natl Acad Sci USA 1999;96:5698–5703.
164. Mosier DE, Picchio GR, Gulizia RJ, Sabbe R, Poignard P, Picard L, Offord RE, Thompson DA, Wilken J. Highly potent RANTES analogues either prevent CCR5-using human immunodeficiency virus type 1 infection in vivo or rapidly select for CXCR4-using variants. J Virol 1999; 73:3544–3550.
165. Michael NL, Moore JP. HIV-1 entry inhibitors: evading the issue. Nat Med 1999;5:740–742.
166. D'Souza MP, Cairns JS, Plaeger SF. Current evidence and future directions for targeting HIV entry — Therapeutic and prophylactic strategies. J Am Med Assoc 2000;284:215–222.
167. Mack M, Luckow B, Nelson PJ, Cihak J, Simmons G, Clapham PR, Signoret N, Marsh M, Stangassinger M, Borlat F, Wells TN, Schlondorff D, Proudfoot AE. Amino-oxypentane-RANTES induces CCR5 internalization but inhibits recycling: a novel inhibitory mechanism of HIV infectivity. J Exp Med 1998;187:1215–1224.
168. Rusconi S, La Seta Catamancio S, Citterio P, Bulgheroni E, Croce F, Herrmann SH, Offord RE, Galli M, Hirsch MS. Combination of CCR5 and CXCR4 inhibitors in therapy of human immunodeficiency virus type 1 infec-

tion: in vitro studies of mixed virus infections. J Virol 2000;74:9328–9332.
169. Tremblay CL, Kollmann C, Giguel F, Chou T-C, Hirsch MS. Strong in vitro synergy between the fusion inhibitor T-20 and the CXCR4 blocker AMD-3100. JAIDS 2000; 25:99–102.

Universes in Delicate Balance: Chemokines and the Nervous System
R.M. Ransohoff, K. Suzuki, A.E.I. Proudfoot, W.F. Hickey and J.K. Harrison, editors

Chapter 5.4

Infections: Meningitis and Encephalitis

Katharina-Susanne Spanaus and Adriano Fontana

Section of Clinical Immunology, Department of Internal Medicine, University Hospital, 8044 Zurich, Switzerland

1. OVERVIEW

The central nervous system (CNS) is protected from invading infectious agents by the blood–brain barrier, which also limits the access of leukocytes and plasma components from the blood to the brain parenchyma, the choroid plexus and to the cerebrospinal fluid in the subarachnoid space. Even though infections of the central nervous system may occur by extension from infections in the neighboring tissue or by communication of the subarachnoid space or the brain parenchyma with the exterior, the majority of cases is caused by hematogenous spread of infectious agents. An inflammatory reaction leads to the breakdown of the blood–brain barrier, which is followed by a migration of leukocytes and leakage of protein. This process involves the meninges and the subarachnoid space in the case of meningitis, the brain parenchyma in encephalitis or both tissues in meningoencephalitis. The course and the symptoms of the CNS inflammation vary greatly in dependence on the triggering agent. While viral meningitis is a disease with rare fatal outcome, the mortality rate in bacterial meningitis is still high and many survivors present severe neurologic sequelae as a result of an irreparable destruction of nervous tissue. At the onset of the disease, bacterial cell wall components, complement factors like C5a, chemokines and other inflammatory cytokines draw large amounts of leukocytes in the subarachnoid space. These cells initiate and propagate brain injury by releasing cytotoxic factors including proteases, glutamate, reactive oxygen intermediates and nitric oxide [1]. Further brain damage results from ischemia due to vasculitis and thrombophlebitis of cerebral vessels in the inflamed area, and from brain edema due to disruption of the blood–brain barrier. In addition, the growing exudate may damage cranial nerves and lead to the obliteration of cerebrospinal fluid (CSF) pathways, thus causing hydrocephalus and increasing intracranial pressure [2]. It is noteworthy that the brain damage seems to be caused mainly by the inflammatory reaction in response to the pathogen, rather than by the pathogen itself. This is illustrated in a mouse model system where solely the expression of the chemokine N51/KC (CXCL1) under an oligodendrocyte-specific promotor caused neutrophil infiltration and brain damage similar to that in infectious encephalitis [3]. Therefore, chemokines, which are produced in the course of the inflammatory response to microbial and viral pathogens, may play a central role in the brain damage in infectious diseases of the CNS.

2. MENINGEAL INFLAMMATION AND ENCEPHALITIS WITH INFECTIOUS ETIOLOGY

2.1. Bacterial Meningitis

From a wide variety of bacterias which can cause bacterial meningitis, *Neisseria meningitidis* and *Streptococcus pneumoniae* are the most common. Less common are Gram-negative meningitides due to infection with *Escherichia coli*, *Klebsiella* or *Enterobacter* sp. The occurrence of meningeal infections with one of the latter bacteria are, like listerial meningitis, often associated with predisposing factors like compromised immunity. Adding to this list, staphylococcal meningitis may occur in patients after penetrating head injuries. In terms of patient ages, *Haemophilus influenzae* type b was the most common cause of disease in young children until vaccination efforts in recent years drastically reduced the incidence rates. Meningitis due to *E. coli* and group B streptococcus are still relatively frequent in children during the first 2 years of life.

Acute bacterial meningitis is characterized by a rapid influx of large numbers of leukocytes into the CSF resulting in a white cell count in the range of 1000–5000/μl with a neutrophil predominance [4]. In patients, as well as in animal models, it was found that the expression of proinflammatory cytokines precedes the cellular inflammatory response. Significant amounts of the proinflammatory cytokines TNF-α, IL-1 and IL-6 could be detected in the CSF of patients with bacterial meningitis [5–7]. In animals, intracisternal injection of LPS or TNF-α and IL-1β mimics infection and subsequent cytokine production and leads to pleocytosis in the CSF [8–10]. In accordance with this invasion of lymphocytes into the CNS, high levels of IFN-γ were found over a prolonged period of time in the CSF of children with tuberculous meningitis [11], whereas TNF-α was detected only at low concentrations. Conversely, meningeal inflammation due to pneumococcal infection can be neutralized by intracisternal injection of anti-IL-1 or anti-TNF-α-antibodies [8]. Meningitis due to infection with *Mycobacterium tuberculosis*, *Borrelia burgdorferi* or *Treponema pallidum*, take up special positions. In these cases, a subacute meningitis evolves over a longer period of time and is characterized by a predominantly lymphocytic pleocytosis.

Among other factors, cytokines induce the expression of adhesion molecules on the cerebrovascular endothelium, an important precondition for the recruitment of leukocytes to the site of inflammation. As outlined in more detail in Section 3, soluble mediators like chemokines are then necessary, to guide the once adherent cells into the subarachnoid space. Proinflammatory cytokines are known to stimulate a variety of cells to secrete chemokines which induce the migration of specific subsets of leukocytes. Interactions between proinflammatory cytokine levels and production of chemokines and their effect on immune cells are discussed further in Section 7.

2.2. Aseptic Meningitis

The most common causes of viral meningitis are enteroviruses and paramyxoviruses. In addition, meningeal inflammations can also be caused by a number of other infectious agents including HIV, protozoans like *Toxoplasma* and funghi. In the following, main focus will be given to the more abundant forms of viral and bacterial meningitis.

The term viral meningitis is often used to describe an aseptic meningitis syndrome of suspected or documented viral etiology. The disease is characterized by a meningeal inflammation which is clinically determined by a mononuclear CSF pleocytosis and an absence of bacteria in CSF culture. In contrast to bacterial meningitis, the CSF infiltrates mainly consist of monocytes and activated T-cells and the cell counts are much lower. Only in the very early phase an influx of neutrophils into the CSF can be observed. As in bacterial meningitis, the proinflammatory cytokines TNF-α and IL-1 are also present in viral meningitis but at much less elevated levels. In contrast, high concentrations of IFN-γ could be detected in viral but not in bacterial meningitis. This pathogen-specific cytokine pattern has consequences for the induction of chemokine expression. Since TNF-α and IL-1 are potent inducers of several chemokines in different cell types, chemokine concentrations might be lower in viral than in bacterial meningitis as a result of the less abundant proinflammatory cytokines. Contrarily, chemokines whose production is induced by IFN-γ should be detectable mainly in viral meningitis. Maybe due to the action of the antiinflammatory cytokines IL-10 and TGF-β, which are found in the CSF over a long period of viral infection, the extent of inflammation is in most cases moderate.

2.3. Encephalitis

Encephalitis, an acute inflammatory disease of the brain parenchyma, is usually the result of a viral infection. The most common viruses causing viral encephalitis are arboviruses, echoviruses, coxsackieviruses, rhabdoviruses and herpesviruses. From such primary encephalitides para- or postinfectious forms have to be distinguished. These develop in succession of viral infections and are supposed to be based on immunologic mechanisms.

The more rare bacterial encephalitides are most often caused by an embolic mechanism in patients with endocarditis, the infectious agents being *Staphylococcus aureus* or *Streptococcus*. Less common are encephalitides due to infections with other bacteria, fungi, protozoans and helminths. In encephalitis, besides a CSF pleocytosis, inflammatory cells are found in CNS parenchyma and may form perivascular inflammatory cuffs. The roles of chemokines and proinflammatory cytokines in encephalitis are less well established than in meningitis. So far, the majority of studies provide clear evidence for a correlation between acuteness and severity of the disease and CSF levels of TNF-α and, in part, of IFN-γ [12–14].

3. RECRUITMENT OF LEUKOCYTES TO THE CSF COMPARTMENT: THE ROLE OF CHEMOKINES AND MATRIX METALLOPROTEINASES

Recruiting leukocytes to the site of infection requires the immune cells to leave the blood vessels in a site-directed manner. This process is thought to comprise four steps: rolling of leukocytes along the vessel walls, activation of integrins on the surface of the leukocytes, activation-dependent arrest of the cells and finally migration through the vessel wall [15–19]. Extravasation of immune cells into the CNS is thought to follow these general steps as well. As the blood–brain barrier represents a very tight network of cells that invading leukocytes have to penetrate, the combined action of different groups of molecules is necessary to enable them to reach the site of infection in the CNS [20]. In the guidance of the migrating leukocytes, chemokines are involved in a central step of the immune reaction: chemokines bound to heparan sulfate side chains on the endothelial cell membranes interact with a subset of seven-transmembrane G-protein-coupled chemokine receptors on the surface of leukocytes. The resulting activation of integrins at the cellular membrane of the leukocytes enhances the ability of the immune cells to adhere to the endothelium. Furthermore, leukocytes are thought to migrate through the endothelial cell layer along a gradient of chemokines and other chemoattractant factors. Besides the large group of chemokines, fMLP (*N*-formyl-L-methionyl-L-leucyl-L-phenylalanine), which is produced during bacterial cell lysis, the complement component C5a, leukotriene B4 and platelet-activating factor possess chemoattractant activity. Penetration of the endothelial cell layer is thought to involve the action of proteolytic enzymes including matrix metalloproteinases (MMPs). While MMP-2 is constitutively expressed in the CSF compartment, MMP-9 is specifically found at high levels in bacterial meningitis [21]. Also in viral meningitis, MMP-9 is detected in up to 30% of the CSF samples, the levels correlating with the number of neutrophils in the CSF [22].

To date, more than 40 chemokines have been identified in humans. A significant subset of these factors have been characterized in CSF, tissues and cell cultures from human patients or have been studied as homologues in experimental animal models. Their action on specific subsets of leukocytes — neutrophils, monocytes, lymphocytes and eosinophils — and their effect on the course of infectious diseases in the CNS will be discussed in the following. For a summary of the chemokines relevant in meningitis and their target cell specificity see Table 1.

4. CHEMOTACTIC ACTIVITY ON NEUTROPHILS IN INFECTIOUS DISEASES OF THE CNS

CXC chemokines, which act strongly on neutrophils are present in the CNS during infectious diseases. In patients with infectious meningitis two members of this subfamily, CXCL8/IL-8 and CXCL1/GROα, were found in the CSF at elevated levels compared to the control CSFs of healthy donors [23–31]. Between bacterial and viral meningitis CSF, a significant difference was found in the concentration of CXCL8/IL-8, which was much higher in bacterial than in viral meningitis samples. The CXC chemokine CXCL1/GROα was found in high concentrations in the bacterial meningitis CSF but only in a third of the viral meningitis CSF samples and was not detected in the controls [28,31]. In the sample group of another study, however, all CSF from patients with viral meningitis were positive for CXCL1/GROα [29].

CXCL8/IL-8 was also found in the CSF of patients with tuberculous meningitis [26]. The measured chemokine levels in these patients were signif-

Table 1. Chemokines, receptors and target cells in infectious diseases of the CNS

Systematic name	Human cytokine	Mouse cytokine	Receptor	Target cell
CXC chemokines				
CXCL1	GROα	GRO/MIP-2	CXCR2	Neutrophils, monocytes, basophils, eosinophils
CXCL8	IL-8	Unknown	CXCR1,2	Neutrophils, monocytes
CXCL9	Mig	Mig	CXCR3	Activated T-cells
CXCL10	IP-10	IP-10/crg-2	CXCR3	Activated T-cells
CXCL11	I-TAC	Unknown	CXCR3	Activated T-cells
CC chemokines				
CCL2	MCP-1	JE	CCR2	Activated T-cells, monocytes, basophils
CCL3	MIP-1α	MIP-1α	CCR1, CCR5	Activated T-cells, monocytes, basophils
CCL4	MIP-1β	MIP-1β	CCR5	Activated T-cells, monocytes, dendritic cells
CCL5	RANTES	RANTES	CCR1, CCR3, CCR5	Activated T-cells, monocytes, eosinophils, basophils, dendritic cells

icantly lower than in CSF of patients with pyogenic meningitis. Nonetheless, CXCL8/IL-8 levels in patients with tuberculous meningitis remained elevated over several weeks. In analogy to viral meningitis, CXCL8/IL-8 is also expressed during viral encephalitis: augmented levels of this chemokine, but no correlation with the neutrophil count, were reported in three patients with herpes simplex type 1 encephalitis [32].

CSF from bacterial as well as viral meningitis exerts chemotactic activity on neutrophils as shown in chemotactic assays in vitro using modified Boyden chambers and peripheral derived polymorphonuclear cells (PMN) as target cells [28,29,31]. Neutralization experiments revealed that in bacterial meningitis CXCL8/IL-8 as well as CXCL1/GROα are active in the CSF and contribute to the chemotactic activity of the CSF on PMN. Due to the small amount of CXCL1/GROα found in the CSF of viral meningitis patients neutralization experiments from viral samples were performed for CXCL8/IL-8 only. In bacterial as in viral meningitis, CXCL8/IL-8 was responsible for the major fraction of chemotactic activity in the CSF and in both cases CXCL8/IL-8 concentration correlated with chemotactic activity of the CSF (Fig. 1). While no correlation was found between the number of neutrophils and the levels of CXCL8/IL-8 in CSF in all bacterial meningitis samples, findings in viral meningitis are controversial: the researchers of three groups detected a correlation between CXCL8/IL-8 levels and neutrophil counts in the CSF [23,24,31], whereas those of another group did not [29].

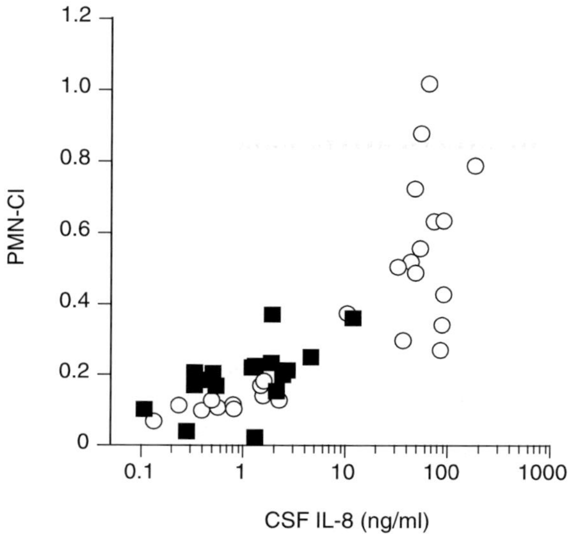

Figure 1. Chemotactic activity on PMN correlates with CXCL8/IL-8 concentrations in the CSF of patients with viral and bacterial meningitis (single patient's values, $P < 0.01$). Chemotactic activity was determined in a modified Boyden chamber. Chemokine concentrations are presented in relation to a chemotactic index (CI). CI results from comparison of CSF chemotactic activity to a positive control (fMLP) [28]. ○, CSF from patients with bacterial meningitis [28]; ■, CSF from patients with viral meningitis [31].

To investigate the time course of CXC chemokine expression in meningeal inflammation, MIP-2, the mouse homolog of CXCL1/GROα, was analyzed in an animal model. In mice, MIP-2 was detected in the CSF compartment as early as 12 h after intracerebral injection of *Listeria monocytogenes* reaching an expression maximum after 72 h and shortly before the animals died [33]. Also in this case, MIP-2 contributed to CSF-mediated chemotaxis on neutrophils in vitro. In other studies, MIP-2 was found in meningitis in infant rats infected with *H. influenzae* type B [34] and in mice with an acute encephalomyelitis after inoculation of mouse hepatitis virus [35].

Besides CXCL8/IL-8 and CXCL1/GROα, other cytokines and chemoattractant factors may also be involved in the migration of PMN into the subarachnoid space. One candidate is granulocyte-colony stimulating factor (G-CSF) which is chemotactically active [36] and has been detected in CSF of patients with viral meningitis [37], the concentrations correlating with PMN counts in the CSF. So far, no data are available on in vitro chemotactic activity of CSF positive for G-CSF. Furthermore, complement C3, a potent attractor of neutrophils, has been identified in the CSF of patients with bacterial meningitis [38].

In summary, in viral meningitis CXCL8/IL-8 is the predominant chemokine for the accumulation of neutrophils in the initial phase of the disease. This contrasts to some extent the situation in bacterial meningitis where in addition to CXCL8/IL-8 also CXCL1/GROα and other molecules including complement factors, leukotriene B4, platelet activating factor, and bacterial cell wall components and toxins have been shown to be present and can lead to CSF pleocytosis.

5. CHEMOTACTIC ACTIVITY ON MONONUCLEAR CELLS IN INFECTIOUS DISEASES OF THE CNS

Monocytes and activated T-cells predominate in viral meningitis. In bacterial meningitis, mononuclear cells are of less importance at early time points but may increase in numbers during the course of the disease. In humans, chemokines active on mononuclear cells include the members of the CC subfamily. Of these chemokines, elevated CCL2/MCP-1 levels were detectable in almost all CSF samples of patients with bacterial or viral meningitis with higher median concentrations in bacterial meningitis [28,29,31]. Furthermore, levels of CCL3/MIP-1α and CCL4/MIP-1β were found to be substantially elevated in bacterial meningitis [28,29,39]. This contrasts with CCL5/RANTES which was only present at low concentrations in 25% of the CSFs from patients with bacterial meningitis [28]. The latter three chemokines were also detected in a minor fraction of the samples of patients with viral meningitis [31], although at very low concentrations. The importance of CCL2/MCP-1 in CNS inflammation is illustrated by the absence of CCL3/MIP-1α, CCL4/MIP-1β, and CCL5/RANTES, but by the presence of CCL2/MCP-1 in the CSF from patients infected with HIV and concomitant cytomegalovirus encephalitis [40]. Also in patients with a herpes simplex type 1 encephalitis, elevated levels of CCL2/MCP-1, CCL3/MIP-1α and CCL5/RANTES were found [32]. Among the three cytokines, CCL2/MCP-1 concentrations were the highest. Moreover, a positive intraindividual correlation was found between CCL2/MCP-1 levels and mononuclear cell count. CCL2/MCP-1 and CCL3/MIP-1α were detectable in the CSF of patients with tuberculous meningitis [26]. In accordance with the course of the disease, which begins subacute and may last for several weeks, the concentrations of both chemokines were found to be elevated over a prolonged period of time and the highest CCL2/MCP-1 levels were significantly lower than in pyogenic meningitis.

Assessment of chemotactic activity by in vitro migration assays using freshly prepared PBMC as target cells revealed that mainly monocytes migrated in response to the CSF of patients with meningitis [28,31]. This is due to the fact that PBMC consist of monocytes and resting T-cells. In contrast to T-cells pre-activated with IL-2, resting T-cells do not migrate in response to CCL2/MCP-1 and CXCL10/IP-10 because they lack the receptors for these chemokines [41,42]. In viral and bacterial meningitis, chemotactic activity on PBMC correlated with the chemokine concentrations in the CSF (Fig. 2). Corresponding to the higher chemokine levels in CSF of bacterial meningitis, also chemotactic activity was significantly higher in bacterial than in viral meningitis. In both meningeal inflammations, anti CCL2/MCP-1 antibodies partially neutralized the chemotactic activity of the CSF. The remaining chemotaxis of PBMC in vitro is thought to reflect the contribution of CCL3/MIP-1α and CCL4/MIP-1β present in the CSF of patients. Although a major part of chemotactic activity of CSF could be abolished by antibody-neutralization of chemokines, neither in viral nor in bacterial meningitis a strict correlation

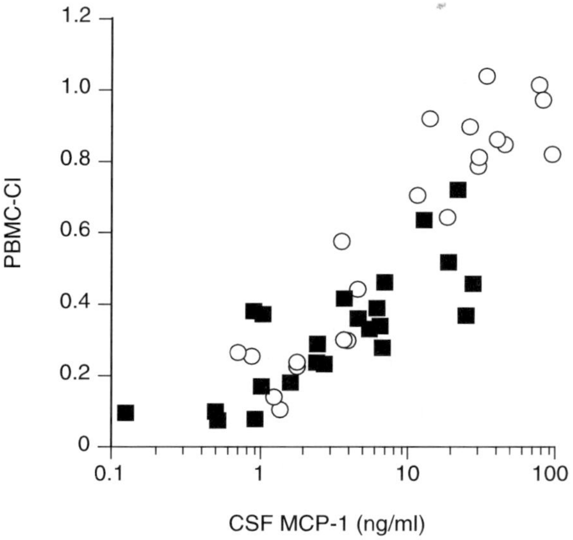

Figure 2. Chemotactic activity on PBMC correlates with the individual CCL2/MCP-1 concentration in the CSF of patients with viral and bacterial meningitis (single patient's values, $P < 0.01$). ○, CSF from patients with bacterial meningitis [28]; ■, CSF from patients with viral meningitis [31].

could be found between chemokine concentrations and mononuclear cell count in the CSF. Adding to this, it is of note that low levels of CCL2/MCP-1 are also found in CSF of controls [28,31]. These CSF samples exert a weak chemotactic activity on mononuclear cells in vitro. This constitutive expression of low levels of CCL2/MCP-1 might explain the presence of small numbers of monocytes and T-cells in the CSF of healthy donors.

The findings in human patients are paralleled by experimental data obtained from different animal model systems. In one of these systems, infant rats infected intraperitoneally with *H. influenzae* type B developed meningitis and showed expression of CCL3/MIP-1α, CCL2/MCP-1 and CCL5/RANTES in the brain [34]. Also in experimental listerial meningoencephalitis in mice CCL3/MIP-1α was found in the CSF. First detection of the chemokine was as early as 12 h after inoculation of bacteria, with maximum levels at the time point 48 h [33]. In situ hybridization analysis revealed that besides the CCL3/MIP-1α gene also the gene of CCL4/MIP-1β was transcribed in an early phase of the disease in infiltrating leukocytes. In the experimental animal system as well, both chemokines contributed to CSF-mediated chemotaxis on mononuclear cells in vitro. Further evidence for the importance of this group of chemokines comes from transgenic mice overexpressing CCL2/MCP-1 which show efficient recruitment of monocytes into the CNS [43].

Taken together, the results listed above suggest a function for CCL2/MCP-1 in the recruitment of monocytes under both physiological as well as under conditions of acute viral and bacterial meningitis. Within the network of chemokine activities, factors can cooperate in synergistic ways as shown for the enhanced chemotactic activity of CCL2/MCP-1 in bacterial meningitis due to the presence of CCL3/MIP-1α and CCL4/MIP-1β. The fundamental role for CCL2/MCP-1 within this set of chemokines and in the host response to brain injury is further underlined by the detection of CCL2/MCP-1 in other acute brain diseases, including experimental brain lesions [44], ischemia [45] and EAE [46].

6. CHEMOTACTIC ACTIVITY ON ACTIVATED T-CELLS IN INFECTIOUS DISEASES OF THE CNS

The chemotactic signaling molecules acting on activated T-cells comprise the CC-chemokines CCL2/MCP-1, CCL3/MIP-1α, CCL4/MIP-1β, and CCL5/RANTES, the CXC-chemokines CXCL9/Mig, CXCL10/IP-10, and CXCL11/I-TAC, and the chemotactically active cytokines IL-15 and IL-16.

Among these factors, CCL2/MCP-1 and CXCL10/IP-10 were detected in the CSF of patients with viral meningitis [31]. CCL5/RANTES, CCL3/MIP-1α, CCL4/MIP-1β, IL-15, and IL-16 were absent or detectable in only a small fraction of the samples. Furthermore, despite the significant detection of CXCL10/IP-10, CXCL9/Mig, a CXC chemokine which also binds to CXCR3 [41], was not found in viral meningitis CSF. Expression of CXCL11/I-TAC was not investigated in the above-mentioned patient studies. As CXCL11/I-TAC, like CXCL10/IP-10 and CXCL9/Mig, belongs to a subfamily of CXC chemokines lacking the ELR motif [47,48] and is induced in response to IFN-γ as well, this T-cell chemoattractant would be of great interest if found to be released in the CSF of patients.

In vitro, T-cells preactivated with IL-2 showed a profound chemotactic response upon stimulation with CSF from patients with viral meningitis

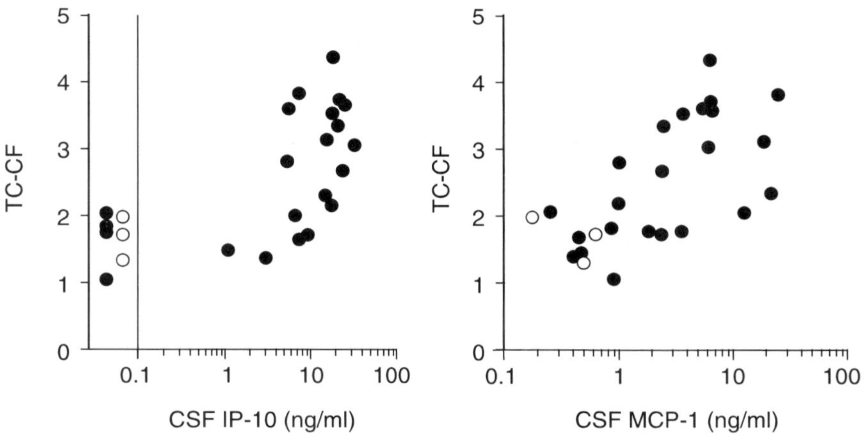

Figure 3. Chemotactic activity on preactivated T-cells correlates with the individual CCL2/MCP-1 and CXCL10/IP-10 concentrations in the CSF from patients with viral meningitis (single patient's values, $P < 0.01$). Chemotactic activity is expressed by a chemotactic factor (CF) [31]. Vertical lines denote detection limit. •, CSF from patients with viral meningitis; ○, control CSF.

[31]. The chemotactic activity correlated with levels of CXCL10/IP-10 and CCL2/MCP-1 in the CSF (Fig. 3). Anti CXCL10/IP-10 antibodies and anti CCL2/MCP-1 antibodies significantly reduced the chemotactic activity of the CSF for activated T-cells with maximum inhibition of chemotaxis being seen upon use of both antibodies in combination. Thus, the chemokines CXCL10/IP-10 and CCL2/MCP-1 were identified as chemotactic signals responsible for chemoattraction of activated T-cells by the CSF of the patients.

CXCL10/IP-10 as well as the other two chemokines which bind to CXCR3 receptors on activated T-cells may be essential for the recruitment of activated CD8 T-cells into the CNS. The essential role of CXCR3-mediated signaling for chemotaxis of CD8 T-cells has been most elegantly shown recently. In CXCR3 gene knockout mice rejection of cardiac allografts is delayed and less severe, the CTL alloresponse being decreased compared to controls [49]. However, when infecting the $CXCR3^{-/-}$ mouse generated by Craig Gerard's group with LCMV, these mice were found to develop lethal meningitis with accumulation of CD8 T-cells in the meninges and CSF. The disease and extent of inflammation was not different from wild-type control mice (A. Fontana, unpublished observation). Thus other chemokines including MCP-1 may compensate for the lack of function of CXCR3-binding chemokines in LCMV disease. As shown in this gene knockout chemokine receptor model, only blockade of several chemokine receptors will mediate sufficient protection in infectious diseases which are in general associated with a wide range of chemotactic signals.

7. CELLULAR SOURCES OF CHEMOKINES IN INFECTIOUS DISEASES OF THE CNS

In the normal brain, chemokines and their receptors are constitutively expressed only at low levels or are not detectable. The production is strongly upregulated in inflammatory diseases in the brain. The cellular sources of the chemokines during the course of CNS infections is complex and not yet fully understood for all participating chemokines. Leukocytes invading the meninges and the choroid plexus are one possible source of chemokines in meningitis. PMN stimulated with outer membrane vesicles (OMV) of *N. meningitidis* B have been shown to produce CXCL8/IL-8, CCL3/MIP-1α and CCL4/MIP-1β. Stimulated with a combination of OMV and IFN-γ, they also released CXCL10/IP-10 [50]. Costimulation of neutrophils with IFN-γ and TNF-α or LPS in vitro led to production of CXCL10/IP-10, CXCL9/Mig, and also CXCL11/I-TAC [51,52]. Activated tissue macrophages together with peripheral blood monocytes were shown to produce CXCL10/IP-10, CCL2/MCP-1, CCL3/MIP-1α and CXCL8/IL-8 during the early course of infection [23,39,53–56]. This has also been shown for macrophages stimulated in vitro with IFN-γ, which induced the production of CCL2/MCP-1 and CXCL10/IP-10 [54,57]. Chemokine production assayed in vivo during the course of listerial meningoencephalitis in mice and meningitis due to infection with *H. influenzae* type B in infant rats showed that neutrophils and macrophages infiltrating the meninges, the ventricular system and the periventricular area contribute to the production of CCL3/MIP-1α, CCL4/MIP-1β, CXCL1/MIP-2 and CCL2/MCP-1 in the CNS [33,34].

In addition to invading immune cells, brain resident cells including microglia and astrocytes as well as endothelial cells are potent producers of chemokines. Cultured microglial cells were shown to produce CCL2/MCP-1 after LPS stimulation [58] and human fetal microglia have been found to produce CCL2/MCP-1, CCL3/MIP-1α and CCL4/MIP-1β after stimulation with LPS and proinflammatory cytokines [58,59]. In vivo, microglia as well as astrocytes were shown to express CCL5/RANTES in rats suffering from experimental meningitis due to the infection with *H. influenzae* type B. In this model system, astrocytes also produced CCL2/MCP-1 [34,58]. Furthermore, stimulation with IL-1β, a cytokine present in the CSF of patients with bacterial and viral meningitis [7,58] induced the production of CXCL8/IL-8 in astrocytes [60]. In a similar manner, proinflammatory factors acting on astrocytes enhanced production of CXCL11/I-TAC (by IFN-γ) and CCL2/MCP-1 (by LPS or TNFα) [48,61]. Sharing induction of expression of CCL2/MCP-1 and CXCL10/IP-10 by activation with IFN-γ with astrocytes [46,61–64], endothelial cells were also shown to be an important source of CXCL8/IL-8 [63,64].

The complex network of proinflammatory cytokines, chemokine-producing cells, and chemokine effects, is illustrated by the role of IFN-γ in chemokine expression during inflammatory brain diseases: as noted above, IFN-γ may play a crucial role in the expression of several chemokines including CCL2/MCP-1 and CXCL10/IP-10 during viral meningitis. Additionally, IFN-γ has been found at high levels in CSF from patients with viral meningitis and from mice with LCMV-induced disease [65–67]. In mice, infected with mouse hepatitis virus, CXCL10/IP-10 is expressed by astrocytes during acute encephalomyelitis, and T-cells invading the CNS express CXCR3. Administration of anti-CXCL10/IP-10 antibodies delayed viral clearance and reduced the number of T-cells and the level of IFN-γ in the CNS. [68]. However, the finding that mice deficient for IFN-γ-receptor also show a strong pleocytosis in LCMV-induced meningitis [69] suggests that in these animals other factors might compensate for the effect of IFN-γ.

8. COMPARISON OF CHEMOKINE LEVELS, CHEMOTACTIC ACTIVITY, AND PATHOGENIC ORIGIN IN MENINGEAL INFLAMMATORY DISEASES

The comparison between bacterial and viral meningitis reveals that in bacterial meningitis a broader range of chemokines is expressed at higher levels than in the viral disease. This is in accordance with more pronounced chemotactic activity of bacterial meningitis CSF compared to viral meningitis CSF. Likewise, other cytokines including TNFα [6], IL-1 [70], IL-6 [71] or G-CSF [72] are much more abundant in bacterial meningitis CSF. This may be due to the presence of more potent inducers of cytokines such as bacterial cell wall components and bacterial toxins.

In acute bacterial meningitis in humans, like in animals, early pleocytosis is dominated by neutrophils, whereas the monocyte counts increase later in the course of the disease. The strong correlation, which was found by comparing individual concentrations of CC and CXC chemokines in the CSF of bacterial meningitis implicates a general upregulation of chemokine production during the early stage of the disease [28]. At a time point with a predominantly polymorphonuclear pleocytosis in vivo, the CSF harbors both CC and CXC chemokines and exerts potent chemotactic activity on PMN and PBMC in Boyden chamber assays. Surprisingly, neither could a correlation be found between CSF leukocyte count and chemokine concentrations in the CSF nor did chemokine levels correlate with the chemotactic activity of CSF on peripheral blood leukocytes. Whether these findings reflect the in vitro assay situation represented by the lack of the vascular endothelium as well as by the use of migrating cells from healthy donors, remains to be shown. Furthermore, the action of cytokines and chemokines on adhesion molecules would be an important parameter to test on the transmigration of leukocytes. This has been shown by the cytokine-mediated upregulation of P- and E-selectin molecules on endothelial cells [73] and by the chemokine-mediated activation of β_2-integrins on neutrophils (by CXCL8/IL-8) [74–76] and on monocytes (by CCL2/MCP-1 and CCL3/MIP-1α) [77,78]. In a different group of patients analyzed, CSF samples also from pyogenic meningitis were investigated. In this study, CXCL8/IL-8 levels declined with the time of disease earlier than CCL2/MCP-1 and CCL3/MIP-1α [26]. The resulting levels of the chemokines would explain the early influx of both, mononuclear cells attracted by CCL2/MCP-1 and CCL3/MIP-1α and, predominantly, neutrophils attracted by CXCL8/IL-8, the decreasing amount of CXCL8/IL-8 at later time points leaving a mononuclear fraction with a decreasing number of neutrophils.

On the role of CXC-chemokines in general and CXCL8/IL-8 in particular in the neutrophilic response in viral and bacterial meningitis, there has to be taken into account that in bacterial meningitis additional chemotactic signals are present which act on neutrophils. This might explain the lack of correlation between the number of neutrophils in the CSF and the chemokine concentrations measured in the CSF in bacterial meningitis in contrast to viral meningitis. Supporting evidence for this theory comes from the limited efficiency of anti-chemokine antibodies to neutralize the chemotactic signals. Neutralization of CXCL8/IL-8 resulted in inhibition of neutrophil chemotaxis to no more than 75% in bacterial meningitis, but to about 95% in viral meningitis [28,31]. Thus, because of the strong correlation between CXCL8/IL-8 concentration in the CSF and the extent of neutrophilic pleocytosis in viral meningitis, this chemokine might play a more dominant role in this disease.

A lack of direct correlation between chemokine levels and cell influx has been described in patients with cryptococcal meningitis. Despite high levels of CXCL8/IL-8 in the CSF only a minimal pleocytosis was reported [79]. In this case, a possible explanation could be the existence of anti-CXCL8/IL-8 antibodies, which could limit the bioavailability of CXCL8/IL-8 in the CSF. However, so far such antibodies have only been described in children with purulent meningitis [80].

That the situation in vivo is complex is illustrated by a report showing that in a murine Listeria infection model the level of CXCL1/MIP-2 reached its maximum in the CSF compartment later than those of CCL3/MIP-1α and CCL4/MIP-1β, contradicting a simple chemokine–cell type correlation of the pleocytosis [33]. Other experimental data show that injection of CCL3/MIP-1 into the CSF compartment of rabbits first leads to an accumulation of PMN followed by PBMC after 4–6 h, whereas MIP-2 injection directly leads to an influx of PMN [8]. The significance of these data is given by the observation in the course of EAE where the expression of CCL5/RANTES, CCL3/MIP-1α, CCL4/MIP-1β, CXCL10/IP-10, CCL2/MCP-1 and CCL7/MCP-3, CXCL1/GROα and CCL1/TCA3 preceded the development of the disease by 1–2 days [81].

9. CHEMOKINES AS POTENTIAL DIAGNOSTIC PARAMETERS AND THERAPEUTIC TARGETS IN INFLAMMATORY BRAIN DISEASES

Reliable parameters to diagnostically distinguish between bacterial and aseptic meningitis are an essential prerequisite for fast and specific treatments of the disease. Because of their elevated levels in bacterial meningitis, TNF-α and IL-1 levels in the CSF can be used to identify the pathogenic origin of the disease in patients with high levels of these cytokines [70,82]. Although the specificity in the prediction of bacterial meningitis of these markers is very high, the sensitivity in many cases is still a problem. To gain additional information, chemokine levels in the CSF could be used for diagnostic purposes. Because of the markedly elevated levels found in bacterial meningitis, CXCL8/IL-8 would be a prime candidate for such use [27]. Other chemokines with diagnostic potential could include CCL3/MIP-1α, CXCL10/IP-10, CXCL9/Mig, and CXCL11/I-TAC the latter three being especially useful in the positive identification of viral meningitis, since all three attract solely activated T-cells. However, the diagnostic value of these parameters has not been proved. As none of the mentioned cytokines or chemokines alone would allow diagnosis of bacterial or viral meningitis with high sensitivity and specificity, established parameters like CSF WBC counts, together with CSF protein and glucose concentrations, have to be included in the differential diagnosis [83,84] together with positive identification tools for single pathogens like PCR [85]. In regard to clinical severity and chemokine levels, to date only one small study including patients with herpes simplex type 1 encephalitis shows a correlation between CCL2/MCP-1 levels and severity of symptoms [32]. In another larger study with bacterial meningitis, the only patient out of a group of 38 who died in the course of the disease neither showed significantly higher or lower chemokine concentrations in the CSF compared to the other patients [28].

Patients who survive a bacterial meningitis often suffer from neurologic sequelae due to irreparable destruction of nervous tissue. Inflammatory cells invading the CNS in response to a pathogen are supposed to play an important role in brain damage. Therefore it might be useful to inhibit the recruitment of immune cells in combination with the established antibiotic therapy. Inhibitors of chemokine activity including neutralizing antibodies and chemokine receptor antagonists may be primary candidates for therapeutic effects. In animal model systems such a strategy has already proven its efficiency: in LPS-induced meningitis in rabbits intravenous administration of neutralizing mononuclear antibodies against CXCL8/IL-8 dramatically reduced LPS dependent pleocytosis [86]. In infant rats suffering from meningitis due to infection with *H. influenzae*, administration of antibodies against CXCL1/MIP-2 or CCL3/MIP-1α resulted in a significant reduction of neutrophil recruitment to the CSF and treatment of the animals with anti-CCL2/MCP-1 antibodies significantly decreased macrophage-infiltration [34]. The results of these studies raise the hopes for a clinical use of chemokine antagonists in the treatment of bacterial meningitis in humans in the future and outline the importance of chemokines as mediators of the accumulation of immune cells in the CNS in response to bacterial pathogens. In viral meningitis, the situation seems to be more complex. Treatment of mice, suffering from acute encephalomyelitis after infection with mouse hepatitis virus, with anti-CCL5/RANTES antisera led to a decreased macrophage infiltration and reduced demyelination [87]. In contrast, in the same experimental system, administration of anti-IP-10 antisera resulted in increased mortality and delayed viral clearance from the CNS [68]. Thus, in viral meningitis inhibition of chemokine activity attracting activated T-cells may be dreadful and illustrates the delicate balance between beneficial and destructive effects of chemokines during inflammatory reactions in the CNS.

REFERENCES

1. Pfister HW, Scheld WM. Brain injury in bacterial meningitis: therapeutic implications. Curr Opin Neurol 1997; 10(3):254–259.
2. Pfister HW, Borasio GD, Dirnagl U, Bauer M, Einhaupl KM. Cerebrovascular complications of bacterial meningitis in adults. Neurology 1992;42(8):1497–1504.
3. Tani M, Fuentes ME, Peterson JW, Trapp BD, Durham SK, Loy JK et al. Neutrophil infiltration, glial reaction, and neurological disease in transgenic mice expressing the chemokine N51/KC in oligodendrocytes. J Clin Invest 1996;98(2):529–539.
4. Tunkel AR, Scheld WM. Acute bacterial meningitis. Lancet 1995;346(8991–8992):1675–1680.
5. Leist TP, Frei K, Kam-Hansen S, Zinkernagel RM, Fontana A. Tumor necrosis factor alpha in cerebrospinal fluid during bacterial, but not viral, meningitis. Evaluation in murine model infections and in patients. J Exp Med 1988;167(5):1743–1748.
6. Nadal D, Leppert D, Frei K, Gallo P, Lamche H, Fontana A. Tumour necrosis factor-alpha in infectious meningitis. Arch Dis Child 1989;64(9):1274–1279.
7. Ramilo O, Mustafa MM, Porter J, Saez-Llorens X, Mertsola J, Olsen KD et al. Detection of interleukin 1 beta but not tumor necrosis factor-alpha in cerebrospinal fluid of children with aseptic meningitis. Am J Dis Child 1990; 144(3):349–352.
8. Saukkonen K, Sande S, Cioffe C, Wolpe S, Sherry B, Cerami A et al. The role of cytokines in the generation of inflammation and tissue damage in experimental Gram-positive meningitis. J Exp Med 1990;171(2):439–448.
9. Waage A, Halstensen A, Shalaby R, Brandtzaeg P, Kierulf P, Espevik T. Local production of tumor necrosis factor alpha, interleukin 1, and interleukin 6 in meningococcal meningitis. Relation to the inflammatory response. J Exp Med 1989;170(6):1859–1867.
10. Ramilo O, Saez-Llorens X, Mertsola J, Jafari H, Olsen KD, Hansen EJ et al. Tumor necrosis factor alpha/cachectin and interleukin 1 beta initiate meningeal inflammation. J Exp Med 1990;172(2):497–507.
11. Donald PR, Schoeman JF, Beyers N, Nel ED, Carlini SM, Olsen KD et al. Concentrations of interferon gamma, tumor necrosis factor alpha, and interleukin-1 beta in the cerebrospinal fluid of children treated for tuberculous meningitis. Clin Infect Dis 1995;21(4):924–929.
12. Navikas V, Haglund M, Link J, He B, Lindqvist L, Fredrikson S et al. Cytokine mRNA profiles in mononuclear cells in acute aseptic meningoencephalitis. Infect Immun 1995;63(4):1581–1586.
13. Ravi V, Parida S, Desai A, Chandramuki A, Gourie-Devi M, Grau GE. Correlation of tumor necrosis factor levels in the serum and cerebrospinal fluid with clinical outcome

in Japanese encephalitis patients. J Med Virol 1997;51(2): 132–136.

14. Tekgul H, Tutuncuoglu S, Kutukculer N, Dizdarer G, Huseyinov A. Lymphocyte subsets and inflammatory mediators in patients with subacute sclerosing panencephalitis. J Child Neurol 1999;14(7):418–421.
15. Bienvenu K, Russell J, Granger DN. Platelet-activating factor promotes shear rate-dependent leukocyte adhesion in postcapillary venules. J Lipid Mediat 1993;8(2):95–103.
16. Bienvenu K, Granger DN. Molecular determinants of shear rate-dependent leukocyte adhesion in postcapillary venules. Am J Physiol 1993;264(5):H1504–H1508.
17. Butcher EC, Picker LJ. Lymphocyte homing and homeostasis. Science 1996;272(5258):60–66.
18. Dunon D, Piali L, Imhof BA. To stick or not to stick: the new leukocyte homing paradigm. Curr Opin Cell Biol 1996;8(5):714–723.
19. Nolte D, Hecht R, Schmid P, Botzlar A, Menger MD, Neumueller C et al. Role of Mac-1 and ICAM-1 in ischemia-reperfusion injury in a microcirculation model of BALB/C mice. Am J Physiol 1994;267(4):H1320–H1328.
20. Springer TA. Traffic signals for lymphocyte recirculation and leukocyte emigration: the multistep paradigm. Cell 1994;76(2):301–314.
21. Paul R, Lorenzl S, Koedel U, Sporer B, Vogel U, Frosch M et al. Matrix metalloproteinases contribute to the blood–brain barrier disruption during bacterial meningitis. Ann Neurol 1998;44(4):592–600.
22. Kolb SA, Lahrtz F, Paul R, Leppert D, Nadal D, Pfister HW et al. Matrix metalloproteinases and tissue inhibitors of metalloproteinases in viral meningitis: upregulation of MMP-9 and TIMP-1 in cerebrospinal fluid. J Neuroimmunol 1998;84(2):143–150.
23. Ishiguro A, Suzuki Y, Inaba Y, Fukushima K, Komiyama A, Koeffler HP et al. The production of IL-8 in cerebrospinal fluid in aseptic meningitis of children. Clin Exp Immunol 1997;109(3):426–430.
24. Lopez-Cortes LF, Cruz-Ruiz M, Gomez-Mateos J, Viciana-Fernandez P, Martinez-Marcos FJ, Pachon J. Interleukin-8 in cerebrospinal fluid from patients with meningitis of different etiologies: its possible role as neutrophil chemotactic factor. J Infect Dis 1995;172(2):581–584.
25. Mastroianni CM, Paoletti F, Rivosecchi RM, Lancella L, Ticca F, Vullo V et al. Cerebrospinal fluid interleukin 8 in children with purulent bacterial and tuberculous meningitis. Pediatr Infect Dis J 1994;13(11):1008–1010.
26. Mastroianni CM, Lancella L, Mengoni F, Lichtner M, Santopadre P, D'Agostino C et al. Chemokine profiles in the cerebrospinal fluid (CSF) during the course of pyogenic and tuberculous meningitis. Clin Exp Immunol 1998;114(2):210–214.
27. Ostergaard C, Benfield TL, Sellebjerg F, Kronborg G, Lohse N, Lundgren JD. Interleukin-8 in cerebrospinal fluid from patients with septic and aseptic meningitis. Eur J Clin Microbiol Infect Dis 1996;15(2):166–169.
28. Spanaus KS, Nadal D, Pfister HW, Seebach J, Widmer U, Frei K et al. C–X–C and C–C chemokines are expressed in the cerebrospinal fluid in bacterial meningitis and mediate chemotactic activity on peripheral blood-derived polymorphonuclear and mononuclear cells in vitro. J Immunol 1997;158(4):1956–1964.
29. Sprenger H, Rosler A, Tonn P, Braune HJ, Huffmann G, Gemsa D. Chemokines in the cerebrospinal fluid of patients with meningitis. Clin Immunol Immunopathol 1996;80(2):155–161.
30. Van Meir E, Ceska M, Effenberger F, Walz A, Grouzmann E, Desbaillets I et al. Interleukin-8 is produced in neoplastic and infectious diseases of the human central nervous system. Cancer Res 1992;52(16):4297–4305.
31. Lahrtz F, Piali L, Nadal D, Pfister HW, Spanaus KS, Baggiolini M et al. Chemotactic activity on mononuclear cells in the cerebrospinal fluid of patients with viral meningitis is mediated by interferon-gamma inducible protein-10 and monocyte chemotactic protein-1. Eur J Immunol 1997; 27(10):2484–2489.
32. Rosler A, Pohl M, Braune HJ, Oertel WH, Gemsa D, Sprenger H. Time course of chemokines in the cerebrospinal fluid and serum during herpes simplex type 1 encephalitis. J Neurol Sci 1998;157(1):82–89.
33. Seebach J, Bartholdi D, Frei K, Spanaus KS, Ferrero E, Widmer U et al. Experimental Listeria meningoencephalitis. Macrophage inflammatory protein-1 alpha and -2 are produced intrathecally and mediate chemotactic activity in cerebrospinal fluid of infected mice. J Immunol 1995; 155(9):4367–4375.
34. Diab A, Abdalla H, Li HL, Shi FD, Zhu J, Hojberg B et al. Neutralization of macrophage inflammatory protein 2 (MIP-2) and MIP-1alpha attenuates neutrophil recruitment in the central nervous system during experimental bacterial meningitis. Infect Immun 1999;67(5):2590–2601.
35. Lane TE, Asensio VC, Yu N, Paoletti AD, Campbell IL, Buchmeier MJ. Dynamic regulation of alpha- and beta-chemokine expression in the central nervous system during mouse hepatitis virus-induced demyelinating disease. J Immunol 1998;160(2):970–978.
36. Wang JM, Chen ZG, Colella S, Bonilla MA, Welte K, Bordignon C et al. Chemotactic activity of recombinant human granulocyte colony-stimulating factor. Blood 1988; 72(5):1456–1460.
37. Fukushima K, Ishiguro A, Shimbo T. Transient elevation of granulocyte colony-stimulating factor levels in cerebrospinal fluid at the initial stage of aseptic meningitis in children. Pediatr Res 1995;37(2):160–164.
38. Stahel PF, Nadal D, Pfister HW, Paradisis PM, Barnum SR. Complement C3 and factor B cerebrospinal fluid concentrations in bacterial and aseptic meningitis. Lancet 1997;349(9069):1886–1887.
39. Inaba Y, Ishiguro A, Shimbo T. The production of macrophage inflammatory protein-1alpha in the cerebrospinal fluid at the initial stage of meningitis in children. Pediatr Res 1997;42(6):788–793.
40. Bernasconi S, Cinque P, Peri G, Sozzani S, Crociati A, Torri W et al. Selective elevation of monocyte chemotactic protein-1 in the cerebrospinal fluid of AIDS patients with cytomegalovirus encephalitis. J Infect Dis 1996;174(5): 1098–1101.
41. Loetscher M, Gerber B, Loetscher P, Jones SA, Piali L, Clark-Lewis I et al. Chemokine receptor specific for IP10

and mig: structure, function, and expression in activated T-lymphocytes. J Exp Med 1996;184(3):963–969.
42. Loetscher P, Seitz M, Baggiolini M, Moser B. Interleukin-2 regulates CC chemokine receptor expression and chemotactic responsiveness in T lymphocytes. J Exp Med 1996; 184(2):569–577.
43. Fuentes ME, Durham SK, Swerdel MR, Lewin AC, Barton DS, Megill JR et al. Controlled recruitment of monocytes and macrophages to specific organs through transgenic expression of monocyte chemoattractant protein-1. J Immunol 1995;155(12):5769–5776.
44. Glabinski AR, Balasingam V, Tani M, Kunkel SL, Strieter RM, Yong VW et al. Chemokine monocyte chemoattractant protein-1 is expressed by astrocytes after mechanical injury to the brain. J Immunol 1996;156(11):4363–4368.
45. Kim JS. Cytokines and adhesion molecules in stroke and related diseases. J Neurol Sci 1996;137(2):69–78.
46. Ransohoff RM, Hamilton TA, Tani M, Stoler MH, Shick HE, Major JA et al. Astrocyte expression of mRNA encoding cytokines IP-10 and JE/MCP-1 in experimental autoimmune encephalomyelitis. FASEB J 1993;7(6):592–600.
47. Cole KE, Strick CA, Paradis TJ, Ogborne KT, Loetscher M, Gladue RP et al. Interferon-inducible T cell alpha chemoattractant (I-TAC): a novel non-ELR CXC chemokine with potent activity on activated T cells through selective high affinity binding to CXCR3. J Exp Med 1998;187(12):2009–2021.
48. Luo Y, Kim R, Gabuzda D, Mi S, Collins-Racie LA, Lu Z et al. The CXC-chemokine, H174: expression in the central nervous system. J Neurovirol 1998;4(6):575–585.
49. Hancock WW, Lu B, Gao W, Csizmadia V, Faia K, King JA et al. Requirement of the chemokine receptor CXCR3 for acute allograft rejection. J Exp Med 2000;192(10): 1515–1520.
50. Lapinet JA, Scapini P, Calzetti F, Perez O, Cassatella MA. Gene expression and production of tumor necrosis factor alpha, interleukin-1beta (IL-1beta), IL-8, macrophage inflammatory protein 1alpha (MIP-1alpha), MIP-1beta, and gamma interferon-inducible protein 10 by human neutrophils stimulated with group B meningococcal outer membrane vesicles. Infect Immun 2000;68(12):6917–6923.
51. Cassatella MA, Gasperini S, Calzetti F, Bertagnin A, Luster AD, McDonald PP. Regulated production of the interferon-gamma-inducible protein-10 (IP-10) chemokine by human neutrophils. Eur J Immunol 1997;27(1):111–115.
52. Gasperini S, Marchi M, Calzetti F, Laudanna C, Vicentini L, Olsen H et al. Gene expression and production of the monokine induced by IFN-gamma (MIG), IFN-inducible T cell alpha chemoattractant (I-TAC), and IFN-gamma-inducible protein-10 (IP-10) chemokines by human neutrophils. J Immunol 1999;162(8):4928–4937.
53. Liebler JM, Kunkel SL, Burdick MD, Standiford TJ, Rolfe MW, Strieter RM. Production of IL-8 and monocyte chemotactic peptide-1 by peripheral blood monocytes. Disparate responses to phytohemagglutinin and lipopolysaccharide. J Immunol 1994;152(1):241–249.
54. Luster AD, Unkeless JC, Ravetch JV. Gamma-interferon transcriptionally regulates an early-response gene containing homology to platelet proteins. Nature 1985;315(6021): 672–676.
55. Wuyts A, Proost P, Put W, Lenaerts JP, Paemen L, Van Damme J. Leukocyte recruitment by monocyte chemotactic proteins (MCPs) secreted by human phagocytes. J Immunol Methods 1994;174(1–2):237–247.
56. Yoshimura T, Yuhki N, Moore SK, Appella E, Lerman MI, Leonard EJ. Human monocyte chemoattractant protein-1 (MCP-1). Full-length cDNA cloning, expression in mitogen-stimulated blood mononuclear leukocytes, and sequence similarity to mouse competence gene JE. FEBS Lett 1989;244(2):487–493.
57. Martin CA, Dorf ME. Differential regulation of interleukin-6, macrophage inflammatory protein-1, and JE/MCP-1 cytokine expression in macrophage cell lines. Cell Immunol 1991;135(1):245–258.
58. Calvo CF, Yoshimura T, Gelman M, Mallat M. Production of monocyte chemotactic protein-1 by rat brain macrophages. Eur J Neurosci 1996;8(8):1725–1734.
59. Ehrlich LC, Hu S, Sheng WS, Sutton RL, Rockswold GL, Peterson PK et al. Cytokine regulation of human microglial cell IL-8 production. J Immunol 1998;160(4): 1944–1948.
60. Aloisi F, Care A, Borsellino G, Gallo P, Rosa S, Bassani A et al. Production of hemolymphopoietic cytokines (IL-6, IL-8, colony-stimulating factors) by normal human astrocytes in response to IL-1 beta and tumor necrosis factor-alpha. J Immunol 1992;149(7):2358–2366.
61. Hayashi M, Luo Y, Laning J, Strieter RM, Dorf ME. Production and function of monocyte chemoattractant protein-1 and other beta-chemokines in murine glial cells. J Neuroimmunol 1995;60(1–2):143–150.
62. Luster AD, Ravetch JV. Biochemical characterization of a gamma interferon-inducible cytokine (IP-10). J Exp Med 1987;166(4):1084–1097.
63. Sica A, Matsushima K, Van Damme J, Wang JM, Polentarutti N, Dejana E et al. IL-1 transcriptionally activates the neutrophil chemotactic factor/IL-8 gene in endothelial cells. Immunology 1990;69(4):548–553.
64. Sica A, Wang JM, Colotta F, Dejana E, Mantovani A, Oppenheim JJ et al. Monocyte chemotactic and activating factor gene expression induced in endothelial cells by IL-1 and tumor necrosis factor. J Immunol 1990;144(8):3034–3038.
65. Chonmaitree T, Baron S. Bacteria and viruses induce production of interferon in the cerebrospinal fluid of children with acute meningitis: a study of 57 cases and review. Rev Infect Dis 1991;13(6):1061–1065.
66. Frei K, Leist TP, Meager A, Gallo P, Leppert D, Zinkernagel RM et al. Production of B cell stimulatory factor-2 and interferon gamma in the central nervous system during viral meningitis and encephalitis. Evaluation in a murine model infection and in patients. J Exp Med 1988; 168(1):449–453.
67. Glimaker M, Olcen P, Andersson B. Interferon-gamma in cerebrospinal fluid from patients with viral and bacterial meningitis. Scand J Infect Dis 1994;26(2):141–147.
68. Liu MT, Chen BP, Oertel P, Buchmeier MJ, Armstrong D, Hamilton TA et al. The T cell chemoattractant IFN-inducible protein 10 is essential in host defense against

viral-induced neurologic disease. J Immunol 2000;165(5): 2327–2330.
69. Huang S, Hendriks W, Althage A, Hemmi S, Bluethmann H, Kamijo R et al. Immune response in mice that lack the interferon-gamma receptor. Science 1993;259(5102): 1742–1745.
70. Lopez-Cortes LF, Cruz-Ruiz M, Gomez-Mateos J, Jimenez-Hernandez D, Palomino J, Jimenez E. Measurement of levels of tumor necrosis factor-alpha and interleukin-1 beta in the CSF of patients with meningitis of different etiologies: utility in the differential diagnosis. Clin Infect Dis 1993;16(4):534–539.
71. Chavanet P, Bonnotte B, Guiguet M, Zeller V, Solary E, Maurice L et al. High concentrations of intrathecal interleukin-6 in human bacterial and nonbacterial meningitis. J Infect Dis 1992;166(2):428–431.
72. Shimoda K, Okamura S, Omori F, Mizuno Y, Hara T, Aoki T et al. Granulocyte colony-stimulating factor in cerebrospinal fluid from patients with meningitis. Blood 1991;77(10):2214–2217.
73. Tang T, Frenette PS, Hynes RO, Wagner DD, Mayadas TN. Cytokine-induced meningitis is dramatically attenuated in mice deficient in endothelial selectins. J Clin Invest 1996;97(11):2485–2490.
74. Detmers PA, Powell DE, Walz A, Clark-Lewis I, Baggiolini M, Cohn ZA. Differential effects of neutrophil-activating peptide 1/IL-8 and its homologues on leukocyte adhesion and phagocytosis. J Immunol 1991;147(12): 4211–4217.
75. Molad Y, Haines KA, Anderson DC, Buyon JP, Cronstein BN. Immunocomplexes stimulate different signalling events to chemoattractants in the neutrophil and regulate L-selectin and beta 2-integrin expression differently. Biochem J 1994;299(3):881–887.
76. Roberts PJ, Pizzey AR, Khwaja A, Carver JE, Mire-Sluis AR, Linch DC. The effects of interleukin-8 on neutrophil fMetLeuPhe receptors, CD11b expression and metabolic activity, in comparison and combination with other cytokines. Br J Haematol 1993;84(4):586–594.
77. Jiang Y, Beller DI, Frendl G, Graves DT. Monocyte chemoattractant protein-1 regulates adhesion molecule expression and cytokine production in human monocytes. J Immunol 1992;148(8):2423–2428.
78. Vaddi K, Newton RC. Regulation of monocyte integrin expression by beta-family chemokines. J Immunol 1994; 153(10):4721–4732.
79. Chaka W, Heyderman R, Gangaidzo I, Robertson V, Mason P, Verhoef J et al. Cytokine profiles in cerebrospinal fluid of human immunodeficiency virus-infected patients with cryptococcal meningitis: no leukocytosis despite high interleukin-8 levels. University of Zimbabwe Meningitis Group. J Infect Dis 1997;176(6):1633–1636.
80. Takasaki J, Ogawa Y. Anti-interleukin-8 auto-antibodies in cerebrospinal fluid of children with purulent meningitis. Pediatr Int 2000;42(2):139–142.
81. Godiska R, Chantry D, Dietsch GN, Gray PW. Chemokine expression in murine experimental allergic encephalomyelitis. J Neuroimmunol 1995;58(2):167–176.
82. Lopez-Cortes LF, Marquez-Arbizu R, Jimenez-Jimenez LM, Jimenez-Mejias E, Caballero-Granado FJ, Rey-Romero C et al. Cerebrospinal fluid tumor necrosis factor-alpha, interleukin-1beta, interleukin-6, and interleukin-8 as diagnostic markers of cerebrospinal fluid infection in neurosurgical patients. Crit Care Med 2000;28(1):215–219.
83. Hoen B, Viel JF, Paquot C, Gerard A, Canton P. Multivariate approach to differential diagnosis of acute meningitis. Eur J Clin Microbiol Infect Dis 1995;14(4):267–274.
84. Spanos A, Harrell FE, Durack DT. Differential diagnosis of acute meningitis. An analysis of the predictive value of initial observations. J Am Med Assoc 1989;262(19): 2700–2707.
85. Chesky M, Scalco R, Failace L, Read S, Jobim LF. Polymerase chain reaction for the laboratory diagnosis of aseptic meningitis and encephalitis. Arq Neuropsiquiatr 2000; 58(3B):836–842.
86. Dumont RA, Car BD, Voitenok NN, Junker U, Moser B, Zak O et al. Systemic neutralization of interleukin-8 markedly reduces neutrophilic pleocytosis during experimental lipopolysaccharide-induced meningitis in rabbits. Infect Immun 2000;68(10):5756–5763.
87. Lane TE, Liu MT, Chen BP, Asensio VC, Samawi RM, Paoletti AD et al. A central role for CD4(+) T cells and RANTES in virus-induced central nervous system inflammation and demyelination. J Virol 2000;74(3):1415–1424.

Universes in Delicate Balance: Chemokines and the Nervous System
R.M. Ransohoff, K. Suzuki, A.E.I. Proudfoot, W.F. Hickey and J.K. Harrison, editors

Chapter 5.5

Chemokines in the Inflamed Peripheral Nervous System

Bernd C. Kieseier and Hans-Peter Hartung

Department of Neurology, Karl-Franzens-University, Auenbruggerplatz 22, 8036 Graz, Austria [1]

1. INTRODUCTION

The peripheral nervous system (PNS), like the central nervous system, has traditionally been considered to be 'immunologically privileged' and not subject to the immune surveillance as found in most other organs within the human body. This concept was based on various assumptions, such as the idea of a strict anatomic separation between the systemic immune compartment and the PNS, a lack of MHC molecules, the absence of lymphatic drainage and T-cell surveillance within the neural tissue. Nowadays our view on these convictions has changed and it appears that differences between the PNS and other organs are more quantitative than qualitative. Although the tissue barrier between blood and neural tissue, the so-called blood–nerve barrier, does restrict access of immunocompetent cells and humoral factors, this restraint is not complete either anatomically or functionally. In fact, circulating lymphocytes can 'traffic' in and out of the PNS provided they are in an activated state, and the peripheral nerve tissue contains large numbers of professional and non-professional antigen-presenting cells, including resident endoneurial and perivascular macrophages and Schwann cells.

Four different cell types can be found in the normal PNS — macrophages, lymphocytes, mast cells, and Schwann cells — which appear to be immunologically relevant to the pathogenesis of immune-mediated neuropathies.

[1] Present address: Department of Neurology, Heinrich-Heine-University, Moorenstrasse 5, 40225 Düsseldorf, Germany

1.1. Macrophages of the Peripheral Nerve

Normal nerves contain large numbers of resident PNS macrophages, which primarily lie near endoneurial vessels, always outside the basal lamina [1,2]. Most of these cells constitutively express major histocompatibility complex (MHC) class II antigens, complement receptor 3 (CR3), the Fc-γ receptor III, and CD4 [3]. The perivascular distribution of these resident macrophages along the blood–nerve interface presumably facilitates their interaction with T-lymphocytes and points to a potential key-role of this cell population in acting as antigen-presenting cells in the PNS [4].

1.2. T-Lymphocytes

In normal nervous tissue T-lymphocytes, unlike macrophages, are rarely detectable as single cells scattered in the endoneurial space. As indicated above, these cells normally traffic into and through the endoneurium, as shown by studies of bone-marrow chimeras [5]. To become activated, these cells require the presentation of an appropriate antigen in the context of MHC class II molecules. Such activation in the course of an inflammatory reaction will increase the number of T-cells within the PNS dramatically, a process in which chemokines appear to play a central role.

1.3. Mast Cells

Although the presence of a large number of mast cells has been described in the PNS, the role of this cell population under normal conditions remains elusive. Abrupt mast cell degranulation can induce endoneurial edema due to damage of the blood–nerve barrier [6].

1.4. Schwann Cells

Besides exhibiting their function as myelinating cells of the peripheral nerve, human and rat Schwann cells, at least in tissue culture, can be induced to express MHC class II antigens by a variety of factors, such as by coculture with T-lymphoblasts, interferon (IFN)-γ, and tumor necrosis factor (TNF)-α [7–9]. Moreover, Schwann cells can present endogenous antigens to autoreactive T-cell lines in vitro [9]. Beyond their potential role as antigen presenters, Schwann cells were found to exert potent immunoregulatory functions: for example via the release of nitric oxide metabolites [10], or the expression of the downregulatory cytokine transforming growth factor (TGF)-β1 [11].

From the perspective of immune-mediated neuropathies, cellular responses in non-immune-mediated models are instructive. Wallerian degeneration of transected nerves represents such a model, which specifies the role of cellular components within the peripheral nerve. In rats, shortly after nerve transection, the axon undergoes dissolution of the cytoskeleton, associated with segmentation of the myelin sheaths, and breakdown of the blood–nerve barrier. A large population of circulating macrophages, but only a few polymorphonuclear leukocytes and no lymphocytes enter the degenerating nerve at the time point of axonal degeneration. This selective invasion of macrophages is mediated, at least in part, by the chemokine MCP-1/CCL2. In a mouse model of Wallerian degeneration of the sciatic nerve, MCP-1 transcripts were found to be synthesized by Schwann cells promptly after axonal breakdown, exhibiting maximum expression levels just prior to the peak of macrophage invasion [12,13]. The recruited macrophages adhere to degenerating nerve fibers and participate in phagocytosis and removal of myelin debris [14]. In vitro studies confirmed that Schwann cells, after stimulation with IFN-γ and TNF-α, can produce MCP-1 [15]. As such, Schwann cells perform a critical role in controlling the cellular composition within the peripheral nerve under non-inflammatory conditions via the release of specific chemokines, such as MCP-1.

2. INFLAMMATORY DEMYELINATION OF THE PERIPHERAL NERVE

Immune-mediated inflammatory disorders of the PNS are characterized by a massive inflammatory reaction, demyelination and axonal loss in the affected part of the nerve.

Most of our current knowledge has been obtained from a model disease, called experimental autoimmune neuritis (EAN), that can be induced in susceptible animals by active immunization with whole peripheral nerve homogenate, myelin, myelin proteins P0 and P2 or peptides thereof [16]. It can also be produced by adoptive transfer of P2, P2 peptide-specific, P0, and P0 peptide-specific T-cell lines [17].

EAN mirrors many of the immunologic, electrophysiologic, and morphologic aspects of the human Guillain–Barré syndrome (GBS) and has therefore been widely used as a model to study disease mechanisms.

The pathological hallmark of EAN is the infiltration of the PNS by lymphocytes and macrophages, which results in multifocal demyelination of axons predominantly around venules. Macrophages actively strip off myelin lamellae from axons, induce vesicular disruption of the myelin sheath, and phagocytose both intact and damaged myelin, as shown by electron microscopy [3]. Macrophages represent the predominant cell population in the inflamed PNS, which can be detected in spinal roots as well as in more distal segments of the affected nerves [18–20]. The fundamental need for this cell group in the pathogenesis of the disease — both in the induction and effector stage — was demonstrated by depletion experiments: macrophage elimination by intraperitoneal injection of silicon dust prevented all clinical, electrophysiologic, and histologic signs of EAN [21].

Crucial to the pathogenesis of inflammatory demyelination is the early invasion of the PNS by leukocytes. Circulating autoreactive T-cells need to be activated in the periphery in order to cross the blood–nerve barrier and to incite a local immuno-inflammatory response. Homing and transmigration depend on a complex interaction of adhesion molecules, chemokines, and matrix metalloproteases [22–24]. Breakdown of the blood–nerve barrier is one of the earliest morphologically demonstrable events in lesion development in EAN. Blood borne T-lymphocytes, once having reached the PNS, are reactivated in situ, expand clonally and release cytokines and chemokines to orchestrate the subsequent local immune response.

3. CHEMOKINES IN EXPERIMENTAL AUTOIMMUNE NEURITIS

The selective entry of macrophages and T-lymphocytes in the inflamed peripheral nervous tissue is a fundamental issue in neurobiology. To elucidate the relevance of chemokines in this process, the expression pattern of various chemokines during the clinical course of EAN has been investigated in recent studies, which yielded similar but also controversial results (Table 1). All studies were performed in Lewis rats; however, gender, age and immunization regimens for all these experiments differed. Therefore, it is difficult to derive distinct conclusions about these given differences.

3.1. IP-10/CXCL10

The restricted expression and selectivity for a single receptor on T-cells suggest that IP-10 is involved in lymphocyte recruitment. In EAN maximum mRNA expression of this α-chemokine was found to precede peak clinical disease severity in the cauda equina in one study [25], whereas in another maximum mRNA levels were observed coincident with peak clinical disease severity in the sciatic nerve of the Lewis rat [26]. Immunohistochemical studies point to endothelial cells as the primary cellular source of IP-10 in the inflamed PNS [26].

Table 1. Chemokine expression in experimental autoimmune neuritis

Chemokine	Expression		References
	mRNA	Protein	
IP-10	▲/•	•	[25,26]
MIP-1α	▲	•	[25–27]
MIP-1β	▲	n.c.	[26]
MIP-2	n.c.	▼	[27]
MCP-1	•	▲/•	[25–27]
RANTES	•	•	[25,26]
Lymphotactin	n.d.	n.c.	[25]

Maximum chemokine expression prior to (▲), at time of (•), or after (▼) peak disease severity. n.d., not detectable; n.c., not conducted.

3.2. MIP-1α/CCL3

Transcript levels of the chemokine MIP-1α were increased in EAN with peak levels preceding clinical disease severity in two studies [25,26], whereas in another, studying protein expression levels, maximal expression was described to coincide with peak of the disease [27]. A cellular source was not defined in any of these studies.

3.3. MIP-1β/CCL4

The expression pattern of this chemokine was the subject of investigation in one study [26]: mRNA levels preceded maximal clinical severity. This increase of transcript was associated with increased protein expression as demonstrated by Western blot. The cellular source, however, remains elusive.

3.4. MIP-2/CXCL2

The expression of the chemokine MIP-2 has been investigated in a single study to date [27]. Increased protein levels were detectable prior to peak disease severity, maximum MIP-2 levels, however, were found during the recovery phase of the disorder. No cellular source has been defined in the inflamed PNS. MIP-2 acts chemotactic for neutrophils, however, EAN is a T-cell mediated disease with a lack of neutrophils. Thus, the functional implication of this observation remains unclear.

3.5. MCP-1/CCL2

In two of three studies [25,26] maximum expression levels of MCP-1, transcript as well as protein, were observed at the time point of peak clinical disease severity. Immunohistochemistry revealed that some macrophages, endothelial cells, and some Schwann cells surrounding mononuclear cellular infiltration, stained positive for MCP-1 [25]. In a different study, MCP-1 protein was found to exhibit peak expression prior to clinical disease severity with a subsequent decline [27].

3.6. RANTES/CCL5

This strong chemoattractant was found to be expressed on the mRNA as well as protein level during the clinical course of EAN with peak expression levels coincident with peak disease severity [25,26]. T-lymphocytes were determined as the primary cellular source of RANTES in the inflamed PNS [26].

3.7. Lymphotactin/XCL1

The expression of the γ-chemokine lymphotactin has been investigated in one study, but no lymphotactin message was observable in the cauda equina at any given time point during the entire course of the disease [25].

Despite these minor discrepancies, this collective evidence supports the concept that a differential expression of individual chemokines orchestrates the infiltration of immunocompetent macrophages and T-lymphocytes into the inflamed peripheral nervous system.

4. CHEMOKINES IN THE INFLAMED PERIPHERAL NERVOUS SYSTEM IN HUMANS

Our current knowledge on the expression of chemokines and especially their receptors is only limited. In one study [28], the chemokine receptor expression pattern was investigated in sural nerve biopsies obtained from patients with Guillain–Barré syndrome, chronic inflammatory demyelinating polyradiculoneuropathy, and various non-inflammatory neuropathies. A consistent chemokine receptor expression pattern was immunohistochemically detected in inflammatory demyelinating neuropathies: CXCR-3, CCR-2, and CCR-4 were primarily expressed by invading T-lymphocytes, whereas CCR-1 and CCR-5 could be localized to endoneurial macrophages. Quantitation of labeled mononuclear cells revealed significantly elevated cell counts compared with controls. The distribution of the two chemokines IP-10 and RANTES mirrored the distribution of their equivalent receptors within the inflamed peripheral nervous system. Further studies are clearly warranted to elucidate the functional implication of these observations.

5. CHEMOKINES AS THERAPEUTIC TARGETS IN INFLAMMATORY DISORDERS OF THE PNS

Similar to studies in experimental autoimmune encephalomyelitis, where the application of anti-MIP-1α antiserum [29] or generation of an immune response to MIP-1α DNA injected as a vaccine [30] prevented the infiltration of mononuclear cells into the central nervous system, the therapeutic application of compounds in an attempt to prevent or ameliorate inflammatory demyelination was performed in EAN as well. Treatment with anti-MIP-1α- and anti-MCP-1 antibodies, respectively, ameliorated the clinical course of myelin induced EAN in the Lewis rat [27]. Whereas animals treated with the anti-MIP-1α antibody exhibited diminished signs of cellular infiltration and demyelination in the sciatic nerve, no clear differences in the number of cellular infiltrates and the degree of demyelination were detectable in the group of animals treated with the anti-MCP-1 antibody. Although these observations support the concept of chemokine inhibition as a promising therapeutic target, further knowledge of specific temporospatial expression patterns of chemokines is clearly needed to specifically define such targets.

6. SUMMARY

Our present knowledge supports the notion that chemokines are critically involved in the composition of cellular infiltrates in immune-mediated demyelinating disorders of the PNS. At least for the chemokines IP-10, MIP-1α, MIP-1β, MIP-2, MCP-1, and RANTES clear evidence supports this assumption.

Results from studies in EAN, describing various maximum chemokine expression levels to be coincident with peak of clinical disease suggest that these chemokines exert an amplifying rather than initiating role in generating PNS inflammation. Moreover it appears that they might also be involved in modulating the immune response during the course of the disease. In a recent report it was demonstrated that the migration and differentiation of dorsal root ganglia cells could be induced by RANTES in vitro [31]. Thus, it appears conceivable that RANTES, and potentially other chemokines, might also exhibit beneficial effects by promoting axonal regeneration after axonal degeneration caused by severe demyelination. Therefore, it can be assumed that the PNS might have its own chemokine ligand–receptor network, the function of which could extend well beyond the regulation of leukocyte trafficking in immuno-inflammatory disease. Further research is needed to shed light on the biological implications of this observation. Increased knowledge of the expression and specific function of individual chemokines and their receptors will hopefully help to design inhibitors that could be administered at critical checkpoints in the evolution of immune-mediated demyelination of the PNS.

ACKNOWLEDGEMENTS

The authors' work is supported in part by the Gemeinnützige Hertie Stiftung and a research grant from the Austrian Federal Ministry of Science and Transport (GZ 70.057/2-Pr/4/99).

REFERENCES

1. Griffin JW, George R, Lobato C, Tyor WR, Li CY, Glass JD. Macrophage responses and myelin clearance during Wallerian degeneration: relevance to immune-mediated demyelination. J Neuroimmunol 1992;40:153–166.
2. Monaco S, Gehrmann J, Raivich G, Kreutzberg GW. MHC-positive, ramified macrophages in the normal and injured rat peripheral nervous system. J Neurocytol 1992; 21:623–634.
3. Griffin JW, George R, Ho T. Macrophage system in peripheral nerves. A review. J Neuropathol Exp Neurol 1993;52:553–560.
4. Hartung H-P, Gold R, Jung S. Local immune responses in the peripheral nervous system. In: Antel J, Birnbaum G, Hartung HP. Clinical Neuroimmunology. Malden, USA: Blackwell Science, 1998:40–54.
5. Vass K, Hickey WF, Schmidt RE, Lassmann H. Bone marrow-derived elements in the peripheral nervous system: an immunohistochemical and ultrastructural investigation in chimeric rats. Lab Invest 1993;69:275–282.
6. Harvey GK, Toyka KV, Hartung H-P. Effects of mast cell degranulation on blood–nerve barrier permeability and nerve conduction in vivo. J Neurol Sci 1994;125:102–109.
7. Gold R, Toyka KV, Hartung H-P. Synergistc effect of IFN gamma and TNF alpha on expression of immune molecules and antigen presentation by Schwann cells. Cell Immunol 1995;165:65–70.
8. Samuel NM, Jessen KR, Grange JM, Mirsky R. Gamma interferon, but not Mycobacterium leprae, induces major histocompatibility class II anitgens on cultured rat Schwann cells. J Neurocytol 1987;16:281–287.
9. Samuel NM, Mirsky R, Grange JM, Jessen KR. Expression of major histocompatibility complex class I and class II antigens in human Schwann cell cultures and effects of infection with *Mycobacterium leprae*. Clin Exp Immunol 1987;68:500–509.
10. Gold R, Zielasek J, Kiefer R, Toyka KV, Hartung H-P. Secretion of nitrite by Schwann cells and its effect on T-cell activation in vitro. Cell Immunol 1996;168:69–77.
11. Kiefer R, Funa K, Schweitzer T, Jung S, Bourde O, Toyka KV, Hartung H-P. Transforming growth factor beta1 in experimental autoimmune neuritis: cellular localization and time course. Am J Pathol 1996;148:211–223.
12. Ho TW, McKhann GM, Griffin JW. Human autoimmune neuropathies. Annu Rev Neurosci 1998;21:187–226.
13. Toews AD, Barret C, Morell P. Monocyte chemoattractant protein 1 is responsible for macrophage recruitment following injury to sciatic nerve. J Neurosci Res 1998;53: 260–267.
14. Stoll G, Griffin JW, Li CY, Trapp BD. Wallerian degeneration in the peripheral nervous system: participation of both Schwann cells and macrophages in myelin degradation. J Neurocytol 1989;18:671–683.
15. Polydefkis MJ, Krivacic K, Tani M, Ransohoff RM, Griffin JW, Ho TW. Rat Schwann cells upregulate MCP-1 synthesis in response to TNF alpha and IFN gamma. Neurology 1998;50:A184.
16. Waksman NH, Adams RD. Allergic neuritis: an experimental disease in rabbits induced by the injection of peripheral nervous tissue and adjuvant. J Exp Med 1955; 102:213–225.
17. Linington C, Izumo S, Suzuki M, Meyermann R, Wekerle H. A permanent rat T cell line that mediates experimental allergic neuritis in the Lewis rat in vivo. J Immunol 1984; 133:1946–1950.
18. Ota K, Irie H, Takahashi K. T cell subsets and Ia-positive cells in sciatic nerve during the course of experimental autoimmune neuritis. J Neuroimmunol 1987;13:283–292.
19. Schmidt B, Toyka KV, Kiefer R, Full J, Hartung H-P, Pollard J. Inflammatory infiltrates in sural nerve biopsies in Guillain–Barré syndrome and chronic inflammatory demyelinating neuropathy. Muscle Nerve 1996;19:474–487.
20. Kiefer R, Kieseier BC, Brück W, Toyka KV, Hartung H-P. Macrophage differentiation antigens in acute and chronic autoimmune polyneuropathies. Brain 1998;121:469–479.
21. Tansey FA, Brosnan CF. Protection against experimental autoimmune neuritis with silicia quartz dust. J Neuroimmunol 1982;3:169–179.
22. Archelos JJ, Previtali SC, Hartung H-P. The role of integrins in immune-mediated diseases of the nervous system. Trends Neurosci 1999;22:30–38.
23. Kieseier BC, Seifert T, Giovannoni G, Hartung H-P. Matrix metalloproteinases in inflammatory demyelination. Targets for treatment. Neurology 1999;53:20–25.
24. Hartung H-P, van der Meché FGA, Pollard JD. Guillain–Barré syndrome, CIDP and other chronic immune-mediated neuropathies. Curr Opin Neurol 1998;11:497–513.

25. Fujioka T, Purev E, Rostami A. Chemokine mRNA expression in the cauda equina of Lewis rats with experimental allergic neuritis. J Neuroimmunol 1999;97:51–59.
26. Kieseier BC, Krivacic K, Jung S, Pischel H, Toyka KV, Ransohoff RM, Hartung H-P. Sequential expression of chemokines in experimental autoimmune neuritis. J Neuroimmunol 2000;110:121–129.
27. Zou LP, Pelidou SH, Abbas N, Deretzi G, Mix E, Schaltzbeerg M, Winblad B, Zhu J. Dynamics of production of MIP-1alpha, MCP-1 and MIP-2 and potential role of neutralization of these chemokines in the regulation of immune responses during experimental autoimmune neuritis in Lewis rats. J Neuroimmunol 1999;98:168–175.
28. Kieseier BC, Tani M, Mahod D, Oka N, Ho T, Woodroofe N, Griffin JW, Toyka KV, Ransohoff RM, Hartung HP. Chemokines and chemokine receptors in inflammatory demyelinating neuropathies – a central role for IP-10. Brain 2002; in press.
29. Karpus WJ, Lukacs NW, McRae BL, Strieter RM, Kunkel SL, Miller SD. An important role for the chemokine macrophage inflammatory protein-1a in the pathogenesis of the T cell-mediated autoimmune disease, experimental autoimmune encephalomyelitis. J Immunol 1995;155: 5003–5010.
30. Youssef S, Wildbaum G, Maor G, Lanir N, Gour-Lavie A, Grabie N, Karin N. Long-lasting protective immunity to experimental autoimmune encephalomyelitis following vaccination with naked DNA encoding C–C chemokines. J Immunol 1998;161:3870–3879.
31. Bolin LM, Murray R, Lukacs NW, Strieter RM, Kunkel SL, Schall TJ, Bacon KB. Primary sensory neurons migrate in response to the chemokines RANTES. J Neuroimmunol 1998;81:49–57.

Subject Index

A

B

C

D

E

F

G

H

I

J

K

L

M

N

O

P

R

S

T

U

V

W

Z